Microalgal Biotechnology

Microalgae Biotechnology: A Water–Soil–Air Nexus Sustainable Perspective offers a comprehensive and integrated overview of microalgal biotechnology, emphasizing its role within the interconnected systems of water, soil, and air. This book brings together cutting-edge research and multidisciplinary perspectives to explore how microalgae contribute to sustainable development, addressing global challenges related to environmental management, resource efficiency, and bio-based innovation.

This volume covers the full spectrum of microalgal biotechnology, from strain bioprospecting, genetic engineering, and synthetic biology to production systems and technological scalability. It examines microbial interactions in the phycosphere, as well as bioethical and risk considerations associated with environmental release and genome editing. This book highlights advances in high-value compounds such as pigments, lipids, exopolysaccharides, and biopolymers, alongside applications in food, feed, agriculture, and functional ingredients. Environmental applications are strongly represented, including wastewater treatment, bioremediation, carbon capture, and odor control. Emerging frontiers—such as sustainable aviation fuels and space life-support systems—are also explored. Finally, this book provides a robust discussion on sustainability assessment, including metrics, life cycle assessment (LCA), and consumer perception, offering readers both scientific depth and practical insight into the viability of microalgal technologies.

This book is designed for researchers, graduate students, and professionals in microalgal biotechnology and related disciplines, including biotechnology, environmental science, food science, and process engineering. It is also a valuable resource for industry stakeholders and decision-makers seeking to better understand the opportunities and challenges associated with microalgae, particularly within the context of the water–energy–food nexus and the growing demand for sustainable technologies.

Microalgal Biotechnology

A Water–Soil–Air Nexus Sustainable Perspective

Edited by
Mariany Costa Deprá,
Eduardo Jacob-Lopes
and Leila Queiroz Zepka

CRC Press
Taylor & Francis Group
Boca Raton London New York

CRC Press is an imprint of the
Taylor & Francis Group, an **informa** business

Designed cover image: Getty

First edition published 2027
by CRC Press
2385 NW Executive Center Drive, Suite 320, Boca Raton FL 33431

and by CRC Press
4 Park Square, Milton Park, Abingdon, Oxon, OX14 4RN

CRC Press is an imprint of Taylor & Francis Group, LLC

ISBN: 978-1-041-00750-0 (hbk)
ISBN: 978-1-041-00751-7 (pbk)
ISBN: 978-1-003-61147-9 (ebk)

DOI: 10.1201/9781003611479

Typeset in Times
by SPi Technologies India Pvt Ltd (Straive)

Contents

Preface viii
About the Editors x
Contributors xi

Chapter 1 Eukaryotic Microalgae Strain Bioprospecting: From Selection and Isolation Approaches to Genetic Engineering and Synthetic Biology 1

Sarah D'Adamo, Sofia Amendola, Julia Bevervanco, Aureo Cutillas Farray, and Igor Vunderink

Chapter 2 Microalgae and Microbes: Understanding Their Interactions in the Phycosphere 38

Udoh Inyang, Akaninyene Jack, Moyosore Adegbeye, Nsikan Ekanem, Ekaette Mbaba, Mona M.M.Y. Elghandour, P.R.K. Reddy, Marie De Angelis, and Abdelfattah Z.M. Salem

Chapter 3 Bioethical and Risk Considerations in Microalgal Biotechnology: Environmental Release and Genome Editing Implications 62

Israel Valencia Quiroz and Yuri Córdoba Campo

Chapter 4 Microalgae Production Systems: Recent Findings to Forward Technological Readiness Level for Microalgal Biotechnology Consolidation 82

Carolina Diaz-Cardenas, Francisco Vera-López Portillo, Gina López, Dulce María Arias, Jorge Daniel Osorio-Márquez, and Carlos Alberto Montenegro-Herrera

Chapter 5 Advancements in Microalgal Exopolysaccharides: Biological Activities, Production Factors, and Industrial Applications 127

Kader Karsavran, Zeynep Elibol Çakmak, and Turgay Çakmak

Chapter 6 Microalgae Biotechnology for Chlorophyll Production: Distribution, Extraction, and Potential Uses 162

Pricila Nass Pinheiro, Tatiele C. Nascimento, Patrícia Acosta Caetano, Luisa Chitolina Schetinger, and Leila Q. Zepka

Chapter 7 Novel Microalgae Metabolites: Production and Industrial Applications 183

Christian Rodrigues Lugcheer, Mimidoo Aver Gaaba, Marcele Leal Nörnberg, Tatiele Casagrande do Nascimento, Eduardo Jacob-Lopes, and Leila Queiroz Zepka

Chapter 8 Optimization Strategies for Enhancing Microalgal Lipid Production toward Sustainable Nutrition 209

Md. Mirazul Islam

Chapter 9 Insights into Microalgae-Based Biopolymer Production: A Technological Survey 235

Anwesha Mondal, Avik Kumar Choudhury, Kristine V. Aleksanyan, and Gour Gopal Satpati

Chapter 10 Microalgae for Sustainable Nutrition: From Functional Benefits to Consumer Adoption 258

Mariany Costa Deprá, Rosangela Rodrigues Dias, Adriane Terezinha Schneider, Richard Luan Silva Machado, Darissa Alves Dutra, and Eduardo Jacob-Lopes

Chapter 11 Microalgae-Based Feed: A Promising and Sustainable Alternative Feedstock for Animal Production Systems 272

Sameh A. Abdelnour and Wael A. Khalil

Chapter 12 Microalgae for Soil Applications: New Perspectives for Plant Growth Additives 292

Iara Barbosa Magalhães, Moacir Caetano do Couto Junior, Diego A. R. Hennessey, Tatiana Cardoso e Búfalo, Joyce Dória, Maria Lúcia Calijuri, Eduardo de Aguiar do Couto, and Paula Assemany

Chapter 13 Microalgae-Based Biotechnology as an Alternative for Industrial Wastewater Treatment 309

Adegoke Isiaka Adetunji and S'fiso Thuthukani Gumbi

Chapter 14 Microalgae as Green Filters: Bioremediation and Odor Neutralization in Wastewaters 340

Nnabueze Darlington Nnaji, Chiemerie Theresa Ekwueme, Job Chinagorom Aleke, Adindu Joy Chinwe, and Helen Onyeaka

Chapter 15 Microalgae in Carbon Capture and Storage: A Clean and Sustainable Approach for Controlling Environmental Pollution...... 383

Zeyuan Wang, Yang Zhou, and Yu Hong

Chapter 16 From Water to Sky: Integrating Microalgal Biomass into Sustainable Aviation Fuel Development... 401

Gabriela F. Ferreira, Fernanda P. Beatriz, and Ingrid L. Motta

Chapter 17 Microalgal Biotechnology in Space Missions: A Potential Strategy for Onboard Sustainable Life-Support Systems................. 435

Giacomo Fais, Giovanni Perra, Upasana Mohanty, Nicola Lai, Giacomo Cao, and Alessandro Concas

Chapter 18 Metrics and Indicators of Sustainability Applied to Microalgal Biotechnology 453

Rosangela Rodrigues Dias and Eduardo Jacob-Lopes

Chapter 19 Life-Cycle Assessment Pathways toward Sustainable Marine Microalgal Biotechnology.. 465

Neha Arora, Elizabeth Pyle, and George P. Philippidis

Chapter 20 Linking Sustainability and Consumer Perceptions: The Nexus Behind Microalgae-Based Food Acceptance 478

Ibrahim Abu-Reidah

Chapter 21 Harnessing the Potential of Microalgae Biotechnology through the Water–Energy–Food Nexus: A System Dynamics Approach to Sustainable Innovation 489

Erika Cristina Francisco

Index... 502

Preface

Sustainability may be understood as the art of living in harmonious relationships. This idea reminds us that no ecosystem exists in isolation: ecological, social, and economic dimensions are interwoven in ways that shape both human well-being and planetary stability. It is precisely within this web of interdependence that contemporary challenges reveal their full scale. Climate change, soil degradation, water scarcity, and declining air quality do not emerge as isolated crises but as interconnected outcomes of how societies produce, consume, and manage resources. In this sense, sustainability invites us to rethink not only individual technologies or sectors but also the relationships among them—seeking pathways that restore balance while respecting the limits of the systems upon which life depends.

A sustainable perspective, therefore, requires moving beyond fragmented approaches and embracing the reality that water, soil, and air form a single, interacting continuum. The water–soil–air nexus underscores how decisions taken in one domain inevitably shape the others: irrigation practices influence erosion and nutrient losses, soil degradation compromises water regulation and amplifies climate vulnerability, and atmospheric emissions alter precipitation regimes, soil fertility, and ecosystem productivity. These feedback mechanisms reveal that sustainability is not achieved by optimizing isolated components but by aligning actions across systems, scales, and institutions. By framing environmental management through a nexus lens, it becomes possible to identify trade-offs transparently, foster synergies, and design strategies that simultaneously protect ecosystems, secure resources, and support human development.

It is within this integrated landscape that microalgal biotechnology emerges not merely as a promising technology but as an illustration of nexus thinking translated into practice. Microalgae harness sunlight, carbon dioxide, nutrients, and water with exceptional efficiency, producing biomass that can serve the food, feed, health, agricultural, and energy sectors. At the same time, they contribute environmental services: capturing atmospheric carbon, recovering nutrients from wastewater, supporting soil improvement, and operating within circular frameworks that transform residues into resources. Situated at the interface of water management, soil health, and air quality, microalgal systems demonstrate how innovation can connect mitigation, regeneration, and resource recovery within a single, integrated platform.

This book invites readers to explore microalgal biotechnology through precisely this nexus lens. Each chapter bridges science, technology, and application, highlighting advances in cultivation systems, bioproduct development, environmental services, and sustainability assessment. More importantly, it demonstrates how interdisciplinary collaboration—between researchers, industry, policymakers, and communities—is essential to move from promising experiments to large-scale, socially just implementation. Ultimately, the path toward sustainable societies will never be built by a single technology. It will emerge from coherent systems that integrate ecology, innovation, and human values. By framing microalgae as a catalyst within the water–soil–air nexus, this volume offers both knowledge and inspiration: an invitation to rethink our relationship with the planet and to transform scientific insight into practical, restorative action.

About the Editors

Dr. Mariany Costa Deprá is currently a researcher at the Department of Food Technology and Science, Federal University of Santa Maria, Brazil. She has published more than 150 scientific publications/communications, including books, book chapters, original research papers, research communications in national and international conferences, and patents. She acts as a reviewer for several international journals. Her research interests include sustainability metrics and indicators with an emphasis on microalgal biotechnology.

Prof. Dr. Eduardo Jacob-Lopes is currently an associate professor at the Department of Food Technology and Science, Federal University of Santa Maria, Brazil. He has more than 15 years of teaching and research experience. He has coordinated/coordinates more than 50 research projects and/or technological developments financed by public funding agencies and private initiatives. He has published more than 800 scientific publications/communications, including books, book chapters, original research papers, research communications in national and international conferences, and patents. He is a member of the editorial board of 10 journals and acts as a reviewer for several national and international journals. His research interests include environmental engineering with an emphasis on microalgal biotechnology.

Prof. Dr. Leila Queiroz Zepka is currently an associate professor at the Department of Food Technology and Science, Federal University of Santa Maria, Brazil. She has more than 15 years of teaching and research experience. She has published more than 700 scientific publications/communications, including books, book chapters, original research papers, research communications in national and international conferences, and patents. She is a member of the editorial board of five journals and acts as a reviewer for several national and international journals. Her research interests include microalgal biotechnology with emphasis on biomolecules and novel products.

Contributors

Sameh A. Abdelnour
Department of Animal Production, Faculty of Agriculture
Zagazig University
Zagazig, Egypt

Ibrahim Abu-Reidah
Memorial University of Newfoundland
St. John's, Canada

Moyosore Adegbeye
Research Centre for Animal Husbandry, National Research and Innovation Agency
Cibinong Science Centre
Bogor, Indonesia

Adegoke Isiaka Adetunji
Institute for Water and Wastewater Technology
Durban University of Technology
Durban, South Africa

Job Chinagorom Aleke
Department of Microbiology
University of Nigeria
Nsukka, Nigeria

Kristine V. Aleksanyan
Semenov Federal Research Center for Chemical Physics
Russian Academy of Sciences
Moscow, Russia

Sofia Amendola
Wageningen University and Research
Wageningen, The Netherlands

Marie De Angelis
Dipartimento di Scienze del Suolo, della Pianta e degli Alimenti (Di.S.S.P.A.)
Università degli Studi di Bari
Bari, Italy

Dulce María Arias
Instituto de Energías Renovables
Universidad Nacional Autónoma de México
Temixco, México

Neha Arora
Department of Biology
Skidmore College
Saratoga Springs, New York, USA
and
Patel College of Global Sustainability
University of South Florida
Tampa, Florida, USA

Paula Assemany
Environmental Engineering Department, Post-Graduate Program in Environmental Engineering
Federal University of Lavras
Lavras, Brazil

Fernanda P. Beatriz
School of Chemical Engineering
University of Campinas
São Paulo, Brazil

Julia Bevervanco
Wageningen University and Research
Wageningen, The Netherlands

Patrícia Acosta Caetano
Department of Food Science and Technology
Federal University of Santa Maria (UFSM)
Santa Maria, Brazil

Turgay Çakmak
Department of Molecular Biology and Genetics
Istanbul Medeniyet University
Istanbul, Türkiye
and

Science and Advanced Technologies Research Center
Istanbul Medeniyet University
Istanbul, Türkiye

Zeynep Elibol Çakmak
Department of Bioengineering
Istanbul Medeniyet University
Istanbul, Türkiye

Maria Lúcia Calijuri
Civil Engineering Department, Post-Graduate Program in Civil Engineering
Federal University of Viçosa
Viçosa, Brazil

Yuri Córdoba Campo
Facultad de Estudios Superiores Iztacala
Universidad Nacional Autónoma de México
Tlalnepantla, México

Giacomo Cao
Interdepartmental Centre of Environmental Science and Engineering (CINSA)
University of Cagliari
Cagliari, Italy
and
Department of Mechanical, Chemical and Materials Engineering
University of Cagliari
Cagliari, Italy

Tatiana Cardoso e Búfalo
Physics Department, Post-Graduate Program in Physics
Federal University of Lavras
Lavras, Brazil

Adindu Joy Chinwe
Department of Applied Microbiology and Brewing
Nnamdi Azikiwe University
Awka, Nigeria

Avik Kumar Choudhury
PG Department of Botany
Ramakrishna Mission Vivekananda Centenary College, Rahara
Khardaha, India

Alessandro Concas
Interdepartmental Centre of Environmental Science and Engineering (CINSA)
University of Cagliari
Cagliari, Italy
and
Department of Mechanical, Chemical and Materials Engineering
University of Cagliari
Cagliari, Italy

Sarah D'Adamo
Wageningen University and Research
Wageningen, The Netherlands

Eduardo de Aguiar do Couto
Environmental Engineering Department, Post-Graduate Program in Environmental Engineering
Federal University of Lavras
Lavras, Brazil

Mariany Costa Deprá
Bioprocess Intensification Group
Federal University of Santa Maria (UFSM)
Santa Maria, Brazil

Rosangela Rodrigues Dias
Bioprocess Intensification Group
Federal University of Santa Maria (UFSM)
Santa Maria, Brazil

Carolina Diaz-Cardenas
Laboratorio de Biotecnología y Recursos Genéticos
Instituto Amazónico de Investigaciones Científicas SINCHI
Bogotá, Colombia

Joyce Dória
Agriculture Department, Post-Graduate Program in Agronomy/Plant Science
Federal University of Lavras
Lavras, Brazil

Darissa Alves Dutra
Bioprocess Intensification Group
Federal University of Santa Maria (UFSM)
Santa Maria, Brazil

Nsikan Ekanem
Division of Animal Science, College of Agriculture, Food and Natural Resources
University of Missouri
Columbia, Missouri, USA

Chiemerie Theresa Ekwueme
School of Health and Life Sciences
Teesside University
Darlington, UK

Mona M.M.Y Elghandour
Facultad de Medicina Veterinaria y Zootecnia
Universidad Autónoma del Estado de México
Toluca, México

Giacomo Fais
Interdepartmental Centre of Environmental Science and Engineering (CINSA)
University of Cagliari
Cagliari, Italy
and
Department of Mechanical, Chemical and Materials Engineering
University of Cagliari
Cagliari, Italy

Aureo Cutillas Farray
Wageningen University and Research
Wageningen, The Netherlands

Gabriela F. Ferreira
Cardiff Catalysis Institute, School of Chemistry
Cardiff University
Cardiff, UK

Erika Cristina Francisco
Swedish University of Agricultural Sciences
Uppsala, Sweden

Mimidoo Aver Gaaba
Department of Food Science and Technology
Federal University of Santa Maria (UFSM)
Santa Maria, Brazil

S'fiso Thuthukani Gumbi
Discipline of Microbiology, School of Life Sciences, College of Agriculture, Engineering, and Science
University of KwaZulu-Natal
Durban, South Africa

Diego A. R. Hennessey
Environmental Engineering Department, Post-Graduate Program in Environmental Engineering
Federal University of Lavras
Lavras, Brazil

Yu Hong
Beijing Key Lab for Source Control Technology of Water Pollution, College of Environmental Science and Engineering
Beijing Forestry University
Beijing, China
and
Engineering Research Center for Water Pollution Source Control & Eco-Remediation, College of Environmental Science and Engineering
Beijing Forestry University
Beijing, China
and
Hebei Key Laboratory for Emerging Contaminants Control and Risk Management, College of Environmental Science and Engineering
Beijing Forestry University
Beijing, China

Udoh Inyang
Department of Animal Science, Faculty of Agriculture
University of Uyo
Uyo, Nigeria

Md. Mirazul Islam
Department of Chemical Engineering and Polymer Science
Shahjalal University of Science and Technology
Sylhet, Bangladesh

Akaninyene Jack
Department of Animal Science, Faculty of Agriculture
University of Uyo
Uyo, Nigeria

Eduardo Jacob-Lopes
Bioprocess Intensification Group
Federal University of Santa Maria (UFSM)
Santa Maria, Brazil

Moacir Caetano do Couto Junior
Agriculture Department, Post-Graduate Program in Agronomy/Plant Science
Federal University of Lavras
Lavras, Brazil

Kader Karsavran
Department of Molecular Biology and Genetics
Istanbul Medeniyet University
Istanbul, Türkiye

Wael A. Khalil
Department of Animal Production, Faculty of Agriculture
Mansoura University
Mansoura, Egypt

Nicola Lai
Interdepartmental Centre of Environmental Science and Engineering (CINSA)
University of Cagliari
Cagliari, Italy
and
Department of Mechanical, Chemical and Materials Engineering
University of Cagliari
Cagliari, Italy

Gina López
Unidad de Saneamiento y Biotecnología Ambiental (USBA), Departamento de Biología
Pontificia Universidad Javeriana
Bogotá, Colombia

Christian Rodrigues Lugcheer
Department of Food Science and Technology
Federal University of Santa Maria (UFSM)
Santa Maria, Brazil

Richard Luan Silva Machado
Bioprocess Intensification Group
Federal University of Santa Maria (UFSM)
Santa Maria, Brazil

Iara Barbosa Magalhães
Civil Engineering Department, Post-Graduate Program in Civil Engineering
Federal University of Viçosa
Viçosa, Brazil

Ekaette Mbaba
Department of Animal Science, Faculty of Agriculture
University of Uyo
Uyo, Nigeria

Upasana Mohanty
Department of Physics
University of Trento
Trento, Italy

Anwesha Mondal
Department of Botany
University of Calcutta
Kolkata, India

Carlos Alberto Montenegro-Herrera
Laboratorio de Biotecnología y Recursos Genéticos
Instituto Amazónico de Investigaciones Científicas SINCHI
Bogotá, Colombia

Ingrid L. Motta
School of Mechanical Engineering
University of Campinas
Campinas, Brazil

Tatiele Casagrande do Nascimento
Department of Food Science and Technology
Federal University of Santa Maria (UFSM)
Santa Maria, Brazil

Pricila Nass Pinheiro
Department of Food Science and Technology
Federal University of Santa Maria (UFSM)
Santa Maria, Brazil

Nnabueze Darlington Nnaji
School of Chemical Engineering
University of Birmingham
Birmingham, UK
and
Department of Microbiology
University of Nigeria
Nsukka, Nigeria

Marcele Leal Nörnberg
Department of Food Science and Technology
Federal University of Santa Maria (UFSM)
Santa Maria, Brazil

Helen Onyeaka
School of Chemical Engineering
University of Birmingham
Birmingham, UK

Jorge Daniel Osorio-Márquez
Facultad de Ciencias Exactas, Naturales y Agropecuarias
Universidad de Santander
Bucaramanga, Colombia

Giovanni Perra
Interdepartmental Centre of Environmental Science and Engineering (CINSA)
University of Cagliari
Cagliari, Italy
and
Department of Mechanical, Chemical and Materials Engineering
University of Cagliari
Cagliari, Italy

George P. Philippidis
Patel College of Global Sustainability
University of South Florida
Tampa, Florida, USA
and
Department of Integrative Biology
University of South Florida
Tampa, Florida, USA

Francisco Vera-López Portillo
Instituto de Energías Renovables
Universidad Nacional Autónoma de México, Privada Xochicalco S/N, Col. Centro
Temixco, México

Elizabeth Pyle
Department of Biology
Skidmore College
Saratoga Springs, New York, USA

Israel Valencia Quiroz
Facultad de Estudios Superiores Iztacala
Universidad Nacional Autónoma de México
Tlalnepantla, México

P.R.K. Reddy
Division of Animal Science, College of Agriculture, Food and Natural Resources
University of Missouri
Columbia, Missouri, USA

Abdelfattah Z.M. Salem
Dipartimento di Scienze del Suolo, della Pianta e degli Alimenti (Di.S.S.P.A.)
Università degli Studi di Bari
Bari, Italy

Gour Gopal Satpati
Department of Botany, Bangabasi Evening College
University of Calcutta
Kolkata, India

Luisa Chitolina Schetinger
Department of Food Science and Technology
Federal University of Santa Maria (UFSM)
Santa Maria, Brazil

Adriane Terezinha Schneider
Bioprocess Intensification Group
Federal University of Santa Maria (UFSM)
Santa Maria, Brazil

Igor Vunderink
Wageningen University and Research
Wageningen, The Netherlands

Zeyuan Wang
Beijing Key Lab for Source Control Technology of Water Pollution, College of Environmental Science and Engineering
Beijing Forestry University
Beijing, China
and
Engineering Research Center for Water Pollution Source Control & Eco-Remediation, College of Environmental Science and Engineering
Beijing Forestry University
Beijing, China
and
Hebei Key Laboratory for Emerging Contaminants Control and Risk Management, College of Environmental Science and Engineering
Beijing Forestry University
Beijing, China

Leila Queiroz Zepka
Department of Food Science and Technology
Federal University of Santa Maria (UFSM)
Santa Maria, Brazil

Yang Zhou
Beijing Key Lab for Source Control Technology of Water Pollution, College of Environmental Science and Engineering
Beijing Forestry University
Beijing, China
and
Engineering Research Center for Water Pollution Source Control & Eco-Remediation, College of Environmental Science and Engineering
Beijing Forestry University
Beijing, China
and
Hebei Key Laboratory for Emerging Contaminants Control and Risk Management, College of Environmental Science and Engineering
Beijing Forestry University
Beijing, China

1 Eukaryotic Microalgae Strain Bioprospecting

From Selection and Isolation Approaches to Genetic Engineering and Synthetic Biology

Sarah D'Adamo, Sofia Amendola, Julia Bevervanco, Aureo Cutillas Farray, and Igor Vunderink

1.1 INTRODUCTION: EUKARYOTIC MICROALGAE AS EMERGING PRODUCTION PLATFORMS FOR MODERN BIOTECHNOLOGY AND SYNTHETIC BIOLOGY

Eukaryotic microalgae constitute a diverse group of unicellular photosynthetic microorganisms that inhabit a wide range of aquatic environments, including both freshwater and marine systems. They account for an estimated 40% of global primary production, thereby serving as pivotal regulators of the biogeochemical carbon cycle and contributing substantially to climate modulation. Their metabolic versatility and ecological ubiquity underscore their integral role in maintaining ecosystem stability and supporting global biogeochemical processes. The diversity within the microalgal kingdom is nothing short of extraordinary. Scientists have catalogued approximately 44,000 distinct species to date (Borowitzka, 2018), yet this represents merely the tip of the iceberg. The vast majority of these microscopic marvels remain undiscovered, with conservative estimates suggesting that between 70,000 and one million species may exist in total (De Clerck et al., 2013). This staggering biodiversity hints at an untapped reservoir of biological potential waiting to be explored.

What also makes microalgae particularly compelling from a biotechnological perspective is their exceptional nutritional profile. These organisms serve as concentrated repositories of essential nutrients, including high-quality proteins, complex carbohydrates, beneficial lipids, vibrant carotenoids, and an array of vitamins and minerals. Often grouped together with photosynthetic bacteria for their similar properties, microalgae have captured the attention of researchers and industry leaders alike for their promising applications in biotechnological production.

DOI: 10.1201/9781003611479-1

The recognition of microalgae as a valuable nutritional resource spans millennia, deeply rooted in indigenous knowledge systems across the globe. Communities in Japan, Hawaii, North and South America, and Africa have harvested and utilized green-blue microalgae, particularly cyanobacteria species. These resourceful societies have developed sophisticated methods for drying and preserving edible cyanobacteria such as *Nostoc*, *Spirulina*, and *Aphanizomenon*, creating vital food reserves for times of scarcity while simultaneously establishing these organisms as valuable commodities in ancient trade networks (Severo et al., 2024).

Building on top of this rich historical foundation, the scientific community began systematic investigation into the potential of cyanobacteria and eukaryotic microalgae. The 1950s marked a pivotal moment in microalgae research, driven by growing concerns about global food security. As scientists predicted that rapidly expanding populations would inevitably lead to protein shortages, microalgae emerged as a promising alternative nutrient source capable of addressing these looming challenges (Spolaore et al., 2006). Concurrent with nutritional research, microalgae's potential as an alternative energy source was investigated already during the oil crisis of the 1970s, when energy security became a pressing global concern (Khan et al., 2018).

The transition from research curiosity to commercial reality began with the establishment of the first commercial microalgae production in South Asia, and by the 1990s, the microalgae industry started its first appearances, with large-scale cultivation facilities operational in many countries across the globe (Spolaore et al., 2006). This geographical expansion reflected both the adaptability of microalgal cultivation technologies and the growing recognition of their commercial versatility.

Today, eukaryotic microalgae are emerging as a platform for applications in synthetic biology, thus moving beyond traditional biomass and nutraceutical applications. Their inherent photosynthetic efficiency, rapid growth, and metabolic diversity provide a flexible platform for engineering value-added compounds, including biofuels, pharmaceuticals, and specialty chemicals. Worth mentioning that microalgae enable the direct conversion of atmospheric CO_2 into high-value metabolites, offering a dual solution for carbon mitigation and sustainable bioproduction (Levasseur et al., 2020).

1.1.1 The Versatility of Eukaryotic Microalgae: How Their Advantages Are Rooted in Their Origins

The advantages of eukaryotic microalgae over traditional terrestrial crops are both numerous and significant. These microscopic organisms demonstrate exceptional photon-to-biomass conversion efficiency, far exceeding that of multicellular plants, while operating independently of seasonal constraints that limit conventional agriculture (Perin et al., 2022).

Most importantly, microalgae cultivation sidesteps the contentious issue of food versus fuel competition by requiring no agricultural land. Thanks to their versatility, microalgae thrive in both controlled systems and various aquatic environments (including freshwater, saltwater, and wastewater), and they can be safely cultivated in photobioreactors with adjustable growth conditions, offering greater adaptability than terrestrial crops (Blanken et al., 2016; Pérez-López et al., 2017).

What truly sets microalgae apart, compared to other biological organisms, is their extraordinary capacity for producing a great array of valuable compounds in one single cell. From lipids and pigments to carbohydrates and proteins, these organisms represent Nature's most versatile biochemical factories, offering unparalleled diversity in biotechnological applications (Barkia et al., 2019; Borowitzka, 2013; Cuellar-Bermudez et al., 2015; de Morais et al., 2015). This remarkable versatility stems from two fundamental factors: the sheer abundance of microalgae species and the evolutionary complexity reflected in the diversity of their metabolic pathways. In fact, the metabolic sophistication and, therefore, adaptability of microalgae reflect their ancient evolutionary heritage and the profound impact of symbiotic relationships throughout their development. Multiple endosymbiotic events over geological time have created intricate metabolic networks that enable different species to produce vastly different compounds and adapt to diverse environmental conditions (Levasseur et al., 2020).

Being among the oldest life forms on Earth, microalgae carry within their cellular machinery the accumulated wisdom of nearly four billion years of evolution, a biological legacy that manifests exceptional metabolic versatility and species diversity (Andersen, 1992).

This evolutionary mastery should position microalgae as uniquely suited for addressing the complex challenges of modern biotechnology, offering sustainable solutions across an unprecedented range of applications while contributing to environmental remediation rather than degradation.

This chapter introduces eukaryotic microalgae as a promising production platform for modern biotechnology and synthetic biology. By looking at their extraordinary biodiversity, ecological diversity, and bioeconomic potential, we will elucidate why the rational selection of strains must be treated as a deliberate design decision in engineering efforts.

From this foundation, the discussion moves from concept to practice, following the journey from wild isolates to bioengineering readiness. Strategies for environmental sampling, niche targeting, and high-throughput phenotypic screening are presented as essential tools for identifying strains with desirable traits, such as productivity of lipids and other products, stress tolerance, and photosynthetic efficiency. This chapter then highlights the role of genomic and molecular characterization as a gateway to deeper engineering, encompassing genome sequencing, genome mining, and the genetic tools and pipeline that enable transformation and gene-expression studies. Case studies on astaxanthin biosynthesis and recombinant vaccine development are presented to exemplify the transition from wild strains to engineered microalgal biofactories, highlighting how well-characterized chassis can be harnessed for the sustainable production of high-value compounds with applications in nutraceuticals and disease control.

1.2 STRAIN SELECTION AS A FOUNDATIONAL DESIGN CHOICE FOR INDUSTRIAL DOMESTICATION

In industrial biotechnology, domestication refers to the systematic adaptation of wild organisms into predictable, controllable, and high-performing production platforms. Just as agriculture transformed wild plants into reliable crops, domestication in biotechnology ensures that microbial or microalgal strains can meet the demands of large-scale, economically viable production. Wild isolates may carry unique metabolic capabilities, but they are often poorly suited for industrial environments: they may grow slowly, be vulnerable to contaminants, or resist genetic modification. Through cycles of strain improvement, cultivation optimization, and genetic engineering, organisms are transformed from ecologically adapted specialists into robust industrial production platforms (Sproles et al., 2021).

For microalgae, domestication is particularly important. While their vast natural diversity holds enormous promise for biofuels, nutraceuticals, pharmaceuticals, and bioplastics, most species remain only partially genetically tractable in laboratory or industrial settings. Historically adopted model species such as *Chlamydomonas reinhardtii* (green algae lineage) and *Phaeodactylum tricornutum* (pennate diatoms, brown algae lineage) demonstrate how domestication can convert a wild-type strain into a versatile biofactory: genomes are sequenced, transformation methods established, and stable phenotypes selected for reproducibility (Butler et al., 2020; Salome & Merchant, 2019; Sproles et al., 2021). Without this domestication process, attempts to scale up biotechnological applications often fail due to instability, contamination, or poor yield.

Thus, domestication is not simply a technical step but a strategic enabler of both modern industrial biotechnology and synthetic biology. It bridges the gap between ecological potential and industrial performance, ensuring that organisms are not only genetically engineerable but also scalable, stable, and economically competitive. In this sense, domestication is the foundation upon which sustainable bio-based industries can be built. To ensure a successful domestication process, the selection of a chassis organism must be recognized not merely as a neutral starting point but as a critical design decision that fundamentally determines the feasibility, efficiency, and scalability of all downstream engineering efforts.

For microalgae, this choice is especially critical given their extraordinary biodiversity, which encompasses vast differences in metabolic pathways, growth rates, stress tolerance, and amenability to genetic manipulation. Strain-specific inherent traits, such as CO_2 fixation efficiency, carbon partitioning, lipid accumulation capacity, pigment diversity, or tolerance to salinity and temperature, will determine not only which products can be synthesized at higher amounts but also how robustly engineered pathways can be expressed under industrial conditions (Work et al., 2013; Work et al., 2012).

Unlike other well-established model microbial chassis such as *Escherichia coli* or *Saccharomyces cerevisiae*, many microalgae remain only partially genetically domesticated. Their genetic intractability, complex cellular architectures (e.g., multiple membranes and rigid cell walls), and diverse lifestyles introduce unique constraints.

For example, strains such as *Chlamydomonas reinhardtii*, *Phaeodactylum tricornutum*, and more recently, *Nannochloropsis* spp. have become model organisms and reference strains, respectively, due to their well-characterized genomes and established transformation protocols, making them ideal for proof-of-concept engineering (Barbosa et al., 2023; Butler et al., 2020; Crozet et al., 2018; Sproles et al., 2021). Yet, focusing solely on these models can overlook other specialized taxa, such as extremophilic microalgae (Rojas-Villalta et al., 2024) that may offer innate resilience for industrial cultivation under adverse conditions (high or fluctuating temperature and evaporation) despite limited genetic accessibility.

In synthetic biology, strain selection is a strategic, front-end design decision, balancing experimental tractability with industrial performance. Incorporating ecological and physiological insights ensures that strain choice aligns with the design–build–test–learn (DBTL) cycle, linking genetic architecture to host biology. This deliberate approach transforms microalgal synthetic biology from opportunistic exploration into systematic, strategic domestication of robust, scalable platforms for sustainable biotechnology.

1.3 MICROALGAE AS INDUSTRIAL PLATFORMS: MATCHING METABOLITES, MARKETS, AND STRAINS

In industrial microalgal biotechnology, the choice of strain should be guided first and foremost by the intended product, whether determined by market demand or by its potential to drive transformative processes and economic development. Microalgae are highly versatile microorganisms capable of producing a diverse spectrum of metabolites of industrial interest, from bulk commodities to high-value products, including novel compounds that can be produced via genetic engineering and synthetic biology (Gangl et al., 2015; Gimpel et al., 2015; Khan et al., 2018).

1.3.1 Industrial Applications of Microalgal Metabolites: Compounds and Process Considerations

The most industrially attractive molecules, which also reflect the unique selling points of eukaryotic microalgae, are oils, pigments, and bioactive secondary metabolites (Gangl et al., 2015). Oils, particularly triacylglycerols (TAGs), are valuable for biofuel and food industries, while specific polyunsaturated fatty acids such as the omega-3 eicosapentaenoic acid, are of interest to nutraceutical and food industries (Ahmmed et al., 2020; Li-Beisson et al., 2019). Additionally, microalgae are explored for their potential in accumulating natural bioactive compounds with antioxidant, anti-inflammatory, and antimicrobial properties; therefore, their extracts are also of interest to the pharmaceutical and functional food industries (Ambati et al., 2019; Barkia et al., 2019; Borowitzka, 2013; Chew et al., 2017; Cuellar-Bermudez et al., 2015). In particular, pigments such as carotenoids (β-carotene, lutein, fucoxanthin, and astaxanthin) find applications in food, feed, cosmetics, and pharmaceuticals, offering antioxidant, coloring, and photoprotective properties (Ambati et al., 2019; Borowitzka, 2013; Chew et al., 2017; Cuellar-Bermudez et al., 2015). Beyond these specialized metabolites, microalgal biomass itself is a valuable product, rich in vitamins (e.g., B12, E, and K), high-quality proteins, and naturally balanced essential amino acids, making it suitable for food and feed applications (Cardozo et al., 2007).

Moreover, microalgae have recently emerged as an attractive production platform for several plant metabolites (such as flavonoids, alkaloids, and terpenoids), which possess potent biological activities but are often difficult or costly to extract in sufficient quantities from their natural sources. By introducing plant biosynthetic pathways into metabolically flexible microalgal hosts, it has been shown how these valuable compounds can be produced often more sustainably and with precise control over yield and quality (Vavitsas et al., 2018).

Collectively, these molecules, whether they are naturally present or produced through genetic engineering, and the biomass itself, highlight the economic and functional versatility that positions microalgae as a uniquely promising platform for industrial biotechnology.

When it comes to production and commercialization, bulk commodities such as biofuels (oil) and animal feeds (protein extracts, full biomass) require strains with high productivity, low-cost cultivation, and resilience under large-scale conditions. Therefore, such types of production may call for low-tech and cheaper cultivation, and the use of open-pond systems could be envisioned and adopted (Qin et al., 2023).

In contrast, the high-value, low-volume products such as nutraceuticals and specialty molecules produced via synthetic biology from plant secondary metabolite pathways may justify the use of more delicate, metabolically specialized strains cultivated in closed photobioreactors to ensure controlled growth and consistent product quality.

Once the product and market scale are defined, the choice of cultivation techniques (open versus closed systems and batch versus continuous operation) may be considered to further narrow down suitable candidates. Finally, the microalgal species can be selected based on its physiological traits, stress tolerance, metabolic profile, and scalability, ensuring that the chosen chassis aligns with both industrial feasibility and the specific economic and functional demands of the target application.

Figure 1.1 illustrates a proposed decision framework that prioritizes the product as the primary factor in the decision-making process.

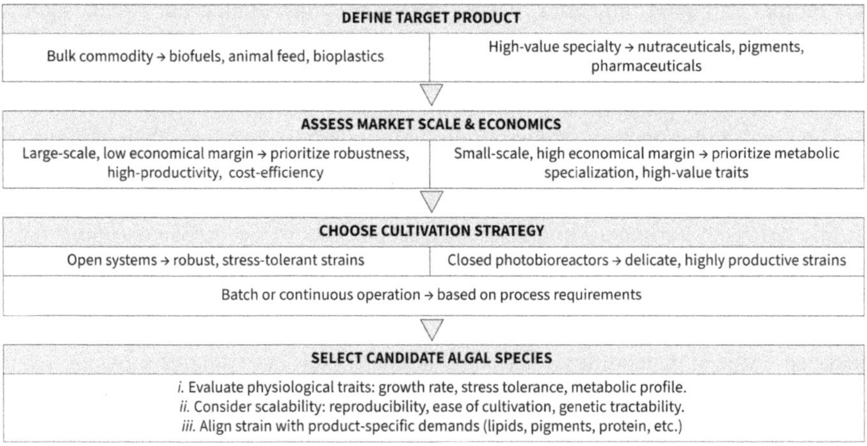

FIGURE 1.1 Proposed decision framework for microalgal strain selection for industrial domestication. In this framework, the first choice of a microalgal species is guided by the target product and its market context.

1.4 FROM THE WILD TO BIOENGINEERING-READINESS: ISOLATION, CULTIVATION, AND MODERN HIGH-THROUGHPUT PHENOMICS

Modern microalgal bioprospecting can integrate targeted environmental sampling, advanced isolation techniques, scalable cultivation, and high-throughput phenotyping to identify strains with industrial potential. Following a forward-thinking approach, there are several examples where strategies focused on environmental sampling in ecological niches that allow selection of species enriched in specific traits, such as hypersaline environments for halotolerance, highlight desert crusts for phototolerance, or CO_2-rich thermal springs for enhanced carbon fixation (D'Elia et al., 2018; Montuori et al., 2023; Rappaport & Oliverio, 2023).

Isolation traditionally relied on enrichment of cultures and serial dilution, which are considered tedious, time-consuming techniques. However, bioprospecting and strain isolation in microalgae are increasingly integrating advanced single-cell techniques such as fluorescence-activated cell sorting (FACS), which enable rapid, functional selection of individual cells while maintaining viability. In recent years, complementary technologies, including microfluidic droplet encapsulation, fluorescence-activated droplet sorting (FADS), and the application of artificial intelligence for data analysis, have also emerged and are considered to become a very powerful asset for strain isolation (Duan et al., 2024; Kim, Pradhan, et al., 2024; Pereira et al., 2018; Zheng et al., 2022).

Likewise, in the last decades, cultivation technologies have evolved from small-scale batch cultures to modular photobioreactors designed to replicate natural light, CO_2 availability, and mixing dynamics. These systems enable the performance of strain selection in a more throughput fashion, and it has the potential to bridge laboratory screening to pilot-scale production. However, it must be stated that several challenges such as self-shading, light regime, low gas transfer, evaporation rates, and hydrodynamics remain significant barriers to the reliability of the results for upscale purposes (Motta et al., 2025).

Once isolated, and even while isolating, the strains may be subjected to phenotyping screening. Traditional phenotyping methods relied on bulk culture measurements, such as growth rate, pigment content, or lipid extraction, which were low-throughput and provided limited insight into heterogeneity within populations (Lee et al., 2013).

High-throughput phenomics is enabling rapid, detailed, and scalable screening of strain diversity for industrial and synthetic biology applications, and it is becoming more frequently adopted in microalgal biotechnology. Modern phenomics platforms assess key traits such as growth dynamics, photosynthetic performance, pigment composition, lipid and carbohydrate accumulation, and stress tolerance. For example, FACS has become a standard technique for screening microalgae based on intrinsic or reporter-linked fluorescence signals (Pereira et al., 2018). Pigment autofluorescence can be used to monitor photosynthetic health, and as a preliminary assessment of their titers, while Nile Red or BODIPY dyes allow rapid identification and isolation of lipid-rich cells (Rumin et al., 2015). FACS enables precise selection and recovery of top-performing individual cells for downstream cultivation or genetic engineering (Südfeld et al., 2021; Sudfeld et al., 2021). Fluorescence-based screening, employing flow cytometry or plate readers, can be utilized to quantitatively assess the expression levels of isolated and genetically engineered mutant lines. This strategy facilitates the identification and selection of mutants with superior protein expression profiles (Baier et al., 2018; Einhaus et al., 2021).

These recent advances in automation, microfluidics, fluorescence-based sorting, and label-free spectroscopic techniques now allow the precise characterization of thousands of individual cells in parallel, bridging the gap between discovery and engineering.

Droplet-based microfluidic assays represent an emerging frontier, with the potential to encapsulate single algal cells in picoliter to nanoliter droplets, each serving as an independent bioreactor (Kim, Pradhan, et al., 2024; Sánchez Barea et al., 2019). In principle, these droplets could be supplemented with nutrients, stressors, or reporter molecules to allow real-time monitoring of metabolic activity. FADS may combine such encapsulation with high-speed sorting to eventually enable the isolation of rare, high-performing cells at kilohertz throughput (Best et al., 2016; Kim et al., 2017; Yu et al., 2021; Zheng et al., 2022). While proof-of-concept studies have demonstrated the possibility of screening for specific traits, such as high lipid content, its broader implementation in microalgal biotechnology remains largely aspirational.

Raman spectroscopy is likewise emerging as a promising tool to streamline analytics. As a label-free technique, it offers the possibility of characterizing cellular biochemistry without staining or cell disruption. Early studies suggest its potential for assessing intracellular lipids, carotenoids, and polysaccharides in microalgae such as *Nannochloropsis* and *Chlorella* (Moudříková et al., 2016). More advanced techniques, including Raman-activated cell sorting (RACS), have also been proposed, and it could in the future integrate optical trapping and Raman analysis to isolate cells based on biochemical signatures, for example, enabling targeted selection of individuals with high TAG or carotenoid content (Wang et al., 2017).

Hybrid platforms like PhenoChip point toward the next generation of experimental tools, integrating microfluidic cell trapping, time-lapse imaging, and environmental perturbation to dynamically monitor phenotypic responses under controlled conditions (Behrendt et al., 2020). When fully realized, the integration of such high-throughput phenotyping approaches with multi-omics datasets (genomics, transcriptomics, proteomics, and metabolomics) could enable a system-level understanding of traits such as CO_2 fixation efficiency, photoprotection, and stress resilience. Although still at an early stage, this envisioned phenomics-to-omics pipeline holds great promise for accelerating the rational selection, improvement, and engineering of microalgal strains for applications ranging from biofuels and nutraceuticals to aquaculture, wastewater treatment, specialty glycoengineering, and carbon capture (Kumar et al., 2023; Sirohi et al., 2021).

However, while high-throughput screening is a powerful first step to preliminarily select strains, it is essential to include a proper biochemical validation stage. Results obtained under controlled, small-scale conditions may not directly translate to larger-scale cultivation, which still represents a true challenge in the field of microalgal biotechnology for industrial applications (Deprá et al., 2025). Microplate or laboratory-scale assays differ fundamentally from industrial photobioreactor conditions in terms of light distribution, mixing, and mass transfer, making simple volumetric productivity measurements unreliable proxies for industrial performance. Therefore, biochemical validation under conditions that closely mimic industrial-scale cultivation, not just high-throughput readouts, is crucial to ensure that candidate strains maintain their desirable traits in other types of cultivation settings and, more importantly, under real-world applications.

1.5 GENOMIC AND MOLECULAR CHARACTERIZATION: A CRITICAL STEP FOR GENETIC ENGINEERING AND SYNTHETIC BIOLOGY

Building upon the initial isolation of wild microalgal strains, subsequent efforts focus on their genetic and phenotypic optimization through advanced synthetic biology techniques. While naturally occurring isolates often exhibit adaptive traits such as enhanced photosynthetic efficiency, nutrient assimilation, stress tolerance, high lipid productivity, or unique pigment profiles, these characteristics may require optimization to meet specific biotechnological goals. Historically, improvement efforts traditionally relied on classical mutagenesis techniques (such as exposure to chemical mutagens or ultraviolet irradiation) to induce random genetic variation and select for phenotypes with enhanced productivity or stress tolerance (Bleisch et al., 2022; Trovão et al., 2022). While these approaches have yielded useful variants, mutations are randomly introduced, often leading to unpredictable pleiotropic effects, and mapping the exact causal mutation can be laborious. This stochastic nature often limits predictability and precision.

Genome sequencing and molecular tool development have undoubtedly expanded the repertoire of possible interventions. In fact, the advent of modern genetic and synthetic biology tools, including CRISPR/Cas-mediated genome editing and modular pathway reconstruction, now enables the precise tailoring of microalgal strains to enhance performance under defined conditions (Fu et al., 2019).

Central to this effort is a detailed understanding and resolution of microalgal genomes. Having a fully sequenced and well-annotated genome is essential for microalgal genetic engineering, providing the blueprint necessary for precise genome editing, functional gene targeting, and rational strain optimization. This host-specific approach provides the foundation for building more predictable and efficient genetic tools tailored to the unique molecular environment of microalgae and allows expanding the molecular tools of different organisms with specific industrially relevant features, beyond the small pool of model organisms.

Within this context, *Chlamydomonas reinhardtii* and *Phaeodactylum tricornutum* have long served as foundational model organisms in microalgal genetic engineering due to their well-characterized genomes and amenability to genetic manipulation (Butler et al., 2020; Li et al., 2025; Salome & Merchant, 2019). The advent of high-throughput sequencing and environmental metagenomics has transformed our ability to discover, annotate, and utilize genes from microalgae and related lineages. Genome mining, broadly encompassing the extraction of biological insight and functional potential from genomic and metagenomic datasets, has emerged as a central strategy for genetic engineering and synthetic biology.

Like other eukaryotes, microalgae possess nuclear genomes organized into linear chromosomes contained within a membrane-bound nucleus. These nuclear genomes encode most genes required for cell growth, metabolism, and organelle function, and they remain the most widely targeted for genetic manipulation. It is worth noting that eukaryotic microalgae also harbor organellar genomes, most notably in chloroplasts and mitochondria (Bowsher & Tobin, 2001). These genomes are of prokaryotic origin, typically circular, and relatively compact, containing mainly essential genes required for organelle function. Importantly, chloroplast genomes are often present in multiple copies per cell, making them attractive engineering targets for high-level expression of transgenes. Their bacterial-like regulatory systems, coupled with their central role in photosynthesis, can provide new and unique opportunities for metabolic and synthetic biology applications. Chloroplast transformations indeed can benefit from the prokaryotic simplicity and high copy number of plastid genomes, often enabling stronger and more stable transgene expression (Kwon et al., 2018). They represent an important extension of the engineering potential available in microalgae, which is still under development and will not be discussed in this chapter.

1.5.1 Genome Mining for Algal Genetic and Metabolic Engineering

As discussed, genome mining is an essential step in genetic engineering, metabolic engineering, and synthetic biology. Public, well-curated databases provide powerful tools to explore thousands of microalgal lineages, enabling identification of regulatory elements (promoters, terminators, and expression enhancers) as well as genes and pathways for targeted engineering. Rich annotations (including gene orthologs, gene-expression profiles, predicted structures, and biogeographies) help pinpoint the most promising genetic targets.

The framework outlined in Table 1.1 supports the systematic discovery and functional characterization of protein-coding genes in microalgae, particularly when working with complex environmental and meta-omics datasets. It integrates multiple, complementary activities: organism-specific portals provide curated genomic references; environmental and meta-"*omics*" catalogs expand the accessible genetic diversity; community protein integration and microalgae-tailored analysis tools enhance functional annotation and biological context; and structure-based discovery approaches employ three-dimensional modeling to reveal remote homologs and previously unrecognized motifs. Collectively, these components constitute a flexible strategy that can be applied independently or in combination, depending on the available data and research objectives.

By incorporating transcriptomic data, the framework can also identify highly expressed genes and their associated promoters or regulatory regions, enabling targeted exploration of transcriptional hotspots for the identification of safe harbors and promoter elements. Altogether, this integrated approach facilitates the discovery of both coding sequences and upstream regulatory regions, providing a comprehensive

TABLE 1.1
Conceptual Framework for Algal Gene and Protein Discovery from Environmental and Meta-Omics Data

Resources	Data Source/Tool Type	Purpose	Key Outcome
1. Organism-specific portals	Reference genomes, transcriptomes, and curated databases (e.g., PhycoCosm and AlgaeBase)	Establish baseline gene and protein annotations for target taxa	High-confidence reference gene sets
2. Environmental/meta-omics catalogs	Metagenomes, metatranscriptomes, and community sequencing repositories	Capture genetic diversity from mixed algal and microbial communities	Expanded pool of potential novel sequences
3. Community protein set integration	Clustered protein families and orthologous group builders (e.g., OrthoFinder and MMseqs2)	Identify conserved and divergent protein families across datasets	Functional and evolutionary context for candidate genes
4. Algae-tailored analysis tools	Specialized prediction tools (e.g., TargetP, PredAlgo, and KEGG pathway mappers)	Annotate algal-specific features and subcellular localizations	Refined functional annotation and localization insight
5. Structure-first discovery	3D modeling and structural homology tools (e.g., AlphaFold, Foldseek, and HHsearch)	Detect remote homologs and motifs undetectable by sequence similarity	Discovery of novel or highly divergent protein genes

The table outlines a modular framework integrating various data sources and computational tools for microalgal gene and protein discovery. Each activity can be applied independently or in combination, depending on available datasets and research objectives.

map of algal functional and regulatory innovation to support precise genetic and metabolic engineering.

Table 1.2 provides a comprehensive list of the principal public databases available for genome-mining approaches applicable to microalgae. These resources have the potential to transform publicly available data into a functional parts pipeline for microalgal synthetic biology.

Collectively, they represent valuable assets for a range of activities, including: (i) identifying lineage-matched promoters, terminators, and transcription factor motifs from high-quality genomes; (ii) selecting candidate genes with ecological and expression context across diverse geographical locations; (iii) cataloguing gene–protein functions across orthogroups; and (iv) bridging sequence-divergence gaps among microalgal lineages through structure-level homology.

1.5.2 Alternative Genome-Mining Sources

While defining genomic safe harbors and improving transformation strategies strengthen the reliability of nuclear engineering, the expansion of molecular tools

TABLE 1.2
Principal Public Databases Available for Genome-Mining Approaches Applicable to Microalgae

Category	Database	Main Features	Limitations	References
Organism-specific	PhycoCosm	>100 algal genomes with comparative tools and multi-omics tracks	Coverage and quality are uneven across clades; many entries remain draft/fragmented assemblies, and annotations differ by project	Grigoriev et al., 2020
	Phytozome	Complements PhycoCosm on the green-lineage side with robust orthology and upstream region pulls	Green-lineage focus only; orthology/promoters inferred within Viridiplantae	Goodstein et al., 2011
	JGI Plant Gene Atlas	Continuously updated RNA-seq compendium for models like Chlamydomonas to guide promoter/terminator choice	n/a	Sreedasyam et al., 2023
	Chlamydomonas Genome Project	Reference assemblies for mating-type plus and minus strains	Lab-strain idiosyncrasies: v6 reveals extensive structural mutations/TE insertions in common laboratory strains	Craig et al., 2022
	Chlamydomonas Resource Center	Strains, protocols, and other resources	n/a	https://www.chlamycollection.org
	PhaeoEpiView	Overlays DNA methylation/histone marks on an updated assembly to pinpoint expression-competent loci	n/a	Wu et al., 2023
	DiatOmicBase	Gene-centered, multi-omics portal built around diatom models, aggregates dispersed public datasets into one interface	Data available for certain species *Phaeodactylum tricornutum*, *Thalassiosira pseudonana*, and *Pseudo-nitzschia multistriata*	Villar et al., 2025

	Nandesyn	Genome-centric portal for *Nannochloropsis* spp.	n/a	Gong et al., 2020
	Ensembl Protists	Genome-centric portal for hundreds of non-vertebrate eukaryotes; browsable genes/transcripts/proteins with comparative genomics (gene trees and homologies)	Coverage is limited to the protist species Ensembl hosts; many microalgal/rare protist genomes aren't there	https://protists.ensembl.org/index.html
Algae-tailored analysis tools	ALGAEFUN with MARACAS	Streamlines RNA-seq/ChIP-seq through to locus- and GO/pathway-level enrichment specifically for microalgae	Species coverage is limited (a small set of microalgal models), and analysis is tuned to RNA-seq/ChIP-seq	Romero-Losada et al., 2022
	AlgaePath 2.0	Aggregates pathway/expression information for model microalgae	Database breadth and updates: focused set of species/datasets; pathway views rely on KEGG-style maps and may miss algal-specific rewiring	Zheng et al., 2014
	AlgaePrAs	Colocate multispecies algal omics for cross-querying	n/a	Kurotani et al., 2017
Meta-omics	Ocean Gene Atlas v2.0	Searches for candidate sequences and immediately map its abundance and phylogeny across Tara Oceans stations	Dataset-restricted: queries map to Tara Oceans catalogs (OM-RGC and MATOU); marine plankton only	Vernette et al., 2022
	MATOU (the Marine Atlas of Tara Oceans Unigenes)	Anchors eukaryotic gene catalogs with expression context—powerful for linking genes or regulatory proteins to niches (light, nutrients, and temperature)	Expression-only unigenes: lacks genomic context/synteny	Carradec et al., 2018
	MGnify	Assembled/metagenomic analyses, MAG catalogs, and uniform functional pipelines. Biome coverage or to reanalyze public datasets with consistent methods	Not algae-specific and pipeline heterogeneity: analysis versions change over time; study metadata quality varies	Richardson et al., 2022

(Continued)

TABLE 1.2 (Continued)
Principal Public Databases Available for Genome-Mining Approaches Applicable to Microalgae

Category	Database	Main Features	Limitations	References
Marker-driven	PR2 18S rRNA PhytoREF plastid 16S	Standards for placing environmental reads and tracing lineage distributions when the genome context is missing	18S (PR2) and plastid 16S (PhytoREF) reference sets are uneven across lineages; primer coverage differences alter perceived community structure	Guillou et al., 2012 Decelle et al., 2015
Community protein sets	EukProt	Compiles predicted proteomes across ~1,000+ diverse eukaryotes	Heterogeneous inputs and QC: proteomes originate from mixed assembly strategies; completeness/contamination varies	Richter et al., 2022
	Ensembl Protists	Protein features precomputed with InterProScan; BioMart for bulk retrieval; regular updates to species taxonomy and homology sets	Protein function annotations largely inherited (e.g., from InterPro/UniProt) rather than deeply curated per protein	https://protists.ensembl.org/index.html
	UniProt	The canonical protein knowledgebase: reviewed Swiss-Prot records with expert curation; unreviewed TrEMBL for breadth and rapid updates; Rich cross-references (GO, InterPro, KEGG, etc.), proteome views, BLAST/ID mapping, and predictable ~8-weekly releases	Vast majority of entries are unreviewed (TrEMBL) and carry automated annotations	Consortium, 2024
	InterPro	Integrates many "signature" databases (Pfam, PROSITE, SMART, etc.) to classify proteins into families/domains and assign GO terms	Signature-based predictions can yield false positives or coarse domain boundaries; GO term propagation reflects domain context, not experiment	Blum et al., 2024

	eggNOG	Hierarchical orthologous groups across thousands of taxa (incl. microbes and eukaryotes) with functional labels (GO/KEGG/COG); optimized for large-scale and metagenomic annotation (eggNOG-mapper)	Automated orthology can mix paralogs at broad levels; group granularity varies across clades (bias toward well-sampled microbes); functional transfers are inferred, not curated	Hernández-Plaza et al., 2022
	OrthoDB	Gene-centric, multi-kingdom orthology with evolutionary context; widely used for single-copy markers (e.g., BUSCO) and pairwise ortholog lists; web/API and bulk downloads	Species selection emphasizes genomes with better assemblies	Tegenfeldt et al., 2024
Structure databases	AlphaFold Database	>200 million predicted structures that can be searched directly	Predictions ≠ experiments: confidence varies (pLDDT/PAE); intrinsically disordered regions, flexible loops, cofactor/ligand states and many complexes are unreliable/absent	Abramson et al., 2024
	AlphaFold Server	Generate your own structures for proteins, which are not available in the main database, as well as multimers and docking with other biomolecules such as RNA and glycans		
	ESM Metagenomic Atlas	Protein dark matter: >600 million metagenomic protein structures	Minimal context: structures for metagenomic proteins often lack clear taxonomy, expression, or function; many predictions are low-confidence and unvalidated	Lin et al., 2023
	Foldseek	Enables ultrafast structure–structure searches to uncover remote homologs or shared active-site geometries; integrates all structure DBs in 1 search	Speed/false-positive trade-off: ultrafast structural hits can include remote analogs with little functional conservation	van Kempen et al., 2024

The summary of major public databases supporting genome-mining approaches in microalgae, highlighting core functionality, primary data types, and known limitations.

for microalgae remains a critical frontier. Many non-model species still lack fully resolved genomes or efficient expression systems, limiting genetic manipulation options. To address this gap, it is valuable to look beyond the host genome itself. Besides microalgae, viruses that naturally infect them are increasingly recognized as an underexplored source of molecular components with enormous potential in algal biotechnology and synthetic biology (D'Adamo et al., 2022). These viruses have evolved mechanisms to efficiently enter host cells and commandeer transcriptional and translational machinery, providing a natural repertoire of promoters, regulatory elements, and expression strategies. Harnessing such viral elements could therefore accelerate the development of versatile, easy-to-establish genetic tools for diverse microalgal systems, including those where classical approaches remain challenging.

The methods and framework described in the previous subsection can also be applied in the mining of viral genomes, and in the last decades, efforts to sample and sequence global aquatic environments have produce massive amounts of data on marine viromes, from which thousands of novel viral draft (meta)genomes and (meta)transcriptomes could be derived and are now publicly available, releasing researchers of the difficult task of isolating and characterizing viruses of interest. Some examples include Global Ocean Viruses (GOV and GOV2), Tara Ocean Virome, Global Ocean RNA Viruses, Pacific Ocean Virome, and Metagenomics of Viral Metagenome (Carradec et al., 2018; Endo et al., 2020; Gregory et al., 2019).

While mainstream web-based protein structure databases do not yet include viral proteins, open-source research tools such as BFVD attempt to make available large numbers of curated viral protein structures currently lacking in the larger repositories (Kim, Levy Karin, et al., 2024). Alternatively, AlphaFold Server and ESM Metagenomic Atlas do allow for in-demand folding of protein sequences of interest, which can be searched against metagenomic databases (ESM) or integrated protein structure databases (Foldseek). A current drawback is that generating and searching large numbers of structures in parallel is not easily automated.

1.5.3 Bridging *In Silico* Discovery and Experimental Validation: The Role of Modular Cloning in Microalgal Synthetic Biology Research

Building on the integrated framework described above, where *in silico* analyses, environmental data mining, and even viral component discovery converge to expand the molecular toolbox, one major challenge becomes apparent. Identifying candidate genes, promoters, or regulatory motifs computationally is now routine, yet translating these discoveries into experimentally testable constructs remains slow and labor-intensive. Because experimental validation is essential to confirm function and compatibility, there is a growing need for efficient cloning and assembly systems that can keep up with the rate of discovery. For this, modular cloning has been introduced, and it is a powerful tool that can be employed in most cases described.

Golden Gate–based systems such as the MoClo, GoldenBraid, and Loop families have transformed the way genetic constructs are built (Crozet et al., 2018; Melero-Cobo et al., 2025; Pollak et al., 2020). These modular frameworks make it possible to "domesticate" individual genetic parts and regulatory elements (coding sequences, leader sequences for translocation, promoters, untranslated regions (UTRs), introns, enhancers, and terminators) into standardized, interchangeable parts. Once in this format, hundreds of combinations of transcriptional units [promoter–coding sequence (CDS)–terminator] or even complete multigene pathways can be assembled hierarchically within a single day. Table 1.3 lists the more advanced modular cloning toolkits currently available for eukaryotic microalgae.

Moreover, the modular frameworks make cloning workflows compatible with low-cost automation and high-throughput assembly for library-scale testing. This speed and flexibility make modular cloning an ideal approach for rapid *in vivo* testing and fine-tuning of genetic designs. This testing can be coupled to high-throughput screening devices for the rapid expression profile analyses and phenotyping, as described in previous subsections.

TABLE 1.3
Most Common Modular Cloning Toolkits for Eukaryotic Microalgae

System	Assembly Logic	Primary Target Organism(s)	Description	Reference
C. reinhardtii MoClo	Hierarchical Golden Gate (BsaI/BpiI)	*Chlamydomonas reinhardtii* (Nuclear)	Large, validated library of parts (119+); standardized syntax; open access; enables rapid, combinatorial assembly of multigene constructs.	Crozet et al., 2018
MoCloro	Golden Gate (BsaI)	*Chlamydomonas reinhardtii* (Chloroplast)	Extends MoClo syntax to the plastome; provides vectors with homologous arms for targeted integration; standardized chloroplast-specific parts.	Melero-Cobo et al., 2025
CyanoGate	Hierarchical Golden Gate (BsaI/BpiI)	Cyanobacteria (e.g., *Synechocystis*)	Adapts the common plant MoClo syntax for prokaryotes; enables engineering of marked and unmarked mutants; interoperable with plant/algal parts.	Vasudevan et al., 2019
Loop/uLoop Assembly	Recursive Golden Gate (alternating BsaI/SapI)/with modular host elements	Originally devised for plants. Universal (demonstrated in diatoms, yeast, and bacteria)	Theoretically unlimited construct size (1, 4, 16+ genes) with a minimal set of eight vectors; highly efficient and versatile. In uLoop, assembly logic is decoupled from host-specific elements, enabling cross-kingdom construct design.	Pollak et al., 2020

Taken together, the modular cloning and part standardization also help to speed up the design–build–test–learn framework, which proposes to advance synthetic biology by enabling rapid, iterative optimization of engineered biological systems.

Specifically, the adoption of the modular cloning directly tackles and enables the Build phase of this strain engineering cycle, going from mined gene candidates to combinatorial assemblies (dozens–hundreds of constructs), screening them *in vivo*, and quickly converging on promoter/UTR/terminator sets or multienzyme layouts that actually perform in microalgal chassis. It becomes a less time-consuming task and permits faster expansion of organism-specific molecular toolboxes that are currently lacking in microalgal research. Similarly, the Design phase is no longer a bottleneck, thanks to computational and systems biology, as well as the unrelenting efforts of many labs broadening the base knowledge and developing genetic tools for different microalgal species.

1.5.4 Overcoming Limitations on Transgene Expression

Currently, nuclear transformation is the gold standard for metabolic engineering, as it enables broad manipulation of cellular functions; however, it often leads to position effects, variable expression, and epigenetic silencing. In fact, the level of gene expression can be profoundly influenced by the local chromatin environment of the integration site: if it is integrated into a region of open transcriptionally active euchromatin, it can result in robust expression (Einhaus et al., 2025). Whereas if it is integrated in tightly packed, transcriptionally silent heterochromatin, expression might be absent. Additionally, eukaryotic microalgae possess sophisticated mechanisms to defend from foreign genetic elements like viruses and transposons, most frequently involving RNA interference pathways (Kim et al., 2015). The combined effects of random integration and gene silencing have made it difficult to rationally design a construct and predict its expression level.

Accordingly, recent advances increasingly rely on endogenous promoter discovery via genome mining, as discussed in the previous section, as well as on high-throughput promoter dissection and synthetic promoter engineering. Historically, many efforts have depended on heterologous promoter elements imported from land plants (or the use of "universal" viral promoters). Yet in many cases, these elements suffer low basal output, high variability among transformants, or rapid silencing. For example, in *Chlamydomonas reinhardtii*, viral or plant promoters such as CaMV35S or SV40 yield poor performance.

Recent works have demonstrated the impact of tuning promoter variants to the regulatory context of a given microalgal host. Native promoter systems (e.g., PSAD, RBCS2), or chimeric fusions like HSP70A-RBCS2 (AR), provide better expression but are still limited: AR reduces transcriptional silencing and leverages heat shock elements, TATA box architecture, etc., yet expression levels vary with chromatin state, promoter spacing, and regulatory motif strength (Baier et al., 2020; Einhaus et al., 2024; Einhaus et al., 2021; Schroda et al., 2000).

One recent study reported that a rationally designed promoter significantly enhanced the heterologous production of a plant secondary metabolite in *Chlamydomonas reinhardtii*, yielding roughly an 18-fold increase of product accumulation, compared to when using the native promoter (Einhaus et al., 2021). Systematic promoter dissection approaches are also being applied to elucidate how features like motif architecture, UTRs, intronic elements, and chromatin accessibility contribute to high, stable expression (Einhaus et al., 2024).

Until now, however, researchers have had to deal with a brute-force screening approach, generating and individually assessing several transformant colonies to identify the few that exhibited both high and stable expression. To overcome these limitations, researchers have begun defining *safe harbour* loci, namely, genomic regions that permit consistent, stable expression of inserted genes without disrupting essential functions (Jacobebbinghaus et al., 2025; Sudfeld et al., 2022).

But also, design choices like using codon harmonization algorithms (Willems et al., 2023), instead of simple codon optimization to overcome codon usage bias, as it has improved the reliable heterologous expression of proteins (Wright et al., 2022). Other approaches involve engineering strains to reduce the silencing mechanism, which in turn results in more stable mutant lines (Einhaus et al., 2025).

Finally, researchers are also exploiting episomes for expression (Karas et al., 2015; Sudfeld et al., 2021). These are extrachromosomal DNA molecules, usually circular, that can replicate and be maintained independently of the host chromosomes, just like plasmids in bacteria (Karas et al., 2015). With such a strategy, one could circumvent position effects, achieving a high and constant copy number resulting in a stable and predictable expression level. While episomal expression is advantageous during the screening phase, genome integration is still regarded as the preferred strategy for developing highly stable industrial strains (Diamond et al., 2023; Trevisan et al., 2025). The combination of all these advances is enabling microalgae as a platform for testing complex, synthetic, genetic circuits and more advanced metabolic engineering approaches. Thus, transforming microalgal nuclear engineering from a stochastic process into a predictable, modular platform for synthetic biology.

Moreover, the rapid advances in molecular biology and modular cloning have also enabled the deployment of increasingly efficient tools for microalgal genome editing and genome engineering techniques. Early innovations were driven by the use of customizable nucleases such as zinc-finger nuclease and TAL effector endonuclease (TALEN), while, more recently, the clustered regularly interspaced palindromic sequences (CRISPR)/Cas9 and Cas12a have expanded the repertoire of possible interventions (Jeon et al., 2017; Jeong et al., 2023; Kroth et al., 2018). In fact, CRISPR–Cas systems now enable targeted gene integration, precise gene deletion, gene interference, and even promoter activation, although the latter capability remains under active development in microalgal research (Tran et al., 2025; Zhang et al., 2019). These tools were initially established in the model species *Chlamydomonas reinhardtii* and *Phaeodactylum tricornutum* and have since been extended to emerging hosts of interest such as *Nannochloropsis* species, particularly *N. gaditana* and *N. oceanica*. The CRISPR/Cas9 and CRISPR/Cas12a have been successfully applied in *Nannochloropsis* species, demonstrating high editing efficiency and multiplexing capability, and making such genome editing toolkits available for the scientific community (Kroth et al., 2018; Naduthodi et al., 2018; Naduthodi et al., 2019; Naduthodi et al., 2021; Poliner, Pulman, et al., 2018; Poliner,

Takeuchi, et al., 2018). These advancements well position *Nannochloropsis* as an emerging model for metabolic engineering and biotechnological applications.

Finally, it is worth mentioning the development and refinement of more powerful GEMs, which allow engineers to simulate the metabolic flux of the entire cell *in silico* (Gudmundsson et al., 2017; Tibocha-Bonilla et al., 2018). Together with improvements in gene delivery and expression, the field has basically transitioned from random insertion to predictable control and developed versatile toolboxes for engineering (Figure 1.2).

1.6 GENETIC TRANSFORMATION WORKFLOW IN MICROALGAE

Genetic engineering in microalgae has matured from isolated proof-of-concept experiments to a reproducible, design-driven discipline that increasingly mirrors workflows in bacteria and yeast. The process generally follows a well-defined pipeline: (i) construct design and assembly in *Escherichia coli*, (ii) DNA delivery into the microalgal host using methods such as particle bombardment, electroporation, glass beads, or bacterial conjugation, (iii) genomic integration or episomal maintenance of the transgene, and (iv) selection and screening for stable transformants.

The genetic engineering of microalgae is very similar to other microbial systems. It implies that foreign DNA is first designed and constructed (this step is performed in the bacterium *Escherichia coli*) and then delivered inside the microalgal cell. The cell integrates the DNA into the genome, and later the transformed cells are identified through selection methods. For DNA delivery, there are several established methods, which vary from species to species. The most commonly used ones are particle bombardment and electroporation. This is due to the fact that microalgae possess a quite strong (and diverse) cell wall that is hard to penetrate (Mosey et al., 2021). A cell-wall-defective mutant has been created for *C. reinhardtii*, which resulted in the development of a simple glass bead method for DNA delivery (Economou et al., 2014; Gutiérrez & Lauersen, 2021). More recently and for certain species only, bacterial conjugation methods have been established. This concerns the incubation of microalgae cells with bacterial cells (typically *E. coli*) carrying a specific conjugative plasmid that transfers the plasmid into the microalgae cells through the formation of a bacterial small channel (*pilus*) (Mosey et al., 2021).

The most commonly used and successful selection method in microalgae research is the use of antibiotics. Microalgae can be naturally resistant to antibiotics, and this varies from species to species; therefore, the right antibiotic needs to be selected. As such, the foreign DNA always contains an expression cassette for an antibiotic resistance gene. Once the foreign DNA is delivered inside the cells, the cells are plated on solid agar medium containing the antibiotic, and only transformant lines that integrated the antibiotic resistance cassette will be able to grow. This process takes roughly 1 month. Figure 1.3 summarizes the overall pipeline from genetic part assembly to transformation and screening. This pipeline effectively embodies the Design, Build, Test, and Learn (DBTL) framework, a key component of synthetic biology, which allows for faster engineering.

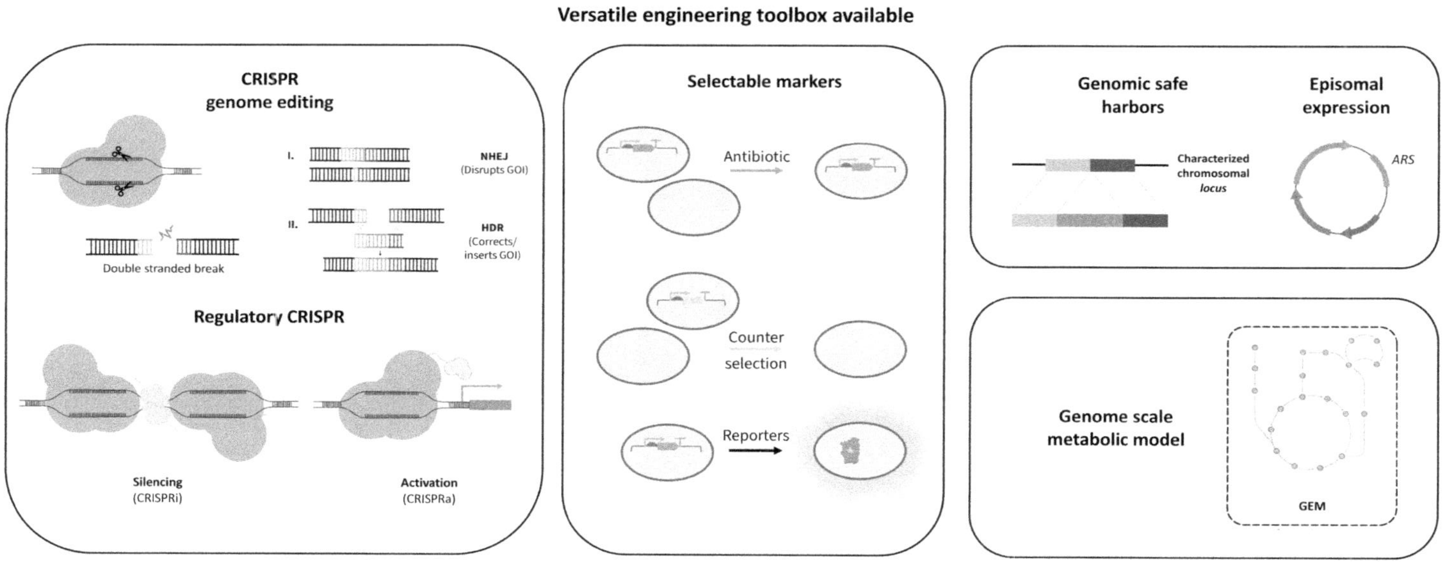

FIGURE 1.2 A versatile genetic toolbox is required to overcome limitations in transgene expression. Ideally, apart from characterized genetic parts (promoters, UTRs, terminators…), such a toolbox should comprise genome editing, gene upregulation/downregulation systems, as well as a wide set of selective markers. Additionally, the characterization of *safe harbours* or the utilization of episomal expression enables a predictable and stable expression of transgenes. More recently, genome-scale metabolic models (GEMs) have also become a powerful tool to perform metabolic simulations and improve predictions.

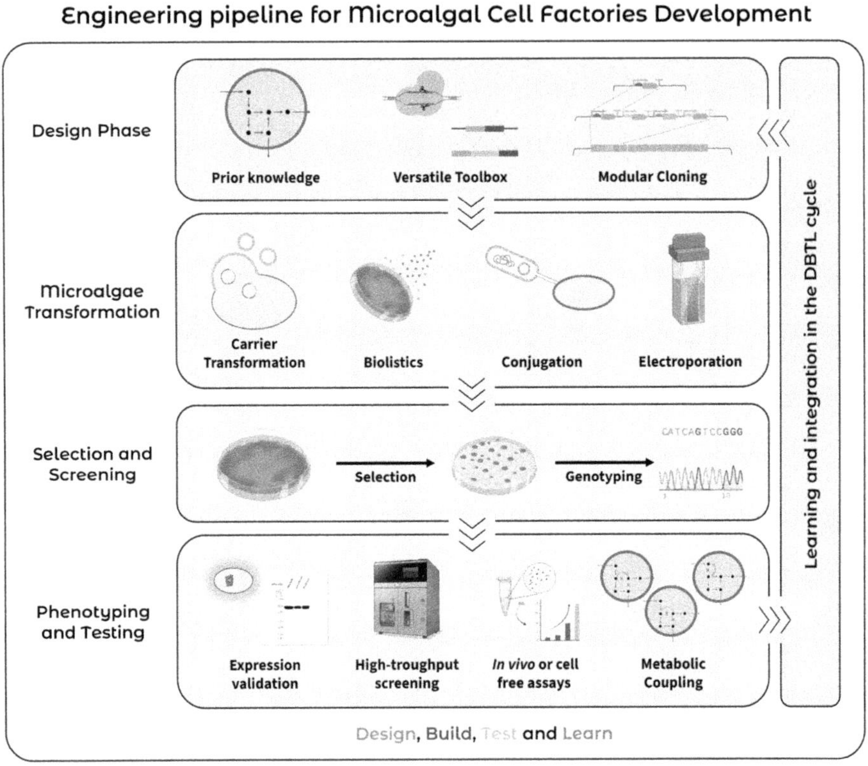

FIGURE 1.3 Overview of the experimental pipeline for the genetic modification of microalgae and the integration into the Design. Build, Test, and Learn (DBTL) cycle for the development of microalgal cell factories.

A typical microalgae genetic modification will start with the construction and amplification of the gene-expression cassette, using *E. coli*. Afterward, DNA will be transferred into the microalgae cells through different methods (such as biolistic, electroporation, and conjugation). Transformed microalgae will be plated on solid medium under antibiotic selection, and transformant colonies will start growing after approximately 2–4 weeks. Afterward, single transformant colonies will be picked and subjected to genetic and physiological screenings.

Once the transformant lines are selected from plates, they are grown in liquid medium to start the genotypic and phenotypic characterization. For nuclear transformation, when the gene integration is random, it is important to screen a relatively large number of lines (on average of 10–30 independent lines), to check for cell fitness, genotype, and to perform phenotype screening. For chloroplast transformation, it is important to transfer the transformant lines to a new plate containing fresh antibiotic for a certain number of generations to reach homoplasmy.

Genotyping is usually performed to check whether the integration event was successful. It is performed by extracting DNA from the cell (by simply boiling the cell for 5 minutes) and then performing a polymerase chain reaction (PCR), using specific primers for amplifying the foreign DNA. It is also possible to determine the location of the insertion in the genome. This method is often called *genome walking*, and it consists of DNA extraction, fragmentation (by an endonuclease), and a set of PCRs using specific and universal primers that amplify (and therefore identify) part of the region flanking the integration site (Gonzalez-Ballester et al., 2011).

Phenotyping really depends on the scope of the engineering, and it often consists in growing the transformant lines in liquid medium and in performing specific analyses, such as determining the cell growth rate, presence of the protein encoded by the foreign DNA, amount of target product, or detecting the expression level of the foreign gene.

Phenotyping can also be carried out in a high-throughput mode, as discussed in Section 1.4, particularly for screening and identifying high gene-expression candidates.

As discussed in the previous sections, the advances in sequencing, bioinformatics, and synthetic biology now enable rational design of promoters, terminators, and coding sequences suited to specific algal species.

1.7 PRACTICAL SUCCESSES IN MICROALGAL BIOTECHNOLOGY: CASE STUDIES IN HIGH-VALUE COMPOUND PRODUCTION

Recent studies and reviews have highlighted several compelling cases of engineering microalgae for high-value products, including plant secondary metabolites such as terpenoids, nutraceuticals, specialty lipids, including novel fatty acid synthesis, pigments, and therapeutic compounds (Cao et al., 2023; Shitanaka et al., 2025). The emphasis on high-value outputs reflects both practical and strategic considerations: genetically modified microalgae require significant investment in design, construction, and regulatory compliance, which is more economically justifiable for high-margin products than for low-value bulk chemicals. By focusing on these targets, researchers are aiming to optimize metabolic pathways, validate engineering strategies, and demonstrate commercial feasibility, providing a scalable blueprint for future applications in modern biotechnology.

For example, enhancing lipid production and specialty lipid engineering, which dominated microalgal research during the first decade of the 2000s, has been extensively reviewed elsewhere; readers are referred to recent comprehensive reviews for detailed insights into the production of omega-3 fatty acids and other high-value lipids (Munoz et al., 2021; Shitanaka et al., 2025). Similarly, the heterologous production of plant terpenoids has attracted scientific interest and has been covered in existing reviews (Costa et al., 2022; Shitanaka et al., 2025; Vavitsas et al., 2018).

While many developments exist and will continue, this chapter considers only the two most relevant examples to illustrate the principles and potential of microalgal engineering. Accordingly, this chapter focuses on only two representative examples: Protein therapeutics and heterologous production of astaxanthin, selected for their commercial relevance and well-documented success in microalgal engineering.

1.7.1 Microalgae-Derived Therapeutic Antigen Production Targeting Human and Animal Diseases

Antigens are a core component of subunit vaccines against human diseases. They are usually produced as recombinant proteins in cell factories using bacteria, yeast, insect cells, and mammalian cells. In the last decades, microalgae have emerged as promising and sustainable alternatives to recombinant antigen production for human and animal vaccines (Barolo et al., 2020). This is achieved by genetically engineering microalgae to express a specific protein from a pathogen, in a process often referred to as "*molecular pharming*". Opposite to commonly used platforms, microalgae are inherently safe, since they don't host human pathogens, are less costly to cultivate, especially if grown photosynthetically, and can grow more rapidly than mammalian cells (Rosales-Mendoza et al., 2020). Therapeutic antigens are produced in microalgae by delivery of a DNA construct containing the sequence of interest, which is then expressed in the host.

Chloroplast transformation can be more advantageous for antigens that do not require glycosylation, as it leads to higher yields compared to nuclear transformation (yield up to 0.02–2% of total soluble protein range (~0.03–3 mg/L of culture medium)) (Rosales-Mendoza et al., 2020). Therapeutic antigen synthesis from microalgae has been successful in the last decades, primarily using chloroplast or nuclear integration, but more recently, transient expression has been explored as well. In many cases, the therapeutic action of recombinant microalgae-made antigens was tested *in vivo* in animal models. Demurtas et al. demonstrated that a recombinant antigen (E7) for the human papillomavirus 16 (HPV16) expressed in the chloroplast of *C. reinhardtii*, injected either as a purified fraction or total cell extract in mice challenged with a tumor cell line expressing the epitope E7, elicited tumor protection in 60% of the tested population (Demurtas et al., 2013). Another promising result is described by Gregory et al. for a treatment against malaria. The authors engineered *C. reinhardtii* to synthesize two antigens (Pfs25 and Pfs28) of *Plasmodium falciparum*, the single-cell parasite responsible for malaria in humans. The genetic construct for the *P. falciparum* antigens was delivered and integrated into the chloroplast, since these antigens don't require any glycosylation. The efficacy of Pfs25 and Pfs28 was tested by feeding mosquitoes a blood meal containing the antigen-producing microalgae that resulted in the disruption of sexual development of the parasites in the mosquito midgut, which prevents human-to-human transmission of malaria (Gregory et al., 2012). A recent study utilized *C. reinhardtii* as a heterologous expression platform to produce a subunit vaccine against Infectious Spleen and Kidney Necrosis Virus (ISKNV). To maximize recombinant protein yield, the ISKNV Major Capsid Protein was expressed in the chloroplast. Immunization of Nile tilapia (*Oreochromis niloticus*) with the soluble protein fraction induced specific anti-ISKNV antibodies and partial *in vitro* neutralization, evidenced by a reduction in cytopathic effects in fish cell cultures.

A particularly interesting feature of using microalgae for therapeutic antigen synthesis relies on the possibility of using them as oral delivery vehicles (Criscuolo et al., 2019). Microalgae have a cell wall that can protect the antigen from the harsh acidic environment of the stomach, enabling the assimilation of the target antigen in the intestinal tract (Kiran & Venkata Mohan, 2021). In addition to this, the European Food Safety Authority and Food and Drug Administration have granted the Generally Recognized as Safe status to several microalgae, making them suitable for human consumption (Ferreira de Oliveira & Bragotto, 2022). The effectiveness of oral delivery of microalgae-synthesized recombinant antigens has been tested in mice with great results. For example, Dreesen et al. engineered *C. reinhardtii* to create the recombinant epitope D2 from *Staphylococcus aureus*. They demonstrated the system's efficacy by feeding the engineered cells to mice for five weeks. They observed that the transgenic microalgae induced specific mucosal and systemic immune responses, significantly reduced the pathogen load in the spleen and intestine of treated mice, and protected 80% of them against a lethal dose of *S. aureus* (Dreesen et al., 2010). Despite this promising result, the strategy of administration (oral delivery and absorption through the intestinal tract) should be carefully considered according to the immune response needed and the biochemical properties of the therapeutic molecule to reach maximum bioavailability (Criscuolo et al., 2019). In another study, authors developed a transgenic *Nannochloropsis* sp. expressing the *OmpK* antigen from *Vibrio harveyi* and evaluated its potential as an oral vaccine against vibriosis in zebrafish. The recombinant microalgae were bioencapsulated in the brine shrimp *Artemia* before delivery to fish. Significant upregulation of immune-related genes (*IgZ, TNF-α, IL-1β*) provided 100% relative percentage survival following bacterial challenge. These results demonstrate that microalgae–shrimp-mediated oral vaccination can effectively enhance fish immunity, offering a sustainable strategy for controlling vibriosis in aquaculture (Ismail et al., 2025). Overall, efforts toward improving microalgae protein production and developing edible vaccines could be extremely beneficial for global immunization efforts, especially in remote or at-risk areas, where access to medical infrastructure can be extremely challenging.

1.7.2 Metabolic Engineering Strategies for Efficient Astaxanthin Synthesis

Astaxanthin is a vibrant red carotenoid naturally produced in plants and some microalgae and has captured significant attention mainly due to its strong antioxidant properties. Astaxanthin finds its biggest market in animal feed, as a natural enhancer of fish coloring and animal health, but in the last years, the interest in this pigment as a nutraceutical has increased due to its beneficial properties for human health, having strong antioxidant and anti-inflammatory properties that prove to be beneficial for skin, eye, cognitive, and cardiovascular health (Debnath et al., 2024; Lu et al., 2021).

Astaxanthin has recently gained attention in the cosmetic sector, with demand currently estimated at over US$550 million and projected to reach US$3.4 billion by 2027 (Debnath et al., 2024). *Haematococcus pluvialis* (recently reclassified as *H. lacustris*) is a freshwater green alga and the richest natural source of astaxanthin, capable of accumulating 3–5% of its dry weight under stress conditions, when it shifts from a vegetative state to a resting cyst form (Fang et al., 2019). Despite this, commercial production is limited by the difficulty of extracting the pigment from its tough cyst walls, making chemical synthesis economically more competitive (Li et al., 2011). These challenges have driven genetic and metabolic engineering efforts to enhance astaxanthin yields in *H. pluvialis* and to enable production in alternative microalgal hosts.

In general, a common strategy to boost astaxanthin production in microalgae is the optimization of the carotenoid pathway by overexpression of the key enzymes. Astaxanthin synthesis is limited by inefficient enzymes or insufficient precursor availability. Specifically, the conversion of beta-carotene into canthaxanthin and zeaxanthin by the β-carotene ketolase (BKT) and the hydroxylation of these pigments into astaxanthin by the β-carotene hydroxylase (CHYB) have been identified as bottlenecks for astaxanthin production (Amendola et al., 2023; Kathiresan et al., 2015). Increasing the pool of source metabolites has been proven to raise the final astaxanthin concentration in the cell.

By introducing the BKT gene into cells that do not naturally accumulate astaxanthin, such as *C. reinhardtii*, *N. oceanica* can be modified to do so (Amendola et al., 2023; Canini et al., 2024; Liu et al., 2024; Perozeni et al., 2020). As observed in native producers, overexpression of bottleneck genes along the carotenoid biosynthetic pathway increases astaxanthin accumulation. In *C. reinhardtii*, introduction of BKT and overexpression of CHYB and PSY lead to a yield of 4.5 mg of astaxanthin per gram of dry cell weight (Amendola et al., 2023). Similarly, the introduction of BKT and PSY genes from *H. pluvialis* in a *C. reinhrdtii* strain resulted in a 5.39-fold increase in β-carotene concentration compared to wild-type and astaxanthin production up to 1.47 ± 0.063 mg/g DCW (Sharma et al., 2024).

Recently, the removal of the competitive pathway to channel the precursor into astaxanthin and increase the yield has been explored. The ε-lycopene cyclase gene (LCYE) was silenced in *C. reinhardtii* via CRISPR/Cas9 technology, disrupting the conversion of lycopene to α-carotene. Reconstruction of the astaxanthin pathway as described above by introduction of BKT, CHYB, and PSY was done in the LCYE-deficient strain, obtaining a volumetric yield of 2.44 mg/L after 72 h of mixotrophic cultivation (Kneip et al., 2024). Codon optimization, synthetic redesign for high transgene expression, and enzyme-engineering have been explored to achieve strong gene expression and efficient enzyme activity in all the cited studies.

Recent studies have focused on various strategies to augment astaxanthin and other ketocarotenoid biosynthesis in *Nannochloropsis* (Canini et al., 2024; Liu et al., 2024). For instance, iterative metabolic engineering efforts have led to a remarkable 133-fold increase in astaxanthin content in *N. oceanica*, achieving titers up to 7.3 mg/g of biomass under nonstress conditions. This enhancement was accomplished by introducing both the β-carotene ketolase (BKT) and β-carotene hydroxylase (CHY) genes, optimizing their compartmentalization and fusion patterns, and employing a fed-batch cultivation approach (Liu et al., 2024). Overall, metabolic and genetic engineering have been employed successfully to increase astaxanthin production. In addition to genetic engineering, chemical mutagenesis has also been employed to develop *N. gaditana* mutant strains, such as the S4 strain, which exhibit increased carotenoid-to-chlorophyll ratios and enhanced ketocarotenoid accumulation, including astaxanthin. These mutants demonstrated improved photosynthetic activity, increased lipid productivity, and the ability to produce up to 1.3 mg/L/day of ketocarotenoids without the need for stressful conditions (Cecchin et al., 2022). Taken together, these studies highlight how metabolic and genetic engineering strategies, ranging from pathway optimization and enzyme overexpression to CRISPR-mediated gene editing and mutagenesis, can substantially enhance astaxanthin production in microalgae, offering promising avenues for sustainable, high-yield biotechnological production of this valuable carotenoid.

1.8 CONCLUSIONS

Microalgae offer unparalleled potential as versatile and sustainable platforms for modern biotechnology, bridging the gap between natural biodiversity and engineered cellular factories. This chapter has highlighted the critical importance of strain bioprospecting, from isolation and high-throughput phenotyping to genomic and molecular characterization, as a foundation for successful industrial domestication. By integrating genome mining, computational design, and modular cloning strategies, researchers can accelerate the design–build–test–learn cycle, enabling targeted genetic modifications and the production of high-value metabolites. Case studies in metabolic engineering and synthetic biology demonstrate that microalgae can be transformed into efficient cell factories for compounds ranging from carotenoids to bioactive lipids and finally to antigen and biopharmaceutical production. This clearly illustrates both the scientific and commercial promise of these organisms. Moving forward, continued advances in high-throughput screening, gene editing, and systems biology will further unlock microalgal potential, establishing them as key players in sustainable biotechnology and expanding their role across diverse industrial applications.

REFERENCES

Abramson, J., Adler, J., Dunger, J., Evans, R., Green, T., Pritzel, A., Ronneberger, O., Willmore, L., Ballard, A. J., Bambrick, J., Bodenstein, S. W., Evans, D. A., Hung, C.-C., O'Neill, M., Reiman, D., Tunyasuvunakool, K., Wu, Z., Žemgulytė, A., Arvaniti, E., … Jumper, J. M. (2024). Accurate structure prediction of biomolecular interactions with AlphaFold 3. *Nature*, *630*(8016), 493–500. https://doi.org/10.1038/s41586-024-07487-w

Ahmmed, M. K., Ahmmed, F., Tian, H. S., Carne, A., & Bekhit, A. E. (2020). Marine omega-3 (n-3) phospholipids: A comprehensive review of their properties, sources, bioavailability, and relation to brain health. *Comprehensive Reviews in Food Science and Food Safety*, *19*(1), 64–123. https://doi.org/10.1111/1541-4337.12510

Ambati, R. R., Gogisetty, D., Aswathanarayana, R. G., Ravi, S., Bikkina, P. N., Bo, L., & Yuepeng, S. (2019). Industrial potential of carotenoid pigments from microalgae: Current trends and future prospects. *Critical Reviews in Food Science and Nutrition*, *59*(12), 1880–1902. https://doi.org/10.1080/10408398.2018.1432561

Amendola, S., Kneip, J. S., Meyer, F., Perozeni, F., Cazzaniga, S., Lauersen, K. J., Ballottari, M., & Baier, T. (2023). Metabolic engineering for efficient ketocarotenoid accumulation in the green microalga Chlamydomonas reinhardtii. *ACS Synthetic Biology*, *12*(3), 820–831. https://doi.org/10.1021/acssynbio.2c00616

Andersen, R. A. (1992). Diversity of eukaryotic algae. *Biodiversity and Conservation*, *1*(4), 267–292. https://doi.org/10.1007/bf00693765

Baier, T., Jacobebbinghaus, N., Einhaus, A., Lauersen, K. J., & Kruse, O. (2020). Introns mediate post-transcriptional enhancement of nuclear gene expression in the green microalga Chlamydomonas reinhardtii. *PLoS Genetics*, *16*(7), e1008944. https://doi.org/10.1371/journal.pgen.1008944

Baier, T., Wichmann, J., Kruse, O., & Lauersen, K. J. (2018). Intron-containing algal transgenes mediate efficient recombinant gene expression in the green microalga Chlamydomonas reinhardtii. *Nucleic Acids Research*, *46*(13), 6909–6919. https://doi.org/10.1093/nar/gky532

Barbosa, M. J., Janssen, M., Sudfeld, C., D'Adamo, S., & Wijffels, R. H. (2023). Hypes, hopes, and the way forward for microalgal biotechnology. *Trends in Biotechnology*, *41*(3), 452–471. https://doi.org/10.1016/j.tibtech.2022.12.017

Barkia, I., Saari, N., & Manning, S. R. (2019). Microalgae for high-value products towards human health and nutrition. *Marine Drugs*, *17*(5). https://doi.org/10.3390/md17050304

Barolo, L., Abbriano, R. M., Commault, A. S., George, J., Kahlke, T., Fabris, M., Padula, M. P., Lopez, A., Ralph, P. J., & Pernice, M. (2020). Perspectives for glyco-engineering of recombinant biopharmaceuticals from microalgae. *Cells*, *9*(3). https://doi.org/10.3390/cells9030633

Behrendt, L., Salek, M. M., Trampe, E. L., Fernandez, V. I., Lee, K. S., Kühl, M., & Stocker, R. (2020). PhenoChip: A single-cell phenomic platform for high-throughput photophysiological analyses of microalgae. *Science Advances*, *6*(36), eabb2754. https://doi.org/10.1126/sciadv.abb2754

Best, R. J., Lyczakowski, J. J., Abalde-Cela, S., Yu, Z., Abell, C., & Smith, A. G. (2016). Label-free analysis and sorting of microalgae and cyanobacteria in microdroplets by intrinsic chlorophyll fluorescence for the identification of fast growing strains. *Analytical Chemistry*, *88*(21), 10445–10451. https://doi.org/10.1021/acs.analchem.6b02364

Blanken, W., Postma, P. R., de Winter, L., Wijffels, R. H., & Janssen, M. (2016). Predicting microalgae growth. *Algal Research*, *14*, 28–38. https://doi.org/10.1016/j.algal.2015.12.020

Bleisch, R., Freitag, L., Ihadjadene, Y., Sprenger, U., Steingröwer, J., Walther, T., & Krujatz, F. (2022). Strain development in microalgal biotechnology—Random mutagenesis techniques. *Life*, *12*(7), 961. https://www.mdpi.com/2075-1729/12/7/961

Blum, M., Andreeva, A., Florentino, Laise C., Chuguransky, Sara R., Grego, T., Hobbs, E., Pinto, Beatriz L., Orr, A., Paysan-Lafosse, T., Ponamareva, I., Salazar, Gustavo A., Bordin, N., Bork, P., Bridge, A., Colwell, L., Gough, J., Haft, Daniel H., Letunic, I., Llinares-López, F., … Bateman, A. (2024). InterPro: The protein sequence classification resource in 2025. *Nucleic Acids Research*, *53*(D1), D444–D456. https://doi.org/10.1093/nar/gkae1082

Borowitzka, M. A. (2013). High-value products from microalgae—Their development and commercialisation. *Journal of Applied Phycology*, *25*(3), 743–756. https://doi.org/10.1007/s10811-013-9983-9

Borowitzka, M. A. (2018). Biology of microalgae. In *Microalgae in Health and Disease Prevention* (pp. 23–72). Academic Press. https://doi.org/10.1016/b978-0-12-811405-6.00003-7

Bowsher, C. G., & Tobin, A. K. (2001). Compartmentation of metabolism within mitochondria and plastids. *Journal of Experimental Botany*, *52*(356), 513–527. https://doi.org/10.1093/jexbot/52.356.513

Butler, T., Kapoore, R. V., & Vaidyanathan, S. (2020). Phaeodactylum tricornutum: A diatom cell factory. *Trends in Biotechnology*, *38*(6), 606–622. https://doi.org/10.1016/j.tibtech.2019.12.023

Canini, D., Martini, F., Cazzaniga, S., Miotti, T., Pacenza, B., D'Adamo, S., & Ballottari, M. (2024). Genetic engineering of Nannochloropsis oceanica to produce canthaxanthin and ketocarotenoids. *Microbial Cell Factories*, *23*(1), 322. https://doi.org/10.1186/s12934-024-02599-4

Cao, K., Cui, Y., Sun, F., Zhang, H., Fan, J., Ge, B., Cao, Y., Wang, X., Zhu, X., Wei, Z., Yao, Q., Ma, J., Wang, Y., Meng, C., & Gao, Z. (2023). Metabolic engineering and synthetic biology strategies for producing high-value natural pigments in microalgae. *Biotechnology Advances*, *68*, 108236. https://doi.org/10.1016/j.biotechadv.2023.108236

Cardozo, K. H. M., Guaratini, T., Barros, M. P., Falcao, V. R., Tonon, A. P., Lopes, N. P., Campos, S., Torres, M. A., Souza, A. O., Colepicolo, P., & Pinto, E. (2007). Metabolites from algae with economical impact. *Comparative Biochemistry and Physiology, Part C: Toxicology & Pharmacology*, *146*(1–2), 60–78. https://doi.org/10.1016/j.cbpc.2006.05.007

Carradec, Q., Pelletier, E., Da Silva, C., Alberti, A., Seeleuthner, Y., Blanc-Mathieu, R., Lima-Mendez, G., Rocha, F., Tirichine, L., Labadie, K., Kirilovsky, A., Bertrand, A., Engelen, S., Madoui, M.-A., Méheust, R., Poulain, J., Romac, S., Richter, D. J., Yoshikawa, G., … Tara Oceans, C. (2018). A global ocean atlas of eukaryotic genes. *Nature Communications*, *9*(1), 373. https://doi.org/10.1038/s41467-017-02342-1

Cecchin, M., Cazzaniga, S., Martini, F., Paltrinieri, S., Bossi, S., Maffei, M. E., & Ballottari, M. (2022). Astaxanthin and eicosapentaenoic acid production by S4, a new mutant strain of Nannochloropsis gaditana. *Microbial Cell Factories*, *21*(1), 117. https://doi.org/10.1186/s12934-022-01847-9

Chew, K. W., Yap, J. Y., Show, P. L., Suan, N. H., Juan, J. C., Ling, T. C., Lee, D. J., & Chang, J. S. (2017). Microalgae biorefinery: High value products perspectives. *Bioresource Technology*, *229*, 53–62. https://doi.org/10.1016/j.biortech.2017.01.006

Consortium, T. U. (2024). UniProt: The universal protein knowledgebase in 2025. *Nucleic Acids Research*, *53*(D1), D609–D617. https://doi.org/10.1093/nar/gkae1010

Costa, J. A. V., Cassuriaga, A. P. A., Moraes, L., & Morais, M. G. (2022). Biosynthesis and potential applications of terpenes produced from microalgae. *Bioresource Technology Reports*, *19*, 101166. https://doi.org/10.1016/j.biteb.2022.101166

Craig, R. J., Gallaher, S. D., Shu, S., Salomé, P. A., Jenkins, J. W., Blaby-Haas, C. E., Purvine, S. O., O'Donnell, S., Barry, K., Grimwood, J., Strenkert, D., Kropat, J., Daum, C., Yoshinaga, Y., Goodstein, D. M., Vallon, O., Schmutz, J., & Merchant, S. S. (2022). The Chlamydomonas Genome Project, version 6: Reference assemblies for mating-type plus and minus strains reveal extensive structural mutation in the laboratory. *The Plant Cell*, *35*(2), 644–672. https://doi.org/10.1093/plcell/koac347

Criscuolo, E., Caputo, V., Diotti, R. A., Sautto, G. A., Kirchenbaum, G. A., & Clementi, N. (2019). Alternative methods of vaccine delivery: An overview of edible and intradermal vaccines. *Journal of Immunology Research*, *2019*(1), 8303648. https://doi.org/10.1155/2019/8303648

Crozet, P., Navarro, F. J., Willmund, F., Mehrshahi, P., Bakowski, K., Lauersen, K. J., Pérez-Pérez, M.-E., Auroy, P., Gorchs Rovira, A., Sauret-Gueto, S., Niemeyer, J., Spaniol, B., Theis, J., Trösch, R., Westrich, L.-D., Vavitsas, K., Baier, T., Hübner, W., de Carpentier, F., … Lemaire, S. D. (2018). Birth of a photosynthetic chassis: A MoClo Toolkit enabling synthetic biology in the microalga Chlamydomonas reinhardtii. *ACS Synthetic Biology*, *7*(9), 2074–2086. https://doi.org/10.1021/acssynbio.8b00251

Cuellar-Bermudez, S. P., Aguilar-Hernandez, I., Cardenas-Chavez, D. L., Ornelas-Soto, N., Romero-Ogawa, M. A., & Parra-Saldivar, R. (2015). Extraction and purification of high-value metabolites from microalgae: Essential lipids, astaxanthin and phycobiliproteins. *Microbial Biotechnology*, *8*(2), 190–209. https://doi.org/10.1111/1751-7915.12167

D'Adamo, S., Kormelink, R., Martens, D., Barbosa, M. J., & Wijffels, R. H. (2022). Prospects for viruses infecting eukaryotic microalgae in biotechnology. *Biotechnology Advances*, *54*, 107790. https://doi.org/10.1016/j.biotechadv.2021.107790

De Clerck, O., Guiry, M. D., Leliaert, F., Samyn, Y., & Verbruggen, H. (2013). Algal taxonomy: A road to nowhere? *Journal of Phycology*, *49*(2), 215–225. https://doi.org/10.1111/jpy.12020

de Morais, M. G., Vaz Bda, S., de Morais, E. G., & Costa, J. A. (2015). Biologically active metabolites synthesized by microalgae. *BioMed Research International*, *2015*, 835761. https://doi.org/10.1155/2015/835761

Debnath, T., Bandyopadhyay, T. K., Vanitha, K., Bobby, M. N., Nath Tiwari, O., Bhunia, B., & Muthuraj, M. (2024). Astaxanthin from microalgae: A review on structure, biosynthesis, production strategies and application. *Food Research International*, *176*, 113841. https://doi.org/10.1016/j.foodres.2023.113841

Decelle, J., Romac, S., Stern, R. F., Bendif, E. M., Zingone, A., Audic, S., Guiry, M. D., Guillou, L., Tessier, D., Le Gall, F., Gourvil, P., Dos Santos, A. L., Probert, I., Vaulot, D., de Vargas, C., & Christen, R. (2015). PhytoREF: A reference database of the plastidial 16S rRNA gene of photosynthetic eukaryotes with curated taxonomy. *Molecular Ecology Resources*, *15*(6), 1435–1445. https://doi.org/10.1111/1755-0998.12401

D'Elia, L., Del Mondo, A., Santoro, M., De Natale, A., Pinto, G., & Pollio, A. (2018). Microorganisms from harsh and extreme environments: A collection of living strains at ACUF (Naples, Italy). *Ecological Questions*, *29*(3), 63–74.

Demurtas, O. C., Massa, S., Ferrante, P., Venuti, A., Franconi, R., & Giuliano, G. (2013). A chlamydomonas-derived human papillomavirus 16 E7 Vaccine induces specific tumor protection. *PLoS One*, *8*(4), e61473. https://doi.org/10.1371/journal.pone.0061473

Deprá, M. C., Dias, R. R., Zepka, L. Q., & Jacob-Lopes, E. (2025). Tackling old challenges in microalgal biotechnology: The role of photobioreactors to advance the technology readiness level. *Processes*, *13*(1), 51. https://www.mdpi.com/2227-9717/13/1/51

Diamond, A., Diaz-Garza, A. M., Li, J., Slattery, S. S., Merindol, N., Fantino, E., Meddeb-Mouelhi, F., Karas, B. J., Barnabé, S., & Desgagné-Penix, I. (2023). Instability of extrachromosomal DNA transformed into the diatom Phaeodactylum tricornutum. *Algal Research*, *70*, 102998. https://doi.org/10.1016/j.algal.2023.102998

Dreesen, I. A. J., Hamri, G. C.-E., & Fussenegger, M. (2010). Heat-stable oral alga-based vaccine protects mice from Staphylococcus aureus infection. *Journal of Biotechnology*, *145*(3), 273–280. https://doi.org/10.1016/j.jbiotec.2009.12.006

Duan, X., Zheng, X., Liu, Z., Dong, T., Luo, Y., Yan, W., Wang, C., & Song, C. (2024). On-chip photoacoustics-activated cell sorting (PA-ACS) for label-free and high-throughput detection and screening of microalgal cells. *Analytical Chemistry*, *96*(3), 1301–1309. https://doi.org/10.1021/acs.analchem.3c04665

Economou, C., Wannathong, T., Szaub, J., & Purton, S. (2014). A simple, low-cost method for chloroplast transformation of the green alga Chlamydomonas reinhardtii. In P. Maliga (Ed.), *Chloroplast Biotechnology: Methods and Protocols* (pp. 401–411). Humana Press. https://doi.org/10.1007/978-1-62703-995-6_27

Einhaus, A., Baier, T., & Kruse, O. (2024). Molecular design of microalgae as sustainable cell factories. *Trends in Biotechnology*, *42*(6), 728–738. https://doi.org/10.1016/j.tibtech.2023.11.010

Einhaus, A., Baier, T., Rosenstengel, M., Freudenberg, R. A., & Kruse, O. (2021). Rational promoter engineering enables robust terpene production in microalgae. *ACS Synthetic Biology*, *10*(4), 847–856. https://doi.org/10.1021/acssynbio.0c00632

Einhaus, A., Krieger, A., Köhne, L., Rautengarten, B., Jacobebbinghaus, N., Saudhof, M., Baier, T., & Kruse, O. (2025). Genome editing of epigenetic transgene silencing in Chlamydomonas reinhardtii. *Trends in Biotechnology*, *43*(8), 1961–1981. https://doi.org/10.1016/j.tibtech.2025.04.019

Endo, H., Blanc-Mathieu, R., Li, Y., Salazar, G., Henry, N., Labadie, K., de Vargas, C., Sullivan, M. B., Bowler, C., Wincker, P., Karp-Boss, L., Sunagawa, S., & Ogata, H. (2020). Biogeography of marine giant viruses reveals their interplay with eukaryotes and ecological functions. *Nature Ecology & Evolution*, *4*(12), 1639–1649. https://doi.org/10.1038/s41559-020-01288-w

Fang, N., Wang, C., Liu, X., Zhao, X., Liu, Y., Liu, X., Du, Y., Zhang, Z., & Zhang, H. (2019). De novo synthesis of astaxanthin: From organisms to genes. *Trends in Food Science & Technology*, *92*, 162–171. https://doi.org/10.1016/j.tifs.2019.08.016

Ferreira de Oliveira, A. P., & Bragotto, A. P. A. (2022). Microalgae-based products: Food and public health. *Future Foods*, *6*, 100157. https://doi.org/10.1016/j.fufo.2022.100157

Fu, W., Nelson, D. R., Mystikou, A., Daakour, S., & Salehi-Ashtiani, K. (2019). Advances in microalgal research and engineering development. *Current Opinion in Biotechnology*, *59*, 157–164. https://doi.org/10.1016/j.copbio.2019.05.013

Gangl, D., Zedler, J. A., Rajakumar, P. D., Martinez, E. M., Riseley, A., Wlodarczyk, A., Purton, S., Sakuragi, Y., Howe, C. J., Jensen, P. E., & Robinson, C. (2015). Biotechnological exploitation of microalgae. *Journal of Experimental Botany*, *66*(22), 6975–6990. https://doi.org/10.1093/jxb/erv426

Gimpel, J. A., Henriquez, V., & Mayfield, S. P. (2015). In metabolic engineering of eukaryotic microalgae: Potential and challenges come with great diversity. *Frontiers in Microbiology*, *6*, 1376. https://doi.org/10.3389/fmicb.2015.01376

Gong, Y., Kang, N. K., Kim, Y. U., Wang, Z., Wei, L., Xin, Y., Shen, C., Wang, Q., You, W., Lim, J.-M., Jeong, S.-W., Park, Y.-I., Oh, H.-M., Pan, K., Poliner, E., Yang, G., Li-Beisson, Y., Li, Y., Hu, Q., … Xu, J. (2020). The NanDeSyn database for nannochloropsis systems and synthetic biology. *The Plant Journal*, *104*(6), 1736–1745. https://doi.org/10.1111/tpj.15025

Gonzalez-Ballester, D., Pootakham, W., Mus, F., Yang, W., Catalanotti, C., Magneschi, L., de Montaigu, A., Higuera, J. J., Prior, M., Galván, A., Fernandez, E., & Grossman, A. R. (2011). Reverse genetics in Chlamydomonas: A platform for isolating insertional mutants. *Plant Methods*, *7*(1), 24. https://doi.org/10.1186/1746-4811-7-24

Goodstein, D. M., Shu, S., Howson, R., Neupane, R., Hayes, R. D., Fazo, J., Mitros, T., Dirks, W., Hellsten, U., Putnam, N., & Rokhsar, D. S. (2011). Phytozome: A comparative platform for green plant genomics. *Nucleic Acids Research*, *40*(D1), D1178–D1186. https://doi.org/10.1093/nar/gkr944

Gregory, A. C., Zayed, A. A., Conceicao-Neto, N., Temperton, B., Bolduc, B., Alberti, A., Ardyna, M., Arkhipova, K., Carmichael, M., Cruaud, C., Dimier, C., Dominguez-Huerta, G., Ferland, J., Kandels, S., Liu, Y., Marec, C., Pesant, S., Picheral, M., Pisarev, S., … Sullivan, M. B. (2019). Marine DNA viral macro- and microdiversity from pole to pole. *Cell*, *177*(5), 1109–1123. https://doi.org/10.1016/j.cell.2019.03.040

Gregory, J. A., Li, F., Tomosada, L. M., Cox, C. J., Topol, A. B., Vinetz, J. M., & Mayfield, S. (2012). Algae-produced Pfs25 elicits antibodies that inhibit malaria transmission. *PLoS One*, *7*(5), e37179. https://doi.org/10.1371/journal.pone.0037179

Grigoriev, I. V., Hayes, R. D., Calhoun, S., Kamel, B., Wang, A., Ahrendt, S., Dusheyko, S., Nikitin, R., Mondo, Stephen J., Salamov, A., Shabalov, I., & Kuo, A. (2020). PhycoCosm, a comparative algal genomics resource. *Nucleic Acids Research*, *49*(D1), D1004–D1011. https://doi.org/10.1093/nar/gkaa898

Gudmundsson, S., Agudo, L., & Nogales, J. (2017). 4 - Applications of genome-scale metabolic models of microalgae and cyanobacteria in biotechnology. In C. Gonzalez-Fernandez & R. Muñoz (Eds.), *Microalgae-Based Biofuels and Bioproducts* (pp. 93–111). Woodhead Publishing. https://doi.org/10.1016/B978-0-08-101023-5.00004-2

Guillou, L., Bachar, D., Audic, S., Bass, D., Berney, C., Bittner, L., Boutte, C., Burgaud, G., de Vargas, C., Decelle, J., del Campo, J., Dolan, J. R., Dunthorn, M., Edvardsen, B., Holzmann, M., Kooistra, W. H. C. F., Lara, E., Le Bescot, N., Logares, R., … Christen, R. (2012). The protist ribosomal reference database (PR2): A catalog of unicellular eukaryote Small Sub-Unit rRNA sequences with curated taxonomy. *Nucleic Acids Research*, *41*(D1), D597–D604. https://doi.org/10.1093/nar/gks1160

Gutiérrez, S., & Lauersen, K. J. (2021). Gene delivery technologies with applications in microalgal genetic engineering. *Biology-Basel*, *10*(4). https://doi.org/10.3390/biology10040265

Hernández-Plaza, A., Szklarczyk, D., Botas, J., Cantalapiedra, Carlos P., Giner-Lamia, J., Mende, D. R., Kirsch, R., Rattei, T., Letunic, I., Jensen, Lars J., Bork, P., von Mering, C., & Huerta-Cepas, J. (2022). eggNOG 6.0: Enabling comparative genomics across 12 535 organisms. *Nucleic Acids Research*, *51*(D1), D389–D394. https://doi.org/10.1093/nar/gkac1022

Ismail, N. F. N., Abidin, A. A. Z., Yusoff, F. M., Karim, M. M. A., Yasin, I. S. M., & Yusof, Z. N. B. (2025). Bioencapsulation of transgenic Nannochloropsis sp. using artemia as a vaccine delivery system to zebrafish as a fish model against vibriosis. *Molecular Biotechnology*. https://doi.org/10.1007/s12033-025-01496-4

Jacobebbinghaus, N., Bigge, F., Saudhof, M., Hübner, W., Kruse, O., & Baier, T. (2025). Transcriptional gene fusions via targeted integration at safe harbors for high transgene expression in Chlamydomonas reinhardtii. *New Phytologist*, *247*(6), 2665–2677. https://doi.org/10.1111/nph.70368

Jeon, S., Lim, J.-M., Lee, H.-G., Shin, S.-E., Kang, N. K., Park, Y.-I., Oh, H.-M., Jeong, W.-J., Jeong, B.-R., & Chang, Y. K. (2017). Current status and perspectives of genome editing technology for microalgae. *Biotechnology for Biofuels*, *10*(1), 267. https://doi.org/10.1186/s13068-017-0957-z

Jeong, B. R., Jang, J., & Jin, E. (2023). Genome engineering via gene editing technologies in microalgae. *Bioresource Technology*, *373*, 128701. https://doi.org/10.1016/j.biortech.2023.128701

Karas, B. J., Diner, R. E., Lefebvre, S. C., McQuaid, J., Phillips, A. P., Noddings, C. M., Brunson, J. K., Valas, R. E., Deerinck, T. J., Jablanovic, J., Gillard, J. T., Beeri, K., Ellisman, M. H., Glass, J. I., Hutchison, C. A., 3rd, Smith, H. O., Venter, J. C., Allen, A. E., Dupont, C. L., & Weyman, P. D. (2015). Designer diatom episomes delivered by bacterial conjugation. *Nature Communications*, *6*, 6925. https://doi.org/10.1038/ncomms7925

Kathiresan, S., Chandrashekar, A., Ravishankar, G. A., & Sarada, R. (2015). Regulation of astaxanthin and its intermediates through cloning and genetic transformation of β-carotene ketolase in Haematococcus pluvialis. *Journal of Biotechnology*, *196–197*, 33–41. https://doi.org/10.1016/j.jbiotec.2015.01.006

Khan, M. I., Shin, J. H., & Kim, J. D. (2018). The promising future of microalgae: Current status, challenges, and optimization of a sustainable and renewable industry for biofuels, feed, and other products. *Microbial Cell Factories*, *17*(1), 36. https://doi.org/10.1186/s12934-018-0879-x

Kim, E.-J., Ma, X., & Cerutti, H. (2015). Gene silencing in microalgae: Mechanisms and biological roles. *Bioresource Technology*, *184*, 23–32. https://doi.org/10.1016/j.biortech.2014.10.119

Kim, H. S., Hsu, S.-C., Han, S.-I., Thapa, H. R., Guzman, A. R., Browne, D. R., Tatli, M., Devarenne, T. P., Stern, D. B., & Han, A. (2017). High-throughput droplet microfluidics screening platform for selecting fast-growing and high lipid-producing microalgae from a mutant library. *Plant Direct*, *1*(3), e00011. https://doi.org/10.1002/pld3.11

Kim, R. S., Levy Karin, E., Mirdita, M., Chikhi, R., & Steinegger, M. (2024). BFVD—A large repository of predicted viral protein structures. *Nucleic Acids Research*, *53*(D1), D340–D347. https://doi.org/10.1093/nar/gkae1119

Kim, T., Pradhan, B., & Ki, J.-S. (2024). Staining to machine learning: An emerging technology for determination of microalgal cell viability. *Journal of Applied Phycology*, *36*(5), 2573–2592.

Kiran, B. R., & Venkata Mohan, S. (2021). Microalgal cell biofactory-therapeutic, nutraceutical and functional food applications. *Plants (Basel)*, *10*(5). https://doi.org/10.3390/plants10050836

Kneip, J. S., Kniepkamp, N., Jang, J., Mortaro, M. G., Jin, E., Kruse, O., & Baier, T. (2024). CRISPR/Cas9-mediated knockout of the lycopene ε-cyclase for efficient astaxanthin production in the green microalga Chlamydomonas reinhardtii. *Plants (Basel), 13*(10). https://doi.org/10.3390/plants13101393

Kroth, P. G., Bones, A. M., Daboussi, F., Ferrante, M. I., Jaubert, M., Kolot, M., Nymark, M., Rio Bartulos, C., Ritter, A., Russo, M. T., Serif, M., Winge, P., & Falciatore, A. (2018). Genome editing in diatoms: Achievements and goals. *Plant Cell Reports*, *37*(10), 1401–1408. https://doi.org/10.1007/s00299-018-2334-1

Kumar, A., Baldia, A., Rajput, D., Kateriya, S., Babu, V., & Dubey, K. K. (2023). Multiomics and optobiotechnological approaches for the development of microalgal strain for production of aviation biofuel and biorefinery. *Bioresource Technology*, *369*, 128457. https://doi.org/10.1016/j.biortech.2022.128457

Kurotani, A., Yamada, Y., & Sakurai, T. (2017). Alga-PrAS (algal protein annotation suite): A database of comprehensive annotation in algal proteomes. *Plant and Cell Physiology*, *58*(1), e6–e6. https://doi.org/10.1093/pcp/pcw212

Kwon, Y. M., Kim, K. W., Choi, T. Y., Kim, S. Y., & Kim, J. Y. H. (2018). Manipulation of the microalgal chloroplast by genetic engineering for biotechnological utilization as a green biofactory. *World Journal of Microbiology and Biotechnology*, *34*(12), 183. https://doi.org/10.1007/s11274-018-2567-8

Lee, T.-H., Chang, J.-S., & Wang, H.-Y. (2013). Current developments in high-throughput analysis for microalgae cellular contents. *Biotechnology Journal*, *8*(11), 1301–1314. https://doi.org/10.1002/biot.201200391

Levasseur, W., Perre, P., & Pozzobon, V. (2020). A review of high value-added molecules production by microalgae in light of the classification. *Biotechnology Advances*, *41*, 107545. https://doi.org/10.1016/j.biotechadv.2020.107545

Li, J., Zhu, D., Niu, J., Shen, S., & Wang, G. (2011). An economic assessment of astaxanthin production by large scale cultivation of Haematococcus pluvialis. *Biotechnology Advances*, *29*(6), 568–574. https://doi.org/10.1016/j.biotechadv.2011.04.001

Li, Y., Deng, L., Walker, E. J. L., Karas, B. J., & Mock, T. (2025). Genetic engineering in diatoms: Advances and prospects. *The Plant Journal*, *121*(6), e70102. https://doi.org/10.1111/tpj.70102

Li-Beisson, Y., Thelen, J. J., Fedosejevs, E., & Harwood, J. L. (2019). The lipid biochemistry of eukaryotic algae. *Progress in Lipid Research*, *74*, 31–68. https://doi.org/10.1016/j.plipres.2019.01.003

Lin, Z., Akin, H., Rao, R., Hie, B., Zhu, Z., Lu, W., Smetanin, N., Verkuil, R., Kabeli, O., Shmueli, Y., dos Santos Costa, A., Fazel-Zarandi, M., Sercu, T., Candido, S., & Rives, A. (2023). Evolutionary-scale prediction of atomic-level protein structure with a language model. *Science*, *379*(6637), 1123–1130. https://doi.org/10.1126/science.ade2574

Liu, M., Yu, L., Zheng, J., Shao, S., Pan, Y., Hu, H., Shen, L., Wang, W., Zhou, W., & Liu, J. (2024). Turning the industrially relevant marine alga Nannochloropsis red: One move for multifaceted benefits. *New Phytologist*, *244*(4), 1467–1481. https://doi.org/10.1111/nph.20114

Lu, Q., Li, H., Zou, Y., Liu, H., & Yang, L. (2021). Astaxanthin as a microalgal metabolite for aquaculture: A review on the synthetic mechanisms, production techniques, and practical application. *Algal Research*, *54*, 102178. https://doi.org/10.1016/j.algal.2020.102178

Melero-Cobo, X., Gallemí, M., Carnicer, M., Monte, E., Planas, A., & Leivar, P. (2025). MoCloro: An extension of the Chlamydomonas reinhardtii modular cloning toolkit for microalgal chloroplast engineering. *Physiologia Plantarum*, *177*(1), e70088. https://doi.org/10.1111/ppl.70088

Montuori, E., Saggiomo, M., & Lauritano, C. (2023). Microalgae from cold environments and their possible biotechnological applications. *Marine Drugs*, *21*(5), 292.

Mosey, M., Douchi, D., Knoshaug, E. P., & Laurens, L. M. L. (2021). Methodological review of genetic engineering approaches for non-model algae. *Algal Research*, *54*. https://doi.org/10.1016/j.algal.2021.102221

Motta, D. N., Torres, V. G. S., de Queiroz Hermino, V. L., Neves, S. M. B., de Oliveira, L. V. V., Carvalho, R. P., & da Silva, E. P. (2025). Challenges of microalgae production: From the laboratory to large scale. In I. A. Severo, J. C. Ordóñez, A. B. Mariano, & J. V. C. Vargas (Eds.), *Microalgae Horizons: Fundamentals, Innovations, and Industrial Applications* (pp. 11–21). Springer Nature Switzerland. https://doi.org/10.1007/978-3-031-84282-5_2

Moudříková, Š., Mojzeš, P., Zachleder, V., Pfaff, C., Behrendt, D., & Nedbal, L. (2016). Raman and fluorescence microscopy sensing energy-transducing and energy-storing structures in microalgae. *Algal Research*, *16*, 224–232. https://doi.org/10.1016/j.algal.2016.03.016

Munoz, C. F., Sudfeld, C., Naduthodi, M. I. S., Weusthuis, R. A., Barbosa, M. J., Wijffels, R. H., & D'Adamo, S. (2021). Genetic engineering of microalgae for enhanced lipid production. *Biotechnology Advances*, *52*, 107836. https://doi.org/10.1016/j.biotechadv.2021.107836

Naduthodi, M. I. S., Barbosa, M. J., & van der Oost, J. (2018). Progress of CRISPR-Cas based genome editing in photosynthetic microbes. *Biotechnology Journal*, *13*(9), e1700591. https://doi.org/10.1002/biot.201700591

Naduthodi, M. I. S., Mohanraju, P., Sudfeld, C., D'Adamo, S., Barbosa, M. J., & van der Oost, J. (2019). CRISPR-Cas ribonucleoprotein mediated homology-directed repair for efficient targeted genome editing in microalgae Nannochloropsis oceanica IMET1. *Biotechnology for Biofuels*, *12*, 66. https://doi.org/10.1186/s13068-019-1401-3

Naduthodi, M. I. S., Sudfeld, C., Avitzigiannis, E. K., Trevisan, N., van Lith, E., Alcaide Sancho, J., D'Adamo, S., Barbosa, M., & van der Oost, J. (2021). Comprehensive genome engineering toolbox for microalgae Nannochloropsis oceanica based on CRISPR-Cas systems. *ACS Synthetic Biology*, *10*(12), 3369–3378. https://doi.org/10.1021/acssynbio.1c00329

Pereira, H., Schulze, P. S. C., Schüler, L. M., Santos, T., Barreira, L., & Varela, J. (2018). Fluorescence activated cell-sorting principles and applications in microalgal biotechnology. *Algal Research*, *30*, 113–120. https://doi.org/10.1016/j.algal.2017.12.013

Pérez-López, P., de Vree, J. H., Feijoo, G., Bosma, R., Barbosa, M. J., Moreira, M. T., Wijffels, R. H., van Boxtel, A. J. B., & Kleinegris, D. M. M. (2017). Comparative life cycle assessment of real pilot reactors for microalgae cultivation in different seasons. *Applied Energy*, *205*, 1151–1164. https://doi.org/10.1016/j.apenergy.2017.08.102

Perin, G., Gambaro, F., & Morosinotto, T. (2022). Knowledge of regulation of photosynthesis in outdoor microalgae cultures is essential for the optimization of biomass productivity. *Frontiers in Plant Science*, *13*, 846496. https://doi.org/10.3389/fpls.2022.846496

Perozeni, F., Cazzaniga, S., Baier, T., Zanoni, F., Zoccatelli, G., Lauersen, K. J., Wobbe, L., & Ballottari, M. (2020). Turning a green alga red: Engineering astaxanthin biosynthesis by intragenic pseudogene revival in Chlamydomonas reinhardtii. *Plant Biotechnology Journal*, *18*(10), 2053–2067. https://doi.org/10.1111/pbi.13364

Poliner, E., Pulman, J. A., Zienkiewicz, K., Childs, K., Benning, C., & Farre, E. M. (2018). A toolkit for Nannochloropsis oceanica CCMP1779 enables gene stacking and genetic engineering of the eicosapentaenoic acid pathway for enhanced long-chain polyunsaturated fatty acid production. *Plant Biotechnology Journal*, *16*(1), 298–309. https://doi.org/10.1111/pbi.12772

Poliner, E., Takeuchi, T., Du, Z. Y., Benning, C., & Farre, E. M. (2018). Nontransgenic marker-free gene disruption by an episomal CRISPR system in the oleaginous microalga, Nannochloropsis oceanica CCMP1779. *ACS Synthetic Biology*, *7*(4), 962–968. https://doi.org/10.1021/acssynbio.7b00362

Pollak, B., Matute, T., Nunez, I., Cerda, A., Lopez, C., Vargas, V., Kan, A., Bielinski, V., von Dassow, P., Dupont, C. L., & Federici, F. (2020). Universal loop assembly: Open, efficient and cross-kingdom DNA fabrication. *Synth Biology (Oxford)*, *5*(1), ysaa001. https://doi.org/10.1093/synbio/ysaa001

Qin, S., Wang, K., Gao, F., Ge, B., Cui, H., & Li, W. (2023). Biotechnologies for bulk production of microalgal biomass: From mass cultivation to dried biomass acquisition. *Biotechnology for Biofuels and Bioproducts*, *16*(1), 131. https://doi.org/10.1186/s13068-023-02382-4

Rappaport, H. B., & Oliverio, A. M. (2023). Extreme environments offer an unprecedented opportunity to understand microbial eukaryotic ecology, evolution, and genome biology. *Nature Communications*, *14*(1), 4959.

Richardson, L., Allen, B., Baldi, G., Beracochea, M., Bileschi, Maxwell L., Burdett, T., Burgin, J., Caballero-Pérez, J., Cochrane, G., Colwell, Lucy J., Curtis, T., Escobar-Zepeda, A., Gurbich, Tatiana A., Kale, V., Korobeynikov, A., Raj, S., Rogers, Alexander B., Sakharova, E., Sanchez, S., … Finn, Robert D. (2022). MGnify: The microbiome sequence data analysis resource in 2023. *Nucleic Acids Research*, *51*(D1), D753–D759. https://doi.org/10.1093/nar/gkac1080

Richter, D. J., Berney, C., Strassert, J. F. H., Poh, Y.-P., Herman, E. K., Muñoz-Gómez, S. A., Wideman, J. G., Burki, F., & de Vargas, C. (2022). EukProt: A database of genome-scale predicted proteins across the diversity of eukaryotes. *Peer Community Journal*, *2*. https://doi.org/10.24072/pcjournal.173

Rojas-Villalta, D., Rojas-Rodriguez, D., Villanueva-Ilama, M., Guillen-Watson, R., Murillo-Vega, F., Gomez-Espinoza, O., & Nunez-Montero, K. (2024). Exploring extremotolerant and extremophilic microalgae: New frontiers in sustainable biotechnological applications. *Biology-Basel*, *13*(9). https://doi.org/10.3390/biology13090712

Romero-Losada, A. B., Arvanitidou, C., de los Reyes, P., García-González, M., & Romero-Campero, F. J. (2022). ALGAEFUN with MARACAS, microALGAE FUNctional enrichment tool for MicroAlgae RnA-seq and Chip-seq AnalysiS. *BMC Bioinformatics*, *23*(1), 113. https://doi.org/10.1186/s12859-022-04639-5

Rosales-Mendoza, S., Solís-Andrade, K. I., Márquez-Escobar, V. A., González-Ortega, O., & Bañuelos-Hernandez, B. (2020). Current advances in the algae-made biopharmaceuticals field. *Expert Opinion on Biological Therapy*, *20*(7), 751–766. https://doi.org/10.1080/14712598.2020.1739643

Rumin, J., Bonnefond, H., Saint-Jean, B., Rouxel, C., Sciandra, A., Bernard, O., Cadoret, J. P., & Bougaran, G. (2015). The use of fluorescent Nile red and BODIPY for lipid measurement in microalgae. *Biotechnology for Biofuels*, *8*, 42. https://doi.org/10.1186/s13068-015-0220-4

Salome, P. A., & Merchant, S. S. (2019). A series of fortunate events: Introducing Chlamydomonas as a reference organism. *Plant Cell*, *31*(8), 1682–1707. https://doi.org/10.1105/tpc.18.00952

Sánchez Barea, J., Lee, J., & Kang, D.-K. (2019). Recent advances in droplet-based microfluidic technologies for biochemistry and molecular biology. *Micromachines*, *10*(6), 412. https://www.mdpi.com/2072-666X/10/6/412

Schroda, M., Blöcker, D., & Beck, C. F. (2000). The HSP70A promoter as a tool for the improved expression of transgenes in chlamydomonas. *The Plant Journal*, *21*(2), 121–131. https://doi.org/10.1046/j.1365-313x.2000.00652.x

Severo, I. A., de Lira, G. S., Ambati, R. R., Gokare, R. A., Vargas, J. V. C., Ordonez, J., & Mariano, A. B. (2024). Disruptive potential of microalgae proteins: Shaping the future of the food industry. *Future Foods*, *9*, 100318. https://doi.org/10.1016/j.fufo.2024.100318

Sharma, A., Nawkarkar, P., Kapase, V. U., Chhabra, M., & Kumar, S. (2024). Engineering of ketocarotenoid biosynthetic pathway in Chlamydomonas reinhardtii through exogenous gene expression. *Systems Microbiology and Biomanufacturing*, *4*(3), 983–995. https://doi.org/10.1007/s43393-024-00240-4

Shitanaka, T., Wang, Y., Do, S., Yuson, J., Khanal, S. K., Zienkiewicz, K., & Du, Z.-Y. (2025). Synthetic biology and metabolic engineering of microalgae for sustainable lipid and terpenoid production: An updated perspective. *Plant Biotechnology Journal*. https://doi.org/10.1111/pbi.70405

Sirohi, R., Joun, J., Choi, H., II, Gaur, V. K., & Sim, S. J. (2021). Algal glycobiotechnology: Omics approaches for strain improvement. *Microbial Cell Factories*, *20*(1), 163. https://doi.org/10.1186/s12934-021-01656-6

Spolaore, P., Joannis-Cassan, C., Duran, E., & Isambert, A. (2006). Commercial applications of microalgae. *Journal of Bioscience and Bioengineering*, *101*(2), 87–96. https://doi.org/10.1263/jbb.101.87

Sproles, A. E., Fields, F. J., Smalley, T. N., Le, C. H., Badary, A., & Mayfield, S. P. (2021). Recent advancements in the genetic engineering of microalgae. *Algal Research*, *53*. https://doi.org/10.1016/j.algal.2020.102158

Sreedasyam, A., Plott, C., Hossain, M. S., Lovell, John T., Grimwood, J., Jenkins, Jerry W., Daum, C., Barry, K., Carlson, J., Shu, S., Phillips, J., Amirebrahimi, M., Zane, M., Wang, M., Goodstein, D., Haas, Fabian B., Hiss, M., Perroud, P.-F., Jawdy, Sara S., … Schmutz, J. (2023). JGI plant gene atlas: An updateable transcriptome resource to improve functional gene descriptions across the plant kingdom. *Nucleic Acids Research*, *51*(16), 8383–8401. https://doi.org/10.1093/nar/gkad616

Südfeld, C., Hubáček, M., D'Adamo, S., Wijffels, R. H., & Barbosa, M. J. (2021). Optimization of high-throughput lipid screening of the microalga Nannochloropsis oceanica using BODIPY 505/515. *Algal Research*, *53*. https://doi.org/10.1016/j.algal.2020.102138

Sudfeld, C., Hubacek, M., Figueiredo, D., Naduthodi, M. I. S., van der Oost, J., Wijffels, R. H., Barbosa, M. J., & D'Adamo, S. (2021). High-throughput insertional mutagenesis reveals novel targets for enhancing lipid accumulation in Nannochloropsis oceanica. *Metabolic Engineering*, *66*, 239–258. https://doi.org/10.1016/j.ymben.2021.04.012

Sudfeld, C., Pozo-Rodriguez, A., Manjavacas Diez, S. A., Wijffels, R. H., Barbosa, M. J., & D'Adamo, S. (2022). The nucleolus as a genomic safe harbor for strong gene expression in Nannochloropsis oceanica. *Molecular Plant*, *15*(2), 340–353. https://doi.org/10.1016/j.molp.2021.11.003

Tegenfeldt, F., Kuznetsov, D., Manni, M., Berkeley, M., Zdobnov, E. M., & Kriventseva, E. V. (2024). OrthoDB and BUSCO update: Annotation of orthologs with wider sampling of genomes. *Nucleic Acids Research*, *53*(D1), D516–D522. https://doi.org/10.1093/nar/gkae987

Tibocha-Bonilla, J. D., Zuniga, C., Godoy-Silva, R. D., & Zengler, K. (2018). Advances in metabolic modeling of oleaginous microalgae. *Biotechnology for Biofuels*, *11*, 241. https://doi.org/10.1186/s13068-018-1244-3

Tran, Q.-G., Le, T. T., Choi, D.-Y., Cho, D.-H., Yun, J.-H., Choi, H. I., Kim, H.-S., & Lee, Y. J. (2025). Progress and challenges in CRISPR/Cas applications in microalgae. *Journal of Microbiology*, *63*(3), e2501028. https://doi.org/10.71150/jm.2501028

Trevisan, N., Fabris, M., van der Oost, J., Barbosa, M., & D'Adamo, S. (2025). Heterologous production of β-pinene in the chloroplast of the marine diatom Phaeodactylum tricornutum. *New Biotechnology*, *90*, 174–184. https://doi.org/10.1016/j.nbt.2025.10.003

Trovão, M., Schüler, L. M., Machado, A., Bombo, G., Navalho, S., Barros, A., Pereira, H., Silva, J., Freitas, F., & Varela, J. (2022). Random mutagenesis as a promising tool for microalgal strain improvement towards industrial production. *Marine Drugs*, *20*(7), 440. https://www.mdpi.com/1660-3397/20/7/440

van Kempen, M., Kim, S. S., Tumescheit, C., Mirdita, M., Lee, J., Gilchrist, C. L. M., Söding, J., & Steinegger, M. (2024). Fast and accurate protein structure search with Foldseek. *Nature Biotechnology*, *42*(2), 243–246. https://doi.org/10.1038/s41587-023-01773-0

Vasudevan, R., Gale, G. A. R., Schiavon, A. A., Puzorjov, A., Malin, J., Gillespie, M. D., Vavitsas, K., Zulkower, V., Wang, B., Howe, C. J., Lea-Smith, D. J., & McCormick, A. J. (2019). CyanoGate: A modular cloning suite for engineering cyanobacteria based on the plant MoClo syntax. *Plant Physiology*, *180*(1), 39–55. https://doi.org/10.1104/pp.18.01401

Vavitsas, K., Fabris, M., & Vickers, C. E. (2018). Terpenoid metabolic engineering in photosynthetic microorganisms. *Genes (Basel)*, *9*(11). https://doi.org/10.3390/genes9110520

Vernette, C., Lecubin, J., Sánchez, P., Coordinators, T. O., Sunagawa, S., Delmont, T. O., Acinas, S. G., Pelletier, E., Hingamp, P., & Lescot, M. (2022). The ocean gene atlas v2.0: Online exploration of the biogeography and phylogeny of plankton genes. *Nucleic Acids Research*, *50*(W1), W516–W526. https://doi.org/10.1093/nar/gkac420

Villar, E., Zweig, N., Vincens, P., Cruz de Carvalho, H., Duchene, C., Liu, S., Monteil, R., Dorrell, R. G., Fabris, M., Vandepoele, K., Bowler, C., & Falciatore, A. (2025). DiatOmicBase: A versatile gene-centered platform for mining functional omics data in diatom research. *The Plant Journal*, *121*(6), e70061. https://doi.org/10.1111/tpj.70061

Wang, X., Ren, L., Su, Y., Ji, Y., Liu, Y., Li, C., Li, X., Zhang, Y., Wang, W., Hu, Q., Han, D., Xu, J., & Ma, B. (2017). Raman-activated droplet sorting (RADS) for label-free high-throughput screening of microalgal single-cells. *Analytical Chemistry*, *89*(22), 12569–12577. https://doi.org/10.1021/acs.analchem.7b03884

Willems, T., Hectors, W., Rombaut, J., De Rop, A.-S., Goegebeur, S., Delmulle, T., De Mol, M. L., De Maeseneire, S. L., & Soetaert, W. K. (2023). An exploratory in silico comparison of open-source codon harmonization tools. *Microbial Cell Factories*, *22*(1), 227. https://doi.org/10.1186/s12934-023-02230-y

Work, V. H., Bentley, F. K., Scholz, M. J., D'Adamo, S., Gu, H., Vogler, B. W., Franks, D. T., Stanish, L. F., Jinkerson, R. E., & Posewitz, M. C. (2013). Biocommodities from photosynthetic microorganisms. *Environmental Progress & Sustainable Energy*, *32*(4), 989–1001. https://doi.org/10.1002/ep.11849

Work, V. H., D'Adamo, S., Radakovits, R., Jinkerson, R. E., & Posewitz, M. C. (2012). Improving photosynthesis and metabolic networks for the competitive production of phototroph-derived biofuels. *Current Opinion in Biotechnology*, *23*(3), 290–297. https://doi.org/10.1016/j.copbio.2011.11.022

Wright, G., Rodriguez, A., Li, J., Milenkovic, T., Emrich, S. J., & Clark, P. L. (2022). CHARMING: Harmonizing synonymous codon usage to replicate a desired codon usage pattern. *Protein Science*, *31*(1), 221–231. https://doi.org/10.1002/pro.4223

Wu, Y., Chaumier, T., Manirakiza, E., Veluchamy, A., & Tirichine, L. (2023). PhaeoEpiView: An epigenome browser of the newly assembled genome of the model diatom Phaeodactylum tricornutum. *Scientific Reports*, *13*(1), 8320. https://doi.org/10.1038/s41598-023-35403-1

Yu, Z., Geisler, K., Leontidou, T., Young, R. E. B., Vonlanthen, S. E., Purton, S., Abell, C., & Smith, A. G. (2021). Droplet-based microfluidic screening and sorting of microalgal populations for strain engineering applications. *Algal Research*, *56*, 102293. https://doi.org/10.1016/j.algal.2021.102293

Zhang, Y.-T., Jiang, J.-Y., Shi, T.-Q., Sun, X.-M., Zhao, Q.-Y., Huang, H., & Ren, L.-J. (2019). Application of the CRISPR/Cas system for genome editing in microalgae. *Applied Microbiology and Biotechnology*, *103*(8), 3239–3248. https://doi.org/10.1007/s00253-019-09726-x

Zheng, G., Gu, F., Cui, Y., Lu, L., Hu, X., Wang, L., & Wang, Y. (2022). A microfluidic droplet array demonstrating high-throughput screening in individual lipid-producing microalgae. *Analytica Chimica Acta*, *1227*, 340322. https://doi.org/10.1016/j.aca.2022.340322

Zheng, H.-Q., Chiang-Hsieh, Y.-F., Chien, C.-H., Hsu, B.-K. J., Liu, T.-L., Chen, C.-N. N., & Chang, W.-C. (2014). AlgaePath: Comprehensive analysis of metabolic pathways using transcript abundance data from next-generation sequencing in green algae. *BMC Genomics*, *15*(1), 196. https://doi.org/10.1186/1471-2164-15-196

2 Microalgae and Microbes

Understanding Their Interactions in the Phycosphere

Udoh Inyang, Akaninyene Jack, Moyosore Adegbeye, Nsikan Ekanem, Ekaette Mbaba, Mona M.M.Y. Elghandour, P.R.K. Reddy, Marie De Angelis, and Abdelfattah Z.M. Salem

2.1 INTRODUCTION

Microalgae, found in diverse environments worldwide, are photosynthetic microbes that significantly contribute to global primary productivity and influence essential biogeochemical processes (Falkowski et al., 1998), especially when in association with bacteria. Microalgae represent a diverse phylogenetic grouping of unicellular photosynthetic organisms, encompassing both eukaryotic and prokaryotic microorganisms that thrive in Oceanic and inland water environments (Fulbright et al., 2018; Krohn-Molt et al., 2013). In aquatic systems, microalgae rarely exist in isolation. Instead, they live in close association with bacteria and other microorganisms, especially in an area referred to as the phycosphere, which serves as the algal counterpart to the plant rhizosphere (Seymour et al., 2017). Seymour et al. (2017) further elucidate this microenvironment to be deepened with dissolved organic carbon, vitamins, and signaling molecules exuded by the algae cell, which supports the growth of a unique and diverse microbial community. Through photosynthesis, these organisms serve a critical function in balancing the elements C (carbon) and O (oxygen) levels in aquatic ecosystems, producing oxygen and sequestering approximately 40% of global carbon (Azam & Ammerman, 1984).

The importance of microalgae cannot be overemphasized with regard to ruminant health and nutrition by incorporating their bioactive compounds into animal-derived products like dairy and beef products and boosting performance (Adegbeye et al., 2025; Jack et al., 2023). They are recognized for their indirect role in climate regulation through the production of dimethylsulfoniopropionate (DMSP), a compound

 DOI: 10.1201/9781003611479-2

with climatic significance. When DMSP is oxidized, it forms tiny sulfate particles that help form clouds and change how sunlight reaches Earth (Falkowski et al., 1998).

Historically, research on microalgae–bacteria interactions has focused on large-scale dynamics. Still, emerging evidence suggests that these associations are primarily driven by microscale processes occurring near the phytoplankton cells (Seymour et al., 2017). As microalgae grow, they release organic compounds known as exudates, which serve as nutrients for bacterial proliferation. Conversely, numerous studies (Amin et al., 2015; Seyedsayamdost et al., 2011) have demonstrated that these associated bacteria significantly influence the growth and overall health of microalgae. Importantly, the accessibility of nutrients is important in shaping the nature of microbial interaction (Azam & Ammerman, 1984). Under certain conditions, bacteria have been reported to vie for resources with microalgae for essential inorganic biologically important compounds, notably nitrogen and phosphorus, and can also produce detrimental substances—including algicidal agents and exoenzymes—that may lead to algal mortality (Cooper & Smith, 2015; Fuentes et al., 2016; Ramanan et al., 2016). Overall, the coexistence between microbes encompasses a spectrum of ecological relationships, which can be broadly categorized into mutualism, commensalism, competition, and parasitism. The interactions in the phycosphere range from mutualism, where bacteria supply micronutrients (e.g., vitamin B_{12}) to fungal (Croft et al., 2005), to commensalism, competition, and parasitism, which can suppress algae blooms or influence community succession (Ramanan et al., 2016). The intimate relationship existing within the phycosphere has led to the development of an intricate web of interactions between different biological kingdoms, fostering potential specialization among different organisms. These relationships are characterized by coordinated nutrient sharing, reciprocal enhancement of growth-promoting indices, and communication through quorum-sensing, all of which contribute to the diverse and complex communities formed by unicellular microalgae and their bacterial partners (Cirri & Pohnert, 2019).

The beneficial influence of microbial interaction within the phycosphere on fungal growth and flocculation has shifted the perception of bacteria from being seen merely as contaminants in algal cultures (Lee et al., 2013). Numerous investigations (Krohn-Molt et al., 2013) have confirmed that the microalgal microbiome is predominantly composed of species belonging to the phyla Bacteroidota, Flavobacteria, Alphaproteobacteriota, Betaproteobacteriota, and Gammaproteobacteriota. Although many uncultivable microorganisms can be identified, the total number of microbial species typically remains around 30 (Krohn-Molt et al., 2013. It remains uncertain whether all these bacterial species are essential for the health and persistence of algae within their communities. Notably, a few potentially invasive or pathogenic species have also been observed during microbial screenings. However, these organisms seem to coexist in equilibrium within the community, likely due to selective pressures that promote stable microalgae–bacteria consortia (Segev et al., 2016; Wang et al., 2014; Wieczorek et al., 2019).

In the context of aquaculture, the interaction between microalgae and microorganisms holds promise for enhancing fungal biomass production and infusing it with valuable biotechnological compounds such as lipids, carbohydrates, and pigments. These symbiotic relationships have been stated by authors (Fuentes et al., 2016), also benefit bacterial growth and photosynthetic performance, facilitated by mechanisms like adhesion, aggregation, motility, chemotaxis, specialized secretion systems, and communication pathways including quorum-sensing (QS) and quorum-quenching (QQ), as well as the synthesis of growth-promoting substances. These relationships profoundly affect microalgal physiology, population dynamics, and ecological outcomes. As such, the phycosphere is increasingly recognized as a hotspot of microbial communication and coevolution, with emerging implications for biotechnology, bioenergy, and ecosystem resilience.

Despite advancements in metagenomics, metabolomics, and microscopy, the spatial–temporal dynamics and mechanistic understanding of microalgae–microbe interactions remain underexplored. Most current research is limited to simplified model systems, lacking ecological relevance or real-time resolution of interactions. Furthermore, the combined influence of microalgal species traits, microbial community composition, and environmental conditions on the functioning of the phycosphere remains poorly understood. Understanding the structure and function of the phycosphere is increasingly important not only for ecological theory but also for applications in biotechnology, aquaculture, and climate change mitigation.

This review aims to bridge the gap by synthesizing recent findings on microalgae–microbe interactions within the phycosphere. It explores key mechanisms such as nutrient exchange, chemical signaling, and physical associations, and discusses how new tools (e.g., microfluidics and single-cell sequencing) are helping to unravel these intricate relationships. The review also identifies potential applications and outlines future directions for leveraging phycosphere interactions in environmental and industrial contexts.

2.2 THE PHYCOSPHERE: A MICROBIAL INTERFACE AROUND MICROALGAE

The phycosphere refers to the localized microenvironment around microalgal cells, enriched with organic compounds they release, which promote intimate interactions with surrounding microbes (Seymour et al., 2017). Functioning as the aquatic counterpart of plant rhizosphere, the phycosphere is a vital microenvironment for nutrient transfer, microbial settlement, and chemically mediated communication (Seymour et al., 2017).

Microalgae are widely recognized for their capacity to form associations with diverse microorganisms across both natural ecosystems and controlled environments. Among these, prokaryotic bacteria constitute the most prevalent and influential partners (Cole, 1982; Seymour et al., 2017). Although axenic algal cultures are preferred for specific experimental and commercial purposes (Vu et al., 2018), huge biomass cultivation systems inherently host bacterial populations and are susceptible to microbial incursions. Early investigations into mass-cultured microalgae acknowledged the ubiquity of algal–bacterial interactions, yet often characterized their influence on algal growth as neutral or detrimental (Fuentes et al., 2016).

Contemporary research has revealed that these interactions are governed by intricate community dynamics, encompassing a spectrum of interdependence (Seymour et al., 2017). In light of these findings, the concept of the phycosphere emerged—analogous to the plant rhizosphere—supported by compelling evidence that bacteria exhibit chemotactic responses to algal exudates, thereby delineating a microenvironment of intense microbial–algal interplay (Seymour et al., 2017). Notably, the phycosphere is frequently inhabited by microbial pathogens capable of inducing measurable biomass loss. Microalgae, whether in marine habitats or industrial cultivation systems, are subject to both indirect growth suppression and direct cell lysis mediated by a range of organisms, including amoebae, and several other microorganisms (Gachon et al., 2010; Wang et al., 2010a; Carney & Lane, 2014; Carney et al., 2014).

Conversely, microbial symbioses with higher plants are well-documented, often enhancing host vitality and reproductive success (Lugtenberg & Kamilova, 2009; Nadeem et al., 2014). Similar mutualistic interactions, duly observed in microalgae, include *Chlorella sorokiniana*, which demonstrated elevated growth rates when cocultured with diazotrophic, indole-3-acetic acid-producing strains of *Azospirillum brasilense* and *Bacillus pumilus* (Amavizca et al., 2017), as well as with certain naturally associated fungal species (Watanabe et al., 2005). Crucially, host–microbiome dynamics are more accurately represented through multispecies abundance models rather than single-taxon analyses, underscoring the systemic interdependence inherent in biological communities (Sze & Schloss, 2018). Within the phycosphere, a finely tuned balance between beneficial and antagonistic interactions governs community structure, mediating outcomes that scale from mutualism to parasitism (Mugnai et al., 2022) (Figure 2.1).

2.3 MICROBIAL COMMUNITIES IN THE PHYCOSPHERE

2.3.1 Microbial Diversity

The phycosphere is colonized by a rich consortium of bacteria, archaea, viruses, and sometimes fungi. These microbes differ significantly from the ambient water column due to elevated levels of organic carbon and other compounds released by algae (Amin et al., 2012). Large multicellular macroalgae, commonly referred to as seaweeds, occupy the shoreline zone and offer extensive surfaces conducive to bacterial colonization. The symbiotic relationships between these macroalgae and their associated bacterial groups have been extensively documented (Singh & Reddy, 2014). Beyond serving as habitats for prokaryotic organisms, certain macroalgae derive direct physiological benefits from these associations. For instance, the green macroalga *Codium decorticatum* hosts cyanobacterial endosymbionts such as *Calothrix* sp., *Anabaena* sp., and *Phormidium* sp., which contribute to nitrogenous fixation for the algal host (Rosenberg & Paerl, 1981). In oligotrophic marine environments, even unicellular algal species can act as localized nutrient sources for bacteria, functioning analogously to particulate organic matter like fecal pellets or marine snow. These algae release substantial quantities of dissolvable organic matter that generates discernible nutrient gradients extending from the algal cell surface into the surrounding

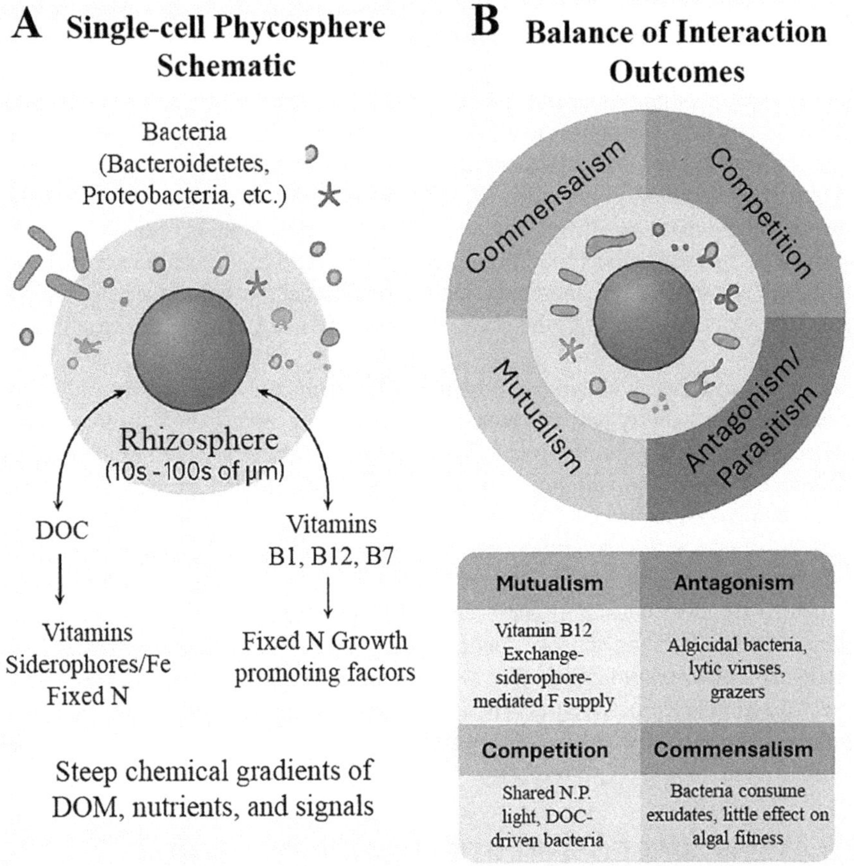

FIGURE 2.1 Conceptual model of the phycosphere as a microbial interface around microalgae. (a) Single-cell phycosphere schematic. (b) Balance of interaction outcomes.

water column (Amin et al., 2012). This microenvironment often harbors bacterial populations that differ markedly from those in the adjacent open water.

Investigations into algal-associated bacterial communities have revealed evidence of coevolutionary dynamics. For example, pyrosequencing of 16S rRNA genes from bacteria associated with *Pseudo-nitzschia* (three observed species) revealed distinct community compositions. Moreover, experimental movement of bacterial communities into non-native diatom hosts occasionally ensued in adverse effects on algal health (Sison-Mangus et al., 2014).

A comprehensive meta-analysis carried out by Goecke et al. (2013), examining bacterial associations across more than 40 macro- and microalgal species identified 101 bacterial isolates, predominantly belonging to six phyla: Bacteroidetes, Proteobacteria, Actinobacteria, Firmicutes, Verrucomicrobia, and Planctomycetes, possessing the following genera assemblage: 42, 32, 8, 7, 3, and 1, respectively. Similarly, broad surveys of diatom-bacteria coexistence have highlighted the dominance of the first two genera of Bacteroidetes and Proteobacteria (Amin et al., 2012). Notably, analogous patterns of microbial enrichment have been observed in terrestrial systems; for instance, the maize rhizosphere exhibits a higher abundance of Proteobacteria compared to bulk soil (Peiffer et al., 2013).

Among the Alphaproteobacteria, the Rhodobacterales family demonstrates a notable adaptation for close associations with phytoplankton (Buchan et al., 2014). Affiliates of this group are frequently detected in natural marine algal communities and exhibit increased abundance during phytoplankton bloom events (Teeling et al., 2012; Mayali et al., 2011). A prominent example is *Dinoroseobacter shibae*, which has been identified as a mutualistic partner to three distinct marine microalgal species—*Amphidinium operculatum*, *Isochrysis galbana*, and *Thalassiosira pseudonana*—indicating its broad symbiotic potential (Paul et al., 2012; Wagner-Döbler et al., 2010).

In freshwater environments, Rhizobiales family, a subset of Alphaproteobacteria, is frequently encountered as a dominant component of xenic green algal cultures, accounting for approximately 50% of the bacterial community in cultures of *Chlamydomonas reinhardtii*, *Chlorella vulgaris*, and *Scenedesmus* spp. (Kim et al., 2014). Beyond their prevalence, Rhizobiales have also been implicated in promoting algal growth; for instance, the *Rhizobium* strain isolated from *C. vulgaris* was shown to significantly enhance the advancement of multiple algal species (Kim et al., 2014). Although rhizobial presence in marine systems appears to be less common, recent evidence has revealed the presence of rhizobial DNA in cultures of 13 strains of *Ostreococcus tauri*, suggesting a broader ecological distribution than previously recognized (Abby et al., 2014) (Figure 2.2).

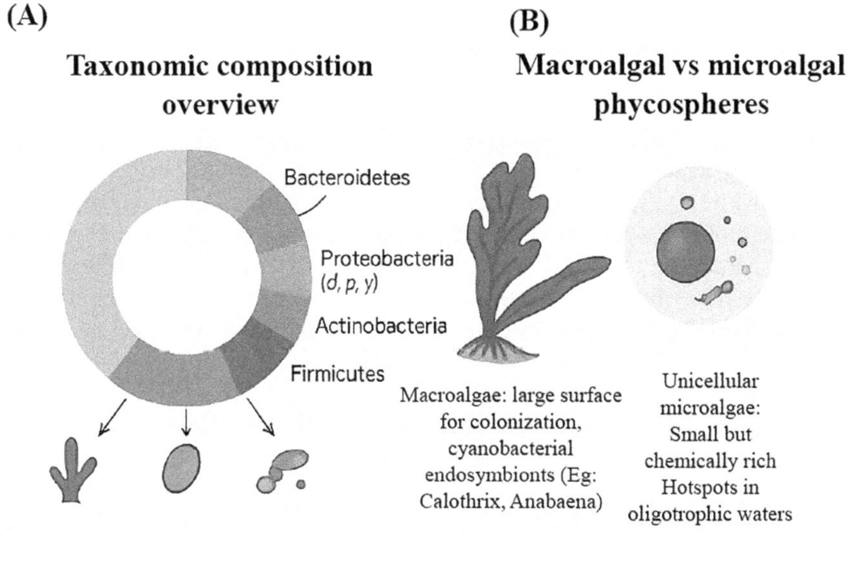

FIGURE 2.2 Taxonomic and habitat diversity of microbial communities in the phycosphere. (a) Taxonomic composition overview. (b) Macroalgal vs microalgal phycospheres.

2.3.2 Community Assembly Factors

Certain algal species form specific associations with distinct bacterial taxa as posited by Sison-Mangus et al. (2014), while another researches had suggested that such relationships may be more stochastic, shaped largely by environmental variables. This perspective is supported by findings such as the environmental modulation of bacterial communities associated with *Leptocylindrus* sp. (Ajani et al., 2018), temporal shifts in microbial composition during algal cultivation (Sapp et al., 2007; Riemann et al., 2000; Bagatini et al., 2014), and the emergence of unique bacterial assemblages in separate laboratory cultures of *Pseudo-nitzschia multiseries* (Kaczmarska et al., 2005). Additionally, non-native bacterial strains, including *Escherichia coli*, have been shown to enhance diatom growth, suggesting that beneficial interactions may occur independently of coevolutionary history (Bruckner et al., 2011).

The dynamic nature of bacterial coexistence with the green algae, *Tetraselmis suecica*, across its growth phases further supports the notion that stable coadaptation is not a prerequisite for functional symbiosis (Park et al., 2021; Stock et al., 2019). Stock et al. (2019) also emphasized the role of environmental filtering in shaping diatom-associated microbiomes, noting that while community assembly may appear random, host-specific microhabitats impose selective constraints on microbial composition. Grossart et al. (2005) similarly observed fluctuations in bacterial populations during algal development, indicating that environmental variability may favor bacteria with flexible adaptive traits. Despite evidence supporting environmentally driven associations, consistent co-occurrence of particular bacterial species with specific algae has been documented in both laboratory and natural settings. For instance, bacterial productivity often correlates with phytoplankton biomass in marine systems (Pinhassi et al., 2004), and algal species composition has been shown to alter bacterial community structure (Kerkhof et al., 1999). These findings suggest a degree of reciprocal adaptation, where algae release organic compounds that enrich the surrounding phycosphere and potentially attract compatible bacterial partners. Signal-mediated interactions may further reinforce these associations, as certain bacteria respond to algal exudates by establishing persistent relationships.

Species-specific bacterial communities have also been identified in cyanobacterial cultures, with intra-species similarity exceeding interspecies variation, supporting the concept of taxon-specific microbial associations (Shi et al., 2009). Sapp et al. (2007) reported that microalgae often harbor distinct bacterial consortia, and Stock et al. (2019) described vastly explicit interactions of *Cylindrotheca* strains and their associated bacteria. While not all algae–bacteria interactions exhibit strict specificity, recurring patterns of co-occurrence suggest that some bacterial taxa preferentially associate with particular algal groups (Trainer et al., 2020). Thus, species specificity should be considered at the taxonomic level rather than generalized across all bacteria or algae, reflecting the nuanced complexity of these ecological relationships.

Several factors may underlie these associations, including nutritional compatibility. For example, the frequent links between algae and Bacteroidetes are predicated on the latter's affinity for environments rich in organic macromolecules (Goecke et al., 2013). Advances in metagenomic and metaproteomic approaches have further clarified these interactions. Zheng et al. (2020) demonstrated that in *Synechococcus* cultures, Flavobacteria dominate early stages by degrading complex organic substrates, while *Oricola* sp. (Alphaproteobacteria) becomes prevalent later, utilizing simpler dissolved carbon compounds.

Synthetic phycosphere models have shown that bacterial community composition can be anticipated based on phytoplankton identity and metabolite profiles, indicating that resource potentials play a central role in microbial recruitment (Fu et al., 2020). This supports the hypothesis that phytoplankton can actively shape their microbiomes by giving forth targeted metabolites that attract beneficial bacteria. Beyond nutritional preferences, mutualistic exchanges may also sustain these associations. For instance, *Methylophaga* sp. and *Donghicola* sp. have been implicated in nitrogen provisioning to the diatom *Phaeodactylum tricornutum* through methylamine degradation, while the diatom supplies organically dissolved carbon to support bacterial growth (Suleiman et al., 2016).

Environmental factors like nutrient availability, water turbulence, salinity, temperature, and pH significantly impact the phycosphere by influencing the composition and activity of the phycosphere. Given that nutrient exchange is central to these relationships, any alteration in environmental conditions that disrupts this exchange equilibrium can lead to shifts in the interaction dynamics. For example, the form of inorganic nitrogen present in the environment (such as nitrate, ammonium, or their combination) has been shown to influence the release of extracellular organic matter by algae, thereby modulating their mutualistic engagement with bacteria (Perera et al., 2022). Supporting this, Cao et al. (2020) reported that elevated phosphorus levels and a balanced nitrogen-to-phosphorus ratio promote cooperative interactions, as bacteria utilize algal-derived dissolved organic carbon. In contrast, an overabundance of carbon can drive competitive behavior, with bacteria sequestering nutrients and suppressing algal proliferation.

Investigations into *Phaeocystis globosa* and its microbial consortia revealed that the presence of bacteria could sustain algal growth under B-vitamin deprivation, whereas nitrogen limitation accelerated culture collapse—highlighting the sensitivity of interaction modes to nutrient shifts. Similarly, Palacios et al. (2018) observed that coexistent cultures of *Chlorella sorokiniana* and *Azospirillum brasilense* produced signaling compounds like tryptophan and indole-3-acetic acid (IAA) under nitrogen-rich conditions, but these molecules were absent when nitrogen or carbon was withheld. Iron scarcity has also been implicated in triggering the manufacture of bioactive metabolites in *Chlorella*, which confer resistance against the pathogen *Vampirovibrio chelorellavorus* (Bagwell et al., 2016).

Beyond nutrient dynamics, thermal conditions significantly influence bacterial physiology and community structure. For instance, *Ruegeria* spp. (Rhodobacteriaceae) exhibited pathogenic behavior toward the coccolithophore *Emiliania huxleyi* at elevated temperatures (25°C) but remained benign at lower temperatures (18°C) (Mayers et al., 2016). Another determinant of interaction mode is the molecular signaling received by bacteria from their algal hosts. A well-documented case involves *Phaeobacter gallaeciensis* (Roseobacter clade), which initially supports *E. huxleyi* growth by producing antimicrobial agents and phytohormones that suppress competing microbes and stimulate algal development. In return, the bacterium

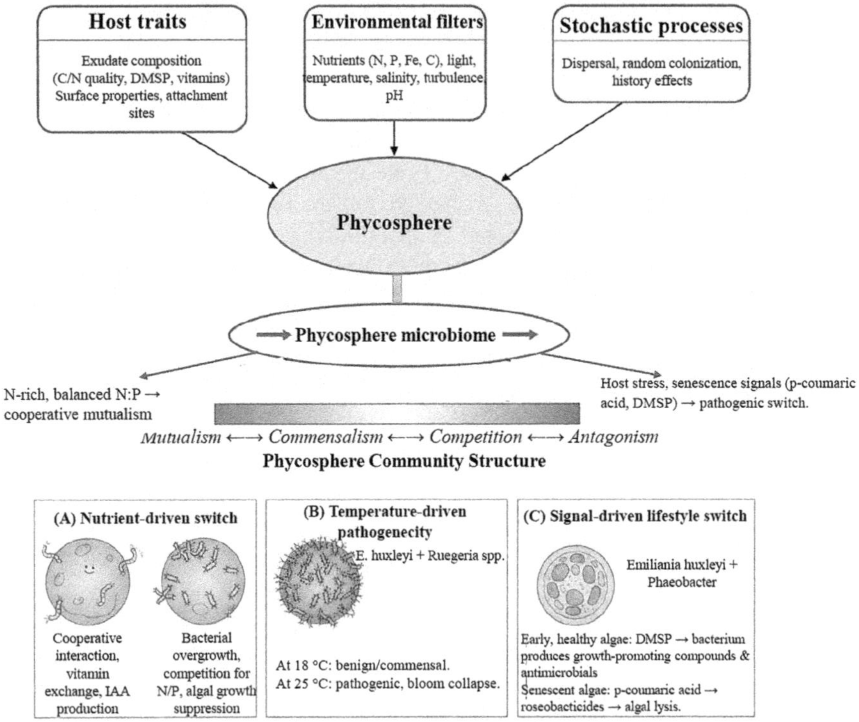

FIGURE 2.3 Drivers of bacterial community assembly and interaction outcomes in the phycosphere.

benefits from algal-derived nutrients and colonization surfaces. However, this mutualism transitions to pathogenicity upon detection of p-coumaric acid, which prompts the synthesis of roseobacticides, potent algicidal compounds (Seyedsayamdost et al., 2011).

An akin shift was observed by Segev et al. (2016), where *Phaeobacter inhibens* transitioned from a growth-promoting symbiont to a lethal pathogen of *E. huxleyi*. More recently, Barak-Gavish et al. (2023) demonstrated that DMSP, released by *E. huxleyi*, modulates the lifestyle of *Sulfitobacter* spp. While DMSP promotes pathogenic behavior, the concurrent production of benzoate by the alga counteracts this effect, enabling stable coexistence between the two organisms (Figure 2.3).

2.4 CHEMICAL ECOLOGY OF THE PHYCOSPHERE

The chemical ecology of the phycosphere involves the complex chemical interactions between phytoplankton and the bacteria surrounding them. Microalgae release a complex mixture of exudates, including dissolved organic carbon, vitamins (e.g., B_{12}), phytohormones, and sulfur compounds (e.g., DMSP). These compounds serve multiple functions, such as feed for heterotrophic bacteria, acting as chemoattractants, and mediating the communication between the microbes (Amin et al., 2012). DMSP produced by algae such as *Emiliania huxleyi* supports sulfur-metabolizing bacteria, while simultaneously modulating bacterial behavior (Durham et al., 2015). Compounds produced by symbiotic bacteria that enhance algal growth encompass a diverse array of bioactive substances, including micronutrients such as vitamins, indispensable minerals, PUFA (polyunsaturated fatty acids), QS mediators, plant hormone analogs like auxins, as well as various carbohydrates and proteins (Krug et al., 2020; Zhang et al., 2020). Numerous microalgal taxa lack the biosynthetic capacity to produce certain essential vitamins, such as cobalamin and thiamin, and thus rely on associated bacterial communities to fulfill these nutritional requirements (Natrah et al., 2014). For instance, *Amphidinium operculatum* acquires cobalamin through its symbiotic relationship with *Halomonas* spp. (Croft et al., 2005). Similarly, *Ruegeria pomeroyi*, which coexists with *Thalassiosira pseudonana*, contributes vitamin B12 to the partnership while utilizing organic carbon compounds released by its associate (Seymour et al., 2017).

Beyond vitamin provisioning, bacteria linked with algal hosts synthesize a variety of bioactive molecules, including plant hormone analogs such as indole-3-acetic acid, iron-chelating siderophores, and antimicrobial agents (Lian et al., 2018; Natrah et al., 2014; Villarreal-Gómez et al., 2010). The phycosphere plays a critical role in regulating the availability of these metabolites and mediating the biochemical interactions between algae and their microbial partners (Seymour et al., 2017). Within this niche, microalgae release substantial quantities of organic compounds, often accounting for up to half of their photosynthetic output, which supports the metabolic activity of beneficial bacteria. These bacteria reciprocate by producing growth-promoting substances that enhance algal physiology and development.

2.5 TYPES OF ALGAL–MICROBIAL INTERACTIONS

2.5.1 Mutualism

Mutualism in the phycosphere involves a mutually beneficial relationship between a microalga and associated bacteria within a chemical exchange zone called the phycosphere. Symbiotic relationships in which distinct species engage in reciprocal benefit are broadly categorized as mutualistic (Fuentes et al., 2016; Cooper & Smith, 2015; Seyedsayamdost et al., 2011; Wang et al., 2010b). Within the phycosphere, mutualistic coexistence of microalgae and bacteria is frequently observed and represents a dominant ecological pattern (Buchan et al., 2014). One well-documented form of this cooperation involves bacterial provision of essential vitamins, such as thiamine (B1), biotin (B7), and cobalamin (B12), to algal species that lack the biological means of pathways for their production. In return, the algae release organic carbon compounds that serve as substrates for bacterial metabolism (Ramanan et al., 2016; Helliwell et al., 2011; Croft et al., 2005). Further evidence suggests that certain bacterial taxa contribute to algal development by synthesizing bioactive compounds, including phytohormones, while simultaneously assimilating organic molecules and amino acids released by their algal hosts (Kim et al., 2014). The complexity of these interactions is exemplified by the dynamic bond between the coccolithophore *Emiliania huxleyi* and the bacterium *Phaeobacter gallaeciensis*. During the early stages of algal growth, when *E. huxleyi* actively secretes DMSP, the bacterium engages in a cooperative role by producing growth-enhancing substances and antimicrobial agents that protect the alga from pathogenic threats. However, as the algal cells enter senescence, they begin to release p-coumaric acid—a compound associated with cellular aging—which acts as a molecular cue for *P. gallaeciensis* to shift its behavior. In response, the bacterium transitions from a mutualistic partner to a parasitic antagonist, synthesizing roseobacticides that induce algal cell death and lysis (Seyedsayamdost et al., 2011). Findings from Suleiman et al. (2016), Amin et al., 2015, and Villa et al. (2014) indicate that microalgae primarily assimilate nitrogen (inorganic forms), as nitrate and ammonium, through two predominant mechanisms, that is engaging in symbiotic or commensal relationships with nitrogen-fixing bacterial species and via forming mutualistic associations with bacteria capable of converting monomethylamine into bioavailable form (ammonium).

Iron, though vital for numerous cellular processes, remains a scarce resource in aquatic environments (Kazamia et al., 2016; Wan et al., 2014). Its limited availability poses a constraint on photosynthetic efficiency and overall algal productivity. A notable example of nutrient exchange within the phycosphere of autotrophic microalgae and heterotrophic bacteria involves the synthesis of the siderophore vibrioferrin by *Marinobacter* sp., which chelates ferric iron (Fe^{3+}), thereby enhancing its accessibility to the dinoflagellate *Scrippsiella trochoidea*. This mutualistic interaction exemplifies a bidirectional exchange, wherein the bacterium supplies iron while receiving carbon compounds from its algal partner (Amin et al., 2009).

2.5.2 Commensalism

Commensalism in the phycosphere describes interactions where one organism, like a bacterium, benefits by living near algae (the host) without harming the host. Distinguishing between ecological interactions such as commensalism, mutualism, and parasitism can be challenging due to their fluid boundaries. Commensalism, in particular, may be conceptualized as an intermediate state within a broader spectrum of interspecies relationships (Zapalski, 2011). Empirical evidence from studies on the phycosphere suggests that environmental variables—especially nutrient concentrations of nitrogen and phosphorus, along with light availability—can drive transitions between mutualistic and parasitic modes of interaction, often passing through a commensal phase (Ramanan et al., 2016).

An illustrative case involves the cocultivation of the green microalga *Neochloris oleoabundans* with the bacterium *Azotobacter vinelandii*. In this association, the microalga accesses nitrogen via the siderophore azotobactin, secreted by the bacterium. *A. vinelandii* is capable of nitrogen fixation under aerobic conditions, provided it receives reduced carbon substrates (sucrose or glycerol), which are supplied by its algal partner. Additionally, the bacterium synthesizes a suite of siderophores that mobilize trace metals from the surrounding medium, further supporting algal growth (Villa et al., 2014). These observations point to a commensalistic interaction, with potential to evolve into a mutualistic partnership under optimized conditions.

2.5.3 Antagonism

Both antagonism and parasitism occur in the phycosphere, where parasitism is a specific type of antagonism. Parasitism involves a parasite harming its host (like an alga), while broader antagonism includes other negative interactions like predation, competition, and antibiosis, which can occur between algae and other microbes, such as bacteria. Copious bacterial taxa identified as detrimental to microalgal populations, prompting increasing interest in their potential relevance as biological agents for managing harmful fungal and cyanobacterial blooms (Lee et al., 2010; Wang et al., 2010b; Kim et al., 2008). For instance, seven algicidal bacterial strains were isolated from the microenvironment surrounding the dinoflagellate *Cochlodinium polykrikoides*, demonstrating the capacity to suppress bloom formation through induction of algal cell lysis (Kim et al., 2008). These bacteria exert their effects via the secretion of hydrolytic enzymes, including β-glucosidase and chitinase, which compromise algal cell integrity and facilitate lysis (Wang et al., 2010b). Following cell disruption, bacterial cells utilize the released intracellular contents as nutrient sources.

Conversely, certain microalgal species exhibit resistance to bacterial antagonism. Although the underlying mechanisms remain poorly characterized, the diatom *Chaetoceros didymus* has shown resilience against lytic activity from *Kordia algicida*, suggesting the presence of innate or inducible defense strategies (Carsten & Pohner, 2011).

2.5.4 Competition

Competition in the phycosphere involves bacteria vying for resources released by phytoplankton, such as dissolved organic matter (DOM). Microbial competition arises when different microorganisms coexist within the same habitat and vie for limited nutrients and resources. This phenomenon was illustrated through a comparative study of axenic and non-axenic cultures of *Scenedesmus obliquus* subjected to phosphate-deficient situations, where the existence of bacteria in non-axenic cultures led to a marked reduction in algal growth, indicating competitive nutrient uptake by the bacterial community (Danger et al., 2007).

Additionally, Bratbak and Thingstad (1985) explored a counterintuitive ecological dynamic involving phytoplankton and bacteria using a theoretical model. Their simulation demonstrated that under conditions of mineral nutrient scarcity, phytoplankton release extracellular organic carbon, inadvertently promoting bacterial proliferation. As bacterial populations expand, their demand for mineral nutrients intensifies, thereby exacerbating the nutrient limitation experienced by the phytoplankton. This feedback loop exemplifies a paradox in which nutrient-stressed algae inadvertently support the growth of their microbial competitors, further intensifying resource competition.

2.6 TOOLS AND TECHNOLOGIES TO STUDY PHYCOSPHERE INTERACTIONS

Recent methodological advances include metagenomics and metatranscriptomics to reveal community structure and function, stable isotope probing (SIP) to trace nutrient flows (Mayali et al., 2011), single-cell omics and microfluidics to study interactions at the microscale (Grossart et al., 2020), and fluorescence in situ hybridization (FISH) to visualize microbial colonization. These tools help overcome the limitations of culture-based studies and allow insights into unculturable taxa.

Recent advancements in high-throughput genome sequencing technologies, including metagenomics and next-generation sequencing, have opened new avenues for exploring the microbial diversity within the phycosphere. Complementary omics approaches—such as proteomics, metabolomics, and transcriptomics—offer deeper discernments into the biochemical roles and ecological significance of metabolites produced by these bacterial groups (Zhang et al., 2020). For example, transcriptomic analysis enabled Cooper and Smith (2015) to elucidate the thiamine biosynthesis pathway in *Pseudoalteromonas* sp. TW7 when cocultured with *Ostreococcus tauri*. These findings underscore the potential of functional metagenomics, especially when paired with optimized cultivation strategies, to drive the discovery of novel bioactive molecules.

The concept of engineering bacterial consortia represents a promising frontier in industrial microbiology. Such consortia, composed of multiple bacterial species, can perform intricate biochemical processes that exceed the capabilities of individual strains (Bhatia et al., 2018). Therefore, a comprehensive understanding of microalgae–bacteria interactions, supported by advanced functional genomics, is essential for designing tailored microbial communities with versatile applications. Genome sequencing plays a pivotal role in this process by enabling precise identification of community members and quantifying their relative abundance. As such, DNA sequencing technologies are indispensable tools in the development and refinement of synthetic consortia, forming the foundation for subsequent functional and ecological analyses.

Transcription, the process by which DNA is converted into RNA, represents a critical regulatory checkpoint in gene expression. Consequently, measuring RNA abundance through metatranscriptomic approaches offers a more precise reflection of functional gene activity. Transcriptomic profiling is particularly valuable for examining how microbial communities respond to varying environmental conditions and developmental stages, a common focus in consortium-based research (Vannier et al., 2023; Wang et al., 2023).

Yi et al. (2020) employed transcriptomic analysis to investigate gene expression dynamics within a wastewater-degrading association composed of *Chlorella* sp. and *Cupriavidus necator*, exposed to phenol. Comparative analysis between the algal monoculture and the consortium revealed differential regulation of 3,529 genes, with notable enrichment in pathways associated with carbon dioxide assimilation and light-harvesting complexes. These transcriptomic findings, coupled with biochemical indicators such as elevated oxygen output in the consortium, support a model in which phenol-degrading bacteria supply CO_2 to the microalgae, which reciprocate by releasing oxygen through photosynthetic activity.

Microbial populations within the phycosphere play integral roles in the origination, preservation, and degeneration of algal blooms through diverse processes such as nutrient turnover, chemical interference, and interspecies communication (Kouzuma & Watanabe, 2015; Shao et al., 2014). Early phases of *Microcystis aeruginosa* proliferation, nitrogen-fixing bacteria often increase in abundance, contributing bioavailable nitrogen to support phytoplankton growth (Yang et al., 2017). Concurrently, these bacteria compete with algae for limited resources, particularly phosphorus (Xu et al., 2022).

As blooms of *Phaeocystis globosa* mature and nutrient levels diminish, microbial succession favors functional groups like denitrifiers and sulfate-reducing bacteria, which enhance nutrient recovery and recycling (Shi et al., 2023). Under these conditions, bacterial utilization of phosphate is exaggerated, resulting in elevated concentrations of inorganic phosphorus in the surrounding medium (Xu et al., 2022). Insights into the molecular and metabolic interactions between *Phaeocystis* and its associated bacterial communities have been significantly advanced through integrative omics approaches, shedding light on the complexity of these ecological relationships (Shi et al., 2023; Gibson et al., 2022; Xu et al., 2022) (Table 2.1).

2.7 ECOLOGICAL AND EVOLUTIONARY IMPLICATIONS

Phycosphere interactions significantly shape the structure and function of aquatic ecosystems. These interactions have ecological consequences that ripple across food webs, biogeochemical cycles, and bloom dynamics. Microbial partners can both promote and suppress phytoplankton blooms. Beneficial bacteria, for instance, may enhance algae growth via nutrient supplementation, while antagonistic microbes release algicidal compounds that trigger bloom collapse (Mayali & Azam, 2004). A notable example is *Ruegeria pomeroyl*, which promotes the growth of *Emiliania huxleyi*, while certain Roseobacter strains contribute to its mortality (Seyedsayamdost et al., 2011).

The phycosphere serves as a crucial interface for nutrient exchange, including carbon cycling—-alga photosynthesis fixes CO_2, some of which is transferred to bacteria as DOC. For nitrogen cycling, diazotrophic bacteria fix atmospheric nitrogen, making it available to algae (Louca et al., 2016). DMSP production and degradation affect global sulfur emissions and climate. Long-term associations have led to metabolic coevolution, as seen in microalgae that have lost the ability to synthesize certain vitamins (e.g., B12) and rely on bacterial partners. This dependency reflects strong co-adaptive pressures in the marine microbial loop (Croft et al., 2005; Helliwell et al., 2011).

2.8 APPLICATIONS FOR BIOTECHNOLOGY AND ENVIRONMENTAL MANAGEMENT

Harnessing the power of phycosphere interactions has opened new frontiers in applied science. Engineered algal–microbial consortia are being developed for various industrial and environmental applications.

2.8.1 Wastewater Treatment

Microalgae–bacteria consortia efficiently remove nutrients and contaminants from wastewater. Algae provide oxygen via photosynthesis, supporting aerobic bacterial degradation of organic pollutants (Li et al., 2019). Engineered algal–bacterial consortia improve nitrogen and phosphorus removal.

TABLE 2.1
Interaction of Microalgae and Microbes

Interaction Type	Description	Examples and Mechanisms	References
Mutualism	Microalgae and microbes benefit each other	• Nutrient exchange: Bacteria break down complex compounds, and microalgae provide oxygen via photosynthesis. • Growth promotion: Bacteria produce growth-promoting hormones (indole-3-acetic acid) in response to fungal exudates like tryptophan, which in turn promotes algal growth. • Vitamin provision: Bacteria like *Variovorax* and *Porphyrobacter* provide vitamins (e.g., B vitamins) to microalgae. Bacteria provide essential vitamins (e.g., B1, B7, and B12) to auxotrophic microalgae, which in turn provide bacteria with fixed organic carbon from photosynthesis. Bacteria also perform nutrient regeneration (nitrogen and phosphorus) in exchange for organic carbon compounds released by microalgae.	Abate et al., 2024a; Iqhrammullah et al., 2024; Astafyeva et al., 2022; Amin et al., 2012, 2015
Parasitism and competition	One partner benefits at the expense of the other, or both compete for resources.	• Competition for nutrients: Bacteria compete with algae for nutrients available in the phycosphere. • Pathogenesis: Some microbes can parasitise microalgae. • Antibiotic production: Some bacteria produce antibiotics, potentially inhibiting competing organisms or establishing a dominant niche.	Abate et al., 2024b; Chakraborty et al., 2020
Commensalism	One partner benefits, while the other is neither harmed nor helped.	• Grazing: Microbes may consume a fraction of the organic matter exuded by microalgae without causing significant harm or benefit to the algae. • Certain bacteria may use the algal surface as a habitat or for defense against predation without significantly impacting the alga's primary functions.	Fuentes et al., 2016
Synergistic interactions in bioremediation	The combined action of microalgae and microbes is greater than the sum of their individual effects.	• Wastewater treatment: Microalgae–bacteria consortia can efficiently degrade organic pollutants, remove nutrients, pathogens, and heavy metals. • Biodegradation of pollutants: For example, a consortium of microalgae and bacteria achieved over 99% biodegradation of motor oil.	Dinh et al., 2022
Antagonism	One organism is harmed while the other benefits or is unaffected.	• Some bacteria produce algicidal compounds or toxins that harm or lyse algal cells, often in later stages of a bloom to access nutrients from decaying biomass. Microalgae can also release allelochemicals to inhibit the growth of competing bacterial species.	Chakraborty et al., 2020; Dow, 2021; Rolland et al., 2016; Zhu et al., 2022

2.8.2 Biofuel and Bioproduct Production

Bacterial partners can enhance algae lipid accumulation and growth rates, boosting yields of biofuels, pigments, antioxidants, and polysaccharides (Ramanan et al., 2016). Some bacteria also degrade inhibitors or supply growth factors that optimize algae productivity. Algal productivity can be enhanced by beneficial microbial partners (Ramanan et al., 2016).

2.8.3 Aquaculture

Algae biomass is a viable feedstock in aquaculture. Probiotic bacteria in the phycosphere can improve nutrient profiles, suppress pathogens, and enhance digestibility.

In aquaculture environments, bacteria associated with microalgae contribute to the suppression of potentially harmful microorganisms, including those pathogenic to fish larvae, through the production of bioactive compounds such as antimicrobials. For instance, cultures of *Tetraselmis* spp. have demonstrated significant antimicrobial efficacy against common aquaculture pathogens (Tulli et al., 2012). This microbial activity may partially account for the improved survival rates observed in fish larvae reared using the "green water" method (Silva, 1999; Reitan et al., 1997). In this approach, microalgae—integral to the larval diet—are cultivated directly within the rearing tanks. The method offers several benefits, including enhanced larval health and survival, which may be attributed to mechanisms such as competitive exclusion, whereby beneficial microbes outcompete pathogenic species for resources and space.

2.8.4 Climate Change Mitigation

Optimized consortia can be used in carbon capture systems, improving both carbon fixation and biomass productivity. The production and degradation of DMSP by these communities may also impact cloud formation and climate feedback loops (Durham et al., 2015). In their review, Jack et al. (2023) noted the effectiveness of microalgae in suppressing methane gas production in ruminants. Phycosphere has been noted to enhance CO_2 capture and sequestration via optimized phycosphere dynamics (Figure 2.4).

2.9 RESEARCH GAPS AND FUTURE DIRECTIONS

Despite significant progress, the study of phycosphere interactions remains limited by several methodological and conceptual gaps. Most studies are conducted under controlled laboratory conditions. Neutral environments present dynamic, multifactorial contexts (including diel cycles, multispecies interactions, and variable nutrient regimes) that are often overlooked. The dynamics of microbial community shifts throughout the developmental stages of algae, as well as their fine-scale spatial variability, remain largely unexplored. Advanced live-cell imaging and real-time metabolomics could address this gap. Few standardized synthetic communities have been

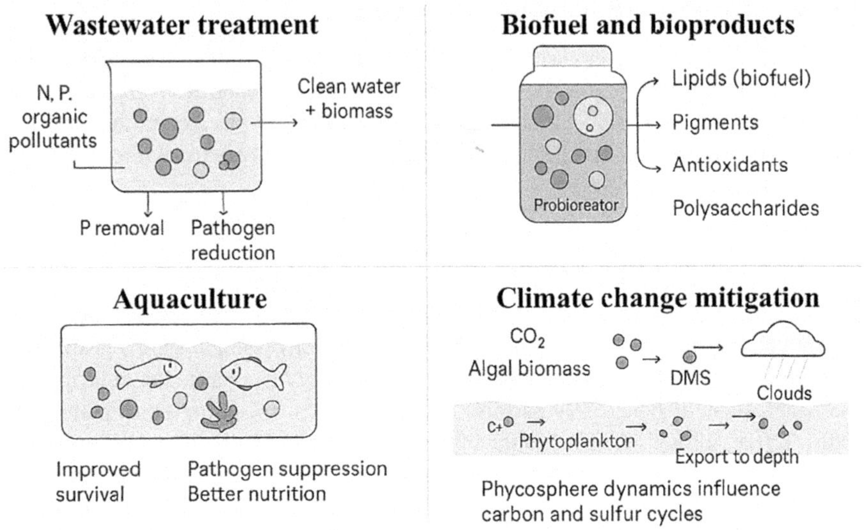

FIGURE 2.4 Applied microalgae–microbe consortia enabled by phycosphere interactions.

developed for mechanistic studies. Creating model phycospheres with defined species and metabolic traits would allow better control and reproducibility. Combining multi-omics (genomics, metabolomics, and transcriptomics) with computational modeling, network analysis, and machine learning can uncover deeper insights into microbial interactions and emergent properties of the phycosphere. Future research must embrace systems biology approaches, linking molecular functions to ecological outcomes (Faust & Raes, 2012).

2.10 CONCLUSION

The phycosphere represents a frontier in microbial ecology—an interface rich with chemical, ecological, and evolutionary interactions. As we move beyond a purely algae or microbe-centric perspective, a more holistic understanding of algae–microbe consortia is emerging, highlighting their central roles in nutrient cycling, bloom regulation, and aquatic productivity.

Ongoing advancements in omics technologies, synthetic biology, and environmental monitoring are rapidly expanding our ability to study and engineer these interactions. Addressing the current research gaps—especially through integrative, interdisciplinary approaches—will unlock new opportunities for sustainable biotechnology, climate mitigation, and ecosystem restoration.

By elucidating the principles governing life in the phycosphere, we not only deepen our ecological knowledge but also pave the way for transformative applications at the interface of biology and the environment.

REFERENCES

Abate, R., Oon, Y., Oon, Y., & Bi, Y. (2024a). Microalgae-bacteria nexus for environmental remediation and renewable energy resources: Advances, mechanisms, and biotechnological applications. *Heliyon 10*(10), e31170. https://doi.org/10.1016/j.heliyon.2024.e31170

Abate, R., Oon, Y., Oon, Y., Bi, Y., Mi, W., Song, G., & Gao, Y. (2024b). Diverse interactions between bacteria and microalgae: A review for enhancing harmful algal bloom mitigation and biomass processing efficiency. *Heliyon 10*(17), e36503. https://doi.org/10.1016/j.heliyon.2024.e36503

Abby, S.S., Touchon, M., De Jode, A., Grimsley, N., & Piganeau, G. (2014). Bacteria in *Ostreococcus 'tauri* cultures: P friends, foes or hitchhikers? *Frontiers in Microbiology 5*, 505.

Adegbeye, M., Jack, A., Sierra-Alarcón, A.M., Inyang, U., Elghandour, M.M.M.Y., Maggiolino, A., De Palo, P., & Salem, A.Z.M. (2025). Microalgae and their impact on ruminant nutrition and health. In: Antonio Pérez-Gálvez, Eduardo Jacob-Lopes, Leila Queiroz Zepka, and María Roca (eds.), *Microalgae and One Health Fundamentals, Biocompounds, and Health and Environmental Applications* (pp. 493–504). Academic Press, Elsevier.

Ajani, P.A., Kahlke, T., Siboni, N., Carney, R.L., Murray, S.A., & Seymour, J.R. (2018). The microbiome of the cosmopolitan diatom *Leptocylindrus* reveals significant spatial and temporal variability, *Frontiers in Microbiology 9*. https://doi.org/10.3389/fmicb.2018.02758

Amavizca, E., Bashan, Y., Ryu, C.-M., Farag, M.A., Bebout, B.M., & de-Bashan, L.E. (2017). Enhanced performance of the microalga *Chlorella sorokiniana* remotely induced by the plant growth-promoting bacteria *Azospirillum brasilense* and *Bacillus pumilus*. *Scientific Report 7*, 41310.

Amin, S.A., Green, D.H., Hart, M.C., Küpper, F.C., Sunda, W.G., & Carrano, C.J. (2009). Photolysis of iron-siderophore chelates promotes bacterial-algal mutualism. *Proceedings of the National Academy of Sciences of the United States of America 106*, 17071–17076.

Amin, S.A., Hmelo, L.R., Van Tol, H.M., Durham, B.P., Carlson, L.T., Heal, K.R., Morales, R.L., Berthiaume, C.T., Parker, M.S., & Djunaedi, B. (2015). Interaction and signalling between a cosmopolitan phytoplankton and associated bacteria. *Nature 522*, 98–101.

Amin, S.A., Parker, M.S., & Armbrust, E.V. (2012). Interactions between diatoms and bacteria. *Microbiology and Molecular Biology Reviews 79*(3), 293–310.

Astafyeva, Y., Gurschke, M., Qi, M., Bergmann, L., Indenbirken, D., de Grahl, I., Katzowitsch, E., Reumann, S., Hanelt, D., Alawi, M., Streit, W.R., & Krohn, I. (2022). Microalgae and bacteria interaction—Evidence for division of diligence in the alga microbiota. *Microbiology Spectrum 10*(4), 1–15.

Azam, F. & Ammerman, J.W. (1984). *Flows of Energy and Materials in Marine Ecosystems: Theory and Practice*, Fasham, M.J.R. (ed.). Springer, pp. 345–360.

Bagatini, I.L., Eiler, A., Bertilsson, S., Klaveness, D., Tessarolli, L.P., & Vieira, A.A.H. (2014). Host-specificity and dynamics in bacterial communities associated with bloom-forming freshwater phytoplankton. *PLoS One 9*, 1–13.

Bagwell, C.E., Abernathy, A., Barnwell, R., Milliken, C.E., Noble, P.A., Dale, T., Beauchesne, K.R., & Moeller, P.D.R. (2016). Discovery of bioactive metabolites in biofuel microalgae that offer protection against predatory bacteria, *Frontiers in Microbiology 7*. https://doi.org/10.3389/fmicb.2016.00516

Barak-Gavish, N., Dassa, B., Kuhlisch, C., Nussbaum, I., Brandis, A., Rosenberg, G., Avraham, R., & Vardi, A. (2023). Bacterial lifestyle switch in response to algal metabolites. *Elife 12*.

Bhatia, S.K., Bhatia, R.K., Choi, Y.K., Kan, E., Kim, Y.G., & Yang, Y.H. (2018). Biotechnological potential of microbial consortia and future perspectives. *Critical Review Biotechnology 38*(8), 1209–1229. https://doi.org/10.1080/07388551.2018.1471445

Bratbak, G., & Thingstad, T.F. (1985). Phytoplankton–bacteria interactions: An apparent paradox? Analysis of a model system with both competition and commensalism. *Marine Ecology Progress Series 25*, 23–30.

Bruckner, C., Rehm, C., Grossart, H.-P., & Kroth, P. (2011). Growth and release of extracellular organic compounds by benthic diatoms depend on interactions with bacteria. *Environmental Microbiology 13*, 1052–1063. https://doi.org/10.1111/j.1462-2920.2010.02411.x

Buchan, A., LeCleir, G.R., Gulvik, C.A., & González, J.M. (2014). Master recyclers: Features and functions of bacteria associated with phytoplankton blooms. *Nature Review Microbiology 12*, 686–698.

Cao, X., Li, H., Zhou, Y., & Song, C. (2020). The shift of mutualistic relationships among algae, free-living and attached bacteria through different nutrient addition modes: A mesocosm study. *Journal of Freshwater Ecology 35*, 535–548.

Carney, L.T., & Lane, T.W. (2014). Parasites in algae mass culture. *Aquatic Microbiology 5*, 278.

Carney, L.T., Reinsch, S.S., Lane, P.D., Solberg, O.D., Jansen, L.S., Williams, K.P., Trent, J.D., & Lane, T.W. (2014). Microbiome analysis of a microalgae mass culture growing in municipal wastewater in a prototype OMEGA photobioreactor. *Algal Research 4*, 52–61.

Carsten, P., & Pohner, G. (2011). Interactions of the algicidal bacterium *Kordia algicida* with diatoms: Regulated protease excretion for specific algal lysis. *PLoS One 6*, e21032.

Chakraborty, K., Kizhakkekalam, V.K., Joy, M., & Chakraborty, R.D. (2020). Moving away from traditional antibiotic treatment: Can macrocyclic lactones from marine macroalga-associated heterotrophs be the alternatives? *Applied Microbiology and Biotechnology 104*, 7117–7130. https://doi.org/10.1007/s00253-020-10658-0

Cirri, E., & Pohnert, G. (2019). Algae-bacteria interactions that balance the planktonic microbiome. *New Phytology 223*, 100–106. https://doi.org/10.1111/nph.15765

Cole, J.J. (1982). Interactions between bacteria and algae in aquatic ecosystems. *Annual Review Ecological Systems 13*, 291–314.

Cooper, M.B., & Smith, A.G. (2015). Exploring mutualistic interactions between microalgae and bacteria in the omics age. *Current Opinion in Plant Biology 26*, 147–153.

Croft, M.T., Lawrence, A.D., Raux-Deery, E., Warren, M.J., & Smith, A.G. (2005). Algae acquire vitamin B12 through a symbiotic relationship with bacteria. *Nature 438*(7064), 90–93.

Danger, M., Oumarou, C., Benest, D., & Lacroix, G. (2007). Bacteria can control the stoichiometry and nutrient limitation of phytoplankton. *Functional Ecology*. https://doi.org/10.1111/j.1365-2435.2006.01222.x

Dinh, N.T., Thu, L.P., Dat, N.T., Toan, N.K., & Le Anh, P. (2022). The roles of microalgae and bacteria in wastewater treatment. *Vietnam Journal of Biotechnology 20*(3), 573–588.

Dow, L. (2021). How do quorum-sensing signals mediate algae–bacteria interactions? *Microorganisms, 9*(7), 1391. https://doi.org/10.3390/microorganisms9071391

Durham, B.P., Sharma, S., Luo, H., Smith, C.B., Amin, S.A., Bender, S.J., Dearth, S.P., Van Mooy, B.A., Campagna, S.R., Kujawinski, E.B., Armbrust, E.V., & Moran, M.A. (2015). Cryptic carbon and sulfur cycling between surface ocean plankton. *Proceedings of the National Academy of Sciences 112*(2), 453–457. https://doi.org/10.1073/pnas.1413137112

Falkowski, P.G., Barber, R.T., & Smetacek, V. (1998). Biogeochemical controls and feedbacks on ocean primary production. *Science 281*(5374), 200–206.

Faust, K., & Raes, J. (2012). Microbial interactions: From networks to models. *Nature Reviews Microbiology 10*(80), 538–550.

Fu, H., Uchimiya, M., Gore, J., & Moran, M.A. (2020). Ecological drivers of bacterial community assembly in synthetic phycospheres. *Proceedings of the National Academy of Sciences of the United States of America 117*(2020), 3656–3662.

Fuentes, J.L., Garbayo, I., Cuaresma, M., Montero, Z., Gonzlez-Del-Valle, M., & Vilchez, C. (2016). Impact of microalgae-bacteria interactions on the production of algal biomass and associated compounds. *Marine Drugs 14*(5), 1–16.

Fulbright, S.P., Robbins-Pianka, A., Berg-Lyons, D., Knight, R., Reardon, K.F., & Chisholm, S.T. (2018). Bacterial community changes in an industrial algae production system. *Algal Research 31*, 147–156. https://doi.org/10.1016/j.algal.2017.09.010

Gachon, C.M.M., Sime-Ngando, T., Strittmatter, M., Chambouvet, A., & Kim, G.H. (2010). Algal diseases: Spotlight on a black box. *Trends in Plant Science 15*, 633–640.

Gibson, K., Song, H., & Chen, N. (2022). Metabarcoding analysis of microbiome dynamics during a *Phaeocystis globosa* bloom in the Beibu Gulf, China. *Harmful Algae 114*, 102217. https://doi.org/10.1016/j.hal.2022.102217

Goecke, F., Thiel, V., Wiese, J., Labes, A., & Imhoff, J.F. (2013). Algae as an important environment for bacteria – phylogenetic relationships among new bacterial species isolated from algae. *Phycologia 52*, 14–24.

Grossart, H.P., Jorgensen, B.B., Arnosti, C., Labadie, J.P., & Joint, I. (2020). Beyond the microbial loop: Complex algal–bacterial interactions. *Nature Review Microbiology 18*(11), 735–745.

Grossart, H.P., Levold, F., Allgaier, M., Simon, M., & Brinkhoff, T. (2005). Marine diatom species harbour distinct bacterial communities. *Environmental Microbiology 7*(6), 860–873. https://doi.org/10.1111/j.1462-2920.2005.00759.x

Helliwell, K.E., Wheeler, G.L., Leptos, K.C., Goldstein, R.E., & Smith, A.G. (2011). Insights into the evolution of vitamin B12 auxotrophy from sequenced algal genomes. *Molecular Biology and Evolution 28*, 2921–2933.

Iqhrammullah, M., Chiari, W., Hudaa, S., Irhamni, I., & Akbar, S.A. (2024). Microalgal—Bacterial interactions: Research trend and updated review. *Heliyon, 10*(15), e35324. https://doi.org/10.1016/j.heliyon.2024.e35324

Jack, A., Adegbeye, M., Ekanem, D., Faniyi, T., Fajemisin, A.N., Elghandour, M.M.M.Y., Salem, A.Z.M., Rivas-Caceres, R.R., Adewumi, K., & Edoh, O. (2023). Microalgae application in feed for ruminants. In: Eduardo Jacob-Lopes, Maria Isabel Queiroz, Mariana Manzoni Maroneze, and Leila Queiroz Zepka (eds.), *Handbook of Food and Feed from Microalgae*, pp. 397–409. Academic Press. https://doi.org/10.1016/B978-0-323-99196-4.00042-5

Kaczmarska, I., Ehrman, J., Bates, S., Green, D., Léger, C., & Harris, J. (2005). Diversity and distribution of epibiotic bacteria on *Pseudo-nitzschia multiseries* (Bacillariophyceae) in culture, and comparison with those on diatoms in native seawater. *Harmful Algae 4*, 725–741. https://doi.org/10.1016/j.hal.2004.10.001

Kazamia, E., Helliwell, K.E., Purton, S., & Smith, A.G. (2016). How mutualisms arise in phytoplankton communities: Building eco-evolutionary principles for aquatic microbes. *Ecological Letters 19*, 810–822.

Kerkhof, L.J., Voytek, M.A., Sherrell, R.M., Millie, D., & Schofield, O. (1999). Variability in bacterial community structure during upwelling in the coastal ocean. *Hydrobiologia 401*, 139–148. https://doi.org/10.1023/A:1003734310515

Kim, B.N.H., Ramanan, R., Cho, N.H., Oh, H.N.M., & Kim, H.N.S. (2014). Role of *Rhizobium*, a plant growth-promoting bacterium, in enhancing algal biomass through mutualistic interaction. *Biomass and Bioenergy 69*, 95–105.

Kim, M.J., Jeong, S.Y., & Lee, S.J. (2008). Isolation, identification, and algicidal activity of marine bacteria against *Cochlodinium polykrikoides*. *Journal of Applied Phycology 20*, 1069–1078.

Kouzuma, A., & Watanabe, K. (2015). Exploring the potential of algae/bacteria interactions. *Current Opinion in Biotechnology 33*, 125–129. https://doi.org/10.1016/j.copbio.2015.02.007

Krohn-Molt, I., Wemheuer B., Alawi, M., Poehlein, A., Güllert, S., Schmeisser, C., Pommerening-Röser, A., Grundhoff, A., Daniel, R., Hanelt, D., & Streit, W.R. (2013). Metagenome survey of a multispecies and alga-associated biofilm revealed key elements of bacterial-algal interactions in photobioreactors. *Applied Environmental Microbiology 79*, 6196–6206. https://doi.org/10.1128/AEM.01641-13

Krug, L., Morauf, C, Donat, C, Müller, H, Cernava, T., & Berg, G. (2020). Plant growth-promoting Methylobacteria selectively increase the biomass of biotechnologically relevant microalgae. *Frontiers in Microbiology 11*, 427. https://doi.org/10.3389/fmicb.2020.00427

Lee, J.K., Park, K.S., Park, M.S., Kim, S.M., Chung, S.Y., & Lee, D.S. (2013). Surgical treatment of lumbar hyperextension injury in ankylosing spondylitis. *Korean Journal of Spine 10*, 195–199. https://doi.org/10.14245/kjs.2013.10.3.195

Lee, Y.K., Ahn, C.Y., Kim, H.S., & Oh, H.M. (2010). Cyanobactericidal effect of Rhodococcus sp. isolated from an eutrophic lake on *Microcystis* sp. *Biotechnology Letters 32*, 1673–1678.

Li, K., Liu, Q., Fang, F., Luo, R., Lu, Q., Zhou, W., Huo, S., Cheng, P., Liu, J., Addy, M., Chen, D., & Ruan, R. (2019). Microalgae-based wastewater treatment for nutrients recovery: A review, *Bioresource Technology 291*, 121934.

Lian, J, Wijffels, R.H., Smidt, H, & Sipkema, D. (2018). The effect of the algal microbiome on industrial production of microalgae. *Microbiology Biotechnology 11*(5), 806–818. https://doi.org/10.1111/1751-7915.13296

Louca, S., Parfrey, L.W., & Doebeli, M. (2016). Decoupling function and taxonomy in the global ocean microbiome, *Science 353*(6305), 1272–1277.

Lugtenberg, B., & Kamilova, F. (2009). Plant-growth-promoting Rhizobacteria. *Annual Review in Microbiology 63*, 541–556.

Mayali, X., & Azam, F. (2004). Algicidal bacteria in the sea and their impact on algal blooms. *Journal of Eukaryotic Microbiology 51*, 139–144.

Mayali, X., Franks, P., & Burton, R. (2011). Temporal attachment dynamics by distinct bacterial taxa during a dinoflagellate bloom. *Aquatic Microbiology Ecology 63*, 111–122.

Mayers, T.J. Bramucci, A.R., Yakimovich, K.M. & Case, R.J. (2016). A bacterial pathogen displaying temperature-enhanced virulence of the microalga *Emiliania huxleyi. Frontiers in Microbiology 7*. https://doi.org/10.3389/fmicb.2016.00892

Mugnai, S., Derossi, M., & Hendlin, Y. (2022). Algae communication, conspecific and interspecific: The concepts of phycosphere and algal-bacteria consortia in a photobioreactor (PBR). *Plant Signaling & Behavior 18*(1), e2148371. https://doi.org/10.1080/15592324.2022.2148371

Nadeem, S.M., Ahmad, M., Zahir, Z.A., Javaid, A., & Ashraf, M. (2014). The role of mycorrhizae and plant growth-promoting rhizobacteria (PGPR) in improving crop productivity under stressful environments. *Biotechnology Advances 32*, 429–448.

Natrah, F.M.I., Bossier, P., Sorgeloos, P., Yusoff, F.M., & Defoirdt, T. (2014). Significance of microalgal-bacterial interactions for aquaculture. *Reviews Aquaculture 6*, 48–61. https://doi.org/10.1111/raq.12024

Palacios, O.A., Lopez, B.R., Bashan, Y., & Bashan, L.E. (2018). Early changes in nutritional conditions affect formation of synthetic mutualism between *Chlorella sorokiniana* and the bacterium *Azospirillum brasilense. Microbiology Ecology 77*, 980–992.

Park, B.S., Choi, W.-J., Guo, R., Kim, H., & Ki, J.-S. (2021). Changes in free-living and particle-associated bacterial communities depending on the growth phases of marine green algae, *Tetraselmis suecica. Journal of Marine Science Engineering 9*(2), 171.

Paul, C., Mausz, M.A., & Pohnert, G. (2012). A co-culturing/metabolomics approach to investigate chemically mediated interactions of planktonic organisms reveals the influence of bacteria on diatom metabolism. *Metabolomics 9*, 349–359.

Peiffer, J.A., Spor, A., Koren, O., Jin, Z., Tringe, S.G., Dangl, J.L., Buckler, E.S., & Ley, R.E. (2013). Diversity and heritability of the maize rhizosphere microbiome under field conditions. *Proceedings of the National Academy of Sciences of the United States of America 110*, 6548–6553.

Perera, I.A., Abinandan, S., Subashchandrabose, S.R., Venkateswarlu, K., Naidu, R., & Megharaj, M. (2022). Combined inorganic nitrogen sources influence the release of extracellular compounds that drive mutualistic interactions in microalgal–bacterial co-cultures. *Journal of Applied Phycology 34*, 1311–1322.

Pinhassi, J., Sala, M.M., Havskum, H., Peters, F., Guadayol, O., Malits, A., & Marrase, C. (2004). Changes in bacterioplankton composition under different phytoplankton regimens. *Applied Environmental Microbiology 70*(11), 6753–6766. https://doi.org/10.1128/aem.70.11.6753-6766.2004

Ramanan, R., Kim, B.H., Cho, D.H., Oh, H.M., & Kim, H.S. (2016). Algae–bacteria interactions: Evolution, ecology and emerging applications. *Biotechnology Advances 34*(1), 14–29.

Reitan, K.I., Rainuzzo, J.R., Øie, G., & Olsen, Y. (1997). A review of the nutritional effects of algae in marine fish larvae. *Aquaculture 155*, 207–221.

Riemann, L., Steward, G.F., & Azam F. (2000). Dynamics of bacterial community composition and activity during a mesocosm diatom bloom. *Applied Environmental Microbiology 66*, 578–587.

Rolland, J.L., Stien, D., Sanchez-Ferandin, S., & Lami, R. (2016). Quorum sensing and quorum quenching in the phycosphere of phytoplankton: A case of chemical interactions in ecology. *Journal of Chemical Ecology 42*(12), 1201–1211. Epub: 2016 Nov 7. PMID: 27822708. https://doi.org/10.1007/s10886-016-0791-y

Rosenberg, G., & Paerl, H.W. (1981). Nitrogen fixation by blue-green algae is associated with the products. *Biology Bullentine 143*, 265–277.

Sapp, M., Schwaderer, A.S., Wiltshire, K.H., Hoppe, H.G., Gerdts, G., & Wichels, A. (2007). Species-specific bacterial communities in the phycosphere of microalgae? *Microbiology Ecology 53*(4), 683–699. https://doi.org/10.1007/s00248-006-9162-5

Segev, E., Wyche, T.P., Kim, K.H., Petersen, J., Ellebrandt, C., Vlamakis, H., Barteneva, N., Paulson, J.N., Chai, L., Clardy, J., & Kolter, R. (2016). Dynamic metabolic exchange governs a marine algal-bacterial interaction. *Elife 5*, e17473. https://doi.org/10.7554/eLife.17473

Seyedsayamdost, M.R., Case, R.J., Kolter, R., & Clardy, J. (2011). The Jekyll-and-Hyde chemistry of *Phaeobacter gallaeciensis*. *Nature Chemistry 3*, 331–335.

Seymour, J.R., Amin, S.A., Raina, J.B., & Stocker, R. (2017). Zooming in on the phycosphere: The ecological interface for phytoplankton–bacteria relationships. *Nature Microbiology 2*, 17065.

Shao, K., Zhang, L., Wang, Y., Yao, X., Tang, X., Qin, B., & Gao, G. (2014). The responses of the taxa composition of particle-attached bacterial community to the decomposition of Microcystis blooms. *Science of the Total Environment 488*, 236–242. https://doi.org/10.1016/j.scitotenv.2014.04.101

Shi, L., Cai, Y., Yang, H., Xing, P., Li, P., Kong, L., & Kong, F. (2009). Phylogenetic diversity and specificity of bacteria associated with *Microcystis aeruginosa* and other cyanobacteria. *Journal of Environmental Science 21*, 1581–1590. https://doi.org/10.1016/S1001-0742(08)62459-6

Shi, R., Qi, Z., Han, T., Dai, M., Zhang, S., & Huang, H. (2023). Responses of bacterioplankton, particle-and colony-attached bacterial communities to *Phaeocystis globosa* blooms in Mirs Bay, China. *Frontiers in Marine Science 10*, 1075059. https://doi.org/10.3389/fmars.2023.1075059

Silva, A. (1999). Effect of the microalga Isochrysis galbana on the early larval culture of Paralichthys adspersus. *Ciencias Marinas 25*, 267–276.

Singh, R.P., & Reddy, C.R.K. (2014). Seaweed microbial interactions: Key functions of seaweed-associated bacteria. *FEMS Microbiology Ecology 88*, 213–230.

Sison-Mangus, M.P., Jiang, S., Tran, K.N., & Kudela, R.M. (2014). Host-specific adaptation governs the Interaction of the marine diatom. *Pseudo-nitzschia*, and its microbiota. *ISME Journal 8*, 63–76.

Stock, W., Blommaert, L., De Troch, M., Mangelinckx, S., Willems, A., Vyverman, W., & Sabbe, K. (2019). Host specificity in diatom-bacteria interactions alleviates antagonistic effects. *FEMS Microbiology Ecology 95*. https://doi.org/10.1093/femsec/fiz171

Suleiman, M., Zecher, K., Yücel, O., Jagmann, N., & Philipp, B. (2016). Interkingdom cross-feeding of ammonium from marine methylamine-degrading bacteria to the diatom *Phaeodactylum tricornutum. Applied Environmental Microbiology 82*, 7113–7122.

Sze, M.A., & Schloss, P.D. (2018). Leveraging existing 16S rRNA gene surveys to identify reproducible biomarkers in individuals with colorectal tumors. *MBio 9*, e00630-18.

Teeling, H., Fuchs, B.M., Becher, D., Klockow, C., Gardebrecht, A., Bennke, C.M., Kassabgy, M., Huang, S., Mann, A.J., Waldmann, J., Weber, M., Klindworth, A., Otto, A., Lange, J., Bernhardt, J., Reinsch, C., Hecker, M., Peplies, J., Bockelmann, F.D., … Amann, R. (2012). Substrate-controlled succession of marine bacterioplankton populations induced by a phytoplankton bloom. *Science 336*, 608–611.

Trainer, V.L., Moore, S.K., Hallegraeff, G., Kudela, R.M., Clement, A., Mardones, J.I., & Cochlan, W.P. (2020). Pelagic harmful algal blooms and climate change: Lessons from nature's experiments with extremes. *Harmful Algae 91*, 101591. https://doi.org/10.1016/j.hal.2019.03.009

Tulli, F., Chini Zittelli, G., Giorgi, G., Poli, B.M., Tibaldi, E., & Tredici, M.R. (2012). Effect of the inclusion of dried *Tetraselmis suecica* on growth, feed utilization, and fillet composition of European sea bass juveniles fed organic diets. *Journal of Aquatic Food Production Technology*, *19*(1), 188–197.

Vannier, N., Mesny, F., Getzke, F., Chesneau, G., Dethier, L., Ordon, J., Thiergart, T., & Hacquard, S. (2023). Genome resolved metatranscriptomics reveals conserved root colonization determinants in a synthetic microbiota. *Nature Communications 14*, 8274. https://doi.org/10.1038/s41467-023-43688-z

Villa, J.A., Ray, E.E., & Barney B.M. (2014). *Azotobacter vinelandii* siderophore can provide nitrogen to support the culture of the green algae *Neochloris oleoabundans* and *Scenedesmus* sp. BA032. *FEMS Microbiology Letters 351*, 70–77.

Villarreal-Gómez, L.J., Soria-Mercado, I.E., Guerra-Rivas, G., & Ayala-Sánchez, N.E. (2010). Antibacterial and anticancer activity of seaweeds and bacteria associated with their surface. *Revista de Biología Marina y Oceanografía 45*, 267–275. https://doi.org/10.4067/S0718-19572010000200008

Vu, C.H.T., Lee, H.-G., Chang, Y.K., & Oh, H.-M. (2018). Axenic cultures for microalgal biotechnology: Establishment, assessment, maintenance, and applications. *Biotechnol Advances 36*, 380–396.

Wagner-Döbler, I., Ballhausen, B., Berger, M., Brinkhoff, T., Buchholz, I., Bunk, B., Cypionka, H., Daniel, R., Drepper, T., Gerdts, G., Hahnke, S., Han, C., Jahn, D., Kalhoefer, D., Kiss, H., Klenk, H.P., Kyrpides, N., Liebi, W., Liesegang, H., … Simon, M. (2010). The complete genome sequence of the algal symbiont *Dinoroseobacter 'shibae*: A hitchhiker's guide to life in the sea. *ISME Journal 4*, 61–77.

Wan, M., Jin, X., Xia, J., Rosenburg, J.N., Yu, G., Nie, Z., Oyler, G.A., & Betenbaugh, M.J. (2014). The effect of iron on growth, lipid accumulation, and gene expression profile of the freshwater microalga *Chlorella sorokiniana. Applied Microbiology Biotechnology 98*, 9473–9481.

Wang, H., Tomasch, J., Jarek, M., & Wagner-Döbler, I. (2014). A dual-species cocultivation system to study the interactions between roseobacters and dinoflagellates. *Frontiers in Microbiology 5*, 311. https://doi.org/10.3389/fmicb.2014.00311

Wang, X., Li, Z., Su, J., Tian, Y., Ning, X., Hong, H., & Zheng, T. (2010a). Lysis of a red-tide causing alga, Alexandrium Siphonous green seaweed *Codium 'decorticatum*: Effects on ammonium uptake. *Marine Biology 61*, 151–158.

Wang, X., Li, Z., Su, J., Tian, Y., Ning, X., Hong, H., & Zheng, T. (2010b). Lysis of a red-tide causing alga, Alexandrium tamarense, caused by bacteria from its phycosphere. *Biological Control 52*, 123–130.

Wang, Y., Jimenez, D.J., Zhang, Z., & Van Elsas, J.D. (2023). Functioning of a tripartite lignocellulolytic microbial consortium cultivated under two shaking conditions: A metatranscriptomic study. *Biotechnology for Biofuels and Bioproducts 16*(1). https://doi.org/10.1186/s13068-023-02289-0

Watanabe, K., Takihana, N., Aoyagi, H., Hanada, S., Watanabe, Y., Ohmura, N., Saiki, H. & Tanaka, H. (2005). Symbiotic association in Chlorella culture. *FEMS Microbiology Ecology 51*, 187–196.

Wieczorek, A.S., Schmidt, O., Chatzinotas, A., von Bergen, M., Gorissen, A., & Kolb, S. (2019). Ecological functions of agricultural soil bacteria and microeukaryotes in chitin degradation: A case study. *Frontiers in Microbiology 10*, 1293. https://doi.org/10.3389/fmicb.2019.01293

Xu, S., He, C., Kang, Z., Song, S., & Li, C. (2022). The ecological responses of bacterioplankton during a *Phaeocystis globosa* bloom in Beibu Gulf, China, were highlighted by integrated metagenomics and metatranscriptomics. *Marine Biology 169*, 36. https://doi.org/10.1007/s00227-022-04021-2

Yang, C., Wang, Q., Simon, P., Liu, J., Liu, L., Dai, X., Zhang, X., Kuang, J., Igarashi, Y., Pan, X., & Luo, F. (2017). Distinct network interactions in particle-associated and free-living bacterial communities during a *Microcystis aeruginosa* bloom in a plateau lake. *Frontiers in Microbiology 8*, 1202. https://doi.org/10.3389/fmicb.2017.01202

Yi, T., Shan, Y., Huang, B., Tang, T., Wei, W., & Quinn, N.W.T. (2020). An efficient chlorella Cupriavidus necator microcosm for phenol degradation and its cooperation mechanism. *Science of the Total Environment 743*, 140775. https://doi.org/10.1016/j.scitotenv.2020.140775

Zapalski, M.K. (2011). Is the absence of proof a proof of absence? Comments on commensalism. *Palaeogeography, Palaeoclimatology, Palaeoecology 302*, 484–488.

Zhang, B., Li, W., Guo, Y., Zhang, Z., Shi, W., Cui, F., Lens, P.N.L., & Tay, J.H. (2020). Microalgal-bacterial consortia: From interspecies interactions to biotechnological applications. *Renewal Sustainable Energy Reviews 118*, 109563. https://doi.org/10.1016/j.rser.2019.109563

Zheng, Q., Wang, Y., Lu, J., Lin, W., Chen, F., & Jiao, N. (2020). Metagenomic and metaproteomic insights into photoautotrophic and heterotrophic interactions in a Synechococcus culture. *mBio 11*, 1–18.

Zhu, Z., Sun, J., Fa, Y., Liu, X., & Lindblad, P. (2022). Enhancing microalgal lipid accumulation for biofuel production. *Frontiers in Microbiology*, *13*, 1024441. https://doi.org/10.3389/fmicb.2022.1024441

3 Bioethical and Risk Considerations in Microalgal Biotechnology

Environmental Release and Genome Editing Implications

Israel Valencia Quiroz and Yuri Córdoba Campo

3.1 INTRODUCTION

Microalgae are microscopic, photosynthetic organisms that have profoundly influenced the Earth's biosphere. Over billions of years, they have contributed to oxygenating the atmosphere and establishing the foundation for aquatic food webs (Falkowski & Raven, 2007). Nowadays, through a polyphasic approach related to the sequencing of the 16S rDNA gene, the following heterocystous families are recognized: *Chlorogloeopsidaceae, Coleospermaceae, Fischerellaceae, Hapalosiphonaceae, Microchaetaceae, Nostocaceae, Nostochopsidaceae, Rivulariaceae, Scytonemataceae, Symphyonemataceae*, and *Tolypothrichaceae*. On the other hand, some genera of filamentous cyanobacteria without defined heterocysts include *Anathece, Cyanobium, Gloeobacter, Halothece, Nodosilinea, Oculatella, Synechococcus*, and *Willmottia*, among others (Fragoso Rodríguez, 2023; Komárek, 2014). Currently, microalgae can convert sunlight and nutrients into biomass with high potential in multiple industries and environmental applications, including bioenergy production, wastewater remediation, and the pharmaceutical and nutraceutical industries (Richmond & Hu, 2013), with a significant impact given the growing nature of climate change, water scarcity, and energy crises, among others (Yu et al., 2024).

Genome editing and genetic engineering allow the modification of microalgae with higher lipid content, growth rates, and nutrient absorption, generating growth efficiencies and additional economic benefits; however, with significant ecological risks, freshwater habitats could be under the greatest threat due to their restricted

 DOI: 10.1201/9781003611479-3

hydrology. Freshwater dynamics pose risks to water stability (Dodds & Whiles, 2020).

This chapter explores ecological risks, bioethical issues, and governance frameworks related to genetically modified microorganisms (GMMs), focusing on rotifers and cladocerans as key consumers and bioindicators of environmental impact.

3.2 ECOLOGICAL RISKS OF GENETICALLY MODIFIED MICROALGAE

3.2.1 Key Freshwater Organisms: Rotifers and Cladocerans

Freshwater ecosystems are particularly vulnerable to biological perturbations due to their relatively closed hydrology, limited dilution capacity, and high sensitivity to nutrient inputs. Zooplankton communities, especially rotifers and cladocerans, occupy central positions in these systems as primary consumers, linking primary producers (phytoplankton) to higher trophic levels, including fish and macroinvertebrates (Dodds & Whiles, 2020; Matsuzaki et al., 2018); their ecological functions are important because they help transfer energy between trophic levels, control phytoplankton biomass, and recycle nutrients through excretion (Choi et al., 2016; Sterner & Elser, 2002).

Rotifers (*phylum Rotifera*) are microscopic freshwater invertebrates with high reproductive adaptability. They reproduce by cyclic parthenogenesis, between clonal and sexual reproduction, in response to different stressors, that is, nutrient limitation, high population density, or changes in water composition. The production of resting eggs enables them to survive desiccation, extreme chemical conditions, or freezing, which preserves genetic diversity and population resilience for decades (Gómez & Carvalho, 2000; Sierra et al., 2019). Because rotifers are sensitive to changes in food availability and composition, they could be used as bioindicators of water quality; for example, in rotifers, growth, fertility, and survival are affected by changes in the content, size, and chemical composition of lipids in water (Choi et al., 2016). They are also useful for monitoring the health of ecosystems in temperate and tropical lakes because they can react quickly to eutrophication, nutrient enrichment, and changes in algae composition (Alcocer et al., 2023).

Cladocerans, particularly *Daphnia* species such as *Daphnia magna* and *Daphnia pulex*, play a key regulatory role in controlling phytoplankton blooms, including species such as *Chlorella vulgaris*, and are among the most extensively studied freshwater zooplankton. They are highly effective herbivores that can quickly reduce algal biomass, stabilizing the nutrient cycle and water clarity (Frost et al., 2025; Sterner & Elser, 2002). For example, there are studies on the consumption of *Chlorella vulgaris* by cladocerans such as *Daphnia pulex* Leydig, *Diaphanosoma birgei* Korínek, and *Moina macrocopa* Goulden (Bravo Hernández, 2018), as well as by rotifers, for example, *Brachionus angularis* (Gama-Flores et al., 2004). However, Paerl and Otten (2013) indicated that cladocerans are vulnerable to environmental stressors such as cyanobacterial toxins—for example, microcystins and nodularins; alkaloid-type compounds such as anatoxin-α, anatoxin-α(S), aplysiatoxins, cylindrospermopsin, lyngbyatoxin, and saxitoxins; and nonprotein amino acids like β-methylamino-L-alanine and lipopolysaccharides (Fragoso Rodríguez, 2023). Temperature fluctuations and changes in nutrient composition also affect them. During cyanobacterial blooms, this sensitivity can cause population collapses, leading to cascading effects on the food web by reducing fish recruitment and impairing nutrient cycling.

In addition, cladocerans allow the dynamics of their populations to reveal the overall health of freshwater systems, and they are significant ecological indicators. The relationship between algae composition and zooplankton resistance is highlighted by long-term studies in European lakes, which show that *Daphnia* populations decline during periods of cyanobacterial dominance (Davidson et al., 2011; Haberman et al., 2007). Also, changes in Latin American lakes from cladoceran-dominated communities to rotifer-dominated communities often indicate ecological stress or nutrient enrichment (Alcocer et al., 2023). The *Daphnia magna* species has been referred to as the norm used by the EPA; however, its use seems not to be adequate in other Latin American countries, such as Mexico, as it is an exotic species and its presence in tropical waters is almost null. Studies have been conducted to propose the use of *Daphnia pulex* instead, as it is naturally present in water bodies in Mexico (Bravo Hernández, 2018).

Although both rotifers and cladocerans are ecologically resilient, they are highly sensitive to changes in algal quality (Figure 3.1), including the following:

- Lipid and protein content: High lipid or protein content can enhance zooplankton growth, but genetic modifications that alter these parameters may disrupt the balance of essential compounds such as long-chain polyunsaturated fatty acids (PUFAs) (Choi et al., 2016).

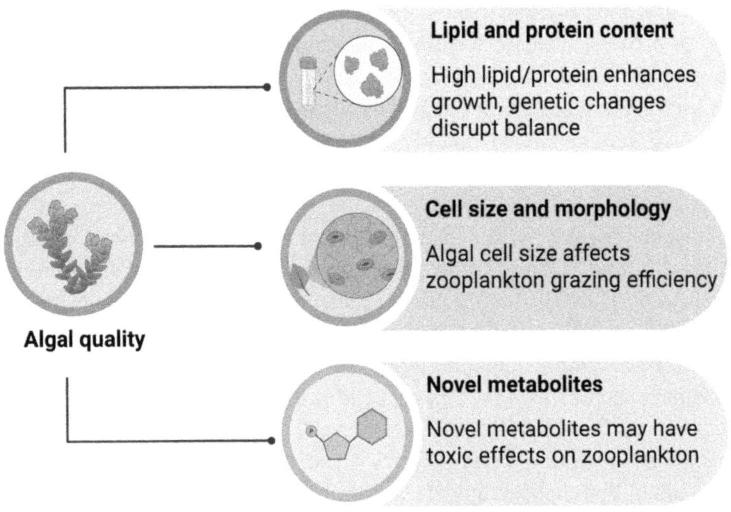

FIGURE 3.1 Changes in algal quality may affect rotifers and cladocerans.

(Created in https://BioRender.com.)

- Cell size and morphology: Changes in algal cell size can affect grazing efficiency, especially for filter feeders such as *Daphnia* (Choi et al., 2016; Cyr & Curtis, 1999).
- Novel metabolites: The production of metabolites by modified microalgae, which are not found in the natural environment, may have toxic or unidentified consequences for zooplankton populations (Bagwell et al., 2016). The production of toxins such as microcystin and its derivatives is also of concern. These derivatives include anatoxin-α, anatoxin-α(S), cyclic peptides (hepatotoxins), cylindrospermopsin, and saxitoxin, which are classified as alkaloids. Of particular interest is the LR-variant of microcystin, which is highly stable under extreme conditions and highly water-soluble (Fragoso Rodríguez, 2023).

Therefore, rotifer and cladoceran populations may be affected by changes in water chemistry, nutrient fluctuations, and the introduction of transgenics, which can alter trophic relationships and ecosystem stability. Cnidarians consume cladocerans, and fish, in turn, prey on cnidarians (Aguirre Rodríguez, 2023). In freshwater environments, the risk assessment and management of GMMs require a thorough understanding of these ecological dynamics (Angeler et al., 2014; Dodds & Whiles, 2020).

3.2.2 Impacts of the GMM on Trophic Dynamics

Before implementing strategies to manage GMM, it is essential to understand the potential ecological consequences of their introduction into freshwater ecosystems. The results are predominantly channeled through modifications in the nutritional aspects of trophic resources, the formation of distinct metabolites, and the competitive exchanges with native microalgae. Such events might interfere with the dynamics of trophic relationships, leading to substantial impacts on ecosystem durability, functional mechanisms, and the array of ecological services (Choi et al., 2016; Dodds & Whiles, 2020; Bagwell et al., 2016).

Changes in nutritional quality: Genetic engineering can increase the amount of lipids or proteins in microalgae, making them more useful in a wider range of fields, such as biotechnology or biofuels. Yet, these changes could introduce a downturn in necessary compounds for zooplankton, especially sterols, significant amino acids, and long-chain PUFAs (Choi et al., 2016; Richmond & Hu, 2013). Key metabolic factors are necessary for the growth, reproductive success, and overall survival of zooplankton, particularly for rotifers and cladocerans such as *Daphnia* spp.

Empirical laboratory investigations suggest that reductions in PUFA concentrations could lead to impaired reproductive success, diminished growth rates, and compromised population resilience, potentially destabilizing primary consumer demographics (Weers & Gulati, 1997; Yan et al., 2024).

Changes in the nutritional quality of algae also affect secondary consumers. Fish larvae and invertebrate predators rely on high-quality zooplankton as a food source; thus, nutritional deficiencies at the base of the food web can propagate upward, influencing fish recruitment, survival, and fisheries productivity (Sterner & Elser, 2002; Dodds & Whiles, 2020).

Novel metabolites: New metabolites, that is, secondary metabolites, signaling compounds, or toxins, which are not present in native microalgae, can be synthesized by GMMs. Sublethal stress and even population decline are the effects that these new metabolites can have on primary and secondary consumers. In some cases, modified microalgae can generate compounds that can prevent rotifers or cladocerans from feeding or digesting, reducing their reproductive success or causing behavioral alterations that change predator–prey interactions (Bagwell et al., 2016; Paerl & Otten, 2013). New metabolite release alters microbial interactions in freshwater ecosystems and affects bacteria–algae ratios, decomposition, and nutrient recycling, influencing biogeochemical cycles and nutrient transfer to higher trophic levels (Dodds & Whiles, 2020). Genetically modified organisms (GMOs) may outcompete native microalgae because they grow faster, produce toxins, and assimilate nutrients more efficiently. In the event that GMMs take the forefront in phytoplankton communities, this event could trigger a drop in biodiversity, transformations in species arrangement, and modifications in how ecosystems function (Scheffer et al., 2001; Choi et al., 2016). Dominance by a single GMM species could also worsen eutrophication dynamics, promote harmful algal blooms, and disrupt traditional food webs by limiting food availability for zooplankton and higher trophic levels. In contrast to those fed with wild-type algae, Bøhn et al. (2008) showed that *Daphnia magna* fed with genetically modified algae with an altered lipid composition exhibited significantly reduced fertility and growth. Extrapolating these laboratory results to natural ecosystems, trophic disruption could hamper fishing, jeopardize ecosystem services, and reduce fish harvests, particularly in semi-enclosed, nutrient-rich freshwater systems.

Ecological implications: There are very important ecological factors to think about when introducing GMOs. These factors stress the need for thorough environmental risk analysis and ongoing ecological monitoring. Regarding microalgae, monitoring activities can typically be conducted around three times weekly, wherein the assessment of nutrients including orthophosphates, nitrites, nitrates, ammonium, and dissolved organic matter serves as a proxy for chemical oxygen demand, alongside the evaluation of total lipid content. In addition, the determination of the chlorophyll concentration and microalgal biomass production should also be conducted. A monitoring procedure has been described for *Nannochloropsis* sp., *Dunaliella* sp., and *Tetraselmis* sp.

The trophic effects extend beyond direct consumer–resource interactions to broader ecosystem functions, including nutrient cycling, energy flow, and tolerance to disturbances. For example, studies have been conducted using microalgae for the removal of nitrogen and phosphorus from swine wastewater. These studies concluded that the characteristics of the wastewater may influence the removal efficiency of contaminants, which may include antibiotics, heavy metals, or other pollutants (Sales Lara, 2024).

Ethical decision-making, adaptive management, and the long-term integration of microalgal biotechnology into aquatic systems depend on understanding the trophic effects of transgenic organisms (Dodds & Whiles, 2020; Choi et al., 2016).

Treatment of contaminants: Some studies have reported a reduction in contaminants, such as malathion, in water bodies containing *Chlorella vulgaris*, although the results are not conclusive (Bravo Hernández, 2018). In this context, genetically modified Chlorella vulgaris may augment its ability to degrade specific pollutants or function as a buffer by elevating the LC_{50} of certain insecticides, such as methyl parathion (Gama-Flores et al., 2004).

Biodegradation of antibiotics and methicillin-resistant *Staphylococcus aureus* (MRSA): Preliminary studies have shown that microalgal bioreactors can help in the decomposition of antibiotics and MRSA present in porcine residual waters. However, although the process has a low yield, the inclusion of the modified microalgae could help increase its effectiveness.

3.2.3 Eutrophication and Regime Shifts

Eutrophication is widely recognized as one of the primary drivers of biodiversity loss and ecosystem degradation in freshwater environments. It happens when too many nutrients, especially nitrogen (N) and phosphorus (P) from agricultural runoff, wastewater, and urban effluents, cause phytoplankton to grow too much. This procedure changes the chemistry of the water, the structure of the community, and the function of the ecosystem (Smith & Schindler, 2009; Dodds & Whiles, 2020).

How eutrophication occurs: High levels of nutrients are good for fast-growing, opportunistic phytoplankton species, such as cyanobacteria, which can produce hazardous toxins such as microcystins and anatoxins (Paerl & Otten, 2013).

These blooms can:

- Reduce light penetration, negatively affecting submerged macrophytes, which are crucial for habitat complexity, nutrient cycling, and water stabilization (Scheffer et al., 2001).
- Cause oxygen depletion through the respiration and decomposition of organic matter, leading to hypoxic or anoxic conditions in deeper waters (Smith & Schindler, 2009).
- Because hypoxia alters trophic interactions and reduces habitat suitability, fish mortality occurs.

The change in the dynamics of the food web, the threat to ecosystem services (fishing, water purification, and recreational value), and the reduction in biodiversity are some of the ecological consequences resulting from these processes.

Changes in freshwater ecosystem regimes: Eutrophic systems are highly vulnerable to critical thresholds, which cause lasting and abrupt changes in ecosystems. Lakes dominated by macrophytes and clear water can rapidly change to turbid conditions dominated by phytoplankton, reflected in low water clarity, loss of benthic vegetation, and characterized by cyanobacteria or other opportunistic algae (Scheffer et al., 2001.

It often takes decades of nutrient reduction, hydrological management, and active restoration initiatives, that is, reintroduction of macrophytes or sediment dredging, to reverse these regime shifts. Reversing these regime shifts is often difficult and time-consuming, requiring decades of nutrient reduction, hydrological management, and active restoration efforts such as macrophyte reintroduction or sediment dredging.

For example, lakes in Europe and North America that experienced long-term nutrient enrichment have required persistent multi-decade management strategies to recover clear-water conditions (Dunalska, 2025; Søndergaard, 2007).

Role of GMM: The introduction of GMM into nutrient-rich freshwater systems could exacerbate eutrophication dynamics in multiple ways (Dodds & Whiles, 2020; Paerl & Otten, 2013; Richmond & Hu, 2013):

- Enhanced phytoplankton dominance: GMM with traits such as faster growth rates or increased nutrient uptake could outcompete native algae, amplifying algal biomass and shifting community composition toward monocultures.
- More organic matter: increased biomass production from GMMs can result in more waste, which increases oxygen depletion in bottom waters and supports microbial respiration, promoting hypoxia.
- Toxin production: New metabolites or secondary compounds produced by modified microalgae could increase the frequency or intensity of toxic algal blooms (Paerl & Otten, 2013).

These factors demonstrate that alterations in zooplankton populations, reductions in fish populations, and alterations in the nutrient cycle may induce cascading effects throughout the ecosystem, undermining its stability (Dodds & Whiles, 2020; Smith & Schindler, 2009).

3.2.4 Case Studies

The sensitivity of freshwater ecosystems to nutrient enrichment and species introduction is evident from multiple case studies. These examples demonstrate the ecological consequences that could arise from the release of genetically modified microalgae. These case studies underscore that even moderate anthropogenic changes can destabilize freshwater ecosystems, reinforcing the need for careful GMM risk assessment.

3.2.4.1 Latin America

- Lake Titicaca (Peru/Bolivia): Cyanobacterial blooms have led to a drop in cladoceran populations, which has hurt local livelihoods and made it harder to capture fish (Montoya, Komárková & Komárek, 2014).
- Lake Chapala (Mexico): It shows zooplankton populations that are mostly made up of rotifers, which means that the water quality is poor and the nutrients are causing stress to the ecosystem (Alcocer et al., 2023).
- Lake Zumpango (Mexico): A laboratory study subjected indigenous cladoceran species—*Daphnia magna, Daphnia laevis, Ceriodaphnia dubia*, and *Simocephalus vetulus*—to heightened levels of the cyanobacterial toxin microcystin-LR, resulting in altered survival and reproductive patterns (Fragoso Rodríguez, 2023).

3.2.4.2 International

Sulejów (Poland): The study conducted from 1999 to 2008 in the Sulejów Reservoir reveals a significant increase in cyanobacterial biomass and a decrease in *Daphnia longispina* densities during the summer months, indicating a strong negative correlation between the density of D. longispina and cyanobacterial biomass during these periods (Wojtal-Frankiewicz et al., 2015).

Mar Menor (Spain): The runoff of agricultural nutrients has caused oxygen levels to drop, which has killed benthic animals (Carstensen et al., 2014; Sandonnini et al., 2024).

Paraná River Basin (Brazil): This study illustrates how invasive species, particularly Eichhornia crassipes (water hyacinth), can modify habitat structures and disturb native biodiversity, thus establishing a basis for the expected effects of genetic modification techniques (Agostinho et al., 2009; Khanna et al., 2012).

3.2.5 Ecological Scenarios and Projections

The potential ecological impacts of genetically modified microalgae (GMMs) can be conceptualized by integrating experimental laboratory data, field observations, and ecological theory. Different scenarios for the introduction of GMMs into freshwater systems could occur, taking into account nutrient availability, hydrological connectivity, species interactions, and environmental resilience (Dodds & Whiles, 2020; Paerl & Otten, 2013; Scheffer et al., 2001; Bøhn et al., 2008). These scenarios demonstrate the need to use specific assessments before any environmental release and to provide a framework for risk assessment, management, and adaptive governance (Figure 3.2).

In a low-risk situation, GMMs should only be kept in controlled environments, such as closed ponds that do not connect to natural bodies of water, photobioreactors, or wastewater treatment plants. This is because the modified microalgae cannot get out into open aquatic systems, so the chance of ecological disruption is low.

- Ecological implications: To prevent accidental trophic disturbances, containment halts interactions with higher trophic levels, zooplankton, and indigenous phytoplankton.
- Management measures: Unintentional release could be reduced by containment in the laboratory, physical barriers, and genetic safeguards such as sterility or the use of fail-safe genetic circuits (Sebesta et al., 2022; Richmond & Hu, 2013).

According to the precautionary principle of environmental biotechnology, this is the safest environment for experimental applications and biomass production (Raffensperger & Tickner, 1999).

Moderate-risk scenario: This involves the unintentional or accidental release of GMMs into small lakes, reservoirs, or ponds rich in nutrients. Due to ecological constraints such as zooplankton predation or seasonal variations in nutrient

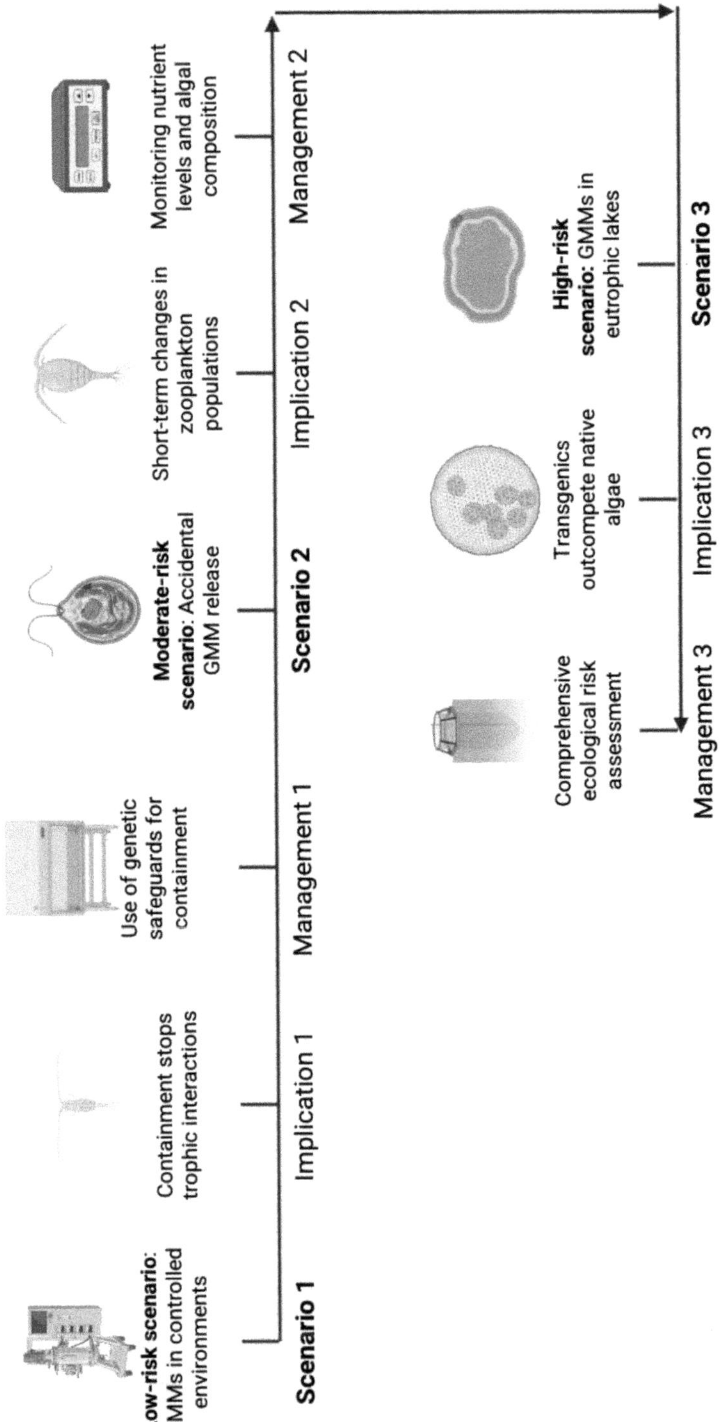

FIGURE 3.2 Different scenarios, management strategies, and implications ranging from low to high risk.

(Created in https://BioRender.com.)

availability, GMMs may temporarily compete with native microalgae in these environments but are unlikely to establish long-term dominance.

- Ecological implications: Changes in algal quality, cell size, or metabolite production could cause short-term changes in zooplankton populations, including declines in rotifer and cladoceran abundance (Choi et al., 2016; Bøhn et al., 2008). Studies on a lake showed that nutrients were added until cyanobacteria bloomed, but when increases in resilience indicators were detected, the addition of nutrients was suspended. Subsequently, some indicators returned to the levels seen when the lake was not enriched (Pace et al., 2017).
- Management measures: There are different strategies that allow ecological impacts to be monitored, that is, monitoring toxins, algae composition, and others. These monitoring efforts will enable faster responses, which will reduce the severity of the damage (Howard et al., 2023).

Moderate-risk scenarios show how important it is to find problems early and manage them in a way that works for everyone. This is because even small changes can have social and ecological effects on systems used for fishing, recreation, or water supply.

High-risk scenario: When GMMs are added to eutrophic or highly productive lakes, they can become dominant due to characteristics such as faster growth rates, more effective nutrient uptake, or the ability to produce toxins. Transgenics can outcompete native algae in these systems, increase cyanobacterial toxins, and cause cascading effects at the trophic level.

- Ecological implications: Clear water states dominated by macrophytes can give way to turbid systems dominated by phytoplankton because of the dominance of transgenics (Scheffer et al., 2001). Fish larvae, benthic invertebrates, and zooplankton are some examples of secondary consumers that may experience decreased growth, fertility, and survival, which could affect ecosystem services and function (Paerl & Otten, 2013; Dodds & Whiles, 2020).
- Management strategies: To reduce the potential negative effects on the environment and economy, high-risk scenarios demonstrate the need for comprehensive ecological risk assessment (ERA), site-specific modeling, and emergency response plans.

3.3 BIOETHICAL CONSIDERATIONS

3.3.1 Ethical Frameworks

Prior to the implementation of any strategy that incorporates genetically modified microorganisms (GMM), it is imperative to scrutinize the ethical implications associated with their utilization. The ecological release of genetically engineered microbes involves considerable uncertainties, including unidentified trophic interactions, potential regime shifts, and enduring alterations in ecosystems (Dodds & Whiles, 2020; Paerl & Otten, 2013). A careful ethical analysis gives a framework for making smart choices that strikes a balance between new technologies, the environment, and civic obligation (Aftab et al., 2022; Martínez-Alier, 2002; Raffensperger & Tickner, 1999). The ethical considerations pertinent to the application of GMMs can be organized around three principal frameworks, each yielding distinct insights and priorities:

Precautionary ethics: This paradigm underscores the necessity of restraint in the face of uncertainty, particularly when actions possess the potential to inflict irreversible damage upon ecosystems or human communities (Raffensperger & Tickner, 1999). In relation to GMMs, precautionary ethics advocates for comprehensive environmental risk assessments prior to their release, which should include site-specific investigations, modeling of trophic interactions, and rigorously controlled laboratory experiments. Researchers investigated the composition of microalgae across various aquatic systems in the State of Mexico and discovered that Presa Valle de Bravo demonstrated an exceptionally high density of microalgal organisms (94,478 org/mL). *Aulacoseira italica, Fragilaria crotonensis, Pediastrum simplex, Anabaena* sp., and *Navicula radiosa* were some of the taxa that were found (Juárez Álvarez, 2005).

The high concentration of orthophosphates in this system allows it to be classified as eutrophic. In contrast, in *Presa Villa Victoria*, a density of 58,961 org/mL was observed, with *Aulacoseira italica, Fragilaria crotonensis, Pediastrum simplex*, and *Scenedesmus quadricauda*, indicating a mesotrophic condition with a tendency toward eutrophication. Hydrological conditions and nutrient concentrations influence these differences (Juárez Álvarez, 2005).

Containment solutions that follow the precautionary principle include using closed bioreactors, fail-safe genetic changes (such as sterility genes), or temporary growth limits to stop inadvertent ecological spread (Sebesta et al., 2022; Richmond & Hu, 2013). Iterative monitoring and adaptive management provide prompt reactions to unexpected ecological effects (Holling, 1978).

The introduction of new species into freshwater ecosystems makes them very fragile, and even small changes can cause cascading impacts or regime shifts. This is why the precautionary approach is so important (Scheffer et al., 2001; Smith & Schindler, 2009). In this regard, a study conducted in the streams of *Parque Nacional Izta-Popo* (State of Mexico) reported a lower microalgal density (26,038 org/mL), attributed to low system productivity and the continuous flow of water (Juárez Álvarez, 2005).

Justice ethics: Discuss how the benefits and dangers of biotechnology are distributed (Martínez-Alier, 2002). The application of transgenics can benefit industrial or research organizations economically, but it can have negative social and environmental effects on vulnerable groups, such as indigenous communities, rural farmers, and people who depend on fishing. Nondiscriminatory access to information, collective decision-making, and contracts for benefit-sharing serve as salient examples of ethical considerations. Empirical studies in Latin America indicate that ethical and justice concerns with the management of GMOs impact rural and indigenous communities, where invasive species and eutrophication threaten freshwater resources vital for survival (Toledo & Barrera-Bassols, 2008).

Justice-focused methodologies: To support inclusive governance that includes communities in making decisions so that biotechnology doesn't make social disparities worse. There has been a report of a specific case in Mexico City, where the Laguna de Xico is a very polluted body of water. Like other bodies of water in the area, it has been used as a dumping ground for solid urban waste and special residues, which has made it easier for pests and diseases to spread. Ironically, the communities surrounding these water bodies often lack access to potable water.

In this context, a study was conducted to develop strategies for cleaning and rehabilitating the lagoon, which also contains a high concentration of microalgae (García Caballero, 2023). The objectives of this study include:

- To calculate the catchment area and runoff volume using Geographic Information Systems and data from the National Meteorological System to obtain key hydrological variables (García Caballero, 2023).
- To determine the steps and processes needed to remove microalgae so that the water is safe to drink, we will use treatability tests in the lab (García Caballero, 2023).
- To create a water treatment train and do a mass balance for making lagoon water drinkable (García Caballero, 2023).

Ethical frameworks grounded in ecological consciousness underscore the intrinsic value of each organism and ecological relationship, irrespective of their utility to humans (Aftab et al., 2022). Within this paradigm, the environmental dissemination of GMOs necessitates the acknowledgment of the rights of nonhuman entities to preserve their ecological functions and evolutionary trajectories, the sanctity of ecosystem processes, encompassing nutrient cycling, trophic dynamics, and the conservation of biodiversity (Dodds & Whiles, 2020), and the prevention of actions that prioritize immediate human advantages at the expense of enduring ecosystem resilience. By recognizing the moral significance of ecosystem health itself, ecocentric ethics encourage strategies that minimize ecological disruption and maintain the complexity and functionality of freshwater systems.

Recognizing that health in an ecosystem must be conceptualized and managed in terms of an integrated and holistic unit, where biophysical and socioeconomic aspects are sustainable and interdependent, with the need to maintain ecological structure, functions, and services, not just human benefits. In addition, the authors recognize the centrality of values in managing the health of aquatic ecosystems and the role of ethics in negotiating and constructively balancing conflicting values to achieve healthy ecosystems in socio-ecological systems (Odume & de Wet, 2019).

The combination of precautionary, judicial, and ecocentric ethics gives us a way to look at the usage of GMMs from different angles. The precautionary principle embodies an epistemic challenge, necessitating a comprehension of the distinctions among ignorance, uncertainty, and risk. When evolution is taken into account, uncertainty is inherent, and probability analysis stops being useful for making predictions. But if there are ecological risks, innovation cannot be stopped. The precautionary principle can be applied with procedural rationality, which helps society find satisfactory solutions (Giampietro, 2002).

In this regard, several examples can be mentioned from the Mexican Official Standards (*Normas Oficiales Mexicanas*) concerning wastewater treatment. For example, these rules set the highest levels of contaminants that can be released into urban and municipal sewage systems, the highest levels of contaminants that can be reused in public services, the highest levels of contaminants that can be present in sludge and biosolids for their use and final disposal, and the requirements for artificial aquifer recharge using treated wastewater (Reyes Landeros, 2022). Such integration supports adaptive governance, where ethical reflection, scientific monitoring, and stakeholder participation guide responsible applications of microalgal biotechnology, aligning with sustainable development goals and environmental stewardship (Dodds & Whiles, 2020; Holling, 1978; Aftab et al., 2022).

In this context, a typical monitoring protocol for wastewater treatment involving microalgae includes the following:

Evaluating water quality in the influent by measuring parameters such as fats and oils, total suspended solids, helminth eggs, biochemical oxygen demand (BOD), and total coliforms to determine the contaminant load entering the system (Reyes Landeros, 2022).

Assessing the quality of the treated effluent with respect to the same parameters—fats and oils, total suspended solids, helminth eggs, BOD, and total coliforms—to determine contaminant removal efficiency during the treatment process (Reyes Landeros, 2022).

Determining whether the treated water meets the standards required for reuse in public services involving direct or occasional human contact (Reyes Landeros, 2022).

3.3.2 Combining Local and Indigenous Knowledge

Local and Indigenous knowledge provides an alternative viewpoint to traditional scientific methodologies. People in the Andes keep an eye on the health of lakes by watching how fish act, how clear the water is, and how well indicator plants are growing (Pohle and Gerique, 2006). Integrating these viewpoints into GMM governance facilitates improved ecological monitoring by conventional indicators, participative decision-making that aligns with community interests, and increased social legitimacy and acceptance of biotechnology initiatives.

3.3.3 Governance and Regulatory Frameworks

Strong regulatory frameworks are essential for reducing the risks associated with the use of transgenics. Freshwater ecosystems often span multiple jurisdictions, and governance approaches vary around the world.

- European Union: before the release of GMOs, Directive 2001/18/EC (European Commission, 2018) requires a thorough environmental risk assessment and public consultation.
- Brazil: The National Technical Commission on Biosafety (CTNBio) evaluates GMO projects, although enforcement varies regionally (Soares, 2015; Velini et al., 2017).
- Mexico: Aquatic biotechnology remains weakly regulated despite strict GMO maize policies (McAfee, 2008).
- Andean countries: Biosafety frameworks are incomplete, challenging transboundary water governance (Soares, 2015; Velini et al., 2017; McAfee, 2008).

3.4 RISK MANAGEMENT AND ADAPTIVE STRATEGIES

3.4.1 Environmental Risk Assessment Strategies

Implementing GMM requires rigorous, site-specific risk assessments. Water chemistry, trophic structure, and populations of sensitive species, particularly rotifers and cladocerans, should be included in the baseline studies. Strategies include laboratory containment, genetic safeguards such as sterility or fail-safe genetic modifications, monitoring of bioindicators to identify early ecological changes, and site-specific ERAs to evaluate nutrient status and connectivity (Sebesta et al., 2022). Therefore, characterizing the target ecosystem is a fundamental step in ERA. This includes:

- Water chemistry: Measurement of nutrient concentrations (nitrogen, phosphorus), pH, dissolved oxygen, and trace elements provides insight into the trophic status of the system and the potential for algal proliferation (Dodds & Whiles, 2020).
- Trophic structure: Identifying primary producers, herbivorous zooplankton (rotifers and cladocerans), benthic invertebrates, and fish populations is critical for understanding the potential cascading effects of GMM introduction (Pace, Cole & Carpenter, 1998). It is also necessary to consider mollusks, birds, reptiles, and higher plants (Pérez-Morales et al., 2016).
- Sensitive species populations: Certain taxa such as rotifers and *Daphnia* spp. are extremely sensitive to changes in metabolite composition, cell morphology, or algal quality. Monitoring these bioindicators provides early warning signs of ecological stress (Bhandarkar & Bansod, 2025; Sao et al., 2024). In addition, *Daphnia spp.* are particularly susceptible to the presence of specific contaminants, such as insecticides such as malathion

(Bravo Hernández, 2018). Furthermore, when a water body is used for domestic consumption, the production of cyanobacterial toxins can also affect domestic animals, livestock, and humans. Depending on their mechanism of action, these toxins can be hepatotoxic, neurotoxic, cytotoxic, dermotoxic, and bioaccumulative. In Mexico, harmful bloom-forming cyanobacterial species have been reported, including *Anabaena, Anabaenopsis, Cylindrospermopsis, Lyngbya, Microcystis, Nodularia, Phormidium, Planktothrix*, and *Pseudanabaena* (Pérez-Morales et al., 2016).

Lake size, connectivity, nutrient loading, and hydrological stability affect the vulnerability of freshwater ecosystems. Site-specific assessments should determine the following:

- Level of nutrition: While oligotrophic systems may restrict or limit the growth of GMMs, cyanobacterial blooms can reduce water clarity, suppress aquatic macrophytes, cause hypoxia or anoxia when cells die, and damage habitats of invertebrates and fish, among other effects (Paerl & Otten, 2013).
- Connectivity: Transgenics can spread more easily through lakes and ponds. This connectivity depends on climate change, metrics, and land use, which could generate and increase adverse ecological effects (Dodds & Whiles, 2020).
- Historical disturbances: The various interactions of transgenics create a more vulnerable ecosystem and alter the trophic or biogeochemical network, which involves changing competition and available niches. The effects of GMOs will depend on the history and status of the affected ecosystem; they will not depend solely on the intrinsic characteristics of the system (Scheffer et al., 2001).

Continuous monitoring of key taxa and environmental parameters is essential for adaptive management. Rotifers and cladocerans are particularly valuable because changes in population density or reproduction rates can indicate reductions in food quality or toxicity of introduced microalgae; shifts in community composition may signal early signs of trophic disruption; and coupled with water chemistry measurements, bioindicator data enable the detection of ecosystem-level impacts before irreversible changes occur (Choi et al., 2016; Dodds & Whiles, 2020; Bøhn et al., 2008).

A series of operational and biotechnological techniques have been suggested to reduce environmental risk, that is, GMMs can be kept out of natural water bodies through the use of wastewater treatment systems, mesocosms, or closed photobioreactors (Richmond & Hu, 2013). GMMs cannot survive outside regulated environments due to engineered characteristics such as nutrient auxotrophy, inducible sterility, or "fail-safe" genetic circuits. According to Sebesta et al. (2022), microalgal strains cannot thrive without artificial nutrients, thus preventing them from spreading throughout the environment. Furthermore, pilot-scale trials in mesocosms or isolated ponds allow the evaluation of ecological interactions under seminatural conditions before full-scale deployment (Dodds & Whiles, 2020).

Contemporary ERAs frequently integrate empirical data with predictive ecological models to simulate the population dynamics of GMOs and native species under varying nutrient loads and hydrological conditions. Scenario analyses estimate the potential outcomes in cases of accidental escape, high-nutrient conditions, or prolonged persistence. Additionally, model outputs guide risk mitigation strategies, monitoring frequency, and containment protocols (Scheffer et al., 2001; Henley et al., 2013).

3.4.2 Adaptive Governance and Participatory Decision-Making

Proper governance calls for an adaptable management technique that is proficient at evolving with fresh empirical discoveries (Holling, 1978). Combining input from the community, professionals, and governing bodies not only makes the process more reliable but also speeds up the process of finding out about environmental effects. Responsible implementation necessitates a thorough strategic framework that encompasses the amalgamation of ethical, ecological, and social factors; the alleviation of unintended ecological consequences through ongoing monitoring and control measures; alignment with sustainable development objectives; and a cyclical assessment process that incorporates both scientific research and local observational data.

REFERENCES

Aftab, J., Abid, N., Sarwar, H., & Venezian, M. (2022). Environmental ethics, green innovation, and sustainable performance: Exploring the role of environmental leadership and environmental strategy. *Journal of Cleaner Production 378*(10), 134639. https://doi.org/10.1016/j.jclepro.2022.134639

Agostinho, A. A., Bonecker, C. C., & Gomes, L. C. (2009). Effects of water quantity on connectivity: The case of the upper Paraná River floodplain. *Ecohydrology & Hydrobiology 9*(1), 99–113.

Aguirre Rodríguez, E. G. (2023). *Crecimiento poblacional y tabla de vida de Hydra sp. (Cnidaria: Hydrozoa) alimentada con diferentes densidades de cladóceros (Ceriodaphnia dubia Richard, 1894 y Daphnia pulex Leydig, 1860)* [Tesis de licenciatura, Universidad Nacional Autónoma de México, Facultad de Estudios Superiores Iztacala]. Recurso en línea.

Alcocer, J., Espinosa-Rodríguez, C. A., Fernández, R., Lugo-Vázquez, A., Macek, M., Maeda-Martínez, A. M., Martínez-Jerónimo, F., Ortega-Mayagoitia, E., & Oseguera, L. A. (2023). The ecology of the zooplankton in Mexican inland waters: What we know so far. *Limnologica 97*, 126084. https://doi.org/10.1016/j.limno.2023.126084

Angeler, D. G., Allen, C. R., Birgé, H. E., Drakare, S., McKie, B. G., & Johnson, R. K. (2014). Assessing and managing freshwater ecosystems vulnerable to environmental change. *Ambio 43*(Suppl 1), 113–125. https://doi.org/10.1007/s13280-014-0566-z

Bagwell, C. E., Abernathy, A., Barnwell, R., Milliken, C. E., Noble, P. A., Dale, T., & Moeller, P. D. (2016). Discovery of bioactive metabolites in biofuel microalgae that offer protection against predatory bacteria. *Frontiers in Microbiology 7*, 516. https://doi.org/10.3389/fmicb.2016.00516

Bhandarkar, S. V., & Bansod, S. (2025). The role of Rotifera as sentinel organism of trophic structure on freshwater ecosystems. *International Journal of Scientific Research in Biological Sciences 12*(2), 41–52. https://doi.org/10.26438/ijsrbs.v12i2.677

Bøhn, T., Primicerio, R., Hessen, D. O., & Traavik, T. (2008). Reduced fitness of *Daphnia magna* fed a Bt-transgenic maize variety. *Archives of Environmental Contamination and Toxicology 55*(4), 584–592. https://doi.org/10.1007/s00244-008-9150-5

Bravo Hernández, E. (2018). *Efectos subletales de malatión sobre variables demográficas de tres especies de cladóceros (Cladocera): Daphnia pulex Leydig, Diaphanosoma birgei Korinek y Moina macrocopa Goulden* [Tesis de maestría, Universidad Nacional Autónoma de México, Programa de Posgrado en Ciencias del Mar y Limnología]. Recurso en línea.

Carstensen, J., Conley, D. J., Bonsdorff, E. Gustafsson, B.G., Hietanen, S., Janas, U., … Reed, D.C. (2014). Hypoxia in the Baltic Sea: Biogeochemical cycles, benthic fauna, and management. *Ambio 43*, 26–36. https://doi.org/10.1007/s13280-013-0474-7

Choi, J.-Y., Kim, S. K., La, G. H., Chang, K. H., Kim, D. K., Jeong, K. Y., … Jeong, K. S. (2016). Effects of algal food quality on sexual reproduction of *Daphnia magna*. *Ecology and Evolution 6*(4), 1003–1012. https://doi.org/10.1002/ece3.2058

Cyr, H., & Curtis, J. M. (1999). Zooplankton community size structure and taxonomic composition affects size-selective grazing in natural communities. *Oecologia 118*(3), 306–315. https://doi.org/10.1007/s004420050731

Davidson, T. A., Bennion, H., Jeppesen, E., Clarke, G. H., Sayer, C. D., Morley, D., Odgaard, B. V., Rasmussen, P., Rawcliffe, R., Salgado, J., Simpson, G. L., & Amsinck, S. L. (2011). The role of cladocerans in tracking long-term change in shallow lake trophic status. *Hydrobiologia 676*, 299–315. https://doi.org/10.1007/s10750-011-0851-9

Dodds, W. K., & Whiles, M. R. (2020). *Freshwater Ecology: Concepts and Environmental Applications* (2nd ed.). Elsevier.

Dunalska, J. (2025). Lake restoration techniques: A review of methods and future pathways. *Science of the Total Environment 979*, 179450. https://doi.org/10.1016/j.scitotenv.2025.179450

European Commission. (2018). Directive 2001/18/EC on the deliberate release into the environment of genetically modified organisms. *Official Journal of the European Union 6*, 1–39.

Falkowski, P. G., & Raven, J. A. (2007). *Aquatic Photosynthesis* (2nd ed.). Princeton University Press. https://doi.org/10.1515/9781400849727

Fragoso Rodríguez, M. E. (2023). *Efecto ecotoxicológico de un florecimiento cianobacterial sobre diferentes especies nativas de cladóceros* [Tesis de licenciatura, Universidad Nacional Autónoma de México, Facultad de Estudios Superiores Iztacala]. Recurso en línea.

Frost, P. C., DaSilva, K., Frost-Xenopoulos, J. A., Burr, W. S., & Jones, C. L. C. (2025). Effects of phosphorus enrichment on Daphnia–algae interactions in laboratory microcosms. *Journal of Plankton Research 47*(2), Article fbaf002. https://doi.org/10.1093/plankt/fbaf002

Gama-Flores, J. L., Sarma, S. S. S., & Nandini, S. (2004). Acute and chronic toxicity of the pesticide methyl parathion to the rotifer *Brachionus angularis* (Rotifera) at different algal (*Chlorella vulgaris*) food densities. *Aquatic Ecology 38*(1), 27–36. https://doi.org/10.1023/B:AECO.0000020986.92471.32

García Caballero, J. C. (2023). *Potabilización de agua con exceso de microalgas: Caso de estudio: Laguna de Xico* [Tesis de licenciatura, Universidad Nacional Autónoma de México, Facultad de Ingeniería]. Recurso en línea.

Giampietro, M. (2002). The precautionary principle and ecological hazards of genetically modified organisms. *Ambio 31*(6), 466–470. http://www.jstor.org/stable/4315287

Gómez, A., & Carvalho, G. R. (2000). Sex, parthenogenesis and genetic structure of rotifers: Microsatellite analysis of contemporary and resting egg bank populations. *Molecular Ecology 9*(2), 203–214. https://doi.org/10.1046/j.1365-294x.2000.00849.x

Haberman, J., Laugaste, R., & Nõges, T. (2007). The role of cladocerans reflecting the trophic status of two large and shallow Estonian lakes. *Hydrobiologia 584*, 157–166. https://doi.org/10.1007/s10750-007-0592-y

Henley, W. J., Litaker, R. W., Novoveská, L., Duke, C. S., Quemada, H. D., & Sayre, R. T. (2013). Initial risk assessment of genetically modified (GM) microalgae for commodity-scale biofuel cultivation. *Algal Research 2*(1), 66–77.

Holling, C. S. (1978). *Adaptive Environmental Assessment and Management.* John Wiley & Sons.

Howard, M. D., Smith, J., Caron, D. A., Kudela, R. M., Loftin, K., Hayashi, K., ... Theroux, S. (2023). Integrative monitoring strategy for marine and freshwater harmful algal blooms and toxins across the freshwater-to-marine continuum. *Integrated Environmental Assessment and Management 19*(3), 586–604. https://doi.org/10.1002/ieam.4651

Juárez Álvarez, O. E. (2005). *Microalgas en tres sistemas acuáticos del Estado de México* [Tesis de licenciatura, Universidad Nacional Autónoma de México, Facultad de Estudios Superiores Zaragoza]. Recurso en línea.

Khanna, S., Santos, M. J., Hestir, E. L., & Ustin, S. L. (2012). Plant community dynamics relative to the changing distribution of a highly invasive species, Eichhornia crassipes: A remote sensing perspective. *Biological Invasions 14*, 717–733. https://doi.org/10.1007/s10530-011-0112-x

Komárek, J. (2014). Modern classification of cyanobacteria. In N. K. Sharma, A. K. Rai, & L. J. Stal (eds.), *Cyanobacteria* (Chapter 2). Wiley-Blackwell. https://doi.org/10.1002/9781118402238.ch2

Martínez-Alier, J. (2002). *The Environmentalism of the Poor: A Study of Ecological Conflicts and Valuation.* Edward Elgar Publishing.

Matsuzaki, S. S., Suzuki, K., Kadoya, T., Nakagawa, M., & Takamura, N. (2018). Bottom-up linkages between primary production, zooplankton, and fish in a shallow, hypereutrophic lake. *Ecology 99*(9), 2025–2036. https://doi.org/10.1002/ecy.2414

McAfee, K. (2008). Beyond techno-science: Transgenic maize in the fight over Mexico's future. *Geoforum 39*(1), 148–160. https://doi.org/10.1016/j.geoforum.2007.06.002

Montoya, H., Komárková, J., & Komárek, J. (2014). Cyanobacterial species, potentially forming water-blooms in the Lake Titicaca (Peru). *Arnaldoa 21*(2), 381–390.

Odume, O. N., & de Wet, C. (2019). A systemic-relational ethical framework for aquatic ecosystem health research and management in social–ecological systems. *Sustainability 11*(19), 5261. https://doi.org/10.3390/su11195261

Pace, M. L. Batt, R. D., Buelo, C. D., Carpenter, S. R., Cole, J. J., Kurtzweil, J. T., & Wilkinson, G. M. (2017). Reversal of a cyanobacterial bloom in response to early warnings. *Proceedings of the National Academy of Sciences of the United States of America 114*(2), 352–357. https://doi.org/10.1073/pnas.1612424114

Pace, M. L., Cole, J. J., & Carpenter, S. R. (1998). Trophic cascades and compensation: Differential responses of microzooplankton in whole-lake experiments. *Ecology 79*(1), 138–152. https://doi.org/10.1890/0012-9658(1998)079[0138:TCACDR]2.0.CO;2

Paerl, H. W., & Otten, T. G. (2013). Harmful cyanobacterial blooms: Causes, consequences, and controls. *Microbial Ecology 65*(4), 995–1010. https://doi.org/10.1007/s00248-012-0159-y

Pérez-Morales, A., Olivos-Ortiz, A., Quijano-Scheggia, S. I., Espinosa-Rodríguez, C. A., & Jiménez-Santos, M. A. 2016. Estado actual del estudio de cianobacterias dulceacuícolas formadoras de florecimientos en el centro de México. In: E. García-Mendoza, S. I. Quijano-Scheggia, A. Olivos-Ortiz, & E. J. Núñez-Vázquez (eds.), *Florecimientos Algales Nocivos en México* (pp. 408–421). Ensenada, CICESE.

Pohle, P., & Gerique, A. (2006). Traditional ecological knowledge and biodiversity management in the Andes of southern Ecuador. *Geographica Helvetica 61*, 275–285. https://doi.org/10.5194/gh-61-275-2006

Raffensperger, C., & Tickner, J. A. (1999). *Protecting Public Health and the Environment: Implementing the Precautionary Principle*. Island Press.

Reyes Landeros, J. (2022). *Evaluación del tratamiento de agua residual mediante cultivo de microalgas para obtener un efluente que cumpla con la NOM-003-SEMARNAT-1997* [Tesis de licenciatura, Universidad Nacional Autónoma de México, Facultad de Ciencias]. Recurso en línea.

Richmond, A., & Hu, Q. (eds.). (2013). *Handbook of Microalgal Culture: Applied Phycology and Biotechnology* (2nd ed.). Wiley-Blackwell.

Sales Lara, G. G. (2024). *Efecto de la intensidad de luz sobre un consorcio de microalgas y bacterias para la remoción de nitrógeno y fósforo en agua residual porcícola* [Informe académico de servicio social, Universidad Nacional Autónoma de México, Escuela Nacional de Estudios Superiores, Unidad Mérida]. Recurso en línea.

Sandonnini, J., Del Pilar-Ruso, Y., Valverde-Urrea, M., & Gimenez-Casalduero, F. (2024). Infauna evolution of the invertebrate community after a severe eutrophic crisis in the coastal lagoon of the Mar Menor (W Mediterranean). *Regional Studies in Marine Science 79*, 103824. https://doi.org/10.1016/j.rsma.2024.103824

Sao, S., Soni, D. K., Singh, R. K., & Pandey, P. (2024). Rotifers and their role in water health: A review. *International Journal for Multidisciplinary Research 6*(4), Article 24907. https://doi.org/10.36948/ijfmr.2024.v06i04.24907

Scheffer, M., Carpenter, S., Foley, J. A., Folke, C., & Walker, B. (2001). Catastrophic shifts in ecosystems. *Nature 413*(6856), 591–596. https://doi.org/10.1038/35098000

Sebesta, J., Xiong, W., Guarnieri, M. T., & Yu, J. (2022). Biocontainment of genetically engineered algae. *Frontiers in Plant Science 13*, 839446. https://doi.org/10.3389/fpls.2022.839446

Sierra, M., García-Reyes, M., Ortells, R., & Carmona, M. (2019). Cyclically parthenogenetic rotifers and the theories of population and evolutionary ecology. *Limnetica 38*(1), 67–93. https://doi.org/10.23818/limn.38.13

Smith, V. H., & Schindler, D. W. (2009). Eutrophication science: Where do we go from here? *Trends in Ecology & Evolution 24*(4), 201–207. https://doi.org/10.1016/j.tree.2008.11.009

Soares, E. (2015, September 6). *Restrictions on Genetically Modified Organisms: Brazil*. Law Library of Congress. Retrieved from https://stuff.coffeecode.net/www.loc.gov/law/help/restrictions-on-gmos/brazil.php

Søndergaard, M. (2007). Lake restoration: Successes, failures and long-term effects. *Journal of Applied Ecology 44*(6), 1095–1105. https://doi.org/10.1111/j.1365-2664.2007.01363.x

Sterner, R. W., & Elser, J. J. (2002). *Ecological Stoichiometry: The Biology of Elements from Molecules to the Biosphere*. Princeton University Press.

Toledo, V. M., & Barrera-Bassols, N. (2008). *La memoria biocultural: La importancia ecológica de los conocimientos tradicionales*. ICARIA.

Velini, E. D., Dagli, M. L. Z., de Souza, G. D., Nascimento, R. J., Pinho, T. F., Paes de Andrade, P., & Carrer, H. (2017). The Brazilian GMO regulatory system: A historical view and perspective. In: A. A. Adenle, E. Jane Morris, & D. J. Murphy (eds.), *Genetically Modified Organisms in Developing Countries: Risk Analysis and Governance* (pp. 258–270). Cambridge University Press. https://doi.org/10.1017/9781316585269.023

Weers, P., & Gulati, R. (1997). Effect of the addition of polyunsaturated fatty acids to the diet on the growth and fecundity of *Daphnia galeata*. *Freshwater Biology 38*(3), 721–729. https://doi.org/10.1046/j.1365-2427.1997.00237.x

Wojtal-Frankiewicz, A., Kruk, A., Frankiewicz, P., Oleksińska, Z., & Izydorczyk, K. (2015). Long-term patterns in the population dynamics of *Daphnia longispina, Leptodora kindtii* and cyanobacteria in a shallow reservoir: A self-organising map (SOM) approach. *PloS One 10*(12), e0144109. https://doi.org/10.1371/journal.pone.0144109

Yan, K., Guo, F., Kainz, M. J., Li, F., Gao, W., Bunn, S. E., & Zhang, Y. (2024). The importance of omega-3 polyunsaturated fatty acids as high-quality food in freshwater ecosystems with implications of global change. *Biological Reviews 99*(1), 200–218. https://doi.org/10.1111/brv.13017

Yu, B. S., Pyo, S., Lee, J., & Han, K. (2024). Microalgae: A multifaceted catalyst for sustainable solutions in renewable energy, food security, and environmental management. *Microbial Cell Factories 23*(1), 308. https://doi.org/10.1186/s12934-024-02588-7

4 Microalgae Production Systems

Recent Findings to Forward Technological Readiness Level for Microalgal Biotechnology Consolidation

Carolina Diaz-Cardenas, Francisco Vera-López Portillo, Gina López, Dulce María Arias, Jorge Daniel Osorio-Márquez, and Carlos Alberto Montenegro-Herrera

4.1 INTRODUCTION

Microalgae have become key biological platforms in modern biotechnology due to their high productivity, metabolic versatility, efficient resource utilization, and capacity to integrate into waste valorization and carbon mitigation strategies. These attributes position microalgal systems as promising sources of bioproducts for sectors such as food, bioenergy, bioplastics, cosmetics, and pharmaceuticals.

Despite this potential, the industrial deployment of microalgal biotechnology remains limited by challenges in energy efficiency, operational costs, culture stability, and scale-up complexity. Addressing these constraints requires more substantial alignment between scientific innovation and technological implementation.

Within this context, Technological Readiness Levels (TRLs) provide a structured framework to assess the development stage of microalgal bioprocesses and their proximity to market application. The TRL approach helps identify technical, economic, and regulatory barriers while guiding strategies to transition from basic research to industrial validation.

This chapter examines recent advances that enhance the scalability and commercial viability of microalgae production systems. It reviews open, closed, and hybrid cultivation configurations; criteria for selecting a system based on strain physiology

 DOI: 10.1201/9781003611479-4

and target products; and key innovations in photobioreactor design and automation. Additionally, it explores the integration of microalgae into industrial processes under a circular-economy perspective, emphasizing the relevance of TRLs for overcoming scale-up barriers. Finally, sustainability considerations and Life-Cycle Assessment (LCA) are highlighted as essential tools for developing efficient, environmentally responsible, and economically competitive bioprocesses.

4.2 OVERVIEW OF MICROALGAE PRODUCTION SYSTEMS

Microalgae exhibit notable photosynthetic capacities, often surpassing those of terrestrial plants in efficiency and growth rate (Khor et al., 2022). Their ability to capture and fix large amounts of CO_2 into biomass has positioned them as a strategic resource in global efforts to mitigate climate change and develop circular bioeconomy models. Microalgal biomass can be transformed into a broad spectrum of products, including biofuels such as biodiesel and biohydrogen, food and feed supplements, fertilizers, pigments, and bioactive compounds with potential uses in nutraceutical, pharmaceutical, and cosmeceutical industries. Under optimal conditions, their productivity may exceed that of other conventional crops, making them among the most promising renewable biological resources of the 21st century.

In particular, the Global Sustainability Goals have intensified interest in these organisms. According to recent Food and Agriculture Organization (FAO) estimations, by 2030, between 638 and 720 million people, equivalent to 7.8 and 8.8 percent of the global population, could face hunger (FAO et al., 2025). In parallel, governments around the world continue to allocate substantial resources to cereal production to ensure food security (Barboza-Rodríguez et al., 2024). At the same time, the growing energy demand driven by urbanization and population growth is forcing a transition away from fossil fuels, whose declining energy return and heavy carbon footprint aggravate biodiversity loss and ocean acidification (Geng et al., 2025).

Within this context, microalgae emerge as a dual opportunity: to supply sustainable energy and high-value bioproducts while contributing to CO_2 capture and nutrient recycling. Their metabolic versatility (rich in polysaccharides, lipids, carotenoids, proteins, and vitamins) makes them valuable as nutritional supplements and sources of functional compounds (Uma et al., 2022). Nevertheless, most studies on the production of microalgae-derived bioactive metabolites have been conducted under laboratory conditions, limiting our understanding of industrial implementation and subsequent commercial applications (Chen et al., 2025).

Cultivation strategy is a critical determinant of microalgal biomass yield and composition. Modern systems seek to optimize light distribution, gas exchange, and nutrient supply while reducing contamination risks and operational costs, making system design and operation central to productivity and economic feasibility. Efficient cultivation platforms are therefore essential for scaling microalgal production from laboratory to industrial levels.

Despite advances in indoor and outdoor configurations, large-scale outdoor cultivation still faces challenges such as environmental variability, biofouling, and elevated energy requirements. To overcome these constraints, current technologies employ three main system types, open, closed, and hybrid, each with distinct advantages and limitations in scalability, productivity, and cost-effectiveness. Understanding their design principles and operational characteristics is key to selecting the appropriate system for a given strain and application. The following section outlines these cultivation strategies and highlights innovations relevant to contemporary microalgal biotechnology.

4.2.1 Classification of Production Systems

Microalgae cultivation systems must be designed to provide optimal conditions for each strain while minimizing energy demand and costs. Each configuration (open, closed, and hybrid) presents unique trade-offs between scalability, productivity, and economic feasibility.

An efficient cultivation system must ensure appropriate light exposure, adequate mixing and gas exchange, sufficient nutrient delivery, minimal contamination risk, and high land-use efficiency (Tan et al., 2020). Therefore, the selection of a particular configuration depends on factors such as the physiological requirements of the species, the desired metabolites, and the intended application.

Open systems, such as raceway ponds and open tanks, are valued for their simplicity and low capital investment, whereas closed systems (photobioreactors) allow precise environmental control and higher product purity. Hybrid systems integrate both approaches to balance productivity and cost-efficiency (Table 4.1).

4.2.1.1 Open System

Open systems include a variety of configurations such as raceway ponds, open tanks, and Bormicroalgae cultivation (Figure 4.1). These systems operate under natural illumination and temperature conditions, relying on continuous agitation to enhance light penetration and gas exchange. Their main appeal lies in their simple design, scalability, and low operational costs (Clagnan et al., 2023).

Raceway ponds, introduced in the 1960s for wastewater treatment with macroalgae (Oswald & Golueke, 1960), remain the most widespread configuration, accounting for more than 90% of global microalgae production (Barceló-Villalobos et al., 2019). Typically, these ponds are shallow (30–70 cm deep) and are arranged in oval or circular channels where a paddlewheel provides circulation, ensuring homogeneous exposure of cells to light and nutrients (Choudhary et al., 2015; Oves et al., 2022). Robust and fast-growing strains such as *Spirulina* (cyanobacteria), *Chlorella*, *Tetraselmis*, and *Dunaliella* are commonly cultivated in this type of reactor (Narala et al., 2016).

Economic scalability in open pond systems benefits from their relatively low capital expenditure and modular design, allowing phased expansion from 10 to 20 ha pilot operations to commercial facilities exceeding 100 ha (Michael et al., 2026). Despite these economic advantages, open systems are highly influenced by environmental fluctuations and typically achieve low biomass concentrations (0.3–1.5 g L^{-1}) (Apel et al., 2017). Modified configurations, such as thin-layer cascade reactors with water depths of 0.01–0.05 m, have demonstrated significantly higher productivities

TABLE 4.1
Comparative Characteristics of Open and Closed Microalgae Cultivation Systems

System Type	Advantages	Disadvantages
Open systems (e.g., raceway ponds and open tanks)	• Low capital and operational costs (Wang et al., 2024). • Simple construction and scalability for mass culture (Michael et al., 2026). • Suitable for robust, fast-growing strains. • Easier CO_2 diffusion and nutrient addition (Sundaram et al., 2025).	• High contamination risk. • Limited control over environmental parameters (light, temperature, and pH) (Villaró et al., 2022). • Water loss by evaporation (Benedetti et al., 2021). • Low biomass concentration and productivity (Borovkov et al., 2020). • Seasonal dependence on environmental conditions (Sundaram et al., 2025).
Closed systems (e.g., tubular, flat-plate, and column photobioreactors)	• Controlled environmental and operational parameters. • Reduced contamination risk (axenic cultures). • High surface-to-volume ratio for efficient light use. • Higher productivity and metabolite yield. • Suitable for high-value compounds.	• High installation and maintenance costs. • Technical complexity and need for skilled operation. • Scale-up limitations due to light attenuation and oxygen accumulation. • Potential overheating under high light exposure.
Hybrid systems (e.g., coupled photobioreactor–raceway configurations and cascade–tubular combinations)	• Combine the high productivity and control of PBRs with the low-cost scalability of open ponds (Sundaram et al., 2025). • Enhanced CO_2 fixation efficiency through sequential cultivation stages. • Reduced contamination risk compared to fully open systems. • Improved resource utilization (light, nutrients, and water). • Suitable for industrial applications requiring large volumes and moderate quality control.	• Higher complexity in system design and operation. • Requires synchronization between cultivation stages. • Potential cross-contamination between open and closed units. • Energy demand is still higher than in open systems. • Optimization parameters vary widely depending on configuration and strain (Rörig et al., 2024).

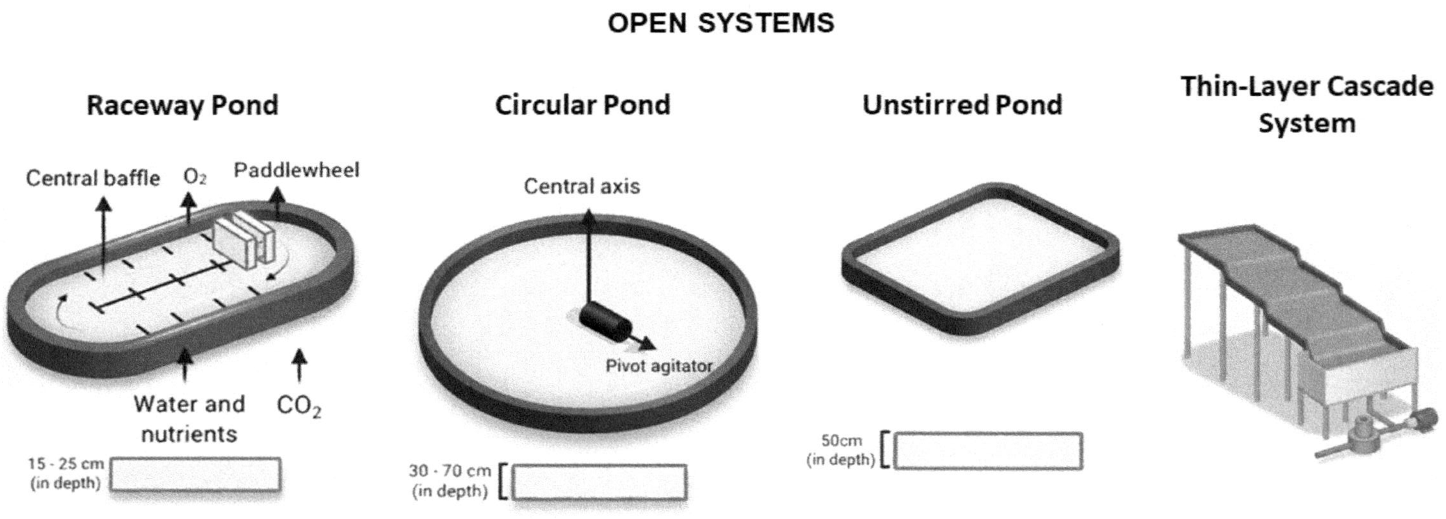

FIGURE 4.1 Schematic representation of common open culture systems.

(Own elaboration adapted from Razzak et al., 2024.)

(30–50 g L^{-1}) in outdoor cultures of *Nannochloropsis*, *Chlorella*, and *Scenedesmus* (Apel et al., 2017; Schädler et al., 2019).

However, open systems remain difficult to operate due to their dependence on natural sunlight and climatic conditions. Conventional raceway designs are often inefficient in light utilization, mainly because of suboptimal culture depth (0.10–0.30 m), which restricts photon availability and limits biomass density (Apel et al., 2017; Barceló-Villalobos et al., 2019).

Several design innovations have aimed to improve mixing, light utilization, and carbon fixation efficiency. For instance, electrostatic field stimulation has been proposed as a strategy to prevent cell sedimentation in unstirred ponds, enhance membrane permeability, and increase the productivity of pigments such as phycocyanin and storage compounds such as triacylglycerols (Ren et al., 2025). Similarly, automated paddle systems and light-supplemented mixers have improved gas–liquid transfer and reduced dead zones, increasing both carbon fixation and biomass yield (Ren et al., 2025).

Unstirred or naturally mixed lagoons remain the simplest and most cost-effective form of open cultivation, relying on wind and convection for mixing (Razzak et al., 2024). Nevertheless, they provide limited control over environmental parameters and are therefore more suitable for low-value biomass production or environmental applications, such as wastewater treatment.

Overall, open systems continue to play a central role in large-scale microalgae cultivation due to their scalability and economic feasibility. However, their performance is often constrained by contamination risk, low volumetric productivity, and strong dependence on climatic conditions, factors that have driven the development of closed and hybrid systems as complementary technologies.

Table 4.2 summarizes representative studies that illustrate the versatility of open cultivation systems across different operational conditions and scales. The examples range from laboratory-scale experiments to pilot and industrial operations, emphasizing the scalability and adaptability of open configurations. These cases highlight the capacity of raceway ponds and thin-layer cascades to support high-value metabolite production while integrating wastewater valorization and process intensification strategies.

4.2.1.2 Closed Systems

While open ponds remain the dominant low-cost technology, closed systems, commonly referred to as photobioreactors (PBRs), have gained increasing attention due to their superior process control (Figure 4.2). These systems provide highly regulated environments with minimal contamination risk, enabling the production of high-quality biomass for food, pharmaceutical, and cosmeceutical applications (Narala et al., 2016; Mathivanan et al., 2023). PBRs can be customized according to strain-specific requirements, offer efficient use of space, and ensure product consistency.

Structurally, PBRs consist of closed vessels that may employ artificial or natural light as an illumination source. Depending on their geometry, they are commonly classified as tubular, flat-panel, column, or bag systems (Razzak et al., 2024; Michael

TABLE 4.2
Overview of Bioreactor Technologies Applied to Microalgae Cultivation and Bioproduct Production

Bioreactor Type	Microalgae	Capacity	Key Variables/Innovation Conditions	Applications/ Metabolite	Biotechnological Application	Ref
Open ponds	*Pavlova granifera* IML2238	6 ton (semi-batch)	Optimization of inoculum size maintained stable biomass and carotenoid productivity under semi-batch conditions. inoculum	Fucoxanthin 4.31 mg L^{-1} day^{-1}	Antioxidant cosmeceutical applications	Chen et al., 2025
Thin-layer cascade	*Tetradesnus almeriensis*	63 m^2	Semicontinuous culture. Influence of culture medium (freshwater, wastewater, and diluted pig slurry) on biomass yield and composition.	Biomass productivity: 30.3 g m^{-2}·day^{-1}	Wastewater valorization, biofertilizers, and feed production	Villaró et al., 2022
Thin-layer cascade	*Microchloropsis salina*	8 m^2	Repeated water recycling with nutrient replenishment maintained stable biomass productivity and reduced nutrient consumption.	Biomass productivity: 14.9–21.3 g m^{-2} d^{-1}; lipid concentration of 6.6 g L^{-1}	Sustainable production through medium and nutrient recycling, minimizing wastewater generation	Schädler et al., 2019
Open in rectangular ponds (batch and semicontinuous)	*Dunaliella salina* IBSS-2	Pilot-scale	Two-stage culture optimizing strain and nutrient medium under variable temperature and irradiance.	β-carotene productivity accumulation up to 8%; biomass productivity: ~6 g m^{-2} day^{-1}; total carotenoids 200–600 mg m^{-2}	Commercial-scale pigment production using stress-induced carotenogenesis	Borovkov et al., 2020
Raceway pond with automated mixing and electrostatic field system	*Limnospira fusiformis*	Laboratory scale—71.4 dm^3	Integration of automated light-supplemented mixers (75° inclined blades, 300 rpm) and electrostatic field treatment improved gas–liquid mass transfer, reducing dead zones.	Biomass productivity >20%; carbon fixation enhanced by 43%; phycocyanin yield is higher 14%.	Process intensification strategy for enhanced productivity and light utilization efficiency	Ren et al., 2025

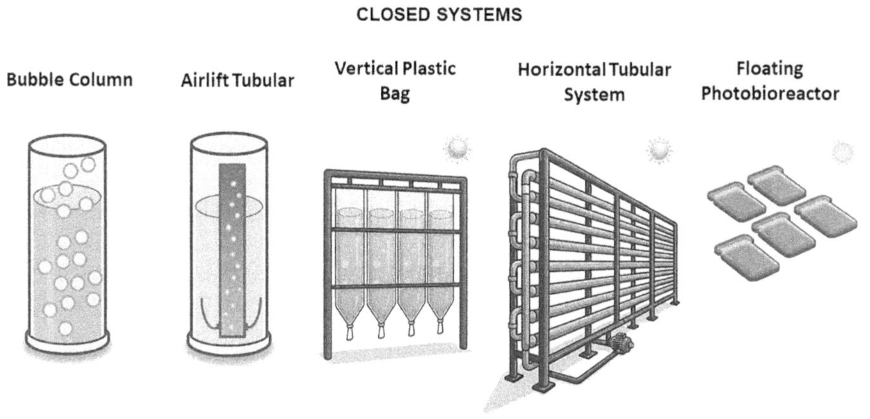

FIGURE 4.2 Schematic representation of common close photobioreactor systems.

(Own elaboration adapted from Razzak et al., 2024; Montenegro-Herrera et al., 2025.)

et al., 2026). Based on the mode of liquid flow dynamics, they can also be grouped into airlift, stirred-tank, and bubble-column configurations (Barboza-Rodríguez, et al., 2024). Among these, horizontal tubular, vertical bubble-column, and flat-plate reactors are the most widely used for large-scale microalgae cultivation due to their high surface-to-volume ratios, which enhance photon utilization and photosynthetic efficiency (Zhang et al., 2025).

Efficient gas exchange and nutrient delivery in these systems are achieved by recirculating the culture medium through pumps or airlift mechanisms under turbulent flow. Bubble-column PBRs are among the simplest and most compact configurations. They improve nutrient distribution, gas transfer, and physicochemical stability, providing favorable growth conditions (Hernandez-de Leon et al., 2025). For instance, *Chlorella sorokiniana* exhibited a growth rate of 0.097 g $DW{\cdot}L^{-1}{\cdot}day^{-1}$ in a bubble-column reactor, outperforming airlift and stirred-tank systems (Uyar et al., 2024). Similarly, studies on *Neochloris oleoabundans* UTEX 1185 demonstrated that biomass concentration and lipid content are highly dependent on reactor configuration.

Airlift PBRs (ALRs) are derived from bubble-column designs and include draft tubes or baffles that enhance gas–liquid mass transfer. Gas is introduced through internal or external spargers, generating improved mixing and CO_2 dissolution (Lara et al., 2018). For example, a ring-sparger bubble-column reactor achieved the highest biomass and lipid productivity at an airflow rate of 600 $mL{\cdot}min^{-1}$, due to enhanced gas transfer and reduced shear stress (Banerjee et al., 2020).

Flat-plate PBRs are another widely adopted configuration, suitable for both indoor laboratory and outdoor pilot-scale applications. Their vertical design maximizes the illuminated surface-to-volume ratio and facilitates temperature regulation. Typical dimensions range between 1 and 2 m in height and ≤10 cm in width, reducing material costs while maintaining optical efficiency (Barboza-Rodríguez et al., 2024). Mixing is achieved through air sparging, pumps, or mechanical impellers. Hybrid designs such as flat-plate airlift photobioreactors (FPA-PBRs) combine the advantages of tubular and flat-panel systems, coupling effective light capture with superior gas–liquid transfer. Under optimized hydrodynamic conditions, the Rectangular Airlift Photobioreactor with Baffles (RAPBR-B) achieved a biomass yield of 3.226 g DW for *Synechococcus* HS-9, demonstrating improved turbulence, bubble dispersion, and overall mass transfer (Rahman et al., 2025).

Hybrid PBRs that integrate both open and closed configurations aim to balance productivity with operational costs. They combine the low-cost scalability of raceway ponds with the controlled conditions of enclosed systems (Sundaram et al., 2025). For instance, *Tetradesmus obliquus* cultivated in a hybrid reactor combining tubular and thin-layer cascade features achieved biomass productivities of 1,216 mg with CO_2 supplementation, compared to 76 mg without enrichment (Rörig et al., 2024). Such designs overcome many limitations of open-air ponds while reducing the energy intensity typical of fully closed systems.

Nonconventional PBRs, including membrane, plastic-bag, and solid-state systems, are being developed to address biomass washout and limited residence times typical of continuous cultures (Marbelia et al., 2014; Zuccaro et al., 2020). Membrane photobioreactors (MPBRs) are particularly effective for wastewater treatment, as membranes decouple the hydraulic retention time from the biomass retention time (MRT), thereby increasing cell density and productivity. MPBRs have demonstrated biomass yields up to 9 times higher than those of conventional systems, with water savings above 75% and energy consumption reductions of approximately 0.75 $kWh{\cdot}m^{-3}$ (Morais et al., 2025). Similarly, plastic-bag reactors offer a low-cost alternative for small-scale operations, using transparent polyethylene bags illuminated by sunlight and aerated from the bottom.

Recently, floating photobioreactors (FPBRs) have gained interest as a cost-effective solution for offshore microalgae cultivation. Installed on ocean surfaces, these systems utilize wave energy for mixing and benefit from natural temperature regulation and sunlight exposure (Khor et al., 2022). They are particularly promising for the aquaculture industry due to their scalability, low-energy input, and high biomass productivity. For example, *Isochrysis zhangjiangensis* FACHB-1750 reached maximum cell densities of $16.1 \pm 0.61 \times 10^6$ $cells{\cdot}L^{-1}$ in an outdoor FPBR with a 5 cm culture depth, outperforming deeper systems with reduced light penetration (Zhu et al., 2019). Despite these advantages, challenges remain, including unpredictable wave dynamics, biofouling, and potential impacts on marine ecosystems.

Collectively, these examples demonstrate the diversity of photobioreactor configurations and their adaptability to a range of production goals, from CO_2 biofixation and wastewater valorization to the synthesis of high-value pigments and lipids. Process optimization through hydrodynamic control, nutrient recycling, and gas–liquid transfer enhancement continues to drive the scalability and economic feasibility of closed cultivation systems (Table 4.3).

TABLE 4.3
Representative Photobioreactor Systems for High-Value Metabolite Production and Resource Valorization

Bioreactor Type	Microalgae	Capacity	Key Variables/ Innovation Conditions	Applications/ Metabolite	Biotechnological Application	Ref
Bubble-column photobioreactor	*Chlorella* sp. S-01, *Chlorella* sp. S-02, and consortium Co-CC	5 L (laboratory scale)	Evaluation of different CO_2 concentrations (0.03%, 5%, and 10%) on monocultures and consortium, obtaining better results at 10% CO_2	Biomass productivity: 1.87 g L^{-1}, and lipid content 19–20%. De-oiled biomass contained 46% carbohydrates and 31% protein.	Integrated biorefinery approach for biodiesel production and valorization of residual biomass	Mathivanan et al., 2023
Bubble-column photobioreactor	*Prymnesium parvum*	10 L (laboratory scale)	Valorization of desalination brine (61 g L^{-1} salinity) by reverse osmosis using a halotolerant strain acclimated to 5–61 g L^{-1}. Optimized nutrients and irradiance. Salinity stress increased saturated and monounsaturated fatty acids and carotenoids.	Biomass yield: 1.9 g L^{-1}. Saturated and monounsaturated fatty acids increase by 83%, and carotenoids enhanced more than 25 %.	Brine valorization for circular bioeconomy; production of antioxidant, antimicrobial, and antitumoral metabolites.	Contreras-Gómez et al., 2026
Controlled PBR	*Chlorella sorokiniana*	0.3 L (Laboratory scale).	Development of a fertilizer-based medium for food-grade *Chlorella*. Optimization of nitrogen source ($NaNO_3$ vs. NH_4Cl) and N:P ratio (21). Semicontinuous recirculation reduced nutrient and water use by 70%.	Biomass productivity: 0.29 ± 0.02 $g{\cdot}L^{-1}{\cdot}day^{-1}$ ($NaNO_3$); 0.10 ± 0.02 $g{\cdot}L^{-1}{\cdot}day^{-1}$ (NH_4Cl).	Nutrient recycling and medium optimization for cost-effective and sustainable large-scale production	Salinas-García et al., 2025

(Continued)

TABLE 4.3 (Continued)
Representative Photobioreactor Systems for High-Value Metabolite Production and Resource Valorization

Bioreactor Type	Microalgae	Capacity	Key Variables/ Innovation Conditions	Applications/ Metabolite	Biotechnological Application	Ref
Airlift bioreactor (internal vs. external sparger)	*Chlorella vulgaris*	Laboratory scale (air–water system, 25 ± 2 °C)	Hydrodynamic comparison between internal and external spargers under varying gas velocities. Highest CO_2 dissolution and mixing achieved with external sparger at 1.88×10^{-3} m·s^{-1}.	Biomass concentration: 26.95×10^6 cells·mL^{-1}; CO_2 removal efficiency: 94%.	Efficient CO_2 biofixation via optimized gas distribution; development of low-energy airlift systems.	Azhand et al., 2020
Rectangular airlift PBR with baffles (RAPBR-B)	*Synechococcus* HS-9	25 L (Laboratory scale)	Optimization of hydrodynamic and mass-transfer parameters (bubble velocity, diameter, gas hold-up, kLa). Baffles improved aeration and mixing efficiency.	Biomass yield: 3.226 g of DW	Optimization for cost-effectiveness enhances biomass productivity in large-scale production	Rahman et al., 2025
Floating photobioreactor (FPBR)	*Dunaliella salina* CCAP 19/18	Pilot-scale (1 m^3 seawater desalination concentrate)	Utilization of desalination brine with carbonate-precipitate pretreatment removing 97.6% Ca^{2+} and 60.3% Mg^{2+}. CO_2 bubbling increased the total inorganic carbon to 0.254 mol·L^{-1}.	300 g of biomass containing 14.3 g of β-carotene per m^3 of medium	β-carotene production and enhanced profitability in the desalination industry.	Zhu et al., 2018.
Stirred-tank photobioreactor (scale-up study)	*Haematococcus pluvialis* EGE MACC-32	2 L and 10 L PBRs	Evaluation of three scale-up criteria. Constant volumetric power input (P/V) was identified as optimal for maintaining high productivity and carotenoid yield.	Total carotenoids: 4.57 mg·g^{-1} (2 L) and 4.77 mg·g^{-1} (10 L); antioxidant activity: 46.91 % inhibition; total phenolics: 11.76 mg GAE·L^{-1}.	Scale-up strategy for commercial antioxidant carotenoid production under controlled PBR conditions.	Deniz, 2020

4.2.2 Challenges

Realizing the full potential of microalgae-based technologies requires overcoming significant technical and economic challenges. Among these, light availability remains one of the most critical limiting factors for both phototrophic and mixotrophic cultivation. Depending on geographic and climatic conditions, microalgal growth may rely on natural sunlight, artificial illumination, or a combination of both. Although sunlight is the most sustainable and cost-effective energy source, its variability, particularly in regions with low or seasonal irradiance, hampers stable, year-round productivity. Conversely, artificial lighting systems enable continuous cultivation and high biomass yields but substantially increase operational costs. As a result, hybrid lighting strategies are being actively developed to balance sustainability and productivity, especially in temperate and high-latitude regions (Zhang et al., 2025; Razzak et al., 2024; Yang et al., 2024).

Nutrient availability and recovery represent another major challenge for future large-scale production. The global recovery of nitrogen and phosphorus from oceans, surface waters, and wastewater streams has become a priority to mitigate fertilizer scarcity and eutrophication. Microalgae offer a promising platform for nutrient recycling and bioremediation, converting inorganic waste into valuable biomass (Michael et al., 2026). However, integrating wastewater-derived nutrients into food and feed production remains controversial due to potential safety concerns regarding heavy metals, pathogens, and residual contaminants (Geng et al., 2025; Wang et al., 2024).

Despite these challenges, microalgae remain exceptional multifunctional biofactories capable of addressing energy, environmental, and resource sustainability crises simultaneously. Their successful industrial exploitation depends on optimizing cultivation systems that ensure both economic feasibility and biosafety. While ideal photobioreactor conditions are theoretically unattainable, advances in reactor engineering, light distribution, automation, and process control continue to bring designs closer to these optimal parameters—enabling efficient, scalable, and sustainable microalgal production (Barboza-Rodríguez et al., 2024; Hernandez-de-Leon et al., 2025; Sundaram et al., 2025).

Looking ahead, the next generation of microalgal biotechnologies will increasingly rely on digitalization and in silico optimization to overcome current scalability barriers. Computational modeling, ranging from computational fluid dynamics (CFD) and metabolic flux analysis to machine learning (ML) algorithms, is now enabling precise simulation of light distribution, gas transfer, and nutrient dynamics in large-scale reactors (Chen et al., 2025). These tools allow researchers to predict growth behavior, identify metabolic bottlenecks, and optimize reactor geometry or operating parameters before experimental implementation, reducing costs and accelerating process development.

Furthermore, digital twins of PBRs, integrating real-time sensor data with predictive algorithms, are transforming microalgae cultivation into an intelligent, self-regulating process. Combined with advances in AI-driven control systems and multi-omics integration, these innovations will pave the way for automated, energy-efficient, and resilient biorefineries capable of producing fuels, nutraceuticals, and specialty metabolites under sustainable circular-economy frameworks (Zhang et al., 2025; Wang et al., 2024).

Ultimately, the convergence of biological insight, engineering innovation, and computational intelligence defines the future of algal biotechnology. As these tools mature, microalgae are poised to move from experimental laboratories to industrial-scale, data-driven platforms that contribute decisively to global sustainability and climate resilience.

4.3 RECENT ADVANCES IN MICROALGAL CULTIVATION SYSTEMS

4.3.1 Photobioreactor Design Optimization

The design of PBRs is a central component in the development of microalgal bioprocesses, as it directly influences light capture, mass transfer, and biomass productivity. Over the past years, several configurations have been developed, including flat-panel, tubular, bubble column, and airlift systems, each offering specific advantages depending on the application context (Section 4.2; Razzak et al., 2024; Rezvani & Rostami, 2023). Their performance depends on physicochemical parameters, including geometry, height-to-diameter ratio, internal fluid dynamics, and bubble dispersion efficiency. A persistent challenge is the control of dissolved oxygen (O_2). This photosynthetic by-product may reach inhibitory concentrations, while carbon dioxide (CO_2) availability is constrained by the system's ability to transfer the gas from the gas to the liquid phase.

To address these challenges, strategies have been developed that focus on mass transfer, including optimizing bubble size and rising velocity, as well as redesigning spargers to generate microbubbles with larger interfacial areas. These modifications increase the volumetric mass-transfer coefficient (k_La), a parameter that combines the intrinsic transfer efficiency (k_L) with the available interfacial area per unit volume (a). k_La provides an integrated measure of the system's ability to move gases between the liquid and gaseous phases and is therefore considered a key indicator of PBR performance. A higher k_La translates into more efficient CO_2 absorption, which is essential to sustain photosynthesis, and into faster removal of accumulated O_2, thereby reducing metabolic inhibition and oxidative stress (Razzak et al., 2024; Rezvani & Rostami, 2023). Notably, bubble column and airlift configurations offer additional advantages, enhancing gas distribution, minimizing dead zones, and improving overall system efficiency, with lower energy consumption than mechanical agitation (Rezvani & Rostami, 2023). Material selection is also critical, where the use of polycarbonate, acrylic, or borosilicate glass enables high light transmission without compromising thermal stability or reactor durability (Sero et al., 2020).

A remarkable advancement has been the incorporation of CFD tools, which enable the prediction of irradiance distributions, mixing patterns, and bubble trajectories. These models have facilitated the internal redesign of PBRs using baffles, compartments, or other geometries that induce controlled turbulence, thereby increasing cell-light interactions and improving overall photosynthetic efficiency (Amanna et al., 2024; Rezvani & Rostami, 2023). Beyond mass transfer, illumination remains a critical factor in the performance of dense cultures, where optical attenuation creates light gradients and dark zones. To counteract these limitations, photonic solutions have been explored to improve light distribution and quality inside the reactor. Among them, optical waveguides stand out, as they can capture external radiation and redirect it in a controlled manner into the culture, enhancing light penetration in shaded regions and reducing irradiance gradients. Adjustable-spectrum LED systems enable modulation of both intensity and wavelength, tailoring light to the physiological needs of microalgae and promoting the activation of specific photosynthetic pigments. Reflective coatings further enhance internal reflection and improve illumination uniformity, while diffusive materials scatter incident light, generating a more homogeneous irradiance throughout the reactor volume (Sero et al., 2020; Razzak et al., 2024).

Beyond engineering strategies, mathematical modeling has played an essential role in predicting and optimizing growth under various irradiance and nutrient conditions. Kinetic models, such as the general growth equation $dX / dt = \mu X$, the Monod formulation adapted for light, the Arrhenius model for temperature, and dual-substrate approaches for nitrogen and phosphorus (Hossain et al., 2020; Razzak et al., 2024), among others, have made it possible to define optimal operating ranges and anticipate limiting phenomena such as photoinhibition or nutrient depletion. Table 4.4 presents these key equations, which integrate the effects of light, temperature, CO_2 , and nutrient availability in predicting microalgal productivity.

Taken together, advances in geometry, materials, simulation tools, and model integration have driven the rational design of more efficient and robust PBRs. These innovations have not only supported the scaling-up of microalgal cultivation systems but have also enabled their adaptation to specific operating conditions, paving the way for industrial applications in bioenergy, nutraceuticals, and bioplastics. These improvements in geometry, materials, and modeling not only increase cultivation productivity but also lay the foundation for their automation and intelligent control. As systems become more complex and scalable, it becomes essential to complement engineering design with digital tools that can ensure stability and resilience.

4.3.2 Automation and Real-Time Monitoring

The introduction of automation tools and real-time monitoring has transformed microalgal cultivation by overcoming the limitations of manual control and periodic sampling. In recent years, the convergence of low-cost sensors, online flow cytometry, artificial intelligence (AI), ML, and the Internet of Things (IoT) has driven the development of smart bioreactors able to adjust parameters, thereby improving both productivity and stability dynamically. These innovations have also reached downstream processing, where digitalization and predictive modeling optimize the harvesting, extraction, and purification of valuable biomolecules (Wang et al., 2022). The complexity of cultivation, influenced by the interaction of factors such as light, pH, temperature, nutrients, dissolved CO_2, and O_2, requires control systems able to integrate heterogeneous data in real time. In this context, hybrid models that combine mechanistic approaches with ML algorithms provide clear advantages over traditional kinetic models by capturing system variability and anticipating responses under changing conditions (Syed et al., 2024; Yang et al., 2024).

TABLE 4.4
Key Equations for Microalgal Growth Modeling and Critical Parameters in PBRs

Factor	Equation	Application	Variables and Meaning
Microalgal growth	$\frac{dX}{dt} = \mu X$	Describes biomass accumulation as a function of the specific growth rate.	X : biomass concentration; μ : specific growth rate
CO_2 influence	$R_{CO_2} = k_L a \cdot \left(C^*_{CO_2} - C_{CO_2}\right)$	Describes CO_2 transfer from the gas to the liquid phase, driven by the difference between equilibrium and actual concentrations	R_{CO_2} : CO_2 transfer rate; $k_L a$: volumetric mass-transfer coefficient; C_{CO_2} *: equilibrium CO_2 concentration (Henry's law); C_{CO_2} : actual CO_2 concentration in the liquid.
Light intensity	$\mu = \mu_{max} \cdot \frac{I}{K_I + I}$	Relates growth rate to light availability, considering saturation	I : light intensity; K_I : half-saturation constant for light; μ_{max} : maximum specific growth rate.
Temperature (Arrhenius)	$\mu = A \cdot e^{-\frac{Ea}{RT}}$	Represents the impact of temperature on enzymatic reaction rates or growth	μ : specific growth rate; A : Arrhenius constant; Ea : activation energy; R : universal gas constant; T : temperature.
Combined light and CO_2	$\mu = \mu_{max} \cdot \frac{I}{K_I - I} \cdot \frac{C_{CO_2}}{K_{CO_2} + C_{CO_2}}$	Integrates light and carbon availability simultaneously to simulate real cultivation conditions.	I : light intensity; C_{CO_2} concentration; K_I and K_{CO_2} : half-saturation constants; μ_{max} : maximum specific growth rate.

Source: Hossain et al. (2020) and Razzak et al. (2024).

Next-generation optical sensors have become key tools for monitoring biomass in both open and closed systems. Multispectral sensors operating in the ultraviolet and near-infrared ranges have demonstrated precise correlations with dry biomass, enabling continuous measurements without manual intervention. Their applications have also expanded to include additional parameters, such as cellular imaging, transcriptomics (RNA-seq), humidity, nutrients, and population dynamics, which require robust multivariate datasets for model training (Oruganti et al., 2023). However, the limited availability of broad, high-quality databases remains a critical constraint for the reliability of these approaches (Syed et al., 2024). Due to their low cost and high versatility, these devices not only facilitate growth control but can also be easily integrated into IoT-based digital architectures, enabling the creation of large-scale distributed monitoring networks.

Online flow cytometry has significantly increased the temporal resolution of monitoring by providing simultaneous data on cell viability, composition, and microbial contaminants. Beyond its role as an analytical tool, it generates large volumes of real-time data that feed ML models, making it an essential source of big data for predictive control. This integration enables early detection of instabilities and supports a proactive management approach by combining cytometry with analytical algorithms to overcome the limitations of manual methods and enhance the resilience of algal cultures. Consequently, online cytometry represents a decisive step toward adaptive management of microalgal ecosystems (Haberkorn et al., 2021).

Automation has been further strengthened by the incorporation of IoT systems into microalgal cultivation and biorefineries. These platforms combine wireless sensors, deep learning algorithms, and low-power microdevices, enabling real-time monitoring and adjustment of critical parameters in large-scale operations. In open cultures, where environmental variability is higher, IoT facilitates early detection of contamination and control of light and nutrients. In closed systems, it enables continuous operation with minimal human intervention. More recently, these technologies have been extended to downstream processing, optimizing harvesting, drying, and the extraction of high-value metabolites (Wang et al., 2022). In addition, their integration with AI has enabled advanced applications, such as automatic strain identification, species classification, toxin detection, and comprehensive biorefinery optimization (Lim et al., 2022). Taken together, these innovations establish a comprehensive digitalization framework for the microalgal industry.

AI and ML play a central role in predicting and optimizing microalgal cultivation. Algorithms such as neural networks, support vector machines, random forests, and deep learning approaches have been applied to model growth under different conditions and to predict the accumulation of target metabolites, such as lipids and carotenoids (Oruganti et al., 2023; Syed et al., 2024). In addition to improving the understanding of cell physiology, their use has enabled the development of adaptive control strategies that can dynamically respond to environmental disturbances or fluctuations in production systems. Although many of these models function as "black boxes" that provide high accuracy but limited interpretability (Syed et al., 2024), hybrid approaches and explainable learning techniques have been developed to combine the robustness of mechanistic models with the predictive power of neural networks, thereby reducing decision-making uncertainty. Complementarily, integrating AI and ML with digital twins enables the creation of virtual replicas of cultivation systems, simulating complex scenarios and optimizing operational decisions in real time (Sheik et al., 2025).

Taken together, these innovations mark the transition toward intelligent microalgal cultivation, where digitalization, automation, and advanced analytics converge to enable precise, sustainable control. Advances in sensors, cytometry, IoT, and AI outline a scenario in which real-time monitoring is no longer an auxiliary tool but a central component of production strategies. Within this framework, digitalization and predictive models enable the management of microalgal systems with greater precision and flexibility. This ability to operate under variable conditions facilitates their integration into real industrial processes, where nutrient availability and environmental conditions constantly fluctuate. In this context, automation opens the door to effective symbiosis with other industries and to the consolidation of circular-economy schemes.

4.3.3 Microalgal–Wastewater Integration in the Circular Economy

Recent advances in PBR design and process automation have laid the groundwork for microalgal cultivation systems to transition beyond the experimental stage and integrate into real industrial schemes. This transition opens the door to circular-economy models, in which waste streams and emissions are transformed into inputs for producing biomass and high-value compounds. An overview of the main technological advances addressed in this section is presented in Figure 4.3.

The integration of microalgal cultivation systems with other industries represents one of the most promising pathways to transform waste management into high-value-added production processes. Unlike conventional treatments, which are typically focused on pollutant removal, algal systems simultaneously enable wastewater treatment, carbon dioxide capture, and biomass generation with applications in energy, agriculture, food, and pharmaceuticals (Abdelfattah et al., 2023; Cheirsilp et al., 2023). In this sense, the industrial integration of microalgae is a clear example of the circular economy, where waste is no longer a liability but an input for new processes.

The use of municipal and industrial effluents as a nutrient source is one of the most extensively studied applications. These effluents, rich in nitrogen and phosphorus, offer a cost-effective alternative to synthetic media and allow for high pollutant removal rates. Reported efficiencies include phosphorus removal above 96% and ammonium removal close to 94%, while chemical oxygen demand (COD) reduction can exceed 70% under controlled conditions. This reflects the substantial reduction

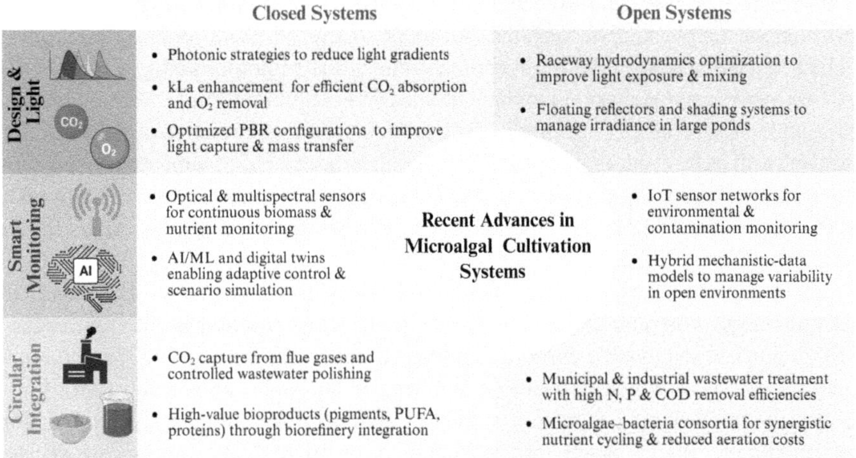

FIGURE 4.3 Overview of recent advances in microalgal cultivation systems.

(Own elaboration.)

in organic matter from the effluent, as COD is a direct indicator of the organic load present in wastewater (Zhang et al., 2025). These results not only ensure compliance with increasingly stringent environmental regulations but also facilitate the production of biotechnologically valuable biomass that can be commercialized.

Many wastewater treatment technologies are based not just on microalgae but also on microalgae–bacteria symbiosis. In these consortia, microalgae capture light and nutrients, fix CO_2, and release oxygen, which bacteria use for respiration and nitrification. In turn, bacteria degrade organic matter and release CO_2, which is reused by the algae in photosynthesis. In addition, both microalgae and bacteria secrete phytohormones (auxins, cytokinins, and gibberellins), vitamins, and secondary metabolites that promote mutual growth, improve resistance to stress conditions, and increase overall system efficiency (Saravanan et al., 2021; Liu et al., 2024). This metabolic exchange significantly reduces the need for mechanical aeration, one of the most energy-intensive operations in conventional treatment plants. By lowering the electricity consumption associated with aeration, these systems not only make the process more efficient but also help reduce indirect greenhouse-gas emissions linked to energy generation. Recent studies show that this coupling can decrease emissions by more than 350 g of CO_2 equivalent per cubic meter of treated wastewater, reinforcing its potential as a carbon-neutral technology (Liu et al., 2024).

Several examples of integration highlight the versatility of these systems. At the municipal level, the incorporation of *Chlorella vulgaris* in a wastewater treatment plant in Portugal not only significantly reduced N and P concentrations but also generated annual gross revenues of around €619,100, derived from the valorization of sludge as soil amendments and the production of bioenergy for self-consumption (Nobre et al., 2024). In the agri-food sector, the cultivation of diatoms in dairy wastewater demonstrated that these effluents not only provide nutrients but also promote lipid accumulation in the biomass, thereby improving the quality of the biodiesel produced and generating a valuable by-product for the energy industry (Singh et al., 2024). Added to this is the integration with industrial gas streams, where microalgae act as highly efficient biological sinks. On average, 1 kg of dry algal biomass requires about 1.8 kg of CO_2, a fixation rate at least ten times higher than that observed in conventional crops (Kumar et al., 2024).

The integral valorization of the obtained biomass represents another key component of this industrial symbiosis. Beyond biofuels, algal biomass can be fractionated to obtain value-added products, including pigments, proteins, polysaccharides, polyunsaturated fatty acids, and bioactive compounds with applications in the food, cosmetic, and pharmaceutical industries (Cheirsilp et al., 2023). This biorefinery approach aims to exploit all biomass fractions, ensuring economic viability while reducing environmental impacts. In this way, treatment plants are conceived not only as purification facilities but also as platforms for diversified production.

Nevertheless, the integration of microalgae with other industries still faces important limitations. The seasonal variability of effluents affects nutrient availability and process stability; certain toxic compounds present in some industrial residues can inhibit algal growth; and downstream processing costs (harvesting, extraction, and purification) remain a significant barrier to scalability (Kumar et al., 2024; Cheirsilp et al., 2023). In addition, most documented experiences remain at pilot or demonstration scales, highlighting the need to advance in TRLs (Section 4.4) and to develop robust business models that ensure economic sustainability at large scale.

In addition to their technical and economic benefits, the integration of microalgae with other industries is strongly driven by regulatory frameworks and international climate commitments. The European Union and the United States have established increasingly stringent limits on nitrogen and phosphorus discharges, compelling industries to adopt more efficient treatment technologies. In parallel, countries such as China and India have set carbon neutrality targets for 2060 and 2070, respectively (Kumar et al., 2024). In this context, algal systems not only provide treatment solutions but also emerge as strategic allies for achieving Global Sustainability Goals.

Integrated microalgal cultivation systems with other industries not only fulfill an environmental role but also constitute a transformative platform toward the circular bioeconomy. Wastewater treatment, carbon capture, and the production of multiproduct biomass converge in an approach that turns environmental liabilities into strategic resources. The future outlook is that treatment plants will evolve into true sustainable biorefineries, capable of simultaneously generating clean water, energy, biofertilizers, and high-value compounds, consolidating microalgae as central actors in the transition toward more resilient, sustainable, and decarbonized economies.

4.4 SCALING AND TRL IN MICROALGAL BIOPROCESSES

The development of microalgal bioprocesses follows a complex trajectory from basic research to commercial implementation. The concept of Technology Readiness Level (TRL), initially developed by NASA to evaluate aerospace technologies (Mankins, 2009), has been adapted to the biotechnological context as a tool to identify gaps between research and industry, providing a valuable means to assess technological maturity, risk, and investment requirements at each stage. In the microalgal sector, TRLs link experimental progress, from strain selection and proof of concept (TRL 1–3) to pilot demonstration and full-scale industrial operation (TRL 7–9), with specific technical, economic, and regulatory milestones. This framework aligns biological advances with engineering feasibility and policy or market requirements, facilitating decision-making for researchers, investors, and regulatory bodies (Vázquez-Romero et al., 2022; Figueroa-Torres & Theodoropoulos, 2023).

The scaling trajectory is constrained by multiple technical, economic, and social barriers. Significant limitations include lower productivity under real-world conditions compared to theoretical yields, the high cost and energy intensity of harvesting and downstream processing, and the difficulty of maintaining strain stability in open cultivation systems. These challenges are consistently reflected in techno-economic analyses and LCAs, which demonstrate that microalgal biofuels remain economically uncompetitive. In contrast, high-value products such as pigments, nutraceuticals, and functional proteins have reached higher readiness levels, approaching commercialization. Recent techno-economic and life-cycle studies confirm that biofuel production remains economically marginal, while high-value compounds exhibit greater resilience and scalability, achieving TRL levels near industrial deployment (Guo et al., 2022; Wang et al., 2024; Quiroz et al., 2023).

Nonetheless, several strategies can accelerate progress through TRLs and bridge the gap between laboratory innovation and industrial application. These include early codesign with industry to address defined market needs, long-term pilot demonstrations that incorporate environmental variability, and public–private partnerships to share infrastructure and scaling costs. Moreover, embedding techno-economic assessment (TEA) and LCA frameworks from the earliest development stages provides quantitative feedback to guide technical optimization and sustainability evaluation (Vázquez-Romero et al., 2022; Figueroa-Torres & Theodoropoulos, 2023).

The contrasting trajectories of Cyanotech in Hawaii, successfully consolidating a profitable niche in nutraceuticals, and Sapphire Energy, whose large-scale biofuel enterprise collapsed despite significant investment, exemplify both the potential and fragility of microalgal innovation.

Integrating TEA and LCA within the TRL Framework

Progression along the TRL scale involves not only demonstrating biological performance but also providing evidence that a process can be economically and environmentally sustainable. TEA examines investment requirements and operational viability through analyses of capital expenditures (CAPEX) and operating expenditures (OPEX), productivity benchmarks, and market competitiveness indicators (Guo et al., 2022; Figueroa-Torres & Theodoropoulos, 2023). LCA complements this by quantifying impacts across energy use, greenhouse-gas emissions, eutrophication, and water demand (Vázquez-Romero et al., 2022). Together, these methodologies provide the factual basis for TRL progression. Early-stage systems (TRL 3–5) typically show high energy consumption and production costs, ranging from €50 to €100 per kg dry biomass, whereas integrated pilot and industrial facilities (TRL 7–9) may lower costs to €2–5 per kg while achieving net-negative carbon balances when coupled with waste–CO_2 capture or nutrient recovery (Wang et al., 2024). These transitions demonstrate that technological maturity in microalgal biotechnology is not only a function of scale but of design integration, iterative validation, and long-term economic realism.

4.4.1 Framework of Technology Readiness Levels (TRLs) Applied to Microalgal Biotechnology

The TRL framework offers a structured perspective for assessing the developmental maturity of emerging biotechnology innovations. In the context of microalgal systems, it facilitates the alignment of fundamental research with pilot-scale validation and potential industrial deployment. Figure 4.4 summarizes this framework as applied to microalgal biotechnology, illustrating the progression from basic research to scalable bioprocesses.

4.4.1.1 TRL 1–3: Discovery, Formulation, and Proof of Concept

The initial stages of technological readiness correspond to fundamental research and the preliminary validation of biological hypotheses. In the case of microalgae, this includes strain prospecting and selection, identification of metabolic pathways, description of biochemical processes, and physiological characterization under controlled laboratory conditions (Dębowski et al., 2023; Deprá et al., 2025). At these

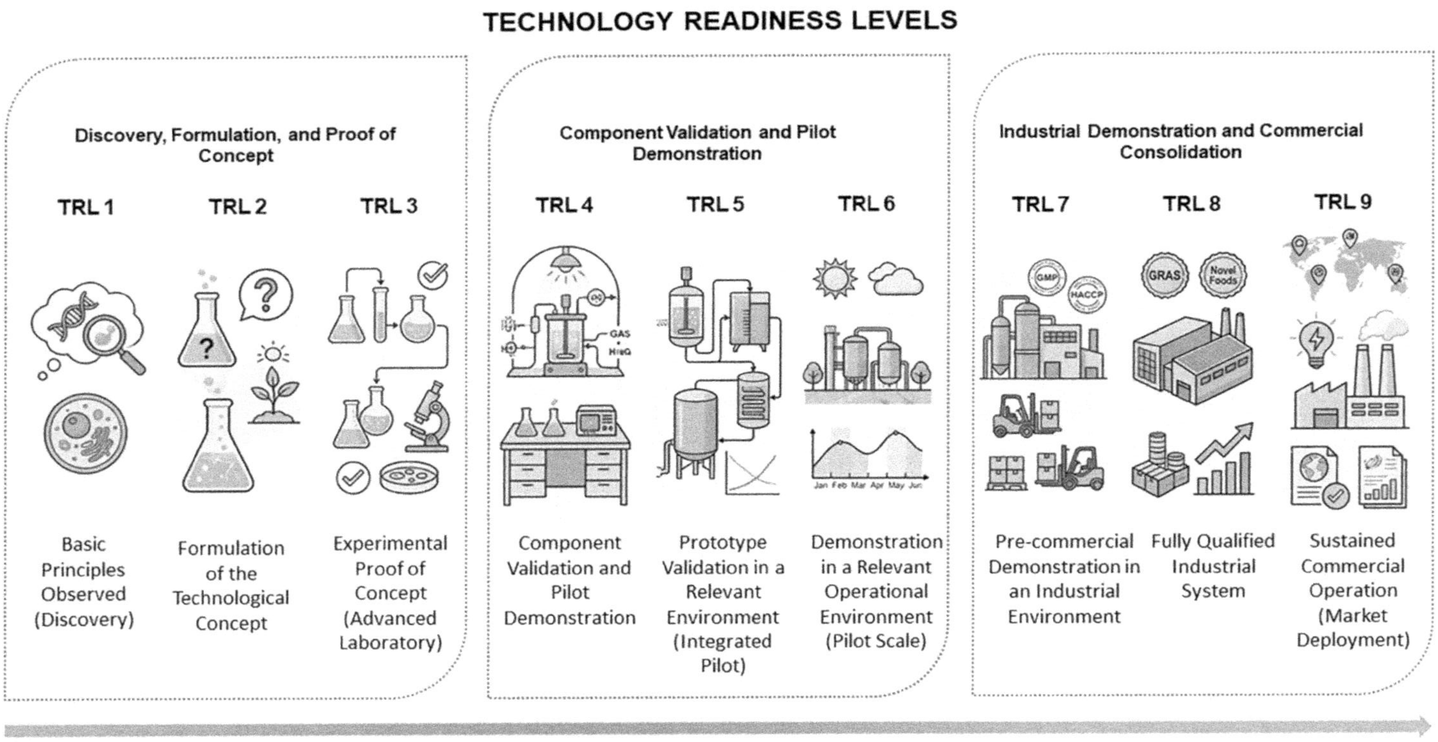

FIGURE 4.4 TRL framework applied to microalgal bioprocess development.

(Own elaboration.)

early TRL stages (1 to 3), laboratory-scale experiments are conducted in small, closed systems such as flasks or PBRs (≤1 L), to measure growth rates, cellular composition, and optimal environmental parameters. These data serve as the basis for modeling initial metabolic balances, which are essential for progressing to subsequent TRL stages (Razzak et al., 2024; Rozenberg et al., 2024).

Critical risks at this phase include biological variability among isolates, loss of phenotypic traits during prolonged culture, lack of reproducibility when transferring conditions beyond the laboratory (Wijffels et al., 2013), and strain loss due to contamination or reduced viability. The reliability of results depends on standardized cultivation protocols, strain conservation and purity, and genetic stability under different experimental conditions (Berthelier et al., 2023). These materials form the foundation for advancing to later stages of technological development.

4.4.1.1.1 Technology Readiness Level 1 (TRL 1): Basic Principles Observed (Discovery)

At TRL 1, scientific inquiry begins. In microalgal biotechnology, this level involves observation, literature review, and documentation of biological phenomena with industrial potential, such as the characterization of biosynthetic pathways for pigment production, salt-induced lipid accumulation, protein and bioactive polysaccharide biosynthesis, and the capacity to generate bioelectricity or biodiesel, among others. The goal is to identify metabolic mechanisms relevant to bioproducts, energy, biomedicine, cosmetics, and phycoremediation (Hu et al., 2008; Borowitzka & Vonshak, 2017; Deprá et al., 2025; Khairuddin et al., 2024; Amouri et al., 2024; Ugya et al., 2021).

This stage does not yet involve reproducible experimentation. Its purpose is to establish the scientific and conceptual basis supporting hypothesis development in later phases through observation, modeling, and critical literature analysis without experimental validation.

4.4.1.1.2 Technology Readiness Level 2 (TRL 2): Formulation of the Technological Concept

At TRL 2, the technological concept is defined, and a working hypothesis is formulated under laboratory conditions. In microalgal biotechnology, this includes the design of cultivation strategies such as photoperiod, CO_2 supply, medium composition, and light intensity, as well as the evaluation of alternative substrates such as wastewater or industrial effluents (Sajjadi et al., 2018; Suparmaniam et al., 2019). Controlled stress conditions may also be designed to induce specific metabolic pathways associated with lipid or pigment accumulation (Figueroa-Torres et al., 2021).

Experimentation remains exploratory, focusing on evidence of biological capacity rather than reproducibility. Typical tests are performed at scales below 1 L, generating preliminary experimental designs and metabolic models that confirm the biological plausibility of the concept. A proof-of-concept (PoC) plan is developed, including hypotheses, variables, controls, success criteria, and experimental design protocols (Figueroa-Torres et al., 2021; Klein & Davis, 2023). Characteristic TRL 2 metrics include determining biomass yield oriented toward the target compound, such as lipids or astaxanthin (Sajjadi et al., 2018; Wang et al., 2024). The information generated provides a quantitative reference point for advancing toward experimental validation (TRL 3).

4.4.1.1.3 Technology Readiness Level 3 (TRL 3): Experimental Proof of Concept (Advanced Laboratory)

At TRL 3, applied research progresses toward reproducible experimental validation of the concept (PoC) under controlled laboratory conditions. This stage confirms the initial technical feasibility of the technology by testing hypotheses and models through bench-scale photobioreactor assays (1–10 L), where pH, temperature, illumination, mixing, and CO_2 supply are precisely controlled. The purpose is to verify kinetic models and experimental outcomes, establishing growth and productivity parameters specific to each strain and system (Brennan & Owende, 2010; Deprá et al., 2025).

Required tests at this stage include reproducible growth curves with biological replicates, biochemical composition profiles (lipid, protein, and pigment content), environmental stress assays (salinity, light intensity, temperature variations), and initial extraction or recovery trials for biomolecules (Brennan & Owende, 2010). Standard operating procedures (SOPs) for cultivation, sampling, and extraction are established to ensure reproducibility and traceability. These experiments verify technical feasibility and consistency, providing the basis for validation in relevant environments at TRL 4.

4.4.1.2 TRL 4–6: Component Validation and Pilot Demonstration

The intermediate TRL stages represent the so-called "technological valley of death" (Deprá et al., 2025), where most microalgal developments encounter economic and technical limitations that prevent successful scaling. At this phase, the main challenge is to integrate and validate cultivation and processing subsystems under conditions that partially represent real environments. Recent literature emphasizes that sustained progress depends on reducing photobioreactor costs, improving the energy efficiency of harvesting and drying processes, and maintaining biomass stability under environmental fluctuations (Deprá et al., 2025; Acién et al., 2012).

4.4.1.2.1 Technology Readiness Level 4 (TRL 4): Laboratory-Scale Validation (Bench to Mini-Pilot)

TRL 4 corresponds to the functional validation of technology under semi-representative environmental conditions. In microalgal biotechnology, this entails the continuous operation of integrated subsystems such as tubular or flat-plate PBRs, or instrumented high-rate algal ponds (HRAPs), typically in the range of 10–100 L, with partially natural light, controlled CO_2 supply, and, when applicable, real input streams such as wastewater. The goal is to demonstrate that the system maintains biomass productivity and operational control beyond closed-laboratory conditions (TRL 3) while withstanding moderate environmental variations (Quiroz et al., 2023).

Required evidence includes documentation of cultivation stability, productivity records expressed as $g{\cdot}L^{-1}{\cdot}day^{-1}$ in closed PBRs or $g{\cdot}m^{-2}{\cdot}day^{-1}$ in open raceways/HRAPs, and measurements of specific energy consumption ($kWh{\cdot}kg^{-1}$ dry biomass) for mixing, aeration, and harvesting. Initial harvesting and concentration trials are expected, including chemical flocculation, centrifugation, or hollow-fiber membrane filtration, along with estimates of biomass recovery efficiency. Validation in wastewater-based systems should also assess environmental co-benefits, such as nitrogen and phosphorus removal (Acién et al., 2012; Tan et al., 2020).

Deliverables at TRL 4 include continuous operation data over at least 8–12 weeks, documenting productivity, energy consumption, operational failures, and corrective measures. SOPs for cultivation, primary harvesting, and effluent management must be established, along with a TEA and a cradle-to-gate LCA based on actual productivity and energy-use data. These analyses serve as inputs for pilot design at TRL 5 (Quinn & Davis, 2015; Figueroa-Torres et al., 2021).

4.4.1.2.2 Technology Readiness Level 5 (TRL 5): Prototype Validation in a Relevant Environment (Integrated Pilot)

TRL 5 involves validating the entire system under relevant environmental conditions, incorporating natural fluctuations in temperature, irradiance, and hydrodynamic mixing. This stage integrates both upstream and downstream processes within pilot facilities of 100–1,000 L or small HRAPs operating continuously for several months (Craggs et al., 2013).

Evidence required to reach intermediate readiness levels (TRL 5–6) includes sustained operation for several months, demonstrating stability and reproducibility; an intermediate TEA with detailed CAPEX and OPEX estimates; and a preliminary LCA identifying key environmental impacts. Product quality validation is also required through analyses of purity, composition, and functionality aligned with target-market specifications, ensuring commercialization potential (Garoma, 2019).

Deliverables include operational reports spanning at least four weeks, recording system failures and adjustments, mass and energy balances, harvesting performance, biomass recovery yields ($kWh \cdot kg^{-1}$), and energy consumption estimates, along with a plan for full-season operation. Risks include unmodeled seasonal variability, product degradation, and energy inefficiencies during drying or concentration.

4.4.1.2.3 Technology Readiness Level 6 (TRL 6): Demonstration in a Relevant Operational Environment (Pilot Scale)

At TRL 6, microalgal biotechnology reaches the pilot-plant scale, typically within the cubic-meter range (10^3–10^4 L), designed for continuous or semicontinuous operation over six to twelve months under real environmental conditions. The main objective is to demonstrate consistent productivity and complete operational integration of cultivation, harvesting, and processing systems while accounting for seasonal variations in solar radiation, temperature, and nutrient availability, as well as variability in effluent quality when these inputs are used as culture media.

Scaling to pilot level involves validating light and CO_2 transfer efficiency, energy use for mixing and aeration, and the feasibility of integrating upstream and downstream stages (harvesting, wet biomass concentration, stabilization or storage, and pretreatment for extraction), which traditionally concentrate most of the process costs and energy demands (Posten & Schaub, 2009).

Evidence at TRL 6 requires sustained operation for at least 12 months, accompanied by comprehensive TEA and LCA analyses, including sensitivity assessments of productivity, costs, and market value. Operational control must be documented regarding biological contamination, natural predators, and climatic variability, demonstrating process stability across complete seasons.

The incorporation of initial quality assurance and control (QA/QC) protocols is a key milestone. The DISCOVR program (Development of Integrated Screening, Cultivar Optimization and Verification Research) and the State of Technology reports compiled by the National Renewable Energy Laboratory (NREL), in collaboration with the US Department of Energy's Bioenergy Technologies Office (DOE/BETO), document this transition to seasonal pilot operation in open environments. These multi-annual reports show progressive improvements in productivity and operational stability through strain rotation, salinity management, and biota control strategies (Klein & Davis, 2023).

At this stage, the first pre-commercial batches or trial contracts with potential customers are usually produced, making it essential to consider regulatory requirements according to the intended product application.

4.4.1.3 TRL 7–9: Industrial Demonstration and Commercial Consolidation

The higher stages of technological maturity (TRL 7–9) mark the transition from technological validation to industrial and commercial consolidation of microalgal bioprocesses. At this phase, attention is focused on operational stability, regulatory compliance, product safety, and long-term economic viability. The technology is demonstrated under real industrial conditions, typically through pilot biorefineries or demonstration-scale facilities (Nguyen et al., 2025).

In addition to the TRL framework, the Commercial Readiness Index is introduced as a complementary metric to evaluate market maturity, consumer acceptance, and business model sustainability, thus linking technical feasibility with commercial competitiveness.

Following this perspective, Table 4.5 presents various success stories of companies and startups that have made substantial progress in their TRLs. These examples demonstrate that the combination of appropriate scaling strategies, innovation in biorefineries, and sustainable business models has enabled microalgae-based technologies to move from experimental phases to established commercial applications. If you'd like, I can also help you integrate this section with the rest of the text or adjust the tone to match the style of the entire document.

4.4.1.3.1 Technology Readiness Level 7 (TRL 7): Pre-commercial Demonstration in an Industrial Environment

At TRL 7, microalgal biotechnology advances from pilot scale (TRL 6) to pre-commercial demonstration in modular plants of 10^4–10^5 L or replicable biorefinery units operating under real industrial conditions. The most notable difference from TRL 6 lies in the full integration of the process under continuous flow, scale-up by one or two orders of magnitude, and the implementation of advanced QA/QC protocols aligned with international standards for Good Manufacturing Practices (GMP) and Hazard Analysis and Critical Control Points (HACCP). The main objective of this phase is to demonstrate that the process can maintain stable yields, comply with intermediate regulatory requirements, and produce pilot batches suitable for commercialization under traceable and sanitary-controlled conditions.

TABLE 4.5
Success Stories in TRL Progression: Examples of Industrial Implementation of Microalgae Biotechnology

Company	Microalga/Process	System/Scale	TRL	Evidence (URL)
Cyanotech – BioAstin® Foundation: 1983 Location: Hawai (EE. UU)	*Haematococcus pluvialis* for astaxanthin production; *Arthrospira platensis*.	Open raceway ponds/Farm area approximately 90 acres.	9	https://www.cyanotech.com/astaxanthin/astaxanthin-process
Corbion–AlgaPrime™ Founded: 1919 (Amsterdam, Netherlands) Location: AlgaPrime™ production facility in Brazil	*Schizochytrium* sp. for omega-3 (DHA) production.	Fermentation in closed tanks.	9	https://www.corbion.com/products/algae-ingredients-products/algaprimedha
Parry Nutraceuticals Founded: 1996 Location: India	Cultivation of *Spirulina* (*Arthrospira*) and *Chlorella* for food and nutritional ingredient applications.	Open raceway ponds are operated under solar illumination.	9	https://parrynutraceuticals.com
AlgaEnergy Founded: 2007 Location: Spain	Multispecies consortia for biostimulant formulations	Photobioreactors (PBRs) + commercial formulation process	9	https://ag.algaenergy.com/news
Microphyt Founded: 2007 Location: France	Various species for high-value bioactive ingredients	Hydro-biomimetic cultivation technology using CAMARGUE® photobioreactors; projected capacity of 100 t $year^{-1}$ (2025)	9	https://www.cbe.europa.eu/news/scales-unique-microalgae-biorefinery-opens-france https://microphyt.eu/exclusive-technology
Algenuity Founded: 2012 Location: United Kingdom	*Chlorella* (Chlorella Colours™ and EdenFlo™ lines), non-photosynthetic strains used for food market applications	Microbial fermentation under dark conditions using glucose as a carbon source; process designed for controlled, heterotrophic biomass production.	9	https://algenuity.com https://www.cambridgenetwork.co.uk/news/algenuitys-chlorella-coloursr-platform-launches-years-vitafoods-europe?utm_source https://www.qi3.co.uk/clients/case-studies/algenuity-ltd/?utm_source

(*Continued*)

TABLE 4.5 (Continued)
Representative Photobioreactor Systems for High-Value Metabolite Production and Resource Valorization

Company	Microalga/Process	System/Scale	TRL	Evidence (URL)
Microalgas Oleas de México Founded: 2015 Location: Mexico	*Arthrospira maxima*, native microalgae, and cyanobacteria for the development of agricultural biostimulants, pigments, and polyunsaturated fatty acid extracts (omega-3 and omega-6) for the food industry.	Pre-pilot operations began in 2016 using raceway pond systems. Since 2016, vertical tubular photobioreactors have been implemented, with a total production capacity exceeding 150 m^3.	~TRL 6–7	https://microalgasoleas.com.mx/?utm_source https://agroorganico.info/la-riqueza-de-las-microalgas-bioestimula-y-nutre-a-las-plantas/?utm_source
Allmicroalgae Founded: 2015 Location: Portugal	*Chlorella vulgaris*, *Arthrospira platensis*, *Tetraselmis* sp., *Nannochloropsis* sp., *Scenedesmus obliquus*, *Phaeodactylum tricornutum*, and *Chlorococcum amblystomatis* for applications in the food and nutraceutical industries, animal	Industrial-scale tubular photobioreactor system.	9	https://www.allmicroalgae.com/en
Grean Stream Farms Founded: 2016 Location: Columbus, New Mexico (USA)	*Nannochloropsis* sp. cultivated at industrial scale for nutraceuticals, animal feed, and other sustainable bioproducts.	Large-scale outdoor raceway ponds covering approximately 96 acres; annual biomass production exceeds 700 tons.	9	https://greenstreamalgae.com/about https://thefishsite.com/articles/algae-farmer-set-for-180-percent-production-increase-new-mexico
Nutré Founded: 2016 Location: Colombia	*Spirulina* (*Arthrospira*), organically certified for human nutrition.	Open raceway ponds operated under greenhouse protection to prevent external contamination in tropical environments; solar illumination; downstream processing includes dehydration and extrusion.	~ TRL 7–8	https://somosnutre.com
Algiecel Founded: 2021 Location: Denmark	Microalgae application: CO_2 capture converted into oils and proteins. Production of high-quality, microalgae-derived ingredients with consistent functionality and full traceability.	High-efficiency, closed photobioreactor (PBR) designed for carbon capture using microalgae, integrated into standard shipping containers. The PBR circulates algae through horizontally aligned, LED-illuminated pipes supplied with pure CO_2 from industrial point-source emissions.	6–7	https://www.algiecel.com/reactor-design-verified-by-cfd-modelling

At this level, cultivation systems (raceways or PBRs) operate under continuous production regimes, with annual areal productivity records (g m^{-2} day^{-1}), energy consumption (kWh kg^{-1}), and product recovery efficiency above 80 percent, along with safety indicators confirming the absence of heavy metals, toxins, or pathogens.

Evidence required for TRL 7 includes continuous operation for at least 6–12 months, auditable QA/QC records, updated TEA and LCA analyses using real plant data, and the establishment of initial supply or pilot sales contracts. The predominant risks relate to regulatory complexity (use of effluents and approval of natural ingredients) and price volatility in high-value markets.

4.4.1.3.2 Technology Readiness Level 8 (TRL 8): Fully Qualified Industrial System

At TRL 8, the technology has achieved a fully functional, certified industrial facility with regular supply contracts. Unlike TRL 7, where operational stability and market confidence are still being consolidated, TRL 8 focuses on achieving full industrial validation through national and international regulatory compliance and demonstrated economic sustainability.

Production operates under certified standards (GMP, ISO 22000, and HACCP) and region-specific regulations for food or cosmetic products, such as Regulation (EU) 2015/2283 on Novel Foods or the GRAS (Generally Recognized as Safe) status granted by the US FDA. Industrial facilities at this level operate continuously throughout the year, maintaining controlled seasonal variability ($<\pm20\%$), product stability, and positive operating margins, as per final TEA evaluations.

4.4.1.3.3 Technology Readiness Level 9 (TRL 9): Sustained Commercial Operation (Market Deployment)

TRL 9 represents the final stage, where the process has been tested, certified, and operated under sustained commercial conditions. The key distinction from TRL 8 is that the technology now competes directly in global markets, maintaining stable performance, consolidated profit margins, and an integrated value chain encompassing production, distribution, and post-sale activities.

At this stage, companies maintain audited financial and environmental reports, certified quality and sustainability systems (ISO 9001/22000, FSSC 22000, GRAS), and continuous improvement strategies (QA/QC). Operations include monitoring of environmental, social, and governance indicators, regular LCA updates, and international expansion plans. A consolidated example is Cyanotech Corporation (Hawaii), whose industrial-scale *Spirulina* production consistently supplies the food, cosmetic, and pharmaceutical sectors.

4.4.2 Barriers to Industrial Scale-Up: Economic, Technical, Regulatory, and Social

The potential of microalgal biotechnology is widely recognized, and numerous studies have documented its applications across the food, cosmetic, biofuel, and bioremediation sectors. However, the TRL of many microalgal processes remains at intermediate stages, creating uncertainty among investors and hindering industrial consolidation (Dębowski et al., 2023; Geng et al., 2025). Achieving viable industrial-scale operations presents multiple technical, economic, regulatory, and social barriers that challenge the transition from laboratory or pilot studies to commercial production.

4.4.2.1 Technical Barriers

From a technical perspective, industrial-scale microalgae cultivation involves multiple interacting variables, including species selection, cultivation mode (autotrophic or heterotrophic), reactor type and design, medium composition, light source and distribution, nutrient balance, CO_2 supply mechanism, agitation system, and the downstream processes of harvesting, drying, cell disruption, and product extraction (Yadav et al., 2020; Dębowski et al., 2023; Geng et al., 2025).

Currently, a limited number of species are cultivated reliably at an industrial scale, mainly from the genera *Chlorella*, *Spirulina* (*Arthrospira*), *Dunaliella*, *Haematococcus*, *Nannochloropsis*, *Isochrysis*, *Tetraselmis*, *Scenedesmus*, and *Porphyridium*. These species are favored for their robustness, fast growth, and high content of value-added compounds such as proteins, lipids, pigments, and fatty acids. However, over 50,000 potential species have been described in natural environments, suggesting an enormous and largely unexplored genetic and metabolic diversity with potential applications in biotechnology, bioenergy, and environmental mitigation (Labara Tirado et al., 2025). The bioprospecting of new strains adapted to real-world conditions could mitigate performance, stability, and adaptability issues that arise when scaling from controlled laboratory environments to open systems (Novoveská et al., 2023).

The transition from PBRs to open raceway ponds introduces challenges related to operational stability and environmental control. Although PBRs provide superior cultivation control and can achieve biomass productivities of 2–8 $g{\cdot}L^{-1}$, their installation and operating costs are 200–300% higher than those of open systems, which typically yield 0.1–0.2 $g{\cdot}L^{-1}$ under natural conditions (Borowitzka & Vonshak, 2017; Ahmad et al., 2021). Conversely, open ponds are more susceptible to biological contamination, which increases operational risks (Quiroz et al., 2025). Regarding cultivation strategies, batch systems are more stable and less prone to contamination, but they require larger inoculum-production infrastructure. In contrast, semicontinuous systems involve lower capital investment but greater exposure to environmental stress and culture crashes (Quiroz et al., 2025).

Another limiting factor is the decline in light efficiency as cell density increases, leading to uneven light distribution and self-shading effects that negatively impact microalgal growth. This phenomenon becomes increasingly challenging to manage as PBRs are scaled up to outdoor pilot systems. Recent designs have attempted to improve light utilization but most rely on artificial illumination, which increases operational costs and thermal load, potentially exacerbating system instability (Chen et al., 2011).

Downstream processes—including centrifugation, drying, and cell disruption—represent major bottlenecks. Centrifugation consumes between 0.8 and 1.2 $kWh{\cdot}m^{-3}$ of culture, and thermal drying exceeds 3 $MJ{\cdot}kg^{-1}$ of biomass. In the case of microalgal biodiesel, even when productivity doubles or energy consumption is reduced by half, production costs (USD 6.27–8.34 per gallon) remain significantly higher than those of fossil fuels. Emerging technologies, such as electroflocculation and microbial bioflocculants, show promise but require further research before large-scale application (Min et al., 2022). Similarly, hybrid systems combining flotation and screw-press harvesting have demonstrated efficiencies of approximately 93% and reduced energy consumption, yet their scalability remains limited for many product types (Yoon et al., 2025).

Operationally, significant fragilities include biofilm formation, seasonal variability, and microbial contamination (Geng et al., 2025). These instabilities reduce reproducibility and reliability at the pilot scale, preventing sustained industrial validation. Overall, the high energy demand of downstream operations, costly cultivation infrastructure, and biological instability remain the dominant technical barriers hindering progress toward microalgal industrialization.

4.4.2.2 Economic Barriers

Despite sustained global market growth, high production costs remain the main economic obstacle to industrial consolidation of microalgal biotechnology. According to the Business Research Insights (2024) report, the global microalgae market is projected to reach USD 0.38 billion in 2025 and USD 0.88 billion by 2034, with a compound annual growth rate of 9.79%. However, 36% of producers report high operational costs and 28% cite inconsistent productivity, affecting profitability and the feasibility of commercial scaling.

Key economic constraints include:

- High infrastructure and energy costs, particularly for PBRs that require thermal regulation, artificial illumination, and costly agitation systems. Deprá et al. (2025) estimate that tubular PBRs, currently operating at TRL 7–9, cost approximately USD 40,000–50,000 per cubic meter of working volume, whereas flat-plate systems (TRL 6–8) reach up to USD 150,000·m^{-3} due to their high surface-to-volume ratio and structural complexity. Hybrid designs (TRL 5–7) and plastic-bag reactors (TRL 4–6) reduce initial capital requirements but have shorter lifespans and lower operational control, resulting in higher long-term costs. In contrast, open raceway ponds or HRAP systems offer low installation costs but suffer from up to 50% lower productivity compared to laboratory conditions, along with biological instability and containment challenges. These differences highlight that the specific cost per illuminated volume remains a structural barrier to industrial profitability.
- Limited price competitiveness relative to conventional plant-based proteins, oils, and fossil-based biofuels, restricting market penetration in price-sensitive sectors such as energy and staple foods.
- High financial risk, driven by low technological maturity (TRL 4–6) and the lack of consolidated business models, which discourages sustained private investment.

- Macroeconomic vulnerability, as recent geopolitical conflicts (COVID-19, the Russia–Ukraine war, and the Israel–Hamas conflict) have increased energy and fertilizer prices, disrupting logistics and large-scale biomass production.

The same report identifies emerging opportunities associated with the circular bioeconomy and high-value markets in nutraceuticals, cosmetics, and functional ingredients. In this context, *Spirulina*, *Chlorella*, and *Dunaliella salina* together account for over 90% of industrial production, with Asia–Pacific leading global output (42%), followed by Europe (26%). However, the production of biotechnological commodities still requires optimization of strains, bioreactors, and industrial scaling to compete with petrochemical products. High capital, energy, and operating costs remain the primary economic barriers to the industrial scale-up of microalgal processes (Business Research Insights, 2024. https://www.businessresearchinsights.com/market-reports/microalgae-market-119735, accessed on 27 October 2025).

4.4.2.3 Regulatory Barriers

Understanding the regulatory framework applicable to microalgal-derived products is essential, as their classification and permissible uses vary across jurisdictions and industrial sectors. Several recent reviews have examined the legal and normative aspects of their incorporation into food, feed, and pharmaceutical products in both the European Union and the United States, emphasizing the importance of compliance with safety, traceability, and contaminant limit standards (Cruz & Vasconcelos, 2023; Su et al., 2023; Chentir et al., 2025). Likewise, technical and regulatory requirements for their use in fertilizers and biostimulants have been described under Regulation (EU) 2019/1009, which defines maximum permissible levels for heavy metals and establishes criteria for the scientific validation of biostimulant claims (Su et al., 2023).

A clear understanding of international regulatory frameworks is necessary to ensure the safe and sustainable commercialization of microalgal products across multiple sectors. In particular, the development of products derived from microalgae cultivated in wastewater or alternative media requires precise knowledge of current regulations on safety, traceability, and contaminant thresholds. While scientific research has advanced in using waste streams as nutrient sources for high-value biomass production, toxicological evaluation and specific regulation of these products remain limited. Currently, neither the United States nor the European Union explicitly authorizes the use of wastewater-grown microalgae for human or animal consumption, due to potential risks of microbial or heavy metal contamination.

In the United States, only a few microalgal species hold GRAS status granted by the Food and Drug Administration (FDA), including *Spirulina*, *Chlorella*, *Dunaliella*, *Haematococcus*, and *Schizochytrium* (FDA GRAS Database, accessed on 17 October 2025). In the European Union, Regulation (EU) 2015/2283 on Novel Foods recognizes 22 safe species for consumption, including *Tetraselmis chuii*, *Chlamydomonas reinhardtii*, and *Arthrospira platensis* (EU Novel Food Catalogue, accessed on 27 October 2025). Both frameworks prohibit the use of biomass obtained from urban wastewater for food or feed purposes, following Regulation (EC) No. 1881/2006 (maximum levels for heavy metals) and Regulation (EC) No. 2073/2005 (microbiological criteria). However, their use in nonfood applications such as cosmetics and biofertilizers is permitted, provided strict quality control is applied to eliminate toxic compounds or pathogenic microorganisms (Su et al., 2023).

In the field of biofertilizers, Regulation (EU) 2019/1009 allows the use of microalgae and recycled materials in fertilizing products, establishing safety limits for cadmium, mercury, and arsenic. For cosmetic and pharmaceutical applications, the European Committee for Standardization (CEN) has developed technical standards (CEN/TR 17611, 17612, 17739) defining uniform analytical methods for assessing algal-derived ingredients and contaminants, ensuring traceability and safety.

In summary, the technological maturity of microalgal biotechnologies must be accompanied by regulatory maturity. Compliance with safety, traceability, and sustainability frameworks not only facilitates market access but is also a prerequisite for generating trust and social acceptance of microalgal-derived products.

4.4.2.4 Social Barriers

The social acceptance of microalgal products depends on public perception of their safety, sustainability, and added value compared to conventional alternatives. While acceptance is increasing in wellness, cosmetic, and supplement markets, skepticism remains in bioenergy and staple food sectors.

4.4.3 Strategies to Advance through TRL Levels

To overcome these barriers and progress toward higher TRL stages, a combination of technological, structural, and regulatory strategies has been proposed. Key strategic actions include:

- Strain improvement and use of alternative substrates: Genetically optimized strains combined with agro-industrial residues or wastewater as nutrient sources can reduce capital and operating inputs while improving process sustainability (Dias et al., 2025).
- Innovation in low-cost photobioreactor design: Development of new or optimized PBRs (tubular, flat-plate, hybrid, or plastic-bag systems) can reduce specific CAPEX per illuminated volume and facilitate scale-up to TRL 7–9. Engineering challenges and maturity levels of these systems are discussed in Deprá et al. (2025).
- Optimization of downstream processes: Improvements in harvesting, drying, and extraction technologies are essential to enhance efficiency and reduce energy consumption.
- Integration of science, industry, and regulation: Collaboration across sectors is required to enable commercial validation of bioproducts. A notable example is the SCALE biorefinery by Microphyt (France), inaugurated in 2025, which demonstrates how strategic partnerships can elevate a microalgal process to TRL 8–9, producing over 100 tons of active ingredients annually for health, nutrition, and cosmetics.
- Regulatory compliance verification in early-stage projects: From TRL 2–3 onward, it is crucial to ensure that final products meet safety, quality, and

sustainability standards defined by regulatory frameworks. Each application, such as nutraceutical, cosmetic, biofertilizer, or bioenergy, has specific legal requirements. Product classification should be defined during the PoC phase, following the main frameworks: FDA (U.S.) GRAS for food ingredients, Regulation (EU) 2015/2283 (Novel Foods), Regulation (EU) 1223/2009 (Cosmetics), and Regulation (EU) 2019/1009 (Fertilizing Products). Proper classification guides the design of safety and efficacy testing.

- Implementation of quality assurance systems: GMP and Good Laboratory Practices should be progressively integrated from TRL 4 onward. In both open (raceway, HRAP) and closed (PBR) systems, strain traceability, operational logs, and substrate documentation are essential. For cultures using effluents, verification of pathogen absence, heavy metal content, and microcystin levels is mandatory before commercialization (Regulations EC 1881/2006; 2073/2005).
- Integrated Techno-Economic and Environmental Assessment (TEA/LCA): Scale-up must be accompanied by LCA and TEA, evaluating carbon footprint, energy efficiency, and biosafety to assess investment scenarios, environmental sensitivity, and economic feasibility (Quiroz et al., 2025).

4.5 SUSTAINABILITY AND LIFE-CYCLE ASSESSMENT APPLIED TO MICROALGAE-BASED TECHNOLOGICAL PRODUCTS

The market for biotechnology products, particularly those derived from microalgae, has become a viable option for various sectors of the economy. Biofuels, bioactive compounds, fertilizers, and even biopolymers can be obtained from the biomass of these microorganisms. Their value lies in several unique characteristics: they produce large quantities of biomass in a short time, efficiently capture CO_2, can grow using different metabolic strategies (autotrophic, heterotrophic, or mixotrophic), which facilitates the utilization of underutilized nutrients and by-products, and can be cultivated in spaces unsuitable for traditional agriculture. All of this makes microalgae a promising alternative to replace natural resources typically obtained through more intensive extraction processes.

While working with biological systems offers significant advantages, this alone does not guarantee that a process will automatically be sustainable. Therefore, it is essential to have scientific methods that allow for measuring, with real data, the impact of the entire production system. In this scenario, LCA has established itself as the most comprehensive and reliable tool for objectively evaluating the environmental performance of microalgae-derived technological products (Zhang et al., 2022; Sousa et al., 2025).

4.5.1 Principles and Methodological Structure of Life-Cycle Assessment

LCA is a methodology regulated by ISO 14040:2006 and ISO 14044:2006, allowing the identification and quantification of potential impacts of a product, process, or service throughout all life-cycle stages, from raw material extraction to end-of-life disposal. The analysis is structured into four interrelated phases.

4.5.1.1 Goal and Scope Definition

The purpose of the study, system boundaries, and functional units are defined. The functional unit acts as a reference parameter for all flows (e.g., "1 MJ of algal biodiesel").

4.5.1.2 Life-Cycle Inventory (LCI)

This phase involves the detailed collection of data on inputs (water, energy, and nutrients) and outputs (emissions, waste, and coproducts) in each stage of the evaluated system.

4.5.1.3 Life-Cycle Impact Assessment (LCIA)

The inventory data are translated into environmental impacts using models that assign burdens to specific categories such as global warming potential, eutrophication, acidification, water use, and cumulative energy demand.

4.5.1.4 Interpretation

This phase integrates the findings, identifies critical points (hotspots), assesses data uncertainty, and proposes environmental improvement or system redesign strategies.

4.5.2 Integration of Environmental Analyses within the Life-Cycle Framework

LCA provides a common framework for integrating different types of environmental analysis. Although each could be applied separately, integrating them into the life-cycle logic provides a much more complete view of the system. Thus, tools such as carbon footprint, water footprint, and energy balance do not function as independent methodologies, but rather as impact categories that form part of the overall LCA analysis.

Thanks to this integration, it is possible to understand not only the magnitude of each impact but also the trade-offs between them. This allows for a more precise identification of the strengths and limitations of each technological configuration. This integration enables understanding not only the magnitude of each impact but also its advantages and disadvantages, revealing the strengths and weaknesses of each technological configuration (Table 4.6).

4.5.2.1 Life-Cycle Assessment in Microalgae-Based Systems

Microalgae-based biotechnological systems have complex production chains, each phase of which has specific environmental implications. LCA allows for a structured examination of the entire process, from cultivation to final product conversion, including transport, use, and disposal (Table 4.7).

TABLE 4.6
Specific Environmental Analyses and Their Integration within the LCA Framework

Specific Environmental Analysis	LCA Category	Typical Unit	Function within LCA
Carbon footprint	Climate change (GWP100)	kg CO_2-eq	Quantification of greenhouse-gas emissions
Energy analysis	Cumulative energy demand (CED, NER)	MJ or MJ/MJ product	Evaluation of efficiency and total energy demand
Water footprint	Water resource use	m^3	Estimation of direct and indirect water consumption
Eutrophication	Nutrient enrichment of water bodies	kg PO_4-eq	Assessment of potential nutrient release
Acidification	Acidifying compound emissions	kg SO_2-eq	Evaluation of impacts on soils and aquatic systems

Source: Adapted from ISO 14040:2006, Sangtanoo et al. (2022), and Bradley et al. (2023).

TABLE 4.7
Life-Cycle Stages of Algal Products

Life-Cycle Stage	Main Inputs	Key Environmental Outputs
Cultivation	Water, nutrients, CO_2, and electricity	Wet biomass, oxygen, and residual heat
Harvesting and drying	Thermal energy, centrifugation, and aeration	Dried biomass, vapor, and liquid waste
Lipid extraction	Solvents, pressure, and temperature	Lipid extract and residual biomass
Conversion to biodiesel	Methanol, catalyst, and energy	Biodiesel, glycerol, and by-products
Use and end of life	Transport and distribution	Emissions and organic residues

Source: Adapted from Sangtanoo et al. (2022) and Bradley et al. (2023).

4.5.2.2 Environmental Indicators and Comparative Performance

Bradley et al. (2023) developed a comprehensive LCA for microalgal biodiesel produced through a hybrid system (PBRs + fermenters) powered by solar energy. Key findings include:

- Carbon footprint: 0.148 kg CO_2-eq/MJ, compared to 0.0884 kg CO_2-eq/MJ for fossil diesel.
- Net energy consumption: 0.99 MJ per MJ of biodiesel produced.
- Hotspots: electrical consumption during fermentation and lipid extraction.
- Valuable coproducts: glycerol, organic fertilizers, and residual biomass.

4.5.2.3 Improvement Strategies: Urban Integration and Circular Approaches

One of the most innovative strategies to enhance the environmental profile of algal systems is their integration within urban environments, particularly through the use of wastewater as a culture medium. This approach transforms municipal waste streams into productive resources, aligning with the principles of the circular bioeconomy.

Sangtanoo et al. (2022) modeled a system installed in an urban wastewater treatment plant, where microalgae captured nitrogen, phosphorus, and CO_2 directly from effluents. The results showed:

- Up to 70% reduction in eutrophication potential.
- Net-negative greenhouse-gas emissions.
- Local and decentralized biomass production for biofuels, biofertilizers, or bioplastics.
- Efficient use of urban space (rooftops, façades, vertical greenhouses).
- Oxygen generation and air quality improvement.

This model simultaneously treats wastewater, captures CO_2, and produces biomass, significantly reducing external inputs and enhancing system sustainability. Its urban integration transforms microalgae from industrial feedstocks into active agents of environmental regeneration.

4.5.2.4 Systemic Sustainability and Technological Maturity

The relationship between LCA and Technology Maturity Level (TML) has become increasingly important in the development of microalgae-based biotechnology (Gurreri et al., 2023). As a process progresses from the initial experimental stages (TML 3–4) to more advanced demonstration and pre-commercial trial phases (TML 6–8), LCA becomes a key tool for guiding decisions. This analysis enables identification of the technological pathways that offer the best environmental and economic performance, avoiding efforts in configurations that, while promising in the laboratory, may not be viable at scale.

In this transitional phase, indicators from LCA, such as the carbon footprint or net energy balance, help adjust and optimize the process: improving energy efficiency, maximizing coproduct recovery, or evaluating how to integrate renewable sources into the system.

Furthermore, incorporating more recent approaches, such as dynamic LCA and multicriteria sustainability analyses (MCA), broadens the scope (Liu et al., 2025; Padri et al., 2025). These frameworks allow simultaneous evaluation of environmental, economic, and social dimensions, enabling more balanced comparisons across different technological designs, including closed PBRs, open ponds, and hybrid models.

All of this contributes to a more complete and realistic understanding of how microalgae technologies are evolving and which have the most significant potential for sustainable market consolidation.

4.6 FINAL CONSIDERATIONS AND RECOMMENDATIONS

Microalgae biotechnology, despite its enormous potential as a versatile platform for producing biocompounds and mitigating CO_2, still faces significant challenges. Aspects such as scalability, operating costs, and cultural stability remain obstacles that have limited its industrial adoption. To overcome these barriers and advance through TMLs, comprehensive strategies are needed that connect basic research with commercial-scale validation. In this process, reducing costs and improving sustainability are not only desirable goals but also fundamental requirements for achieving industrial maturity.

A key part of this progress lies in optimizing both the strains and the substrates. The development of improved strains, combined with the use of agro-industrial waste or wastewater as nutrient sources, reduces both initial investment and operating costs. In parallel, innovation in PBR design is equally crucial. Lower-cost systems, such as tubular, flat-plate, hybrid, or even plastic-bag fiber-reinforced resins (FBRs), need to be developed or refined to reduce investment per unit of illuminated volume and facilitate reliable scaling up to TRLs 7–9.

Another critical point is the post-harvest stage, which represents one of the biggest energy bottlenecks in the process. Continuous improvement of harvesting, drying, and extraction technologies is essential to increase system efficiency and significantly reduce total energy consumption.

To achieve solid progression through the TRLs, it is crucial to rely on clear metrics and rigorous technical, economic, and environmental feasibility analyses. Therefore, integrating both TEA and LCA from early stages (TRLs 2–3) provides a reliable quantitative basis for decision-making. These analyses guide the design, establish verifiable milestones, and reduce technological uncertainty.

During the intermediate TRLs (4–6), a more detailed ATE (Technological Assessment of Enabling) is necessary, carefully evaluating both OPEX and CAPEX. In parallel, an LCA helps identify critical environmental points in the process. The goal at this stage is to substantially reduce production costs.

To overcome this phase, simply increasing scale is not enough; it requires validating each component and subsystem in relevant environments, demonstrating continuous and reproducible operation in pilot plants, and rigorously documenting mass and energy balances, including failures and adjustments. It is also essential to verify the quality, purity, and functionality of the final bioproduct from the outset, ensuring that it meets the standards of the target market, whether nutraceutical, cosmetic, agricultural, or energy-related.

Finally, to consolidate microalgae-based technology in the industrial environment, close collaboration between science, industry, and regulation is essential. Strategic alliances have proven to be an effective way to bring these technologies to high levels of maturity. Furthermore, regulatory compliance must be considered from the early stages of development (TRL 2–3) to ensure products meet the safety, quality, and sustainability requirements established by each sector's regulatory frameworks, thereby facilitating a much smoother transition to commercial application.

REFERENCES

Abdelfattah, A., Ali, S. S., Ramadan, H., El-Aswar, E. I., Eltawab, R., Ho, S.-H., Elsamahy, T., Li, S., El-Sheekh, M. M., Schagerl, M., Kornaros, M., & Sun, J. (2023). Microalgae-based wastewater treatment: Mechanisms, challenges, recent advances, and future prospects. *Environmental Science and Ecotechnology 13*, 100205. https://doi.org/10.1016/j.ese.2022.100205

Acién, F. G., Fernández, J. M., Magán, J. J., & Molina, E. (2012). Production cost of a real microalgae production plant and strategies to reduce it. *Biotechnology Advances 30*(6), 1344–1353. https://doi.org/10.1016/j.biotechadv.2012.02.005

Ahmad, I., Abdullah, N., Koji, I., Yuzir, A., & Muhammad, S. E. (2021). Evolution of photobioreactors: A review based on microalgal perspective. *IOP Conference Series: Materials Science and Engineering 1142*, 012004. https://doi.org/10.1088/1757-899X/1142/1/012004

Amanna, M., Kumar, S., Singh, N., & Reddy, S. (2024). Application of computational fluid dynamics in optimizing microalgal photobioreactors. *Algal Research 78*, 103718. https://doi.org/10.1016/j.algal.2024.103718

Amouri, M., Aziza, M., Kaidi, F., Abert Vian, M., Chemat, F., Amrane, A., Assunção, M. F. G., Santos, L. M. A., Ounnar, A., Zitouni, D., & Berrached, A. (2024). Indigenous microalgae strains characterization for a sustainable biodiesel production. *Biotechnology Journal 19*(1), e2300096. https://doi.org/10.1002/biot.202300096

Apel, A. C., Pfaffinger, C. E., Basedahl, N., Mittwollen, N., Göbel, J., Sauter, J., Brück, T., & Weuster-Botz, D. (2017). Open thin-layer cascade reactors for saline microalgae production evaluated in a physically simulated Mediterranean summer climate. *Algal Research 25*, 381–390. https://doi.org/10.1016/j.algal.2017.06.004

Azhand, N., Sadeghizadeh, A., & Rahimi, R. (2020). Effect of superficial gas velocity on CO_2 capture from air by *Chlorella vulgaris* microalgae in an airlift photobioreactor with external sparger. *Journal of Environmental Chemical Engineering 8*(4), 104022. https://doi.org/10.1016/j.jece.2020.104022

Banerjee, S., Dasgupta, S., Das, D., & Atta, A. (2020). Influence of photobioreactor configuration on microalgal biomass production. *Bioprocess and Biosystems Engineering 43*(8), 1487–1497. https://doi.org/10.1007/s00449-020-02342-4

Barboza-Rodríguez, R., Rodríguez-Jasso, R. M., Rosero-Chasoy, G., Rosales Aguado, M. L., & Ruiz, H. A. (2024). Photobioreactor configurations in cultivating microalgae biomass for biorefinery. *Bioresource Technology 394*, 130208. https://doi.org/10.1016/j.biortech.2023.130208

Barceló-Villalobos, M., Fernández-Del Olmo, P., Guzmán, J. L., Fernández-Sevilla, J. M., & Acién Fernández, F. G. (2019). Evaluation of photosynthetic light integration by microalgae in a pilot-scale raceway reactor. *Bioresource Technology 280*, 404–411. https://doi.org/10.1016/j.biortech.2019.02.032

Benedetti, M., Vecchi, V., Barera, S., & Bauer, A., & Minceva, M. (2021). Techno-economic analysis of a new downstream process for the production of astaxanthin from the microalgae *Haematococcus pluvialis*. *Bioresources and Bioprocessing 8*(1), 111. https://doi.org/10.1186/s40643-021-00463-6

Berthelier, J., Saint-Jean, B., Casse, N., Bougaran, G., & Carrier, G. (2023). Phenotype stability and dynamics of transposable elements in a strain of the microalga *Tisochrysis lutea* with improved lipid traits. *PloS One 18*(4), e0284656. https://doi.org/10.1371/journal.pone.0284656

Borovkov, A. B., Gudvilovich, I. N., & Avsiyan, A. L. (2020). Scale-up of Dunaliella salina cultivation: From strain selection to open ponds. *Journal of Applied Phycology 32*, 1545–1558. https://doi.org/10.1007/s10811-020-02104-5

Borowitzka, M. A., & Vonshak, A. (2017). Scaling up microalgal cultures to commercial scale. *European Journal of Phycology 52*(4), 407–418. https://doi.org/10.1080/09670262.2017.1365177

Bradley, T., Rajaeifar, M. A., Kenny, A., Hainsworth, C., del Pino, V., del Valle Inclán, Y., Povoa, I., Mendonça, P., Brown, L., Smallbone, A., & Roskilly, A.P. (2023). Life cycle assessment of microalgae-derived biodiesel. *International Journal of Life Cycle Assessment 28*, 590–609. https://doi.org/10.1007/s11367-023-02140-6

Brennan, L., & Owende, P. (2010). Biofuels from microalgae—A review of technologies for production, processing, and extractions of biofuels and co-products. *Renewable and Sustainable Energy Reviews 14*(2), 557–577. https://doi.org/10.1016/j.rser.2009.10.009

Business Research Insights. (2024). *Microalgae Market Size, Share, Growth, and Forecast 2025–2034*. https://www.businessresearchinsights.com/market-reports/microalgae-market-119735

Cheirsilp, B., Maneechote, W., Srinuanpan, S., & Angelidaki, I. (2023). Microalgae as tools for bio-circular-green economy: Zero-waste approaches for sustainable production and biorefineries of microalgal biomass. *Bioresource Technology 387*, 129620. https://doi.org/10.1016/j.biortech.2023.129620

Chen, C.-Y., Chang, Y.-H., Leong, Y. K., & Chang, J.-S. (2025). Developing an efficient open-pond microalgae cultivation system for producing biomass and fucoxanthin using *Pavlova granifera* IML2238. *Biochemical Engineering Journal 220*, 109762. https://doi.org/10.1016/j.bej.2025.109762

Chen, C.-Y., Yeh, K.-L., Aisyah, R., Lee, D.-J., & Chang, J.-S. (2011). Cultivation, photobioreactor design and harvesting of microalgae for biodiesel production: A critical review. *Bioresource Technology 102*(1), 71–81. https://doi.org/10.1016/j.biortech.2010.06.159

Chentir, I., Gnouma, A., & Markou, G. (2025). Legislative aspects of microalgae-based products. In: Antonio Pérez-Gálvez, Eduardo Jacob-Lopes, Leila Queiroz Zepka, and María Roca (eds.), *Microalgae and One Health: Fundamentals, Biocompounds, and Health and Environmental Applications* (pp. 613–624). Elsevier. https://doi.org/10.1016/B978-0-443-22080-7.00013-2

Choudhary, P., Bhattacharya, A., Prajapati, S. K., Kaushik, P., & Malik, A. (2015). Phycoremediation-coupled biomethanation of microalgal biomass. In: S.-K. Kim (ed.), *Handbook of Marine Microalgae* (pp. 483–499). Academic Press. https://doi.org/10.1016/B978-0-12-800776-1.00032-7

Clagnan, E., Dell'Orto, M., Štěrbová, K., Grivalský, T., Câmara Manoel, J. A., Masojídek, J., D'Imporzano, G., Acién-Fernández, F. G., & Adani, F. (2023). Impact of photobioreactor design on microalgae–bacteria communities grown on wastewater: Differences between thin-layer cascade and thin-layer raceway ponds. *Bioresource Technology 374*, 128781. https://doi.org/10.1016/j.biortech.2023.128781

Contreras-Gómez, A., López-Rosales, L., Martínez-Burgos, J., Molina-Miras, A., Cerón-García, M. C., Sánchez-Mirón, A., & García-Camacho, F. (2026). Valorisation of seawater desalination brine via cultivation of a halotolerant microalga in bubble column photobioreactors. *Bioresource Technology*, *439*, 133261.

Craggs, R. J., Lundquist, T. J., & Benemann, J. (2013). Wastewater treatment and algal biofuel production. In: M. A. Borowitzka & N. R. Moheimani (eds.), *Algae for Biofuels and Energy* (pp. 153–163). Springer. https://doi.org/10.1007/978-94-007-5479-9_9

Cruz, J. D., & Vasconcelos, V. (2023). Legal aspects of microalgae in the European food sector. *Foods (Basel, Switzerland) 13*(1), 124. https://doi.org/10.3390/foods13010124

Dębowski, M., Swica, I., Kazimierowicz, J., & Zieliński, M. (2023). Large scale microalgae biofuel technology—Development perspectives in light of the barriers and limitations. *Energies 16*(1), 81. https://doi.org/10.3390/en16010081

Deniz, I. (2020). Scaling-up of *Haematococcus pluvialis* production in stirred tank photobioreactor. *Bioresource Technology 310*, 123434. https://doi.org/10.1016/j.biortech.2020.123434

Deprá, M. C., Dias, R. R., Zepka, L. Q., & Jacob-Lopes, E. (2025). Tackling old challenges in microalgal biotechnology: The role of photobioreactors to advance the technology readiness level. *Processes 13*(1), 51. https://doi.org/10.3390/pr13010051

Dias, R. R., Deprá, M. C., de Menezes, C. R., Zepka, L. Q., & Jacob-Lopes, E. (2025). Microalgae cultivation in wastewater: How realistic is this approach for value-added product production? *Processes 13*(7), 2052. https://doi.org/10.3390/pr13072052

FAO, IFAD, UNICEF, WFP, & WHO. (2025). *The State of Food Security and Nutrition in the World 2025: Addressing High Food Price Inflation for Food Security and Nutrition*. Food and Agriculture Organization of the United Nations (FAO). https://doi.org/10.4060/cd6008en

Figueroa-Torres, G. M., Pittman, J. K., & Theodoropoulos, C. (2021). Optimisation of microalgal cultivation via nutrient-enhanced strategies: The biorefinery paradigm. *Biotechnology for Biofuels 14*(1), 64. https://doi.org/10.1186/s13068-021-01912-2

Figueroa-Torres, G. M., & Theodoropoulos, C. (2023). Techno-economic analysis of a microalgae-based biorefinery network for biofuels and value-added products. *Bioresource Technology Reports 23*, 101524. https://doi.org/10.1016/j.biteb.2023.101524

Garoma, T. (2019). *Energy-Efficient and Cost-Effective Microalgae Disruption for Extraction of Lipids for Biodiesel Production*. California Energy Commission. https://www.energy.ca.gov (Publication No. CEC-600-2019-015).

Geng, Y., Shaukat, A., Azhar, W., Raza, Q. U., Tahir, A., Abideen, M. Z. U., Zia, M. A. B., Bashir, M. A., & Rehim, A. (2025). Microalgal biorefineries: A systematic review of technological trade-offs and innovation pathways. *Biotechnology for Biofuels and Bioproducts 18*(1), 93. https://doi.org/10.1186/s13068-025-02694-7

Guo, B., Hornung, U., Zhang, S., & Dahmen, N. (2022). Techno-economic assessment of a microalgae biorefinery. *Chemie Ingenieur Technik 94*(12), 1–6. https://doi.org/10.1002/cite.202200007

Gurreri, L., Calanni Rindina, M., Luciano, A., Lima, S., Scargiali, F., Fino, D., & Mancini, G. (2023). Environmental sustainability of microalgae-based production systems: Roadmap and challenges towards the industrial implementation. *Sustainable Chemistry and Pharmacy 35*, 101191. https://doi.org/10.1016/j.scp.2023.101191

Haberkorn, I., Off, C. L., Besmer, M. D., Buchmann, L., & Mathys, A. (2021). Automated online flow cytometry advances microalgal ecosystem management as in situ, high-temporal resolution monitoring tool. *Frontiers in Bioengineering and Biotechnology 9*, 642671. https://doi.org/10.3389/fbioe.2021.642671

Hernandez-de Leon, H. R., Martinez-Morgan, J. C., Fragoso-Mandujano, J. A., Rodriguez-Rangel, H., Amabilis-Sosa, L. E., & Flores-Guirao, S. K. (2025). A comprehensive dataset of biomass and critical variables for *Verrucodesmus verrucosus* culture in bubble column photobioreactors. *Data in Brief 62*, 112003. https://doi.org/10.1016/j.dib.2025.112003

Hossain, S. M. Z., Al-Bastaki, N., Alnoaimi, A. M. A., Ezuber, H., Razzak, S. A., & Hossain, M. M. (2020). Mathematical modeling of temperature effect on algal growth for biodiesel application. In: A. Sayigh (Ed.), *Renewable Energy and Sustainable Buildings* (pp. 517–528). Springer Nature Switzerland AG. https://doi.org/10.1007/978-3-030-18488-9_41

Hu, Q., Sommerfeld, M., Jarvis, E., Ghirardi, M., Posewitz, M., Seibert, M., & Darzins, A. (2008). Microalgal triacylglycerols as feedstocks for biofuel production: Perspectives and advances. *The Plant Journal: For Cell and Molecular Biology*, *54*(4), 621–639. https://doi.org/10.1111/j.1365-313X.2008.03492.x

ISO. (2006). *ISO 14040:2006 – Environmental Management – Life Cycle Assessment – Principles and Framework*. https://www.iso.org/standard/37456.html

ISO. (2006). *ISO 14044:2006 – Environmental Management – Life Cycle Assessment – Requirements and Guidelines*. https://www.iso.org/standard/38498.html

Khairuddin, F., Mohd Fuzi, S. F. Z., Ahmad, A., Low, K. O., Bokhari, A., Dailin, D. J., Habila, M. A., Nawaz, A., & Chuah, L. F. (2024). Evaluation on microalgae for the production of bio-chemicals and electricity. *Chemosphere 350*, 141007. https://doi.org/10.1016/j.chemosphere.2023.141007

Khor, W. H., Kang, H.-S., Lim, J.-W., Iwamoto, K., Tang, C. H.-H., Goh, P. S., Quen, L. K., Shaharuddin, N. M. R. B., & Lai, N. Y. G. (2022). Microalgae cultivation in offshore floating photobioreactor: State-of-the-art, opportunities and challenges. *Aquacultural Engineering 98*, 102269. https://doi.org/10.1016/j.aquaeng.2022.102269

Klein, B., & Davis, R. (2023). *Algal Biomass Production Via Open Pond Algae Farm Cultivation: 2022 State of Technology and Future Research* (NREL/TP-5100-85661). National Renewable Energy Laboratory. https://www.nrel.gov/docs/fy23osti/85661.pdf

Kumar, R., Chung, W. J., Khan, M. A., Son, M., Park, Y. K., Lee, S. S., & Jeon, B. H. (2024). Breakthrough innovations in carbon dioxide mineralization for a sustainable future. *Reviews in Environmental Science and Bio/Technology*, *23*(3), 739–799.

Labara Tirado, J., Herdean, A., & Ralph, P. J. (2025). The need for smart microalgal bioprospecting. *Natural Products and Bioprospecting 15*(1), 7. https://doi.org/10.1007/s13659-024-00487-3

Lara, G. A., Moreno, L., Ramírez, Y., & Cisternas, L. A. (2018). Modeling an airlift reactor for the growing of microalgae. *The Open Chemical Engineering Journal 12*, 80–94. https://doi.org/10.2174/1874123101812010080

Lim, H. R., Khoo, K. S., Chia, W. Y., Chew, K. W., Ho, S.-H., & Show, P. L. (2022). Smart microalgae farming with Internet-of-Things for sustainable agriculture. *Biotechnology Advances 57*, 107931. https://doi.org/10.1016/j.biotechadv.2022.107931

Liu, Y., Xia, R., Wang, Y., Zhao, Y., & Ju, M. (2025). Life cycle environmental impact assessment of microalgae biodiesel derived from different processes. *Industrial Crops and Products 233*, 121381. https://doi.org/10.1016/j.indcrop.2025.121381

Liu, Z., Feng, L., & Chenxi, L. (2024). Effect of bacteria-algae ratio on treatment of anaerobic digested wastewater by symbiotic coupling of bacteria and algae under the background of carbon neutralization. *Environmental Research 251*, 118771. https://doi.org/10.1016/j.envres.2024.118771

Mankins, J. C. (2009). Technology readiness assessments: A retrospective. *Acta Astronautica 65*(9–10), 1216–1223. https://doi.org/10.1016/j.actaastro.2009.03.058

Marbelia, L., Bilad, M. R., Passaris, I., Discart, V., Vandamme, D., Beuckels, A., Muylaert, K., & Vankelecom, I. F. (2014). Membrane photobioreactors for integrated microalgae cultivation and nutrient remediation of membrane bioreactors effluent. *Bioresource Technology 163*, 228–235. https://doi.org/10.1016/j.biortech.2014.04.012

Mathivanan, K., Ameen, F., Zhang, R., Ravi, G., & Beduru, S. (2023). Bubble column photobioreactor (BCPR) for cultivating microalgae and microalgal consortium (Co-CC) with additional CO_2 supply for enhancing biomass, lipid, and preferable fatty acids production. *Environmental Research 238*(Pt 2), 117284. https://doi.org/10.1016/j.envres.2023.117284

Michael, M., Vanness, V., Henry, H., Tjandra, T. M., Kristian, K., Nasution, J. A., Masytha, M., Sidabutar, R., Trisakti, B., Irvan, I., Dalimunthe, N. F., Al Fath, M. T., Nabilah, Y., Kobayashi, T., Daimon, H., & Takriff, M. S. (2026). Microalgae for sustainable biodiesel and omega-3: A comprehensive review of production, processing, and implementation. *Renewable and Sustainable Energy Reviews 226*(Part C), 116327. https://doi.org/10.1016/j.rser.2025.116327

Min, K. H., Kim, D. H., Ki, M. R., & Pack, S. P. (2022). Recent progress in flocculation, dewatering, and drying technologies for microalgae utilization: Scalable and low-cost harvesting process development. *Bioresource Technology 344*(Pt B), 126404. https://doi.org/10.1016/j.biortech.2021.126404

Montenegro-Herrera, C. A., Sangregorio-Soto, V., & Maroneze, M. M. (2025). Heterotrophic microalgal bioreactors. In: Eduardo Jacob-Lopes, Leila Queiroz Zepka, and Mariany Costa Deprá (eds.), *Algal Bioreactors: Science, Engineering and Technology of Upstream Processes* (Woodhead Series in Bioenergy, pp. 331–344). Elsevier. https://doi.org/10.1016/B978-0-443-14058-7.00029-4

Morais, M. G., Rosa, G. M., Moraes, L., Lopes, L. C., & Costa, J. A. V. (2025). Membrane technologies for bioengineering microalgae: Sustainable applications in biomass production, carbon capture, and industrial wastewater valorization. *Membranes 15*(7), 205. https://doi.org/10.3390/membranes15070205

Narala, R. R., Garg, S., Sharma, K. K., Thomas-Hall, S. R., Deme, M., Li, Y., & Schenk, P. M. (2016). Comparison of microalgae cultivation in photobioreactor, open raceway pond, and a two-stage hybrid system. *Frontiers in Energy Research 4*, Article 29. https://doi.org/10.3389/fenrg.2016.00029

Nguyen, T. D., Le, H. T. H., Khoo, K. S., & Bui, H. M. (2025). Algae windows: A novel approach towards sustainable building design and energy conservation. *Polish Journal of Environmental Studies, 34*(3), 3243–3255.

Nobre, M. L. F., Tavares, D., Fraga, C., Oliveira, B., Dias, M., Mesquita, S., Oliveira, C. M., & Pires, J. C. M. (2024). Techno-economic analysis of a circular microalgal approach for enhanced wastewater treatment and resource recovery in Northern Portugal. *Journal of Cleaner Production 434*, 140389. https://doi.org/10.1016/j.jclepro.2023.140389

Novoveská, L., Nielsen, S. L., Eroldoğan, O. T., Haznedaroglu, B. Z., Rinkevich, B., Fazi, S., Robbens, J., Vasquez, M., & Einarsson, H. (2023). Overview and challenges of large-scale cultivation of photosynthetic microalgae and cyanobacteria. *Marine Drugs 21*(8), 445. https://doi.org/10.3390/md21080445

Oruganti, R. K., Biji, A. P., Lanuyanger, T., Show, P. L., Sriariyanun, M., Upadhyayula, V. K. K., Gadhamshetty, V., & Bhattacharyya, D. (2023). Artificial intelligence and machine learning tools for high-performance microalgal wastewater treatment and algal biorefinery: A critical review. *Science of the Total Environment 876*, 162797. https://doi.org/10.1016/j.scitotenv.2023.162797

Oswald, W. J., & Golueke, C. G. (1960). Biological transformation of solar energy. In: W. W. Umbreit (ed.), *Advances in Applied Microbiology* (Vol. 2, pp. 223–262). Academic Press. https://doi.org/10.1016/S0065-2164(08)70127-8

Oves, M., Qari, H. A., & Ismail, I. M. (2022). Biofuel formation from microalgae: A renewable energy source for eco-sustainability. *Current World Environment 17*(1), 4–19. https://doi.org/10.12944/CWE.17.1.2

Padri, M., Amdah, M., Latief, M. M., & Fuentes-Grünewald, C. (2025). Site suitability assessment for microalgae plant deployment in Saudi Arabia using multi-criteria decision making and the analytic hierarchy process: A spatial approach. *Sustainability 17*(23), 10480. https://doi.org/10.3390/su172310480

Posten, C., & Schaub, G. (2009). Microalgae and terrestrial biomass as source for fuels--A process view. *Journal of Biotechnology 142*(1), 64–69. https://doi.org/10.1016/j.jbiotec.2009.03.015

Quinn, J. C., & Davis, R. (2015). The potentials and challenges of algae based biofuels: A review of the techno-economic, life cycle, and resource assessment modeling. *Bioresource Technology 184*, 444–452. https://doi.org/10.1016/j.biortech.2014.10.075

Quiroz, D., Greene, J. M., Limb, B. J., & Quinn, J. C. (2023). Global life cycle and techno-economic assessment of algal-based biofuels. *Environmental Science & Technology 57*(31), 11541–11551. https://doi.org/10.1021/acs.est.3c02892

Quiroz, D., McGowen, J. A., & Quinn, J. C. (2025). Techno-economic analysis of microalgae cultivation strategies: Batch and semi-continuous approaches. *Algal Research 90*, 104109. https://doi.org/10.1016/j.algal.2025.104109

Rahman, A., Prihantini, N. B., Oktaufik, M. A. M., Irwansyah, R., Kholiq, M.A., Levison, A.K., Rahadiyan, L., Aziz, M., Wahyono, Y., Ariyanti, D., & Nasruddin, N. (2025). Effect of hydrodynamics and mass transfer processes on biomass production of *Synechococcus* HS-9 cultivated using rectangular airlift photobioreactors with baffles. *Scientific Reports 15*, 32070. https://doi.org/10.1038/s41598-025-13711-y

Razzak, S. A., Bahar, K., Oajedul Islam, K. M., Haniffa, A. K., Faruque, M. O., Hossain, S. M. Z., & Hossain, M. M. (2024). Microalgae cultivation in photobioreactors: Sustainable solutions for a greener future. *Green Chemical Engineering 5*, 418–439. https://doi.org/10.1016/j.gce.2023.10.004

Ren, S., Shao, C., Zhu, F., Schagerl, M., Hu, X., Sobhi, M., Xu, L., Qian, J., & Huo, S. (2025). Optimization and synergistic enhancement of microalgae productivity in laboratory raceway ponds via co-regulation of automated light-supplemented mixers and electric field system. *Biotechnology for Biofuels and Bioproducts 18*, 63. https://doi.org/10.1186/s13068-025-02658-x

Rezvani, F., & Rostami, K. (2023). Photobioreactors for utility-scale applications: Effect of gas-liquid mass transfer coefficient and other critical parameters. *Environmental Science and Pollution Research 30*, 76263–76282. https://doi.org/10.1007/s11356-023-27644-4

Rörig, L. R., Gressler, P. D., Tramontin, D. P., de Souza Schneider, R. C., Derner, R. B., de Oliveira Bastos, E., de Souza, M. P., & Oliveira, C. Y. B. (2024). Biomass productivity and characterization of *Tetradesmus obliquus* grown in a hybrid photobioreactor. *Bioprocess and Biosystems Engineering 47*(3), 367–380. https://doi.org/10.1007/s00449-024-02969-7

Rozenberg, J. M., Sorokin, B. A., Mukhambetova, A. N., Emelianova, A. A., Kuzmin, V. V., Belogurova-Ovchinnikova, O. Y., & Kuzmin, D. V. (2024). Recent advances and fundamentals of microalgae cultivation technology. *Biotechnology Journal 19*(3), e2300725. https://doi.org/10.1002/biot.202300725

Sajjadi, B., Chen, W.-Y., Abdul Raman, A. A., & Ibrahim, S. (2018). Microalgae lipid and biomass for biofuel production: A comprehensive review on lipid enhancement strategies and their effects on fatty acid composition. *Renewable and Sustainable Energy Reviews 97*, 200–232. https://doi.org/10.1016/j.rser.2018.07.050

Salinas-García, M., Calatrava-Arrizabalaga, P., Ciardi, M., Villaró-Cos, S., & Lafarga, T. (2025). Development and reutilisation of a fertiliser-based culture medium for the commercial production of *Chlorella sorokiniana*. *Scientific Reports 15*, 23891. https://doi.org/10.1038/s41598-025-08320-8

Sangtanoo, P., Songserm, P., Watsuntorn, W., & Karnchanatat, A. (2022). Microalgae-based wastewater treatment for developing economic and environmental sustainability: Current status and future prospects. *Frontiers in Bioengineering and Biotechnology 10*, 904046. https://doi.org/10.3389/fbioe.2022.904046

Saravanan, A., Senthil Kumar, P., Varjani, S., Jeevanantham, S., Yaashikaa, P. R., & Abirami, B. (2021). A review on algal-bacterial symbiotic system for effective treatment of wastewater. *Chemosphere 271*, 129540. https://doi.org/10.1016/j.chemosphere.2021.129540

Schädler, T., Caballero Cerbon, D., de Oliveira, L., Garbe, D., Brück, T., & Weuster-Botz, D. (2019). Production of lipids with *Microchloropsis salina* in open thin-layer cascade photobioreactors. *Bioresource Technology 289*, 121682. https://doi.org/10.1016/j.biortech.2019.121682

Sero, E. T., Siziba, N., Bunhu, T., Shoko, R., & Jonathan, E. (2020). Biophotonics for improving algal photobioreactor performance: A review. *International Journal of Energy Research 44*(7), 5071–5092. https://doi.org/10.1002/er.5059

Sheik, A. G., Sireesha, M., Kumar, A., Dasari, P. R., Patnaik, R., Bagchi, S. K., Ansari, F.A., & Bux, F. (2025). The role of Industry 4.0 enabling technologies for predicting and managing algal blooms: Bridging gaps and unlocking potential. *Marine Pollution Bulletin 205*, 117493. https://doi.org/10.1016/j.marpolbul.2024.117493

Singh, P. K., Marella, T. K., Bhattacharjya, R., Tyagi, R., Plaha, N. S., Kaushik, N., & Tiwari, A. (2024). Marine diatom algae cultivation in simulated dairy wastewater and biomass valorization. *Environmental Science and Pollution Research 31*, 57466–57477. https://doi.org/10.1007/s11356-023-31531-3

Sousa, V., Novais, R., Mata, T. M., Martins, A. A., & Pereira, R. N. (2025). Life cycle assessment of protein extraction from microalgae biomass using ohmic heating. *Algal Research 86*, 103962. https://doi.org/10.1016/j.algal.2025.103962

Su, M., Bastiaens, L., Verspreet, J., & Hayes, M. (2023). Applications of microalgae in foods, pharma and feeds and their use as fertilizers and biostimulants: Legislation and regulatory aspects for consideration. *Foods (Basel, Switzerland) 12*(20), 3878. https://doi.org/10.3390/foods12203878

Sundaram, T., Kamalesh, R., Saravanan, A., Yaashikaa, P. R., Vickram, A. S., Teena, R. A., & Thiruvengadam, S. (2025). Harnessing algal biomass for renewable energy and biofuel production: Current strategies and future insights. *International Journal of Biological Macromolecules 319*(Part 2), 145487. https://doi.org/10.1016/j.ijbiomac.2025.145487

Suparmaniam, U., Lam, M. K., Uemura, Y., Lim, J. W., Lee, K. T., & Shuit, S. H. (2019). Insights into the microalgae cultivation technology and harvesting process for biofuel production: A review. *Renewable and Sustainable Energy Reviews 115*, 109361. https://doi.org/10.1016/j.rser.2019.109361

Syed, T., Krujatz, F., Ihadjadene, Y., Mühlstädt, G., Hamedi, H., Mädler, J., & Urbas, L. (2024). A review on machine learning approaches for microalgae cultivation systems. *Computers in Biology and Medicine 172*, 108248. https://doi.org/10.1016/j.compbiomed.2024.108248

Tan, J. S., Lee, S. Y., Chew, K. W., Lam, M. K., Lim, J. W., Ho, S.-H., & Show, P. L. (2020). A review on microalgae cultivation and harvesting, and their biomass extraction processing using ionic liquids. *Bioengineered 11*(1), 116–129. https://doi.org/10.1080/21655979.2020.1711626

Ugya, Y. A., Hasan, D. B., Tahir, S. M., Imam, T. S., Ari, H. A., & Hua, X. (2021). Microalgae biofilm cultured in nutrient-rich water as a tool for the phycoremediation of petroleum-contaminated water. *International Journal of Phytoremediation 23*(11), 1175–1183. https://doi.org/10.1080/15226514.2021.1882934

Uma, V. S., Usmani, Z., Sharma, M., Diwan, D., Sharma, M., Guo, M., Tuohy, M. G., Makatsoris, C., Zhao, X., Thakur, V. K., & Gupta, V. K. (2022). Valorisation of algal biomass to value-added metabolites: Emerging trends and opportunities. *Phytochemistry Reviews*, 1–26. https://doi.org/10.1007/s11101-022-09805-4

Uyar, B., Ali, M. D., & Uyar, G. E. O. (2024). Design parameters comparison of bubble column, airlift and stirred tank photobioreactors for microalgae production. *Bioprocess and Biosystems Engineering 47*(2), 195–209. https://doi.org/10.1007/s00449-023-02952-8

Vázquez-Romero, B., Perales, J. A., Pereira, H., Barbosa, M., & Ruiz, J. (2022). Techno-economic assessment of microalgae production, harvesting and drying for food, feed, cosmetics, and agriculture. *Science of the Total Environment 837*, 155742. https://doi.org/10.1016/j.scitotenv.2022.155742

Villaró, S., Sánchez-Zurano, A., Ciardi, M., Alarcón, F. J., Clagnan, E., Adani, F., Morillas-España, A., Álvarez, C., & Lafarga, T. (2022). Production of microalgae using pilot-scale thin-layer cascade photobioreactors: Effect of water type on biomass composition. *Biomass and Bioenergy 163*, 106534. https://doi.org/10.1016/j.biombioe.2022.106534

Wang, K., Khoo, K. S., Leong, H. Y., Nagarajan, D., Chew, K. W., Ting, H. Y., Selvarajoo, A., Chang, J. S., & Show, P. L. (2022). How does the Internet of Things (IoT) help in microalgae biorefinery? *Biotechnology Advances 54*, 107819. https://doi.org/10.1016/j.biotechadv.2021.107819

Wang, M., Ye, X., Bi, H., & Shen, Z. (2024). Microalgae biofuels: Illuminating the path to a sustainable future amidst challenges and opportunities. *Biotechnology for Biofuels and Bioproducts 17*(1), 10. https://doi.org/10.1186/s13068-024-02461-0

Wijffels, R. H., Kruse, O., & Hellingwerf, K. J. (2013). Potential of industrial biotechnology with cyanobacteria and eukaryotic microalgae. *Current Opinion in Biotechnology 24*(3), 405–413. https://doi.org/10.1016/j.copbio.2013.04.004

Yadav, G., Dubey, B. K., & Sen, R. (2020). A comparative life cycle assessment of microalgae production by CO_2 sequestration from flue gas in outdoor raceway ponds under batch and semi-continuous regime. *Journal of Cleaner Production 258*, 120703. https://doi.org/10.1016/j.jclepro.2020.120703

Yang, C. T., Kristiani, E., Leong, Y. K., & Chang, J. S. (2024). Machine learning in microalgae biotechnology for sustainable biofuel production: Advancements, applications, and prospects. *Bioresource Technology 413*, 131549. https://doi.org/10.1016/j.biortech.2024.131549

Yoon, H. J., Lee, J. S., Min, K. H., Kim, D. H., Sim, S. J., & Pack, S. P. (2025). Economic and demonstrative pilot-scale harvesting of microalgae biomass via novel combined process of dissolved air flotation and screw-press filtration. *Bioresource Technology 418*, 131892. https://doi.org/10.1016/j.biortech.2024.131892

Zhang, D., An, S., Yao, R., Fu, W., Han, Y., Du, M., Chen, Z., Lei, A., & Wang, J. (2022). Life cycle assessment of auto-tropically cultivated economic microalgae for final products such as food, total fatty acids, and bio-oil. *Frontiers in Marine Science 9*, 990635. https://doi.org/10.3389/fmars.2022.990635

Zhang, H., Liu, Z., Chen, Y., Liu, G., Bai, H., & Wu, J. (2025). Universal temperature model for large-scale outdoor horizontal tubular microalgae photobioreactor plants. *International Journal of Thermal Sciences 214*, 109921. https://doi.org/10.1016/j.ijthermalsci.2025.109921

Zhu, C., Han, D., Li, Y., Zhai, X., Chi, Z., Zhao, Y., & Cai, H. (2019). Cultivation of aquaculture feed Isochrysis zhangjiangensis in low-cost wave driven floating photobioreactor without aeration device. *Bioresource Technology 293*, 122018. https://doi.org/10.1016/j.biortech.2019.122018

Zhu, C., Zhai, X., Jia, J., Wang, J., Han, D., Li, Y., Tang, Y., & Chi, Z. (2018). Seawater desalination concentrate for cultivation of *Dunaliella salina* with floating photobioreactor to produce β-carotene. *Algal Research 35*, 319–324. https://doi.org/10.1016/j.algal.2018.08.035

Zuccaro, G., Yousuf, A., Pollio, A., & Steyer, J.-P. (2020). Microalgae cultivation systems. In: A. Yousuf (ed.), *Microalgae Cultivation for Biofuels Production* (pp. 11–29). Academic Press. https://doi.org/10.1016/B978-0-12-817536-1.00002-3

5 Advancements in Microalgal Exopolysaccharides

Biological Activities, Production Factors, and Industrial Applications

Kader Karsavran, Zeynep Elibol Çakmak, and Turgay Çakmak

5.1 INTRODUCTION

Microalgae constitute an ancient group of photosynthetic microorganisms that thrive in diverse microbiomes. They are always at the base of the food chain in aquatic ecosystems such as lakes, seas, and oceans in the pelagic zone. Moreover, they spread over all terrestrial ecosystems, including polar regions, deserts, and caves, where enough sunlight and moisture are present. Their morphological characteristics and bioactive compounds vary depending on the biotic and abiotic factors. Bioactive compounds such as pigments, proteins, lipids, and saccharides produced by microalgae have attracted the food, cosmetics, and medical industries (Magierek & Krzemińska, 2018). Apart from the intracellular bioactive compounds, some microalgae produce extracellular polymeric substances in response to changing environmental conditions. These substances are synthesized inside the cell and later are secreted out (Delattre et al., 2016; Xiao & Zheng, 2016). Extracellular polymeric substances may contain large amounts of polysaccharides, lower proteins, and trace amounts of nucleic acids, lipids, and pigments. Since most of these molecules are composed of polysaccharides, they are generally referred to as exopolysaccharides (EPSs). Microalgal EPSs are synthesized in the Golgi apparatus, while prokaryotes synthesize them in the cytoplasm (Delattre et al., 2016). EPSs have a high molecular weight of 10–1000 kDa, depending mainly on the homogeneity and the number of monosaccharides they include (Pierre et al., 2019).

EPS production allows microorganisms to better adapt to changing environmental conditions (Cruz et al., 2021). Due to their potential use as bioactive materials, bacterial, fungal, or microalgal EPSs are appropriate for biotechnological exploitation

DOI: 10.1201/9781003611479-5

(Rossi & De Philippis, 2016). However, EPSs from bacteria or fungi are preferred over microalgal EPSs in industrial applications. The differences in the EPS production yields of these microorganisms are considered one of the main reasons for this situation. *Cyanothece* sp. is known in the literature as the highest EPS-producing cyanobacterium, with a 22.34 g/L EPS yield (Chi et al., 2007). *Botrycoccus braunii* is one of the highest EPS-producing microalgae, with an EPS yield of 2.5 g/L (Lupi et al., 1994). However, it is known in the literature that the average EPS production yield of microalgae and cyanobacteria is 0.5–1 g/L. The bacterium *Leuconostoc mesenteroides* has the potential for industrial use with an EPS yield of 53.79 g/L, and the EPS of this strain is commercially available as dextran (Xing et al., 2018). The bacterium *Xanthomonas campestris* has the potential for industrial use with an EPS yield of 11.19 g/L, and the EPS of this species is available for sale in the market as xanthan gum (Demirci et al., 2017). *Aureobasidium pullulans* fungus produces a widely used EPS known commercially as pullulan and can yield 37.55 g/L EPS (Hamidi et al., 2019). Some bacteria and fungi can yield almost 20 times more EPS than most EPS-producing microalgae species.

Nevertheless, it seems possible to consider microalgae as renewable EPS-producing resources, making them a green production platform. Since microalgae are sensitive to environmental stimuli, their metabolite production profiles can be enriched by manipulating environmental conditions or culture media. For example, in a valuable study conducted with *Arthrospira platensis* (*Spirulina*), EPS yield could be increased from 20 mg/L to 60 mg/L with 600 mM NaCl application (Liu et al., 2016). Moreover, microalgal EPS production is superior to others due to its rich monosaccharide composition and its evaluation within the scope of green production (Naveed et al., 2019; Xiao & Zheng, 2016). Microalgal EPSs can contain 9–12 types of monosaccharides, while bacterial EPSs can contain a maximum of four monosaccharide types. In addition to microalgal EPSs, cyanobacterial EPSs are rich in monosaccharides and considered renewable, photosynthetic microorganisms. Although both cyanobacterial and microalgal EPSs are presented in this chapter, microalgal EPSs are primarily considered.

Microalgal EPSs exhibit structural and compositional differences according to the microalgae species from which they are produced (Table 5.1). Sulfated polysaccharides synthesized by various diatoms show similar monosaccharide profiles, including glucose, galactose, arabinose, rhamnose, mannose, and xylose (Nur et al., 2019). Many red microalgae, especially *Porphyridium* (1–9%) and *Rhodella* (4.7%), are known for their capacity to produce EPSs with high sulfate content (Drira et al., 2021; Villay et al., 2013). *Chlorella, Chlamydomonas*, and *Botryococcus* species are the most pronounced chlorophytes for EPS production. EPS of *Botryococcus braunii* contains a high proportion of galactose, glucose, and arabinose (Wang et al., 2021). In addition, uronic acid (7–8%), protein (2–4%), and sulfate groups (1.5–1.9%) were detected in the composition of EPS. When the EPS contents of the species are reviewed, sulfate groups are more common in red microalgae.

This chapter outlines the promising characteristics and describes factors affecting the composition and production of microalgal EPSs. Moreover, it evaluates recent developments in EPSs and their potential applications in food, cosmetics, and medical technology.

TABLE 5.1
Rich Chemical Composition of Microalgal EPS

	Arabinose	Rhamnose	Galactose	Glucose	Xylose	Mannose	Fucose	Ribose	Uronic Acid	Sulfate	Protein	References
Achnanthes longipes (Bac)	✓	✓		✓	✓	✓			✓		✓	Wustman et al., 1997
Amphora coffeaeformis (Bac)	✓	✓	✓	✓	✓		✓			✓	✓	Wustman et al., 1997
Amphora nostrata (Bac)	✓	✓	✓	✓	✓	✓					✓	Patil & Anil, 2005
Amphora holsatica (Bac)	✓	✓		✓	✓		✓		✓			Leandro et al., 2003
Amphora rostrata (Bac)		✓	✓	✓	✓	✓	✓		✓	✓	✓	Khandeparker & Bhosle, 2001
Aulacoseira granulata (Bac)			✓			✓		✓	✓			Gouvêa et al., 2005
Botryococcus braunii (Tre)	✓	✓	✓	✓	n.d.	n.d.	✓	✓	✓	✓	✓	Wang et al., 2021
Chaetoceros decipiens (Bac)		✓		✓	✓	✓	✓	✓				Urbani et al., 2012
Chaetoceros socialis (Bac)		✓	✓	✓	✓	✓	✓		✓		✓	Haug & Myklestad, 1976
Chlamydomonas mexicana (Chl)	✓	✓	✓	✓	✓	✓	✓	✓	✓			Barclay & Lewin, 1985
Chlamydomonas reinhardtii (Chl)	✓	✓	✓	✓	✓	n.d.	n.d.	✓			n.d.	Bafana, 2013

(Continued)

TABLE 5.1 (Continued)
Rich Chemical Composition of Microalgal EPS

	Arabinose	Rhamnose	Galactose	Glucose	Xylose	Mannose	Fucose	Ribose	Uronic Acid	Sulfate	Protein	References
Chlamydomonas sajao (Chl)	✓	✓	✓	✓	✓	✓	✓	✓	✓			Barclay & Lewin, 1985
Chlorella pyrenoidosa (Tre)	✓	✓	✓	✓		✓	✓					Naveed et al., 2019
Chlorella sorokiniana (Tre)			✓	✓		✓						Koçer et al., 2021
Chlorella vulgaris (Tre)	✓	✓	✓					✓	n.d.		✓	Barboríková et al., 2019
Chlorella zofingiensis (Tre)			✓	✓		✓			✓			Zhang, Liu, & Chen, 2019
Chlorococcum sp. (Chl)	✓	✓	✓	✓	✓	✓	✓				✓	Harun et al., 2011
Coccomyxa sp. (Tre)		✓	✓	✓	✓	✓	✓			✓		Concórdio-Reis et al., 2023
Cosmarium sp. (Zyg)	✓	✓	✓		✓	✓	✓	✓	✓	✓	✓	Kiemle et al., 2007
Cylindrotheca closterium (Bac)		✓	✓		✓	✓	✓		✓	✓	✓	Staats et al., 1999
Cylindrotheca fusiformis (Bac)	✓	✓		✓	✓	✓	✓	✓	✓	✓		Magaletti et al., 2004
Cymbella cistula (Bac)	✓	✓		✓	✓	✓	✓		✓		✓	Wustman et al., 1997

Desmococcus olivaceus (Chl)	✓	✓	✓	✓	✓	✓		✓	✓			Hokputsa et al., 2003
Dixoniella grisea (Rho)	✓		✓	✓	✓	✓				✓	✓	Arad & Levy-Ontman, 2010
Dunaliella salina (Chl)			✓	✓	✓						✓	Mishra & Jha, 2009
Dunaliella sp. (Chl)	✓	✓	✓	✓	✓	✓			✓	✓		Concórdio-Reis et al., 2023
Dunaliella tertiolecta (Chl)	✓	✓	✓	✓	✓	✓	✓		✓		✓	Goo et al., 2013
Graesiella sp. (Tre)	✓		✓	✓	✓	✓	✓					Trabelsi et al., 2016
Nannochloropsis sp. (Eus)	✓	✓	✓	✓	✓	✓	✓	✓	✓			Templeton et al., 2012
Navicula curvilineata (Bac)		✓	✓	✓	✓				✓			Tesson & Hildebrand, 2013
Navicula directa (Bac)	✓	✓	✓	✓	✓		✓	✓	✓			Leandro et al., 2003
Navicula phyllepta (Bac)	✓	✓		✓	✓	✓	✓			✓		Bellinger et al., 2005
Navicula salinarum (Bac)		✓	✓	✓	✓	✓			✓	✓	✓	Staats et al., 1999
Navicula subinflata (Bac)	✓	✓	✓	✓	✓	✓	✓	✓			✓	Patil & Anil, 2005
Palmella mucosa (Chl)	✓			✓			✓		✓			Tischer & Moore, 1964

(*Continued*)

TABLE 5.1 (Continued)
Rich Chemical Composition of Microalgal EPS

	Arabinose	Rhamnose	Galactose	Glucose	Xylose	Mannose	Fucose	Ribose	Uronic Acid	Sulfate	Protein	References
Penium margaritaceum (Zyg)	✓		✓	✓		✓	✓		✓			Domozych et al., 2005
Pinnularia viridis (Bac)	✓	✓	✓	✓	✓	✓	✓		✓	✓	✓	Chiovitti et al., 2003
Phaeodactylum tricornutum (Bac)	✓	✓	✓	✓	✓	✓	✓	✓	✓		✓	Willis et al., 2013
Pleurotaenium trabecula (Bac)	✓	✓	✓	✓		✓	✓	✓	✓	✓	✓	Kiemle et al., 2007
Porphyridium cruentum (Por)	✓		✓	✓					✓	✓	✓	de Jesus Raposo et al., 2014
Porphyridium marinum (Por)			✓	✓	✓		✓		✓	✓	✓	Roussel et al., 2015
Porphyridium sordidum (Por)		✓	✓	✓	✓	✓				✓	✓	Nikolova et al., 2019
Rhodella maculata (Rho)	✓	✓	✓	✓	✓				✓	✓	✓	Roussel et al., 2015
Rhodella violacea (Rho)	✓	✓	✓	✓	✓				✓	✓	✓	Villay et al., 2013
Scenedesmus sp. (Chl)		✓	✓	✓		✓			✓		n.d.	Zhang, Liu, Ren, et al., 2019
Tetraselmis suecica (Chd)				✓		✓		✓	✓			Parra-Riofrío et al., 2020

Note: Bac, Bacillariophyceae; Chd, Chlorodendrophyceae; Chl, Chlorophyceae, Eus, Eustigmatophyceae; Por, Porphyridiophyceae; Rho, Rhodophyceae; Tre, Trebouxiophyceae; Zyg, Zygnematophyceae; n.d., not determine.

5.2 CHARACTERISTIC FEATURES OF EPS

According to their location and physiological properties, EPSs can be categorized as cell-bound or released polysaccharides (Figure 5.1). Released polysaccharides are dissolved in the medium. In contrast, cell-bound polysaccharides can be found in capsular, sheath, and slime forms according to their attachment strength to the cell (Tiwari et al., 2020). Slime disperses around the cell as a mucilaginous layer without representing the cell shape, and extraction of this form is more straightforward than that of the other two forms. The sheath form loosely attaches to the cell surface as a thin layer and is visible under light microscopy without staining, while the capsule form is a thick layer around the cell and firmly adheres to the cell surface (Kumar et al., 2018).

Monosaccharide content influences the biological activity of EPSs. Glucose, galactose, mannose, arabinose, rhamnose, fucose, and xylose are known to be the most common monosaccharides for microalgal EPSs (Delattre et al., 2016; Kumar et al., 2018). Arabinose improves cell aggregation; rhamnose and fucose can influence flocculation and cell adhesion, while mannose removes heavy metals (Delattre et al., 2016; Zhang, Liu, & Chen, 2019). Proteins, the second macromolecule found in large proportions, play a role in the stability of EPS structure, cell attachment, and disintegration (Xiao & Zheng, 2016). Sulfate groups, pyruvate, O-acetyl, and O-methyl groups are the main non-sugar compounds in the EPS structure, and some of these groups, especially sulfate groups, give the anionic nature to the EPS and affect the solubility (Delattre et al., 2016). The sulfate groups and uronic acid make EPSs valuable for the biomedical and pharmaceutical industries (Rossi & De Philippis, 2016). Sulfate content is significant in the biomedical field as it provides hydrophilicity in gel structures such as hydrogels. Polysaccharides obtained from marine microalgae attract attention because of their high sulfate content (Xiao &

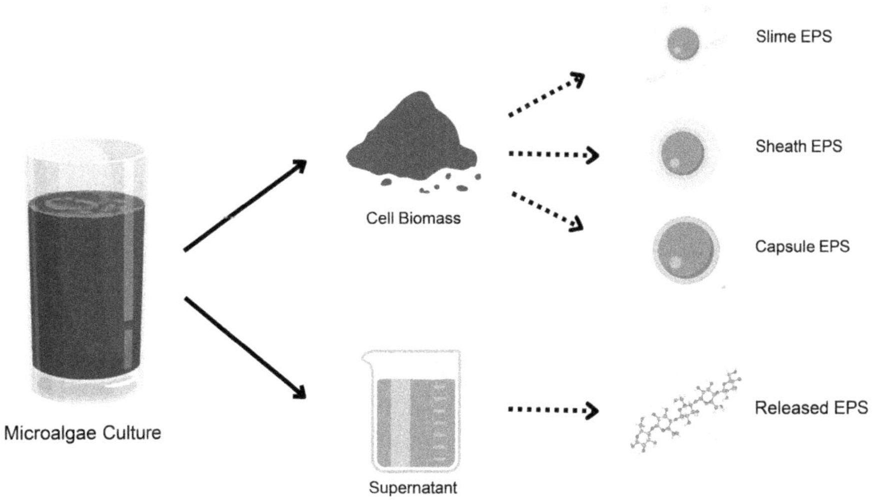

FIGURE 5.1 Categorization of EPSs according to the location and attachment strength.

Zheng, 2016). EPSs can be further categorized based on the configuration of their constituent monosaccharide chains. Specifically, EPSs with straight monosaccharide chains are designated linear polysaccharides, while those with bends and arms are classified as branched polysaccharides. Despite the surge in research endeavors concerning EPSs in recent years, their comprehensive structural characterization remains elusive. This is primarily attributable to the intricate nature of microalgal polysaccharides. The intricacy of microalgal EPSs stems from several factors, including the presence of a diverse array of monosaccharides within their structure, the existence of non-sugar groups, and the conspicuous absence of repeating units (Laroche, 2022). The incomplete structural characterization and the lack of a clear relationship between structural properties and biological activities may be related to the fact that microalgal EPSs are not as widespread in industrial use as bacterial EPSs.

5.3 BIOLOGICAL ACTIVITIES OF MICROALGAL EPS

A wide range of publications have addressed the biological activities of microalgal EPSs, including those involving antioxidants, antiviral, antibacterial, anti-inflammatory, antitumor, flocculant, and cryoprotective roles (Table 5.2).

TABLE 5.2
Biological Activities of Microalgal EPS

Species	Biological Activity	References
Picochlorum sp. *Chlorella sorokiniana* *Chlorella* sp.	Antioxidant and Anticoagulant	Mousavian et al., 2022
Chlorella vulgaris *Chlorella zofingiensis*	Anticancer and Antioxidant	Zhang, Liu, & Chen, 2019
Chlorella sp.	Antioxidant	Kokarakis et al., 2022
Chlorella pyrenoidosa *Scenedesmus* sp. *Chlorococcum* sp.	Antioxidant and Antitumor	Zhang, Liu, Ren, et al., 2019
Graesiella sp.	Antioxidant	Gongi et al., 2022
Porphyridium purpureum	Antibacterial, Antibiofilm, and Anticancer	Li et al., 2023
Porphyridium cruentum	Antiviral and Antibacterial	de Jesus Raposo et al., 2014
Porphyridium cruentum	Antioxidant	Sun et al., 2009
Porphyridium cruentum	Antitumor and Immunomodulatory	Sun et al., 2012
Chlamydomonas reinhardtii	Antioxidant	Bafana, 2013
Tetraselmis suecica *Porphyridium cruentum*	Antiviral	Parra-Riofrio et al., 2023
Botryococcus braunii	Antioxidant	Wang et al., 2021
Graesiella sp.	Antioxidant and Antiproliferative	Trabelsi et al., 2016

(*Continued*)

TABLE 5.2 (Continued)
Biological Activities of Microalgal EPS

Species	Biological Activity	References
Auxenochlorella protothecoides	Anti-inflammatory	Zhou et al., 2023
Neochloris oleoabundans	Immunostimulatory	Li, Wang et al., 2020
Dunaliella salina	Plant Biostimulant	Rachidi et al., 2021
Phaeodactylum tricornutum		
Porphyridium sp.		
Desmodesmus sp.		
Chlorella vulgaris	Plant Biostimulant	El-Naggar et al., 2020
Phaeodactylum tricornutum	Anticancer	Yang et al., 2019
Porphyridium sp.	Antibacterial	Liberman et al., 2016
Tribonema sp.	Anticancer and Immunomodulation	Chen et al., 2019
Chlorella vulgaris	Anti-inflammatory	Barboríková et al., 2019
Rhodella reticulata	Antibacterial	Najdenski et al., 2013
Porphyridium sp.	Antiviral	Huheihel et al., 2002
Gyrodinium impudicum	Antiviral	Yim et al., 2003
Porphyridium sp.	Antioxidant	Tannin-Spitz et al., 2005
Phaeodactylum tricornutum *Chlorella stigmatophora*	Anti-inflammatory and Immunomodulatory	Guzmán et al., 2003
[illegible]	Antioxidant	Chen et al., 2010
Porphyridium sp.	Antiviral	Huleihel et al., 2001
Synedropsis sp.	Cryoprotective	Aslam et al., 2012
Fragilariopsis curta	Cryoprotective	Aslam et al., 2012
Fragilariopsis cylindrus	Cryoprotective	Aslam et al., 2012
Coccomyxa subellipsoidea	Cryoprotective	Blanc et al., 2012
Chlamydomonas reinhardtii	Bioflocculant	Zhu et al., 2012
Neochloris oleoabundans	Bioflocculant	Paredes-Molina et al., 2025
Chlamydomonas reinhardtii	Wastewater Treatment	Li et al., 2025
Chlorella vulgaris	Wastewater Treatment	Joelyna et al., 2024

Antioxidant activity of EPS has been reported for *Chlorella pyrenoidosa, Scenedesmus* sp., *Chlorococcum* sp., *B. braunii*, and *Porphyridium* sp. (Wang et al., 2021; Zhang, Liu, Ren, et al., 2019). EPSs are of interest as natural antioxidants. The study's findings demonstrate that the EPSs derived from these microalgae species exhibit remarkable efficacy in scavenging free radicals. While the precise mechanism underlying the antioxidant activity remains to be elucidated, the study provides evidence that these EPSs possess antioxidant properties, as evidenced by their ability to scavenge free radicals. An evaluation of the antioxidant activity of EPS derived from *Porphyridium aerugineum* revealed a concentration-dependent response, indicating significant antioxidant potential (Ivanova et al., 2024). Due to their high antioxidant activity, EPSs exhibit potential for use in food, cosmetics, pharmaceuticals, and medical fields. In a valuable study, an attempt was made to develop a body cream from the EPS of the cyanobacterium Spirulina (Sed et al., 2017).

Microalgal EPSs are also pronounced with their high antibacterial and antifungal properties, making them valuable feedstock for various applications (Gongi et al., 2022). EPSs produced by *B. braunii* and *C. pyrenoidosa* showed potent antibacterial activity against *Escherichia coli* and *Staphylococcus aureus* (Gallón et al., 2019). The effective antibacterial activity of the EPS extract of *Pavlova lutheri* was proven against *Bacillus subtilis* (Bashir et al., 2018). On the other hand, EPSs released by cyanobacterium *Gloeaocapsa* sp. exhibited antifungal activity against *Candida albicians* (Najdenski et al., 2013). Researchers analyzed the different extracts of various microalgae species against *S. aureus, Streptococcus pyogenes, Pseudomonas aeruginosa*, and *C. albicians*. According to the results, most microalgal EPSs showed antibacterial activity, while only EPS from *Gloeaocapsa* sp. showed antifungal activity. The antibacterial activity of the EPS produced by *Neochloris oleoabundans* was examined in vitro against gram-negative bacteria (*Klebsiella pneumoniae* and *Pseudomonas aeruginosa*) and gram-positive bacteria (*S. aureus* and *Nocardia brasiliensis*) (Paredes-Molina et al., 2025). The study demonstrated that the extract exhibited antibacterial properties against *S. aureus* at all concentrations examined. However, its efficacy against *N. brasiliensis* was concentration-dependent, with the highest levels of activity observed at 20 and 30 mg/mL.

Studies show that sulfated polysaccharides may be used as natural virucides. The antiviral activity of EPSs has been reported primarily on sulfated polysaccharides obtained from marine microalgae (Kanekiyo et al., 2005; Kumar et al., 2018). EPSs with high sulfate content have been reported to exert growth-suppressive effects on some human pathogens, such as HIV (Yim et al., 2004). The antiviral activity of cyanobacterial EPSs was demonstrated against HIV-1 and HIV-2, influenza, and some other viruses by inhibiting the interaction of the virus with cells (Delattre et al., 2016). The infection-inhibitory activity of EPSs produced from *A. platensis* and *P. purpureum* was demonstrated on *Vaccinia* and *Ectromelia* virus strains (Radonić et al., 2011). The sulfated EPSs are evaluated as a potential therapeutic agent blocking the pathogen adhesion.

Some microalgal EPSs have anti-inflammatory activities, as demonstrated in animal models. The cyanobacterium *Mastigocladus luminous* produces capsular EPSs, and these EPSs show anti-inflammatory activity (Gloaguen et al., 2007). The anti-inflammatory activity of EPSs from *Phormidium* sp. ETS05 was demonstrated on Zebrafish embryos, and the results showed that the EPS has anti-inflammatory activity by decreasing inflammation (Zampieri et al., 2020). EPSs obtained from *C. vulgaris* show anti-inflammatory activity by reducing cytokine levels and providing a potential way for asthma treatment (Barboríková et al., 2019). The anti-inflammatory activity of EPSs secreted by *Cyanobacterium aponium*, a dominating organism in the Blue Lagoon, is presented in detail in human dendritic cells (Gudmundsdottir et al., 2019a). The researchers hypothesized that the EPS released by this cyanobacterium might contribute to the beneficial effects of regular bathing in the Blue Lagoon. Later, they studied the immunoregulatory activity of the EPS released by *C. aponinum*, and they reported a decrease in T cell and keratinocyte secretion, indicating that EPS may be involved in the positive effects that psoriasis patients get from bathing in the Blue Lagoon (Gudmundsdottir et al., 2019b).

The antitumor activity of EPS was also investigated in various studies, and some researchers have shown that EPS can be used as a feedstock for potential treatments for cancer cases. The inhibitory effect of the EPS obtained from *Spirulina platensis* on tumor invasion was demonstrated in previous studies (Dewi et al., 2018). The EPSs were isolated from *C. vulgaris* and *Chlorella zofingiensis*, and later, the antitumor activity was analyzed in the colon cancer cell line (Naveed et al., 2019). The results showed that the cell viability of cancer cells decreased when the EPSs were applied to the cell lines. The anticancer activity of EPSs on colon cancer was tested for *C. pyrenoidosa, Scenedesmus* sp., and *Chlorococcum* sp. species. Results showed a decreasing effect on colon cancer cell line cell viability, so the anticancer activity potential of EPSs in further studies gained attention (Zhang, Liu, Ren, et al., 2019). A research study demonstrated that the *Coelastrella* sp. BGV EPSs triggered apoptosis in HeLa cervical cancer cells (Toshkova-Yotova et al., 2024). The anticancer activity of EPSs isolated from *Thraustochytriidae* sp. has been confirmed in ovarian, breast, and colon cancer cell lines. Subsequent examination of the mechanisms underlying this anticancer activity revealed alterations in the expression of cell cycle-related proteins (Park et al., 2017). A study evaluated the anticancer effect of EPS from *P. aerugineum* (Ivanova et al., 2024). The results indicated that EPS exhibited cell type-specific antiproliferative activity on breast cancer cells while demonstrating no significant cytotoxic effect on normal and cervical cancer cells.

EPSs are secreted from cells to provide a defense against changing environmental conditions. It was shown that some diatoms living in sea ice protect themselves against freezing via EPS secretion (Aslam et al., 2012). The positive correlation between cell survival and EPS availability showed the cryoprotective effect of EPS released by the sea-ice diatoms *Synedropsis* sp., *Fragilariopsis curta*, and *F. cylindrus*. The cold adaptation activity of EPSs and antifreeze proteins against cold temperatures was reported for an Antarctic chlorophyte, *Coccomyxa subellipsoidea* (Blanc et al., 2012). EPS's cryoprotective role is also called the osmotic balance inside and outside the cell. It was shown that the EPS released by the mesophilic cyanobacterium *Nostoc commune* makes the colony highly desiccation-tolerant (Tamaru et al., 2005). When the *N. commune* was covered with EPS, the cells became heat-tolerant, and their resistance to freeze-thaw cycles improved. Thus, EPS is produced as an effective solid barrier for balancing cellular response against environmental fluctuations.

In recent years, EPSs have been evaluated as natural bio-flocculants that can reduce the harvesting cost, which is a limiting factor for using microalgae on an industrial scale (Li et al., 2021; Moreira et al., 2022). The flocculant activity of EPS provides a more effortless and cheaper harvesting procedure. In another study, researchers described the optimum physical conditions for bioflocculant EPS production by *Chlamydomonas reinhardtii* as a temperature range between 15 and 25°C, pH 6–10, and low-light illumination of 40–60 μmol photons m^{-2} s^{-1} (Zhu et al., 2012). EPSs produced by *C. reinhardtii* have been confirmed for their potential application as bio-flocculants. The efficacy of EPS from *N. oleoabundans* as a bio-flocculant has been demonstrated at a concentration of 25 mg/L in microalgae harvesting (Paredes-Molina et al., 2025).

The interest in biofertilizers and biostimulants has been increasing in recent years due to the harmful effects of chemical biofertilizers on soil and the environment. Microalgal exopolysaccharides can be used as biostimulants in agriculture due to their protective activity against biotic and abiotic stresses on crops (Braun & Colla, 2023). Salt stress is one of the principal abiotic stresses affecting plants' physiology and biochemical mechanisms. In a research study, the biostimulant activity of *Dunaliella salina* EPS was studied in tomatoes for salt stress tolerance (El Arroussi et al., 2018). The results showed that applying EPS under salt stress improved shoot and root length compared to the group in which EPSs were not used under salt stress. In another study, EPSs extracted from *Spirulina* sp. were evaluated as a plant biostimulant (Chaudhuri & Balasubramanian, 2023). *Spirulina* sp. EPS was applied to two plants (*Ocimum basilicum* and *Spinacia oleracea*) at the 0.25–1.5 mg/ml concentration ranges. The results showed that maximum growth for *Ocimum basilicum* was observed at 1.5 mg/ml, while for *S. oleracea*, 0.5 mg/ml EPS concentration was verified as optimal. The biostimulant activity of EPSs extracted from three different microorganisms (*A. platensis, D. salina*, and *Porphyridium* sp.) was well documented on tomato plants (Rachidi et al., 2020). Considering that EPSs can be used as feedstock for hydrogel production, it is assumed that these biomolecules are valuable compounds that can be used in agriculture as biostimulants and in the production of controlled biological release systems.

EPSs play an instrumental role in the treatment of wastewater, given their capacity to enhance biological treatment processes and facilitate the removal of pollutants. EPSs of *C. vulgaris* have been demonstrated to effectively remove microplastics by the flocculation method through electrostatic interaction with microplastics, thanks to the functional groups present in their structure (Joelyna et al., 2024). The removal of microplastic particles by the hetero-aggregation method was found to be significantly enhanced by the presence of EPSs derived from three distinct species of microalgae: *Tetraselmis chuii*, *C. vulgaris*, and *D. salina* (Hadiyanto et al., 2025). A recent study has demonstrated that EPS secreted by the photosynthetic organism *C. reinhardtii* plays a significant role in removing heavy metals from the environment (Li et al., 2025). In the context of bioremediation, the role of EPS is of particular significance in the process of metal pollution cleanup. The fundamental mechanism of this process involves the electrostatic interaction between EPS and metal cations, facilitated by the hydrophilic functional groups present in EPS, including carboxyl, hydroxyl, and amine groups.

5.4 FACTORS AFFECTING EPS PRODUCTIVITY AND COMPOSITION

EPSs are secreted as a mechanism of protection and adaptation, especially in adverse environmental conditions. Thereby, EPSs play a role as a defense mechanism against unsuitable conditions such as ultraviolet radiation (UV), oxidation, temperature, and desiccation. The role of EPSs in the cell attachment process was demonstrated in cyanobacteria *Gloeothece* sp., *Calothrix* sp., and *Anabaena* sp., showing that EPS protects nitrogenase from the deleterious effect of oxygen (Kallas et al., 1983). EPSs, formed as a gel structure, protect the cell against desiccation stress by controlling water uptake or loss (Kumar et al., 2018). The yield and content of EPSs vary in species-specific environmental conditions. Previous studies have shown that abiotic factors such as nutrient starvation, temperature, light intensity, aeration, heavy metals, pH, and biotic stress conditions such as bacterial and fungal contamination are practical on EPS productivity (Costa, Lucas et al., 2021).

5.4.1 Light

While light is essential for the growth of photosynthetic microorganisms, EPS production also varies according to the type and intensity of light. EPS production from *Porphyridium cruentum* was investigated at different light intensities, and maximum EPS productivity was obtained at 80 μm^{-2} s^{-1} light intensity (Sun et al., 2008). On the contrary, increasing light intensity caused a decrease in the EPS production for *C. vulgaris* (Clément-Larosière et al., 2014). The light period also affects the EPS synthesis, so different light/dark cycles have been evaluated for *P. cruentum* to determine the most suitable light period for maximizing the EPS productivity, and the 18:6 h light/dark cycle was determined as the most suitable light period (Babiak & Krzemińska, 2021). In addition to light density and period, the wavelength also influences EPS productivity. Blue light (400–500 nm) and red light (600–700 nm) were shown to increase EPS production in *P. cruentum* (You & Barnett, 2004). Microalgal EPS production seems to be associated with photosynthesis, which may be another factor differentiating it from bacterial and fungal EPS.

5.4.2 Temperature

Optimum growth temperature values in microalgae may show species-specific differences, and EPS productivity may show variation in temperature ranges (De Philippis et al., 1991; Pereira et al., 2009). The optimum temperature range to obtain the highest EPS yield is 25–30°C for *B. braunii* (Lupi et al., 1991). The EPS yield increased with rising temperature in cyanobacterium *Anabaena* sp. (Moreno et al., 1998). However, in another study, the cyanobacterium *Spirulina* sp. reported a decrease in EPS production with increasing temperature (Nicolaus et al., 1999). Therefore, the temperature effect may be assessed as a species-specific parameter for EPS production. Elevated temperatures can induce oxidative stress in microalgae (Xing et al., 2022). In response, microalgae may augment the synthesis of EPS, which binds water and protects cellular components while concurrently exhibiting antioxidant properties. The ability of EPS to assist in the maintenance of osmotic balance suggests the potential for an increase in its production as a strategic adaptation to temperature fluctuations. This phenomenon has been particularly observed in species adapted to extreme environments, such as Antarctic microalgae, where increased EPS production has been documented at low temperatures (Peng et al., 2021).

5.4.3 pH

Studies address the fact that alkali pH values are favorable for microalgal EPS production. A recent study showed that the maximum EPS synthesis occurs at the stationary phase and the pH value of 8–9 due to the aging of the culture (Babiak & Krzemińska, 2021). On the other hand, *C. reinhardtii* was studied to optimize EPS production conditions, and the highest EPS production was observed at pH 7 (Bafana, 2013). However, in *A. platensis*, the highest EPS production was obtained at a higher pH, accompanied by growth inhibition (Li et al., 2022). In summary, photosynthesis and carbon metabolism function at optimal pH levels with greater efficiency, and EPS production may be elevated under these conditions.

5.4.4 Salinity

Generally, high salinity levels increase EPS production, and the salt tolerance observed in some extremophilic microorganisms is related to increased carbohydrate synthesis (Rossi & De Philippis, 2016). Cyanobacteria such as *Spirulina* sp. and *Nostoc* sp., as well as eukaryotic microalgae such as *D. salina*, showed an increase in EPS production in proportion to the increase in salinity (Mishra & Jha, 2009; Nicolaus et al., 1999; Yoshimura et al., 2012). Salt stress has been demonstrated to exert an influence on the production of EPS by microalgae, which can be attributed to three principal mechanisms: osmotic stress, metabolic adaptation, and cellular protection mechanisms (Ferreira et al., 2021).

5.4.5 Aeration

Aeration of culture is an essential parameter for growth due to its effect on photosynthetic mechanisms. Aeration influences photosynthetic activity by affecting the availability of light and nutrient uptake. Aeration is assessed as a factor that influences EPS production (Pereira et al., 2009). A research study on the heterocystous cyanobacteria *Anabaena torulose* and *Phormidium* sp. demonstrated the aeration effect on EPS production (Nicolaus et al., 1999). Results showed that biomass production decreased by 95% when the aeration stopped, while *Phormidium* sp. recorded a 90 % decrease in EPS production per cell.

On the other hand, *A. torulosa* exhibited a 50% decrease in growth accompanied by a 47 % increase in EPS production when the aeration stopped. A two-stage EPS production strategy may include the growth of microalgae or cyanobacteria and settling the dense cultures without aeration to induce EPS production for a certain period. In another study conducted with *P. cruentum, the effect of the* aeration rate on growth and EPS production was investigated (Iqbal & Zafar, 1993). Different aeration rates (125–1000 $mg.ml^{-1}.L^{-1}$) were evaluated, and results showed that the maximum growth and EPS productivity (1.9 g/L) were obtained with a 500 $mg.ml^{-1}.L^{-1}$ aeration rate.

5.4.6 Nutrients

Mineral nutrients are generally assimilated in inorganic ion forms and cycled from one organism to another in nature. Hydrogen, oxygen, carbon, nitrogen, phosphorus, and sulfur are essential elements in the structure of microalgae and are vital for their growth. Therefore, remarkable changes in metabolism and morphology, such as EPS production, are observed when microalgae are limited or deprived of essential elements (Sivaramakrishnan et al., 2020). In a research study, the ability of three different *Porphyridium* species (P-271, P-293, and P-519) to produce EPS in response to the chemical composition of the nutrient medium was compared (Boychenko et al., 2024). The findings demonstrated that certain species exhibited enhanced EPS production under nutrient deficiencies.

Mixotrophic and heterotrophic growth conditions change EPS productivity, and carbon in the medium increases intracellular and extracellular polysaccharide production. Different trophic conditions were experienced in a study to maximize EPS productivity (Trabelsi et al., 2009). According to the results, the mixotrophic growth conditions showed the highest EPS yield in *A. platensis*. Positive effects of mixotrophic growth on EPS production were demonstrated in *C. vulgaris* (Naveed et al., 2019). The carbon/nitrogen ratio influences EPS synthesis, and an increase in this ratio will enhance the production of EPS. The C/N ratio effect on EPS synthesis was studied for *Porphyridium purpureum*, and results indicated that the C/N ratio may be directly proportional to EPS synthesis. The mixotrophic cultivation of *P. purpureum* gave the highest EPS yield (Li, Ji et al., 2020; Rossi & De Philippis, 2016). Adding bicarbonate (190 mM) to *C. vulgaris* culture resulted in a fivefold increase in EPS production (Tripathi et al., 2025). Bicarbonate has been demonstrated to indirectly promote the production of EPS by providing a carbon source. Furthermore, bicarbonate functions as a stress factor at elevated concentrations, prompting cells to synthesize increased amounts of protective EPS. Additionally, it is postulated that bicarbonate-induced pH alterations may positively influence the process of EPS biosynthesis.

Nitrogen (N) is the only macronutrient available for microalgae in oxidized and reduced forms. Among mineral nutrients, N is the most important and, in most cases, a limiting factor for the growth and development of photosynthetic organisms. Microalgae need N in almost all developmental stages. Nitrogen forms approximately 2% of microalgae dry weight. It plays significant roles in all microalgae metabolic functions from metabolism to source sharing, as it is a principal constituent of the nucleoside and amino acids of proteins, nucleic acids, and several cofactors (Yaakob et al., 2021). Several reports showed that EPS production is influenced by the N availability (Babiak & Krzemińska, 2021). N limitation was reported to induce EPS production in oil-rich microalga *B. braunii* (Díaz Bayona & Garcés, 2014) and *Parachlorella* sp. (Sasaki et al., 2020). A recently isolated microalgae species, *Chlamydomonas* sp. Xylok, has demonstrated both elevated levels of EPS production and substantial biomass production (Kim et al., 2024). In this study, researchers have demonstrated that nitrogen deprivation conditions significantly augment EPS production by 49%.

Along with concentration, the nitrogen source is an important parameter, as the available nitrogen sources show variation in nature, such as organic or inorganic oxidized or reduced forms. In a study investigating the impact of differences in nitrogen sources on EPS production in *B. braunii*, potassium nitrate was reported as superior to ammonium carbonate and urea (Lupi et al., 1994). In the methodology of the study, 2 mM of nitrogen source was applied, and the effect of different nitrogen sources was evaluated after 10- and 14-day cultivation. According to the results, all nitrogen sources showed a similar impact on EPS productivity after 10 days of cultivation. On the other hand, potassium nitrate showed higher EPS productivity than urea and ammonium carbonate after 14 days of cultivation. Researchers showed that urea and ammonium induce acidification of the intracellular space, causing decreased cellular activity stemming from deformation in cellular homeostasis and EPS production; nitrate was inferred as an appropriate N source in *B. braunii* for elevated EPS production.

Studies show that phosphate (P) limitation or deprivation induces EPS production in many organisms (Rossi & De Philippis, 2015). Valuable research conducted with *Cylindrotheca fusiformis* investigated the effects of P limitation on EPS yield and composition (Magaletti et al., 2004). Results showed that the P limitation increases the EPS yield and affects the distribution of its content by increasing the galactose percentage. Cyanobacteria may give a differential response to phosphate limitation. *Spirulina subsalsa* was reported to increase EPS production due to phosphate limitation or deprivation (Chakraborty et al., 2015). On the other hand, P deficiency did not considerably impact EPS production in the cyanobacterium *A. torulosa* (Nicolaus et al., 1999).

Sulfur is one of the nutrients that influence the growth and development of microalgae, and sulfur deficiency causes dramatic fluctuations in the metabolism and morphology of cells (Pereira et al., 2009; Rossi & De Philippis, 2015). Sulfur availability was noted as an essential parameter for EPS production. The effect of sulfur limitation or depletion on EPS productivity was studied for *Anabaena* sp. (Kharwar et al., 2021). Researchers applied different sulfur concentrations to *Anabaena* sp. for 8 days. According to the results, the highest EPS yield was obtained in sulfur-limited conditions. In general, an increase in EPS production as a response to a significant change in element concentration in the growth medium may also address the role of EPS as a unique stress signal in EPS-producing microalgae. Treatment of *Auxenochlorella protothecoides* with sodium bisulfite (1.25 mmol/L) resulted in a 122% increase in EPS yield (Zhou et al., 2025). The researchers hypothesize that the observed effect is a result of sodium bisulfite reducing protein synthesis, thereby diverting carbon to EPS synthesis

5.4.7 Growth Phase and Cultivation Methods

Previous studies showed that EPS yield increases at the stationary growth phase, and researchers concluded that nutrient limitation or starvation may cause this effect (Badireddy et al., 2010; Penna et al., 1999). Twenty different *Chlamydomonas* strains were evaluated for their EPS production in different growth stages, and the results indicated that the highest EPS productivity was obtained at the stationary phase (Katona et al., 2018). In another study, aging stress effectively increased EPS production by 1.8-fold in cyanobacterium *S. subsalsa* (Chakraborty et al., 2015).

The cultivation system influences EPS yield because when microalgae are starved of nutrients, EPS synthesis increases, but biomass productivity and cell growth slow down. A two-stage cultivation system, which involves growing the culture to the exponential phase in the first stage and applying stress parameters in the second stage, is advantageous for large-scale EPS production (Medina-Cabrera et al., 2020).

The antibacterial and antifungal activities of EPSs are seen as defense mechanisms developed against various microorganisms, including bacteria and fungi, which are considered biotic stress factors (Bell, 1983). Consequently, an increase in EPS production under biotic stress is anticipated. Cocultivating bacteria or fungi with microalgae may enhance EPS production and increase diversity in its composition (Angelis et al., 2012).

5.4.8 Heavy Metals

In a research study, Cd(II), Pb(II), and Cr(VI) stresses were applied to *Synechocystis* sp. for 48 hours, and their effects on EPS production were investigated (Cui et al., 2021). The results demonstrated a 14.2% increase in EPS production for Cd(II), a 29.1% increase for Pb(II), and a 30.9% increase for Cr(VI). The exposure of *B. braunii* to elevated concentrations of cobalt resulted in alterations to its EPS's composition and antimicrobial activity (Cheng et al., 2022). A study was conducted to examine the dynamic responses of *C. reinhardtii* to different heavy metal exposures, including cadmium, copper, lead, and zinc (Li et al., 2025). The study's results demonstrated how metal stress alters the composition (monosaccharides and proteins) and structure of EPS and increases its amount.

5.4.9 Microplastics

In the presence of polypropylene (PP) and polyethylene terephthalate (PET) microplastics, *C. vulgaris* exhibited a 1.14- and 1.41-fold increase in EPS production, respectively, in comparison to the control group (Joelyna et al., 2024). Another study revealed that the species *T. chuii*, *C. vulgaris*, and *D. salina* exhibited an increase in EPS production yield when polypropylene microplastic was applied (Hadiyanto et al., 2025).

5.5 EXTRACTION AND CHARACTERIZATION OF EPSS

5.5.1 Extraction

The culture conditions and species selection are essential parameters to obtain maximum EPS productivity, yet the extraction method is significant in maximizing the yield and composition of the extracted EPSs. The extraction procedures differ for released and cell-bound polysaccharides (Delattre et al., 2016; Rossi & De Philippis, 2016). While the released polysaccharides can be obtained via centrifugation or filtration, some physical or chemical treatments must be performed for cell-bound polysaccharides. Released polysaccharides can generally be extracted by the alcoholic precipitation method; the ultrafiltration method can also be used as an alternative (Kumar et al., 2018). A study compared the efficacy of various methods for extracting EPS from *P. cruentum*. The study evaluated dialysis, solvent precipitation, and membrane separation techniques. The study results indicated that dialfiltration (DF) was the most effective method for EPS extraction (Patel et al., 2013). In a subsequent study, the EPS yield of alcoholic precipitation and ultrafiltration techniques was compared, and the results indicated that ultrafiltration yielded a 1.2-fold higher EPS yield (Ziadi et al., 2018).

Furthermore, EPSs are separated using an ultrafiltration method to obtain higher purity EPSs by removing low-molecular-weight impurities. However, the use of this method is constrained by the requirement of expensive membrane systems, which limits its scalability. In contrast, alcoholic precipitation is preferred due to its simplicity and economy. However, this method provides lower purity due to the precipitation of components such as proteins and salts. In the alcoholic precipitation method, the cost can be reduced by recovering the alcohol (Cruz et al., 2021).

In physical and chemical methods used to extract cell-bound polysaccharides, cell rupture during the removal of EPS must be prevented to avoid contamination with intracellular polysaccharides. Formaldehyde, EDTA, and NaOH are used for chemical treatment, while sonication, physical agitation, and heating are used for physical treatments (Comte et al., 2006; Katona et al., 2018; Takahashi et al., 2009).

5.5.2 Characterization Techniques for Chemical Composition and Structure of EPSs

The contents and characteristics of EPS can vary depending on the species, cultivation conditions, and extraction methods. In the initial stage of the characterization process, the quantity of sulfate and uronic acid groups, present in addition to the essential components like saccharides and proteins, is measured in the composition of EPS (Delattre et al., 2016). The presence of nucleic acids in very low proportions in EPS structure can be tested for purity by measuring them using UV–visible spectrophotometry (Costa, Lucas et al., 2021). The monosaccharide content of EPS varies depending on the producing species and can also impact their biological activities. For chromatographical analysis of EPS, commonly accepted gas chromatography or high-pressure liquid chromatography are the most reliable methods, and coupling this analysis with mass spectrometry increases the efficacy of the results (Decamp et al., 2021). Fourier transform infrared (FTIR) spectroscopy is used to determine the biological molecule fingerprint of EPSs, and some characteristic peaks that are seen commonly in all polysaccharides are obtained (Castro et al., 2014; Rossi & De Philippis, 2016). Another difference observed among EPSs is their molecular weights, which can be characterized using gel permeation chromatography (Cristofoli et al., 2023). Given the established correlation between the variation in EPS viscosity and its biological activity, investigating EPS viscoelastic properties through rheometry and viscosity measurements is imperative (Gargouch et al., 2021). X-ray diffraction (XRD) can determine the crystalline or amorphous structure of EPSs, while thermogravimetric analysis and differential scanning calorimetry can evaluate their thermal stability. Furthermore, the binding type or the degree of branching of EPSs may also be subject to variation. A variety of analytical methods, including spectroscopic (nuclear magnetic resonance (NMR), circular dichroism, FTIR, Raman), XRD, microscopic (atomic force microscopy, scanning electron microscopy), and computational (molecular dynamics simulations), can be employed to analyze the secondary, tertiary, and helical structure of EPS (Liu et al., 2017). The detection of fibrous, helical, or straight-chain structures is facilitated using atomic force microscopy and scanning electron microscopy, which enables the examination of three-dimensional surface morphology. At the same time, the presence of acetyl and methyl groups that may not be determined via colorimetric analysis can be analyzed by ^{1}H NMR (Costa, Lucas et al., 2021). Thus, after the characterization of EPS, which can vary according to species and environmental conditions, suitable routes for the biomaterial production of EPS can be defined.

5.6 APPLICATION AND USE OF EPSS IN BIOMATERIAL PRODUCTION

Microalgal EPSs are suitable for use in various areas due to their diverse biological activities, which include antioxidant, antiviral, antimicrobial, and anti-inflammatory properties. As represented in Figure 5.2, microalgal EPSs are also considered potential natural and biodegradable products in the cosmetic, food, pharmaceutical, and biomedical industries. In recent years, EPSs have been evaluated for their potential to create biocompatible, biodegradable, and nontoxic biomaterials, such as hydrogels, films, and fibers, particularly for use in the food and medical fields.

5.6.1 Cosmetic

The interest in biocompatible cosmetic products is rapidly increasing as consumers prefer new, natural, and safe products over synthetic chemicals. In this sense, EPSs attract researchers in cosmetic applications due to their nontoxic and biodegradable

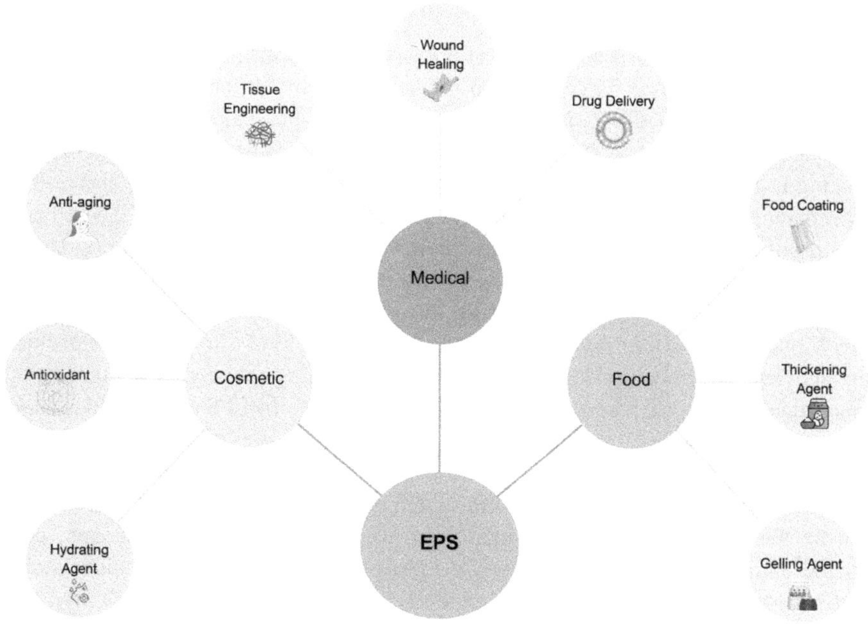

FIGURE 5.2 The application areas of EPSs.

nature and their antibacterial and antioxidant activities. Antioxidant compounds help prevent free radicals caused by UV radiation, so cream or lotion formulations with antioxidant activity are of interest in cosmetics. In a study, a lotion formulation was developed with EPS extracted from *Spirulina* sp., and no side effects were found to prevent its use (Winahyu & Primadiamanti, 2021). The antiaging potential of EPSs increases the attention to cosmetic applications. For instance, the antiwrinkle activity was demonstrated for EPSs extracted from *Trimella fuciformis* (Pintathong et al., 2021).

Currently, several patented products of EPSs in cosmetic applications are available. EPS from *P. cruentum* is commercially used as a hydrating agent in the cosmetic field (Costa, Lucas et al., 2021). This product, known as Hydrintense™, has applications as moisturizing care, long-lasting hydration cream, and antiaging serum. EPS Revive is another cosmetic product formulated with sulfated polysaccharides of *P. cruentum*. It attracts attention due to its intense antiaging and anti-inflammatory activities. Silidine® is a cosmetic product formulated with oligo-elements and EPS of *P. purpureum* and used for redness and heavy leg syndrome. Epsiline® is another commercial product that was formulated with EPS of *P. purpureum*. These two cosmetic products sell for around €100/kg (Pierre et al., 2019). Astaderm™ is one of the commercial products of microalgal EPSs and is obtained from *P. cruentum*. Another commercial product is Acqualift, obtained from *Porphyridium*, and it can be used in cosmetics due to its antiaging activity. Porphyderm® is another patented product obtained from the EPS of *Porphyridium* species and is used in the cosmetic field (Morais et al., 2022).

5.6.2 Food

EPSs have been used as food additives for years, and industrial applications are available. EPSs can be used as gelling and thickening agents in the food industry, substituting synthetic food additives with natural and non-toxic alternatives (Nechita & Roman, 2020; Nešić et al., 2019). Bacterial EPSs have broad application potential for coating food in the food industry due to their antibacterial, antioxidant, and protective characteristics (Zannini et al., 2016). On the other hand, microalgal EPSs emerge as a prominent feedstock for creating edible films due to their film-forming ability and bioactivities, such as antibacterial and antioxidant properties. Released EPSs of *Graesiella* sp. were investigated for film formation that can be used in food packaging (Gongi et al., 2022). The results showed that EPS films showed better packaging activity than polyvinylchloride used in the packaging process. In another study, microalgal EPSs were incorporated with *Gracilaria gracilis* EPS to create film material for edible coatings, and their suitability for shrimp preservation was assessed (Balti et al., 2020). They reported that the EPS-based coating treatments improved the quality and extended the shelf life of refrigerated shrimp. EPS as a raw material for the preparation of biopolymeric film was reported in a research study conducted with cyanobacterium *N. commune* (Angel et al., 2017). In a valuable study, EPSs of *Nostoc* sp. and *P. purpureum* were evaluated for potential bioactive film formation, and results showed that transparent and flexible biofilms can be obtained

(Morales-Jiménez et al., 2020). Therefore, microalgal EPSs show potential for use in food packaging applications as an alternative to commercial packaging materials.

5.6.3 Pharmaceutical and Biomedicine

EPSs demonstrate potential for use in the pharmacology industry due to their antiviral, anticancer, and anti-inflammatory activities. The novel alginate polymer produced from the cyanobacterium *Synechocystis algini* was evaluated as a new alternative antibacterial medication against resistant bacteria due to its superior antibacterial activity (Saad et al., 2023). In another study, the sulfated polysaccharide of red microalgae *Porphyridium* sp. was used to form transparent and stable hydrogels by mixing fluorenyl methoxycarbonyl diphenylalanine peptide (Halperin-Sternfeld et al., 2022). In addition to their ability to form hydrogels, sulfated polysaccharides' anti-inflammatory and antioxidant properties increase their potential usage in the biomedical field.

In recent years, the pharmaceutical industry's shift toward natural polymers as excipients for drug delivery has increased interest in EPS (Asgher et al., 2021). Algal polysaccharides gain attention with nanoparticle formation that can be used as nanocarriers in a drug delivery system (Zhang et al., 2022). Nanotechnology is a rapidly developing field that promotes the production of materials at the nanometer scale. The potential of EPSs to be used in this field adds significant importance to EPSs. EPSs can be utilized for nanoencapsulation owing to their biocompatibility, biodegradability, and nontoxic characteristics. Moreover, EPSs can be used in the production and stabilization of nanoparticles. EPSs obtained from *B. braunii* and *C. pyrenoidosa* were used to produce the silver nanoparticles (AgNPs), and the antibacterial activity of nanoparticles was evaluated. Results showed that EPS-based nanoparticles show antibacterial activity against *E. coli* and *S. aureus* (Gallón et al., 2019). The potential of microalgal EPSs as a stabilizing agent in nanomaterial formation has been reported in several studies. The literature study showing that the stability of gold nanoparticles coated with EPSs obtained from *Chlorella* sp. increased can be considered an example of this situation (Jakhu et al., 2021). Nanomaterials produced from EPSs can be used not only in the pharmacology field but also in cleaning polluted water. The microsphere formed by polymerizing EPS obtained from *Chlorella* sp. with alginate can efficiently remove sulfamethoxazole from water (Roy et al., 2024).

Besides the pharmacological use of EPSs, they can be considered potential candidates for forming hydrogel or fiber materials used in biomedical and tissue engineering applications. In recent years, EPSs have gained popularity for their ability to create hydrogel structures with potential applications in various fields, such as wound healing, tissue engineering, and drug delivery systems. The EPS secreted by *Navicula* sp. was used for hydrogel production, and gelation was achieved with $FeCl_3$ (Fimbres-Olivarría et al., 2016). The sulfated polysaccharides were extracted from the biomass of *Navicula* sp. and characterized. The sulfated polysaccharides (1 % w/v) were mixed with $FeCl_3$ (0.4 % w/v) for hydrogel formation by an ionic crosslinking mechanism. The same gelation reaction was used to produce hydrogels from EPSs of the *Nostoc* strain and was found to be effective when the wound-healing potential of the hydrogel was evaluated (Alvarez et al., 2021). Sulfated polysaccharides produced by the unicellular cyanobacterium *Crocosphaera chwakensis*, called Cyanoflan, have wound-healing activity (Costa, Costa et al., 2021). The released EPS obtained from *Trichormus variabilis* was evaluated to produce a photo-polymerizable hydrogel, and the potential application of this EPS for producing enzyme or cell carrier systems was demonstrated (Bellini et al., 2018). In a remarkable study, EPSs obtained from *Porphyridium* were mixed with chitosan at varying ratios to form a microgel structure (Ruiz-Dávila et al., 2023). The suitability of the obtained microgel structure for use as a drug or vaccine delivery system has been confirmed by characterization studies. EPS released into the media by *Arthrospira* forms stable gel beads with the addition of Al^{3+} and Fe^{3+} (Cuellar-Bermudez et al., 2023). In a valuable study, a liquid-crystal gel structure was formed with EPSs (known as sacran) obtained from *Aphanothece sacrum* cyanobacterium (Okeyoshi et al., 2021). Thus, with its high-water absorption capacity, the potential for using sacran in forming biomaterials that can be used in tissue engineering and biomedical fields has been reported. In a valuable study, a microcarrier system based on marine EPSs was developed to carry signaling proteins that induce cartilage tissue regeneration (Zykwinska et al., 2019). In another research study, EPSs produced by *Cyanothece* sp. cyanobacteria were evaluated for the controlled delivery of functional proteins, and the results showed that EPSs may be a suitable candidate for use as carrier systems (Leite et al., 2017).

Many studies showing that microalgal EPSs can be used in biomedical and tissue engineering with their rich content and valuable biological activities are presented in this chapter. Moreover, the commercial use of EPSs from macroalgae confirms these theories. As evaluated in the previous sections, it is likely to develop products that can be used commercially with increased EPS yield if the production of microalgal EPSs is optimized.

5.7 MARKETING VALUE OF EPS

Lactic acid bacteria are the most popular microorganisms employed for industrial EPS production. Some bacterial EPSs, such as dextran, xanthan, and hyaluronan, are sold in the food and cosmetics market (Moscovici, 2015). On the other hand, cyanobacteria and microalgae have gained interest as new potential EPS sources under the green production concept (Pierre et al., 2019). The characteristic properties of microalgal EPSs offer some advantages over bacterial EPSs. Microalgal EPSs have higher molecular weight and can contain more than ten different monosaccharides, while bacterial EPSs have low molecular weight and can contain less than four types of monosaccharides (Naveed et al., 2019).

Even if cyanobacteria and algae are defined as potential sources, most of the industrial EPS production is mainly based on bacterial fermentation. The main reason for this is the abundance of production despite its limited variety in content. To overcome this limitation, selecting suitable microalgae for EPS production and optimizing culture conditions and downstream processes are the main parameters to focus on. Considerably high cost and sustainability are the main obstacles for microalgal EPS production. However, depending on the type of microalgae and the desired output production, specific growth techniques reduce the production cost for growing microalgae in bioreactors and open ponds, popularizing microalgal EPS production (Naveed et al., 2019). In recent years, with the development of algal growth techniques, the interest in EPSs of microalgae has increased linearly. Moreover, using various low-cost nutrient sources, such as wastewater, also reduces the cost of microalgae production (Morais et al., 2022).

Microalgae marketing comprises essential ingredients such as carotenoids, fatty acid derivatives, natural pigments, and polysaccharides. The increasing attention to microalgal polysaccharides stems from the higher diversity of their compositions and biological functions. Microalgal polysaccharides can be marketed as moisturizing agents, immune stimulants, and hydrating agents. The selling price of microalgal polysaccharides ranges from €2.5k to €200k/ton (Pierre et al., 2019). Much of the selling price belongs to immune stimulant purposes as €200k/ton (Patel et al., 2023). It is predicted that green EPS production will reach approximately US$ 1.5 billion in 2027, while the market value of xanthan gum was nearly US$ 1 billion in 2019 (Nadzir et al., 2021). EPS production from microalgae and cyanobacteria will increase in popularity in the coming years within the scope of the green science concept.

5.8 CONCLUSIONS

Microalgae are considered a suitable production platform for several value-added metabolites, and microalgal EPSs are one of these metabolites that have attracted attention in recent years. However, production cost and lack of knowledge about the synthesis mechanisms are the most common limitations for producing and using microalgal EPSs in industrial areas. These limitations have been mitigated recently using wastewater as a nutrient medium and improved culture systems. With their ability to produce high amounts of EPS, bacteria have been intensively evaluated in this context; however, microalgae emerged as green production platforms for EPS production. Their potential comes from the diversity in their EPS composition, particularly monosaccharide content. Monosaccharide content is one of the most critical factors affecting the functions of EPSs, and it is essential to clarify the relationship between content and function so that EPSs can be used in appropriate application fields.

In addition to the industrial use of EPSs in the food and cosmetic industries, these EPSs create biomaterials that can be used in the biomedical and food sectors, providing an innovative perspective. Using EPSs in nanotechnology will open up new insights into the utilization of EPSs produced by different bio sources. EPSs are evaluated as alternative polymer sources against synthetic-based polymers with their biodegradable and nontoxic natures. Moreover, the antibacterial and antioxidant nature of EPSs ensures the additive needs for these activities in the biomaterial. Their rich composition makes microalgal EPSs more profitable and superior to other bio sources. Once the production steps are optimized, microalgal EPS can be utilized at a large scale to produce specific biomaterials such as biofilms, nanofibers, and hydrogels.

AUTHORS CONTRIBUTIONS

T.Ç., K.K., and Z.E.Ç. designed and coordinated the study. T.Ç., K.K., and Z.E.Ç. wrote the manuscript. All authors reviewed and approved the final manuscript.

FUNDING

This study was supported by TÜBİTAK – The Scientific and Technological Research Council of Turkiye under Grant Number 119Y399. The authors thank TUBITAK for their support.

AVAILABILITY OF DATA AND MATERIALS

All data generated or analyzed during this study are included in this published article.

ETHICAL APPROVAL AND CONSENT TO PARTICIPATE

We all declare that manuscript reporting studies do not involve any human participants, human data, or human tissue. So, it is not applicable. The study protocol must comply with relevant institutional, national, and international guidelines and legislation.

CONSENT TO PUBLISH

Not applicable.

COMPETING INTERESTS

The authors declare no competing interests.

REFERENCES

Alvarez, X., Alves, A., Ribeiro, M. P., Lazzari, M., Coutinho, P., & Otero, A. (2021). Biochemical characterization of Nostoc sp. exopolysaccharides and evaluation of potential use in wound healing. *Carbohydrate Polymers*, *254*, 117303.

Angel, S., Torres, F. G., & López, D. (2017). Preparation and characterization of polysaccharide films from the cyanobacteria Nostoc commune. *Polymers from Renewable Resources*, *8*(4), 133–150.

Angelis, S., Novak, A. C., Sydney, E. B., Soccol, V. T., Carvalho, J. C., Pandey, A., Noseda, M. D., Tholozan, J. L., Lorquin, J., & Soccol, C. R. (2012). Co-culture of microalgae, cyanobacteria, and macromycetes for exopolysaccharides production: Process preliminary optimization and partial characterization. *Applied Biochemistry and Biotechnology*, *167*(5), 1092–1106.

Arad, S. M., & Levy-Ontman, O. (2010). Red microalgal cell-wall polysaccharides: Biotechnological aspects. *Current Opinion in Biotechnology*, *21*(3), 358–364.

Asgher, M., Qamar, S. A., & Iqbal, H. M. (2021). Microbial exopolysaccharide-based nanocarriers with unique multi-functionalities for biomedical sectors. *Biologia*, *76*(2), 673–685.

Aslam, S. N., Cresswell-Maynard, T., Thomas, D. N., & Underwood, G. J. (2012). Production and characterization of the intra- and extracellular carbohydrates and polymeric substances (Eps) of three sea-ice diatom species, and evidence for a cryoprotective role for Eps. *Journal of Phycology*, *48*(6), 1494–1509.

Babiak, W., & Krzemińska, I. (2021). Extracellular polymeric substances (EPS) as microalgal bioproducts: A review of factors affecting EPS synthesis and application in flocculation processes. *Energies*, *14*(13), 4007).

Badireddy, A. R., Chellam, S., Gassman, P. L., Engelhard, M. H., Lea, A. S., & Rosso, K. M. (2010). Role of extracellular polymeric substances in bioflocculation of activated sludge microorganisms under glucose-controlled conditions. *Water Research*, *44*(15), 4505–4516.

Bafana, A. (2013). Characterization and optimization of production of exopolysaccharide from Chlamydomonas reinhardtii. *Carbohydrate Polymers*, *95*(2), 746–752.

Balti, R., Mansour, M. B., Zayoud, N., Le Balc'h, R., Brodu, N., Arhaliass, A., & Massé, A. (2020). Active exopolysaccharides based edible coatings enriched with red seaweed (Gracilaria gracilis) extract to improve shrimp preservation during refrigerated storage. *Food Bioscience*, *34*, 100522.

Barboríková, J., Šutovská, M., Kazimierová, I., Jošková, M., Fraňová, S., Kopecký, J., & Capek, P. (2019). Extracellular polysaccharide produced by Chlorella vulgaris–chemical characterization and anti-asthmatic profile. *International Journal of Biological Macromolecules*, *135*, 1–11.

Barclay, W. R., & Lewin, R. A. (1985). Microalgal polysaccharide production for the conditioning of agricultural soils. *Plant and Soil*, *88*(2), 159–169.

Bashir, K. M. I., Lee, J. H., Petermann, M. J., Shah, A. A., Jeong, S. J., Kim, M. S., & Cho, M. G. (2018). Estimation of antibacterial properties of chlorophyta, rhodophyta and haptophyta microalgae species. *Microbiology and Biotechnology Letters*, *46*(3), 225–233.

Bell, W. H. (1983). Bacterial utilization of algal extracellular products. 3. The specificity of algal-bacterial interaction. *Limnology and Oceanography*, *28*(6), 1131–1143.

Bellinger, B. J., Abdullahi, A. S., Gretz, M. R., & Underwood, G. J. C. (2005). Biofilm polymers: Relationship between carbohydrate biopolymers from estuarine mudflats and unialgal cultures of benthic diatoms. *Aquatic Microbial Ecology*, *38*, 169–180.

Bellini, E., Ciocci, M., Savio, S., Antonaroli, S., Seliktar, D., Melino, S., & Congestri, R. (2018). Trichormus variabilis (Cyanobacteria) biomass: From the nutraceutical products to novel EPS-cell/protein carrier systems. *Marine Drugs*, *16*, 298.

Blanc, G., Agarkova, I., Grimwood, J., Kuo, A., Brueggeman, A., Dunigan, D. D., Salamov, A., Schmutz, J., Weeks, D., Yamada, T., Lomsadze, A., Borodovsky, M., Claverie, J., Grigoriev, I. V., & Van Etten, J. L. (2012). The genome of the polar eukaryotic microalga Coccomyxa subellipsoidea reveals traits of cold adaptation. *Genome Biology*, *13*(5), 1–12.

Boychenko, Y. D., Budenkova, E., Kostyushina, N., Babich, O., & Kashinskikh, E. (2024). Porphyridium sp. Microalgae as a source of polysaccharides. *Journal of Applied Biology and Biotechnology*, *12*(3), 74–79. https://doi.org/10.7324/JABB.2024.165844

Braun, J. C., & Colla, L. M. (2023). Use of microalgae for the development of biofertilizers and biostimulants. *Bioenergy Research*, *16*(1), 289–310.

Castro, L., Zhang, R., Muñoz, J. A., González, F., Blázquez, M. L., Sand, W., & Ballester, A. (2014). Characterization of exopolymeric substances (EPS) produced by Aeromonas hydrophila under reducing conditions. *Biofouling*, *30*(4), 501–511.

Chakraborty, T., Sen, A. K., & Pal, R. (2015). Stress induced extra cellular polysaccharide of Spirulina subsalsa Stress induced enhancement in exo-polysaccharide production in Spirulina subsalsa and its chemical characterization. *Journal of Algal Biomass Utilization*, *6*(3), 24–38.

Chaudhuri, R., & Balasubramanian, P. (2023). Evaluating the potential of exopolysaccharide extracted from the spent cultivation media of Spirulina sp. as plant biostimulant. *Biomass Conversion and Biorefinery*, *1*, 1–10.

Chen, B., You, W., Huang, J., Yu, Y., & Chen, W. (2010). Isolation and antioxidant property of the extracellular polysaccharide from Rhodella reticulata. *World Journal of Microbiology and Biotechnology*, *26*(5), 833–840.

Chen, X., Song, L., Wang, H., Liu, S., Yu, H., Wang, X., Li, R., Liu, T., & Li, P. (2019). Partial characterization, the immune modulation and anticancer activities of sulfated polysaccharides from filamentous microalgae tribonema sp. *Molecules*, *24*(2), 322.

Cheng, P., Chang, T., Wang, C., Yao, C., Zhou, C., Liu, T., Wang, G., Yan, X., & Ruan, R. (2022). High cobalt exposure facilitates bioactive exopolysaccharides production with a novel molecular structure in Botryococcus braunii. *Chemical Engineering Journal*, *442*. https://doi.org/10.1016/j.cej.2022.136294

Chi, Z., Su, C. D., & Lu, W. D. (2007). A new exopolysaccharide produced by marine Cyanothece sp. 113. *Bioresource Technology*, *98*(6), 1329–1332.

Chiovitti, A., Higgins, M. J., Harper, R. E., Wetherbee, R., & Bacic, A. (2003). The complex polysaccharides of the raphid diatom pinnularia viridis (BACILLARIOPHYCEAE)1. *Journal of Phycology*, *39*(3), 543–554.

Clément-Larosière, B., Lopes, F., Gonçalves, A., Taidi, B., Benedetti, M., Minier, M., & Pareau, D. (2014). Carbon dioxide biofixation by Chlorella vulgaris at different CO2 concentrations and light intensities. *Engineering in Life Sciences*, *14*(5), 509–519.

Comte, S., Guibaud, G., & Baudu M. (2006). Relations between extraction protocols for activated sludge extracellular polymeric substances (EPS) and EPS complexation properties: Part I. Comparison of the. *Enzyme and Microbial Technology*, *38*(1–2), 237–245.

Concórdio-Reis, P., David, H., Reis, M. A., Amorim, A., & Freitas, F. (2023). Bioprospecting for new exopolysaccharide-producing microalgae of marine origin. *International Microbiology*, *26*(4), 1123–1130.

Costa, J. A. V., Lucas, B. F., Alvarenga, A. G. P., Moreira, J. B., & de Morais, M. G. (2021). Microalgae polysaccharides: An overview of production, characterization, and potential applications. *Polysaccharides*, *2.4*, 759–772.

Costa, R., Costa, L., Rodrigues, I., Meireles, C., Soares, R., Tamagnini, P., & Mota, R. (2021). Biocompatibility of the biopolymer cyanoflan for applications in skin wound healing. *Marine Drugs*, *19*(3), 147.

Cristofoli, N. L.; Lima, A. R., da Costa, A. M. R., Evtyugin, D., Silva, C., Varela, J., & Vieira, M. C. (2023). Structural characterization of exopolysaccharides obtained from Porphyridium cruentum exhausted culture medium. *Food and Bioproducts Processing*, *138*, 162–171.

Cruz, D., Vasconcelos, V., Pierre, G., Michaud, P., & Delattre, C. (2021). Exopolysaccharides from cyanobacteria: Strategies for bioprocess development. *Applied Sciences*, *10*(11), 3763. https://doi.org/10.3390/app10113763

Cuellar-Bermudez, S. P., Bleus, D., Mees, M. A., Struyf, T., Lama, S., Vandamme, D., & Muylaert, K. (2023). Isolation, purification, and metal-induced gelation of released polysaccharides from spent culture medium of Arthrospira. *Elsevier*, *70*, 103010.

Cui, L., Fan, L., Li, Z., Wang, J., Chen, R., Zhang, Y., Cheng, J., Wu, X., Li, J., Yin, H., Zeng, W., & Shen, L. (2021). Characterization of extracellular polymeric substances from Synechocystis sp. PCC6803 under Cd (II), Pb (II) and Cr (VI) stress. *Journal of Environmental Chemical Engineering*, *9*(4). https://doi.org/10.1016/j.jece.2021.105347

de Jesus Raposo, M. F., de Morais, A. M. M. B., & de Morais, R. M. S. C. (2014). Influence of sulphate on the composition and antibacterial and antiviral properties of the exopolysaccharide from Porphyridium cruentum. *Life Sciences*, *101*(1–2), 56–63.

De Philippis, R., Sili, C., Tassinato, G., Vincenzini, M., & Materassi, R. (1991). Effects of growth conditions on exopolysaccharide production by Cyanospira capsulata. *Bioresource Technology*, *38*(2–3), 101–104.

Decamp, A., Michelo, O., Rabbat, C., Laroche, C., Grizeau, D., Pruvost, J., & Gonçalves, O. (2021). A new, quick, and simple protocol to evaluate microalgae polysaccharide composition. *Marine Drugs*, *19*(2), 101.

Delattre, C., Pierre, G., Laroche, C., & Michaud, P. (2016). Production, extraction and characterization of microalgal and cyanobacterial exopolysaccharides. *Biotechnology Advances*, *34*(7), 1159–1179.

Demirci, A. S., Palabiyik, I., Altan, D. D., Apaydın, D., & Gumus, T. (2017). Yield and rheological properties of exopolysaccharide from a local isolate: Xanthomonas axonopodis pv. vesicatoria. *Electronic Journal of Biotechnology*, *30*, 18–23.

Dewi, I. C., Falaise, C., Hellio, C., Bourgougnon, N., & Mouget, J. L. (2018). Anticancer, antiviral, antibacterial, and antifungal properties in microalgae. In: Ira A. Levine and Joël Fleurence (eds.), *Microalgae in Health and Disease Prevention* (pp. 235–261). Academic Press.

Díaz Bayona, K. C., & Garcés, L. A. (2014). Effect of different media on exopolysaccharide and biomass production by the green microalga Botryococcus braunii. *Journal of Applied Phycology*, *26*(5), 2087–2095.

Domozych, D. S., Kort, S., Benton, S., & Yu, T. (2005). The extracellular polymeric substance of the green alga Penium margaritaceum and its role in biofilm formation. *Biofilms*, *2*(2), 129–144.

Drira, M., Elleuch, J., Hlima, H. B., Hentati, F., Gardarin, C., Rihouey, C., & Fendri, I. (2021). Optimization of exopolysaccharides production by Porphyridium sordidum and their potential to induce defense responses in Arabidopsis thaliana against Fusarium oxysporum. *Biomolecules*, *11*(2), 282.

El Arroussi, H., Benhima, R., Elbaouchi, A., Sijilmassi, B., El Mernissi, N., Aafsar, A., & Smouni, A. (2018). Dunaliella salina exopolysaccharides: A promising biostimulant for salt stress tolerance in tomato (Solanum lycopersicum). *Journal of Applied Phycology*, *30*(5), 2929–2941.

El-Naggar, N. E. A., Hussein, M. H., Shaaban-Dessuuki, S. A., & Dalal, S. R. (2020). Production, extraction and characterization of Chlorella vulgaris soluble polysaccharides and their applications in AgNPs biosynthesis and biostimulation of plant growth. *Scientific Reports*, *10*(1), 3011.

Ferreira, A. S., Mendonça, I., Póvoa, I., Carvalho, H., Correia, A., Vilanova, M., Silva, T. H., Coimbra, M. A., & Nunes, C. (2021). Impact of growth medium salinity on galactoxylan exopolysaccharides of Porphyridium purpureum. *Algal Research*, *59*. https://doi.org/10.1016/j.algal.2021.102439

Fimbres-Olivarría, D., López-Elías, J. A., Carvajal-Millán, E., Márquez-Escalante, J. A., Martínez-Córdova, L. R., Miranda-Baeza, A., Enríquez-Ocaña, F., Valdéz-Holguín, J. E., & Brown-Bojórquez, F. (2016). Navicula sp. sulfated polysaccharide gels induced by Fe(III): Rheology and microstructure. *International Journal of Molecular Sciences*, *17*(8), 1238.

Gallón, S. M. N., Alpaslan, E., Wang, M., Larese-Casanova, P., Londoño, M. E., Atehortúa, L., & Webster, T. J. (2019). Characterization and study of the antibacterial mechanisms of silver nanoparticles prepared with microalgal exopolysaccharides. *Materials Science and Engineering*, *99*, 685–695.

Gargouch, N., Elleuch, F., Karkouch, I., Tabbene, O., Pichon, C., Gardarin, C., Rihouey, C., Picton, L., Abdelkafi, S., Fendri, I., & Laroche, C. (2021). Potential of exopolysaccharide from Porphyridium marinum to contend with bacterial proliferation, biofilm formation, and breast cancer. *Marine Drugs*, *19*(2), 66. https://doi.org/10.3390/MD19020066

Gloaguen, V., Morvan, H., Hoffmann, L., Sainte Catherine, O., Kraemer, M., & Krausz, P. (2007). Bioactive capsular polysaccharide from the thermophilic cyanophyte/cyanobacterium Mastigodadus laminosus - Cytotoxic properties. *Planta Medica*, *73*(13), 1402–1406.

Gongi, W., Pinchetti, J. L. G., Cordeiro, N., Sadok, S., & Ouada, H. B. (2022). Characterization of biodegradable films based on extracellular polymeric substances extracted from the thermophilic microalga Graesiella sp. *Algal Research*, *61*, 102565.

Goo, B. G., Baek, G., Choi, D. J., Park, Y. I., Synytsya, A., Bleha, R., & Park, J. K. (2013). Characterization of a renewable extracellular polysaccharide from defatted microalgae Dunaliella tertiolecta. *Bioresource Technology*, *129*, 343–350.

Gouvêa, S. P., Vieira, A. A. H., & Lombardi, A. T. (2005). Copper and cadmium complexation by high molecular weight materials of dominant microalgae and of water from a eutrophic reservoir. *Chemosphere*, *60*(9), 1332–1339.

Gudmundsdottir, A. B., Brynjolfsdottir, A., Olafsdottir, E. S., Hardardottir, I., & Freysdottir, J. (2019a). Exopolysaccharides from Cyanobacterium aponinum induce a regulatory dendritic cell phenotype and inhibit SYK and CLEC7A expression in dendritic cells. *International Immunopharmacology*, *69*, 328–336.

Gudmundsdottir, A. B., Brynjolfsdottir, A., Olafsdottir, E. S., Hardardottir, I., & Freysdottir, J. (2019b). Exopolysaccharides from Cyanobacterium aponinum induce a regulatory dendritic cell phenotype and inhibit SYK and CLEC7A expression in dendritic cells, T cells and keratinocytes. *International Immunopharmacology*, *69*, 328–336.

Guzmán, S., Gato, A., Lamela, M., Freire-Garabal, M., & Calleja, J. M. (2003). Anti-inflammatory and immunomodulatory activities of polysaccharide from Chlorella stigmatophora and Phaeodactylum tricornutum. *Phytotherapy Research*, *17*(6), 665–670.

Hadiyanto, H., Khoironi, A., Dianratri, I., Joelyna, F. A., Christwardana, M., Sabhira, A. I., & Baihaqi, R. A. (2025). Microplastic removal in aquatic systems using extracellular polymeric substances (EPS) of microalgae. *Sustainable Environment*, *11*(1). https://doi.org/10.1080/27658511.2025.2454756

Halperin-Sternfeld, M., Netanel Liberman, G., Kannan, R., Netti, F., Ma, P. X., Arad, S. M., & Adler-Abramovich, L. (2022). Thixotropic red microalgae sulfated polysaccharide-peptide composite hydrogels as scaffolds for tissue engineering. *Biomedicines*, *10*(6), 1388.

Hamidi, M., Kennedy, J. F., Khodaiyan, F., Mousavi, Z., & Hosseini, S. S. (2019). Production optimization, characterization and gene expression of pullulan from a new strain of Aureobasidium pullulans. *International Journal of Biological Macromolecules*, *138*, 725–773.

Harun, R., Jason, W. S. Y., Cherrington, T., & Danquah, M. K. (2011). Exploring alkaline pre-treatment of microalgal biomass for bioethanol production. *Applied Energy*, *88*(10), 3464–3467.

Haug, A., & Myklestad, S. (1976). Polysaccharides of marine diatoms with special reference to Chaetoceros species. *Marine Biology*, *34*(3), 217–222.

Hokputsa, S., Hu, C., Paulsen, B. S., & Harding, S. E. (2003). A physico-chemical comparative study on extracellular carbohydrate polymers from five desert algae. *Carbohydrate Polymers*, *54*(1), 27–32.

Huheihel, M., Ishanu, V., Tal, J., & Arad, S. (2002). Activity of Porphyridium sp. polysaccharide against herpes simplex viruses in vitro and in vivo. *Journal of Biochemical and Biophysical Methods*, *50*(2–3), 189–200.

Huleihel, M., Ishanu, V., Tal, J., & Arad, S. (2001). Antiviral effect of red microalgal polysaccharides on herpes simplex and varicella zoster viruses. *Journal of Applied Phycology*, *13*(2), 127–134.

Iqbal, M., & Zafar, S. I. (1993). Effects of photon flux density, CO2, aeration rate, and inoculum density on growth and extracellular polysaccharide production by Porphyridium cruentum. *Folia Microbiologica*, *38*(6), 509–514.

Ivanova, J. G., Toshkova-Yotova, T. S., Toshkova, R. A., Deleva, V. R., Georgieva, A. K., & Gigova, L. G. (2024). Antioxidant and anticancer potential of extracellular polysaccharide from Porphyridium aerugineum (rhodophyta). *Fermentation, 10*(5). https://doi.org/10.3390/fermentation10050259

Jakhu, S., Sharma, Y., Sharma, K., Vaid, K., Dhar, H., Kumar, V., Singh, R. P., Shekh, A., & Kumar, G. (2021). Production and characterization of microalgal exopolysaccharide as a reducing and stabilizing agent for green synthesis of gold-nanoparticle: A case study with a Chlorella sp. from Himalayan high-altitude psychrophilic habitat. *Journal of Applied Phycology, 33*(6), 3899–3914.

Joelyna, F. A., Hadiyanto, Khoironi, A., & Pratama, W. D. (2024). The utilization of exopolysaccharide (EPS) from microalgae Chlorella vulgaris in microplastic removal. In *IOP Conference Series: Earth and Environmental Science* (Vol. 1414, No. 1, p. 012017). https://doi.org/10.1088/1755-1315/1414/1/012017

Kallas, T., Rebitre, M. C., Rippka, R., & Tandeau de Marsac, N. (1983). The structural nif genes of the cyanobacteria Gloeothece sp. and Calothrix sp. share homology with those of Anabaena sp., but the gloeothece genes have a different arrangement. *Journal of Bacteriology, 155*(1), 427–431.

Kanekiyo, K., Lee, J. B., Hayashi, K., Takenaka, H., Hayakawa, Y., Endo, S., & Hayashi, T. (2005). Isolation of an antiviral polysaccharide, nostoflan, from a terrestrial cyanobacterium, Nostoc flagelliforme. *Journal of Natural Roducts, 68*(7), 1037–1041.

Katona, S., Horváth, N., Molnár, Z., & Ördög, V. (2018). Extracellular polysaccharides in twenty Chlamydomonas strains of the Mosonmagyaróvár algal culture collection. *Acta Agronomica Óváriensis, 59*(1), 62–81.

Khandeparker, R. D., & Bhosle, N. B. (2001). Extracellular polymeric substances of the marine fouling diatom Amphora rostrata Wm.Sm. *Biofouling, 17*(2), 117–127.

Kharwar, S., Bhattacharjee, S., & Mishra, A. K. (2021). Disentangling the Impact of Sulfur Limitation on Exopolysaccharide and Functionality of Alr2882 by In Silico Approaches in Anabaena sp. PCC 7120. *Applied Biochemistry and Biotechnology, 193*(5), 1447–1468.

Kiemle, S. N., Domozych, D. S., & Gretz, M. R. (2007). The extracellular polymeric substances of desmids (Conjugatophyceae, Streptophyta): Chemistry, structural analyses and implications in wetland biofilms. *Phycologia, 46*(6), 617–627.

Kim, H. J., Yun, Y. S., & Kim, D. G. (2024). Production of xylose polysaccharides using newly isolated Chlamydomonas sp. Xylok with nitrogen depletion. *Algal Research, 77*. https://doi.org/10.1016/j.algal.2023.103374

Koçer, A. T., İnan, B., Kaptan Usul, S., Özçimen, D., Yılmaz, M. T., & Işıldak, İ. (2021). Exopolysaccharides from microalgae: Production, characterization, optimization and techno-economic assessment. *Brazilian Journal of Microbiology, 52*(4), 1779–1790.

Kokarakis, E. J., Nazos, T. T., Mavroudakis, L., Stratigakis, N. C., Sfendourakis, G. P., Lioudaki, S., Spyros, A., Pergantis, S. A., & Ghanotakis, D. F. (2022). Structural and physicochemical characterization of an aminosugar-rich exopolysaccharide isolated from a Chlorella sp. *Algal Research, 68*, 102881.

Kumar, D., Kaštánek, P., & Adhikary, S. P. (2018). Current science association exopolysaccharides from cyanobacteria and microalgae and their commercial application. *Current Science, 115*(2), 234–241.

Laroche, C. (2022). Exopolysaccharides from microalgae and cyanobacteria: Diversity of strains, production strategies, and applications. *Marine Drugs, 20*(5), 336. https://doi.org/10.3390/md20050336

Leandro, S. M., Gil, M. C., & Delgadillo, I. (2003). Partial characterisation of exopolysaccharides exudated by planktonic diatoms maintained in batch cultures. *Acta Oecologica, 24*, 49–55.

Leite, J. P., Mota, R., Durão, J., Neves, S. C., Barrias, C. C., Tamagnini, P., & Gales, L. (2017). Cyanobacterium-derived extracellular carbohydrate polymer for the controlled delivery of functional proteins. *Macromolecular Bioscience*, *17*(2), 1600206.

Li, C., Li, P., Fu, H., She, Z., Zhang, C., Li, Y., Zhang, M., & Ge, Y. (2025). Dynamic responses and adsorption mechanisms of Chlamydomonas reinhardtii extracellular polymeric substances under Cd, Cu, Pb, and Zn exposure. *Environmental Pollution*, *368*. https://doi.org/10.1016/j.envpol.2025.125747

Li, Q., Chen, Y., Liu, X., Li, Y., Xu, J., Li, T., Xiang, W., & Li, A. (2023). Effect of salinity on the biochemical characteristics and antioxidant activity of exopolysaccharide of Porphyridium purpureum FACHB 806. *Frontiers in Marine Science*, *9*, 1097200.

Li, S., Ji, L., Chen, C., Zhao, S., Sun, M., Gao, Z., & Fan, J. (2020). Efficient accumulation of high-value bioactive substances by carbon to nitrogen ratio regulation in marine microalgae Porphyridium purpureum. *Bioresource Technology*, *309*, 123362.

Li, T., Hu, J., & Zhu, L. (2021). Self-flocculation as an efficient method to harvest microalgae: A mini-review. *Water*, *13*(18), 2585.

Li, Y., Wang, C., Liu, H., Su, J., Lan, C. Q., Zhong, M., & Hu, X. (2020). Production, isolation and bioactive estimation of extracellular polysaccharides of green microalga Neochloris oleoabundans. *Algal Research*, *48*, 101883.

Li, Z., Liu, Y., Zhou, T., Cao, L., Cai, Y., Wang, Y., Cui, X., Yan, H., Ruan, R., & Zhang, Q. (2022). Effects of culture conditions on the performance of Arthrospira platensis and its production of exopolysaccharides. *Foods*, *11*(14), 2020.

Liberman, G. N., Ochbaum, G., Arad, S. M., & Bitton, R. (2016). The sulfated polysaccharide from a marine red microalga as a platform for the incorporation of zinc ions. *Carbohydrate Polymers*, *152*, 658–664.

Liu, C., Li, L. J., Wu, C. Y., Guo, K. N., & Li, J. H. (2016). Growth and antioxidant production of Spirulina in different NaCl concentrations. *Biotechnology Letters*, *38*(7), 1089–1096.

Liu, J., Wang, X., Pu, H., Liu, S., Kan, J., & Jin, C. (2017). Recent advances in endophytic exopolysaccharides: Production, structural characterization, physiological role and biological activity. *Carbohydrate Polymers*, *157*, 1113–1124. https://doi.org/10.1016/j.carbpol.2016.10.084

Lupi, F. M., Fernandes, H. M., Sá-Correia, I., & Novais, J. M. (1991). Temperature profiles of cellular growth and exopolysaccharide synthesis by Botryococus braunii Kütz. UC 58. *Journal of Applied Phycology*, *3*(1), 35–42.

Lupi, F. M., Fernandes, H. M. L., Tomé, M. M., Sá-Correia, I., & Novais, J. M. (1994). Influence of nitrogen source and photoperiod on exopolysaccharide synthesis by the microalga Botryococcus braunii UC 58. *Enzyme and Microbial Technology*, *16*(7), 546–550.

Magaletti, E., Urbani, R., Sist, P., Ferrari, C. R., & Cicero, A. M. (2004). Abundance and chemical characterization of extracellular carbohydrates released by the marine diatom Cylindrotheca fusiformis under N- and P-limitation. *European Journal of Phycology*, *39*(2), 133–142.

Magierek, E., & Krzemińska, I. (2018). Effect of stress conditions on improvement of lipid and carbohydrate accumulation under photoautotrophic cultivation of Chlorophyta. *Phycologia*, *57*(6), 601–618.

Medina-Cabrera, E. V., Rühmann, B., Schmid, J., & Sieber, V. (2020). Optimization of growth and EPS production in two Porphyridum strains. *Bioresource Technology Reports*, *11*, 100486.

Mishra, A., & Jha, B. (2009). Isolation and characterization of extracellular polymeric substances from micro-algae Dunaliella salina under salt stress. *Bioresource Technology*, *100*(13), 3382–3386.

Morais, M. G., Santos, T. D., Moraes, L., Vaz, B. S., Morais, E. G., & Costa, J. A. V. (2022). Exopolysaccharides from microalgae: Production in a biorefinery framework and potential applications. *Bioresource Technology Reports*, *18*, 101006.

Morales-Jiménez, M., Gouveia, L., Yáñez-Fernández, J., Castro-Muñoz, R., & Barragán-Huerta, B. E. (2020). Production, preparation and characterization of microalgae-based biopolymer as a potential bioactive film. *Coatings*, *10*(2), 120.

Moreira, J. B., Kuntzler, S. G., Bezerra P. Q. M., Cassuriaga, A. P. A., Zaparoli, M., da Silva, J. L. V., & de Morais, M. G. (2022). Recent advances of microalgae exopolysaccharides for application as bioflocculants. *Polysaccharides*, *3*(1), 264–276.

Moreno, J., Vargas, M. A., Olivares, H., Rivas, J., & Guerrero, M. G. (1998). Exopolysaccharide production by the cyanobacterium Anabaena sp. ATCC 33047 in batch and continuous culture. *Journal of Biotechnology*, *60*(3), 175–182.

Moscovici, M. (2015). Present and future medical applications of microbial exopolysaccharides. *Frontiers in Microbiology*, *6*, 137616.

Mousavian, Z., Safavi, M., Azizmohseni, F., Hadizadeh, M., & Mirdamadi, S. (2022). Characterization, antioxidant and anticoagulant properties of exopolysaccharide from marine microalgae. *AMB Express*, *12*(1). https://doi.org/10.1186/S13568-022-01365-2

Nadzir, M. M., Nurhayati, R. W., Idris, F. N., & Nguyen, M. H. (2021). Biomedical applications of bacterial exopolysaccharides: A review. *Polymers*, *13*, 530.

Najdenski, H. M., Gigova, L. G., Iliev, I. I., Pilarski, P. S., Lukavský, J., Tsvetkova, I. V., Ninova, M. S., & Kussovski, V. K. (2013). Antibacterial and antifungal activities of selected microalgae and cyanobacteria. *International Journal of Food Science & Technology*, *48*(7), 1533–1540.

Naveed, S., Li, C., Lu, X., Chen, S., Yin, B., Zhang, C., & Ge, Y. (2019). Microalgal extracellular polymeric substances and their interactions with metal(loid)s: A review. *Critical Reviews in Environmental Science and Technology*, *49*(19), 1769–1802.

Nechita, P., & Roman, M. (2020). Review on polysaccharides used in coatings for food packaging papers. *Coatings*, *10*(6), 566.

Nešić, A., Cabrera-Barjas, G., Dimitrijević-Branković, S., Radovanović, N., & Delattre, C. (2019). Prospect of polysaccharide-based materials as advanced food packaging. *Molecules*, *25*(1), 135.

Nicolaus, B., Panico, A., Lama, L., Romano, I., Manca, M. C., De Giulio, A., & Gambacorta, A. (1999). Chemical composition and production of exopolysaccharides from representative members of heterocystous and non-heterocystous cyanobacteria. *Phytochemistry*, *52*(4), 639–647.

Nikolova, B., Semkova, S., Tsoneva, I., Antov, G., Ivanova, J., Vasileva, I., Kardaleva, P., Stoineva, I., Christova, N., Nacheva, L., & Kabaivanova, L. (2019). Characterization and potential antitumor effect of a heteropolysaccharide produced by the red alga Porphyridium sordidum. *Engineering in Life Sciences*, *19*(12), 978–985.

Nur, M. M. A., Swaminathan, M. K., Boelen, P., & Buma, A. G. J. (2019). Sulfated exopolysaccharide production and nutrient removal by the marine diatom Phaeodactylum tricornutum growing on palm oil mill effluent. *Journal of Applied Phycology*, *31*(4), 2335–2348.

Okeyoshi, K., Okajima, M. K., & Kaneko, T. (2021). The cyanobacterial polysaccharide sacran: Characteristics, structures, and preparation of LC gels. *Polymer Journal*, *53*(1), 81–91.

Paredes-Molina, D. M., Cervantes-López, M. A., Orona-Tamayo, D., Lozoya-Pérez, N. E., Beltrán-Ramírez, F. I., Vázquez-Martínez, J., Macias-Sánchez, K. L., Alonso-Romero, S., & Quintana-Rodríguez, E. (2025). Lactic whey as a potential feedstock for exopolysaccharide production by microalgae strain Neochloris oleoabundans UTEX 1185. *Biotechnology for Biofuels and Bioproducts*, *18*(1). https://doi.org/10.1186/s13068-024-02595-1

Park, G. T., Go, R. E., Lee, H. M., Lee, G. A., Kim, C. W., Seo, J. W., Hong, W. K., Choi, K. C., & Hwang, K. A. (2017). Potential anti-proliferative and immunomodulatory effects of marine microalgal exopolysaccharide on various human cancer cells and lymphocytes in vitro. *Marine Biotechnology*, *19*(2), 136–146. https://doi.org/10.1007/s10126-017-9735-y

Parra-Riofrío, G., García-Márquez, J., Casas-Arrojo, V., Uribe-Tapia, E., & Abdala-Díaz, R. T. (2020). Antioxidant and cytotoxic effects on tumor cells of exopolysaccharides from Tetraselmis suecica (Kylin) butcher grown under autotrophic and heterotrophic conditions. *Marine Drugs*, *18*(11), 534.

Parra-Riofrio, G., Moreno, P., García-Rosado, E., Alonso, M. C., Uribe-Tapia, E., Abdala-Diaz, R. T., & Bejar, J. (2023). Tetraselmis suecica and Porphyridium cruentum exopolysaccharides show anti-VHSV activity on RTG-2 cells. *Aquaculture International*, *31*(6), 3145–3157.

Patel, A. K., Laroche, C., Marcati, A., Ursu, A. V., & Jubeau, S. (2013). Separation and fractionation of exopolysaccharides from Porphyridium cruentum. *Bioresource Technology*, *145*, 345–350.

Patel, A. K., Vadrale, A. P., Singhania, R. R., Michaud, P., Pandey, A., Chen, S. J., Chen, C. W., & Dong, C. D. (2023). Algal polysaccharides: Current status and future prospects. *Phytochemistry Reviews*, *22*(4), 1167–1196.

Patil, J. S., & Anil, A. C. (2005). Influence of diatom exopolymers and biofilms onmetamorphosis in the barnacle Balanus amphitrite. *Marine Ecology Progress Series*, *301*, 231–245.

Peng, Z., Liu, G., & Huang, K. (2021). Cold adaptation mechanisms of a snow alga Chlamydomonas nivalis during temperature fluctuations. *Frontiers in Microbiology*, *11*. https://doi.org/10.3389/FMICB.2020.611080

Penna, A., Berluti, S., Penna, N., & Magnani, M. (1999). Influence of nutrient ratios on the in vitro extracellular polysaccharide production by marine diatoms from the Adriatic Sea. *Journal of Plankton Research*, *21*(9), 1681–1690.

Pereira, S., Zille, A., Micheletti, E., Moradas-Ferreira, P., De Philippis, R., & Tamagnini, P. (2009). Complexity of cyanobacterial exopolysaccharides: Composition, structures, inducing factors and putative genes involved in their biosynthesis and assembly. *FEMS Microbiology Reviews*, *33*(5), 917–941.

Pierre, G., Delattre, C., Dubessay, P., Jubeau, S., Vialleix, C., Cadoret, J. P., Probert, I., & Michaud, P. (2019). What is in store for EPS microalgae in the next decade? *Molecules*, *24*(23), 4296.

Pintathong, P., Chomnunti, P., Sangthong, S., Jirarat, A., & Chaiwut, P. (2021). The feasibility of utilizing cultured Cordyceps militaris residues in cosmetics: Biological activity assessment of their crude extracts. *Journal of Fungi*, *7*(11), 973.

Rachidi, F., Benhima, R., Kasmi, Y., Sbabou, L., & Arroussi, H. E. (2021). Evaluation of microalgae polysaccharides as biostimulants of tomato plant defense using metabolomics and biochemical approaches. *Scientific Reports*, *11*(1), 930.

Rachidi, F., Benhima, R., Sbabou, L., & El Arroussi H. (2020). Microalgae polysaccharides bio-stimulating effect on tomato plants: Growth and metabolic distribution. *Biotechnology Reports*, *25*, e00426.

Radonić, A., Thulke, S., Achenbach, J., Kurth, A., Vreemann, A., König, T., Walter, C., Possinger, K., & Nitsche, A. (2011). Anionic polysaccharides from phototrophic microorganisms exhibit antiviral activities to Vaccinia virus. *Journal of Antivirals & Antiretrovirals*, *2*(4). https://doi.org/10.4172/jaa.1000023

Rossi, F., & De Philippis, R. (2015). Role of cyanobacterial exopolysaccharides in phototrophic biofilms and in complex microbial mats. *Life*, *5*(2), 1218–1238).

Rossi, F., & De Philippis, R. (2016). Exocellular polysaccharides in microalgae and cyanobacteria: Chemical features, role and enzymes and genes involved in their biosynthesis. In: M. Borowitzka, J. Beardall, and J. Raven (eds.), *The Physiology of Microalgae. Developments in Applied Phycology*, vol. 6. Springer. https://doi.org/10.1007/978-3-319-24945-2_21

Roussel, M., Villay, A., Delbac, F., Michaud P., Laroche, C., Roriz, D., & Diogon, M. (2015). Antimicrosporidian activity of sulphated polysaccharides from algae and their potential to control honeybee nosemosis. *Carbohydrate Polymers*, *133*, 213–220.

Roy, N., Roy, S., Debroy, A., & Mukherjee, A. (2024). Chlorella sp. EPS loaded alginate microsphere: A novel bionanomaterial for adsorptive removal of sulfamethoxazole from the aqueous solutions. *Environmental Technology & Innovation*, *33*, 103513.

Ruiz-Dávila, C. E., Solís-Andrade, K. I., Olvera-Sosa, M., Palestino, G., & Rosales-Mendoza, S. (2023). Core-shell chitosan/Porphyridium-exopolysaccharide microgels: Synthesis, properties, and biological evaluation. *International Journal of Biological Macromolecules*, *246*, 125655.

Saad, M. H., Sidkey, N. M., & El-Fakharany, E. M. (2023). Identification and statistical optimization of a novel alginate polymer extracted from newly isolated Synechocystis algini MNE ON864447 with antibacterial activity. *Microbial Cell Factories*, *22*(1), 229.

Sasaki, M., Takagi, A., Ota, S., Kawano, S., Sasaki, D., & Asayama, M. (2020). Coproduction of lipids and extracellular polysaccharides from the novel green alga Parachlorella sp. BX1. 5 depending on cultivation conditions. *Biotechnology Reports*, *25*, e00392.

Sed, G., Cicci, A., & Bravi, M. (2017). Extraction and purification of exopolysaccharides from exhausted Arthrospira platensis (Spirulina) culture systems. *Chemical Engineering Transactions*, *57*, 211–216.

Sivaramakrishnan, R., Suresh, S., Pugazhendhi, A., Pauline, J. M. N., & Incharoensakdi, A. (2020). Response of Scenedesmus sp. to microwave treatment: Enhancement of lipid, exopolysaccharide and biomass production. *Bioresource Technology*, *312*, 123562.

Staats, N., De Winder, B., Stal, L. J., & Mur, L. R. (1999). Isolation and characterization of extracellular polysaccharides from the epipelic diatoms Cylindrotheca closterium and Navicula salinarum. *European Journal of Phycology*, *34*(2), 161–169.

Sun, L., Wang, C., & Lei, S. (2008). Effects of light regime on extracellular polysaccharide production by Porphyridium cruentum cultured in flat plate photobioreactors. In *2nd International Conference on Bioinformatics and Biomedical Engineering*, Shanghai, China (pp. 1488–1491). https://doi.org/10.1109/ICBBE.2008.701

Sun, L., Wang, C., Shi, Q., & Ma, C. (2009). Preparation of different molecular weight polysaccharides from Porphyridium cruentum and their antioxidant activities. *International Journal of Biological Macromolecules*, *45*(1), 42–47.

Sun, L., Wang, L., & Zhou, Y. (2012). Immunomodulation and antitumor activities of different-molecular-weight polysaccharides from Porphyridium cruentum. *Carbohydrate Polymers*, *87*(2), 1206–1210.

Takahashi, E., Ledauphin, J., Goux, D., & Orvain, F. (2009). Optimising extraction of extracellular polymeric substances (EPS) from benthic diatoms: Comparison of the efficiency of six EPS extraction methods. *Marine and Freshwater Research*, *60*(12), 1201–1210.

Tamaru, Y., Takani, Y., Yoshida, T., & Sakamoto, T. (2005). Crucial role of extracellular polysaccharides in desiccation and freezing tolerance in the terrestrial cyanobacterium Nostoc commune. *Applied and Environmental Microbiology*, *71*(11), 7327–7333.

Tannin-Spitz, T., Bergman, M., Van-Moppes, D., Grossman, S., & Arad, S. (2005). Antioxidant activity of the polysaccharide of the red microalga Porphyridium sp. *Journal of Applied Phycology*, *17*(3), 215–222.

Templeton, D. W., Quinn, M., Van Wychen, S., Hyman, D., & Laurens, L. M. (2012). Separation and quantification of microalgal carbohydrates. *Journal of Chromatography A*, *1270*, 225–234.

Tesson, B., & Hildebrand, M. (2013). Characterization and localization of insoluble organic matrices associated with diatom cell walls: Insight into their roles during cell wall formation. *PLoS One*, *8*(4), e61675.

Tischer, R. G., & Moore, B. G. (1964). An extracellular polysaccharide produced by Palmella mucosa Kütz. *Archiv Für Mikrobiologie*, *49*(2), 158–166.

Tiwari, O. N., Muthuraj, M., Bhunia, B., Bandyopadhyay T. K., Annapurna, K., Sahu, M., & Indrama, T. (2020). Biosynthesis, purification and structure-property relationships of new cyanobacterial exopolysaccharides. *Polymer Testing*, *89*, 106592.

Toshkova-Yotova, T., Sulikovska, I., Djeliova, V., Petrova, Z., Ognyanov, M., Denev, P., Toshkova, R., & Georgieva, A. (2024). Exopolysaccharides from the green microalga strain Coelastrella sp. BGV—Isolation, characterization, and assessment of anticancer potential. *Current Issues in Molecular Biology*, *46*(9), 10312–10334. https://doi.org/10.3390/cimb46090614

Trabelsi, L., Ben Ouada, H., Bacha, H., & Ghoul, M. (2009). Combined effect of temperature and light intensity on growth and extracellular polymeric substance production by the cyanobacterium Arthrospira platensis. *Journal of Applied Phycology*, *21*(4), 405–412.

Trabelsi, L., Chaieb, O., Mnari, A., Abid-Essafi, S., & Aleya, L. (2016). Partial characterization and antioxidant and antiproliferative activities of the aqueous extracellular polysaccharides from the thermophilic microalgae Graesiella sp. *BMC Complementary and Alternative Medicine*, *16*(1), 210.

Tripathi, S. D., Mitra, R., Kulshrestha, S., Rabiya, R., & Sen, R. (2025). Bicarbonate induced enhanced production of microalgal extracellular polymeric substance and its characterization. *Bioresource Technology*, *423*. https://doi.org/10.1016/j.biortech.2025.132232

Urbani, R., Sist, P., Pletikapić, G., Radić, T. M., Svetličić, V., & Žutic, V. (2012). Diatom polysaccharides: Extracellular production, isolation and molecular characterization. *The Complex World of Polysaccharides*, 345–370.

Villay, A., Laroche, C., Roriz, D., El Alaoui, H., Delba, F., & Michaud, P. (2013). Optimisation of culture parameters for exopolysaccharides production by the microalga Rhodella violacea. *Bioresource Technology*, *146*, 732–735.

Wang, W. N., Li, T., Li, Y., Zhang, Y., Wu, H. L., Xiang, W. Z., & Li, A. F. (2021). Exopolysaccharides from the energy microalga strain Botryococcus braunii: Purification, characterization, and antioxidant activity. *Foods*, *11*(1), 110.

Willis, A., Chiovitti, A., Dugdale, T. M., & Wetherbee, R. (2013). Characterization of the extracellular matrix of Phaeodactylum tricornutum (Bacillariophyceae): Structure, composition, and adhesive characteristics. *Journal of Phycology*, *49*(5), 937–949.

Winahyu, D. A., & Primadiamanti, A. (2021). Formulation and evaluation of the exopolysaccharide compound extract lotion from the microalgae Spirulina sp. *Journal of Physics: Conference Series*, *1882*, 12107.

Wustman, B. A., Gretz, M. R., & Hoagland, K. D. (1997). Extracellular matrix assembly in diatoms (Bacillariophyceae) (I. A model of adhesives based on chemical characterization and localization of polysaccharides from the marine diatom Achnanthes longipes and other diatoms). *Plant Physiology*, *113*(4), 1059–10069.

Xiao, R., & Zheng, Y. (2016). Overview of microalgal extracellular polymeric substances (EPS) and their applications. *Biotechnology Advances*, *34*(7), 1225–1244.

Xing, C., Li, J., Yuan, H., & Yang, J. (2022). Physiological and transcription level responses of microalgae Auxenochlorella protothecoides to cold and heat induced oxidative stress. *Environmental Research*, *211*. https://doi.org/10.1016/j.envres.2022.113023

Xing, H., Du, R., Zhao, F., Han, Y., Xiao, H., & Zhou, Z. (2018). Optimization, chain conformation and characterization of exopolysaccharide isolated from Leuconostoc mesenteroides DRP105. *International Journal of Biological Macromolecules*, *112*, 1208–1216.

Yaakob, M. A., Mohamed, R. M. S. R., Al-Gheethi, A., Aswathnarayana Gokare, R., & Ambati, R. R. (2021). Influence of nitrogen and phosphorus on microalgal growth, biomass, lipid, and fatty acid production: An overview. *Cells*, *10*(2), 393.

Yang, S., Wan, H., Wang, R., & Hao, D. (2019). Sulfated polysaccharides from Phaeodactylum tricornutum: Isolation, structural characteristics, and inhibiting HepG2 growth activity in vitro. *PeerJ*, *7*, e6409.

Yim, J. H., Kim, S. J., Ahn, S. H., Lee, C. K., Rhie, K. T., & Lee, H. K. (2004). Antiviral effects of sulfated exopolysaccharide from the marine microalga Gyrodinium impudicum strain KG03. *Marine Biotechnology*, *6*(1), 17–25.

Yim, J. H., Kim, S. J., Ahn, S. H., & Lee, H. K. (2003). Optimal conditions for the production of sulfated polysaccharide by marine microalga Gyrodinium impudicum strain KG03. *Biomolecular Engineering*, *20*(4–6), 273–280.

Yoshimura, H., Kotake, T., Aohara, T., Tsumuraya, Y., Ikeuchi, M., & Ohmori, M. (2012). The role of extracellular polysaccharides produced by the terrestrial cyanobacterium Nostoc sp. strain HK-01 in NaCl tolerance. *Journal of Applied Phycology*, *24*(2), 237–243.

You, T., & Barnett, S. M. (2004). Effect of light quality on production of extracellular polysaccharides and growth rate of Porphyridium cruentum. *Biochemical Engineering Journal*, *19*(3), 251–258.

Zampieri, R. M., Adessi, A., Caldara, F., Codato, A., Furlan, M., Rampazzo, C., & Dalla Valle, L. (2020). Anti-inflammatory activity of exopolysaccharides from Phormidium sp. ETS05, the most abundant cyanobacterium of the therapeutic euganean thermal muds, using the zebrafish model. *Biomolecules*, *10*(4), 582.

Zannini, E., Waters, D. M., Coffey, A., & Arendt, E. K. (2016). Production, properties, and industrial food application of lactic acid bacteria-derived exopolysaccharides. *Applied Microbiology and Biotechnology*, *100*(3), 1121–1135.

Zhang, J., Liu, L., & Chen, F. (2019). Production and characterization of exopolysaccharides from Chlorella zofingiensis and Chlorella vulgaris with anti-colorectal cancer activity. *International Journal of Biological Macromolecules*, *134*, 976–983.

Zhang, J., Liu, L., Ren, Y., & Chen, F. (2019). Characterization of exopolysaccharides produced by microalgae with antitumor activity on human colon cancer cells. *International Journal of Biological Macromolecules*, *128*, 761–767.

Zhang, S., Qamar, S. A., Junaid, M., Munir, B., Badar, Q., & Bilal, M. (2022). Algal polysaccharides-based nanoparticles for targeted drug delivery applications. *Starch - Stärke*, *74*(7–8), 2200014.

Zhou, R., Qian, Ya., Lei, Z., Tang, Y., & Li, Y. (2023). Production and characterization of exopolysaccharides from salinity-induced Auxenochlorella protothecoides and the analysis of anti-inflammatory activity. *International Journal of Biological Macromolecules*, *240*, 124217.

Zhou, R., Zuo, W., Xie, Y., Wang, Y., Li, C., Tang, Y., Wang, Z., & Li, Y. (2025). Sodium bisulfite boosted exopolysaccharide production by Auxenochlorella protothecoides: Potential mechanisms harnessing H2O2 signaling and carbon reallocation. *Bioresource Technology*, *420*. https://doi.org/10.1016/j.biortech.2025.132121

Zhu, C., Chen, C., Zhao, L., Zhang, Y., Yang, J., Song, L., & Yang, S. (2012). Bioflocculant produced by Chlamydomonas reinhardtii. *Journal of Applied Phycology*, *24*(5), 1245–1251.

Ziadi, M., Bouzaiene, T., M'Hir, S., Zaafouri, K., Mokhtar, F., Hamdi, M., & Boisset-Helbert, C. (2018). Evaluation of the efficiency of ethanol precipitation and ultrafiltration on the purification and characteristics of exopolysaccharides produced by three lactic acid bacteria. *BioMed Research International*, *2018*. https://doi.org/10.1155/2018/1896240

Zykwinska, A., Marquis, M., Godin, M., Marchand, L., Sinquin, C., Garnier, C., & Cuenot, S. (2019). Microcarriers based on glycosaminoglycan-like marine exopolysaccharide for TGF-β1 long-term protection. *Marine Drugs*, *17*(1), 65.

6 Microalgae Biotechnology for Chlorophyll Production

Distribution, Extraction, and Potential Uses

Pricila Nass Pinheiro, Tatiele C. Nascimento, Patrícia Acosta Caetano, Luisa Chitolina Schetinger, and Leila Q. Zepka

6.1 INTRODUCTION

Consumers are increasingly aware of the potential long-term health risks of synthetic compounds commonly used in the food industry. This awareness has driven a growing demand for natural alternatives and a preference for food components that not only fulfill basic nutritional and energy requirements but also offer added physiological benefits (Mohammadi et al., 2022; Galanakis, 2021).

At the same time, there is an urgent need to identify natural sources and develop more sustainable processes to produce these components. Within this context, natural pigments have gained considerable attention as viable replacements for synthetic dyes. Microalgae have emerged as a highly sustainable solution, functioning as biofactories capable of producing a wide range of biologically active compounds, including pigments such as chlorophylls while supporting photosynthetic carbon sequestration (Nabi et al., 2023; Manzoor et al., 2021).

These microorganisms are valuable sources of chlorophyll due to their high content of this pigment in their structure (approx. 7 wt%) (Rosa et al., 2023; Sarkar et al., 2020). These biomolecules have been associated with various health benefits, including antimicrobial properties, antimutagenic and antigenotoxic effects, enhanced antioxidant activity to neutralize free radicals, and the prevention of lipid oxidation (Perez-Galvez, Viera, & Roca, 2018; Zepka, Jacob-Lopes, & Roca, 2019).

As a result, the global market for natural food coloring grew from USD 1.64 billion in 2021 and is projected to reach a market size of USD 3.66 billion by 2030, reflecting a compound annual growth rate (CAGR) of 9.31% during the forecast period (2022–2030) (Straits Research, 2024).

DOI: 10.1201/9781003611479-6

Despite this promising growth, the commercial production of microalgal pigments faces significant challenges. These include high production costs, stringent culture conditions, susceptibility to contamination, and the difficulty of balancing high pigment accumulation with optimal cell growth. To address these challenges, advancing biotechnological strategies to develop microalgal cell factories with enhanced pigment production capacity is essential (Hu et al., 2018).

This chapter explores the importance of chlorophyll production by microalgae, focusing on the processes involved in pigment production using algae bioreactors and the latest extraction techniques. Additionally, it examines the global market of high-value microalgal pigments and provides a concise analysis of key patented inventions relevant to this study.

6.2 CULTIVATION OPTIMIZATION FOR CHLOROPHYLL PRODUCTION

Chlorophyll-a, the main pigment of green plants, serves as a universal ecological and physiological marker of biomass, photosynthetic activity, and production capabilities of microalgae. Microalgae are the basis of the trophic pyramid in the freshwater ecosystem (Mineeva & Semadeni, 2023). The main ecological role of microalgae is oxygen generation and photosynthetic production of autochthonous organic matter, which forms the energy basis for organisms of higher trophic levels and for all subsequent stages of the production process in large lakes and reservoirs (Peng et al., 2024).

Microalgae cells contain photosynthetic pigments that are among the most common indicators used in the study of algocenoses. Chlorophyll is an optically active component of the aquatic environment due to its ability to absorb light in a narrow wavelength range. The amount of chlorophyll is closely related to the biomass of microalgae, which makes it possible to express biomass in units of this important component of the plant cell. A detailed review of studies on the relationship between cultivation and microalgal biomass will be provided (Bourgougnon, Burlot, & Jacquin, 2021).

The prominence of microalgae cultivation methods lies in the fact that, as photoautotrophic microorganisms, microalgae thrive under diverse environmental conditions. They utilize nitrogen (N), phosphorus (P), and carbon dioxide (CO_2) as nutrients and carbon sources, absorbing light energy through photosynthesis to enhance growth and biomass production (Yaqoubnejad, Rad, & Taghavijeloudar, 2021).

There are various techniques for cultivating microalgae, which can be categorized as photoautotrophic, heterotrophic, or facultative cultures. Depending on the species and the peculiarities of microalgal cells, the cultivation approaches can include open ponds (Figure 6.1), closed systems (photobioreactors (PBRs)) (Figure 6.2), and hybrid or integrated systems, each presenting unique advantages and drawbacks (Yang et al., 2024).

FIGURE 6.1 Raceway pond.

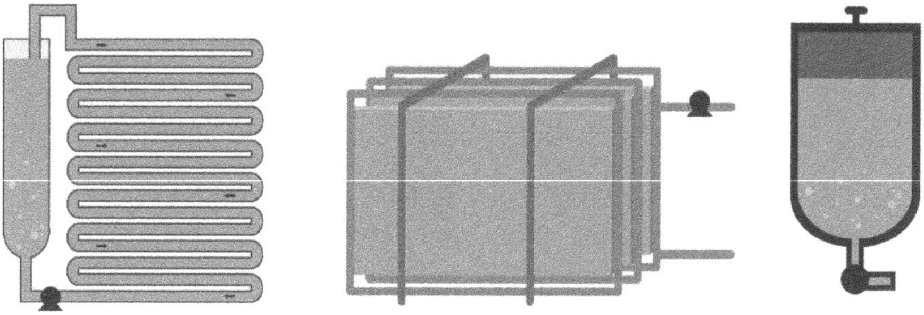

FIGURE 6.2 Styles of bioreactors used to produce pigments.

6.2.1 Open Cultivation Systems

Open ponds have historically been the most commonly employed approach for industrial-scale microalgae production, primarily due to low construction costs and simple scalability. These systems can be subdivided into natural (lakes, lagoons) and artificial (raceway ponds) configurations. Among artificial ponds, high-rate algal ponds (HRAPs) have garnered attention for wastewater treatment applications, often integrating paddle wheels that enhance circulation, reduce sedimentation, and maintain favorable growth conditions (Yaqoubnejad, Rad, & Taghavijeloudar, 2021).

Despite their cost-effectiveness and relative ease of operation, open ponds are generally constrained by shallow water depths (commonly 0.2–0.4 m), which can result in elevated CO_2 losses and reduced overall productivity. Other issues include susceptibility to contamination by competing microorganisms or predators, as well as substantial land requirements. These limitations have prompted efforts to refine open cultivation systems. For instance, high-rate microalgal/bacterial ponds (HRABPs) integrate algae and bacteria to achieve higher biomass yields and improved pollutant removal in wastewater scenarios (Torres-Franco et al., 2021).

Raceway ponds, in particular, operate as closed-loop channels, typically propelled by paddle wheels for continuous circulation. Constructed from materials like concrete, fiberglass, or liners, they remain a popular choice for large-scale microalgae cultivation because of their low investment and straightforward operation. Nonetheless, raceway ponds still face hurdles such as limited biomass densities and exposure to environmental fluctuations that can compromise both productivity and product quality (Otálora et al., 2023).

6.2.2 Closed Cultivation Systems

To address the challenges posed by open ponds, especially where strict quality control is essential (human-grade biomass, bioactive compounds), closed PBRs have been developed. In these systems, key parameters such as temperature, pH, dissolved gases, and nutrient concentrations (N, P) are managed precisely to support consistent microalgal growth. Depending on process objectives, one can choose from different PBR configurations, including tubular, flat plate, column, bubble column, airlift, and hybrid designs (Deng et al., 2022).

Because they are locked environments, PBRs reduce the risks of microbial contamination and enable high cell-density cultures. However, supplying adequate light becomes a critical engineering challenge as light penetration decreases with rising cell concentration (Sartori et al., 2022). Moreover, gas exchange can be complicated by the accumulation of oxygen produced via photosynthesis, necessitating effective degassing strategies. The design, dimensioning, and high initial capital investment of these systems often underscore their complexity, even though they promise more controlled conditions and superior bioproduct quality (Casagli et al., 2023).

6.2.3 Fermenters and Heterotrophic Approaches

Another closed system option involves fermenters, originally adapted from the chemical, food, and biochemical industries. When applied to microalgae, fermenters support heterotrophic growth by supplying organic carbon sources (glucose or other sugars). This approach can lead to high cell productivities but demands carefully optimized mixing and aeration to ensure consistent culture conditions. Any malfunction in the mixing or aeration units risks cell damage or death due to excessive shear stress or oxygen deficiency. Despite these challenges, fermenters remain an attractive alternative, particularly for microalgae that thrive under heterotrophic or mixotrophic regimes, supporting the production of valuable metabolites (Saumyakanti Khanra et al., 2018).

6.2.4 Hydric or Integrated Systems and Emerging Designs

Beyond conventional open and closed suspension cultures, hydric and integrated approaches provide alternative methods that address both operational efficiency and economic viability. In these systems, microalgae may be immobilized on solid supports or combined with membrane-based technologies, thereby increasing cell density and simplifying biomass harvesting. This strategy, exemplified by biofilm PBRs, often proves beneficial in wastewater treatment, as it improves nutrient transfer and gas exchange (Padrón & Mérida, 2025).

Moreover, hybrid membrane processes have emerged as effective means of advancing both nutrient recovery and bioenergy production. By coupling membrane units with anaerobic reactors, researchers can simultaneously treat wastewater and generate energy (Gao et al., 2021).

Meanwhile, novel reactor configurations such as disposable wave-induced bioreactors, dynamic membrane aeration technology, and traveling wave bioreactors are undergoing laboratory and pilot-scale trials. Despite their diverse benefits, these systems often rely heavily on electrical power for mixing, aeration, and temperature control, making energy demands a key consideration in both economic and environmental assessments. Consequently, optimizing resource consumption while maintaining high productivity remains critical for the successful implementation of hydric or integrated microalgal systems (Silva et al., 2024).

Finally, microalgae cultivation can be implemented through various systems, ranging from open raceway ponds to closed PBRs and specialized fermenters. Each strategy presents a unique balance of benefits (cost, scalability, and product purity) and limitations (contamination, productivity, or energy demand). As the microalgae industry expands, ongoing research focuses on hybrid and emerging cultivation strategies that merge the best features of open and closed systems while minimizing operational costs and environmental impacts. Ultimately, the choice of cultivation system must align with the target application, be it wastewater treatment, biofuel production, or high-value bioproducts, taking into account the need for consistent quality, operational reliability, and sustainability.

6.3 EXTRACTION OF CHLOROPHYLLS FROM MICROALGAE

Among microalgal pigments, chlorophylls are tetrapyrrolic compounds with a hydrophobic nature. As shown in Figure 6.3, the structure consists of a cyclic base with a characteristic five-membered isocyclic ring, a central magnesium (Mg) atom, and a phytol chain (which imparts hydrophobicity to the molecule) (Zepka et al., 2019). Also in Figure 6.3, the main chlorophyll structures found in most microalgae are displayed, including chlorophylls a, b, c, d, and f, with the first two being the most abundant. In addition to these primary structures, various derivatives are found, such as pheophorbides (without the central Mg and without phytol), pheophytins (without central Mg), chlorophyllides (without phytol), hydroxychlorophylls (with additional hydroxyl groups), epimers, and lactones (Maroneze et al., 2019).

These structures are accumulated inside the microalgal cell, and consequently, their extraction requires methods capable of promoting the release of this intracellular compound. Various approaches have been explored to extract microalgal chlorophyll, as no universal protocol applies to all species (Viera & Roca, 2020). Therefore, Table 6.1 presents the most relevant mechanical/physical, chemical, and biological methods (categorized for didactic purposes) for promoting the extraction of this microalgal pigment. Additionally, promising approaches are also highlighted. Although based on different principles, all methods aim to primarily disrupt the multilayer cellulose structure of the microorganism's cell wall.

Among the physical methods, the most commonly used are: grinding with a mortar and pestle, bead-beating, ultrasound-assisted extraction (UAE), and microwave-assisted extraction (MAE). Grinding with a mortar and pestle is based on the mixing of biomass and an organic solvent with a polarity similar to that of chlorophyll (typically acetone, ethyl acetate, or methanol), followed by homogenization, centrifugation, and repeated collection of the pigmented extract (Simon & Helliwell, 1998). In this method, the friction between the biomass and the mortar causes cell rupture, exposing the chlorophyll, which is then transferred to the solvent. Typically,

MICROALGAE CHLOROPHYLLS

	R1	R2	R3
Chlorophyll *a*	CH_3	$H_2C{=}CH_2$	CH_3
Chlorophyll *b*	CHO	$H_2C{=}CH_2$	CH_3
Chlorophyll *d*	CH_3	COH	CH_3
Chlorophyll *f*	CH_3	$H_2C{=}CH_2$	COH

	R4	R5	R6
Chlorophyll c_1	CH_3	CH_2CH_3	CH_2CH_2COOH
Chlorophyll c_2	CH_3	$H_2C{=}CH_2$	CH_2CH_2COOH
Chlorophyll c_3	$COOCH_3$	$H_2C{=}CH_2$	CH_2CH_2COOH
Chlorophyll c_4	$COOCH_3$	CH_2CH_3	CH_2CH_2COOH

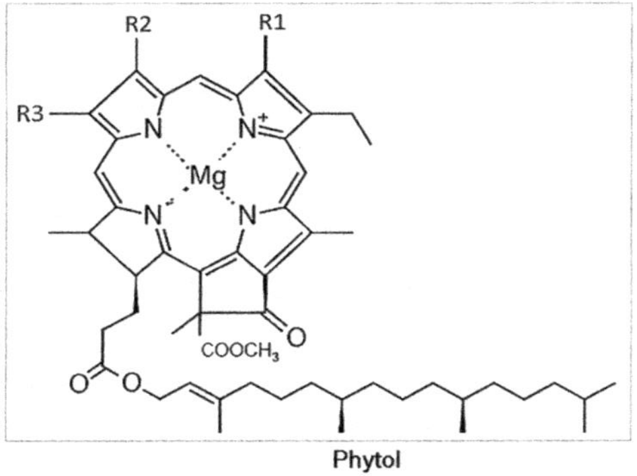

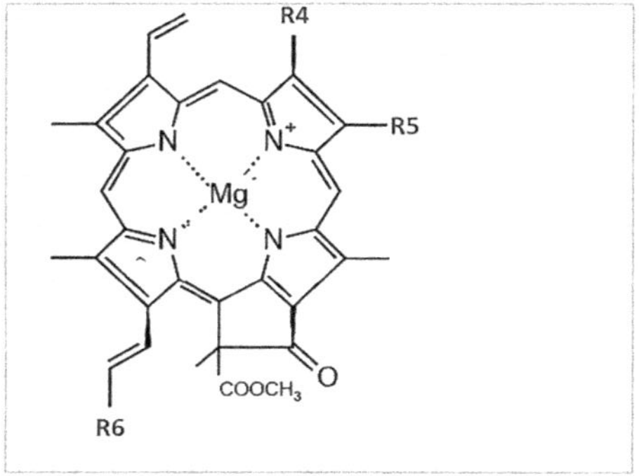

FIGURE 6.3 Structure of main microalgae chlorophylls.

TABLE 6.1
Main Methods of Microalgal Chlorophyll Extraction

Methods	Principle	Advantages	Disadvantages	Microalgae	References
			Physical/Mechanical		
Maceration	The biomass is macerated/crushed in a mortar and pestle together with an organic solvent of similar polarity to the pigment	Simple to use; effective quantitative	Longer extraction time; degradation of compounds; use of toxic solvents	*Phormidium; Scenedesmus obliquus; Chlorella vulgaris; Arthrospira platensis; Chaetoceros calcitrans*	Rodrigues et al., 2015; Vendruscolo et al., 2019; Fernandes et al., 2021; Nass et al., 2022; Nascimento, Pinheiro, et al., 2022
Bead-beating	Damage to the cells caused by the high-speed spinning of the biomass slurry with fine beads of a specific frequency to destroy the cell wall	Applicable to wet microalgae	Sample heating	*Chlorella vulgaris; Arthrospira platensis*	Hamed, 2016; Águila-Carricondo et al., 2023
UAE	Rupture of the cell wall of microalgae through the propagation of ultrasonic waves	Not heating; more economical	Generation of cell-damaging heat when prolonged	*Nannochloropsis gaditana, Dunaliella salina, Cylindrotheca closterium, Phormidium autumnale, Scenedesmus obliquus, and Arthrospira platensis*	Macías-Sánchez et al., 2005; Macías-Sánchez et al., 2009; Pasquet et al., 2011; Fernandes et al., 2016; Fernandes et al., 2024; Pérez & Albero, 2023
MAE	Absorption of microwave radiation converted into thermal energy – cell rupture	Instantaneous heating; reduced energy; effective; higher yield; solvent volumes and time-reduced	Generates heat; high maintenance cost	*Cylindrotheca closterium; Dunaliella tertiolecta; Chlorella vulgaris*	Pasquet et al., 2011; Al Hattab et al., 2015; Thirugnanasambandham et al., 2015; Georgiopoulou et al., 2023
			Chemical		
Solvents	polarity and/or solubility with the compound	Simple to use (Traditional method); industrial scale	Requires prior cell rupture; toxic; not eco-friendly	*Phormidium autumnale; Scenedesmus obliquus; Phaeodactylum tricornutum; Chlorella; Scenedesmus quadricauda; Selenastrum capricornutum; Microcystis aeruginosa; Dunaliella salina*	Fernandes et al., 2016; Maroneze et al., 2019; Mantoura et al., 1983; Schumann et al., 2005; Sartory & Grobbelaar, 1984; Macías-Sánchez et al., 2009

SPE	Is based on the use of solvents at temperatures and pressures above their critical points	Effective; nontoxic; Eco-friendly	Prior optimization of process conditions is required	*Nannochloropsis sp.; Dunaliella salina*	Macías-Sánchez et al., 2008; Nobre et al., 2013; Esquivel-Hernández et al., 2017
			Biological		
EAE	Specific action – lysis of the cell/membrane wall	Rapid; effective; specific; eco-friendly	Costly; limited industrial scale	*Phaeodactylum tricornutum**	Zuorro et al., 2016; Gille et al., 2019; Pagels et al., 2021
			Promising Approach		
ILs	Disruption of the hydrogen bond network of the microalgae cell wall	Effective; nontoxic; eco-friendly; biocompatibility; thermal stability; low boiling point; high conductivity; diverse electrochemical and polar characteristics	High cost; challenging manufacturing processes; low biodegradability	*Neochloris oleoabundans; Arthrospira platensis; Scenedesmus obliquus*	Ünlü et al., 2019; Suarez Ruiz et al., 2020; Assunção et al., 2022; Zhang et al., 2023; Fernandes et al., 2024
DESs	Interaction of the eutectic solvent with target solutes affects the structure of the cell membrane and releases intracellular compounds	Effective; nontoxic; eco-friendly; ease of manufacture; cheap components; biodegradable	Synergistic effect of its components may result in fewer ecological impacts; exploratory studies	*Scenedesmus dimorphus*	Cicci et al., 2017; Boateng, 2023; Bashir et al., 2023; Amesho et al., 2022; Fatima et al., 2024

UAE, ultrasound-assisted extraction; MAE, microwave-assisted extraction; SPE, solid-phase extraction; EAE, enzyme-assisted extraction; ILs, ionic liquids; DESs, deep eutectic solvents.

diatomaceous earth or glass microspheres are used to increase the biomass–mortar–solvent contact surface, which consequently enhances the extraction efficiency.

The advantage of this method lies in its simplicity and the possibility of exhaustive extraction, allowing for a more effective quantitative yield. On the other hand, the multiple steps result in longer extraction times, exposing the structures to environmental factors (light, oxygen, and the solvent itself), which can interfere with qualitative determination (leading to the formation of degradation compounds). Additionally, although this method is applicable to dry biomass, repeated extractions relying on organic solvents lead to a high accumulation of toxic waste. Despite these drawbacks, the use of mortar and pestle has been widely employed in the extraction of chlorophylls from *Phormidium* (Rodrigues et al., 2015), *Scenedesmus obliquus* (Fernandes et al., 2021; Vendruscolo et al., 2019), *Chlorella vulgaris, Arthrospira platensis* (Nass et al., 2022), and *Chaetoceros calcitrans* (Nascimento, Pinheiro, et al., 2022).

Another physical/mechanical method used to extract microalgal metabolites is bead-beating. This method damages the microalgal cells through high-speed centrifugation of a biomass slurry with fine beads, oscillating at a specific frequency to break the cell wall. These beads vary depending on the type of cells, with those of 0.5 mm in diameter being ideal for microalgae (Hamed, 2016). A significant advantage is the possibility of applying the method to wet biomass, eliminating drying costs. On the other hand, this approach increases temperature due to the high rotation speed, necessitating a cooling jacket to prevent chlorophyll degradation. These factors hinder the scalability of the process and increase energy consumption (Kumar et al., 2015). Bead-beating has proven effective for chlorophyll extraction from *Chlorella vulgaris* and *Arthrospira platensis* (Águila-Carricondo et al., 2023).

Unlike the methods previously mentioned, UAE is considered a green technique, or eco-friendly (Somwongin et al., 2023). This method allows the rupture of microalgal cell walls through the propagation of ultrasonic waves in the extraction solvent. Its efficiency is associated with the cavitation effect, which is responsible for increasing the mass transfer rate, where bubbles may collapse, generating localized pressure, causing cell rupture, and enhancing the release of intracellular substances into the solvent (Kurd & Samavati, 2015; Zhao et al., 2018). The ultrasonic waves can be propagated in a bath or through a probe.

UAE is the most widely used recent extraction technique for solid samples, a fact attributed to its simplicity and the ability to use a wide range of solvents with different polarities that are less toxic (Pérez & Albero, 2023). Moreover, it does not generate heat (unless exposure to ultrasonic waves is prolonged), thus causing less thermal degradation of the compounds. From an economic perspective, it is considered less costly because it is faster and does not require additives or chemicals that would need to be removed afterward (Al Hattab et al., 2015). When combined with the use of green solvents, it becomes even more sustainable (Fernandes et al., 2024). Macías-Sánchez et al. (2005), Macías-Sánchez et al. (2009), Pasquet et al. (2011), Fernandes et al. (2016), and Fernandes et al. (2024) used this technology for the extraction of chlorophylls from *Nannochloropsis gaditana*, *Dunaliella salina*, *Cylindrotheca closterium*, *Phormidium autumnale*, *Scenedesmus obliquus*, and *Arthrospira platensis*, respectively.

Another physical method in line with the principles of green chemistry is MAE. MAE involves the rapid increase in temperature of the biomass (conversion of microwave radiation into thermal energy) through molecular rotation (dipole–dipole interactions) and increased ionic movement, resulting in the rupture of the outer layer of the cell structure (Thirugnanasambandham et al., 2015). The advantages of this unconventional method include reduced thermal gradients, along with the instantaneous heating of the sample, reduced energy consumption, higher extraction yield, reduced solvent volumes, and faster processing compared to traditional methods (Georgiopoulou et al., 2023).

On the other hand, the heat generated can degrade biomolecules, and the cost of maintaining the process is high (Al Hattab et al., 2015). Another drawback of this process is that the solid/liquid ratio, microwave power, time, and temperature can significantly affect the extraction efficiency (Silva et al., 2018). Although MAE has been extensively studied for the extraction of apolar compounds, there are few studies focused on the extraction of microalgal chlorophylls. Notable examples of applications include a study by Pasquet et al. (2011) with *Cylindrotheca closterium* and *Dunaliella tertiolecta*, as well as a study by Georgiopoulou et al. (2023) with *Chlorella vulgaris*.

Among the chemical extraction methods, the use of organic solvents such as dimethylformamide, ethanol, methanol, acetone, or their mixtures is more common (traditional method), with the first solvent being the most effective; however, its use is avoided due to its high toxicity (Jorge et al., 2024). The polarity or solubility of the compound determines the best extraction solvent, where the organic solution must be capable of penetrating the cell membrane, dissolving lipids and lipoproteins from the chloroplast, and thus facilitating the extraction of the pigments (Viera & Roca, 2020). Since it is an intracellular compound (part of a solid biomass), this type of extraction is characterized as solid–liquid extraction (biomass–solvent), requiring prior treatments aimed at breaking the microalgal cell to facilitate the release of chlorophylls, which will then be carried by the extracting solvent (Jorge et al., 2024). As mentioned in the earlier paragraphs (maceration method), the combination of mechanical extraction and solvent is inseparable for microalgae, often resulting in physical–chemical approaches in most cases.

The main advantage of this method is the simplicity of use and industrial scalability. However, the need for pretreatment, along with the high toxicity to humans and the environment, makes the use of organic solvents a disadvantageous approach (Hamed, 2016). *Phormidium autumnale* (Fernandes et al., 2016), *Scenedesmus obliquus* (Maroneze et al., 2019), *Phaeodactylum tricornutum* (Mantoura et al., 1983), *Chlorella* (Schumann et al., 2005), *Scenedesmus quadricauda*, *Selenastrum capricornutum*, *Microcystis aeruginosa* (Sartory & Grobbelaar, 1984), and *Dunaliella salina* (Macías-Sánchez et al., 2009) are examples of the diverse applications in phytoplankton.

Unlike the traditional chemical method, supercritical fluid extraction (SFE) is a chemical approach considered a green technology, which uses carbon dioxide under supercritical conditions as a nontoxic extracting agent to separate lipophilic compounds from cells (Esquivel-Hernández et al., 2017). SFE offers several advantages, including high purity, fewer processing steps, the use of moderate temperatures that reduce compound degradation, and greater safety compared to traditional extraction methods. However, since the extraction of microalgal chlorophyll by SFE depends on the fluid density (a function of pressure and temperature), it is necessary to optimize the process conditions for each microalgal species (Viera & Roca, 2020). Nobre et al. (2013) and Macías-Sánchez et al. (2008) applied SFE to the biomass of *Nannochloropsis sp.* and *Dunaliella salina*, respectively.

In this same perspective, biological methods are ecologically friendly alternatives, such as enzyme-assisted extraction (EAE), which is based on the specific action of enzymes. This action (hydrolysis) aims to break down the structure of the cell wall or membrane, releasing intracellular compounds such as chlorophyll (Wijesinghe & Jeon, 2012). The specificity of the enzymes (cell wall hydrolysis, in the case of cellulase) is the main advantage of this methodology. Likewise, the mild conditions of pH and temperature, absence of corrosion, higher yield, and the ability to use wet biomass are significant benefits (Pagels et al., 2021). However, the high cost of enzymes, the need for process control and stability, coupled with the low reaction speed, limit their large-scale application (Zuorro et al., 2016). To date, most EAE methods in microalgae are used as a pretreatment, as demonstrated in a study by Gille et al. (2019) with *Phaeodactylum tricornutum.*

In addition to the aforementioned methods, current global trends increasingly drive the adoption of promising approaches grounded in the principles of green chemistry and sustainability. This leads to a new era of developing more ecologically friendly and sustainable extraction processes (Assunção et al., 2022). Within this trend, ionic liquids (ILs) and deep eutectic solvents (DESs) have been explored to obtain more nontoxic and eco-friendly microalgal extracts.

ILs are considered promising options for the fractionation of microalgal molecules as they disrupt the hydrogen bond network of the cell wall, exposing intracellular components (Zhang et al., 2023). The prominence of ILs stems from their ability to enhance extraction efficiency, biocompatibility, and recyclability, which minimizes environmental and economic impact, as well as their thermal stability, low volatility, high conductivity, and diverse electrochemical and polarity characteristics (Mesquita et al., 2019; Fernandes et al., 2024). However, the high cost, challenging manufacturing processes, and low biodegradability are some disadvantages of these liquids (Bashir et al., 2023; Ünlü et al., 2019). Recently, this approach has been used to extract chlorophyll and other bioactive compounds from *Scenedesmus obliquus*, *Arthrospira platensis*, and *Neochloris oleoabundans* (Fernandes et al., 2024; Suarez Ruiz et al., 2020).

Considered analogs of ILs, DESs differ in that they result from the mixture of two or more liquid components that act as hydrogen bond acceptors or donors (Boateng, 2023). A hydrogen bond network formed by the components reduces the liquefaction of solid mixtures. Thus, the extraction mechanism is based on the interaction of the eutectic solvent with target solutes, which affects the cell membrane structure and releases intracellular compounds (Bashir et al., 2023). It is considered an economic, eco-friendly extraction method, easy to manufacture, with inexpensive and biodegradable components at elevated temperatures (Fatima et al., 2024). On the other hand, some studies suggest that the synergistic effects of certain components may result in less eco-friendly impacts (Martínez et al., 2022). Additionally, the high viscosity and hygroscopicity of some DESs, the difficulty in regenerating and separating the reactive components, as well as the lack of structural information, make this type of approach less attractive (Amesho et al., 2022). DESs composed of 1,2-propanediol, choline chloride, and water (1:1:1) were used to extract chlorophylls and other cellular compounds from *Scenedesmus dimorphus* biomass (Cicci et al., 2017).

Finally, there is no universal method for microalgal chlorophyll extraction; all the approaches mentioned above are applicable for extracting this valuable bioactive compound. However, it is important to consider that chlorophyll is subject to structural transformations due to its instability in response to environmental factors such as light, temperature, and pH. Therefore, to ensure unequivocal identification later, these factors need to be taken into account when choosing the extraction method. Another consideration arises from the simultaneous extraction of compounds with similar chemical characteristics, such as carotenoids. Both pigments are valuable bioactive, so exploring strategies to ensure efficient, eco-friendly extraction that allows for rapid fractionation without structural damage to both molecules would be an ambitious goal.

6.4 INNOVATIONS AND TRENDS

6.4.1 Global Microalgae Chlorophyll Market

Growing awareness of the beneficial functional compounds found in microalgae and their potential to offer more sustainable alternatives to traditional food and energy production methods is driving increased interest in this area. Amid the COVID-19 pandemic in 2020, the global microalgae market was valued at approximately USD 3.4 billion. It is expected to grow at a CAGR of 4.3%, potentially reaching USD 4.6 billion by 2027 (Curtin et al., 2019; Loke Show, 2022). Recent applications of microalgae include wastewater treatment, natural food colorants and dyes, biofuel production, dietary supplements, antioxidants, and innovative uses in cosmetics and pharmaceuticals (Jacob-Lopes et al., 2019; Nascimento, Nass, et al., 2022). With the expansion of the microalgae pigment market, there has been a substantial increase in the number of registered patents over the past few decades. This surge reflects the growing interest and innovation in harnessing microalgae for various applications, ranging from sustainable food production to advanced biotechnological solutions (Silva et al., 2020).

Like many natural pigments, chlorophylls serve multiple purposes in the food additive industry, with their most widespread use being as colorants. Due to concerns over the potential side effects of synthetic pigments, food manufacturers are increasingly opting for natural green colorants to enhance, intensify, or standardize the color of foods, delivering optimal color intensity and hue to meet consumer expectations (García et al., 2021; Sun et al., 2023). Microalgae have emerged as a sustainable source for producing natural food-grade colorants with low allergenicity, toxicity, and carcinogenicity (Nabi et al., 2023). Chlorophylls, some of the most abundant green pigments, are gaining significant attention as colorants and functional food supplements in the food, feed, cosmetic, and pharmaceutical industries. In addition to their role as colorants, microalgal pigments also offer auxiliary benefits, such as preservation and antiseptic properties in food storage, thanks to their physiological activities. Chlorophylls have also been proven to be effective as food deodorizers. Chlorophylls, with their vibrant green hue, have been widely applied to various products such as beverages, fruit juices, pasta, dairy, sweeteners, and soups (Carvalho et al., 2011; Silva et al., 2020).

However, maintaining the stability of natural pigments like chlorophyll remains a significant technological challenge, as chlorophyll tends to lose its antioxidant properties when exposed to factors such as oxygen, elevated temperatures, or light. One promising approach to enhance their stability is through stabilization in ethanolic solutions for food applications. Moreover, studies have shown that soybean protein isolation can effectively improve the thermal stability of chlorophyll. This improvement is attributed to the interactions between the chlorophyll molecules and the hydrophobic cavity of the protein, which helps protect the pigment under thermal stress (Cao et al., 2021; Ferreira et al., 2023; Sanna & Fadda, 2022). Furthermore, in some bioproducts, heat treatment can be used to deactivate the chlorophyllase enzyme and reduce the conversion of chlorophyll to pheophytin. Additionally, the formation of a vibrant green metallo-chlorophyll complex can be achieved by adding Cu^{2+} or Zn^{2+} salts. Controlling process conditions such as temperature, pH, and ionic strength, as well as incorporating surface-active agents, can further enhance chlorophyll stability (Indrasti et al., 2018; Zheng et al., 2014).

6.4.2 Recent Patents of Chlorophylls from Microalgae

The patent landscape plays a crucial role in identifying innovation opportunities and guiding financial investors in their decision-making processes for the commercialization of disruptive and novel technologies (Kim & Lee, 2019). Thus, patents provide a valuable framework for assessing the technological advancements within a particular field. In the case of microalgae-derived products, producers are actively forming strategic partnerships to enhance commercialization efforts and capture larger market shares. These collaborations often include comprehensive plans to advance research, development, and innovation (R&D&I) initiatives. Within the Food Technology and Engineering sector, alternative products derived from microalgae are receiving increased attention. A wealth of cutting-edge research and patents is being dissected, although many of these innovations remain in experimental phases and have yet to be fully validated. Moreover, numerous studies focus on technological innovations that still require further investigation before being widely adopted or commercialized. This evolving landscape highlights the significant potential for growth and the need for continued research in the microalgae field (Naik et al., 2024; Severo et al., 2024).

In terms of chlorophyll production, the microalgae *Chlorella Vulgaris* stands out as a great source of this pigment (Khairunnisa et al., 2024). From 2009 to 2019, research predominantly focused on the pigments phycocyanin, chlorophylls, β-carotene, and astaxanthin. The main microalgae sources identified for these pigments were *Chlorella vulgaris* for chlorophylls (Khairunnisa et al., 2024), *Spirulina platensis* for phycocyanin (Van Hieu et al., 2021), *Haematococcus pluvialis* for astaxanthin (He et al., 2018), and *Dunaliella salina* for β-carotene (Pourkarimi et al., 2020). Bibliometric studies revealed that China, Brazil, and France were the most prolific in publishing research in this field. *Haematococcus pluvialis* stood out in terms of patent filings, suggesting that its production technology is the most advanced. There has also been a rise in both European projects and company investments aimed at microalgae cultivation and pigment extraction. *Spirulina platensis* has gained particular attention for its phycocyanin production, an area where the biorefinery approach is increasingly being applied. Some examples of patents that aim to improve its rentability include: "Method for increasing content of Chlorella chlorophyll" (CN109355194A), "Chlorella with high content of chlorophyll and carotenoid" (JP2016067313A), "Method for producing Chlorella powder with increased chlorophyll content and antioxidant activity" (KR101389471B1), "Green color-based beverage composition" (JP2008161094A), and "'Chlorophyll-rich and salt-resistant Chlorella" (CA2355289A1) (Silva et al., 2020).

Additionally, recent patents have introduced methods aimed at boosting overall biomass yield, which in turn enhances the industrial extraction of chlorophyll and other valuable compounds. Such as patent RO137350A2, titled "Biotechnological process for stimulation of growth of *chlorella sorokiniana utex* 1230 biomass quantity" (Ardelean et al., 2021). Moreover, patent WO2024130271A2 proposes a method to control contaminants in large-scale microalgae production, by balancing the growth of contaminant fermentative bacteria. By controlling the metals and urea present in the medium, a better growth scale can be reached. This patent was published in 2024 (Hamrick, 2024). Interestingly, there is a recent patent that showcases a strain of Chlorella deficient in chlorophyll production, which owns 50% of protein by its dry weight, and the patent number is WO2024161108A1 (Spicer et al., 2024). Thus, the patent landscape for microalgae is evolving rapidly, with innovations spanning various aspects of production, extraction, and diverse applications.

6.5 CONCLUSION

Microalgae pigments are poised for success in the global market. They help capture carbon dioxide, contributing to the fight against climate change. Ongoing research is refining production techniques and expanding their uses. As we strive for a more sustainable and healthier future, microalgae pigments are providing innovative solutions to transform industries. With continuous advancements in science, production methods will improve, leading to new applications and benefits. The future looks promising, with fresh discoveries and opportunities driving us toward a more sustainable world.

REFERENCES

Águila-Carricondo, P., de la Roche Cadavid, J. P., Galán, P. L., Bautista, L. F., & Vicente, G. (2023). New green biorefineries from cyanobacterial-microalgal consortia: Production of chlorophyll-rich extracts for the cosmetic industry and sustainable biogas. *Journal of Cleaner Production*, *429*, 139652. https://doi.org/10.1016/J.JCLEPRO.2023.139652

Al Hattab, M., Ghaly, A., & Hammoud, A. (2015). Microalgae oil extraction pre-treatment methods: Critical review and comparative analysis. *Journal of Fundamentals of Renewable Energy and Applications*, *5*(4). https://doi.org/10.4172/2090-4541.1000172

Amesho, K. T. T., Lin, Y. C., Mohan, S. V., Halder, S., Ponnusamy, V. K., & Jhang, S. R. (2022). Deep eutectic solvents in the transformation of biomass into biofuels and fine chemicals: A review. *Environmental Chemistry Letters*, *21*(1), 183–230. https://doi.org/10.1007/S10311-022-01521-X

Ardelean, I., Moisescu, C., Ardelean, A. V., & Negruț, C. D. (2021). Biotechnological process for stimulation of growth of Chlorella sorokiniana UTEX 1230 biomass quantity (Patent RO137350A2). Romania.

Assunção, J., Amaro, H. M., Malcata, F. X., & Guedes, A. C. (2022). Factorial optimization of ultrasound-assisted extraction of phycocyanin from Synechocystis salina: Towards a biorefinery approach. *Life (Basel, Switzerland)*, 12(9). https://doi.org/10.3390/LIFE12091389

Bashir, I., Dar, A. H., Dash, K. K., Pandey, V. K., Fayaz, U., Shams, R., Srivastava, S., & Singh, R. (2023). Deep eutectic solvents for extraction of functional components from plant-based products: A promising approach. *Sustainable Chemistry and Pharmacy*, *33*, 101102. https://doi.org/10.1016/J.SCP.2023.101102

Boateng, I. D. (2023). Evaluating the status quo of deep eutectic solvent in food chemistry. Potentials and limitations. *Food Chemistry*, *406*, 135079. https://doi.org/10.1016/J.FOODCHEM.2022.135079

Bourgougnon, N., Burlot, A. S., & Jacquin, A. G. (2021). Algae for global sustainability? In Jean-Pierre Jacquot (ed.), *Advances in Botanical Research* (Vol. 100, pp. 145–212). Academic Press. https://doi.org/10.1016/bs.abr.2021.01.003

Cao, J., Li, F., Li, Y., Chen, H., Liao, X., & Zhang, Y. (2021). Hydrophobic interaction driving the binding of soybean protein isolate and chlorophyll: Improvements to the thermal stability of chlorophyll. *Food Hydrocolloids*, *113*. https://doi.org/10.1016/j.foodhyd.2020.106465

Carvalho, A. P., Silva, S. O., Baptista, J. M., & Malcata, F. X. (2011). Light requirements in microalgal photobioreactors: An overview of biophotonic aspects. *Applied Microbiology and Biotechnology*, *89*(5). https://doi.org/10.1007/s00253-010-3047-8

Casagli, F., Beline, F., Ficara, E., & Bernard, O. (2023). Optimizing resource recovery from wastewater with algae-bacteria membrane reactors. *Chemical Engineering Journal*, *451*, 138488. https://doi.org/10.1016/j.cej.2022.138488

Cicci, A., Sed, G., & Bravi, M. (2017). Potential of choline chloride – Based natural deep eutectic solvents (NaDES) in the extraction of microalgal metabolites. *Chemical Engineering Transactions*, *57*, 61–66. https://doi.org/10.3303/CET1757011

Curtin, J., McInerney, C., Gallachóir, B. Ó., Hickey, C., Deane, P., & Deeney, P. (2019). Quantifying stranding risk for fossil fuel assets and implications for renewable energy investment: A review of the literature. *Renewable and Sustainable Energy Reviews*, *116*, 109402. https://doi.org/10.1016/j.rser.2019.109402

Deng, Z., Khatiwada, J. R., Fan, L., & Qin, W. (2022). Algae for wastewater treatment and production of biofuels and bioproducts. *International Journal of Environmental Research*, *16*(3), 34. https://doi.org/10.1007/s41742-022-00412-x

Esquivel-Hernández, D. A., Ibarra-Garza, I. P., Rodríguez-Rodríguez, J., Cuéllar-Bermúdez, S. P., Rostro-Alanis, M. D. J., Alemán-Nava, G. S., García-Pérez, J. S., & Parra-Saldívar, R. (2017). Green extraction technologies for high-value metabolites from algae: A review. *Biofuels, Bioproducts and Biorefining*, *11*(1), 215–231. https://doi.org/10.1002/BBB.1735

Fatima, I., Munir, M., Qureshi, R., Hanif, U., Gulzar, N., & Sheikh, A. A. (2024). Advanced methods of algal pigments extraction: A review. *Critical Reviews in Food Science and Nutrition*, *64*(27). https://doi.org/10.1080/10408398.2023.2216782

Fernandes, A., Nogara, G. P., Zepka, L. Q., Jacob-Lopes, E., Menezes, C. R., Cichoski, A. J. S., & Mercadante, A. Z. (2016). Identification of chlorophyll molecules with peroxyl radical scavenger capacity in microalgae Phormidium autumnale using ultrasound-assisted extraction. *Food Research International*, *99*, 1036–1041. https://doi.org/10.1016/j.foodres.2016.11.011

Fernandes, A. S., Caetano, P. A., Jacob-Lopes, E., Zepka, L. Q., & de Rosso, V. V. (2024). Alternative green solvents associated with ultrasound-assisted extraction: A green chemistry approach for the extraction of carotenoids and chlorophylls from microalgae. *Food Chemistry*, *455*, 139939. https://doi.org/10.1016/J.FOODCHEM.2024.139939

Fernandes, A. S., Nascimento, T. C., Pinheiro, P. N., de Rosso, V. V., de Menezes, C. R., Jacob-Lopes, E., & Zepka, L. Q. (2021). Insights on the intestinal absorption of chlorophyll series from microalgae. *Food Research International*, *140*, 110031. https://doi.org/10.1016/J.FOODRES.2020.110031

Ferreira, A. S., Pereira, L., Canfora, F., Silva, T. H., Coimbra, M. A., & Nunes, C. (2023). Stabilization of natural pigments in ethanolic solutions for food applications: The case study of *Chlorella vulgaris*. *Molecules*, *28*(1). https://doi.org/10.3390/molecules28010408

Galanakis, C. M. (2021). Functionality of food components and emerging technologies. *Foods*, *10*(1), 1–26. https://doi.org/10.3390/foods10010128

Gao, F., Yang, Z. Y., Zhao, Q. L., Chen, D. Z., Li, C., Liu, M., … Chen, J. M. (2021). Mixotrophic cultivation of microalgae coupled with anaerobic hydrolysis for sustainable treatment of municipal wastewater in a hybrid system of anaerobic membrane bioreactor and membrane photobioreactor. *Bioresource Technology*, *337*, 125457. https://doi.org/10.1016/j.biortech.2021.125457

García-Vaquero, M., Brunton, N., & Lafarga, T. (2021). Microalgae as a source of pigments for food applications. In Tomás Lafarga and Gabriel Acién (eds.), *Cultured Microalgae for the Food Industry: Current and Potential Applications*. Academic Press. https://doi.org/10.1016/B978-0-12-821080-2.00014-9

Georgiopoulou, I., Tzima, S., Louli, V., & Magoulas, K. (2023). Process optimization of microwave-assisted extraction of chlorophyll, carotenoid and phenolic compounds from *Chlorella vulgaris* and comparison with conventional and supercritical fluid extraction. *Applied Sciences*, *13*(4), 2740. https://doi.org/10.3390/APP13042740

Gille, A., Hollenbach, R., Trautmann, A., Posten, C., & Briviba, K. (2019). Effect of sonication on bioaccessibility and cellular uptake of carotenoids from preparations of photoautotrophic *Phaeodactylum tricornutum*. *Food Research International*, *118* (July 2017), 40–48. https://doi.org/10.1016/j.foodres.2017.12.040

Hamrick, E. B. (2024). Contamination control when growing green algae. U.S. Patent Application n. 18/640,396, 14 ago. 2025.

Hamed, I. (2016). The evolution and versatility of microalgal biotechnology: A review. *Comprehensive Reviews in Food Science and Food Safety*, *15*(6), 1104–1123. https://doi.org/10.1111/1541-4337.12227

He, B., Hou, L., Dong, M., Shi, J., Huang, X., Ding, Y., Cong, X., Zhang, F., Zhang, X., & Zang, X. (2018). Transcriptome analysis in *Haematococcus pluvialis*: Astaxanthin induction by high light with acetate and Fe2+. *International Journal of Molecular Sciences*, *19*(1). https://doi.org/10.3390/ijms19010175

Hu, J., Nagarajan, D., Zhang, Q., Chang, J. S., & Lee, D. J. (2018). Heterotrophic cultivation of microalgae for pigment production: A review. *Biotechnology Advances*, 36(1), 54–67. https://doi.org/10.1016/j.biotechadv.2017.09.009

Indrasti, D., Andarwulan, N., Purnomo, E. H., & Wulandari, N. U. R. (2018). Stability of chlorophyll as natural colorant: A review for suji (dracaena angustifolia (medik.) roxb.) leaves' case. *Current Research in Nutrition and Food Science*, *6*(3). https://doi.org/10.12944/CRNFSJ.6.3.04

Jacob-Lopes, E., Maroneze, M. M., Deprá, M. C., Sartori, R. B., Dias, R. R., & Zepka, L. Q. (2019). Bioactive food compounds from microalgae: An innovative framework on industrial biorefineries. *Current Opinion in Food Science*, *25*. https://doi.org/10.1016/j.cofs.2018.12.003

Jorge, A. M. S., Pedroso, P. R. M., & Pereira, J. F. B. (2024). Sustainable extraction and utilization of chlorophyll from microalgae for eco-friendly wool dyeing. *Journal of Cleaner Production*, *451*, 142009. https://doi.org/10.1016/J.JCLEPRO.2024.142009

Khairunnisa, K., Hartati, R., & Widowati, I. (2024). Chlorophyll content of Chlorella vulgaris (Beijerinck, 1890) on different light intensity. *Buletin Oseanografi Marina*, *13*(1). https://doi.org/10.14710/buloma.v13i1.59218

Kim, S. Y., & Lee, H. J. (2019). The effect of patent acquisition on subsequent patenting activity. *World Patent Information*, *59*. https://doi.org/10.1016/j.wpi.2019.101933

Kumar, R. R., Rao, P. H., & Arumugam, M. (2015). Lipid extraction methods from microalgae: A comprehensive review. *Frontiers in Energy Research*, *3*(January), 125610. https://doi.org/10.3389/FENRG.2014.00061

Kurd, F., & Samavati, V. (2015). Water soluble polysaccharides from *Spirulina platensis*: Extraction and in vitro anti-cancer activity. *International Journal of Biological Macromolecules*, *74*, 498–506. https://doi.org/10.1016/j.ijbiomac.2015.01.005

Loke Show, P. (2022). Global market and economic analysis of microalgae technology: Status and perspectives. *Bioresource Technology*, *357*. https://doi.org/10.1016/j.biortech.2022.127329

Macías-Sánchez, M. D., Mantell, C., Rodríguez, M., Martínez De La Ossa, E., Lubián, L. M., & Montero, O. (2005). Supercritical fluid extraction of carotenoids and chlorophyll a from *Nannochloropsis gaditana*. *Journal of Food Engineering*, *66*(2), 245–251. https://doi.org/10.1016/J.JFOODENG.2004.03.021

Macías-Sánchez, M. D., Mantell, C., Rodríguez, M., Martínez de la Ossa, E., Lubián, L. M., & Montero, O. (2009). Comparison of supercritical fluid and ultrasound-assisted extraction of carotenoids and chlorophyll a from *Dunaliella salina*. *Talanta*, *77*(3), 948–952. https://doi.org/10.1016/J.TALANTA.2008.07.032

Macías-Sánchez, M. D., Serrano, C. M., Rodríguez, M. R., de la Ossa, E. M., Lubián, L. M., & Montero, O. (2008). Extraction of carotenoids and chlorophyll from microalgae with supercritical carbon dioxide and ethanol as cosolvent. *Journal of Separation Science*, *31*(8), 1352–1362. https://doi.org/10.1002/JSSC.200700503

Mantoura, R. F. C., Llewellyn, C. A., Mantoura, R. F. C., & Llewellyn, C. A. (1983). The rapid determination of algal chlorophyll and carotenoid pigments and their breakdown products in natural waters by reverse-phase high-performance liquid chromatography. *AcAC*, *151*(C), 297–314. https://doi.org/10.1016/S0003-2670(00)80092-6

Manzoor, M., Singh, J., Gani, A., & Noor, N. (2021). Valorization of natural colors as health-promoting bioactive compounds: Phytochemical profile, extraction techniques, and pharmacological perspectives. *Food Chemistry*, *362*(May), Article 130141. https://doi.org/10.1016/j.foodchem.2021.130141

Maroneze, M. M., Zepka, L. Q., Lopes, E. J., Pérez-Gálvez, A., & Roca, M. (2019). Chlorophyll oxidative metabolism during the phototrophic and heterotrophic growth of *Scenedesmus obliquus*. *Antioxidants (Basel, Switzerland)*, *8*(12). https://doi.org/10.3390/ANTIOX8120600

Martínez, G. M., Townley, G. G., & Martínez-Espinosa, R. M. (2022). Controversy on the toxic nature of deep eutectic solvents and their potential contribution to environmental pollution. *Heliyon*, *8*(12), e12567. https://doi.org/10.1016/J.HELIYON.2022.E12567

Mesquita, L. M., Murador, D. C., & de Rosso, V. V. (2019). Application of Ionic Liquid Solvents in the Food Industry. *Encyclopedia of Ionic Liquids*, 1–16. https://doi.org/10.1007/978-981-10-6739-6_8-1

Mineeva, N., & Semadeni, I. (2023). Chlorophyll content and photosynthetic activity of phytoplankton in reservoirs of the Volga River (Russia). *Phycology*, *3*(3), 368–381. https://doi.org/10.3390/phycology3030024

Mohammadi, M. A., Ahangari, H., Mousazadeh, S., Hosseini, S. M., & Dufossé, L. (2022). Microbial pigments as an alternative to synthetic dyes and food additives: A brief review of recent studies. *Bioprocess and Biosystems Engineering*, *45*(1), 1–12. https://doi.org/10.1007/s00449-021-02621-8

Nabi, B., Mukhtar, K., Ahmed, W., Faisal, M., Modassar, M., Nawaz, A., … Muhammad, R. (2023). Natural pigments: Anthocyanins, carotenoids, chlorophylls, and betalains as colorants in food products. *Food Bioscience*, *52*(January), Article 102403. https://doi.org/10.1016/j.fbio.2023.102403

Naik, B., Mishra, R., Kumar, V., Mishra, S., Gupta, U., Rustagi, S., Gupta, A. K., Preet, M. S., Bhatt, S. C., & Rizwanuddin, S. (2024). Micro-algae: Revolutionizing food production for a healthy and sustainable future. *Journal of Agriculture and Food Research*, *15*. https://doi.org/10.1016/j.jafr.2023.100939

Nascimento, T. C., Nass, P. P., Fernandes, A. S., Nörnberg, M. L., Zepka, Q. Z., & Jacob-Lopes, E. (2022). Microalgae carotenoids: An overview of biomedical applications. In Ashfaq Ahmad, Fawzi Banat, and Hanifa Taher (eds.), *Algal Biotechnology: Integrated Algal Engineering for Bioenergy, Bioremediation, and Biomedical Applications*. Elsevier. https://doi.org/10.1016/B978-0-323-90476-6.00013-3

Nascimento, T. C., Pinheiro, P. N., Fernandes, A. S., Caetano, P. A., Jacob-Lopes, E., & Zepka, L. Q. (2022). Insights on the bioaccessibility of natural pigments from diatom Chaetoceros calcitrans. *Molecules*, *27*(10), 3305. https://doi.org/10.3390/MOLECULES27103305

Nass, P. P., do Nascimento, T. C., Fernandes, A. S., Caetano, P. A., de Rosso, V. V., Jacob-Lopes, E., & Zepka, L. Q. (2022). Guidance for formulating ingredients/products from *Chlorella vulgaris* and *Arthrospira platensis* considering carotenoid and chlorophyll bioaccessibility and cellular uptake. *Food Research International*, *157*, 111469. https://doi.org/10.1016/J.FOODRES.2022.111469

Nobre, B. P., Villalobos, F., Barragán, B. E., Oliveira, A. C., Batista, A. P., Marques, P. A. S. S., Mendes, R. L., Sovová, H., Palavra, A. F., & Gouveia, L. (2013). A biorefinery from *Nannochloropsis* sp. microalga – Extraction of oils and pigments. Production of biohydrogen from the leftover biomass. *Bioresource Technology*, *135*, 128–136. https://doi.org/10.1016/J.BIORTECH.2012.11.084

Otálora, P., Guzmán, J. L., Acién, F. G., Berenguel, M., & Reul, A. (2023). An artificial intelligence approach for identification of microalgae cultures. *New Biotechnology*, *77*, 58–67. https://doi.org/10.1016/j.nbt.2023.07.003

Padrón, R. A. R., & Mérida, L. G. R. (2025). Hybrid and nonconventional photobioreactors applied to microalgae production. *Algal Bioreactors*, 483–491. https://doi.org/10.1016/B978-0-443-14058-7.00035-X

Pagels, F., Pereira, R. N., Vicente, A. A., & Guedes, A. C. (2021). Extraction of pigments from microalgae and cyanobacteria-A review on current methodologies. *Applied Sciences*, *11*(11), 5187. https://doi.org/10.3390/APP11115187

Pasquet, V., Chérouvrier, J. R., Farhat, F., Thiéry, V., Piot, J. M., Bérard, J. B., Kaas, R., Serive, B., Patrice, T., Cadoret, J. P., & Picot, L. (2011). Study on the microalgal pigments extraction process: Performance of microwave assisted extraction. *Process Biochemistry*, *46*(1), 59–67. https://doi.org/10.1016/J.PROCBIO.2010.07.009

Peng, S., Bi, R., Liu, J., Cui, J., Fu, X., Xiao, X., … Zhao, M. (2024). Spatial distribution of bacteria in response to phytoplankton community and multiple environmental factors in surface waters in *Sanggou Bay. Marine Environmental Research*, 106912. https://doi.org/10.1016/j.marenvres.2024.106912

Pérez, R. A., & Albero, B. (2023). Ultrasound-assisted extraction methods for the determination of organic contaminants in solid and liquid samples. *Trends in Analytical Chemistry, 166*, 117204. https://doi.org/10.1016/J.TRAC.2023.117204

Perez-Galvez, A., Viera, I., & Roca, M. (2018). Chemistry in the bioactivity of chlorophylls: An overview. *Current Medicinal Chemistry*, 24(40), 4515–4536. https://doi.org/10.2174/0929867324666170714102619

Pourkarimi, S., Hallajisani, A., Nouralishahi, A., Alizadehdakhel, A., & Golzary, A. (2020). Factors affecting production of beta-carotene from *Dunaliella salina* microalgae. *Biocatalysis and Agricultural Biotechnology, 29.* https://doi.org/10.1016/j.bcab.2020.101771

Rodrigues, D. B., Menezes, C. R., Mercadante, A. Z., Jacob-Lopes, E., & Zepka, L. Q. (2015). Bioactive pigments from microalgae *Phormidium autumnale. Food Research International, 77*, 273–279. https://doi.org/10.1016/j.foodres.2015.04.027

Rosa, M. E., Ferreira, A. M., Neves, C. M. S. S., Almeida, M. R., Barros, R., Cristovão, A. C., Sousa, A. C. A., Reis, P. M., Rebelo, L. P. N., Esperança, J. M. S. S., Coutinho, J. A. P., & Freire, M. G. (2023). Valorisation of red beet waste: One-step extraction and separation of betalains and chlorophylls using thermoreversible aqueous biphasic systems. *Green Chemistry, 25*(5), 1852–1864. https://doi.org/10.1039/D2GC04480C

Sanna, D., & Fadda, A. (2022). Waste from food and agro-food industries as pigment sources: Recovery techniques, stability and food applications. *Nutraceuticals, 2*(4). https://doi.org/10.3390/nutraceuticals2040028

Sarkar, S., Manna, M. S., Bhowmick, T. K., & Gayen, K. (2020). Extraction of chlorophylls and carotenoids from dry and wet biomass of isolated *Chlorella thermophila*: Optimization of process parameters and modelling by artificial neural network. *Process Biochemistry, 96*, 58–72. https://doi.org/10.1016/j.procbio.2020.05.025

Sartori, R. B., Vendruscolo, R. G., Ribeiro, S. R., Furlan, V. J. M., Wagner, R., Zepka, L. Q., & Jacob-Lopes, E. (2022). The role of photo-cycles in the modulation of growth and biochemical profile of microalgae: Part I-Food interest compounds. *Life, 12*(3), 462. https://doi.org/10.3390/life12030462

Sartory, D. P., & Grobbelaar, J. U. (1984). Extraction of chlorophyll a from freshwater phytoplankton for spectrophotometric analysis. *Hydrobiologia, 114*(3), 177–187. https://doi.org/10.1007/BF00031869

Saumyakanti Khanra, S. K., Madhumanti Mondal, M. M., Gopinath Halder, G. H., Tiwari, O. N., Kalyan Gayen, K. G., & Bhowmick, T. K. (2018). Downstream processing of microalgae for pigments, protein and carbohydrate in industrial application: A review. *Food and Bioproducts Processing, 110*, 60–84. https://doi.org/10.1016/j.fbp.2018.02.002

Schumann, R., Häubner, N., Klausch, S., & Karsten, U. (2005). Chlorophyll extraction methods for the quantification of green microalgae colonizing building facades. *International Biodeterioration & Biodegradation, 55*(3), 213–222. https://doi.org/10.1016/J.IBIOD.2004.12.002

Severo, I. A., de Lira, G. S., Ambati, R. R., Gokare, R. A., Vargas, J. V. C., Ordonez, J., & Mariano, A. B. (2024). Disruptive potential of microalgae proteins: Shaping the future of the food industry. *Future Foods, 9*. https://doi.org/10.1016/j.fufo.2024.100318

Silva, A. D. S., de Magalhães, W. T., Moreira, L. M., Rocha, M. V. P., & Bastos, A. K. P. (2018). Microwave-assisted extraction of polysaccharides from *Arthrospira* (*Spirulina*) platensis using the concept of green chemistry. *Algal Research, 35*(March), 178–184. https://doi.org/10.1016/j.algal.2018.08.015

Silva, S. C., Ferreira, I. C. F. R., Dias, M. M., & Filomena Barreiro, M. (2020). Microalgae-derived pigments: A 10-year bibliometric review and industry and market trend analysis. *Molecules*, *25*(15). https://doi.org/10.3390/molecules25153406

Silva, T. A., de Paula Pereira, A. S. A., Ferreira, J., Lorentz, J. F., de Assis, M. L., Assemany, P. P., … Calijuri, M. L. (2024). Enhancing microalgae biomass production: Exploring improved scraping frequency in a hybrid cultivation system. *Journal of Environmental Management*, *355*, 120505. https://doi.org/10.1016/j.jenvman.2024.120505

Simon, D., & Helliwell, S. (1998). Extraction and quantification of chlorophyll a from freshwater green algae. *Water Research*, *32*(7), 2220–2223. https://doi.org/10.1016/S0043-1354(97)00452-1

Somwongin, S., Sirilun, S., Chantawannakul, P., Anuchapreeda, S., Yawootti, A., & Chaiyana, W. (2023). Ultrasound-assisted green extraction methods: An approach for cosmeceutical compounds isolation from Macadamia integrifolia pericarp. *Ultrasonics Sonochemistry*, *92*, 106266. https://doi.org/10.1016/J.ULTSONCH.2022.106266

Spicer, A., Pudney, A., Taleb, A., Schiano di Visconte, G., Ransom-Jones, E., Smith, R., & Dickson, A. (2024). *Chlorella microalgae* (Patent WO2024161108A1). Algenuity Holdings Limited. https://patents.google.com/patent/WO2024161108A1/en

Straits Research. (2024). *Natural Food Colorants Market Size, Share & Trends Analysis Report By Type (Carmine, Anthocyanin, Caramel, Carotenoid, Curcumin, Paprika, Spirulina, Lycopene, Betalain, Others), By Application (Beverage, Dairy-based Products, Bakery, Confectionery, Snacks & Cereals, Nutraceutical, Processed Food Products, Others), By Source (Plant, Animal), By Form (Liquid, Powder) and By Region (North America, Europe, APAC, Middle East and Africa, LATAM) Forecasts, 2022–2030*. Retrieved from https://straitsresearch.com/report/natural-food-colorants-market

Suarez Ruiz, C. A., Kwaijtaal, J., Peinado, O. C., Van Den Berg, C., Wijffels, R. H., & Eppink, M. H. M. (2020). Multistep fractionation of microalgal biomolecules using selective aqueous two-phase systems. *ACS Sustainable Chemistry and Engineering*, *8*(6), 2441–2452. https://doi.org/10.1021/ACSSUSCHEMENG.9B06379/ASSET/IMAGES/LARGE/SC9B06379_0005.JPEG

Sun, H., Wang, Y., He, Y., Liu, B., Mou, H., Chen, F., & Yang, S. (2023). Microalgae-derived pigments for the food industry. *Marine Drugs*, *21*(2). https://doi.org/10.3390/md21020082

Thirugnanasambandham, K., Sivakumar, V., & Maran, J. P. (2015). Microwave-assisted extraction of polysaccharides from mulberry leaves. *International Journal of Biological Macromolecules*, *72*, 1–5. https://doi.org/10.1016/J.IJBIOMAC.2014.07.031

Torres-Franco, A. F., Figueredo, C. C., Barros, L., Gücker, B., Boëchat, I. G., Muñoz, R., & Mota, C. R. (2021). Assessment of a deep, LED-enhanced high-rate algal pond for the treatment of digestate. *Algal Research*, *59*, 102444. https://doi.org/10.1016/j.algal.2021.102444

Ünlü, A. E., Arlkaya, A., & Takaç, S. (2019). Use of deep eutectic solvents as catalyst: A mini-review. *Green Processing and Synthesis*, *8*(1), 355–372. https://doi.org/10.1515/GPS-2019-0003/ASSET/GRAPHIC/J_GPS-2019-0003_FIG_018.JPG

Van Hieu, H., Quang-Tuong, L., & Cong Doan, B. (2021). The production of high phycocyanin by applications of red light-emitting diodes (LEDs) *in vitro* algae growth on *Spirulina platensis*. *Journal of Nano- and Electronic Physics*, *13*(3). https://doi.org/10.21272/jnep.13(3).03034

Vendruscolo, R. G., Fagundes, M. B., Maroneze, M. M., do Nascimento, T. C., de Menezes, C. R., Barin, J. S., Zepka, L. Q., Jacob-Lopes, E., & Wagner, R. (2019). Scenedesmus obliquus metabolomics: Effect of photoperiods and cell growth phases. *Bioprocess and Biosystems Engineering*, *42*(5). https://doi.org/10.1007/s00449-019-02076-y

Viera, I., & Roca, M. (2020). Analytical protocols in chlorophyll analysis. *Pigments from Microalgae Handbook*, 127–149. https://doi.org/10.1007/978-3-030-50971-2_6

Wijesinghe, W. A. J. P., & Jeon, Y. J. (2012). Enzyme-assistant extraction (EAE) of bioactive components: A useful approach for recovery of industrially important metabolites from seaweeds: A review. *Fitoterapia*, *83*(1), 6–12. https://doi.org/10.1016/J.FITOTE.2011.10.016

Yang, K., Wang, J., Zheng, J., & Cai, W. (2024). Microalgae-based wastewater treatment: Advances and challenges in membrane harvesting technologies. *Separation and Purification Technology*, 130805. https://doi.org/10.1016/j.seppur.2024.130805

Yaqoubnejad, P., Rad, H. A., & Taghavijeloudar, M. (2021). Development a novel hexagonal airlift flat plate photobioreactor for the improvement of microalgae growth that simultaneously enhance CO_2 bio-fixation and wastewater treatment. *Journal of Environmental Management*, *298*, 113482. https://doi.org/10.1016/j.jenvman.2021.113482

Zepka, L. Q., Jacob-Lopes, E., & Roca, M. (2019). Catabolism and bioactive properties of chlorophylls. *Current Opinion in Food Science*, *26*, 94–100. https://doi.org/10.1016/j.cofs.2019.04.004

Zhang, X., Zhu, Y., & Zhao, Q. (2023). Green extraction of bioactive compounds from microalgae by ionic liquids. *Algal Biotechnology*, 115–129. https://doi.org/10.1079/9781800621954.0008/ASSET/710CDE06-EC1F-4354-BCAF-99BBF782CE2B/ASSETS/GRAPHIC/C08FIG0002.JPG

Zhao, W., Duan, M., Zhang, X., & Tan, T. (2018). A mild extraction and separation procedure of polysaccharide, lipid, chlorophyll and protein from *Chlorella* spp. *Renewable Energy*, *118*, 701–708. https://doi.org/10.1016/j.renene.2017.11.046

Zheng, Y., Shi, J., Pan, Z., Cheng, Y., Zhang, Y., & Li, N. (2014). Effect of heat treatment, pH, sugar concentration, and metal ion addition on green color retention in homogenized puree of Thompson seedless grape. *LWT*, *55*(2). https://doi.org/10.1016/j.lwt.2013.10.011

Zuorro, A., Maffei, G., & Lavecchia, R. (2016). Optimization of enzyme-assisted lipid extraction from *Nannochloropsis* microalgae. *Journal of the Taiwan Institute of Chemical Engineers*, *67*, 106–114. https://doi.org/10.1016/J.JTICE.2016.08.016

7 Novel Microalgae Metabolites

Production and Industrial Applications

Christian Rodrigues Lugcheer, Mimidoo Aver Gaaba, Marcele Leal Nörnberg, Tatiele Casagrande do Nascimento, Eduardo Jacob-Lopes, and Leila Queiroz Zepka

7.1 INTRODUCTION

Microalgae are an ample group of unicellular, photosynthetic microorganisms, known for their bioactive composition and growth versatility, being able to be grown in open systems for larger harvests, such as ponds or lakes, and in closed systems for ease of culture control such as bioreactors. It's natural pigments, peptides, polyunsaturated fatty acids (PUFAs), polysaccharides and volatile compounds are widely known and used for the sake of preventive care for cancer, cardiovascular diseases, among other diseases, due to their antioxidative, anti-inflammatory and antibacterial properties (Nashath et al., 2025).

Microalgae's primary and secondary metabolite profile and biomass yield at the time of harvest vary from each species and may be improved or modified according to necessity through exposure to different physical conditions during growth, such as changes in salinity, oxidative stress, temperature and pH (Rani et al., 2026). Such capabilities turn microalgae into highly adaptable ingredients for different food matrices, health supplements and cosmetics.

Despite the known long-term health benefits derived from the inclusion of microalgae in the human diet, an overarching obstacle that impedes its inclusion in mainstream nutritional trends stems from their foreign sensory characteristics that may directly clash with the sensory profiles of supplemented foods. Traits such as their earthy fish-like taste or varying green colouration depending on concentration may discourage consumers from interacting with products supplemented with microalgae. Among strategies to minimize negative sensory characteristics in microalgae-supplemented foods, spray drying, encapsulation and the use of secondary ingredients such as herbs and spices are popular methods that can aid in masking unwanted effects (Ang et al., 2025; de Souza et al., 2025).

DOI: 10.1201/9781003611479-7

Microalgae cultivation is remarkably more compact than conventional agriculture, requiring no arable land, as well as showing faster growth and consequently more frequent harvests, while simultaneously removing pollutants such as heavy metals, greenhouse gases, pesticides and other toxins. Aside from the previously mentioned growth medium physical conditions that affect metabolites and biomass yields in harvests, the culture's scale and configuration are of extremely importance as well, as they define points of caution and the need for special care as the culture grows. Open systems such as raceway ponds and lakes are cheaper and easier to maintain and allow for large-scale biomass production, at the expense of increased vulnerability, as the lack of a controlled environment means the possibility of external interferents as well as loss of product through volatilization. In contrast, closed systems such as photobioreactors (PBRs) limit biomass production scale and present an increase in maintenance costs but are smaller in size and offer more control on growth conditionals and safety, ideal for more sensitive microalgae and for the synthesis of specific metabolites (Rani et al., 2026).

7.2 MICROALGAL SPECIES FOR HIGH-VALUE BIOPRODUCT PRODUCTION

Despite microalgae's positive response to controlled physical stimuli for the sake of metabolite synthesis, it is paramount that the standard productivity of the selected strain is taken into consideration to maximize the production of target compounds, as different strains present different structures and reactions to the medium and its environment, which may result in the production of different bioproducts in varying concentrations (Beirne et al., 2023; Piligaev et al., 2018).

7.2.1 *Spirulina*

A popular cyanobacterium, currently holding the GRAS label by the FDA (Food and Drug Administration). *Spirulina* is seen as an efficient source of various nutraceuticals, phytonutrients and probiotics, being a highly absorbable supplement rich in essential amino acids (47% total protein weight) and proteins (60–70% dry weight), surpassing the nutritional value of common animal and vegetal foods such as meat, fish, milk and soy, alongside the presence of many vitamins (A, B, C, D and E) and carbohydrates (poly/mono/disaccharides) (Marjanović et al., 2024).

Spirulina also presents high concentrations of different natural photosynthetic pigments such as chlorophylls, carotenoids and especially phycocyanin, which are responsible for its notable antioxidative, anticancer and anti-inflammatory effects through the neutralization of free radicals and reduction of oxidative stress, while also being useful as a source of natural pigmentation for sensory purposes on food (Marjanović et al., 2024; Nova et al., 2024).

As a collective, *Spirulina*'s known bioactivity and bioaccessibility turn it into a functional ingredient of ample applications, such as baked goods, pastas, drinks and even reaching towards the cosmetics industry through skincare products. Currently, most *Spirulina* cultivation focuses solely on the production of protein, specifically phycocyanin. Current *Spirulina* production is dominated by China, with the species *Arthrospira platensis*, contributing approximately 66% of a global yearly output of about 10,000 tons (Marjanović et al., 2024; Ragusa et al., 2021).

7.2.2 Haematococcus pluvialis

H. pluvialis, a freshwater species of green alga, is popularly cultivated for *astaxanthin* production, with a resulting yearly biomass yield of approximately 300 tons. *Astaxanthin* is a natural pigment also found in different aquatic species such as shrimps and crabs, sought after by the pharmaceutical, cosmetic and food industries for its exceptional antioxidative capabilities, far surpassing those of carotenoids, vitamins and many phenolic compounds. With a market value of approximately $2,000/kg^{-1}, and most of the market available *astaxanthin* being obtained through synthetic means, the demand for natural *astaxanthin* grows due to increasing health concerns (Ren, Deng, et al., 2021).

Astaxanthin is a xanthophyll carotenoid with multiple geometric isomer forms, its all-*E*-isomer form being the most abundantly found in nature, and its *Z*-isomer form presenting higher bioavailability and antioxidative activity, considered one of the strongest naturally occurring antioxidants, due to its long-conjugated polyene chain, which aids in delocalizing electrons and enabling the scavenging of reactive oxygen species (Ghosh & Honda, 2026; Ren, Deng, et al., 2021).

A popular production method involving *H. pluvialis* relies on the accumulation of *astaxanthin* via regulatory action to multiple environmental factors, a previously mentioned capability of other microalgal strains. Its main factors, such as light intensity, nutrient concentration, medium pH and temperature, are key factors for production optimization, which must be adapted according to the culture's two different stages: green, for biomass productivity, and red for astaxanthin production, if following a two-stage model. Whereas during the green stage, low intensity lighting, nutrient abundance and low salinity are necessary for culture development and biomass production, the red stage involves subjecting the culture to a stressful environment through high intensity lighting and an increase in salinity to stimulate *astaxanthin* production for the sake of antioxidative activity and photoprotection, though lack of production control may cause photoinhibition directly affecting cell mortality (Sung & Sim, 2022) (Figure 7.1).

7.2.3 Chlorella

Chlorella currently stands as one of the most globally cultivated microalgae, for its uses as a biofuel, and metabolite extraction, with a production rate of biomass between 0.08 and 0.42 g/L per day when cultivated in BG-11 medium, leading to increased bioproduct production, though not sufficient for industrial applications, in which an additional supply of nitrogen, carbon and sodium nitrate must be added to the culture to boost metabolite synthesis (Fang et al., 2024; Ferreira de Oliveira & Bragotto, 2022).

Aside from an increased production rate of proteins, carbohydrates and lipids in comparison to other algae, *Chlorella* also contains natural pigments, including

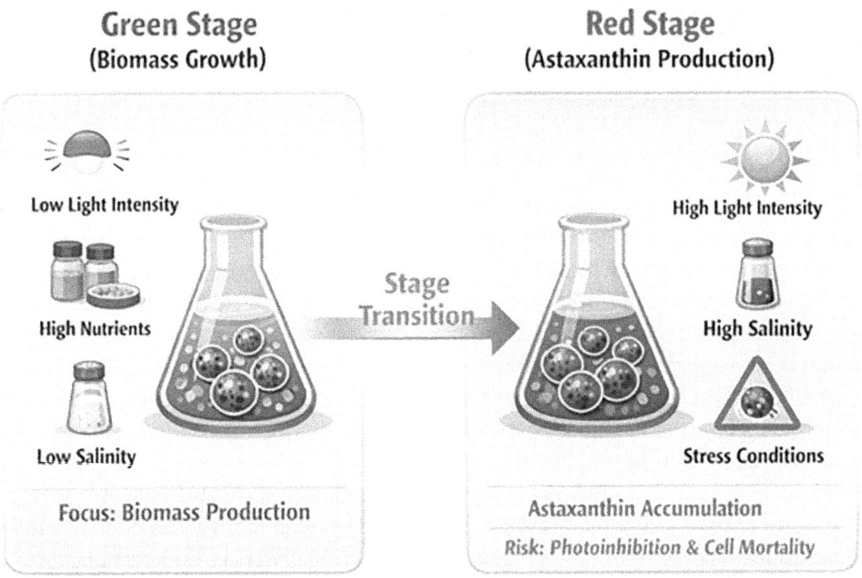

FIGURE 7.1 Two-stage model of *H. pluvialis* cultivation for *Astaxanthin* production.

carotenoids and chlorophylls, further boosting its nutraceutical potential and industry usage. One of *Chlorella's* downsides, however, is the low bioavailability of metabolites, due to the presence of cellulose in its cell walls, requiring additional steps in the extraction process, through physical, chemical or enzymatic means (Fang et al., 2024; Ferreira de Oliveira & Bragotto, 2022).

7.3 MICROALGAE AS GREEN BIOFACTORIES OF BIOACTIVE METABOLITES

As stated previously, different microalgae produce a multitude of bioactive metabolites which are subsequently extracted from biomass, amidst these, *carotenoids*, *chlorophylls*, peptides, polysaccharides, PUFAs and volatile compounds are seen as metabolites of high interest for their nutritional and pharmaceutical value, centring microalgae as a potential for technological expansion for multiple industries (Kaur et al., 2025; Oslan et al., 2025).

7.3.1 Carotenoids

Carotenoids make up a diverse family of terpenoid pigments formed by structures of forty carbon atoms (C40). It is currently estimated that more than 1,200 carotenoids have been identified in different organisms, including species from the plant kingdom, animal kingdom and various microorganisms, such as microalgae. Chemically, carotenoids are classified according to the carbon numbers in their molecules and may have C30, C40, C45 or C50 skeletons. However, most of these pigments have C40 structures (Ren, Sun, et al., 2021; Fernandes et al., 2021).

Another form of classification considers the presence of oxygen atoms in the molecule, dividing them into xanthophylls, which have oxygenated groups, and carotenes, formed only from carbon and hydrogen atoms. This chemical variation directly influences their properties, such as stability, polarity and biological role in microalgae, especially as photoprotective agents. Amongst these compounds, astaxanthin and canthaxanthin stand out, secondary carotenoids that help dissipate excess light energy and prevent oxidative stress (Mostafa & Hashem, 2024; Sirohi et al., 2022).

In addition to their functional importance in the microorganism, several carotenoids have high commercial relevance. Amongst them are carotenes, β-carotene, lycopene, as well as the xanthophylls astaxanthin, canthaxanthin, lutein and zeaxanthin, all already found in microalgae and explored on the global market. The growing demand for compounds that have biologically active functional groups, such as those present in the structure of carotenoids, combined with their nutritional and technological value, is reflected in a market that is expected to exceed three billion dollars by 2030 (Patel et al., 2022; Patwary et al., 2025).

From this perspective, microalgae stand out as superior sources of these pigments, due to their high photosynthetic efficiency, reaching 10 to 20%, a value above that observed in terrestrial plants (1–2%). Furthermore, environmental factors such as temperature, light, nutrient availability and salinity directly influence microalgal metabolism. When modulated in a controlled way, they can redirect to antioxidant pathways, resulting in increased carotenoid synthesis. This ability to convert light and nutrients into high-value metabolites quickly and sustainably reinforces their role as true green biofactories to produce carotenoids (Barreiros et al., 2024; Chen et al., 2022).

Carotenoids in microalgae are synthesized from isopentenyl pyrophosphate (IPP) and dimethylallyl pyrophosphate (DMAPP), produced mainly via the 2-C-methyl-D-erythritol 4-phosphate (MEP) pathway. These precursors are condensed to form geranylgeranyl pyrophosphate (GGPP), which is converted into phytoene synthase (CrtB), a step considered limiting in the pathway. From phytoene, successive desaturation and cyclization reactions give rise to lycopene and, subsequently, carotenes and xanthophylls through the actions of specific enzymes, such as lycopene β-cyclase (LCYB), lycopene ε-cyclase (LCYE) and β-carotene ketolase (CrtW). The regulation of this pathway is highly sensitive to environmental conditions, light, oxidative stress, salinity and nutrients (Su et al., 2023).

In a study with the microalgae *Scenedesmus Obliquus*, it was found that the application of saline stress during cultivation significantly increased the carotenoid content, reaching up to 2.08% of dry biomass after 14 days, compared to 0.68% under normal conditions, showing that environmental factors can modulate the production of these pigments (da Silva et al., 2021). Similarly, adjustments to photoperiods in *S. Obliquus* have been shown to be effective in maximizing photoperiod efficiency, with short light and dark cycles providing greater accumulation of (31.84 mg/L), while longer cycles favour the synthesis of other pigments (do Nascimento et al., 2025).

Carotenoids isolated from microalgae, such as α-carotene, β-carotene, lycopene, phytoene, lutein, β-cryptoxanthin, zeaxanthin, astaxanthin and fucoxanthin, have several bioactive activities, including the potential to be bio-transformed by the intestinal microbiota into metabolites capable of strengthening the intestinal barrier, stimulating beneficial microorganisms and inhibiting harmful ones (Liu et al., 2025). Furthermore, these carotenoids have high antioxidant capacity, acting as scavengers of reactive oxygen species and contributing to cellular protection against oxidative damage. Experimental evidence also indicates neuroprotective effects, with emphasis on fucoxanthin, associated with the inhibition of acetylcholinesterase and the attenuation of processes related to Alzheimer's disease (Oliyaei et al., 2023).

Amongst oxygenated carotenoids, astaxanthin has anti-inflammatory activity in models of metabolic diseases, such as diabetes mellitus, acting in the modulation of signalling pathways associated with oxidative stress, including NF-κB, MAPK and PI3K/Akt. Furthermore, this carotenoid has demonstrated efficacy in complications associated with diabetes, such as cardiovascular disease (Kohandel et al., 2022).

Taken together, carotenoids produced by microalgae constitute a promising class of bioactive compounds. The high biosynthetic capacity of these microorganisms, combined with the possibility of metabolic modulation, makes these pigments extremely valuable for industrial applications. In addition to their recognized antioxidant properties, carotenoids such as astaxanthin and fucoxanthin have demonstrated neuroprotective and anti-inflammatory potential, with applications in degenerative and metabolic diseases. In this context, microalgae are consolidated as true green biofactories, capable of generating carotenoids sustainably, meeting the technological and functional demands of modern industry

7.3.2 Chlorophylls

Chlorophylls belong to one of the three main classes of natural pigments and stand out as the most widely distributed and abundant pigments in nature. Their relevance is directly associated with the role they play, as they are the pigments responsible for photosynthesis, a vital process that converts light energy into chemical energy. During this metabolic stage, in the presence of light, carbohydrates and other compounds are synthesized from carbon dioxide and water, with the release of molecular oxygen. In higher plants, two predominant types of chlorophyll are found, chlorophyll *a* and *b*, with chlorophyll *b* acting as an accessory pigment, while chlorophyll *a* is generally the primary natural photosynthetic pigment (Gross, 1991).

Most photosynthetic organisms, ranging from cyanobacteria to higher plants, contain chlorophyll, differing mainly in the amount present in each organism. Currently, approximately five types of chlorophyll are known: *a*, *b*, *c*, *d* and *f*. Chlorophyll *b* differs from chlorophyll *a* in the presence of a formyl group instead of a methyl group at the C7 position. Chlorophyll *c* differs by lacking a reduced pyrrolic ring, exhibiting a structure closer to that of a porphyrin ring, and is also referred to as chlorophyllide due to the presence of acrylic side chains instead of long phytol chains. Chlorophyll *d* contains a formyl group at the C3 position instead of a vinyl group, and finally, chlorophyll *f* represents the most recently discovered chlorophyll type (Sun et al., 2023)

Beyond their fundamental role in photosynthesis, chlorophylls *a* and *b* exhibit chemical instability and sensitivity to environmental factors. Exposure of these chlorophyll molecules to acidic conditions, oxygen, and light can accelerate their degradation process, leading to the formation of secondary degradation products (Portillo & Jiménez, 2021). A thorough understanding of these mechanisms and associated challenges is essential for industrial applications and future physiological studies. In this context, the use of chlorophyll in various products is evident, with these pigments being key ingredients in pharmaceutical formulations and widely applied as natural colourants in the food industry (Sun et al., 2023). Understanding these degradation mechanisms is therefore crucial for the development of industrial applications and advances in physiological research.

In this context, chlorophyll compounds have attracted scientific interest due to their bioactive properties, highlighting their potential health benefits. Recent studies indicate their role in wound healing, support of cellular respiration, and increased hemoglobin levels in the blood (Ferreira et al., 2021). In addition, scientific interest extends to their broader bioactive properties, with studies reporting health benefits associated with chlorophylls and their derivatives, including antimutagenic, anti-inflammatory and antigenotoxic activities, as well as strong antioxidant capacity capable of scavenging free radicals and preventing lipid oxidation (Fernandes et al., 2021). Furthermore, when compared to synthetic pigments, chlorophylls exhibit greater resistance to ascorbic acid, heat and freezing processes, demonstrating high efficiency even when added in small quantities.

Within the food industry, a promising scenario is observed for the acceptance of chlorophyll as a natural food colourant. However, this acceptance depends on a balance between consumer interest in clean-label products and resistance towards plant-based foods. In parallel, recently investigated ultrasound-assisted extraction methods have proven capable of providing higher purity levels of pigments, particularly chlorophylls *a* and *b*, further enhancing the potential use of these compounds in the pharmaceutical industry and as nutritional supplements (Nikolova et al., 2024).

Finally, understanding the potential of microalgae as sources of these bioactive compounds, including proteins, lipids, carbohydrates, pigments and polysaccharides, among others, has emerged as an increasingly viable alternative (Conceição & Varandas, n.d.)

7.3.3 Bioactive Peptides

Recently, due to their high protein content, rapid growth, metabolic versatility and the possibility of cultivation under controlled conditions, microalgae—including cyanobacteria—have attracted increasing interest as alternative and sustainable sources of bioactive peptides. Genera such as *Chlorella*, *Spirulina* (*Arthrospira*), *Scenedesmus*, *Dunaliella* and *Nannochloropsis* exhibit amino acid profiles favourable for the generation of functionally active peptides (Chitthan et al., 2025).

These secondary metabolites correspond to protein fragments composed of 2 to 50 amino acid residues, capable of exerting beneficial effects on human metabolism (Guo et al., 2024). However, within the intact protein sequence, these compounds remain biologically inactive, and their release is required for bioactivity to be expressed. This process occurs mainly through protein hydrolysis, mediated by digestive processes (pH variations), fermentative processes (the action of bacteria, yeasts and fungi) or enzymatic processes via specific proteases (Fernando et al., 2024).

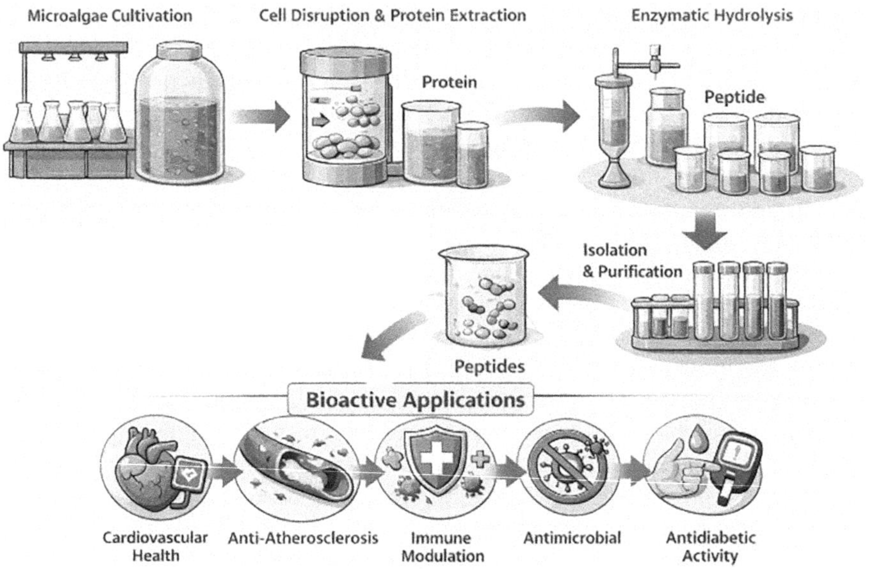

FIGURE 7.2 Biopeptide production and potential applications.

As illustrated in Figure 7.2, the production of biopeptides from microalgae begins with biomass cultivation, followed by the release of intracellular proteins through the disruption of the microalgal cell wall. At this stage, physical, chemical or mechanical methods may be employed, depending on the composition and robustness of the cell wall of the species in question (Chitthan et al., 2025). After protein release, the molecular structure of proteins is broken down through hydrolysis, resulting in the generation of amino acid residues and peptides of lower molecular mass (Siddhnath et al., 2024).

Hydrolysis consists of the cleavage of peptide bonds responsible for linking amino acids within the protein chain (Pereira et al., 2023). Amongst the available methods, enzymatic hydrolysis is the most widely used due to its high specificity, greater control over cleavage sites and lower formation of undesirable by-products. To a lesser extent, fermentative processes and chemical methods may also be employed (Fernando et al., 2024).

During hydrolysis, other cellular constituents, such as lipids, carbohydrates and pigments, may be co-extracted, making the isolation and purification of peptides a necessary step. For this purpose, fractionation techniques such as centrifugation, precipitation, ultrafiltration, electrophoresis and chromatography are employed to obtain purer and more homogeneous peptide fractions (Guo et al., 2025; O'Connor et al., 2022).

Once isolated, peptides can be characterized in terms of their chemical composition and evaluated for their bioactivity. Recent studies indicate that microalgae-derived peptides exhibit a wide range of biological properties, including antioxidant, antihypertensive, anti-inflammatory, antimicrobial, immunomodulatory and antidiabetic activities (Kumar et al., 2022; Qoms et al., 2025). These bioactivities are often associated with amino acid sequence, molecular mass, net charge and degree of hydrophobicity of the peptides.

In a study conducted by Masoumifeshani et al. (2025), enzymatic hydrolysis of microalgal proteins predominantly generated peptides with molecular masses below 3 kDa, which exhibited high antioxidant activity, with values of 78% for 2,2-diphenyl-1-picrylhydrazyl (DPPH), 82% for 2,2′-azinobis(3-ethylbenzothiazoline-6-sulfonic acid (ABTS) and 74% for Ferric Reducing Antioxidant Power (FRAP), highlighting an IC_{50} of 23.44 μg/mL for the ABTS assay in *Chlorella vulgaris*. These peptides also showed marked inhibition of enzymes associated with skin aging, with 84% for elastase, 90% for collagenase and 66% for tyrosinase. Mass spectrometry analysis predominantly identified di- and tripeptides, such as Lys–Val, Val–Arg, His–Ile, and Leu–Gly–Leu, which may be responsible for the observed bioactivities.

Microalgal peptides with antiatherosclerotic activity exert anti-inflammatory and antithrombotic effects, in addition to reducing the expression of endothelial adhesion molecules and pro-inflammatory mediators. Peptides isolated from microalgae such as *Chlorella*, *Spirulina* and *Isochrysis* have demonstrated the ability to modulate inflammatory pathways, reduce oxidative stress and protect endothelial cells (Qoms et al., 2025).

Antioxidant and antidiabetic activities were also demonstrated by Villaro et al., who reported that enzymatic hydrolysates of phycobiliproteins from *Arthrospira platensis* generated bioactive peptides capable of inhibiting markers associated with glycemic control, including DPP-IV, as well as the enzymes α-amylase and α-glucosidase.

Using *Aureococcus anophagefferens* as a model, Zhang et al., (2024) conducted a study based on de novo transcriptomics and in silico screening, identifying 471 putative antimicrobial peptides. Amongst these, three short peptides (P1, P6 and P7) exhibited antimicrobial activity against *Escherichia coli*, *Staphylococcus aureus*, *Micrococcus luteus* and *Pichia pastoris*.

Regarding immunomodulatory activities, the microalga *Nannochloropsis gaditana* demonstrated relevant effects on the human gut microbiota. The results indicated that *N. gaditana* digests significantly modulated gut microbiota composition, promoting an increase in beneficial bacterial genera such as *Akkermansia*, *Butyricicoccus*, *Eisenbergiella*, *Lachnoclostridium* and *Marvinbryantia* when compared with inulin, in addition to increasing the production of short-chain fatty acids (Paterson et al., 2025).

Finally, microalgae have recently been recognized as promising alternative sources to produce biopeptides; however, the still limited availability of confirmatory studies highlights the need for further in-depth research in this field to support and consolidate their use as bioactive peptides in food, nutraceutical and pharmaceutical applications.

7.3.4 Polysaccharides – b-glucans

Depending on the species and cultivation conditions, microalgae can accumulate up to approximately 50% of their dry weight in the form of carbohydrates, highlighting their potential as promising sources of bioactive polysaccharides (Nörnberg,

Schetinger, et al., 2025; Souza et al., 2019). These compounds, predominantly associated with the cell wall and reserve polymers, exhibit a well-recognized capacity to modulate the immune system and inflammatory responses, thereby expanding their potential application in functional foods and as natural therapeutic agents (Nörnberg, de Aguiar, et al., 2025; Nörnberg et al., 2024; Souza et al., 2019).

Amongst the most relevant polysaccharides, β-glucans stand out as polymers composed of D-glucose units linked by β-glycosidic bonds and distributed across plants, fungi, bacteria and microalgae. These compounds physiologically function as soluble dietary fibres and are associated with reductions in low-density lipoprotein cholesterol and cardiovascular disease risk, in addition to exhibiting anti-inflammatory, immunomodulatory, antioxidant, anticancer, prebiotic and glycemic-regulatory properties (Colarte et al., 2025; Li et al., 2025; Nörnberg et al., 2022; Torres-Tiji, Fields & Mayfield, 2020). Green microalgae are important producers of these polymers, with *Chlorella* sp. being particularly noteworthy due to its high β-glucan content (Souza et al., 2019).

In an integrated manner, Figure 7.3 illustrates the microalgal origin of β-glucans, their main structural features, the underlying mechanisms of bioactivity and the associated physiological effects, as well as their potential applications in functional foods, nutraceuticals and health-oriented products, thereby facilitating a systemic understanding of the role of these bioactive biopolymers.

Numerous studies reinforce the high functional value of microalgal polysaccharides. Niccolai et al. (2019) demonstrated that the total dietary fibre content of microalgae can range from 4.4% to 17%, significantly exceeding that of conventional foods such as cereals, fruits and vegetables. This characteristic contributes to physiological benefits related to intestinal health, lipid and glucose metabolism and modulation of gut microbiota. Experimental evidence further supports these observations, as demonstrated by Manzke et al. (2025), who evaluated the effects of dietary supplementation with microalgae-derived β-glucans in an animal model. The authors reported significant improvements in intestinal integrity and mucosal morphology, evidenced by increased villus height and a more favourable villus-to-crypt ratio, indicating enhanced absorptive capacity and gastrointestinal functionality. Additionally, supplementation promoted positive modulation of the intestinal immune response, associated with reduced inflammatory processes and improved intestinal environment, resulting in enhanced zootechnical performance. Collectively, these findings confirm that microalgal β-glucans act as functional bioactive polysaccharides with immunomodulatory and prebiotic potential, reinforcing their applicability as functional ingredients in animal nutrition and, with appropriate caution, as a basis for extrapolation to functional foods targeting human health.

Beyond metabolic and immunological effects, β-glucans and other microalgae-derived polysaccharides exhibit additional therapeutic bioactivities, including antioxidants, wound-healing and antiviral properties, as demonstrated by Vasilakis et al. (2025) through *in vitro* and *ex vivo* assays. In this study, the authors isolated and characterized polysaccharide fractions rich in structural polysaccharides, including β-glucans, and systematically evaluated their biological activities. The results revealed high antioxidant capacity associated with the neutralization of reactive

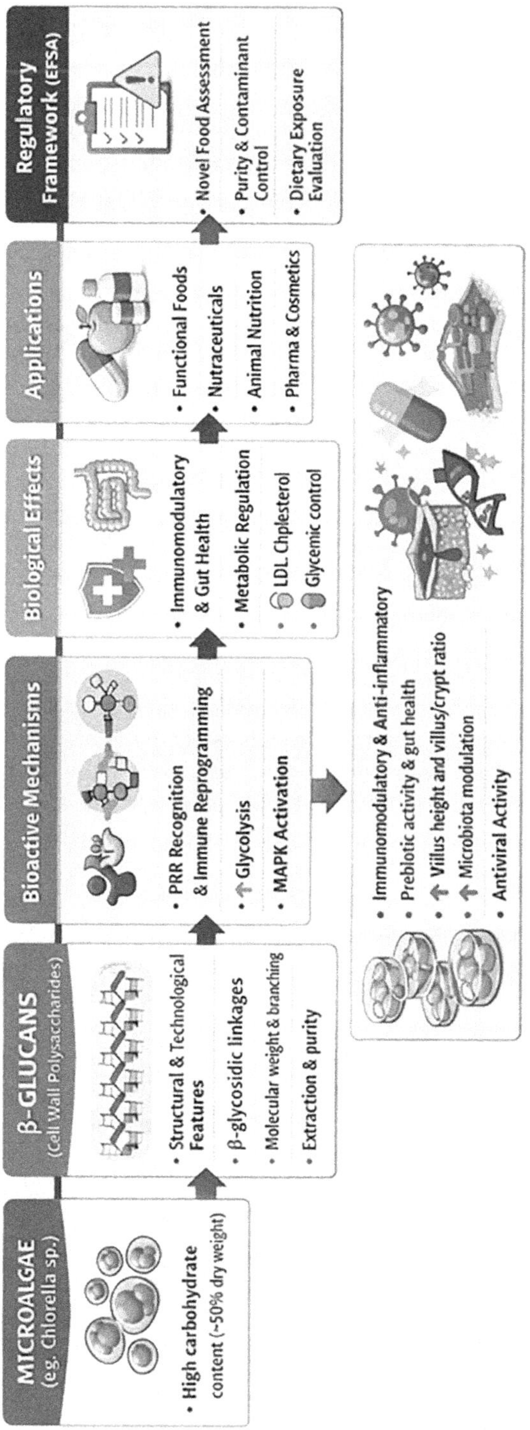

FIGURE 7.3 Microalgal β-glucans: bioactivity and applications.

oxygen species, as well as positive effects on cell migration and tissue regeneration, indicating potential application in wound-healing processes. Furthermore, significant antiviral activity was observed, with inhibition of viral replication in cellular models, possibly related to interactions between the polysaccharides and the cell surface or viral particles. The authors also demonstrated that bioactivity was directly associated with the chemical composition and molecular structure of the polysaccharides, reinforcing the role of β-glucans and other heteropolysaccharides as multifunctional bioactive molecules.

From a technological and regulatory perspective, recent reviews emphasize that the molecular structure, extraction method and degree of purity of β-glucans directly influence their biological functionality, technological performance and safety of use (Ahmad et al., 2025; Prabhakar et al., 2025). Ahmad et al. (2025) presented a comprehensive review of algae-derived β-glucans, highlighting their relevance as strategic bioactive ingredients in the development of nutraceuticals and functional foods. According to the authors, these β-glucans exhibit structural variations depending on the biological source and extraction method, factors that directly impact their bioactivity and technological properties. The review compiles evidence associating these molecules with hypocholesterolemic, immunomodulatory, anti-inflammatory, antioxidant, glycemic-regulatory and prebiotic effects while also emphasizing important advantages over conventional sources, such as sustainability, lack of competition with food crops and greater industrial versatility.

Complementarily, Prabhakar et al. (2025) described β-glucans as microbial polymers with high nutraceutical potential, addressing their chemistry, structure and functionality in an integrated manner. The authors highlight that characteristics such as glycosidic linkage type, degree of branching and molecular weight directly influence the bioactivity and bioavailability of these polymers. The study also discusses microbial biosynthesis mechanisms, as well as the main methods for extraction, identification and industrial production, emphasizing that technologically controlled processes are essential to preserve the biological functionality of β-glucans. Based on evidence from the literature, the authors associate these molecules with immunomodulatory, hypocholesterolemic, antioxidant and metabolic effects, reinforcing their role as key bioactive ingredients for the development of next-generation functional foods and nutraceuticals.

In this context, the European Food Safety Authority (EFSA) establishes that microalgae-derived products, including ingredients rich in β-glucans, must undergo case-by-case safety evaluation for authorization as novel foods, considering aspects such as detailed chemical characterization, stability, contaminant control and estimated dietary exposure (Garciarena et al., 2025). Garciarena et al. (2025) highlighted that microalgal polysaccharides, such as cell wall β-glucans, are generally considered safe when adequately characterized and when they meet EFSA regulatory criteria for purity, composition and quality. However, the authors emphasize that regulatory approval depends on the producing species, the extraction process, the molecular structure of the compound and the proposed consumption level, reinforcing the need for an integrated scientific and regulatory approach. Thus, microalgal β-glucans are consolidated as innovative high-value-added bioproducts, combining proven bioactivity, technological potential and regulatory feasibility, supporting their safe and strategic application in functional foods, nutraceuticals and products aimed at human health.

7.3.5 Polyunsaturated Fatty Acids – EPA and DHA

Microalgae are relevant sources of bioactive compounds with high added value, due to their high lipid production capacity, which can be up to ten times higher than that observed in terrestrial plants. Amongst these compounds, PUFAs from the omega-3 series stand out, such as docosahexaenoic acid (DHA, 22:6n-3) and eicosapentaenoic acid (EPA, 20:5n-3). Chemically, these compounds have a stable structure and high bioavailability, and their intake contributes to the control of cholesterol levels in the diet, helping to prevent cardiovascular diseases (Arenas Colarte et al., 2025; Hassan et al., 2022; Linares-Maurizi et al., 2023).

From a biological point of view, PUFAs from the omega-3 series, such as DHA and EPA, exert regulatory functions in the human body, acting in the modulation of anti-inflammatory processes, the functioning of the nervous system and the stability of the cardiovascular system, in addition to participating in the integrity of cell membranes (Zakaria et al., 2025).

Currently, marine fish, especially from cold waters, constitute the main dietary source of long-chain omega-3 polyunsaturated fatty acids. On the other hand, the increase in environmental pollution and the seasonality of ocean fishing result in the instability of their production and fluctuation in the price of these compounds. In this context, microalgae emerge as sustainable and scalable alternatives to produce DHA and EPA, since their biosynthesis can be modulated by controlled cultivation conditions, such as nutrient availability and environmental factors (Swetha & Mathanghi, 2024; Udayan et al., 2022).

Furthermore, comparative studies in adults indicate that DHA and EPA derived from fish oil and microalgae have equivalent bioavailability effects, reinforcing the potential of these sources as viable substitutes in human nutrition. In this context, omega-3 fatty acids, from microalgae, such as *Pavlova lutheri*, have been evaluated for specific food applications, such as infant formulas. These compounds play an essential role in children's neurological development, justifying their interest in formulation (Bailey et al., 2025; Dhandwal et al., 2025).

Recent evidence indicates that lipids derived from microalgae have specific bioactive potential in cellular models. Extracts of *Microchloropsis gaditana* and *Tisochrysis lutea* showed the ability to modulate the integrity of the intestinal barrier in vitro, highlighting the protective effect of *M. gaditana* lipids against disruption induced by dietary xenobiotics. Notably, a clinical trial evaluated supplementation with oil from the microalgae *Schizochytrium sp.* enriched with DHA/EPA, in patients with rheumatoid arthritis, reporting beneficial effects on joint health and reducing levels of inflammatory markers (Crudo et al., 2025; Cutolo et al., 2023).

Microalgae represent sustainable and promising sources of polyunsaturated fatty acids from the omega-3 series, especially EPA and DHA and bioactive metabolites of nutritional and functional relevance. Experimental and clinical evidence indicates that lipids derived from microalgae have bioavailability and biological effects comparable to those of fish oil, confirming their potential for food and nutritional applications. In this scenario, the advancement of biotechnological and formulation strategies will be essential to expand the use of these compounds in products intended for human health.

7.3.6 Volatile Compounds – Limonene

Limonene, also known as dipentene, is a natural monocyclic monoterpene characterized as a colourless liquid that presents two optical isomers: D-limonene or (+)-limonene, which is dextrorotatory, and L-limonene or (−)-limonene, which is levorotatory. These two isomers exhibit distinct bioactive properties, which can be attributed to differences in their spatial configuration, molecular structures and specific interactions with active sites (Lin et al., 2025).

From a chemical perspective, terpenes can be considered 'natural alkenes', as they contain at least one carbon–carbon double bond and are therefore unsaturated hydrocarbons. When oxygen atoms are incorporated into their structure, they are referred to as terpenoids. Despite the structural differences between terpenes and terpenoids, they share fundamental characteristics, particularly their basic structure formed by five-carbon units known as isoprene units (C_5H_8). These units are typically linked in a 'head-to-tail' arrangement (1–4 linkage), which forms the basis of the so-called 'terpene rule'. However, exceptions to this rule exist, known as irregular terpenes, such as β-carotene, which exhibits 'tail-to-tail' linkages (4–4 linkage). In addition, cyclic terpenes, such as limonene, arise due to distinct biochemical pathways. All carbon atoms in these molecules originate from IPP or its isomer DMAPP. The mevalonate pathway and the 1-deoxy-D-xylulose-5-phosphate (DXP) pathway are responsible for the biosynthesis of the wide diversity of terpenes found in nature, allowing their classification according to the number of isoprene residues present.

Limonene is one of the most abundant monoterpenes in nature and may account for approximately 16% of the components of essential oils, being widely used in the aroma and fragrance industries. Its antioxidant and anticancer properties are well documented in scientific literature, indicating its potential use in food fortification strategies aimed at cancer prevention. Furthermore, the derivatization of limonene into other compounds opens opportunities for future commercial applications (Kousar et al., 2017). In this context, derivatization processes and innovative strategies aimed at improving the extraction of these compounds are highly relevant for future technological advancements.

These innovations are supported by studies that increasingly reveal that limonene, or 4-isopropenyl-1-methylcyclohexene, possesses not only broad industrial applications as an antimicrobial agent in the food sector but also significant relevance in the pharmaceutical field, where it acts in cholesterol dissolution and serves as a chemoprotective agent against various types of cancer. In addition, its applications extend to the flavour and fragrance industry, in which limonene is widely used as a precursor for the synthesis of a range of other terpenes. This evidence highlights that not only limonene itself but also its derivatives, represent promising investments from an economic perspective.

In addition to higher plants, microalgae have been investigated as alternative sources of lipid and volatile metabolites due to their high sustainability and productivity. Recent studies have shown that the use of terpenes obtained from renewable raw materials as alternative solvents represents an efficient approach for the extraction of lipids from microalgae when compared to petroleum-based solvents. The latter are not only toxic to users but also environmentally harmful, contributing negatively to greenhouse gas emissions (Labua & Kenkhunthot, 2020).

7.4 INNOVATIVE FOOD APPLICATIONS OF MICROALGAE-DERIVED BIOPRODUCTS

Malnutrition remains a global health challenge that contributes to over 3.5 million deaths annually and accounts for 45% of all child mortalities. Hence, microalgae are being reputed as a promising solution composed of bioactive compounds capable of meeting the global demand for sustainable, nutrient-dense functional foods (Guil-Guerrero & Prates, 2025). Notably, they can function as nutraceutical additives, whole-food ingredients, natural food stabilizers, natural colourants, flavour compounds and contribute to stability and sensory aspects due to their biochemical flexibility (García et al., 2021). The food industry has shifted towards clean-label, natural and health-promoting ingredients; hence, microalgae-derived bioproducts perfectly align with this trend, serving as functional ingredients, natural colourants, flavour enhancers, stabilizers and sensory modifiers (Chettri et al., 2026).

In recent decades, the global scientific and public health communities have highly recognized the pivotal importance of bioactive compounds in supporting health, growth and disease prevention. Furthermore, these compounds have been demonstrated to exert significant physiological effects and are found naturally in diverse plant and marine foods (Zhou et al., 2022). Common classes include phenolics, carotenoids, phytosterols, flavonoids and PUFAs, each linked with mechanisms that modulate oxidative stress, inflammation, immune response and metabolic regulation (Ortega & Campos, 2019). Hence, owing to their high protein content and bioactive components, it has been proven that the inclusion of microalgal biomass into functional foods increases their functional properties (Camacho et al., 2019; Jacob-Lopes et al., 2019; Ampofo & Abbey, 2022; Hernández et al., 2022; Vieira et al., 2023). They can be incorporated into functional foods in a manner comparable to traditional food products or formulated as nutraceuticals (Vieira et al., 2023). Hence, integrating microalgae into conventional food products can improve their nutritional profile and preserve essential textural properties, including firmness and fracturability, without altering their taste (Caporgno & Mathys, 2018). Also, these microalgae-enriched foods have demonstrated potential health benefits, including reduced serum cholesterol, lower cardiovascular disease risk, improved diabetes symptoms, decreased blood pressure, cancer prevention and enhanced gastric health and nutrient absorption (Chaves et al., 2019; Dhandwal et al., 2025; Salehipour-Bavarsad et al., 2024).

7.4.1 Microalgae-Derived Bioproducts as Natural Colourants

According to the EFSA, food colourants are a category of food additives used in the food industry to make processed foods more attractive, restore colour lost during processing and add colour to otherwise colourless foods (EFSA, 2024). Also, these colourants are in two forms, including synthetic and natural colourants (Singh et al., 2023). However, these natural colourants have gained more interest among consumers who demand natural products. With the growing focus on healthier lifestyles, consumers are seeking cleaner-label foods, prompting research into natural ingredient sources that can substitute synthetic food additives (Mesías et al., 2021). Hence, the presence of pigments in microalgae necessitates their application in various industries, including food, nutraceuticals, pharmaceuticals, aquaculture and cosmetics (Ahmad et al., 2022). Notably, some of these pigments, including phycobiliproteins (PBPs), carotenoids and chlorophylls, are increasingly recognized as promising natural alternatives to synthetic food colourants, offering multifunctional properties, with reduced risks while supporting sustainable, clean-label food systems (Sun et al., 2023; Chettri et al., 2026).

Using spectroscopy and colorimetry methods, *B-phycoerythrin* extract from the microalgae *Porphyridium cruentum* tested as a food colourant in milk-based products for 11 days showed that the pigment exhibited favourable colouring properties and remained stable across a pH range between 4.0 and 9.0, with low pigment levels effectively reproducing the pink colour of strawberry-flavoured dairy products, and remained stable under cold storage (García et al., 2021).

7.4.2 Microalgae as Flavouring Agents in Foods

Several microalgal species, including *Chlorella, Spirulina* (*Arthrospira*), *Nannochloropsis* and *Porphyridium*, are recognized for their high content of umami-related compounds such as glutamic and aspartic acids, as well as 5′-ribonucleotides, which enhance palatability and flavour depth (Lafarga et al., 2020). In addition, lipid-derived volatiles from PUFAs can generate aroma-active compounds that resemble marine or seafood flavours, offering potential for plant-based seafood analogues (Luo et al., 2021). Beyond their sensory contributions, microalgae-derived flavouring agents align well with current consumer demand for clean-label, natural and sustainable ingredients. Compared with synthetic flavour compounds, microalgal flavours offer multifunctional benefits, including nutritional value, antioxidant activity and reduced environmental impact (Guil-Guerrero & Prates, 2025; Baldemir & Çaklı, 2024; Chettri et al., 2026). However, challenges remain, particularly related to off-flavours, bitterness and colour–flavour interactions, which can limit consumer acceptance if not properly controlled through strain selection, cultivation conditions and downstream processing (Lafarga, 2019). In contrast to seaweed, there is a paucity of studies about the flavour of microalgae, even though these have received attention in previous years as a novel source of food. Some studies have investigated the addition of microalgae in food products such as bread, cookies, pasta, soup and yogurt and have reported fishy off-flavours, even at very low additions (Batista et al., 2019; Lafarga, 2019; Nunes et al., 2020).

7.4.3 Microalgae-Derived Bioproducts in Food Stability

One of the most significant contributions of microalgae to food stability lies in their high content of antioxidant compounds, including carotenoids, phycobiliproteins, phenolic compounds and PUFAs (Guil-Guerrero & Prates, 2025). These compounds effectively scavenge free radicals, delay lipid oxidation and preserve sensory attributes such as colour, flavour and aroma, particularly in lipid-rich foods (Luo et al., 2021). In addition, microalgal pigments not only function as natural colourants but also improve oxidative stability due to their inherent antioxidant activity (Chettri et al., 2026; Singh et al., 2023). Furthermore, metabolites, such as phenolics, flavonoids and tocopherols, have been proven to effectively attenuate oxidative stress and lower the risk of chronic diseases (Torres-Tiji et al., 2020).

7.4.4 Sensory Aspects

One of the most notable sensory contributions of microalgae is colour (Qazi et al., 2022). Pigments such as chlorophylls, carotenoids and phycobiliproteins confer vibrant green, yellow-orange, red and blue hues, making them attractive natural alternatives to synthetic food colourants (Singh et al., 2023). Beyond visual appeal, these pigments may influence consumer perception of freshness and naturalness, which are key drivers of acceptance for clean-label foods. Food, health and technology have received immense attention in recent times.

Microalgae affect flavour and aroma through their content of free amino acids, peptides, fatty acids and volatile compounds (Urlass et al., 2023). Umami-related amino acids can improve savory taste, while lipid-derived volatiles may impart marine, vegetal features, making microalgae particularly relevant for plant-based meat and seafood alternatives (Coleman et al., 2022; Hossain et al., 2024). However, excessive incorporation can lead to undesirable off-flavours, bitterness or earthy and fishy aromas, which remain major challenges for consumer acceptance (Lafarga, 2019). Texture and mouthfeel are additional sensory dimensions influenced by microalgae-derived bioproducts. Polysaccharides and proteins from species such as *Chlorella*, *Spirulina* and *Porphyridium* exhibit thickening, gelling and emulsifying properties capable of improving viscosity, creaminess and structural stability in foods (Netanel Liberman et al., 2021; Fortuin et al., 2024; Magnabosco et al., 2025). These functional attributes can positively impact sensory perception when carefully optimized within food formulations. Hence, understanding consumer acceptance and concern is essential as it determines their response to the product.

7.5 CHALLENGES AND TECHNOLOGICAL BOTTLENECKS

At a large scale, cultivation inefficiencies represent a major bottleneck. Although open raceway ponds are cheap, this method has low productivity and a high contamination risk. However, PBRs are expensive, with excessive energy consumption, along with light and mass-transfer limitations (Novoveská et al., 2023). These constraints restrict the consistent production of high-density biomass required for downstream processing (Geng et al., 2025). Also, downstream operations, particularly harvesting and dewatering, further complicate scale-up. Furthermore, large-scale cultivation typically yields dilute biomass streams, necessitating energy-intensive separation processes that significantly increase operational costs and reduce overall system efficiency (Wiatrowski et al., 2022).

Regulatory frameworks are a significant barrier to the commercialization of microalgae-derived biomass in the European food industry. Hence, edible microalgae species and derived compounds are subject to the European Union Novel Food Regulation, which requires comprehensive safety assessments, toxicological data and detailed characterization before market authorization (EFSA, 2024). Although essential for ensuring consumer safety, the approval process is often lengthy, costly and administratively complex, while regulatory uncertainty persists due to limited harmonization of standards for species classification, production methods, labelling and quality specifications across EU member states (Girotto & Scapini, 2025). Collectively, these regulatory challenges delay market entry, reduce consumer exposure and constrain the broader adoption of microalgae-derived biomass as a sustainable food ingredient in Europe (Girotto & Scapini, 2025).

Limited consumer familiarity and concerns about naturalness and processing intensity further constrain market uptake (Raverta et al., 2025). Sensory characteristics represent a major technological bottleneck influencing acceptance (Nunes et al., 2023). Variability in sensory quality due to scale-up and inconsistent processing further exacerbates acceptance challenges (Guil-Guerrero & Prates, 2025; Lin et al., 2025).

Despite their strong potential as renewable and circular bioresources, the sustainability performance of microalgae-derived biomass systems remains highly context-dependent. Recent techno-economic and life-cycle assessment (LCA) studies indicate that large-scale microalgal biorefineries often face sustainability bottlenecks associated with high energy demand, intensive nutrient use and water consumption, particularly during cultivation and downstream processing (Brentner et al., 2011; Zhang et al., 2016). However, LCA-based evaluations show that the environmental performance of such systems is susceptible to system boundaries, allocation methods and process integration levels, underscoring the need for standardized sustainability assessment frameworks (Geng et al., 2025; Mahmod et al., 2024).

7.6 FINAL CONSIDERATIONS

Microalgae stands as the key ingredient for technological expansion on the production line of various industries, with its bioactive metabolites being an invaluable factor for offering both financial value for the product and health for the consumer. Though its positives are known, few strains are properly researched and generally accepted by legislative organs. Current consumer opinions also stand as an obstacle for the proper industrialization of microalgal biomass, alongside logistical issues regarding its culture, harvest and processing. It is paramount that future researchers explore further possibilities found in lesser-known species, bringing to light more positives regarding microalgae, so that it may eventually be known accepted by the general population.

REFERENCES

Ahmad, A., Hassan, S. W., & Banat, F. (2022). An overview of microalgae biomass as a sustainable aquaculture feed ingredient: Food security and circular economy. *Bioengineered*, *13*, 9521–9547. https://doi.org/10.1080/21655979.2022.2061148

Ahmad, S., Jan, K., Sahu, J. K., Habib, M., Jan, S., & Bashir, K. (2025). A comprehensive review on recent trends and utilization of algal β-glucan for the development of nutraceuticals and functional foods. *Food Reviews International*, *41*(2), 469–490. https://doi.org/10.1080/87559129.2024.2404222

Ampofo, J., & Abbey, L. (2022). Microalgae: Bioactive composition, health benefits, safety and prospects as potential high-value ingredients for the functional food industry. *Foods*, *11*(12), 1744. https://doi.org/10.3390/foods11121744

Ang, W. S., Kee, P. E., Lan, J. C. W., Chen, W. H., Chang, J. S., & Khoo, K. S. (2025). Unveiling the rise of microalgae-based foods in the global market: Perspective views and way forward. *Food Bioscience*, *66*, 105390. https://doi.org/10.1016/J.FBIO.2024.105390

Arenas Colarte, C., Balic, I., Díaz, Ó., Moreno, A. A., Amenabar, M. J., Bruna Larenas, T., & Caro Fuentes, N. (2025). High-value bioactive molecules extracted from microalgae. *Microorganisms*, *13*(9). https://doi.org/10.3390/microorganisms13092018

Bailey, E., Wojcik, J., Rahn, M., Roos, F., Spooren, A., & Koshibu, K. (2025). Comparative bioavailability of DHA and EPA from microalgal and fish oil in adults. *International Journal of Molecular Sciences*, *26*(19). https://doi.org/10.3390/ijms26199343

Baldemir, P., & Çaklı, Ş. (2024). Sustainable food ingredients: Micro-algae as source bioactive compounds. *Food Bulletin*, *3*(2), 34–40. https://doi.org/10.61326/foodb.v3i2.315

Barreiros, M., Maciel, F., Pereira, R. S., Vicente, A. A., & Pereira, R. N. (2024). Electric fields to support microalgae growth with a differentiated biochemical composition. *Innovative Food Science and Emerging Technologies*, *97*. https://doi.org/10.1016/j.ifset.2024.103829

Batista, P., Niccolai, A., Bursic, I., Sousa, I., Raymundo, A., Rodolfi, L., … Tredici, M. R. (2019). Microalgae as functional ingredients in savory food products: Application to wheat crackers. *Foods*, *8*(12), 611. https://doi.org/10.3390/foods8120611

Beirne, N., Edmundson, S., Gao, S., Freeman, J., & Huesemann, M. (2023). A streamlined approach to characterize microalgae strains for biomass productivity under dynamic climate simulation conditions. *Algal Research*, *72*, 103099. https://doi.org/10.1016/J.ALGAL.2023.103099

Brentner, L. B., Eckelman, M. J., & Zimmerman, J. B. (2011). Combinatorial life cycle assessment to inform process design of industrial production of algal biodiesel. *Environmental Science & Technology*, *45*(16), 7060–7067. https://doi.org/10.1021/es2006995

Camacho, F., Macedo, A., & Malcata, F. (2019). Potential industrial applications and commercialization of microalgae in the functional food and feed industries: A short review. *Marine Drugs*, *17*(6), 312. https://doi.org/10.3390/md17060312

Caporgno, M. P., & Mathys, A. (2018). Trends in microalgae incorporation into innovative food products with potential health benefits. *Frontiers in Nutrition*, *5*, 58. https://doi.org/10.3389/fnut.2018.00058

Chaves, H., Singh, R. B., Khan, S., & Wilczynska, A. (2019). High omega-6/omega-3 fatty acid ratio diets and risk of noncommunicable diseases. In *The Role of Functional Food Security in Global Health* (pp. 217–259). Elsevier. https://doi.org/10.1016/B978-0-12-813148-0.00014-1

Chen, C., Tang, T., Shi, Q., Zhou, Z., & Fan, J. (2022). The potential and challenge of microalgae as promising future food sources. *Trends in Food Science and Technology*, *126*, 99–112. https://doi.org/10.1016/j.tifs.2022.06.016

Chettri, S., Altemimi, A. B., Das, P., Singh, R., Said, P. P., Nath, P. C., Pandey, V. K., & Rustagi, S. (2026). Recent advances in microalgal pigments as a source of natural colors and their application in next-generation functional foods. *Sustainable Food Technology*. https://doi.org/10.1039/D5FB00537J

Chitthan, S. V., Rasheed, A. S., Muniyandi, M., Dhanasekaran, D., & Thajuddin, N. (2025). Bioactive peptides from microalgae and cyanobacteria and their possible mechanisms of action. *International Journal of Peptide Research and Therapeutics*, *31*(5), 86. https://doi.org/10.1007/S10989-025-10746-7

Colarte, C. A., Balic, I., Díaz, Ó., Moreno, A. A., Amenabar, M. J., Bruna Larenas, T., & Caro Fuentes, N. (2025). High-value bioactive molecules extracted from microalgae. *Microorganisms*, *13*(9), 2018. https://doi.org/10.3390/microorganisms13092018

Coleman, B., Van Poucke, C., Dewitte, B., Ruttens, A., Moerdijk-Poortvliet, T., Latsos, C., … Robbens, J. (2022). Potential of microalgae as flavoring agents for plant-based seafood alternatives. *Future Foods*, *5*, 100139. https://doi.org/10.1016/j.fufo.2022.100139

Crudo, F., Dedić, D., Del Favero, G., & Castejón, N. (2025). Potential of microalgal lipids to prevent intestinal barrier disruption: A case study exploring the response of Caco-2 cells to the food contaminant deoxynivalenol. *Journal of Functional Foods*, *135*. https://doi.org/10.1016/j.jff.2025.107125

Cutolo, E. A., Caferri, R., Campitiello, R., & Cutolo, M. (2023). The clinical promise of microalgae in rheumatoid arthritis: From natural compounds to recombinant therapeutics. *Marine Drugs*, 21(12), 630. https://doi.org/10.3390/md21120630

da Silva, M. E. T., Martins, M. A., de Oliveira Leite, M., Milião, G. L., & Coimbra, J. S. D. R. (2021). Microalga Scenedesmus obliquus: Extraction of bioactive compounds and antioxidant activity. *Revista Ciencia Agronomica*, *52*(2), 1–8. https://doi.org/10.5935/1806-6690.20210036

de Souza, T. S. P., Lemos Junior, W. J. F., Cipolat-Gotet, C., & Sant'Ana, A. S. (2025). Innovations in sodium reduction: Integrating technologies in food reformulation. *Food and Humanity*, *5*, 100817. https://doi.org/10.1016/J.FOOHUM.2025.100817

Dhandwal, A., Bashir, O., Malik, T., Salve, R. V., Dash, K. K., Amin, T., Shams, R., & Wani, A. W. (2025). Sustainable microalgal biomass as a potential functional food and its applications in food industry: A comprehensive review. *Environmental Science and Pollution Research International*, *32*(32), 19110–19128. https://doi.org/10.1007/s11356-024-33431-6

do Nascimento, T. C., Pinheiro, P. N., Deprá, M. C., Sartori, R. B., Roca, M., Jacob-Lopes, E., & Zepka, L. Q. (2025). Photoperiod strategies to increase microalgae-pigment content: A new level-up to natural food colorants production. *Algal Research*, *90*. https://doi.org/10.1016/j.algal.2025.104185

EFSA Panel on Nutrition, Novel Foods and Food Allergens (NDA), Turck, D., Bohn, T., Castenmiller, J., de Henauw, S., Hirsch-Ernst, K. I., … Knutsen, H. K. (2024). Guidance on the scientific requirements for an application for authorisation of a novel food in the context of Regulation (EU) 2015/2283. *EFSA Journal*, *22*(9), 8961. https://doi.org/10.2903/j.efsa.2024.8961

Fang, Y., Cai, Y., Zhang, Q., Ruan, R., & Zhou, T. (2024). Research status and prospects for bioactive compounds of Chlorella species: Composition, extraction, production, and biosynthesis pathways. *Process Safety and Environmental Protection*, *191*, 345–359. https://doi.org/10.1016/J.PSEP.2024.08.114

Fernandes, A. S., Jacob-Lopes, E., & Zepka, L. Q. (2021). An overview on microalgae carotenoids and chlorophylls: Focus in the bioaccessibility / Uma visão geral dos carotenoides e clorofilas microalgais: Foco na bioacessibilidade. *Brazilian Journal of Development*, *7*(8), 82727–82760. https://doi.org/10.34117/bjdv7n8-470

Fernando, R., Sun, X., & Rupasinghe, H. P. V. (2024). Production of bioactive peptides from microalgae and their biological properties related to cardiovascular disease. *Macromol*, *4*(3), 582–596. https://doi.org/10.3390/MACROMOL4030035

Ferreira, A. M., Leite, A. C., Coutinho, J. A., & Freire, M. G. (2021). Chlorophylls extraction from spinach leaves using aqueous solutions of surface-active ionic liquids. *Sustainable Chemistry*, *2*(4), 764–777.

Ferreira de Oliveira, A. P., & Bragotto, A. P. A. (2022). Microalgae-based products: Food and public health. *Future Foods*, *6*, 100157. https://doi.org/10.1016/J.FUFO.2022.100157

Fortuin, J., Hellebois, T., Iken, M., Shaplov, A. S., Fogliano, V., & Soukoulis, C. (2024). Stabilising and functional effects of Spirulina (Arthrospira platensis) protein isolate on encapsulated Lacticaseibacillus rhamnosus GG during processing, storage and gastrointestinal digestion. *Food Hydrocolloids*, *149*, 109519. https://doi.org/10.1016/j.foodhyd.2023.109519

García, A. B., Longo, E., Murillo Cruz, M. C., & Bermejo, R. (2021). Using a B-phycoerythrin extract as a natural colorant: Application in milk-based products. *Molecules*, *26*(2), 297. https://doi.org/10.3390/molecules26020297

Garciarena, I. N., Ackerl, R., Ruiz, E. G., Glymenaki, M., Mendes, V., Muñoz-González, A., & Kass, G. E. (2025). The safety assessment of microalgae-derived products as novel foods by the European food safety authority. *Future Foods*, *11*, 100661. https://doi.org/10.1016/j.fufo.2025.100661

Geng, Y., Shaukat, A., Azhar, W., Raza, Q. U., Tahir, A., Abideen, M. Z. U., Zia, M. A. B., Bashir, M. A., & Rehim, A. (2025). Microalgal biorefineries: A systematic review of technological trade-offs and innovation pathways. *Biotechnology for Biofuels and Bioproducts*, *18*(1), 93. https://doi.org/10.1186/s13068-025-02694-7

Ghosh, A., & Honda, M. (2026). A comprehensive review of recent advances and future perspectives on astaxanthin geometrical isomers: Biological activities, isomerization techniques, analytical methods, physicochemical characteristics, and application potential. *Food Research International*, *225*, 118036. https://doi.org/10.1016/J.FOODRES.2025.118036

Girotto, F., & Scapini, T. (2025). Microalgal biomass in the European food industry: Navigating regulation, technological innovation, and consumer acceptance. *Algal Research*, *91*, 104288. https://doi.org/10.1016/j.algal.2025.104288

Gross, J. (1991). Carotenoids. In *Pigments in Vegetables: Chlorophylls and Carotenoids* (pp. 75–278). Springer US.

Guil-Guerrero, J. L., & Prates, J. A. M. (2025). Microalgae bioactives for functional food innovation and health promotion. *Foods*, *14*(12), 2122. https://doi.org/10.3390/foods14122122

Guo, C., Ling, N., Tian, H., Wang, Z., Gao, M., Chen, Y., & Ji, C. (2025). Comprehensive review of extraction, purification, structural characteristics, pharmacological activities, structure-activity relationship and application of seabuckthorn protein and peptides. *International Journal of Biological Macromolecules*, *294*. https://doi.org/10.1016/J.IJBIOMAC.2024.139447

Guo, X., Wang, Q., Wu, Y., Liu, X., & Gong, Z. (2024). Comprehensive insights into microalgae proteins: Nutritional profiles and innovative applications as sustainable alternative proteins in health and food sciences. *Food Hydrocolloids*, *154*. https://doi.org/10.1016/J.FOODHYD.2024.110112

Hassan, S., Meenatchi, R., Pachillu, K., Bansal, S., Brindangnanam, P., Arockiaraj, J., Kiran, G. S., & Selvin, J. (2022). Identification and characterization of the novel bioactive compounds from microalgae and cyanobacteria for pharmaceutical and nutraceutical applications. *Journal of Basic Microbiology*, *62*(9), 999–1029. https://doi.org/10.1002/jobm.202100477

Hernández, H., Nunes, M. C., Prista, C., & Raymundo, A. (2022). Innovative and healthier dairy products through the addition of microalgae: A review. *Foods*, *11*(5), 755. https://doi.org/10.3390/foods11050755

Hossain, M. J., Alam, A. N., Lee, E. Y., Hwang, Y. H., & Joo, S. T. (2024). Umami characteristics and taste improvement mechanism of meat. *Food Science of Animal Resources*, *44*(3), 515–532. https://doi.org/10.5851/kosfa.2024.e29

Jacob-Lopes, E., Maroneze, M. M., & Deprá, M. C. (2019). Bioactive food compounds from microalgae: An innovative framework on industrial biorefineries. *Current Opinion in Food Science*, *25*, 1–7. https://doi.org/10.1016/j.cofs.2018.12.003

Kaur, S., Sharma, A., Bala, S., Satheesh, N., Nile, A. S., & Nile, S. H. (2025). Microalgae in the food-health nexus: Exploring species diversity, high-value bioproducts, health benefits, and sustainable market potential. *Bioresource Technology*, *427*, 132424. https://doi.org/10.1016/J.BIORTECH.2025.132424

Kohandel, Z., Farkhondeh, T., Aschner, M., Pourbagher-Shahri, A. M., & Samarghandian, S. (2022). Anti-inflammatory action of astaxanthin and its use in the treatment of various diseases. *Biomedicine and Pharmacotherapy*, *145*, 112179. https://doi.org/10.1016/j.biopha.2021.112179

Kousar, S., Nadeem, F., Khan, O., & Shahzadi, A. (2017). Chemical synthesis of various limonene derivatives—A comprehensive review. *International Journal of Chemical and Biochemical Sciences*, *11*, 102–112.

Kumar, R., Hegde, A. S., Sharma, K., Parmar, P., & Srivatsan, V. (2022). Microalgae as a sustainable source of edible proteins and bioactive peptides – Current trends and future prospects. *Food Research International*, *157*, 111338. https://doi.org/10.1016/J.FOODRES.2022.111338

Labua, S., & Kenkhunthot, T. (2020). The extraction of lipid from microalgae found in brackish water by terpenes. *Journal of Current Science and Technology*, *10*(1), 35–40.

Lafarga, T. (2019). Effect of microalgal biomass incorporation into foods: Nutritional and sensory attributes. *Algal Research*, *41*, 101566. https://doi.org/10.1016/j.algal.2019.101566

Lafarga, T., Fernández-Sevilla, J. M., González-López, C. V., & Acién-Fernández, F. G. (2020). Spirulina for the food and functional food industries. *Food Research International*, *137*, 109356. https://doi.org/10.1016/j.foodres.2020.109356

Li, C., Du, M., Han, Y., Sun, W., Chen, Z., Liu, Q., & Wang, J. (2025). Microalgae in health care and functional foods: β-glucan applications, innovations in drug delivery and synthetic biology. *Frontiers in Pharmacology*, *16*, 1557298. https://doi.org/10.3389/fphar.2025.1557298

Lin, X., Maran, N., Lim, A., Ng, S. B., & Teo, P. S. (2025). Current challenges, and potential solutions to increase acceptance and long-term consumption of cultured meat and edible insects – A review. *Future Foods*, *11*, 100544. https://doi.org/10.1016/j.fufo.2025.100544

Linares-Maurizi, A., Reversat, G., Awad, R., Bultel-Poncé, V., Oger, C., Galano, J. M., Balas, L., Durbec, A., Bertrand-Michel, J., Durand, T., Pradelles, R., & Vigor, C. (2023). Bioactive oxylipins profile in marine microalgae. *Marine Drugs*, *21*(3). https://doi.org/10.3390/md21030136

Liu, Y., Wang, M., Castagnini, J. M., Barba, F. J., & Asensio-Grau, A. (2025). The impact of microalgae-derived bioactive compounds on gut microbiota modulation. *Current Opinion in Food Science*, *66*, 101365. https://doi.org/10.1016/j.cofs.2025.101365

Luo, A., Hu, B., Feng, J., Lv, J., & Xie, S. (2021). Preparation, and physicochemical and biological evaluation of chitosan-Arthrospira platensis polysaccharide active films for food packaging. *Journal of Food Science*, *86*(3), 987–995. https://doi.org/10.1111/1750-3841.15639

Magnabosco, C., Santaniello, G., & Romano, G. (2025). Microalgae: A promising source of bioactive polysaccharides for biotechnological applications. *Molecules*, *30*, 2055. https://doi.org/10.3390/molecules30092055

Mahmod, S., Al-Rajabi, M., Abdul, P., Ding, G. T., Kamarudin, K., Nagoor Gunny, A. A., Tan, J. P., & Takriff, M. (2024). Microalgae biomass: A multi-product biorefinery solution for sustainable energy, environmental remediation, and industrial symbiosis. *Algal Research*, *85*, 103839. https://doi.org/10.1016/j.algal.2024.103839

Manzke, N. E., Gormley, A. R., Kim, Y. I., Parnsen, W., & Kim, S. W. (2025). Efficacy of β–glucan from microalgae on the intestinal health and growth of nursery pigs. *Animal Science Journal*, *96*(1), e70056. https://doi.org/10.1111/asj.70056

Marjanović, B., Benković, M., Jurina, T., Cvetnić, T. S., Valinger, D., Kljusurić, J. G., Tušek, A. J., Marjanović, B., Benković, M., Jurina, T., Cvetnić, T. S., Valinger, D., Kljusurić, J. G., & Tušek, A. J. (2024). Bioactive compounds from Spirulina spp.—Nutritional value, extraction, and application in food industry. *Separations*, *11*(9). https://doi.org/10.3390/SEPARATIONS11090257

Masoumifeshani, B., Abedian Kenari, A., Sottorff, I., Crüsemann, M., & Amiri Moghaddam, J. (2025). Identification and evaluation of antioxidant and anti-aging peptide fractions from enzymatically hydrolyzed proteins of Spirulina platensis and Chlorella vulgaris. *Marine Drugs*, *23*(4), 162. https://doi.org/10.3390/MD23040162

Mesías, F. J., Martín, A., & Hernández, A. (2021). Consumers' growing appetite for natural foods: Perceptions towards the use of natural preservatives in fresh fruit. *Food Research International*, *149*, 110749. https://doi.org/10.1016/j.foodres.2021.110749

Mostafa, H. S., & Hashem, M. M. (2024). Microalgae as a source of carotenoids in foods, obstacles and solutions. *Phytochemistry Reviews*. https://doi.org/10.1007/s11101-024-10039-9

Nashath, F. Z., Ng, Y. J., Khoo, K. S., & Show, P. L. (2025). Current contributions of microalgae in global biofuel production; cultivation techniques, biofuel varieties and promising industrial ventures. *Biomass and Bioenergy*, *201*, 108092. https://doi.org/10.1016/J.BIOMBIOE.2025.108092

Netanel Liberman, G., Ochbaum, G., Bitton, R., & Arad, S. (2021). Antimicrobial hydrogels composed of chitosan and sulfated polysaccharides of red microalgae. *Polymer*, *215*, 123353. https://doi.org/10.1016/j.polymer.2020.123353

Niccolai, A., Chini Zittelli, G., Rodolfi, L., Biondi, N., & Tredici, M. R. (2019). Microalgae of interest as food source: Biochemical composition and digestibility. *Algal Research*, *42*, 101617. https://doi.org/10.1016/j.algal.2019.101617

Nikolova, K., Gentscheva, G., Gyurova, D., Pavlova, V., Dincheva, I., Velikova, M., … & Gergov, G. (2024). Metabolomic profile of Arthrospira platensis from a Bulgarian bioreactor—A potential opportunity for inclusion in dietary supplements. *Life*, *14*(2), 174.

Nörnberg, M. L., Bortolotti, C. M., Minella, E., & Nörnberg, J. L. (2022) Use of barley flour as a source of biocompounds in bakery products. *Brazilian Journal of Development*, *8*(2), 10334–10353. https://doi.org/10.34117/bjdv8n2-125

Nörnberg, M. L., de Aguiar, K. L., do Nascimento, T. C., Jacob-Lopes, E., Nörnberg, J. L., & Zepka, L. Q. (2025). Functional enrichment of Minas Frescal cheese with Chlorella sorokiniana: Integration of carotenoids and bioactive fatty acids. *Observatório de LaEconomía Latinoamericana*, *23*(12), e12618. https://doi.org/10.55905/oelv23n12-084

Nörnberg, M. L., Nass, P. P., Caetano, P. A., Schetinger, L. C., & Zepka, L. Q. (2024). Microalgae as food source. In Eduardo Jacob Lopes, Leila Queiroz Zepka, and Mariany Costa Deprá (eds.), *Smart Food Industry: The Blockchain for Sustainable Engineering* (pp. 173–184). CRC Press. https://doi.org/10.1201/9781003231172-12

Nörnberg, M. L., Schetinger, L. C., do Nascimento, T. C., Pinheiro, P. N., Caetano, P. A., Jacob-Lopes, E., & Zepka, L. Q. (2025). Current status of microalgae-based food products: Future trends of functional ingredients. *Recent Patents on Biotechnology*. https://doi.org/10.2174/0118722083380758250802021828

Nova, M., Citterio, S., Martegani, E., & Colombo, S. (2024). Unraveling the anti-aging properties of phycocyanin from the cyanobacterium spirulina (Arthrospira platensis). *International Journal of Molecular Sciences*, *25*(8). https://doi.org/10.3390/IJMS25084215

Novoveská, L., Nielsen, S. L., Eroldoğan, O. T., Haznedaroglu, B. Z., Rinkevich, B., Fazi, S., Robbens, J., Vasquez, M., & Einarsson, H. (2023). Overview and challenges of large-scale cultivation of photosynthetic microalgae and cyanobacteria. *Marine Drugs*, *21*(8), 445. https://doi.org/10.3390/md21080445

Nunes, M. C., Fernandes, I., Vasco, I., Sousa, I., & Raymundo, A. (2020). Tetraselmis chuii as a sustainable and healthy ingredient to produce gluten-free bread: Impact on structure, colour and bioactivity. *Foods*, *9*(5), 579. https://doi.org/10.3390/foods9050579

Nunes, M. C., Ferreira, J., & Raymundo, A. (2023). Volatile fingerprint impact on the sensory properties of microalgae and development of mitigation strategies. *Current Opinion in Food Science*, *51*, 101040. https://doi.org/10.1016/j.cofs.2023.101040

O'Connor, J., Garcia-Vaquero, M., Meaney, S., & Tiwari, B. K. (2022). Bioactive peptides from algae: Traditional and novel generation strategies, structure-function relationships, and bioinformatics as predictive tools for bioactivity. *Marine Drugs*, *20*(5), 317. https://doi.org/10.3390/MD20050317

Oliyaei, N., Moosavi-Nasab, M., Tanideh, N., & Iraji, A. (2023). Multiple roles of fucoxanthin and astaxanthin against Alzheimer's disease: Their pharmacological potential and therapeutic insights. *Brain Research Bulletin*, *193*, 11–21. https://doi.org/10.1016/j.brainresbull.2022.11.018

Ortega, A., & Campos, M. (2019). Bioactive compounds. In Maira Rubi Segura Campos (ed.), *Bioactive Compounds as Therapeutic Alternatives* (pp. 247–264). Springer Nature.

Oslan, S. N. H., Xuan, N. J., Khodzori, F. A., Hakim, B. N. A., Huda, N., Julmohamad, N., & Barzkar, N. (2025). Comprehensive review of bioactive compounds from microalgae as promising source for industrial applications. *Algal Research*, *92*, 104420. https://doi.org/10.1016/J.ALGAL.2025.104420

Patel, A., Rova, U., Christakopoulos, P., & Matsakas, L. (2022). Microalgal lutein biosynthesis: Recent trends and challenges to enhance the lutein content in microalgal cell factories. *Frontiers in Marine Science*, *9*, 1015419. https://doi.org/10.3389/fmars.2022.1015419

Paterson, S., Majchrzak, M., Gómez-Garre, D., Ortega-Hernández, A., Sánchez-González, S., de la Fuente, M. Á., Gómez-Cortés, P., & Hernández-Ledesma, B. (2025). Role of simulated Nannochloropsis gaditana digests in shaping gut microbiota and short-chain fatty acid levels. *Nutrients*, *17*(1), 99. https://doi.org/10.3390/NU17010099

Patwary, M. M., Kim, Y. E., George, S., & Ignea, C. (2025). Integrating biosynthesis insights with biotechnological potential of bioactive carotenoids in food systems. *Trends in Food Science and Technology*, *166*, 105410. https://doi.org/10.1016/j.tifs.2025.105410

Pereira, R. N., Rodrigues, R. M., Madalena, D. A., & Vicente, A. (2023). Tackling food allergens—The role of food processing on proteins' allergenicity. *Advances in Food and Nutrition Research*, *106*, 317–351. https://doi.org/10.1016/BS.AFNR.2023.04.001

Piligaev, A. V., Sorokina, K. N., Shashkov, M. V., & Parmon, V. N. (2018). Screening and comparative metabolic profiling of high lipid content microalgae strains for application in wastewater treatment. *Bioresource Technology*, *250*, 538–547. https://doi.org/10.1016/J.BIORTECH.2017.11.063

Portillo, F., & Jiménez, A. M. (2021). Pigmentos en microalgas: funciones, aplicaciones y técnicas de sobreproducción. *Biotecnología: Revista de la Sociedad Mexicana de Biotecnología y Bioingeniería*, *25*, 35–51.

Prabhakar, P., Verma, D. K., & Banerjee, M. (2025). Beta (β)-glucan as microbial polymer with nutraceutical potential: Chemistry, biosynthesis, extraction, identification, and industrial production of bioactive compound for human health. *Microbial Nutraceuticals: Products and Processes*, 231–256.

Qazi, M. W., de Sousa, I. G., Nunes, M. C., & Raymundo, A. (2022). Improving the nutritional, structural, and sensory properties of gluten-free bread with different species of microalgae. *Foods*, *11*(3), 397.

Qoms, M. S., Wong, S. K., Mohd Fauzi, N., Husain, K., Makpol, S., & Tan, J. K. (2025). Microalgae-derived peptides targeting lifestyle-related diseases: Discovery, mechanisms, structure–activity relationships, and structural modifications. *Antioxidants*, *14*(10), 1170. https://doi.org/10.3390/ANTIOX14101170

Ragusa, I., Nardone, G. N., Zanatta, S., Bertin, W., Amadio, E., Ragusa, I., Nardone, G. N., Zanatta, S., Bertin, W., & Amadio, E. (2021). Spirulina for skin care: A bright blue future. *Cosmetics*, *8*(1), 1–19. https://doi.org/10.3390/COSMETICS8010007

Rani, S., Senapati, V., Senapati, V., Saxena, A., Cheah, W. Y., & Tiwari, A. (2026). Assessment of growth dynamics, metabolite production phasic analysis in microalgae Scenedesmus rotundus under varying pH conditions. *Bioresource Technology*, *441*, 133533. https://doi.org/10.1016/J.BIORTECH.2025.133533

Raverta, P., Sandi, I., Martin, B., & Loera, B. (2025). Unfamiliar familiarity: A scoping review on the role of familiarity in consumer acceptance of cultivated meat. *Appetite*, *196*, 108000. https://doi.org/10.1016/j.appet.2025.108000

Ren, Y., Deng, J., Huang, J., Wu, Z., Yi, L., Bi, Y., & Chen, F. (2021). Using green alga Haematococcus pluvialis for astaxanthin and lipid co-production: Advances and outlook. *Bioresource Technology*, *340*, 125736. https://doi.org/10.1016/J.BIORTECH.2021.125736

Ren, Y., Sun, H., Deng, J., Huang, J., & Chen, F. (2021). Carotenoid production from microalgae: Biosynthesis, salinity responses and novel biotechnologies. *Marine Drugs*, *19*(12), 713. https://doi.org/10.3390/md19120713

Salehipour-Bavarsad, F., Nematollahi, M. A., Pistocchi, R., & Pezzolesi, L. (2024). Algal food safety: Possible contaminations, challenges of harmonized quality assessments, and suggested recommendations for the nascent industry of microalgae-based products. *Algal Research*, *81*, 103579. https://doi.org/10.1016/j.algal.2024.103579

Siddhnath, Surasani, V. K. R., Singh, A., Singh, S. M., Hauzoukim, Murthy, L. N., & Baraiya, K. G. (2024). Bioactive compounds from micro-algae and its application in foods: A review. *Discover Food*, *4*(1), 27. https://doi.org/10.1007/S44187-024-00096-6

Singh, T., Pandey, V. K., Dash, K. K., Zanwar, S., & Singh, R. (2023). Natural bio-colorant and pigments: Sources and applications in food processing. *Journal of Agriculture and Food Research*, *13*, 100628. https://doi.org/10.1016/j.jafr.2023.100628

Sirohi, P., Verma, H., Singh, S. K., Singh, V. K., Pandey, J., Khusharia, S., Kumar, D., Kaushalendra, Teotia, P., & Kumar, A. (2022). Microalgal carotenoids: Therapeutic application and latest approaches to enhance the production. *Current Issues in Molecular Biology*, *44*(12), 6257–6279. https://doi.org/10.3390/cimb44120427

Souza, M. P., Hoeltz, M., Gressler, P. D., Benitez, L. B., & Schneider, R. C. (2019). Potential of microalgal bioproducts: General perspectives and main challenges. *Waste and Biomass Valorization*, *10*(8), 2139–2156. https://doi.org/10.1007/s12649-018-0253-6

Su, B., Deng, M. R., & Zhu, H. (2023). Advances in the discovery and engineering of gene targets for carotenoid biosynthesis in recombinant strains. *Biomolecules*, *13*(12), 1747. https://doi.org/10.3390/biom13121747

Sun, H., Wang, Y., He, Y., Liu, B., Mou, H., Chen, F., & Yang, S. (2023). Microalgae-derived pigments for the food industry. *Marine Drugs*, *21*(2), 82. https://doi.org/10.3390/md21020082

Sung, Y. J., & Sim, S. J. (2022). Multifaceted strategies for economic production of microalgae Haematococcus pluvialis-derived astaxanthin via direct conversion of CO2. *Bioresource Technology*, *344*, 126255. https://doi.org/10.1016/J.BIORTECH.2021.126255

Swetha, N., & Mathanghi, S. K. (2024). Towards sustainable omega-3 fatty acids production – A comprehensive review on extraction methods, oxidative stability and bio-availability enhancement. *Food Chemistry Advances*, *4*, 100603. https://doi.org/10.1016/j.focha.2023.100603

Torres-Tiji, Y., Fields, F. J., & Mayfield, S. P. (2020). Microalgae as a future food source. *Biotechnology Advances*, *41*(107536), 1–13. https://doi.org/10.1016/j.biotechadv.2020.107536

Udayan, A., Sreekumar, N., & Arumugam, M. (2022). Statistical optimization and formulation of microalga cultivation medium for improved omega 3 fatty acid production. *Systems Microbiology and Biomanufacturing*, *2*(2), 369–379. https://doi.org/10.1007/s43393-021-00069-1

Urlass, S., Wu, Y., Nguyen, T. T., Winberg, P., Turner, M. S., & Smyth, H. (2023). Unravelling the aroma and flavour of algae for future food applications. *Trends in Food Science & Technology*, *138*, 370–381.

Vasilakis, G., Marka, S., Ntzouvaras, A., Zografaki, M. E., Kyriakopoulou, E., Kalliampakou, K. I., & Flemetakis, E. (2025). Wound healing, antioxidant, and antiviral properties of bioactive polysaccharides of microalgae strains isolated from Greek coastal lagoons. *Marine Drugs*, *23*(2), 77. https://doi.org/10.3390/md23020077

Vieira, M. V., Pastrana, L. M., & Fuciños, P. (2023). Encapsulation of microalgae-based products for food and feed applications. In Eduardo Jacob-Lopes, Maria Isabel Queiroz, Mariana Manzoni Maroneze, and Leila Queiroz Zepka (eds.), *Handbook of Food and Feed from Microalgae: Production, Application, Regulation, and Sustainability* (p. 371). Elsevier. https://doi.org/10.1016/B978-0-323-99196-4.00013-9

Wiatrowski, M., Klein, B. C., Davis, R. W., Quiroz-Arita, C., Tan, E. C. D., Hunt, R. W., & Davis, R. E. (2022). Techno-economic assessment for the production of algal fuels and value-added products: Opportunities for high-protein microalgae conversion. *Biotechnology for Biofuels and Bioproducts*, *15*(1), 8. https://doi.org/10.1186/s13068-021-02098-3

Zakaria, M. F., Mat Zaidi, F. N. A., Ahmad Kamal, A. H., Aleng, N. A., Abd Wahid, M. E., Katayama, T., & Jusoh, M. (2025). PUFA from microalgae: Challenges, factors affecting high production and industrial application. *Aquaculture and Fisheries*, *10*(4), 545–555. https://doi.org/10.1016/j.aaf.2025.01.004

Zhang, K., Yin, X., Huang, Y., Liu, C., Zhang, Q., Liu, Q., Wang, S., Fei, W., Shi, Q., & Qiu, L. (2024). A Potent Antibacterial Peptide (P6) from the De Novo Transcriptome of the Microalga Aureococcus anophagefferens. *International Journal of Molecular Sciences*, *25*(24), 13736. https://doi.org/10.3390/IJMS252413736

Zhang, T.-Y., Hu, H.-Y., Wu, Y., Zhuang, L.-L., Xu, X.-Q., Wang, X.-X., & Dao, G.-H. (2016). Promising solutions to solve the bottlenecks in the large-scale cultivation of microalgae for biomass/bioenergy production. *Renewable and Sustainable Energy Reviews*, *60*, 1602–1614. https://doi.org/10.1016/j.rser.2016.02.008

Zhou, L., Li, K., Duan, X., Hill, D., Barrow, C., Dunshea, F., Martin, G., & Suleria, H. (2022). Bioactive compounds in microalgae and their potential health benefits. *Food Bioscience*, *48*, 101932. https://doi.org/10.1016/j.fbio.2022.101932

8 Optimization Strategies for Enhancing Microalgal Lipid Production toward Sustainable Nutrition

Md. Mirazul Islam

8.1 INTRODUCTION

Microalgae have emerged as a highly promising and sustainable alternative source for the production of lipids and high-value fatty acids essential to human health and industrial applications (Michael et al., 2026). These photosynthetic microorganisms offer several advantages over conventional lipid sources, including rapid growth rates, high lipid accumulation capacity, and minimal requirements for arable land and freshwater. Consequently, microalgae have gained increasing attention as a viable solution to global challenges related to sustainable energy production, food security, and environmental preservation (Ferreira et al., 2019; Lari et al., 2016; Michael et al., 2026).

The growing global demand for healthy dietary lipids—particularly long-chain omega-3 polyunsaturated fatty acids (PUFAs) such as eicosapentaenoic acid (EPA) and docosahexaenoic acid (DHA)—has highlighted the limitations and sustainability concerns associated with conventional sources (Bagchi et al., 2025; Kumari et al., 2023). Traditionally, fish and fish oils have served as the primary sources of EPA and DHA. However, the long-term sustainability of marine-derived omega-3 fatty acids is increasingly threatened by declining wild fish stocks, overfishing, environmental pollution, and the risk of contamination with heavy metals and persistent organic pollutants. In parallel, shifting consumer preferences toward plant-based and vegan diets have further intensified the search for alternative omega-3 sources (Bagchi et al., 2025; Kumari et al., 2023).

Although vegetable oils provide alpha-linolenic acid (ALA), their nutritional efficacy is limited by the low conversion efficiency of ALA into biologically active EPA and DHA in the human body. As a result, direct dietary sources of long-chain omega-3 fatty acids are required to achieve optimal health benefits (Bagchi et al., 2025; Kumari et al., 2023). In this context, microalgae represent a direct, sustainable, and environmentally friendly source of EPA and DHA, circumventing the ecological, ethical, and supply-chain limitations associated with fish oil production. Moreover, microalgal-derived lipids align with sustainability frameworks in human nutrition

DOI: 10.1201/9781003611479-8

and land-use efficiency, reinforcing their potential as a cornerstone of future green diets (Kumari et al., 2023).

The economic and technological feasibility of microalgae-based lipid production is closely linked to cultivation efficiency and biomass productivity (Ferreira et al., 2019). Microalgal species exhibit substantial variability in lipid composition and accumulation capacity, with certain strains capable of accumulating lipids exceeding 20% of their dry cell weight under optimized conditions (Rehman et al., 2022). Notably, *Pleurastrum insigne* has been reported to achieve lipid contents of approximately 57.06% of dry cell weight, highlighting the remarkable potential of strain selection and process optimization (Chhandama & Satyan, 2021).

Environmental conditions and nutrient availability play a pivotal role in regulating both biomass growth and lipid biosynthesis in microalgae. Optimized control of cultivation parameters—including nutrient supply, light intensity, temperature, pH, and CO_2 concentration—can significantly influence lipid yield and fatty acid composition (Elshobary et al., 2025; Hang Nguyen, 2026; Han et al., 2021). However, cultivation strategies often require a delicate balance between promoting cell growth and inducing lipid accumulation, as these processes are frequently antagonistic (Pourbakhtiar et al., 2022).

Nutrient limitation, particularly nitrogen and phosphorus deprivation, has been extensively investigated as an effective strategy to enhance lipid accumulation in various microalgal species, including *Chlorella* sp., *Chaetoceros calcitrans*, *Phaeodactylum tricornutum*, and *Dunaliella salina* (Dani et al., 2021; Elshobary et al., 2025). While nitrogen deficiency can significantly increase lipid content—reaching up to 46.32% in *Chaetoceros muelleri* and 54.15% in *Dunaliella salina*—it often leads to a concomitant reduction in biomass productivity (Gao et al., 2013). To overcome this limitation, two-stage cultivation strategies have been developed to decouple the growth phase from the lipid induction phase, thereby maximizing overall lipid yields (Dani et al., 2021; Pourbakhtiar et al., 2022). In this approach, favorable growth conditions are applied during the first stage to enhance biomass accumulation, followed by the introduction of stress factors—such as nutrient deprivation or temperature shifts in the second stage to stimulate lipid biosynthesis (Pourbakhtiar et al., 2022).

Beyond conventional cultivation parameters, the use of waste streams—such as municipal wastewater and agri-food by-products—as low-cost nutrient sources has emerged as a promising strategy to improve the economic and environmental sustainability of microalgal lipid production. This approach not only reduces cultivation costs but also contributes to waste valorization and circular resource management (Ferreira et al., 2019; Han et al., 2021; Rafiq et al., 2023).

The aim of this chapter is to critically evaluate and synthesize current strategies for optimizing microalgal lipid production, with a particular focus on high-value omega-3 fatty acids for sustainable human nutrition. Advanced cultivation approaches, including nutrient management, photobioreactor (PBR) design, two-stage cultivation systems, genetic engineering, and metabolic pathway manipulation, are discussed as key tools to enhance lipid yield and tailor fatty acid profiles. Furthermore, this chapter explores the integration of microalgal lipid production into a circular bioeconomy framework, emphasizing the use of waste streams and industrial by-products to improve economic feasibility and environmental performance. Ultimately, this work provides a holistic perspective on the role of microalgae as a sustainable, scalable, and health-promoting source of lipids capable of meeting growing global nutritional demands while minimizing environmental impacts.

8.2 NUTRITIONAL BENEFITS OF MICROALGAL LIPIDS FOR HUMAN CONSUMPTION

Microalgal lipids are a very rich and rapidly renewable source of vital fatty acids for the human diet, with significant nutritional value, particularly in the sustainable diet and in the plant-based diet. This section will focus on the direct health advantages of the microalgal lipids, including the essential fatty acid composition, purity, tailor-made lipid profiles, and sustainability as a source of nutrition.

8.2.1 Essential Fatty Acids (EPA, DHA, and ALA)

Among the most prominent ones are omega-3 PUFAs, which are vital to human health and contain ALA, EPA, and DHA (Kumari et al., 2023; Muthu, 2025). ALA (18:3n-3) is an EPA (20:5n-3) and DHA (22:6n-3) precursor in human beings, although the rate of conversion is significantly low and may be as low as less than 1% of EPA and approximately 10% of DHA (Kumari et al., 2023) (Figure 8.1). These omega-3 fatty acids are long-chain omega-3 fatty acids, which are essential to cardiovascular health, neurocognitive functioning, immune regulation, and fetal and infant development (Calder, 2021; Muthu, 2025). In particular, DHA is important for brain development and vision, especially in infancy (Calder, 2021). The physiological actions of such fatty acids are due to their roles in the structure of cell membranes and as precursors of signaling molecules such as eicosanoids, which control inflammation, blood clotting, and other important body processes (Calder, 2021).

In aquatic food webs, microalgae are primary producers of EPA and DHA (Adarme-Vega et al., 2012; Muthu, 2025). Historically, the primary sources of EPA and DHA in the human diet have been marine organisms, especially fatty fish (Adarme-Vega et al., 2012; Bagchi et al., 2025). Nevertheless, fish also obtain these positive fatty acids through eating microalgae or zooplankton that feed on microalgae (Adarme-Vega et al., 2012). Plant-based oils, including flaxseed, walnuts, and canola oil, contain large quantities of ALA but do not tend to contain large quantities of preformed EPA and DHA (Kumari et al., 2023; Muthu, 2025). Due to the inefficiency of ALA conversion, the increasing demand for EPA and DHA globally, and the limited supply of marine fish oil, which has issues of sustainability and contamination, extensive research has been conducted to genetically enhance oilseed crops to offer more promising and effective land-based sources of omega-3 (Deckelbaum & Torrejon, 2012; Nazir & Zahoor, 2021; Zhou et al., 2023).

The accumulation of stearidonic acid (SDA, 18:4n-3), the product of the $\Delta 6$ desaturase step, has been one of the strategies. SDA bypasses the rate-limiting $\Delta 6$ desaturase in humans and thus is better converted to EPA than ALA, so SDA-rich oils (both naturally SDA-rich and genetically modified plants) are considered an effective

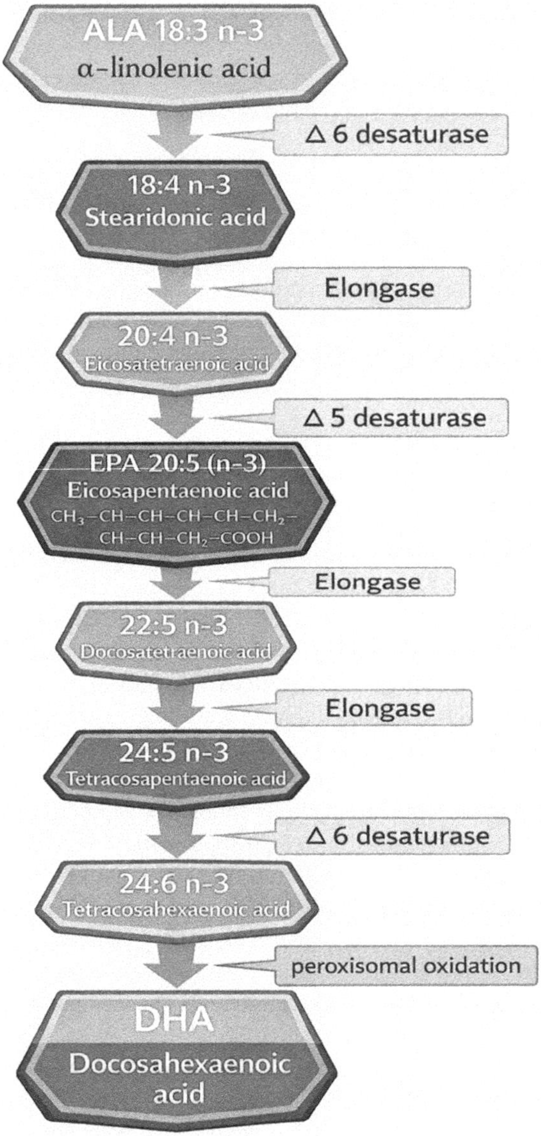

FIGURE 8.1 Biosynthesis of eicosapentaenoic acid (EPA) and docosahexaenoic acid (DHA) from α-linolenic acid (ALA).

plant precursor to increasing EPA status (Baker et al., 2025; Prasad et al., 2021; Walker et al., 2013). A second, more advanced approach is the introduction of complete microalgal-derived biosynthetic pathways for EPA and DHA in oilseed crops like camelina and canola, which allows seeds to store fish oil-like amounts of long-chain omega-3 PUFAs (up to 30% EPA and large quantities of DHA) and therefore a scalable terrestrial substitute to marine oils (Napier et al., 2018; Zhou et al., 2023).

Although genetically engineered land-based oilseed crops yielding SDA, EPA, and DHA provide promising plant sources of long-chain omega-3 fatty acids (Bagchi et al., 2025; Kitessa et al., 2014), microalgae are a direct source of preformed EPA and DHA, bypassing the inefficient human conversion pathway and requiring multi-gene engineering in crops, and thus, a powerful vegetarian alternative to fish oils (Adarme-Vega et al., 2012; Bagchi et al., 2025; Kitessa et al., 2014).

8.2.2 High-Purity Lipids without Contaminants

One of the considerable benefits of microalgae as a source of omega-3 is their ability to produce high-purity lipids without the presence of many pollutants that can contaminate traditional fish oils or marine sources. Although fish and fish oils are excellent sources of EPA and DHA, there are concerns about their potential contamination with heavy metals, especially methylmercury, polychlorinated biphenyls, and persistent organic pollutants, which bioaccumulate in marine food chains (Kumari et al., 2023). Microalgae, in particular, when grown in a controlled bioreactor, provide an alternative that is cleaner and poses a lower risk of contamination (Bagchi et al., 2025; Saini et al., 2023). This contributes to the fact that microalgal lipids are especially appealing to the sensitive markets, including infant nutrition, where the safety and purity are the most important factors (Kumari et al., 2023; Saini et al., 2023). Lacking environmental toxins makes microalgae-derived omega-3s a very viable product in terms of functional foods and nutraceuticals, as it responds to consumer needs of safe, pure, and sustainably sourced nutritional supplements (Saini et al., 2023).

8.2.3 Adjustable Lipid Profile through Cultivation Conditions

The ideal conditions of growth allow adjusting the lipid composition of microalgae, such as the percentage of saturated fatty acids (SFAs), monounsaturated fatty acids (MUFAs), and PUFAs (Veerasamy et al., 2024). Light intensity, temperature, nutrient availability (including nitrogen, phosphate, and carbon dioxide), pH, salinity, and even stress conditions may have a significant effect on the quantity and quality of generated lipids (Kumari et al., 2023; Papapanagiotou et al., 2024; Veerasamy et al., 2024). The most common strategy, such as nitrogen limitation, is often used, and it may often result in the accumulation of lipids in microalgae and, in certain species, may sometimes lead to a change in the fatty acid composition, whether to more neutral lipids or to particular PUFAs (Kumari et al., 2023; Papapanagiotou et al., 2024; Veerasamy et al., 2024). Some authors state that similar light regimes, including intense light, can also increase the lipid content or change the composition of fatty acids (Papapanagiotou et al., 2024). As Veerasamy et al. (2024) observe, UV-C light has been shown to increase EPA synthesis in some microalgal strains, such as *Nannochloropsis sp.*

This capability to tailor the fatty acid content is highly valuable in the process of tailoring microalgal lipids to specific food formulations (Papapanagiotou et al., 2024). To meet some specific dietary or technological requirements, such as oxidative stability or emulsifying properties, microalgal oils, for example, can be designed to be used in beverages, nutritional bars, infant formulas, or encapsulated omega-3 supplements (Liu et al., 2024). The ability to produce lipids with specific SFA, MUFA, and PUFA ratios enables the production of products with optimal nutritional content and preferred physical characteristics (e.g., texture, emulsion stability, and shelf life) (Liu et al., 2024). As an illustration, changing the cultivation conditions can allow the production of microalgal oil that is high in DHA (neurological health products) or has a balanced EPA/DHA ratio (cardiovascular support) to offer a range of flexibility in product production (Calder, 2021).

8.2.4 Sustainable Nutritional Source

The sustainability of microalgae as a source of omega-3 is a key factor in the increase in its consumption (Bagchi et al., 2025; Muthu, 2025). There is also a concern about the sustainability and ecological impact of traditional sources under the pressure of overfishing, habitat degradation, and climate change, that is, marine fish stocks (Adarme-Vega et al., 2012; Bagchi et al., 2025). Microalgal cultivation, with its numerous environmental advantages, provides a solution to these sustainability issues.

8.2.4.1 Reduced Land and Freshwater Footprint

Microalgae can be cultivated on nonarable land (e.g., deserts and coastal areas) and in any type of water source, such as brackish water, seawater, or even wastewater, unlike traditional crop production of oilseeds (e.g., soybean, rapeseed, and sunflower), which consumes a lot of arable land and freshwater (Bagchi et al., 2025). This significantly decreases the need for freshwater and fertile land and competition with food crops (Bagchi et al., 2025). There are also methods of microalgae growth that can be connected to the model of the circular economy using industrial effluent or CO_2 emissions as a source of fertilizers, which enhances their environmental image (Yang et al., 2024).

8.2.4.2 Continuous, Nonseasonal Production

Microalgae may be grown all year round either in regulated bioreactors or in open pond systems, which offer a steady and reliable source of biomass, regardless of the seasonal agricultural patterns or the changing fish populations (Bagchi et al., 2025). This permanent manufacturing power assists in fulfilling the increasing world demand of omega-3 fatty acids without being limited by seasons, which provides a consistent supply chain (Bagchi et al., 2025).

8.2.4.3 Support for Plant-Based Diets and Functional Food Development

As vegan and vegetarian diets become more popular around the world due to ethical, environmental, and health-related reasons, there is a rising need to identify sources of the necessary nutrients, such as EPA and DHA, in plants (Saini et al., 2023). Microalgae are a direct and high-quality vegan source of fish oil, which addresses a large nutritional gap in these populations because of the inefficient conversion of ALA (Bagchi et al., 2025; Saini et al., 2023). Moreover, microalgae are rich in proteins, and that is why they are considered to develop sustainable and nutritious functional food (Yang et al., 2024).

8.2.4.4 Industrial Applications

The microalgal lipids are under investigation in a broad range of industrial applications, which can result in sustainable food systems. These are the fortification of plant-based milks, nutritional bars, infant formulas, enriched omega-3 supplements in capsules, and lipid fortification in new protein foods, including cultured meat (Saini et al., 2023; Wang, 2025). The potential of microalgae as a source of omega-3 supplementation to increase the n-3/n-6 ratio in products like cultured beef has been shown to be a valuable tool for improving nutritional profiles within new food categories and amending current unbalanced nutritional content in Western diets (Wang, 2025). They may be incorporated into food to enhance the human health outcomes like cancer prevention by maintaining a balance in lipid intake (Hao et al., 2024).

The microalgal lipids, therefore, have a significant role in the development of sustainable nutrition, as they are a renewable, environmentally friendly, and nutritionally diverse source of essential fatty acids to meet a wide variety of human nutritional needs. Additional studies on microalgal cultivation and extraction technologies optimization are enhancing their economic feasibility and increasing their applications in the food and nutraceutical sectors (Veerasamy et al., 2024).

8.3 CULTIVATION STRATEGIES TO ENHANCE LIPID PRODUCTION

The cultivation conditions strongly affect the amount and composition of lipids in microalgae. These situations are of great importance to maximize lipid production. Some of the emerging strategies analyzed in this section are light, temperature, salinity, CO_2, and nutrient availability modulation, along with two-stage cultivation processes and other approaches aimed improving microalgal lipid production.

8.3.1 Light Conditions

Light is a very stringent controller of microalgal growth and metabolism that has a direct effect on the photosynthetic efficiency and biomass gain (Brindhadevi et al., 2021; Rao & Ravishankar, 2019). Besides acting as the source of energy to facilitate growth, light conditions may also work as a stressor that leads to lipid accretion in certain species (Kim et al., 2018; Okazaki et al., 2019).

8.3.1.1 Light Intensity and Wavelength

An increase in light intensity tends to stimulate the rate of photosynthesis and biomass growth until it reaches a saturation point, after which photoinhibition occurs (Lv et al., 2010; Tripathi et al., 2025). Nonetheless, when light is too intense, the formation of lipids is a preventive measure against photooxidative stress (Zhu et al., 2022). As an example, Kumar et al. (2018) researched the optimal light intensity of *Dunaliella tertiolecta* to achieve the highest growth of algal biomass, lipid production, and CO_2 fixation. The influence of CO_2 concentration and irradiance on cell growth, the content of chlorophyll a, and lipid accumulation in *Chlorella vulgaris* has been widely studied (Lv et al., 2010). The attachment culture of *Ettlia sp.* was found to be optimized at the best light intensity that enhanced biomass and lipid surface production (Kim et al., 2018).

The quality of light or wavelength is also a critical factor. Dissimilar wavelengths influence metabolic pathways in a different manner. Indicatively, red light and the phytohormone indole-3-acetic acid (IAA) have been observed to exhibit synergistic outcomes in the growth of mixotrophic microalgae (Chang et al., 2022). A two-stage system of *Scenedesmus sp. GTAF 01 IU* reached a maximum biomass of 1.65 g at the light intensity of 27 μmol m^{-2} s^{-1} (Tripathi et al., 2025). *Chlorella vulgaris* grown under certain conditions of light and salt stress reached its optimal lipid productivity at 140 μmol photons m^{-2} s^{-1} (Ali et al., 2020).

8.3.1.2 High Light as a Lipid-Inducing Stressor

Triacylglycerol (TAG) accumulation can be greatly increased by high light intensity, especially when the stressor is other factors such as nutrient limitation (Okazaki et al., 2019). This effect occurs because when light energy is not fully utilized during growth due to nutrient deficiency, the excess energy is redirected toward lipid production to prevent cell damage (Zhu et al., 2022). Nevertheless, the resulting lipid enrichment can frequently be at the cost of biomass growth, so a trade-off will be necessary, which can be resolved by strategies such as two-stage cultivation (Aziz et al., 2020).

8.3.2 Temperature Modulation

One of the key environmental parameters that affects microalgal physiology, such as the growth rate, metabolic pathways, and enzyme activities, and, accordingly, the lipid accumulation and fatty acid profiles, is temperature (Abdelkarim et al., 2024; Brindhadevi et al., 2021). The optimal range of temperatures in which each microalgal species can grow and produce lipids is determined (Cheirsilp et al., 2024; Rao & Ravishankar, 2019).

8.3.2.1 Effects of Temperature on Metabolism and Fatty Acid Profiles

Both low and high deviations of optimal temperatures may cause stress reactions that change lipid metabolism. As an example, a certain amount of research suggests that a lower temperature may improve the accretion of unsaturated fatty acids, which are essential to keep membranes fluid in cold temperatures (X. Wang et al., 2024). On the other hand, the rise in temperature could result in more SFAs or a general lipid level as a protective mechanism (Brindhadevi et al., 2021; Cheirsilp et al., 2024). The terrestrial oleaginous microalga *Vischeria sp. WL1* was cultivated using a two-stage cultivation strategy that was temperature-dependent to separate the accumulation of biomass at optimal temperatures and the accumulation of lipids at low temperatures, leading to improved lipid and fatty acid production (Wang et al., 2024). *Chlamydomonas sp. HP59*, which was isolated in hot springs, showed robustness in temperatures between 30 and 40°C, which means that it could be cultivated in subtropical and tropical areas (Cheirsilp et al., 2024).

8.3.3 Salinity and Osmotic Stress

Salinity or salt concentration is one of the most important environmental stressors of microalgae that leads to physiological modifications that favor lipid deposition, especially TAGs (Nyoni et al., 2024; Singh, Yadav, Kumar, Hashem, et al., 2023).

8.3.3.1 Salinity as an Inducer of Triacylglycerol (TAG) Accumulation

Salinity causes osmotic stress that causes cell dehydration and ion toxicity. Microalgae react with the means of defensive processes, which often include the enhanced production of appropriate solutes and energy storage molecules, including TAGs (Pandey et al., 2024; Singh, Yadav, Kumar, Hashem, et al., 2023). This is a process that utilizes metabolic resources to produce lipids instead of growing (Singh, Yadav, Kumar, Hashem, et al., 2023). The salinity stress may raise lipid levels in *Scenedesmus sp. BHU1*, but this is often coupled with a reduction in the growth and biomass production (Singh, Yadav, Kumar, Hashem, et al., 2023). It was identified that the highest lipid production was observed under 150 mM NaCl salinity stress in a gradient technique of salt addition in *Chlamydomonas reinhardtii* (Pandey et al., 2024). After one round of growth in the presence of 400 mM NaCl, *Scenedesmus sp. CCNM 1077* yielded 33.13% lipid (Pancha et al., 2015). Optimal salinity of *Dunaliella* species has been found to enhance the biomass and the lipid production (Kumar et al., 2018). Nyoni et al. (2024), in their microalgae experiment, discovered the best salinity concentration at which the microalgae had high biomass (2480 ± 75 mg L^{-1}) and lipid content (18.00%).

8.3.3.2 Salt-Tolerant Species and Their Advantages

Dunaliella and certain strains of *Nannochloropsis* are already adapted to salinity and are especially beneficial for lipid production at high salinity (Abdelkarim et al., 2024; Kumar et al., 2018). Their natural tolerance enables them to maintain relatively high growth rates even in conditions that would significantly suppress other species, resulting in a superior biomass-to-lipid storage ratio (Kumar et al., 2018). *Scenedesmus obliquus* showed improved biomass and lipid production under salinity stress in olive processing wastes, reaching 0.15 g L^{-1} d^{-1} biomass productivity and 34.1 mg L^{-1} d^{-1} lipid productivity with a C/N ratio of 40/1 (Nicodemou et al., 2024). In another experiment on *Chlorella vulgaris*, the optimal lipid productivity (15.59 ± 0.10 mg L^{-1} day^{-1}) was obtained at 5 g L^{-1} $MgCl_2$ of a certain light intensity (Ali et al., 2020).

8.3.4 CO_2 Supplementation

Photoautotrophic microalgae use carbon dioxide (CO_2) as the main source of carbon that is vital to photosynthesis and biomass production (Seyed Hosseini et al., 2018). The addition of CO_2 to the cultivation systems can substantially improve the growth rates and lipid productivity (Pourbakhtiar et al., 2022; Seyed Hosseini et al., 2018).

The CO_2 levels up to a certain level of saturation enhance the efficiency of photosynthesis, resulting in an enhanced fixation of carbon (Seyed Hosseini et al., 2018). Such a large amount of fixed carbon can then be directed to other metabolic routes, such as the production of lipids (Chu et al., 2019; Pourbakhtiar et al., 2022). The open microalgae cultivation systems have been successfully used to use industrial off-gas CO_2 to reduce greenhouse gas emissions and increase lipid production (Seyed Hosseini et al., 2018). Under nitrogen starvation, *Chlorella PY-ZU1* had 1.3 times higher biomass concentration and 2.2 times higher lipid productivity with the supplement of 15% CO_2 in a continuous two-step cultivation process, particularly when phosphorus was sufficiently added (Chu et al., 2019). A new two-step procedure with *Chlorella vulgaris* emphasized the concentration of CO_2 as one of the three important parameters to enhance the lipid productivity (Pourbakhtiar et al., 2022). Also, it was demonstrated that *Chlorella vulgaris* grown in a membrane-sparged PBR increased lipid accumulation with optimal CO_2 concentration (Lv et al., 2010). pH control in semicontinuous cultivation was also reported to promote lipid productivity in *Chlorella pyrenoidosa* (Han et al., 2013).

8.3.5 Two-Stage Cultivation

Two-step cultivation is one of the most commonly accepted strategies as a highly successful method to overcome the trade-off inherent in the maximization of biomass growth and high lipid accumulation in microalgae (Aziz et al., 2020; Nayak et al., 2019; Singh et al., 2025).

8.3.5.1 Stage I (Maximal Biomass Growth)

During the first step, the microalgae are grown in the best growth conditions (nutrient-enriched, adequate light, temperature, and CO_2) to reach the highest cell density and biomass (Aziz et al., 2020; Singh et al., 2025). At this stage, the metabolic resources are mainly allocated to cellular growth and production of proteins (Singh et al., 2025).

8.3.5.2 Stage II (Stress Induction for Maximal Lipid Accumulation)

When a high concentration of biomass is reached, the culture conditions are deliberately changed to create stress, which diverts metabolism to lipid production, in particular, TAGs (Nayak et al., 2019; Singh et al., 2025; Udayantha et al., 2025). Nitrogen starvation, phosphorus limitation, high light intensity, or salinity stress are examples of common stressors (Nayak et al., 2019; Udayantha et al., 2025). This method enables the synergistic effect where high biomass in Stage I forms the basis of high lipid in Stage 2 (Aziz et al., 2020). In the case of *Chlamydomonas reinhardtii*, Stage I was associated with biomass increases in the absence of nitrogen, then lipid synthesis in the presence of nitrogen deficiency in Stage II, with positive impacts on the growth kinetics and enzymatic activity (Singh et al., 2025). The optimal lipid content (36.7%) and productivity (216.9 mg L^{-1} d^{-1}) of *Chlorella sp. HS2* were obtained after five days of nitrogen-replete and four days of nitrogen-depleted conditions (Nayak et al., 2019). *Scenedesmus obliquus* showed the highest lipid accumulation in the case of a nitrogen-replete Stage I and a nitrogen-deprived Stage II (Trivedi et al., 2022). A two-step method of combining nitrogen deprivation and glucose supplementation increased the production of lipids in *Tetraselmis sp.* (Udayantha et al., 2025). The two-stage strategy has proven its efficiency, which is why it is widely used in research and possible commercial arrangements (Aziz et al., 2020). Two-step strategies have been continuously created to prevent the harvesting step between stages that may reduce the cost of production (Xie et al., 2021). One example is the cocultivation of microalgae and xylanolytic bacteria as part of a continuous two-step approach to optimize the production of algal lipids (Xie et al., 2021). Other strains of microalgae, including *Chlorella vulgaris*, *Scenedesmus obliquus*, and *Navicula phyllepta*, have demonstrated substantial enhancements in lipid production in two-stage systems through nutrient deprivation, temperature regulation, or salinity stress (Singh, Yadav, Kumar, Solanki, et al., 2023). A comparison also noted the advantages of the two-stage cultivation strategy over the one-stage strategy when using selenite as a lipid inducer in *Dunaliella tertiolecta* (Perečinec et al., 2021).

The researchers have discussed various approaches to maximize lipid productivity, noting that these approaches can sometimes have varying impacts on both the microalgal strain and the specific cultivation parameters used. Other research studies are investigating nutrient starvation, whereas others are investigating the impact of light intensity, temperature, and salinity stress (Aziz et al., 2020; Rafiq et al., 2023). The issue is that it is required to raise lipid content and reduce a decline in the overall production of biomass, which is often a trade-off (Seyed Hosseini et al., 2018). Table 8.1 compares lipid productivity obtained with different cultivation strategies and indicates the variety of results and methods employed with different microalgal species.

TABLE 8.1
Comparison of Lipid Productivity under Different Cultivation Strategies

Microalgal Species	Cultivation Strategy	Key Parameters	Lipid Content (%)	Lipid Productivity (mg/L/d)	Reference
Chlorella sp. HS2	Two-stage (N+ followed by N-)	5 days N+/4 days N-	36.7	216.9	Aziz et al., 2020
Nannochloropsis oculata	Two-stage (N-depleted)	Alkaline flocculation for separation	Not specified	Enhanced	Rezaei et al., 2024
Ankistrodesmus sp. OR119838	Two-stage (N-replete then N-limited)	Doubled nitrate, increased light (180 to 360 μmol/m²/s)	Not specified	0.61 (specific growth rate)	Chen et al., 2014
Scenedesmus obtusus XJ-15	Two-stage (NaCl addition)	NaCl in Stage II	47.7	212.1 (biomass)	Parveen & Patidar, 2022
Coelastrella sp.	Two-stage (1% NaCl, light intensity 4200 Lux)	B3N medium + 1% NaCl	30	77.83	Pancha et al., 2015

(Continued)

TABLE 8.1 (Continued)
Comparison of Lipid Productivity under Different Cultivation Strategies

Microalgal Species	Cultivation Strategy	Key Parameters	Lipid Content (%)	Lipid Productivity (mg/L/d)	Reference
Scenedesmus sp. CCNM 1077	Single-stage (salinity stress)	400 mM NaCl	33.13	Not specified	Rafiq et al., 2023
Desmodesmus sp. F2	Optimized batch cultivation	Thermotolerant conditions	High	263	Ho et al., 2014
Chlorella pyrenoidosa	Continuous culture (variable nitrate)	Optimized nitrate concentration	High	Not specified	S. Wang & Diao, 2024
Auxenochlorella pyrenoidosa	Two-stage (optimal stage conversion time)	72 h in stage I	Not specified	Higher after transfer	Padhi & Nayak, 2025
Scenedesmus abundans	Single-stage (optimized lipid medium)	Photoautotrophic	32	Not specified	Nayak et al., 2019
Nannochloropsis sp. SCSIO-45217	Phosphate adjustment strategy	Optimized phosphorous concentration	Not specified	Enhanced	Singh et al., 2016

8.4 NUTRIENT-BASED APPROACHES TO INCREASE LIPID PRODUCTIVITY

The manipulation of nutrients is one of the foundations of microalgal culture techniques to increase lipid productivity because it directly affects the metabolic pathways of lipid synthesis (Aziz et al., 2020; Brindhadevi et al., 2021).

8.4.1 Nitrogen Limitation or Starvation

Nitrogen is an essential macronutrient in the growth of microalgae, which is used in protein, nucleic acid, and chlorophyll synthesis (Aziz et al., 2020). A well-known and highly effective method of achieving lipid accumulation is to limit or starve microalgae of nitrogen (Abdelkarim et al., 2024; Singh et al., 2025; Trivedi et al., 2022).

The metabolism of cells in nitrogen-limited or starved conditions alters toward non-nitrogen compounds such as proteins and nucleic acids. Rather, the surplus carbon that is fixed during photosynthesis is diverted to the production and storage of carbon-rich compounds, mostly TAGs, as an energy store (Rafiq et al., 2023; Zhu et al., 2022). Although nitrogen limitation is an effective way to induce lipid accumulation, it always results in a decrease in the rate of biomass growth and the total biomass yield since nitrogen is needed to divide cells and synthesize proteins (Abdelkarim et al., 2024; Trivedi et al., 2022). It requires applying such strategies as two-stage cultivation, when the biomass is initially grown in nitrogen-rich conditions and then starved of nitrogen to induce lipids (Aziz et al., 2020; Trivedi et al., 2022). The example is *Chlorella sp. HS2*, in which the optimal duration of the nitrogen-replete and -deplete phases maximized the lipid productivity (Nayak et al., 2019). In the case of *Scenedesmus obliquus*, the maximum lipid content was achieved in the two-stage process under a particular concentration of nitrate (Trivedi et al., 2022). Optimal nitrogen level (25 mg NL^{-1}) and low-frequency ultrasound treatment demonstrated an increase in biomass and lipid accumulation in *Tetradesmus obliquus* (Wei et al., 2022). *Chlorella regularis* has shown better lipid production under nitrogen starvation when grown in glucose heterotrophic conditions, according to a research study by Fu et al. (2017). In the same way, the strains of *Nannochloropsis* and *Microchloropsis* show a tendency of higher lipid accumulation during nitrogen starvation, adversely affecting photosynthesis and growth (Abdelkarim et al., 2024). Nitrogen modulation has also been found to influence the antenna pigments, biomass, and lipid production of *Chlorella vulgaris* during chemostat continuous cultivation (Cho et al., 2019).

8.4.2 Phosphorus Limitation

Phosphorus is another macronutrient required in large amounts for energy metabolism (adenosine triphosphate, ATP), nucleic acids, and membrane formation (Chu et al., 2019). Phosphorus is found in phospholipids, cell membranes, and ATP, which is the cell's primary energy currency (Chu et al., 2019). As a result, microalgae may have a significant impact on the phosphorus restriction metabolism (Fu et al., 2017).

Phosphorus limitation can lead to the accumulation of lipids, notably TAGs, as a survival response, similar to nitrogen limitation (Chu et al., 2019). In the case of phosphorus limitation, microalgae can shift the metabolic flow to carbon storage products. Nevertheless, phosphorus limitation may have species-specific effects that are less severe than the effects of nitrogen limitation on lipid accumulation (Mou et al., 2022). In the case of *Chlorella PY-ZU1*, supplementation with sufficient phosphorus in the second phase and nitrogen starvation significantly enhanced the biomass concentration and lipid productivity (Chu et al., 2019). Excessive phosphorus (maximum 818% in BG11 medium) increased lipid production in nitrogen starvation and glucose heterotrophic growth in *Chlorella regularis* (Fu et al., 2017). *Scenedesmus quadricauda* grew in phosphorus-limited conditions with nitrogen starvation and accumulated lipids, but it achieved the highest lipid level (41% of dry weight) under certain conditions of xylose mixotrophic/heterotrophic growth (Mou et al., 2022).

8.4.3 Iron and Other Trace Elements

Micronutrients like trace elements, including iron, are cofactors of many enzymes that play a role in the most important metabolic processes, including photosynthesis and lipid biosynthesis (Elshobary et al., 2025). Iron is an essential part of proteins involved in the photosynthetic electron transport chain and enzymes such as fatty acid desaturases, which are used to determine the fatty acid composition (Teh et al., 2024). Thus, the availability of iron has a significant impact on the biomass production and lipid quality.

Iron deficiency may disrupt photosynthesis and general growth, which might cause a change in carbon distribution to storage lipids (Teh et al., 2024). However, severe deficiency can also impair the efficiency of lipid biosynthesis enzymes, leading to decreased lipid production or a skewed fatty acid composition. Conversely, vigorous development and successful lipid synthesis require optimal levels of iron. The other trace elements that have not been studied as extensively as nitrogen and phosphorus in terms of lipid induction are also involved in microalgae metabolism (Elshobary et al., 2025). For example, the *Phaeodactylum tricornutum* experiments aimed to maximize biomass, lipid, and fucoxanthin production while considering nutrient proportions, including nitrogen, phosphorus, and iron (Elshobary et al., 2025).

8.4.4 Carbon/Nitrogen Ratio (C:N)

The carbon-to-nitrogen (C:N) ratio is one of the most efficient ways to regulate microalgae metabolism and trigger lipid biosynthesis (Nicodemou et al., 2024; Rao & Ravishankar, 2019). The high C:N ratio that is typically achieved by limiting nitrogen but keeping the carbon content high is a signal to the cell to store the excess carbon as lipids (Rao & Ravishankar, 2019). The ratio allows the greatest fixation of carbon and its further conversion to lipids rather than to nitrogen-rich compounds, including proteins. An optimization of C:N and C:P ratios in synthetic media in the case of *Scenedesmus obliquus* gave high biomass and lipid productivity with a 40/1 C:N ratio, resulting in 0.15 g L^{-1} d^{-1} biomass and 34.1 mg L^{-1} d^{-1} lipid productivity (Nicodemou et al., 2024). The interaction of the C:N ratio with other environmental factors, including light intensity, is critical as far as the maximum lipid production is concerned (Chin et al., 2023).

A comparative analysis of the nutrient manipulation techniques and their effects on lipid accumulation and growth in the biomass of the various microalgal species, which focuses on the compound effects of the above methods, is shown in Table 8.2.

8.5 BIOPROCESS ENGINEERING FOR LARGE-SCALE LIPID PRODUCTION

To implement such methods in the areas of sustainable nutrition and biofuels, it is necessary to devise effective and cost-efficient strategies to produce large-scale microalgal lipids (Singh, Yadav, Kumar, Solanki, et al., 2023). It demands a holistic approach to bioprocess engineering, which incorporates the design and optimization of bioreactors, the effective harvesting and downstream processing, and the biorefinery concepts (Karemore & Sen, 2016).

8.5.1 Bioreactor Design and Optimization

Scaling laboratory findings to the industrial-scale production of microalgal lipids has faced serious issues, particularly the maintenance of optimal conditions of growth and lipid productivity (Gaurav et al., 2024). The open pond systems that are

TABLE 8.2
Nutrient Manipulation Techniques and Their Effects on Lipid Accumulation

Nutrient Manipulation Technique	Mechanism of Action	Effect on Lipid Accumulation	Effect on Biomass Growth	Associated Microalgal Species	Reference
Nitrogen limitation/ starvation	Redirects assimilated carbon from protein synthesis and growth toward neutral lipid (TAG) storage.	• Significantly increases lipid content. Optimal coproduction of hydrogen and lipids achieved at 5% N in *C. protothecoides* (49.6% DW lipids). • Up to 39.93% ± 0.44% lipid accumulation in *Tetradesmus obliquus* KMC24	• Generally, reduces biomass growth rate, creating a trade-off. • However, continuous nitrogen limitation (CNL) can maintain high biomass (4.61 g/L in *C. vulgaris*) near batch nitrogen starvation levels (4.88 g/L).	*Chlorella vulgaris*, *Nannochloropsis* spp., *Schizochytrium* sp., *C. protothecoides*, *Tetradesmus obliquus*, *Scenedesmus obliquus*	Chang et al., 2022; Roy et al., 2021
Phosphorus limitation	Shifts metabolism toward carbon storage in lipids due to deficiency in components like phospholipids and ATP.	• Increases lipid content and productivity. • Enhanced lipid productivity in *Nannochloropsis sp. SCSIO-45217* through phosphate adjustment. • More effective on fatty acid profiles than total lipid in *Spirulina*.	• Can decrease biomass growth. • Slight excess of phosphorus (≤45 mg L^{-1}) under N-limitation can stimulate growth while promoting lipid accumulation in *Chlorella regularis*.	*Chlorella* sp., *Rhodosporidium toruloides*, *Nannochloropsis sp. SCSIO-45217*, *Spirulina*, *Chlorella regularis*	Fu et al., 2019; Wu et al., 2023

(*Continued*)

TABLE 8.2 (Continued)
Nutrient Manipulation Techniques and Their Effects on Lipid Accumulation

Nutrient Manipulation Technique	Mechanism of Action	Effect on Lipid Accumulation	Effect on Biomass Growth	Associated Microalgal Species	Reference
Iron and trace elements	Cofactors for enzymes in photosynthesis and lipid synthesis; affects overall metabolic efficiency.	• Specific concentrations can enhance lipid and hydrocarbon production (e.g., high cobalt in *Botryococcus braunii*). • Myoinositol enrichment increased lipid content by 33% in *Dunaliella* sp.	• Critical for metabolic function and growth; deficiency can severely impair growth.	*Botryococcus braunii*, *Dunaliella* sp.	Patel et al., 2024
High carbon/nitrogen ratio (C:N)	Excess carbon channeled into lipid storage when nitrogen is scarce, leading to increased carbon flow to fatty acid synthesis.	• Promotes significant lipid accumulation. • Optimized C:N ratios (e.g., 20, 70, 120) led to maximal lipid accumulation in *R. toruloides*-1588.	• Can support biomass growth initially if carbon is abundant, but subsequent N-limitation can reduce growth during the lipid accumulation phase.	*Chlorella pyrenoidosa*, *Rhodosporidium toruloides*, *Chlorella vulgaris*	Paranjape et al., 2016; Saini et al., 2023

traditional and used in large areas are prone to contamination, evaporation, and a lack of control of the environmental parameters (Gaurav et al., 2024). PBRs provide a more controlled condition and entirely control the intensity of light, as well as the concentration of nutrients, pH, and the supply of CO_2, which is significant to improve the growth rate of algal biomass and lipid accumulation (Gaurav et al., 2024; Islam et al., 2024; Kumar et al., 2024). As an example, biomass growth was increased by a quarter with the optimization of bioreactor parameters (Joshi et al., 2024). New bioreactor designs are under investigation to address the weaknesses of the conventional designs. It has been demonstrated that multi-cultivator airlift PBRs, such as one, are effective at improving components of the media and light levels to catalyze the production of lipids and hydrocarbons in microalgae such as *Botryococcus braunii* (Patel et al., 2024). Combinations of computational methods, including artificial neural network modeling and multi-objective optimization, can be used to optimize bioreactor design in order to identify process parameters that enhance both biomass and lipid productivities, despite the fact that these two factors are often negatively related (Kumar et al., 2024).

8.5.2 Harvesting and Downstream Processing

The most important in microalgal lipid production is efficient and cost-efficient harvesting technologies because harvesting may constitute the greater part of the overall production costs (Cheirsilp et al., 2020). The harvesting techniques can be divided into physical, chemical, biological, and magnetic techniques to extract microalgae (Gaurav et al., 2024) (Figure 8.2).

Physical techniques are centrifugation, filtration, gravity sedimentation, and flotation. Centrifugation, although useful, is energy-consuming and therefore is not

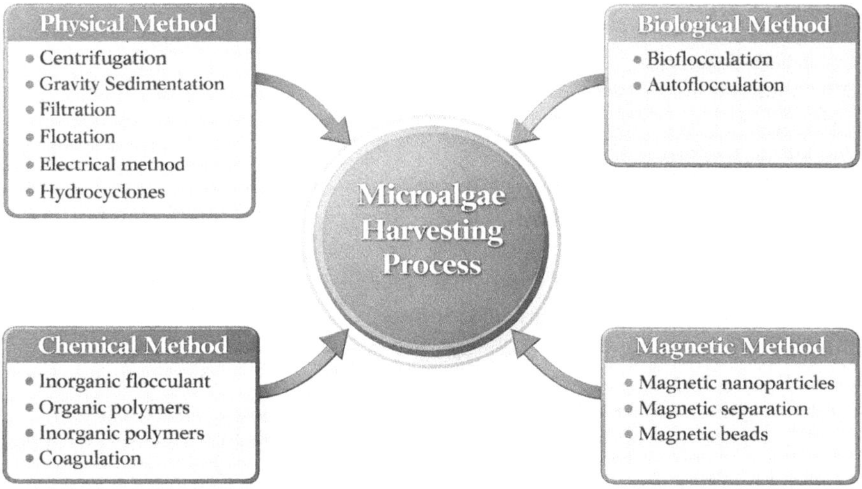

FIGURE 8.2 Emerging and conventional processes for microalgae harvesting.

applicable in large-scale and low-cost processes (Cheirsilp et al., 2020; Gaurav et al., 2024). With the help of the filters and different membrane or cartridge filters, microalgae are separated according to their size (Gaurav et al., 2024). Gravity sedimentation is slow but can be utilized in large-scale applications, whereas flotation exploits the use of air bubbles to bring microalgae to the surface (Gaurav et al., 2024). Chemical methods for flocculating microalgae cells are inorganic flocculants (e.g., aluminum sulfate and ferric chloride), coagulants, or organic/inorganic polymers (Matter et al., 2019). Biological processes like bioflocculation and autoflocculation are based on the usage of microorganisms or inherent flocculation tendency to reduce the cost of harvesting (Gaurav et al., 2024). The least popular is the magnetic method, and microalgae are isolated with the assistance of magnetic particles (Gaurav et al., 2024).

Lipid extraction is an important and usually costly process, which has a significant impact on the efficiency of the whole process and the economic viability of the same process once harvested (Karim et al., 2020). Microalgae have thick cell walls that prevent recovery of lipids, which need to be broken before or during extraction (Correa-Aguado et al., 2023; Oh et al., 2022). A number of extraction methods have been developed, and each of them has its advantages and disadvantages in terms of efficiency, cost, and environmental impact (Chaos-Hernández et al., 2023; Islam & Dixit, 2024). Table 8.3 provides a comparison of the energy and resource demand of different lipid extraction processes.

8.6 CONCLUSION

Microalgae have become one of the most useful and sustainable sources of nutritional lipids and a renewable alternative to satisfy the increasing global demand for essential fatty acids without straining terrestrial or marine ecosystems. The knowledge introduced in this chapter confirms the fact that the economic viability of large-scale production of microalgal lipids is determined, first of all, by the accurate optimization of the cultivation and nutrient environment, which directly determines the productivity and quality. The ability to control the parameters of light, temperature, salinity, and nutrient availability, along with the sophisticated bioprocess engineering, offers the basis of scalable and cost-effective systems of production. Moreover, the use of waste as a source of nutrients and effective downstream processing supports the sustainability and circularity of the whole process. In the future, automation and artificial intelligence should be the main drivers of future developments, as they will allow real-time monitoring, adaptive control, and predictive optimization of microalgal bioprocesses. Selection and development of lipid-specific, high-yield, and robust strains will also increase the industrial potential of this biotechnology. As innovations mature, microalgae will transition from a research focus to a foundation of sustainable nutrition. Microalgae not only support human health with high-quality, lipid-rich products but also protect the environment by serving as a strong, efficient, and eco-friendly model for future biomanufacturing.

TABLE 8.3
Energy and Resource Requirements for Various Lipid Extraction Methods

Method	Description	Energy Requirement	Solvent Type	Advantages	Reference
Dry oil extraction	Chemical process (using organic solvents)	Relatively lower	Hexane (most common), chloroform, and methanol	High selectivity for lipids; lower cost than some alternatives; effective for nonpolar lipids	Gorgich et al., 2020; Shin et al., 2018
Wet oil extraction	Chemical process (using an organic solvent on wet biomass)	More than the dry method	Ethyl acetate (best), chloroform:methanol mixture	Avoids the drying process, potentially saving energy and cost	Sánchez-Bayo et al., 2020
Wet oil extraction using sonication	Physical process (utilizing sound waves to burst cell walls for improved extraction efficiency)	Moderate	ethyl acetate (best), chloroform:methanol mixture	Enhanced lipid extraction efficiency from wet biomass; lower energy input than some mechanical methods	Sánchez-Bayo et al., 2020
Pressure liquid extraction (ethanol)	Green process; utilizes ethanol as a solvent	Moderate	Ethanol	Increased extraction efficiency and speed; decreased solvent use	Višnjevec et al., 2024
Infrared (NIR) extraction	Utilizes near-infrared radiation to disrupt cell walls and enhance lipid release with greener solvents	Moderate (potentially lower due to efficient cell wall breakage)	Nonpolar/polar solvent systems (e.g., hexane, ethyl acetate, and ethanol)	Environmentally friendly due to greener solvents; efficient cell disruption; high lipid recovery	Correa-Aguado et al., 2023
Supercritical CO_2 Extraction	Nontoxic, environmentally benign alternative to organic solvents; effective for lipid extraction due to high diffusivity and low viscosity	Moderate to high (requires specialized equipment)	Supercritical CO_2	Nontoxic, environmentally friendly; tunable selectivity for different lipid classes; solvent-free product	Cheng et al., 2011
Enzyme-assisted extraction	Utilizes enzymes to break down cell walls, facilitating lipid release	Moderate	Green solvents like ethanol	Enhanced lipid extraction efficiency reduced harsh chemical usage, suitability for biorefinery integration	He et al., 2020; Oh et al., 2022

REFERENCES

Abdelkarim, O. H., Wijffels, R. H., & Barbosa, M. J. (2024). Microalgal lipid production: A comparative analysis of Nannochloropsis and Microchloropsis strains. *Journal of Applied Phycology*, *37*(1), 15–34. https://doi.org/10.1007/S10811-024-03318-7

Adarme-Vega, T. C., Lim, D. K. Y., Timmins, M., Vernen, F., Li, Y., & Schenk, P. M. (2012). Microalgal biofactories: A promising approach towards sustainable omega-3 fatty acid production. *Microbial Cell Factories*, *11*. https://doi.org/10.1186/1475-2859-11-96

Ali, H. E. A., El-Fayoumy, E. A., Rasmy, W. E., Soliman, R. M., & Abdullah, M. A. (2020). Two-stage cultivation of Chlorella vulgaris using light and salt stress conditions for simultaneous production of lipid, carotenoids, and antioxidants. *Journal of Applied Phycology*, *33*(1), 227–239. https://doi.org/10.1007/S10811-020-02308-9

Aziz, M. M. A., Kassim, K. A., Shokravi, Z., Jakarni, F. M., Lieu, H. Y., Zaini, N., Tan, L. S., Islam, S., & Shokravi, H. (2020). Two-stage cultivation strategy for simultaneous increases in growth rate and lipid content of microalgae: A review. *Renewable and Sustainable Energy Reviews*, *119*, 109621. https://doi.org/10.1016/J.RSER.2019.109621

Bagchi, S. K., Patnaik, R., Rawat, I., & Bux, F. (2025). Innovative strategies for augmenting Omega-3-fatty acid production in microalgae: Sustainable approaches for vegan food applications. *Bioresource Technology*, *437*. https://doi.org/10.1016/j.biortech.2025.133176

Baker, E. J., Cumberford, G., & Hanaway, P. (2025). Recognizing the health benefits of plant-sourced Omega-3 stearidonic acid: Exploring its complementary role to preformed EPA/DHA. *Lipids*, *60*(6), 435–442. https://doi.org/10.1002/LIPD.12452

Brindhadevi, K., Mathimani, T., Rene, E. R., Shanmugam, S., Chi, N. T. L., & Pugazhendhi, A. (2021). Impact of cultivation conditions on the biomass and lipid in microalgae with an emphasis on biodiesel. *Fuel*, *284*, 119058. https://doi.org/10.1016/J.FUEL.2020.119058

Calder, P. C. (2021). Health benefits of omega-3 fatty acids. *Omega 3 Delivery Systems*, 25–53. https://doi.org/10.1016/B978-0-12-821391-9.00006-5

Chang, W., Li, Y., Qu, Y., Liu, Y., Zhang, G., Zhao, Y., & Liu, S. (2022). Mixotrophic cultivation of microalgae to enhance the biomass and lipid production with synergistic effect of red light and phytohormone IAA. *Renewable Energy*, *187*, 819–828. https://doi.org/10.1016/J.RENENE.2022.01.108

Chaos-Hernández, D., Reynel-Ávila, H. E., Bonilla-Petriciolet, A., & Villalobos-Delgado, F. J. (2023). Extraction methods of algae oils for the production of third generation biofuels – A review. *Chemosphere*, *341*, 139856. https://doi.org/10.1016/J.CHEMOSPHERE.2023.139856

Cheirsilp, B., Sa-Nga, K. C., Baojanya, J., & Maneechote, W. (2024). Effects of temperature and nitrogen limitation on growth and lipid production of oleaginous microalgae from hot springs. *Trends in Sciences*, *21*(12), 8596. https://doi.org/10.48048/tis.2024.8596

Cheirsilp, B., Srinuanpan, S., & Mandik, Y. I. (2020). Efficient harvesting of microalgal biomass and direct conversion of microalgal lipids into biodiesel. *Microalgae Cultivation for Biofuels Production*, 83–96. https://doi.org/10.1016/B978-0-12-817536-1.00006-0

Chen, C. Y., Yeh, K. L., Chang, H. Y., & Chang, J. S. (2014). Strategies to improve oil/lipid production of microalgae in outdoor cultivation using vertical tubular-type photobioreactors. *Energy Procedia*, *61*, 2755–2758. https://doi.org/10.1016/J.EGYPRO.2014.12.298

Cheng, C. H., Du, T. B., Pi, H. C., Jang, S. M., Lin, Y. H., & Lee, H. T. (2011). Comparative study of lipid extraction from microalgae by organic solvent and supercritical CO2. *Bioresource Technology*, *102*(21), 10151–10153. https://doi.org/10.1016/J.BIORTECH.2011.08.064

Chhandama, V. L. M., & Satyan, B. K. (2021). Optimization of lipid accumulation in Pleurastrum insigne for biodiesel production. *Research Journal of Biotechnology*, *16*(10), 144–155.

Chin, G. J. W. L., Andrew, A. R., Abdul-Sani, E. R., Yong, W. T. L., Misson, M., & Anton, A. (2023). The effects of light intensity and nitrogen concentration to enhance lipid production in four tropical microalgae. *Biocatalysis and Agricultural Biotechnology*, *48*, 102660. https://doi.org/10.1016/J.BCAB.2023.102660

Cho, K., Cho, D. H., Heo, J., Kim, U., Lee, Y. J., Choi, D. Y., & Kim, H. S. (2019). Nitrogen modulation under chemostat cultivation mode induces biomass and lipid production by Chlorella vulgaris and reduces antenna pigment accumulation. *Bioresource Technology*, *281*, 118–125. https://doi.org/10.1016/J.BIORTECH.2019.02.063

Chu, F., Cheng, J., Zhang, X., Ye, Q., & Zhou, J. (2019). Enhancing lipid production in microalgae Chlorella PY-ZU1 with phosphorus excess and nitrogen starvation under 15% CO2 in a continuous two-step cultivation process. *Chemical Engineering Journal*, *375*, 121912. https://doi.org/10.1016/J.CEJ.2019.121912

Correa-Aguado, H. C., Cerrillo-Rojas, G. V., Aguilera Flores, M. M., Castillo, S. Z., & Morales-Domínguez, J. F. (2023). Optimized infrared-assisted extraction to obtain total lipid from microalgae Scenedesmus obliquus: A green approach. *International Journal of Chemical Reactor Engineering*, *21*(4), 549–563. https://doi.org/10.1515/IJCRE-2022-0107

Dani, N., Zare, D., Assadi, M. M., Irani, S., & Soltani, N. (2021). Isolation, screening and medium optimization of native microalgae for lipid production using nutritional starvation strategy and statistical design. *International Journal of Environmental Science and Technology*, *18*(10), 2997–3012. https://doi.org/10.1007/S13762-020-03037-9

Deckelbaum, R. J., & Torrejon, C. (2012). The Omega-3 fatty acid nutritional landscape: Health benefits and sources. *The Journal of Nutrition*, *142*(3), 587S–591S. https://doi.org/10.3945/JN.111.148080

Elshobary, M. E., Abo-Shanab, W. A., Ende, S. S. W., Alquraishi, M., & El-Shenody, R. A. (2025). Optimizing Phaeodactylum tricornutum cultivation: Integrated strategies for enhancing biomass, lipid, and fucoxanthin production. *Biotechnology for Biofuels and Bioproducts*, *18*(1). https://doi.org/10.1186/S13068-024-02602-5

Ferreira, G. F., Ríos Pinto, L. F., Maciel Filho, R., & Fregolente, L. V. (2019). A review on lipid production from microalgae: Association between cultivation using waste streams and fatty acid profiles. *Renewable and Sustainable Energy Reviews*, *109*, 448–466. https://doi.org/10.1016/j.rser.2019.04.052

Fu, L., Cui, X., Li, Y., Xu, L., Zhang, C., Xiong, R., Zhou, D., & Crittenden, J. C. (2017). Excessive phosphorus enhances Chlorella regularis lipid production under nitrogen starvation stress during glucose heterotrophic cultivation. *Chemical Engineering Journal*, *330*, 566–572. https://doi.org/10.1016/J.CEJ.2017.07.182

Fu, L., Li, Q., Yan, G., Zhou, D., & Crittenden, J. C. (2019). Hormesis effects of phosphorus on the viability of Chlorella regularis cells under nitrogen limitation. *Biotechnology for Biofuels*, *12*(1), 121. https://doi.org/10.1186/S13068-019-1458-Z

Gao, Y., Yang, M., & Wang, C. (2013). Nutrient deprivation enhances lipid content in marine microalgae. *Bioresource Technology*, *147*, 484–491. https://doi.org/10.1016/j.biortech.2013.08.066

Gaurav, K., Neeti, K., & Singh, R. (2024). Microalgae-based biodiesel production and its challenges and future opportunities: A review. *Green Technologies and Sustainability*, *2*(1), 100060. https://doi.org/10.1016/J.GRETS.2023.100060

Gorgich, M., Mata, T. M., Martins, A. A., Branco-Vieira, M., & Caetano, N. S. (2020). Comparison of different lipid extraction procedures applied to three microalgal species. *Energy Reports*, *6*, 477–482. https://doi.org/10.1016/J.EGYR.2019.09.011

Han, F., Huang, J., Li, Y., Wang, W., Wan, M., Shen, G., & Wang, J. (2013). Enhanced lipid productivity of Chlorella pyrenoidosa through the culture strategy of semi-continuous cultivation with nitrogen limitation and pH control by CO2. *Bioresource Technology*, *136*, 418–424. https://doi.org/10.1016/J.BIORTECH.2013.03.017

Han, W., Jin, W., Ding, W., Lu, S., Song, K., Chen, C., Qin, C., Chen, Y., Tu, R., & Zhou, X. (2021). Effects of nutrient composition, lighting conditions, and metal ions on the growth and lipid yield of the high-lipid-yielding microalgae (Chlorella pyrenoidosa) cultivated in municipal wastewater. *Journal of Environmental Chemical Engineering*, *9*(6). https://doi.org/10.1016/j.jece.2021.106491

Hang Nguyen, T. T. (2026). Diverse cultivation temperatures and nitrogen supplementation levels alter biomass, lipid, and fatty acid productivity in the native microalgae for biodiesel application. *Renewable Energy*, *256*. https://doi.org/10.1016/j.renene.2025.123988

Hao, S., Ge, P., Su, W., Wang, Y., Abd El-Aty, A. M., & Tan, M. (2024). Steady-state delivery and chemical modification of food nutrients to improve cancer intervention ability. *Foods*, *13*(9). https://doi.org/10.3390/FOODS13091363

He, Y., Zhang, B., Guo, S., Guo, Z., Chen, B., & Wang, M. (2020). Sustainable biodiesel production from the green microalgae Nannochloropsis: Novel integrated processes from cultivation to enzyme-assisted extraction and ethanolysis of lipids. *Energy Conversion and Management*, *209*, 112618. https://doi.org/10.1016/J.ENCONMAN.2020.112618

Ho, S. H., Nakanishi, A., Ye, X., Chang, J. S., Hara, K., Hasunuma, T., & Kondo, A. (2014). Optimizing biodiesel production in marine Chlamydomonas sp. JSC4 through metabolic profiling and an innovative salinity-gradient strategy. *Biotechnology for Biofuels*, *7*(1), 97. https://doi.org/10.1186/1754-6834-7-97

Islam, M. M., Alam, H., Acharjee, A., Salatul, M., & Mozumder, I. (2024). Establishment of an effective photobioreactor for growing microalgae: A review. *International Journal of Agricultural Research, Innovation and Technology*, *14*(2), 153–162. https://doi.org/10.3329/IJARIT.V14I2.79511

Islam, M. M., & Dixit, S. (2024). An overview study of lipid extraction methods from microalgae. *SSRN Electronic Journal*. https://doi.org/10.2139/SSRN.4920440

Joshi, S. K., Madhavi, A., Reddy, V. B., Raj, N., Kumar, M., Bhalla, A., Kumar, Y., & Ali, H. M. (2024). Optimization of bioenergy production from algae using bioreactor design. *E3S Web of Conferences*, *581*, 01033. https://doi.org/10.1051/E3SCONF/202458101033

Karemore, A., & Sen, R. (2016). Downstream processing of microalgal feedstock for lipid and carbohydrate in a biorefinery concept: A holistic approach for biofuel applications. *RSC Advances*, *6*(35), 29486–29496. https://doi.org/10.1039/C6RA01477A

Karim, A., Amirul Islam, M., Khalid, Z. Bin, Faizal, C. K. M., Khan, M. M. R., & Yousuf, A. (2020). Microalgal cell disruption and lipid extraction techniques for potential biofuel production. *Microalgae Cultivation for Biofuels Production*, 129–147. https://doi.org/10.1016/B978-0-12-817536-1.00009-6

Kim, S., Moon, M., Kwak, M., Lee, B., & Chang, Y. K. (2018). Statistical optimization of light intensity and CO2 concentration for lipid production derived from attached cultivation of green microalga Ettlia sp. *Scientific Reports*, *8*(1), 15390. https://doi.org/10.1038/S41598-018-33793-1

Kitessa, S. M., Abeywardena, M., Wijesundera, C., & Nichols, P. D. (2014). DHA-containing oilseed: A timely solution for the sustainability issues surrounding fish oil sources of the health-benefitting long-chain Omega-3 oils. *Nutrients*, *6*(5), 2035. https://doi.org/10.3390/NU6052035

Kumar, A., Guria, C., & Pathak, A. K. (2018). Optimal cultivation towards enhanced algae-biomass and lipid production using Dunaliella tertiolecta for biofuel application and potential CO2 bio-fixation: Effect of nitrogen deficient fertilizer, light intensity, salinity and carbon supply strategy. *Energy*, *148*, 1069–1086. https://doi.org/10.1016/J.ENERGY.2018.01.042

Kumar, R. R., Sarkar, D., & Sen, R. (2024). Simultaneously maximizing microalgal biomass and lipid productivities by machine learning driven modeling, global sensitivity analysis and multi-objective optimization for sustainable biodiesel production. *Applied Energy*, *358*, 122597. https://doi.org/10.1016/J.APENERGY.2023.122597

Kumari, A., Pabbi, S., & Tyagi, A. (2023). Recent advances in enhancing the production of long chain omega-3 fatty acids in microalgae. *Critical Reviews in Food Science and Nutrition*, *64*(29), 10564–10582. https://doi.org/10.1080/10408398.2023.2226720

Lari, Z., Moradi-Kheibari, N., Ahmadzadeh, H., Abrishamchi, P., Moheimani, N. R., & Murry, M. A. (2016). Bioprocess engineering of microalgae to optimize lipid production through nutrient management. *Journal of Applied Phycology*, *28*(6), 3235–3250. https://doi.org/10.1007/S10811-016-0884-6

Liu, Q., Liu, Y., Zhao, J., Qiao, W., Hou, J., Wang, Y., Zhang, M., Jia, G., Fan, X., Li, Z., Jia, H., Zhao, X., & Chen, L. (2024). Impact of manufacturing processes on glycerolipid and polar lipid composition and ultrastructure in infant formula. *Food Chemistry*, *444*. https://doi.org/10.1016/j.foodchem.2024.138623

Lv, J. M., Cheng, L. H., Xu, X. H., Zhang, L., & Chen, H. L. (2010). Enhanced lipid production of Chlorella vulgaris by adjustment of cultivation conditions. *Bioresource Technology*, *101*(17), 6797–6804. https://doi.org/10.1016/j.biortech.2010.03.120

Matter, I. A., Bui, V. K. H., Jung, M., Seo, J. Y., Kim, Y.-E., Lee, Y.-C., Oh, Y.-K., Matter, I. A., Bui, V. K. H., Jung, M., Seo, J. Y., Kim, Y.-E., Lee, Y.-C., & Oh, Y.-K. (2019). Flocculation harvesting techniques for microalgae: A review. *Applied Sciences*, *9*(15). https://doi.org/10.3390/APP9153069

Michael, M., Vanness, V., Henry, H., Tjandra, T. M., Kristian, K., Nasution, J. A., Masytha, M., Sidabutar, R., Trisakti, B., Irvan, I., Dalimunthe, N. F., Al Fath, M. T., Nabilah, Y., Kobayashi, T., Daimon, H., & Takriff, M. S. (2026). Microalgae for sustainable biodiesel and omega-3: A comprehensive review of production, processing, and implementation. *Renewable and Sustainable Energy Reviews*, *226*. https://doi.org/10.1016/j.rser.2025.116327

Mou, Y., Liu, N., Su, K., Li, X., Lu, T., Yu, Z., & Song, M. (2022). The growth and lipid accumulation of Scenedesmus quadricauda under nitrogen starvation stress during xylose mixotrophic/heterotrophic cultivation. *Environmental Science and Pollution Research*, *30*(44), 98934–98946. https://doi.org/10.1007/S11356-022-24579-0

Muthu, A. (2025). A perspective of microalga-derived omega-3 fatty acids: Scale up and engineering challenges. *Critical Reviews in Food Science and Nutrition*, *65*(32), 8173–8187. https://doi.org/10.1080/10408398.2025.2494060

Napier, J. A., Olsen, R. E., & Tocher, D. R. (2018). Update on GM canola crops as novel sources of omega-3 fish oils. *Plant Biotechnology Journal*, *17*(4), 703. https://doi.org/10.1111/PBI.13045

Nayak, M., Suh, W. I., Chang, Y. K., & Lee, B. (2019). Exploration of two-stage cultivation strategies using nitrogen starvation to maximize the lipid productivity in Chlorella sp. HS2. *Bioresource Technology*, *276*, 110–118. https://doi.org/10.1016/J.BIORTECH.2018.12.111

Nazir, S., & Zahoor, I. (2021). Oilseed crops as the alternate source of omega fatty acids: A paradigm shift. *Oil Crop Genomics*, 341–352. https://doi.org/10.1007/978-3-030-70420-9_15

Nicodemou, A., Konstantinou, D., & Koutinas, M. (2024). Enhanced biomass and lipid production from olive processing wastewater using Scenedesmus obliquus in a two-stage cultivation strategy under salt stress. *Biochemical Engineering Journal*, *205*, 109290. https://doi.org/10.1016/J.BEJ.2024.109290

Nyoni, K., Kelebopile, L., & Lebogang, L. (2024). Microalgal lipid production for fatty acid methyl esters conversion: A response surface methodology optimisation of media nutrient in batch reactors. *Results in Chemistry*, *9*, 101668. https://doi.org/10.1016/J.RECHEM.2024.101668

Oh, Y. K., Kim, S., Ilhamsyah, D. P. A., Lee, S. G., & Kim, J. R. (2022). Cell disruption and lipid extraction from Chlorella species for biorefinery applications: Recent advances. *Bioresource Technology*, *366*, 128183. https://doi.org/10.1016/J.BIORTECH.2022.128183

Okazaki, H., Takabe, Y., Masuda, T., & Hoshikawa, Y. (2019). Biochemical response of indigenous microalgal consortia to variations in nitrogen concentration of treated effluent. *Bioresource Technology*, *280*, 118–126. https://doi.org/10.1016/j.biortech.2019.02.014

Padhi, D., & Nayak, M. (2025). Impact of surplus P with N and S modulation for enhanced microalgal lipid productivity in closed reactor system. *Chemosphere*, *385*, 144529. https://doi.org/10.1016/J.CHEMOSPHERE.2025.144529

Pancha, I., Chokshi, K., Maurya, R., Trivedi, K., Patidar, S. K., Ghosh, A., & Mishra, S. (2015). Salinity induced oxidative stress enhanced biofuel production potential of microalgae Scenedesmus sp. CCNM 1077. *Bioresource Technology*, *189*, 341–348. https://doi.org/10.1016/J.BIORTECH.2015.04.017

Pandey, S., Visavadiya, P., Patil, V., Jain, J., Archana, G., & Bagchi, D. (2024). Salinity stress-mediated lipid droplet production and liquid–liquid phase separation in Chlamydomonas reinhardtii. *The European Physical Journal Special Topics*, *233*(21), 3105–3121. https://doi.org/10.1140/EPJS/S11734-024-01186-3

Papapanagiotou, G., Charisis, A., Samara, C., Kalogianni, E. P., & Chatzidoukas, C. (2024). Linking cultivation conditions to the fatty acid profile and nutritional value of Chlorella sorokiniana lipids. *Processes*, *12*(12). https://doi.org/10.3390/PR12122770

Paranjape, K., Leite, G. B., & Hallenbeck, P. C. (2016). Effect of nitrogen regime on microalgal lipid production during mixotrophic growth with glycerol. *Bioresource Technology*, *214*, 778–786. https://doi.org/10.1016/J.BIORTECH.2016.05.020

Parveen, S., & Patidar, S. K. (2022). Revisiting algal lipids and cellular stress-causing strategies for ameliorating the productivity of suitable lipids of microalgae for biofuel applications. *Sustainable Energy & Fuels*, *6*(17), 3907–3925. https://doi.org/10.1039/D2SE00574C

Patel, A., Rantzos, C., Krikigianni, E., Rova, U., Christakopoulos, P., & Matsakas, L. (2024). A bioprocess engineering approach for the production of hydrocarbons and fatty acids from green microalga under high cobalt concentration as the feedstock of high-grade biofuels. *Biotechnology for Biofuels and Bioproducts*, *17*(1), 64. https://doi.org/10.1186/S13068-024-02512-6

Perečinec, M. G., Babić, S., Čižmek, L., Selmani, A., Popović, N. T., Sikirić, M. D., Strunjak-Perović, I., & Čož-Rakovac, R. (2021). Selenite as a lipid inductor in marine microalga Dunaliella tertiolecta: Comparison of one-stage and two-stage cultivation strategies. *Applied Biochemistry and Biotechnology*, *194*(2), 930–949. https://doi.org/10.1007/S12010-021-03659-W

Pourbakhtiar, A., Tavakoli, O., & Ahmadi, B. (2022). Design and optimization of a two-stage microalgae-assisted lipid production. *BioEnergy Research*, *16*(1), 565–578. https://doi.org/10.1007/S12155-022-10413-5

Prasad, P., Anjali, P., & Sreedhar, R. V. (2021). Plant-based stearidonic acid as sustainable source of omega-3 fatty acid with functional outcomes on human health. *Critical Reviews in Food Science and Nutrition*, *61*(10), 1725–1737. https://doi.org/10.1080/10408398.2020.1765137

Rafiq, S., Bhat, M. I., Sofi, S. A., Muzzafar, K., Majid, D., Dar, B. N., & Makroo, H. A. (2023). Bioconversion of agri-food waste and by-products into microbial lipids: Mechanism, cultivation strategies and potential in food applications. *Trends in Food Science & Technology*, *139*, 104118. https://doi.org/10.1016/J.TIFS.2023.07.015

Rao, A. R., & Ravishankar, G. A. (2019). Microalgal biomass, lipids, and fatty acids production through open or closed cultivation systems: Challenges and future perspectives. In Gokare Ravishankar and Ranga Rao Ambati (eds.), *Handbook of Algal Technologies and Phytochemicals: Volume II: Phycoremediation, Biofuels and Global Biomass Production* (pp. 91–99). Academic Press. https://doi.org/10.1201/9780429057892-8

Rehman, M., Kesharvani, S., Dwivedi, G., & Gidwani Suneja, K. (2022). Impact of cultivation conditions on microalgae biomass productivity and lipid content. *Materials Today: Proceedings*, *56*, 282–290. https://doi.org/10.1016/j.matpr.2022.01.152

Rezaei, A., Cheniany, M., Ahmadzadeh, H., & Vaezi, J. (2024). A new isolate cold-adapted Ankistrodesmus sp. OR119838: Influence of light, temperature, and nitrogen concentration on growth characteristics and biochemical composition using the two-stage cultivation strategy. *Bioprocess and Biosystems Engineering*, *47*(3), 341–353. https://doi.org/10.1007/S00449-023-02964-4

Roy, M., Bera, S., & Mohanty, K. (2021). Nutrient starvation-induced oxidative stress-mediated lipid accumulation in Tetradesmus obliquus KMC24. *Journal of Applied Phycology*, *33*(6), 3617–3635. https://doi.org/10.1007/S10811-021-02614-W

Saini, R. K., Ravishankar, G. A., & Keum, Y. S. (2023). Microalgae and Thraustochytrids are Sustainable Sources of Vegan EPA and DHA with Commercial Feasibility. *IJSIM*, *63*(1), 155–158. https://doi.org/10.1007/S12088-023-01059-8

Sánchez-Bayo, A., López-Chicharro, D., Morales, V., Espada, J. J., Puyol, D., Martínez, F., Astals, S., Vicente, G., Bautista, L. F., & Rodríguez, R. (2020). Biodiesel and biogas production from Isochrysis galbana using dry and wet lipid extraction: A biorefinery approach. *Renewable Energy*, *146*, 188–195. https://doi.org/10.1016/J.RENENE.2019.06.148

Seyed Hosseini, N., Shang, H., & Scott, J. A. (2018). Biosequestration of industrial off-gas CO2 for enhanced lipid productivity in open microalgae cultivation systems. *Renewable and Sustainable Energy Reviews*, *92*, 458–469. https://doi.org/10.1016/J.RSER.2018.04.086

Shin, H. Y., Shim, S. H., Ryu, Y. J., Yang, J. H., Lim, S. M., & Lee, C. G. (2018). Lipid extraction from Tetraselmis sp. microalgae for biodiesel production using hexane-based solvent mixtures. *Biotechnology and Bioprocess Engineering*, *23*(1), 16–22. https://doi.org/10.1007/S12257-017-0392-9

Singh, A., Singh, S., Prasad, N., Singh, S., Mohanta, A., Kayastha, A. M., & Asthana, R. K. (2025). An exploration of lipid remodeling by microalga Chlamydomonas reinhardtii in autotrophic and mixotrophic cultivation for bioenergy prospects. *Biomass and Bioenergy*, *200*, 108015. https://doi.org/10.1016/J.BIOMBIOE.2025.108015

Singh, P., Kumari, S., Guldhe, A., Misra, R., Rawat, I., & Bux, F. (2016). Trends and novel strategies for enhancing lipid accumulation and quality in microalgae. *Renewable and Sustainable Energy Reviews*, *55*, 1–16. https://doi.org/10.1016/J.RSER.2015.11.001

Singh, R. P., Yadav, P., Kumar, A., Hashem, A., Avila-Quezada, G. D., Abd_Allah, E. F., & Gupta, R. K. (2023). Salinity-induced physiochemical alterations to enhance lipid content in oleaginous microalgae Scenedesmus sp. BHU1 via two-stage cultivation for biodiesel feedstock. *Microorganisms*, *11*(8), 2064. https://doi.org/10.3390/MICROORGANISMS11082064

Singh, R. P., Yadav, P., Kumar, I., Solanki, M. K., Roychowdhury, R., Kumar, A., Gupta, R. K., Singh, R. P., Yadav, P., Kumar, I., Solanki, M. K., Roychowdhury, R., Kumar, A., & Gupta, R. K. (2023). Advancement of abiotic stresses for microalgal lipid production and its bioprospecting into sustainable biofuels. *Sustainability*, *15*(18). https://doi.org/10.3390/SU151813678

Teh, M. R., Armitage, A. E., & Drakesmith, H. (2024). Why cells need iron: A compendium of iron utilisation. *Trends in Endocrinology & Metabolism*, *35*(12), 1026–1049. https://doi.org/10.1016/J.TEM.2024.04.015

Tripathi, G., Dubey, P., Osama, K., Ahmad, S., Irum, Singh, V., Farooqui, A., Vamanu, E., Nand Rai, S., & Mishra, V. (2025). Two-stage cultivation of microalga for sustainable biofuel production. *Green Chemistry Letters and Reviews*, *18*(1). https://doi.org/10.1080/17518253.2025.2543909

Trivedi, J., Atray, N., Agrawal, D., & Ray, A. (2022). Enhanced lipid production in Scenedesmus obliquus via nitrogen starvation in a two-stage cultivation process and evaluation for biodiesel production. *Fuel*, *316*, 123418. https://doi.org/10.1016/J.FUEL.2022.123418

Udayantha, H. M. V., Kim, S. H., Chen, Y., Long, J., Bathige, S. D. N. K., & Park, K. I. (2025). Enhancing lipid accumulation in Tetraselmis sp.: Integrating nitrogen deprivation and glucose supplementation for biofuel production. *Biotechnology for Biofuels and Bioproducts*, *18*(1), 57. https://doi.org/10.1186/S13068-025-02654-1

Veerasamy, V., Neethirajan, V., Singarayar, M. S., Balasundaram, D., Dharmar, P., & Thilagar, S. (2024). Microalgal biomass and lipid synergy for omega fatty acid enrichment: A sustainable source for food supplements & nutraceuticals. *Algal Research*, *80*. https://doi.org/10.1016/j.algal.2024.103514

Višnjevec, A. M., Barp, L., Lucci, P., & Moret, S. (2024). Pressurized liquid extraction for the determination of bioactive compounds in plants with emphasis on phenolics. *TrAC Trends in Analytical Chemistry*, *173*, 117620. https://doi.org/10.1016/J.TRAC.2024.117620

Walker, C. G., Jebb, S. A., & Calder, P. C. (2013). Stearidonic acid as a supplemental source of ω-3 polyunsaturated fatty acids to enhance status for improved human health. *Nutrition*, *29*(2), 363–369. https://doi.org/10.1016/J.NUT.2012.06.003

Wang, S., & Diao, Y. (2024). Printed electronics for cultivating plants in space. *Nature Reviews Materials*, *9*(11), 762–763. https://doi.org/10.1038/s41578-024-00742-6

Wang, X., Li, Z., Wang, D., Yuan, X., Guo, X., Xue, H., & Gao, X. (2024). Temperature-dependent two-stage cultivation: A strategy for lipid and fatty acids production in the terrestrial oleaginous microalga Vischeria sp. WL1. *Journal of Applied Phycology 2024 37:1*, *37*(1), 97–107. https://doi.org/10.1007/S10811-024-03383-Y

Wang, Z. (2025). Microalgae-derived Omega-3 fatty acids improve nutritional and flavor profiles of cultured meat. *Highlights in Science Engineering and Technology*, *153*, 294–303. https://doi.org/10.54097/8J04XC40

Wei, Q., Yao, J., Chen, R., Yang, S., Tang, Y., & Ma, X. (2022). Low-frequency ultrasound and nitrogen limitation induced enhancement in biomass production and lipid accumulation of Tetradesmus obliquus FACHB-12. *Bioresource Technology*, *358*, 127387. https://doi.org/10.1016/J.BIORTECH.2022.127387

Wu, J., Chen, C., Wu, H., Li, T., Chen, X., Wu, H., & Xiang, W. (2023). Enhancement in lipid productivity of the marine microalga Nannochloropsis sp. SCSIO-45217 through phosphate adjustment strategies. *Journal of Applied Phycology*, *35*(3), 1023–1035. https://doi.org/10.1007/S10811-023-02935-Y

Xie, Z., Lin, W., & Luo, J. (2021). Co-cultivation of microalga and xylanolytic bacterium by a continuous two-step strategy to enhance algal lipid production. *Bioresource Technology*, *330*, 124953. https://doi.org/10.1016/J.BIORTECH.2021.124953

Yang, S., Wang, Y., Wang, J., Cheng, K., Liu, J., He, Y., Zhang, Y., Mou, H., & Sun, H. (2024). Microalgal protein for sustainable and nutritious foods: A joint analysis of environmental impacts, health benefits and consumer's acceptance. *Trends in Food Science and Technology*, *143*. https://doi.org/10.1016/j.tifs.2023.104278

Zhou, X.-R., Yao, Z. J., Benedicto, K., Nichols, P. D., Green, A., Singh, S., Zhou, X.-R., Yao, Z. J., Benedicto, K., Nichols, P. D., Green, A., & Singh, S. (2023). New sustainable oil seed sources of Omega-3 long-chain polyunsaturated fatty acids: A journey from the ocean to the field. *Sustainability*, *15*(14). https://doi.org/10.3390/SU151411327

Zhu, Z., Sun, J., Fa, Y., Liu, X., & Lindblad, P. (2022). Enhancing microalgal lipid accumulation for biofuel production. *Frontiers in Microbiology*, *13*, 1024441. https://doi.org/10.3389/FMICB.2022.1024441

9 Insights into Microalgae-Based Biopolymer Production

A Technological Survey

Anwesha Mondal, Avik Kumar Choudhury, Kristine V. Aleksanyan, and Gour Gopal Satpati

9.1 INTRODUCTION

The increasing global concern regarding environmental sustainability and the depletion of fossil fuel resources has accelerated the search for alternative materials to conventional petrochemical-based plastic. The mounting accumulation of nondegradable plastics in terrestrial and marine ecosystems has become one of the defining environmental challenges of the 21st century. In this context, biopolymers derived from renewable sources have gained considerable attention as next-generation materials capable of balancing industrial functionality with ecological responsibility (Halley & Dorgan, 2011). Among various biological resources, microalgae have emerged as a particularly promising feedstock owing to their remarkable growth efficiency, carbon fixation capacity, and metabolic versatility (Madadi et al., 2021).

Microalgae are microscopic photosynthetic organisms capable of thriving in diverse environments, including non-arable lands and nutrient-rich wastewaters, without competing with agricultural crops for freshwater or arable soil (Satpati et al., 2023, 2024). Compared with terrestrial plants, they exhibit superior photosynthetic efficiency and faster biomass accumulation, coupled with minimal nutrient requirements (Mondal et al., 2024a, 2024b). This makes them ideal candidates for large-scale, eco-friendly biopolymer production. These microorganisms synthesize a broad spectrum of value-added biopolymers such as polyhydroxyalkanoates (PHAs), polysaccharides, proteins, and lipids. Many of these polymers are biodegradable, biocompatible, and suitable for a wide range of applications. Among them, polyhydroxybutyrate (PHB), a major type of PHA, has attracted significant industrial interest for its potential to replace petroleum-based plastics due to its excellent biodegradability and mechanical properties (Koller & Mukherjee, 2022). The utility of microalgae-derived polymers extends far beyond packaging, encompassing sectors such as pharmaceuticals, agriculture, and wastewater remediation, all of which align with the principles of a circular bioeconomy (Devadas et al., 2021). Despite

DOI: 10.1201/9781003611479-9

their promise, the commercialization of microalgae-based biopolymers remains hindered by challenges associated with cost-effective cultivation, harvesting, and downstream processing. Ongoing research efforts are therefore directed toward integrating low-cost substrates, optimizing culture parameters, and developing novel biorefinery models that enhance both productivity and sustainability (Mal et al., 2022).

This chapter provides a comprehensive overview of recent technological advancements in microalgae-based biopolymer production. It surveys key trends, cultivation strategies, and process innovations, highlighting how microalgae could serve as the cornerstone of future biopolymer value chains that merge environmental stewardship with industrial innovation.

9.2 MICROALGAE AS RENEWABLE RESOURCES

Microalgae refers to a large group of unicellular to multicellular photosynthetic thalloid organisms that are unobservable with our naked eyes. They are mainly aquatic organisms, although they often tend to be cosmopolitan in distribution and have been reported from diverse habitats ranging from alpine regions to halophilic conditions. Microalgae are dominated by the members of cyanobacteria and green algae. Although cyanobacteria belong to Gram-negative archaebacteria, eukaryotic green algae are closest to viridiplantae, the modern-day land plants. The primary similarity between green algae and modern-day land plants is the presence of double-membrane-bound chloroplasts. There are several variations of the chloroplast morphology among algae, where the presence of chloroplast endoplasmic reticulum (CER) is an important characteristic, as observed in classes like Bacillariophyceae, Eustigmatophyceae, Prymnesiophyceae, and other members of Heterokontophyta. The other microalgae are dinoflagellates and euglenophytes, although they are highly habitat-specific (Lee, 2008). These microalgae are getting increasing attention as a renewable resource due to their ability to thrive in non-arable lands (Devadas et al., 2021), utilize wastewater as a habitat, and sequester CO_2 from the atmosphere, yet show high growth rates under both *in situ* and *in vitro* conditions. In an era with high carbon emissions, the efficiency of microalgae to capture atmospheric CO_2 from diverse habitats makes them an excellent biological carbon sink. The ability to utilize non-arable water like wastewater, saltwater, and brackish water reduces the conflict of "food versus fuel" with first-generation biofuels (Antony & Ramanan, 2021; Devadas et al., 2021; Venkata Mohan et al., 2020). Microalgae are also rich sources of molecules with high nutraceutical properties like omega-3 fatty acids, carotenoids, and other antioxidant molecules. Different microalgae like *Dunaliella* sp. And *Haematococcus* sp. have been commercially used as sources of high-value pigment molecules like carotenoids and astaxanthin. Microalgae have been considered as a third-generation resource of biofuels due to their high lipid production potential, which can be further processed as biodiesel, bioethanol, or biohydrogen. The other aspect that scores microalgae as a renewable resource is the versatility of mass cultivation, as they can be cultivated under the simplest conditions like a pond or advanced setups like open raceway ponds and photobioreactors (PBRs) (Dey et al., 2025a). The ability of microalgae, especially cyanobacteria, to grow in high organic pollution makes them score over other plant forms as a renewable plant

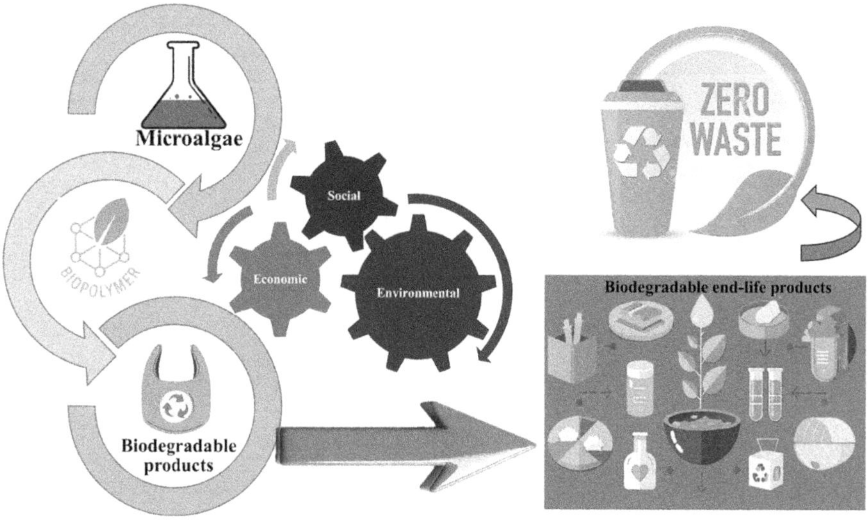

FIGURE 9.1 Microalgae as a source of biopolymers.

resource. In recent years, research on algae has also focused on utilizing their energy storage and conversion ability to produce bioelectricity and fuel cells that may see multifarious applications in future (Dey et al., 2025b). Thus, as arable land for cultivation reduces with time, such inherent abilities of microalgal forms as renewable resources make them an excellent choice for sustainable development in the future (Figure 9.1).

9.3 MICROALGAE-BASED BIOPOLYMER PRODUCTION AND SCALE-UP

Microalgae have gained significant interest as alternative resources for biopolymer synthesis, offering flexible and sustainable platforms to generate various biodegradable plastics and related materials (Figure 9.2). Their ability to produce PHAs, starch, cellulose, proteins, and lipids under stress conditions has contributed to the rise of algae as "green factories" with promising industrial potential (Koller & Mukherjee, 2022).

9.3.1 Diversity of Biopolymers and Sources

Microalgae-derived biopolymers encompass PHB, polyhydroxyvalerate (PHV), polysaccharides (e.g., alginate, cellulose, and starch), and functional proteins, each with specific ecological and industrial values. PHB stands as the most prominent microbial polyester, accumulating in certain cyanobacteria and green algal species up to 85% of biomass under suitable culture conditions. Polysaccharides such as alginate and carrageenan are conventionally sourced from macroalgae but can have analogues from select microalgal strains (Mondal et al., 2021).

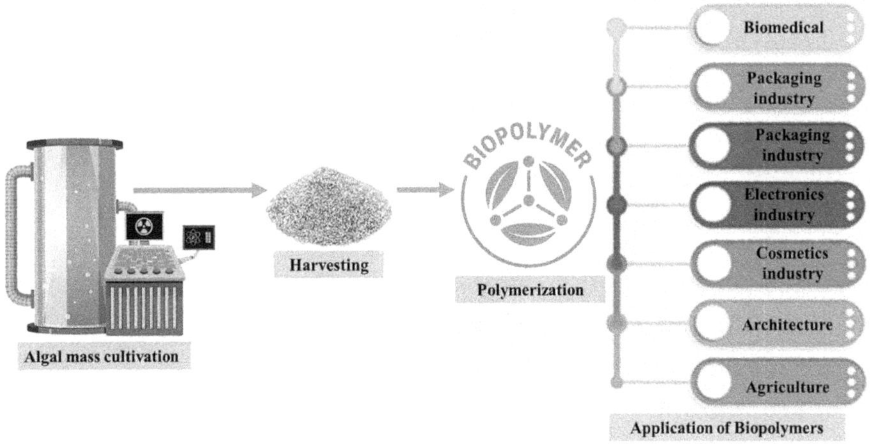

FIGURE 9.2 Algae-based biopolymer production and application.

9.3.2 Cultivation Technologies and Scale-Up Challenges

Large-scale cultivation represents a critical step for viable biopolymer production. Open ponds are cost-effective for mass culture but suffer from contamination risks, variable environmental control, and batch-to-batch variation. Tubular and flat-panel PBRs allow finer control of environmental parameters but require higher capital investment. Wastewater reuse in cultivation not only offers nutrient-rich environments but also integrates environmental remediation with polymer production, reinforcing the circular bioeconomy concept for sustainability (Duan & Shi, 2014). Scale-up demands optimization of nutrient supply, CO_2 sources, and culture medium selection. The use of industrial wastes, flue gases, and agricultural runoff as input streams has been demonstrated to improve cost efficiency and environmental compatibility of large-scale production.

9.3.3 Harvesting, Biomass Processing, and Extraction

Algal biomass harvesting is a major bottleneck in economic scale-up due to the small size and low density of cultures. Cost-effective harvesting strategies include autoflocculation, chemical and bioflocculation, floatation, centrifugation, and membrane filtration; each presents unique trade-offs in efficiency, cost, and product purity. Advanced methods such as magnetic separation and electrocoagulation are emerging, though their economic feasibility requires further exploration (Kumar et al., 2023). After harvesting, drying, whether by solar, convective, or freeze-drying, prepares biomass for polymer extraction but adds substantial cost and energy requirements. The extraction and purification of target biopolymers, whether PHB, proteins, or polysaccharides, involve mechanical, chemical, or enzymatic processing, with solvent systems and value-added by-product recovery gaining industrial momentum (Mal et al., 2022). Once the purified polymer is obtained, various fabrication and processing methods are used to convert it into functional materials (Figure 9.3). The injection molding technique enables rapid shaping of molten biopolymer into complex forms with high precision, suitable for large-scale industrial production. The twin-screw extrusion technique facilitates homogeneous mixing and compounding of biopolymers with additives or reinforcing agents, enhancing mechanical performance. In the solvent casting technique, the polymer is dissolved in a suitable solvent

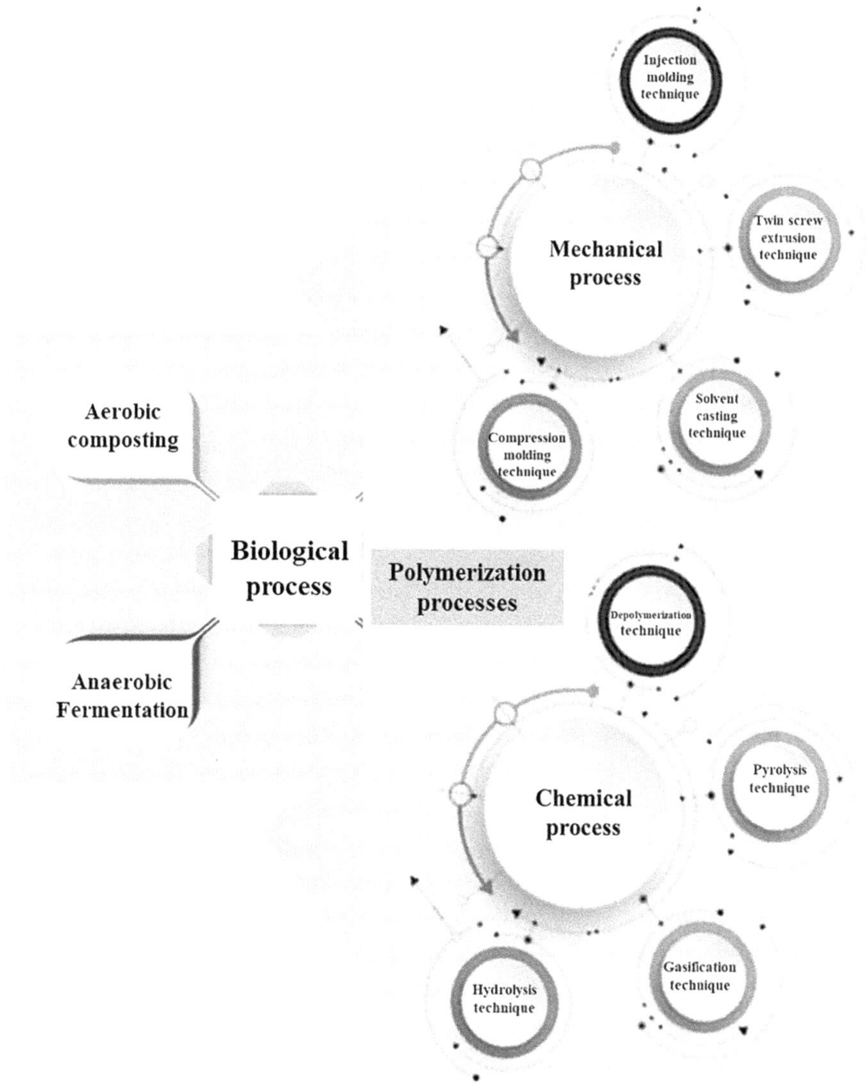

FIGURE 9.3 Different processes of bio-polymerization from algae.

and then evaporated to form thin films or coatings, widely used for packaging and biomedical applications. Meanwhile, the compression molding technique applies heat and pressure to mold solid polymer sheets or components, offering a simple and efficient route for thermoplastic biopolymers (Adetunji & Erasmus, 2024). The chemical processes play a pivotal role in refining or depolymerizing complex macromolecules into valuable monomers or fuel precursors. The depolymerization technique involves the chemical breakdown of polymer chains into smaller units or monomers through catalysts or specific reaction conditions, facilitating recycling and reuse of bioplastics. The pyrolysis technique converts algal biomass into bio-oil, syngas, and char through thermal decomposition in an oxygen-limited environment, providing an efficient route for energy and chemical production. The gasification technique operates at higher temperatures to generate a synthesis gas composed primarily of carbon monoxide and hydrogen, which serves as a versatile feedstock for fuel and polymer synthesis. The hydrolysis technique utilizes acid, alkali, or enzymatic catalysts to cleave complex polysaccharides and proteins into simpler sugars and amino acids, forming the basis for further fermentation or chemical transformation (Devadas et al., 2021). The biological processes contribute to the degradation and conversion of algal biopolymers into useful by-products. Aerobic composting utilizes microorganisms in the presence of oxygen to decompose organic matter, transforming algal residues and biodegradable polymers into nutrient-rich compost that can be used as biofertilizer. In contrast, anaerobic fermentation occurs in oxygen-free conditions, where microbial consortia convert organic substrates into biogas, primarily composed of methane and carbon dioxide, along with valuable digestate. Both processes offer environmentally friendly routes for recycling algal biomass, closing the loop within the biopolymer production cycle, and supporting the principles of a circular bioeconomy.

Together, these processes bridge the gap between laboratory-scale synthesis and commercial-scale product development, underscoring the technological maturity of microalgae-derived biopolymers (Devadas et al., 2021).

9.3.4 Integration and Process Optimization

A biorefinery approach, integrating biopolymer production with simultaneous recovery of biofuels, high-value products (nutraceuticals, pigments), and bioremediation functions, is recognized as key to economic viability. Multimodal cultivation (photoautotrophic, heterotrophic, and mixotrophic) and advanced downstream processing technologies are being refined to drive down costs while maximizing yield and sustainability (Mal et al., 2022). Continued research is directed toward strain improvement (through genetic engineering and adaptive selection), development of novel reactor designs, improved harvesting and drying technologies, and holistic process optimization to meet industrial demand. Life-cycle assessments (LCAs) and techno-economic analyses provide further validation and strategic guidance to position microalgae-based biopolymers as practical, eco-friendly solutions for the global plastic challenge.

9.4 TYPES OF BIOPOLYMERS OBTAINED FROM MICROALGAE

Microalgae-based carbon biopolymers can be divided into three basic types based on their precursors.

9.4.1 Polyhydroxyalkanoates (PHA)

Under conditions of nutrient limitations (namely nitrogen and phosphorus starvation) and excess carbon, many microalgae and cyanobacterial forms produce polyesters. These polymers have properties like conventional plastics, although being derived from biomaterials, they are biodegradable (García et al., 2021). The examples can be PHB and copolymers like poly(3-hydroxybutyrate-co-3-hydroxyvalerate) (PHBV). Chemically, these are aliphatic polyesters of hydroxy fatty acid monomers with ester bonds between the hydroxyl (–OH) group of one monomer and the carboxyl (–COOH) group of the next monomer. These PHAs can be of short-chain length (scl-PHA) with monomers of 3–5 carbon atoms or medium chain length (mcl-PHA) with monomers of 6–14 carbon atoms. The polymerization process involves a family of enzymes called Pha synthases (*Pha C*) as key enzymes with the condensation of multiple acetyl-CoA molecules (Rehm & Steinbüchel, 1999).

9.4.2 Polysaccharides (Carbohydrate-Based Polymers)

A variety of polysaccharides are produced by microalgae either as energy storage molecules or as cell wall components. They have multifarious applications like gelling and thickening agents, stabilizing agents, or precursors for bioplastics (Satpati & Pal., 2018). They can be further categorized in Table 9.1.

9.4.2.1 Starch-Based Polymers

Obtained from green microalgae like *Chlorella* and *Chlamydomonas*, where conventional starch is the main storage polysaccharide. Other than that, hosts of brown and red algae are also important sources of starch-based biopolymers. Starch is a highly polar, hydrophilic group of molecules composed of D-glucose units, amylose and amylopectin. Some of the common biopolymers are alginate and carrageenan. Alginate is an anionic linear copolymer of D-mannuronic acid and L-guluronic acid obtained from brown algae like *Chondrus crispus*. Carrageenan is a polysaccharide of D-galactose and 3,6-anhydro-D-galactose, and the presence of sulfate groups makes it anionic. However, microalgal sources of alginate or carrageenan are largely absent (Mathiot et al., 2019).

9.4.2.2 Cellulose-Based Polymers

Cellulose is a major component of many microalgae, which when extracted can act as a suitable biopolymer. These are generally extracted from green algal cell walls as β-(1,4) linked D-glucose structural polymers (Blockx et al., 2019).

TABLE 9.1
Chemical Nature of Different Biomolecules as Potential Source of Biopolymers and Their Microalgal Sources with Insights into Percentage Content of Target Biomolecule in Algal Biomass

Biomolecule/ Biopolymer	Microalgal Source	Biomolecule Content as Percent of Total Biomass	References
Cellulose, hemicellulose	*Chlorella vulgaris*		Zanchetta et al., 2021
	Nanochloropsis gaditana		Scholz et al., 2014
Cellulose nanocrystals	*Dunaliella tertiolecta*	–	Mondal et al., 2021
Cellulose	*Cladophora* sp.	≤ 95	Ross et al., 2021
Starch	*Tetraselmis subcordiformis*	54.3	Yao et al., 2012
	Chlorella sorokiniana	27	Maia et al., 2020
Polyhydroxybutyrate (PHB)	*Synechococcus* sp. MA19	55 (dry weight)	Nishioka et al., 2001
Polyhydroxybutyrate-co-hydroxyvalerate (PHB-cPV)	*Oscillatoria okeni* TISTR	42 (dry weight)	Madadi et al., 2021; Roja et al., 2019
Polyhydroxyalkanoate (PHA)	*Scenedesmus* sp.	29.9	Garcia et al., 2021
Polyurethane	Defatted *Chlorella* biomass	–	Saha et al., 2021

9.4.3 Proteins

Microalgae have high protein contents, which can be extracted by various methods. On being extracted, these high-molecular-weight biomolecules can be used as biopolymers with application as packaging material. Polymerization or cross-linking using inter- or intra-molecular bonds between polypeptide chains is the main process for forming protein-based biopolymers. However, as both hydrophobic and hydrophilic interactions occur in protein-based biopolymers, the flexibility of the product is often reduced. To solve this issue, thermomechanical methods or the addition of plasticizers like glycerol become essential (Madadi et al., 2021).

9.4.4 Lipid-Based Polymers and Precursors

Microalgae can accumulate significant amounts of lipids mainly as triglycerides. The length of fatty acid side chains and the level of saturation are important parameters to be considered when deriving lipid-based biopolymers. Algal lipids can be extracted and used directly, or they can be blended with other polymers as plasticizers or hydrophobic agents to prepare water-resistant biocomposites. The derived monomeric forms of these lipids can be chemically converted into polymers like polyurethane (PU). The PU formation involves polymerization by the urethane linkage formed by the reaction between a polyol and an isocyanate to provide the necessary softness to the biopolymer. However, the challenge lies in isocyanate production, which is chemically derived from toxic phosgene. Residual fatty acids from algae can be a good source of diisocyanates, which can be used to produce chemically nontoxic polyurethane polymers (Barla et al., 2023). Microalgal biomass can also be processed or added to other bio-based or petroleum-based polymers to create biocomposites or bioplastics like polylactic acid (PLA) or polybutylene succinate (PBS), where the carbon feedstock for production of the biocomposite is provided by microalgae. PLA is a third-generation, sustainable bioplastic production concept where the precursor lactic acid is not obtained from food-crop feedstocks like corn and sugarcane and utilizes non-arable resource options like microalgae (Ibrahim et al., 2023). Such non-arable forms are advantageous as they consume CO_2 and induce carbon neutrality as well as can be grown in wastewater for better wastewater management. PBS is obtained from the polymerization of succinic acid (SA) and 1,4-butanediol (BDO), where fermentable sugars like glucose and mannitol from algae are hydrolyzed and fermented. On obtaining SA and BDO, they are purified and polymerized by esterification and polycondensation to obtain high-molecular-weight PBS as a biocomposite material with high mechanical strength (Madadi et al., 2021).

9.5 COST-EFFECTIVE CULTIVATION

Commercialization of microalgae-based biopolymers will be successful only when the productivity and harvesting of microalgae biomass are cost-effective. Thus, cost-effective cultivation of microalgae is central to the success of microalgae-derived biopolymers. Other than that, as microalgae is primarily aquatic, drying of biomass after harvesting is an important criterion for the success of microalgae-based biopolymers (Mondal et al., 2024a). Thus, the strategies to make cost-effective cultivation mainly involve the reduction in the cost of input resources like nutrients, water, CO_2, and energy along with the selection of the right microalgal cultivation system.

9.5.1 Low-Cost Cultivation System

9.5.1.1 Open Ponds

These are mainly natural lentic systems, which are economical options as the construction, maintenance, and operation costs are minimal, which makes them suitable for scaling up of microalgal biomass. However, they are highly vulnerable to contamination and environmental fluctuations, thereby making the production of unialgal biomass a major challenge. Moreover, optimization of light and temperature intensity is essential to get optimum growth of the target microalgae.

9.5.1.2 Low-Cost Photobioreactors

Simple containers like plastic bag PBRs can be a good option for low-cost biomass production although upscaling of the same is difficult in such containers.

9.5.1.3 Photobioreactors

These are specialized closed systems which can be tubular and are of three basic types: bubble column, flat-panel and airlift reactors. In bubble column and airlift reactors, air is pumped through the base of the reactor in a vertical manner thereby creating an upward current that allows mass transfer and mixing of biomass (Duan & Shi, 2014). In flat-panel bioreactors, there is a membrane that attaches the biomass and the media is stirred by aeration or agitation of the whole system.

9.5.2 Use of Wastewater as a Source of Nutrients and Growth Medium

A major challenge for microalgal growth is the selection of appropriate medium for microalgal cultivation under in vitro conditions. The use of specific nutrients as recommended for culture media and pure CO_2 as carbon source is not only expensive but requires a high level of knowhow and efficiency. The use of wastewater from agriculture runoffs, sewage, or industrial effluents as a natural resource of nitrogen and phosphorus is a highly cost-effective option for microalgal cultivation. This utilization of wastewater is a mutually exclusive option, which not only allows microalgal cultivation but bioremediates the wastewater for other applications also. Thus, coupling of biopolymer and biofuel production in biorefineries has been proposed in recent times (Kumar et al., 2010). Such an innovative and sustainable approach not only bioremediates wastewater but also drawdown GHGs, leading toward carbon neutrality.

9.5.3 Flue Gas and Other Industrial Emissions as CO_2 Source

Being photosynthetic organisms, microalgae require CO_2 as a natural source of carbon for carbon fixation. Under *in vitro* conditions, CO_2 is often needed in pure form to cultivate, thereby making the process expensive. Carbon emissions from industrial and anthropogenic sources like flue gas and other greenhouse gases (GHGs) have caused havoc on the climate of the Earth. This has led to phenomena like global warming, ocean acidification as both the terrestrial and aquatic systems work as sinks of such emissions. As microalgae have sophisticated carbon capture mechanisms, they can easily utilize flue gas and other emissions for their benefit during photosynthesis. This not only helps in enhancing the carbon content of microalgae but reduces the carbon footprints on the Earth as well. As cost-effective sources of other nutrients like nitrogen, organic nitrogen sources of animal origin have been reported in mass culture of microalgae, especially cyanobacteria (Kumar et al., 2010).

9.5.4 Integration and Optimization of the Processes

The biorefinery concept is central to process optimization, where value-added products (VAPs) like astaxanthin, carotenoids, proteins, pharmaceutical, and nutraceutical products, as well as bulk products like biofuels and biopolymers are produced through mass cultivation of microalgae. The use of low-cost resources and methods as mentioned above, as well as production of VAPs, makes it a more economical and commercially viable option. Although extraction and purification techniques can be demanding, the low cost of resource generation and VAP production further underlines the cost-effective nature of microalgae. Heterotrophic mode of microalgae cultivation in the absence of light with inexpensive organic nitrogen sources may yield higher biomass, which may also reduce the cost of PBRs and media requirements (Devadas et al., 2021; Mal et al., 2022).

9.5.5 Harvesting and Dewatering of the Biomass

This is one of the most challenging aspects of biomass availability for biopolymer extraction. Due to the small size and low concentration of microalgae, traditional methods do not yield enough biomass. To overcome this issue, methods like autoflocculation, application of low-cost natural flocculants, and gravity settlements are often implemented (Table 9.2). Microalgal cells are mostly negatively charged on the outside, so they repel each other in solution. The addition of flocculants (e.g., alum and ferric chloride) neutralizes this charge, thereby allowing cell aggregation that settles easily and separates out from the liquid medium. Another nonchemical method of flocculation is electrolytic flocculation, where the application of electrical current in the algal cultures settles the algal cells to the bottom (Uduman et al., 2010; Kim et al., 2012). Autoflocculation can be induced by increasing the culture pH due to higher CO_2 consumption or nutrient depletion. Another contrasting method concerning the process of floatation is where air bubbles are used to float out algal cells in suspension, which can then be segregated and separated as liquid sludge (Teixeira & Rosa, 2006).

Once flocculated, biomass is harvested and dewatered to enhance the solid concentrations of microalgae to 5–25% from their initial culture concentrations (0.02–0.06%). The harvesting process is separated into two phases, where there is an initial bulk harvesting (pre-concentration) phase in which biomass is concentrated to 2–7% total suspended solids (TSS), followed by dewatering. The dewatering process can be mechanical, chemical, or biological. Centrifugation is one of the best-suited methods with a dewatering efficiency of more than 90%, although it is an energy-intensive process. Moreover, a specific instrumentation facility is needed for this process. Filtration through porous membrane (PVC, PVDF, PES, PAN, PTEE, and PES-PVP) is also a well-suited method where water flows through and biomass is retained (Baerdemaeker et al., 2013; Drexler & Yeh, 2014). Microfiltration is a more efficient method where specialized membranes are used for filtration. Recovery efficiency (RE) and concentration factors (CF) are central to the dewatering process.

RE = mass of cells recovered/mass of cells in initial culture
CF = concentration of algae in the final product/initial concentration of algae in the culture

More modern techniques of dewatering include electrocoagulation or magnetic separation. An electrical field is applied at an anode (e.g., Al or Fe), where metal ions are released that act as flocculants to destabilize and aggregate microalgal cells. This is

TABLE 9.2
Different Methods of Harvesting and Dewatering of Microalgal Biomass with Details of Advantages and Disadvantages of Each Method

Method	Principle	Advantages	Disadvantages
	Chemical and Biological Methods (Primary Harvesting)		
Flocculation	The cells' surface negative charge is neutralized to form aggregates of small, unstable cells that settle or float.	Highly efficient and scalable, cost-effective.	High chances of contamination, reuse, or product purity may be compromised.
Chemical flocculation	Addition of inorganic (e.g., alum and ferric chloride) or organic (e.g., polymers) as flocculants	High recovery efficiency	May not be cost-effective, possibility of chemical contamination
Autoflocculation	Enhanced pH induces spontaneous aggregation of microalgal cells	Economic and cost-effective, environmentally friendly	Selective for specific species of microalgae, high pH may induce cellular damage
Bioflocculation	Coculturing with other microbial forms (e.g., bacteria), where biopolymers secreted from microbes induce flocculation	Cost-effective, nontoxic, energy efficient	Slow process, possibility of contamination from cocultured microbes
	Mechanical Methods (Dewatering/Secondary Harvesting)		
Centrifugation	Use of centrifugal force to induce sedimentation	Fast, reliable, and high efficiency in concentrating biomass	High energy consumption can be expensive
Floatation	Use of rising air or gas bubbles for the floating of microalgal cells to the surface for skimming	Effective, scalable, and economic	High operation cost, specific equipment is necessary
Filtration	Separation of biomass by passing through a permeable membrane or filters	Effective and efficient, low chemical use.	Susceptible to membrane fouling and costly due to advanced membrane systems.
Gravity sedimentation	Separation by gravity	Cost-effective	Low efficiency

Source: Shen et al. (2009), Acien-Fernandez et al. (2012), and Saratale et al. (2022).

an environmentally favorable method, which can be cost-effective based on the cost of electricity. Magnetic separation is a more advanced, sustainable option, where magnetic nanoparticles are attached to microalgal cells that separate the cells in a magnetic field. Although highly efficient, the cost of magnetic nanoparticles and their removal and reuse may not be cost-effective (Sharma et al., 2014).

9.5.6 Drying of Biomass

In the final phase, biomass is dried to get a TSS concentration of 90–95%, which is the most energy-intensive process. The biomass can be spray-dried rapidly by putting the concentrated slurry in a hot air stream, which may cause deterioration of biomass quality due to loss of heat-sensitive molecules. It is a cost-intensive process as well as a high-energy-requiring process. Solar energy drying is often preferred as a low-cost alternative, although the rate of drying is low and requires a large area (Saratale et al., 2022). The other more advanced drying options are convective drying and freeze-drying. These are contrasting processes, where freeze-drying involves sublimation of frozen water in vacuum, whereas convective drying involves exposure of the biomass to hot air at controlled temperatures.

Thus, it is our opinion that a combination of factors is to be considered for the processing of microalgae as a resource for biopolymer production. The choice of techniques to be used will depend on the algal species, cell size, and desired quantity and quality of the product. If rapid biomass production is the primary objective, then traditional methods are best suited as they are most cost-effective. However, if both quality and quantity are desired along with an eye for VAPs, then chemical or mechanical as well as advanced methods should be included for harvesting and dewatering of biomass to get the optimum outputs (Acien-Fernandez et al., 2012, 2017).

9.6 BIOPOLYMER PRODUCTION AND SCALE-UP

The capacity of microalgae to synthesize biopolymers such as PHAs, polysaccharides, and proteins is rooted in their highly adaptable metabolic networks and physiological responses to environmental stimuli (Koller & Mukherjee, 2022). Under conditions of nutrient limitation, particularly when nitrogen, phosphorus, or sulfur availability decreases, microalgal cells redirect the carbon fixed during photosynthesis toward the formation of storage compounds. This metabolic shift serves as a protective mechanism to maintain cellular redox equilibrium and energy balance, leading to the intracellular accumulation of polymer reserves. Enzymes including β-ketothiolase, acetoacetyl-CoA reductase, and PHA synthase play pivotal catalytic roles in the transformation of acetyl-CoA into PHB and other related polymers (Ramanna et al., 2017). Similarly, the biosynthesis of extracellular polysaccharides is often triggered by stress-induced secretion processes that help shield algal cells from oxidative or osmotic damage. Enhancement of polymer yield is frequently achieved through controlled modulation of environmental conditions such as nutrient concentration, illumination, and carbon supplementation (Cheng et al., 2015). Advances in metabolic and molecular biology have enabled targeted manipulation of polymer biosynthetic pathways. The overexpression of key enzymes and the redirection of carbon flux toward polymer accumulation have been explored to improve both productivity and polymer quality. The integration of transcriptomic, proteomic, and metabolomic approaches has provided valuable insights into these complex regulatory processes and facilitated the selection of superior algal strains for industrial development.

The transition from laboratory research to large-scale production presents significant engineering and economic challenges. Cultivation systems commonly employed include open raceway ponds and closed PBRs, each offering distinct advantages in terms of cost, contamination control, and operational efficiency. Hybrid configurations that combine the scalability of open systems with the precision of enclosed reactors are increasingly being adopted to optimize performance. Continuous and semicontinuous cultivation strategies, supported by automated control of light, nutrient delivery, and carbon dioxide supply, contribute to maintaining stable biomass growth and consistent polymer accumulation (Devadas et al., 2021).

Through the convergence of metabolic engineering, process optimization, and system integration, microalgae are poised to become a scalable and sustainable platform for the commercial production of high-value biopolymers. There are some major optimization techniques that underpin efficient bioplastic production from microalgae (Figure 9.4). These include metabolic and genetic engineering to enhance biosynthetic pathways, the use of PBRs for controlled cultivation, and the employment of microalgal consortia to improve productivity through synergistic

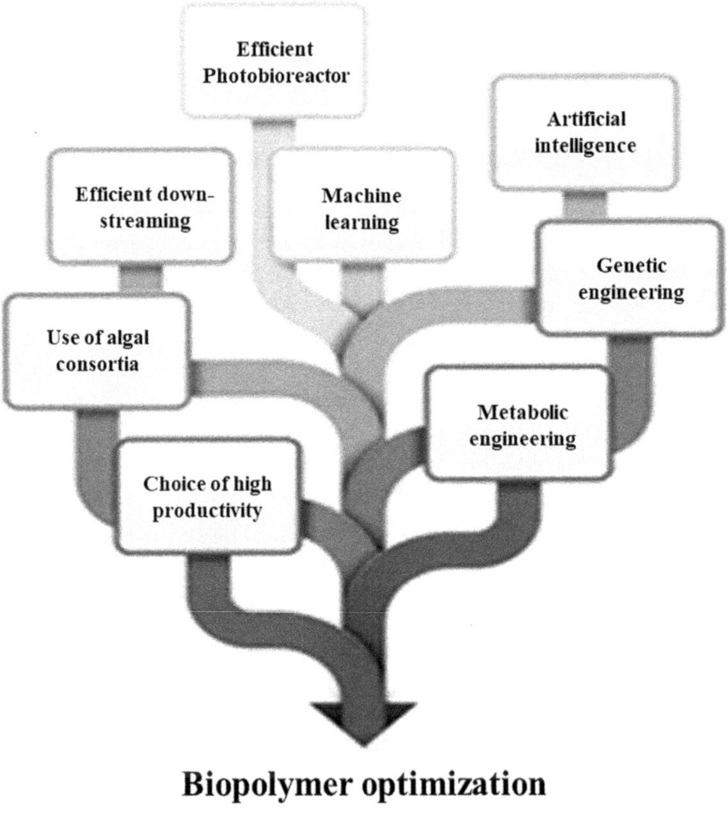

FIGURE 9.4 Different processes of algae-based biopolymer production optimization.

interactions. Statistical tools such as response surface methodology aid in refining process parameters, while machine learning and artificial intelligence offer predictive and real-time control of complex bioprocesses. Collectively, these interdisciplinary strategies provide an integrated framework for optimizing yield, quality, and scalability in microalgae-based bioplastic production. Equally important are downstream operations such as harvesting, extraction, and polymer processing, which must be integrated into a unified biorefinery framework to reduce production costs and environmental impact. Future efforts should focus on improving light utilization efficiency, minimizing energy consumption, and developing closed-loop systems that allow recycling of water and nutrients.

9.7 TECHNO-ECONOMIC ANALYSIS AND LIFE-CYCLE ASSESSMENT

In literature, the techno-economic analysis of the biopolymers obtained from microalgae considers different production methods, as well as different strains, methods of cultivation, and so on. Interest in microalgae as a source for biopolymers is connected with the following advantages compared to traditional crops: fast growth rates, high photosynthetic efficiency, ability to isolate CO_2 (Pignolet et al., 2013). In common sense, considering different crucial factors affecting the economic feasibility, it was concluded that the capital expenditure and operating costs are distributed throughout the bioplastic production process. Microalgae cultivation emerges as a significant contributor to operating costs, emphasizing its pivotal role. Despite the great potential of the microalgae as a source of valuable biopolymers, the literature contains limited exact examples. For instance, Alavianghavanini et al. (2024) analyzed the microalgal protein production features in terms of techno-economic. The authors consider microalgae cultivated on anaerobically digested abattoir effluent. Their results declare that the protein concentrate and protein isolate from anaerobic digestion effluents have a minimum total cost of 28 USD/kg and 33 USD/kg products, respectively, without taking into account such factors as the absence of a centralized plant to treat effluents and transportational expenses. Techno-economic analysis of mixotrophic algal-protein production (only for cell cultivation and fermentative process step) showed that the total cost of protein production from *S. platensis* (using supplemented media) comprised 1,984.87 USD/year at culture medium capacity 5,657.14 L/year (Ferreira et al., 2021). But the model considered for analysis is only partially complete; it is probably underestimating the real cost of the microalgae production.

LCA is effectively used to estimate the item's economic and environmental performance. Besides, LCA is a valuable tool in assessing the impact on the final product by different environmental factors (greenhouse gas emission, energy consumption, etc.) (Ali et al., 2023). Furthermore, the microalgae have a high surface area-to-volume ratio that also facilitates efficient gas exchange with the surrounding environment, enhancing CO_2 uptake. Righi et al. (2017) assessed the environmental impacts related to the new protocol PHA extraction based on dimethyl carbonate. It was demonstrated that dried biomass application followed by solvent evaporation is beneficial and promising in terms of environmental sustainability performance. Beckstrom (2019) discovered that in the case of bioplastic feedstock, impact assessment led to a decrease in greenhouse gas emission ranging between 67 and 116% compared to a petroleum-based plastic feedstock. Bussa et al. (2019) performed a comparative study for PLA production from different sources (microalgae and plant-derived). It was indicated that there were two main factors determining the pronounced impact on the environment: (1) the optimal growing conditions for the microalgae, (2) the effects of the source on the properties of PLA and the resulting need for additives. If these factors are solved, it can be concluded that microalgae can considerably improve the environmental impact in terms of terrestrial ecotoxicity and land use. In contrast to it, Draaisma et al. (2013) stated that despite improvement in the solution of territorial problems via using microalgae biopolymers for solving problems with food items, there is a disadvantage as well, namely, the consumption of freshwater.

9.8 APPLICATIONS OF MICROALGAE-BASED BIOPOLYMERS

Taking into account all valuable characteristics known about microalgae as a source of biopolymers, there are numerous application fields for the materials based on the latter. The need for exploration of microalgal biopolymers is dictated by the increasing interest in the creation of new bioplastics for the future sustainable world. One of the most widespread microalgae-based biopolymers is starch. This polysaccharide can be used in medicine, packaging, food industries, etc. Di Caprio et al. (2024) used a non-axenic strain of *Tetradesmus obliquus* to isolate starch, then mixed it with chitosan to obtain new systems with another polysaccharide, chitosan. The film samples were obtained by a solvent casting technique, and their physicomechanical properties were assessed. The proposed systems can be used for the production of biodegradable packaging materials with improved toughness. But it is well known that, for example, *Chlorella* sp. may consist of 2–11% dw of starch (Cheng et al., 2015). The fraction of starch in the majority of microalgae strains is small. This means that there is a serious limitation for the industrial application of microalgal starch. Despite attempts to increase the starch content up to 50% for the *Chlamydomonas reinhardtii* strain, at pilot scale, this factor did not exceed 25% (Mathiot et al., 2019). Besides, the polysaccharide biomaterials (alginate, cellulose, and agar) from microalgae can be applied as anionic organic linkers in metal–organic framework crystalline structures with the aim of removing the contaminants from wastewater by an eco-friendly and sustainable approach. Ulvan is a sulfated, water-soluble polysaccharide obtained from *Ulva* spp. This polysaccharide is applied in a lot of application fields, namely, for medical materials, drug delivery, regenerative medicine, and tissue engineering. So, only partially purified ulvan from *U. lactuca* demonstrated potential antimicrobial activity against some widespread pathogenic microbes (Ibrahim et al., 2022). Besides, the ulvan-chitosan hydrogel produced had significantly increased antimicrobial activity. Manikandan and Lens (2022) illustrated that cross-linking of ulvan with citric acid allowed one to obtain a hydrogel, while the hydrogels at different content of sodium carboxymethyl cellulose with ulvan led to the formation of systems used in health care and agriculture. Besides, ulvans can be interesting not only for medical but also cosmetic applications, thanks to their mechanical, chemical, and biological characteristics (Fournière et al., 2021). Not only can chitosan be mixed with ulvans to produce potential bioplastics but also cellulose (Gomaa et al., 2022). It was demonstrated that cellulose and ulvan extracted from microalgae with polyvinyl alcohol and glycerol as plasticizers were mixed to develop edible films with natural antioxidant properties. The films obtained had increased water solubility and water vapor permeability, along with decreased oxygen permeability. Not only ulvan but also carrageenan and fucoidan extracted from macroalgae have a wide range of applications due to their nutritional value and therapeutic potential. So, Wang et al. (2020) demonstrated that sulfated polysaccharides of *Hizikia fusiformis* exhibited anti-inflammatory and anti-melanogenesis effects. According to the results, the polysaccharides extracted substantially reduced the production of inflammatory molecules by suppressing the expression of iNOS and COX-2 in LPS-stimulated RAW 264.7 macrophages.

PHB is one of the most promising biopolymers in a lot of fields, but its production is expensive and resources are limited. These disadvantages stimulate it to look for new raw materials for its production. That is why the microalgae-based PHB is attracting more and more interest among researchers worldwide. Under optimized conditions, Chlorella sp. accumulated 80% of PHB. Thanks to its biodegradability and biocompatibility, PHB can be used in industrial, medicinal, agricultural, and other fields. The PHB films with predictable degradation under the action of soil microorganisms proposed to be used as packages for urea fertilizers and insecticides, which, when used in the agricultural land, can release the components in a controlled mode (Aguilar & San Román, 2014). For PHB, one of the major application fields is medicine, namely, PHB is successfully used as drug carriers, scaffolds, bone plates, and surgical sutures. Analysis of the application of PHB from marine sources in tissue engineering, as well in other fields, allowed one to understand the prospectivity in developing the industrial production process of microalgae PHB (Kavitha et al., 2018).

One more valuable component obtained from microalgae is proteins, which can be used in either of these forms: the whole cell, concentrate, isolate, or hydrolysate. Developing a new technique in the production of microalgal proteins, it was shown that the resulting crude protein extract displayed excellent surface activity and gelation behavior (Suarez Garcia, et al., 2018). Thus, protein concentrate and isolate can be used in food applications for their functional properties, such as emulsifying and gelation capabilities, while hydrolysate can be a component of functional food and beverages; purified hydrolysate can yield bioactive peptides with pharmaceutical properties. It is well known that some strains are extremely rich in protein; for example, *Chlorella* and *Arthrospira* (*Spirulina*) contain more than 50% protein that allows one to use these microalgae in food or nutraceutical applications (Andrade et al., 2018). Teuling et al. (2017) proposed a technique allowing one to produce RuBisCo, N-methylpyrrolidone, and N-vinylpyrrolidone from microalgae protein, valuable nitrogen-containing chemicals. However, it is known that proteins are brittle. That is why they can be used in combination with plasticizers. It was proposed to use hot-press processing to convert the microalgae into protein-based bioplastics. Unique protocols for the extraction of proteins from microalgae, such as *Spirulina* (*Arthrospira platensis*) and *Chlorella* (*Chlorella vulgaris*), via an ultrasound-assisted extraction technique offer a promising solution for large-scale, safe microalgae processing and facilitate the production of bioavailable peptides from microalgae species (Guadalupi et al., 2025). By the way, the conversion of microalgal proteins to high-value industrial products (such as polyurethane) by a relatively simple process greatly improves the value of proteins extracted from algae (Kumar et al., 2014).

9.9 CHALLENGES AND FUTURE DIRECTIONS

Although microalgae-based biopolymer production offers great potential, several challenges limit its commercial viability. High costs associated with large-scale cultivation, harvesting, and drying remain the primary bottlenecks, as maintaining optimal light, nutrients, and temperature conditions is energy-intensive. The low cell density of algal cultures also makes harvesting and downstream processing costly, often contributing a major share of total production expenses (Madadi et al., 2021). Variability in polymer yield and composition among algal strains further complicates process standardization. Genetic engineering and metabolic optimization have shown promise in enhancing polymer productivity, yet these approaches are still developing. Similarly, solvent-based extraction and purification methods must be refined to reduce toxicity, solvent loss, and environmental impact.

Future progress depends on integrating biopolymer production with wastewater treatment, CO_2 capture, or other resource-recovery systems to improve overall sustainability. Innovations in PBR design, continuous processing, and green extraction technologies can enhance efficiency and reduce costs. Policy incentives, carbon credit frameworks, and public–private collaborations will also be crucial to scale these technologies. As interdisciplinary research in synthetic biology, materials science, and process engineering advances, microalgae-derived biopolymers are poised to become a viable component of the global bioeconomy.

9.10 CONCLUSION

In the quest for sustainable materials, microalgae stand out as nature's miniature green factories—capable of transforming sunlight, carbon dioxide, and nutrients into high-value biopolymers. These photosynthetic powerhouses offer a renewable route to replace conventional petrochemical plastics while addressing global challenges of pollution and resource depletion. Their remarkable adaptability, rapid growth, and ability to utilize wastewater and non-arable lands position them at the heart of a truly circular bioeconomy.

The diversity of algal-derived biopolymers, including PHAs, proteins, and polysaccharides, provides a broad platform for developing biodegradable and biocompatible materials. Advances in cultivation, harvesting, and downstream processing, supported by modern techniques such as extrusion, molding, and solvent casting, have steadily improved process efficiency. However, large-scale commercialization remains constrained by production costs and technological complexity. Future progress will depend on integrating algal polymer production with waste valorization, CO_2 mitigation, and biorefinery approaches to enhance both sustainability and profitability. Continued innovations in strain engineering, green extraction, and reactor design, coupled with supportive policies and industrial collaboration, can accelerate the transition from experimental systems to market-ready solutions. As research converges across biotechnology, materials science, and environmental engineering, microalgae-derived biopolymers hold the potential to reshape the landscape of sustainable materials and lead the next wave of eco-innovation.

CONFLICT OF INTEREST

Authors declare that there is no conflict of interests.

REFERENCES

Acien Fernandez, F. G., Fernandez-Sevilla, J. M., & Molina Grima, E. (2017). Microalgae: The basis of mankind sustainability. In S. K. Khan (Ed.), *Algae: The green gold*. IntechOpen. https://doi.org/10.5772/67930

Acien-Fernandez, F. G., Fernandez, J. M., Magan, J. J., & Molna, E. (2012). Production cost of a real microalgae production plant and strategies to reduce it. *Biotechnology Advances*, *30*(6), 1344–1353. https://doi.org/10.1016/j.biotechadv.2012.02.005

Adetunji, A. I., & Erasmus, M. (2024). Green synthesis of bioplastics from microalgae: A state-of-the-art review. *Polymers*, *16*(10), 1322. https://doi.org/10.3390/polym16101322

Aguilar, M. R., & San Román, J. (2014). *Smart Polymers and Their Applications*. Woodhead Publishing.

Alavianghavanini, A., Moheimani, N. R., & Bahri, P. A. (2024). Producing protein-based products from microalgae cultivated on anaerobically digested abattoir effluent: Process integration and techno-economic analysis. *Bioresource Technology Reports*, *25*, 101734. https://doi.org/10.1016/j.biteb.2023.101734

Ali, S. S., Abdelkarim, E. A., Elsamahy, T., Al-Tohamy, R., Li, F., Kornaros, M., … Sun, J. (2023). Bioplastic production in terms of life cycle assessment: A state-of-the-art review. *Environmental Science and Ecotechnology*, *15*, 100254. https://doi.org/10.1016/j.ese.2023.100254

Andrade, L. M., Andrade, C. J., Dias, M., Nascimento, C. A. O., & Mendes, M. A. (2018). Chlorella and Spirulina microalgae as sources of functional foods, nutraceuticals, and food supplements: An overview. *MOJ Food Processing & Technology*, *6*(1), 45–58. https://doi.org/10.15406/mojfpt.2018.06.00144

Antony, A. R., & Ramanan, R. (2021). Genetic engineering of microalgae for the production of high-value metabolites: Status and prospects. In Ajam Shekh, Peer Schenk, and R. Sarada (eds.), *Microalgal Biotechnology: Recent Advances, Market Potential, and Sustainability* (pp. 209–253). The Royal Society of Chemistry. https://doi.org/10.1039/9781839162473-00209

Baerdemaeker, T., Lemmens, B., Dotremont, C., Fret, J., Roef, L., Goiris, K., & Diels, L. (2013). Benchmark study on algae harvesting with backwashable submerged flat panel membranes. *Bioresource Technology*, *129*, 582–591. https://doi.org/10.1016/j.biortech.2012.10.153

Barla, F., Lega, M., & Kumar, S. (2023). Process integration opportunities applied to microalgae biomass production. In Eduardo Jacob-Lopes, Rosangela Rodrigues Dias, and Leila Queiroz Zepka (eds.), *Microalgae-Based Systems: Process Integration and Process Intensification Approaches* (p. 271). De Gruyter.

Beckstrom, B. D. (2019). Bioplastic production from microalgae with fuel co-products: A techno-economic and life-cycle assessment [Master's thesis, Colorado State University].

Blockx, J., Verfaillie, A., Eyley, S., Deschaume, O., Bartic, C., Muylaert, K., & Thielemans, W. (2019). Cationic cellulose nanocrystals for flocculation of microalgae: Effect of degree of substitution and crystallinity. *ACS Applied Nano Materials*, *2*(6), 3394–3403. https://doi.org/10.1039/D1NA00102G

Bussa, M., Eisen, A., Zollfrank, C., & Röder, H. (2019). Life cycle assessment of microalgae products: State of the art and their potential for the production of polylactic acid. *Journal of Cleaner Production*, *213*, 1299–1312. https://doi.org/10.1016/j.jclepro.2018.12.048

Cheng, Y.-S., Labavitch, J. M., & VanderGheynst, J. S. (2015). Elevated CO_2 concentration impacts cell wall polysaccharide composition of green microalgae of the genus *Chlorella*. *Letters in Applied Microbiology*, *60*(1), 1–7. https://doi.org/10.1111/lam.12320

Devadas, V. V., Khoo, K. S., Chia, W. Y., Chew, K. W., Munawaroh, H. S. H., Lam, M. K., Lim, J. W., Ho, Y. C., Lee, K. T., & Show, P. L. (2021). Algae biopolymer towards sustainable circular economy. *Bioresource Technology*, *325*, 124702. https://doi.org/10.1016/j.biortech.2021.124702

Dey, I., Mondal, A., & Pal, R. (2025a). Algae-based systems for removal of emerging pollutants from sewage sludge. In Satarupa Dey and Maulin P. Shah (eds.), *Biotechnological Removal of Emerging Pollutants from Wastewater Systems* (pp. 109–133). Springer Nature Singapore. https://doi.org/10.1007/978-981-96-3945-8_5

Dey, I., Mondal, A., Satpati, G. G., & Pal, R. (2025b). Tannery wastewater remediation potential of cyanobacteria and algae with nutrient recovery, ecological monitoring and biomass valorization: A circular economy. *Next Chemical Engineering*, *1*, 100009. https://doi.org/10.1016/j.nxcen.2025.100009

Di Caprio, F., Pedram, N., Brugnoli, B., Francolini, I., Altimari, P., & Pagnanelli, F. (2024). Exploring different processes for starch extraction from microalgae and synthesis of starch-chitosan plastic films. *Bioresource Technology*, *413*, 131516. https://doi.org/10.1016/j.biortech.2024.131516

Draaisma, R. B., Wijffels, R. H., Slegers, P. M. E., Brentner, L. B., Roy, A., & Barbosa, M. J. (2013). Food commodities from microalgae. *Current Opinion in Biotechnology*, *24*, 169–177. https://doi.org/10.1016/j.copbio.2012.09.012

Drexler, I. L. C., & Yeh, D. H. (2014). Membrane applications for microalgae cultivation and harvesting: A review. *Reviews in Environmental Science and Biotechnology*, *13*(4), 487–504. https://doi.org/10.1007/s11157-014-9350-6

Duan, Y., & Shi, F. (2014). Bioreactor design for algal growth as a sustainable energy source. In Fan Shi (ed.), *Reactor and Process Design in Sustainable Energy Technology* (pp. 27–60). Elsevier Inc. https://doi.org/10.1016/B978-0-444-59566-9.00002-8

Ferreira, P. S., Santos, G. A., de Souza, I. T., da Silva Neiro, S. M., Cardoso, V. L., & Batista, F. R. X. (2021). Mixotrophic algal-protein production and techno-economic analysis based on medium formulation. *Research, Society and Development*, *10*(4), e54110414186.

Fournière, M., Bedoux, G., Lebonvallet, N., Leschiera, R., Le Goff-Pain, C., Bourgougnon, N., & Latire, T. (2021). Poly- and oligosaccharide *Ulva* sp. fractions from enzyme-assisted extraction modulate the metabolism of extracellular matrix in human skin fibroblasts: Potential in anti-aging dermo-cosmetic applications. *Marine Drugs*, *19*. https://doi.org/10.3390/md19030156

García, G., Sosa-Hernandez, J. E., Rodas-Zuluaga, L. I., Castillo-Zacarías, C., Iqbal, H., & Parra-Saldívar, R. (2021). Accumulation of PHA in the microalgae *Scenedesmus* sp. under nutrient-deficient conditions. *Polymers*, *13*(1), 131. https://doi.org/10.3390/polym13010131

Gomaa, M., Al-Badaani, A. A., Awatief, A. F., Hifney, F., & Adam, M. S. (2022). Utilization of cellulose and ulvan from the green seaweed *Ulva lactuca* in the development of composite edible films with natural antioxidant properties. *Journal of Applied Phycology*, *34*, 2615–2626. https://doi.org/10.1007/s10811-022-02786-z

Guadalupi, L. S., Bianco, M., Cataldi, T. R. I., Ravnsborg, T., Jensen, O. N., & Calvano, C. D. (2025). Ultrasound-assisted protein extraction for deep proteome analysis of *Spirulina* and *Chlorella* microalgae. *LWT – Food Science and Technology*, *222*, 117647. https://doi.org/10.1016/j.lwt.2025.117647

Halley, P. J., & Dorgan, J. R. (2011). Next-generation biopolymers: Advanced functionality and improved sustainability. *MRS Bulletin*, *36*(9), 687–691. https://doi.org/10.1557/mrs.2011.180

Ibrahim, M. I. A., Amer, M. S., Ibrahim, H. A. H., & Zaghloul, E. H. (2022). Considerable production of ulvan from *Ulva lactuca* with special emphasis on its antimicrobial and anti-fouling properties. *Applied Biochemistry and Biotechnology*, *194*, 3097–3118. https://doi.org/10.1007/s12010-022-03867-y

Ibrahim, N. N., Ahmad, I., Abdullah, N., Koji, I., Mohamad, S. E., & Akhir, F. N. B. M. (2023). Biocomposites from microalgae. In *Composites from the Aquatic Environment* (pp. 99–138). Springer Nature Singapore. https://doi.org/10.1007/978-981-19-5327-9_5

Kavitha, G., Rengasamy, R., & Inbakandan, D. (2018). Polyhydroxybutyrate production from marine source and its application. *International Journal of Biological Macromolecules*, *111*, 102–108. https://doi.org/10.1016/j.ijbiomac.2017.12.155

Kim, J., Ryu, B. G., Kim, B. K., Han, J. I., & Yang, J. W. (2012). Continuous microalgae recovery using electrolysis with polarity exchange. *Bioresource Technology*, *111*, 268–275. https://doi.org/10.1016/j.biortech.2012.01.104

Koller, M., & Mukherjee, A. (2022). A new wave of industrialization of PHA biopolyesters. *Bioengineering*, *9*(2), 74. https://doi.org/10.3390/bioengineering9020074

Kumar, A., Ergas, S., Yuan, X., Sahu, A., Zhang, Q., Dewulf, J., Malcata, F. X., & van Langenhove, H. (2010). Enhanced CO_2 fixation and biofuel production via microalgae: Recent developments and future directions. *Trends in Biotechnology*, *28*(7), 371–380. https://doi.org/10.1016/j.tibtech.2010.04.004

Kumar, N., Banerjee, C., Negi, S., & Shukla, P. (2023). Microalgae harvesting techniques: Updates and recent technological interventions. *Critical Reviews in Biotechnology*, *43*(3), 342–368. https://doi.org/10.1080/07388551.2022.2031089

Kumar, S., Hablot, E., Moscoso, J. L. G., Obeid, W., Hatcher, P. G., DuQuette, B. M., et al. (2014). Polyurethanes preparation using proteins obtained from microalgae. *Journal of Materials Science*, *49*(22), 7824–7833. https://doi.org/10.1007/s10853-014-8493-8

Lee, R. E. (2008). *Phycology*. Cambridge University Press.

Madadi, R., Maljaee, H., Serafim, L. S., & Ventura, S. P. M. (2021). Microalgae as contributors to produce biopolymers. *Marine Drugs*, *19*(8), 466. https://doi.org/10.3390/md19080466

Maia, J. L., Cardoso, J. S., Mastrantonio, D. J., Bierhals, C. K., Moreira, J. B., Costa, J. A. V., & de Morais, M. G. (2020). Microalgae starch: A promising raw material for bioethanol production. *International Journal of Biological Macromolecules*, *165*, 2739–2749. https://doi.org/10.1016/j.ijbiomac.2020.10.159

Mal, N., Satpati, G., Raghunathan, S., & Davoodbasha, M. (2022). Current strategies on algae-based biopolymer production and scale-up. *Chemosphere*, *289*, 133178. https://doi.org/10.1016/j.chemosphere.2021.133178

Manikandan, N. A., & Lens, P. N. L. (2022). Green extraction and esterification of marine polysaccharide (ulvan) from green macroalgae *Ulva* sp. using citric acid for hydrogel preparation. *Journal of Cleaner Production*, *366*, 132952. https://doi.org/10.1016/J.JCLEPRO.2022.132952

Mathiot, C., Ponge, P., Gallard, B., Sassi, J. F., Delrue, F., & Le Moigne, N. (2019). Microalgae starch-based bioplastics: Screening of ten strains and plasticization of unfractionated microalgae by extrusion. *Carbohydrate Polymers*, *208*, 142–151. https://doi.org/10.1016/j.carbpol.2018.12.057

Mondal, A., Dey, I., Mukherjee, A., Ismail, A., Satpati, G. G., Banerjee, S., Paul, S., Paul, S., & Pal, R. (2024b). Spirulina biomass loaded with iron nanoparticles: A novel biofertilizer for the growth and enrichment of iron content in rice plants. *Biocatalysis and Agricultural Biotechnology*, *61*, 103387. https://doi.org/10.1016/j.bcab.2024.103387

Mondal, A., Mukherjee, A., & Pal, R. (2024a). Phycosynthesis of nanoiron particles and their applications: A review. *Biocatalysis and Agricultural Biotechnology*, *55*, 102986. https://doi.org/10.1016/j.bcab.2023.102986

Mondal, K., Sakurai, S., Okahisa, Y., Goud, V. V., & Katiyar, V. (2021). Effect of cellulose nanocrystals derived from *Dunaliella tertiolecta* marine green algae residue on crystallization behaviour of poly(lactic acid). *Carbohydrate Polymers*, *261*, 117881. https://doi.org/10.1016/j.carbpol.2021.117881

Nishioka, M., Nakai, K., Miyake, M., Asada, Y., & Taya, M. (2001). Production of poly-β-hydroxybutyrate by thermophilic cyanobacterium *Synechococcus* sp. MA19 under phosphate-limited conditions. *Biotechnology Letters*, *23*(14), 1095–1099. https://doi.org/10.1023/A:1010551614648

Pignolet, O., Jubeau, S., Vaca-Garcia, C., et al. (2013). Highly valuable microalgae: Biochemical and topological aspects. *Journal of Industrial Microbiology & Biotechnology*, *40*, 781–796. https://doi.org/10.1007/s10295-013-1281-7

Ramanna, L., Rawat, I., & Bux, F. (2017). Light enhancement strategies improve microalgal biomass productivity. *Renewable and Sustainable Energy Reviews*, *80*, 765–773. https://doi.org/10.1016/j.rser.2017.05.202

Rehm, B. H., & Steinbüchel, A. (1999). Biochemical and genetic analysis of PHA synthases and other proteins required for PHA synthesis. *International Journal of Biological Macromolecules*, *25*(1–3), 3–19. https://doi.org/10.1016/s0141-8130(99)00010-0

Righi, S., Baioli, F., Samorì, C., Galletti, P., Tagliavini, E., Stramigioli, C., … Fantke, P. (2017). A life cycle assessment of poly-hydroxybutyrate extraction from microbial biomass using dimethyl carbonate. *Journal of Cleaner Production*, *168*, 692–707. https://doi.org/10.1016/j.jclepro.2017.08.227

Roja, K., Ruben Sudhakar, D., Anto, S., & Mathimani, T. (2019). Extraction and characterization of polyhydroxyalkanoates from marine green alga and cyanobacteria. *Biocatalysis and Agricultural Biotechnology*, *22*, 101358. https://doi.org/10.1016/j.bcab.2019.101358

Ross, I. L., Shah, S., Hankamer, B., & Amiralian, N. (2021). Microalgal nanocellulose: Opportunities for a circular bioeconomy. *Trends in Plant Science*, *26*(9), 924–939. https://doi.org/10.1016/j.tplants.2021.05.004

Saha, P., Aloui, H., Yun, J. H., Kim, H. S., & Kim, B. S. (2021). Development of a novel composite film based on polyurethane and defatted *Chlorella* biomass: Physical and functional characterization. *Journal of Applied Polymer Science*, *138*(14), 50152. https://doi.org/10.1002/app.50152

Saratale, G. D., Cho, S. K., Shinde, S. K., Saratale, R. G., Kadam, A. A., Ghodake, G., Mulla, S. I., Kumar, M., Syed, A., Bahkali, A. H., Kim, D. S., & Shin, H. S. (2022). Microalgae cultivation strategies using cost-effective nutrient sources: Recent updates and progress towards biofuel production. *Bioresource Technology*, *361*, 127691.

Satpati, G. G., Devi, A., Kundu, D., Dikshit, P. K., Saravanabhupathy, S., Banerjee, R., Rajak, R. C., Kamli, M. R., Lee, S.-Y., Kim, J.-W., & Davoodvasha, M. (2024). Synthesis, delineation and technological advancements of algae biochar for sustainable remediation of the emerging pollutants from wastewater-a review. *Environmental Research*, 258, 119408. https://doi.org/10.1016/j.envres.2024.119408

Satpati, G. G., Gupta, S., Biswas, R. K., Choudhury, A. K., Kim, J.-W., & Davoodbasha, M. (2023). Microalgae mediated bioremediation of polycyclic aromatic hydrocarbons: Strategies, advancement and regulations. *Chemosphere*, 344, 140337. https://doi.org/10.1016/j.chemosphere.2023.140337

Satpati, G. G., & Pal, R. (2018). Microalgae-biomass to biodiesel: A review. *Journal of Algal Biomass Utilization*, *9*(4), 11–37.

Scholz, M. J., Weiss, T. L., Jinkerson, R. E., Jing, J., Roth, R., Goodenough, U., Posewitz, M. C., & Gerken, H. G. (2014). Ultrastructure and composition of the *Nannochloropsis gaditana* cell wall. *Eukaryotic Cell*, *13*(11), 1450–1464. https://doi.org/10.1128/EC.00183-14

Sharma, V., Sharma, S., & Sharda, A. (2014). A comprehensive overview on various methods of harvesting microalgae according to Indian perspective. *International Journal of Research in Science & Technology (IJRST)*, *1*(Special Issue 7), 313–317.

Shen, Y., Yuan, W., Pei, Z. J., Wu, Q., & Mao, E. (2009). Microalgae mass production methods. *Transactions of the ASABE*, *52*(4), 1275–1287. https://doi.org/10.13031/2013.27771

Suarez Garcia, E., van Leeuwen, J., Safi, C., Sijtsma, L., Eppink, M. H. M., Wijffels, R. H., et al. (2018). Selective and energy efficient extraction of functional proteins from microalgae for food applications. *Bioresource Technology*, *268*, 197–203. https://doi.org/10.1016/j.biortech.2018.07.131

Teixeira, M. R., & Rosa, M. J. (2006). Comparing dissolved air flotation and conventional sedimentation to remove cyanobacterial cells of *Microcystis aeruginosa*: Part I: The key operating conditions. *Separation and Purification Technology*, *52*(1), 84–94. https://doi.org/10.1016/j.seppur.2006.03.017

Teuling, E., Wierenga, P. A., Schrama, J. W., & Gruppen, H. (2017). Comparison of protein extracts from various unicellular green sources. *Journal of Agricultural and Food Chemistry*, *65*(36), 7989–8002. https://doi.org/10.1021/acs.jafc.7b01788

Uduman, N., Qi, Y., Danquah, M. K., Forde, G. M., & Hoadley, A. (2010). Dewatering of microalgal cultures: A major bottleneck to algae-based fuels. *Journal of Renewable Energy and Sustainable Energy*, *2*(1), 012701. https://doi.org/10.1063/1.3294480

Venkata Mohan, S., Hemalatha, M., Chakraborty, D., Chatterjee, S., Ranadheer, P., & Kona, R. (2020). Algal biorefinery models with self-sustainable closed loop approach: Trends and prospects for blue bioeconomy. *Bioresource Technology*, *295*, 122128. https://doi.org/10.1016/j.biortech.2019.122128

Wang, L., Oh, J. Y., Jayawardena, T. U., Jeon, Y. J., & Ryu, B. (2020). Anti-inflammatory and anti-melanogenesis activities of sulfated polysaccharides isolated from *Hizikia fusiforme*. *International Journal of Biological Macromolecules*, *142*, 545–550. https://doi.org/10.1016/j.ijbiomac.2019.09.128

Yao, C., Ai, J., Cao, X., Xue, S., & Zhang, W. (2012). Enhancing starch production of a marine green microalga *Tetraselmis subcordiformis* through nutrient limitation. *Bioresource Technology*, *118*, 438–444. https://doi.org/10.1016/j.biortech.2012.05.030

Zanchetta, E., Damergi, E., Patel, B., Borgmeyer, T., Pick, H., Pulgarin, A., & Ludwig, C. (2021). Algal cellulose production and potential use in plastics: Challenges and opportunities. *Algal Research*, *56*, 102288. https://doi.org/10.1016/j.algal.2021.102288

10 Microalgae for Sustainable Nutrition
From Functional Benefits to Consumer Adoption

Mariany Costa Deprá, Rosangela Rodrigues Dias, Adriane Terezinha Schneider, Richard Luan Silva Machado, Darissa Alves Dutra, and Eduardo Jacob-Lopes

10.1 INTRODUCTION

Food plays such an essential role in human existence that we rarely stop to reflect on its true purpose. They constitute an essential substrate for life, transcending the mere provision of energy and nutrients to assume complex and multifaceted functions in the social, cultural, and economic spheres (Gowder, 2024). In addition to nourishing, foods act as vectors of interaction, celebration, and strengthening of social bonds, as a structuring of health and individual and collective well-being, intrinsically reflecting socioeconomic dynamics and the management of natural resources in contemporary society (Andersen and Hyldig, 2015).

The centrality of food in human life has driven profound transformations in the ways it is produced throughout history. From the consolidation of agriculture—over 10,000 years ago—to the technological advances of the Industrial Revolution, humanity has expanded its capacity for sustenance and built social systems increasingly capable of intervening in global ecosystems (Funabashi, 2018). However, these achievements have been accompanied by significant environmental paradoxes: the intensification of agricultural production, marked by the extensive use of synthetic agrochemicals, coupled with high energy demands and the exploitation of nonrenewable resources, has compromised the planet's biological and biochemical integrity. Within this context, the sustainability of conventional food production models has been increasingly questioned, as the global supply chain often exceeds planetary boundaries and threatens the maintenance of ecosystem services essential to life (Khatri et al., 2024).

These impacts, however, are not restricted to the natural environment but also reverberate through human health and the global economy. Inadequate diets, strongly shaped by conventional dietary systems, are associated with a greater risk of

 DOI: 10.1201/9781003611479-10

non-transmissible chronic diseases, such as cardiovascular diseases, diabetes, obesity, and some types of cancer (Fresán et al., 2025). In parallel, the processes of food production and consumption contribute to environmental pollution and intensify pressure on natural resources, configuring a complex trilemma among diet, health, and sustainability. Overcoming this challenge requires a profound reform of the food systems, involving the entire chain of production, distribution, and consumption, in order to promote schools that reconcile individual health with environmental preservation (Iqbal et al., 2025).

In response to these demands, an exponential transformation in food standards and societal expectations has been observed, characterized by the convergence between the pursuit of healthy nutritional habits and environmental sustainability criteria, with an emphasis on reducing ecological impacts and strengthening resilient food systems. In this context, microalgae have emerged as promising microorganisms capable of providing innovative ingredients for the food industry and enhancing the nutritional quality of products (do Nascimento et al., 2025). Rich in proteins, carbohydrates, and lipids—macronutrients of recognized importance—as well as various bioactive compounds, they exhibit a versatile and robust biochemical profile that varies according to species and cultivation conditions. Thus, microalgae represent a strategic alternative to address global trends in functional and sustainable foods, directly aligning with the Sustainable Development Goals (SDG), particularly those related to food security, health, industrial innovation, and climate change (Iñiguez-Moreno et al., 2025).

Despite the recognized nutritional and functional potential of microalgal biomass and the growing appeal of its use in sustainable foods, the widespread adoption of such innovations still faces significant challenges. Among the main barriers are cultural constraints, consumer sensory acceptance, production costs, and the resulting price gap compared to conventional products (Lucas and Brunner, 2024). This economic disparity is particularly relevant, as it directly influences purchasing intentions: even consumers who are health- and sustainability-conscious tend to choose more affordable alternatives when microalgae-based products are marketed at higher prices. Therefore, overcoming these obstacles—especially those related to cost and accessibility—will be essential for microalgae-based foods to fulfill their role in promoting healthier and more sustainable diets (Wassmann et al., 2024).

Given this context, this chapter proposes an in-depth analysis of the factors that influence consumer intention and acceptance regarding the incorporation of microalgae-derived ingredients and products into the diet. The focus lies on the health benefits and the perceived value of the bioactive compounds present in microalgal biomass, considering their relevance in shaping informed food choices. To this end, this chapter reviews the main nutritional attributes of microalgae—including proteins, lipids, polysaccharides, and functional pigments—as well as sensory aspects such as flavor, texture, and acceptance across different food matrices. Furthermore, potential applications of microalgal products in distinct food categories are discussed, providing insights into how consumer preferences and expectations may guide formulation strategies and the market introduction of these innovative ingredients.

10.2 MICROALGAL BIOMASS: NUTRITIONAL COMPOSITION AND FUNCTIONAL BENEFITS

Microalgae are microscopic photosynthetic organisms capable of biosynthesizing a wide range of nutrients and bioactive compounds within their cells. This remarkable metabolic capacity has earned them the designation of "unicellular biofactories," consolidating their role as a promising source of functional and nutritious ingredients for human consumption (Rajamanickam et al., 2025).

Although naturally occurring in diverse aquatic ecosystems—such as rivers, lakes, and oceans—where they play essential ecological roles in biogeochemical cycles and primary production, the exploitation of microalgae for food and commercial purposes requires controlled cultivation conditions. In nature, factors such as light availability, nutrient concentration, and temperature are highly variable, leading to unpredictable nutritional composition of the biomass (Guerra et al., 2025). To ensure consistent yields and a desirable biochemical profile, cultivation is typically performed in open systems, such as raceway ponds, or in closed systems, such as photobioreactors, which allow for the maximization of growth and biomass production (Deprá et al., 2025b; Shirai et al., 2025).

Technological improvements, however, do not eliminate the influence of species-specific biological factors. The nutritional profile of microalgal biomass remains sensitive to multiple interdependent variables, including strain selection, medium composition—whether freshwater, marine, or wastewater-based—and nutrient sources such as fertilizers or biological residues. Environmental conditions also directly influence metabolism and the synthesis of target compounds (Esteves et al., 2024). Biotechnological parameters, including light regime and intensity, temperature, pH, cell density, and harvesting time, further modulate biochemical composition. Stress conditions, for instance, may favor the accumulation of carotenoids or lipids, whereas optimized regimes can enhance protein yield (Aditi et al., 2025; do Nascimento et al., 2025). This metabolic flexibility enables the production of biomass tailored to specific food applications, reinforcing microalgae as a strategic functional ingredient.

The appeal of microalgal cultivation extends beyond its nutritional value. It also aligns with sustainability imperatives through practices such as the use of industrial exhaust gases as a carbon source, wastewater recovery, and the implementation of closed systems that reduce water and land use compared to conventional cultures. Nonetheless, large-scale photobioreactors are associated with high energy consumption and substantial infrastructure, maintenance, and monitoring costs, which pose challenges to both perceived sustainability and economic feasibility. In this regard, the scientific field has advanced toward hybrid solutions and industrial integration strategies aimed at balancing productivity, economic viability, and environmental responsibility to achieve a feasible trade-off (Deprá et al., 2025a).

Despite these barriers, the nutritional value of microalgal biomass remains a central argument for its exploitation. Rich in high-quality proteins, essential lipids, bioactive pigments, and micronutrients—and combined with its technological versatility that enables incorporation into a wide range of food matrices—microalgal biomass stands out as a strategic ingredient for functional and sustainable foods (Table 10.1).

TABLE 10.1
Overview of the Macronutritional Composition of the Main Microalgal Strains Intended as a Food Resource

Microalgae Strains	Protein (% DCW)	Carbohydrates (% DCW)	Lipids (% DCW)
Aphanizomenon flos-aquae	**65**	**20**	**6**
Chlamydomonas reinhardtii	48	17	21
Chlorella pyrenoidosa	57	26	2
Chlorella sorokiniana	29–56	17–35	19–55
Chlorella vulgaris	30–58	12–23	10–22
Dunaliella salina	57	32	6–25
Haematococcus pluvialis	68	9	25–26
Scenedesmus obliquus	20–56	10–28	12–24
Spirulina sp.	46–71	8–16	4–9

Source: Adapted from Vendruscolo et al. (2022), Esteves et al. (2024), and Águila-Carricondo et al. (2025).

The balance between its high nutritional potential, industrial applicability, and ongoing efforts to reduce production costs and environmental impacts underscores the role of microalgae as a key resource in the innovation of modern food systems capable of simultaneously addressing the demands for health, nutritional diversity, and economic sustainability.

It is important to highlight that thousands of microalgal strains are preserved in culture collections worldwide, yet only a fraction of them have been thoroughly investigated (Dutra et al., 2025). This limitation in research scope implies that, even among the species already examined, significant differences can be observed in the composition of the main biochemical pools across distinct phyla. Consequently, defining an average composition for microalgal biomass becomes a complex task (Lafarga et al., 2020).

Even so, among the primary constituents of microalgal biomass, proteins stand out as the most valuable compounds, representing up to 65–70% of the dry biomass in certain species (Wang et al., 2021). This remarkable concentration, coupled with a balanced profile of essential amino acids, has justified the early interest in utilizing microalgae as an alternative protein source. Historically, species such as *Arthrospira platensis* (Spirulina), *Chlorella vulgaris* (Chlorella), and *Aphanizomenon flos-aquae* were domesticated and cultivated at scale for human consumption, making the dietary use of microalgae a long-standing practice embedded in several traditional cultures. With the advancement of biochemical analyses, the amino acid composition of microalgal proteins has been elucidated, revealing significant levels of isoleucine, lysine, valine, and leucine—amino acids that ensure high nutritional quality and broaden their applicability in food supplements, animal feed formulations, and nutraceutical development (Kaur et al., 2025). More recently, the extraction of microalgal proteins has enabled the production of bioactive peptides—short chains of 2–20 amino acids that, once released from their parent proteins, can act similarly to hormones in metabolic and biochemical processes (Puri et al., 2022). These peptides exhibit a broad spectrum of functional properties, including immunomodulatory, antioxidant, anticancer, antibacterial, antihypertensive, and hypocholesterolemic effects. In this context, the combination of high nutritional value, bioavailability, and multiple physiological benefits positions microalgal proteins as strategic ingredients for innovative food formulations, capable of meeting the growing demand for vegetarian, sports, and clinical diets while promoting the sustainable modernization of food systems.

Although proteins currently dominate the market narrative, microalgae also harbor another underexplored component: the carbohydrate fraction, which may account for 10–30% of the dry biomass (Van Wychen et al., 2021). This fraction consists of a mixture of structural and extracellular polysaccharides (EPS), oligosaccharides, and low-molecular-weight sugars, each with distinct functions and properties. Among the polysaccharides, fucoidans, carrageenans, alginates, laminarans, and rhamnans stand out due to their bioactivity, including prebiotic, immunomodulatory, antioxidant, anti-inflammatory, and antiviral effects. Low-molecular-weight sugars, such as mannitol and sorbitol, play osmoprotective, antioxidant, and metabolic regulatory roles, contributing to the adaptation of microalgae to environmental stress conditions and potentially offering functional benefits when consumed (Hosseini et al., 2024). Therefore, the nutritional value of microalgal carbohydrates lies not in their caloric contribution but in their content of bioactive molecules capable of modulating physiological processes, reinforcing the role of microalgae as functional and strategic ingredients for the development of innovative, sustainable, and health-promoting foods (Akter et al., 2025).

Completing the macronutrient triad, the lipid fraction represents an essential component of the human diet. Although contemporary dietary trends have historically stigmatized carbohydrates and fats as metabolic "villains," modern nutritional science has repositioned lipids as indispensable molecules. Beyond serving as the main form of energy storage, lipids are crucial for building and protecting cell membranes and are essential for the absorption of fat-soluble vitamins (Vieira et al., 2025). Among them, polyunsaturated fatty acids (PUFAs) are of particular importance due to their key roles in cardiovascular health, neurological and retinal development, and immune response. Considered essential because they cannot be synthesized endogenously, these lipids must be obtained through the diet. Robust scientific evidence demonstrates that regular consumption of n-3 and n-6 PUFAs contributes not only to overall well-being but also to the prevention and management of noncommunicable chronic diseases, such as cardiovascular and respiratory disorders, certain cancers, and diabetes (Djuricic and Calder, 2021). This growing body of evidence has been mirrored in consumer behavior, increasing demand for foods rich in functional lipids, such as fish, fish oils, nuts, and vegetable oils (Katiyar and Arora, 2020). However, marine overfishing and deforestation associated with oilseed cultivation pose serious environmental and economic challenges, threatening global food security and the sustainability of these production chains. Within this context, microalgae have emerged as alternative sources of high-value lipids. Recognized for their ability to accumulate bioactive fatty acids—particularly PUFAs—they have become strategic candidates for use in functional foods and nutraceuticals (Zakaria et al., 2025).

Certain species provide significant levels of n-3 fatty acids, including α-linolenic acid (18:3 n-3), stearidonic acid (18:4 n-3), eicosapentaenoic acid (20:5 n-3, EPA), and even small amounts of docosahexaenoic acid (22:6 n-3, DHA). Moreover, they also supply relevant n-6 fatty acids, such as linoleic acid (18:2 n-6) and arachidonic acid (20:4 n-6). Recently, advances in lipid extract analysis have expanded our understanding of algal-derived bioactivity, revealing promising antioxidant and anti-inflammatory properties (Marques et al., 2024).

Beyond classical macronutrients, microalgae biosynthesize a wide variety of bioactive compounds that confer functional singularity and enhance their relevance as food resources. The fraction of bioactive molecules—particularly microalgal pigments—significantly complements the nutritional mosaic of these organisms. Among these pigments, carotenoids are traditionally the most studied, primarily due to their role in preventing and treating chronic diseases (Fernandes et al., 2021). Structurally, they are tetraterpenes with 40 carbon atoms and extended chains of conjugated double bonds responsible for free radical neutralization and quenching of reactive oxygen species. This molecular configuration underlies their antioxidant potential and the protection of lipids, proteins, and DNA against oxidative damage. In the case of xanthophylls, such as lutein and zeaxanthin, the presence of hydroxyl groups increases polarity and favors interactions with cell membranes and proteins, resulting in anti-inflammatory effects and retinal protection against macular degeneration. β-Carotene is also noteworthy for serving as a precursor of vitamin A, whose endogenous synthesis is absent in humans, making dietary intake essential (Fathi, 2025; Kołodziejczyk and Karwowski, 2025). Industrially, certain microalgal species are recognized for their remarkable pigment productivity, including *Dunaliella salina*, a major producer of β-carotene, and *Haematococcus pluvialis*, renowned for its high astaxanthin yield (Sarkar and Ray, 2025).

Despite the predominance of carotenoids, other bioactive pigments have gained increasing attention. Chlorophylls, for example, though still relatively underexplored, exhibit remarkable functional diversity. These molecules are tetrapyrrolic structures containing a central magnesium atom, which can be replaced by other metals, such as copper or zinc, thereby altering both structural stability and tissue transport kinetics. The phytol chain confers hydrophobicity, promoting aggregation and membrane interactions, whereas peripheral modifications—such as oxidations on the isocyclic ring—modulate light absorption and specific bioactivity (Herrera and Roca, 2023). Such characteristics explain the wide range of reported effects, including antioxidant, antimutagenic, and xenobiotic metabolism–modulating activities, as well as apoptosis induction in tumor cells and antimicrobial properties (Fernandes et al., 2025; Jaén and Roca, 2025). Clinically, chlorophyll derivatives such as chlorophyllin—in which magnesium is replaced by copper or sodium—are used as supplements with antimutagenic and anticancer effects. Industrially, *Chlorella* is recognized as a rich chlorophyll source applied in supplements, cosmetics, and dentifrices, whereas *Arthrospira* contains lower levels but remains relevant for its simultaneous production of other functional pigments (Khanra et al., 2018; Martínez-Ruiz et al., 2025).

The pigment diversity of microalgae extends beyond tetrapyrrolic and terpenoid compounds. Another group of major biotechnological importance is the phycobiliproteins, among which phycocyanin—the main pigment of *Arthrospira platensis*—stands out. Unlike lipophilic carotenoids and hydrophobic chlorophylls, phycobiliproteins are water-soluble, a property that enhances their applicability as natural colorants in foods and beverages. However, their value extends beyond technological appeal: numerous studies have demonstrated their anticancer, anti-inflammatory, immunomodulatory, neuroprotective, and antioxidant properties (García-Vaquero et al., 2021). The antioxidant activity of phycocyanin arises from the presence of the phycocyanobilin chromophore, structurally similar to biliverdin, allowing it to act as an electron donor and reactive radical scavenger. This chemical configuration also explains the reported protective effects in animal models against neurodegenerative diseases and the enhancement of the immune response. Combined with the growing regulatory and market demand for natural colorants, phycocyanin positions *Arthrospira* as a strategic source of high-value pigments, broadening the nutraceutical and functional applications of microalgae.

Despite their remarkable nutritional and functional advantages, the full exploitation of microalgae as food resources requires a critical perspective that extends beyond biochemical characterization. While their pigments and macronutrients provide documented benefits, the effective incorporation of these compounds into the human diet depends on factors such as sensory acceptability, sanitary regulation, technological feasibility, and market competitiveness. Therefore, acknowledging not only the biotechnological potential but also the practical barriers to application is crucial to transforming scientific value into tangible food solutions. In this regard, understanding consumer perception and identifying strategies to promote the adoption of microalgae-based foods emerge as the next key discussion axis, aligning technological innovation, sustainability, and market dynamics.

10.3 INDUSTRIAL APPLICATIONS OF MICROALGAE-DERIVED FOOD INGREDIENTS

The growing global focus on sustainable food systems has driven interest in alternative ingredients, with microalgae recognized as a promising solution within this context. The incorporation of microalgae into food products extends beyond the pursuit of nutritional value; it also encompasses the capacity to meet contemporary consumer expectations regarding health, flavor, aesthetics, and environmental responsibility (Fernandes et al., 2021).

Indeed, the global market already includes a representative share of consolidated microalgae-based products, reflecting diverse strategies for incorporating either whole biomass or specific compounds. Among dietary supplements, powders, capsules, and tablets derived from *Arthrospira* and *Chlorella* are particularly noteworthy, valued for their protein content, vitamins, minerals, and bioactive pigments (Torky et al., 2023). In processed foods, whole microalgal biomass has been incorporated into breads, pastas, cookies, snacks, and energy bars, enriching conventional formulations with high-quality proteins and bioactive compounds (Raymundo et al., 2023). Additionally, microalgal extracts such as phycocyanin, carotenoids, and PUFAs (EPA and DHA) have been applied in functional beverages, meat and dairy alternatives, natural colorants, and condiments, establishing microalgae as a cornerstone of innovation in sustainable food development (Sidari and Tofalo, 2019).

More specifically, proteins derived from microalgae exhibit a broad range of promising applications, spanning from whole-cell biomass utilization to the production of highly purified concentrates and isolates. In the former case, whole-cell proteins provide not only essential amino acids but also benefit from the nutritional synergy of other biomass components, constituting a complete nutritional source. Protein concentrates and isolates, in turn, are obtained through purification, filtration, or precipitation processes, allowing the selective recovery of functional protein fractions (Timira et al., 2022). These derivatives display valuable techno-functional properties—such as emulsifying capacity, water absorption, foaming, and gelling behavior—making them particularly suitable for use in soups, sauces, pasta, baked goods, and biscuits, where they contribute to improved texture, stability, and sensory acceptance (Ang et al., 2025; Sivaramakrishnan et al., 2025). Nevertheless, despite these advantages, significant constraints still limit large-scale commercial expansion. The use of whole biomass, for example, may compromise the sensory properties of the final product due to the distinct "marine-like" flavor and aroma, as well as the intense coloration associated with microalgal pigments. Moreover, the presence of a rigid cell wall can reduce the digestibility and bioavailability of proteins and amino acids. To overcome these challenges, recent research has explored advanced extraction and refining technologies. Some studies have shown that the combination of high-pressure homogenization, fractionation, and lyophilization enables the production of protein extracts with lower pigment content while maintaining protein concentrations comparable to those of skim milk powder (Grossmann et al., 2022). These advances demonstrate that, although microalgal proteins hold undeniable nutritional and functional potential, overcoming technological and sensory barriers remains crucial for their consolidation in the food market (Mosibo et al., 2024).

In parallel, microalgae-derived polysaccharides constitute a class of high-value biopolymers widely recognized for their rheological, emulsifying, and gelling properties. These polymers can be classified as EPS, secreted to the cell surface and forming a protective mucilaginous sheath, or intracellular polysaccharides (IPS), which serve primarily as storage reserves within the cytoplasm (Laroche, 2022). Their composition, degree of sulfation, and relative abundance vary with species, environmental conditions, and growth stage, directly influencing functional and technological potential in food systems (Colusse et al., 2022). Among EPS, those produced by unicellular red microalgae such as *Porphyridium* and *Rhodosorus* are particularly notable for their high sulfation levels and gel-forming capacity. These polymers share structural and functional features with agars and carrageenans traditionally extracted from multicellular red macroalgae, including sulfated galactose residues that confer viscosity, gel strength, and stability in aqueous systems. This similarity underlies their suitability as thickening, gelling, and stabilizing agents in food formulations, such as emulsions, desserts, coatings, and edible films. The functional performance of these polysaccharides is closely linked to their rheological behavior, exemplified by carrageenans and agar. κ-Carrageenan forms rigid gels in the presence of K^+ ions, while λ-carrageenan, which lacks gel-forming capacity, acts primarily as a thickener. Divalent cations, particularly Ca^{2+}, enhance gel firmness through crosslinking between polysaccharide chains. Moreover, polymer concentration, pH, and ionic strength modulate chain conformation and intermolecular interactions, affecting viscosity and gel stability. Importantly, microalgal polysaccharides retain their rheological properties across a wide range of ionic strengths, indicating that common food salts such as sodium chloride (NaCl) and calcium chloride ($CaCl_2$) do not significantly interfere with their functionality (Liberman et al., 2020). This robustness is particularly relevant for food formulations, enabling reliable performance in diverse industrial applications. In practical terms, sulfated polysaccharides from red microalgae act as multifunctional natural additives, providing structure, stabilizing emulsions, forming gels, and serving as carriers for antimicrobial and antioxidant compounds. They also reduce moisture loss and gas exchange, extending the shelf life of perishable products (Felicia et al., 2022). However, their use can negatively affect texture and mouthfeel, especially in foods with neutral or delicate flavors. These sensory challenges are often exacerbated by residual proteins, pigments, or salts remaining after extraction, highlighting the need for optimized purification protocols (van de Velde et al., 2015; Srivastava et al., 2025). Hence, while microalgal polysaccharides offer substantial technological and functional potential, their variable composition and the lack of standardized extraction methods remain challenges for consistent industrial application, sensory quality, and commercial consolidation.

Microalgal pigments represent another group of compounds of high industrial relevance, particularly due to their potential to replace synthetic colorants and add functional value to foods. However, large-scale use of these pigments still faces challenges related to chemical and sensory stability during processing and storage. Chlorophylls, for instance, are highly susceptible to environmental factors such as light, oxygen, pH variation, and temperature. Under such conditions, chlorophyll degradation occurs through the replacement of the central magnesium ion with protons (forming pheophytin) and cleavage of the phytol chain, leading to undesirable color changes—from the characteristic green to olive-brown hues—and the formation of degradation compounds that may impart herbaceous, metallic, or bitter flavors to foods. Although thermal degradation can be mitigated by selecting appropriate food matrices and controlling cooking conditions, photodegradation and instability under acidic conditions (pH 3.5–5.0) continue to restrict their use as food colorants. After extraction, the central magnesium ion becomes unstable and can be readily displaced by weak acids. One strategy to overcome this limitation involves substituting magnesium with copper ions, forming bluish-green complexes with enhanced stability; however, the purity and composition of commercial preparations still yield variable results in the literature. Similarly, phycobiliproteins such as phycocyanin and phycoerythrin display limited stability compared with synthetic colorants. Under normal storage conditions, these proteins are relatively stable across a wide temperature range (4–40°C) and near-neutral pH values (6.0–7.0). Nevertheless, high temperatures (>60°C) and extreme pH conditions significantly reduce their fluorescence and functional potential (Reimão et al., 2025). Carotenoids, though less prone to photodegradation than chlorophylls, are lipophilic in nature, which complicates their incorporation into aqueous food matrices and requires strategies such as encapsulation, emulsification, or protein complexation to maintain stability and bioavailability. Furthermore, being associated with lipid phases, carotenoids are susceptible to oxidation, producing volatile aldehydes and ketones responsible for rancid or oily off-flavors that negatively affect the aromatic profile of foods. These alterations, combined with the loss of color brightness and uniformity, highlight the need for adequate technological protection to ensure commercial viability and sensory acceptance of these natural pigments. Nevertheless, these pigments have been successfully applied as ingredients in a broad range of food formulations. For example, phycocyanins are used to impart red coloration to gelatin desserts and dairy products, whereas carotenoids are widely used as natural food colorants at concentrations ranging from 2 to 50 ppm in products such as juices, beverages, butter, margarine, and cheese (Di Salvo et al., 2023).

Given this brief scenario, the application of microalgal proteins, polysaccharides, and pigments reflects an integrated trend of innovation in the food industry, where functionality, naturalness, and sustainability converge to meet emerging market demands. These biocompounds not only show potential to replace synthetic ingredients but also provide unique technological, nutritional, and sensory properties to food products. However, the consolidation of microalgae-based foods will depend on technological advances aimed at process standardization, stability control, and consumer sensory acceptance. As techno-economic barriers are gradually overcome, microalgae are poised to assume a central role in the transition toward more circular, resilient, and environmentally responsible food systems.

10.4 CONCLUSION AND FUTURE PERSPECTIVES

The incorporation of microalgae-derived ingredients into the food industry presents a promising route for the development of sustainable, nutritious, and functional products. Proteins, polysaccharides, and pigments from microalgae offer unique technological, nutritional, and sensory properties that enable applications ranging from protein-enriched foods and functional beverages to gelling agents, emulsifiers, natural colorants, and protective coatings. These compounds contribute not only to food functionality and nutritional value but also to sensory appeal, aligning with contemporary consumer expectations for health, naturalness, and sustainability.

Despite their potential, several challenges remain. Technological limitations such as extraction efficiency, compositional variability, stability during processing and storage, and sensory issues—including off-flavors, color degradation, or texture changes—need to be addressed. Advances in high-pressure homogenization, enzymatic treatments, encapsulation, and formulation strategies show promise, yet standardization and scalable production remain key hurdles. From a consumer perspective, acceptance depends heavily on sensory quality, visual appeal, and perceived health and sustainability benefits. Addressing these factors is essential for successful market penetration.

Looking forward, a multidisciplinary approach that integrates food science, biotechnology, sensory analysis, and process engineering will be crucial to unlock the full potential of microalgae-based ingredients. Targeted innovations in extraction, purification, stabilization, and formulation, combined with strategic product development and consumer education, can accelerate adoption and foster resilient, sustainable food systems.

REFERENCES

Aditi, Bhardwaj, R., Yadav, A., Swapnil, P., & Meena, M. (2025). Characterization of microalgal β-carotene and astaxanthin: Exploring their health-promoting properties under the effect of salinity and light intensity. *Biotechnology for Biofuels and Bioproducts*, *18*(1), 18. DOI: 10.1186/s13068-025-02612-x

Águila-Carricondo, P., García-García, R., de la Roche, J. P., Galán, P. L., Bautista, L. F., Espada, J. J., & Vicente, G. (2025). Aphanizomenon flos-aquae: A biorefinery for health and energy—Unleashing Phycocyanin's power and biogas potential. *Marine Drugs*, *23*(6), 225. DOI: 10.3390/md23060225

Akter, F., Songsomboon, K., Ralph, P. J., & Kuzhiumparambil, U. (2025). A comprehensive overview of microalgae-and cyanobacteria-derived polysaccharides: Extraction, structural chemistry, techno-functional and bioactive properties. *Bioresource Technology Reports*, 102280. DOI: 10.1016/j.biteb.2025.102280

Andersen, B. V., & Hyldig, G. (2015). Consumers' view on determinants to food satisfaction. A qualitative approach. *Appetite*, *95*, 9–16. DOI: 10.1016/j.appet.2015.06.011

Ang, W. S., Kee, P. E., Lan, J. C. W., Chen, W. H., Chang, J. S., & Khoo, K. S. (2025). Unveiling the rise of microalgae-based foods in the global market: Perspective views and way forward. *Food Bioscience*, *66*, 105390. DOI: 10.1016/j.fbio.2024.105390

Colusse, G. A., Carneiro, J., Duarte, M. E. R., Carvalho, J. C. D., & Noseda, M. D. (2022). Advances in microalgal cell wall polysaccharides: A review focused on structure, production, and biological application. *Critical Reviews in Biotechnology*, *42*(4), 562–577. DOI: 10.1080/07388551.2021.1941750

Deprá, M. C., Dias, R. R., & Jacob-Lopes, E. (2025a). Sustainability metrics and indicators in microalgae-based processes and products. In Pérez-Gálvez, Antonio, Jacob-Lopes, Eduardo, Zepka, Leila Queiroz, Roca, María (eds.), *Microalgae and One Health* (pp. 579–585). Academic Press. DOI: 10.1016/B978-0-443-22080-7.00010-7

Deprá, M. C., Dias, R. R., Zepka, L. Q., & Jacob-Lopes, E. (2025b). Tackling old challenges in microalgal biotechnology: The role of photobioreactors to advance the technology readiness level. *Processes*, *13*(1), 51. DOI: 10.3390/pr13010051

Di Salvo, E., Lo Vecchio, G., De Pasquale, R., De Maria, L., Tardugno, R., Vadalà, R., & Cicero, N. (2023). Natural pigments production and their application in food, health and other industries. *Nutrients*, *15*(8), 1923. DOI: 10.3390/nu15081923

Djuricic, I., & Calder, P. C. (2021). Beneficial outcomes of omega-6 and omega-3 polyunsaturated fatty acids on human health: An update for 2021. *Nutrients*, *13*(7), 2421. DOI: 10.3390/nu13072421

do Nascimento, T. C., Pinheiro, P. N., Deprá, M. C., Sartori, R. B., Roca, M., Jacob-Lopes, E., & Zepka, L. Q. (2025). Photoperiod strategies to increase microalgae-pigment content: A new level-up to natural food colorants production. *Algal Research*, 104185. DOI: 10.1016/j.algal.2025.104185

Dutra, D. A., Dias, R. R., Deprá, M. C., Zepka, L. Q., Jacob-Lopes, E. (2025). The beginning to integration: Evolution and context of microalgae biotechnology. In: Santos Ballardo, D. U., Rossi, S. (eds.), *Microalgae as Promising Source of Commercial Bioproducts. Developments in Applied Phycology* (Vol. 14). Springer. DOI: 10.1007/978-3-031-86433-9_1

Esteves, A. F., Goncalves, A. L., Vilar, V. J., & Pires, J. C. (2024). Is it possible to shape the microalgal biomass composition with operational parameters for target compound accumulation? *Biotechnology Advances*, 108493. DOI: 10.1016/j.biotechadv.2024.108493

Fathi, Z. (2025). Microbial production of β-carotene. In: Jafari, S. M., Harzevili, F. D., Karaca, A. C. (eds.), *Microbial Production of Food Bioactive Compounds*. Springer. DOI: 10.1007/978-3-030-81403-8_12-1

Felicia, W. X. L., Rovina, K., Nur'Aqilah, M. N., Vonnie, J. M., Erna, K. H., Misson, M., & Halid, N. F. A. (2022). Recent advancements of polysaccharides to enhance quality and delay ripening of fresh produce: A review. *Polymers*, *14*(7), 1341. DOI: 10.3390/polym14071341

Fernandes, A. S., Jacob-Lopez, E., Zepka, L. Q., de Rosso, V. V., & Roca, M. (2025). Bioactive compound-loaded food-grade bigels: (II) Delivery systems focusing on the bioaccessibility of chlorophylls and the effects of microstructure and 3D printing. *Food Hydrocolloids*, 111596. DOI: 10.1016/j.foodhyd.2025.111596

Fernandes, A. S., Nascimento, T. C., Pinheiro, P. N., Vendruscolo, R. G., Wagner, R., de Rosso, V. V., … Zepka, L. Q. (2021). Bioaccessibility of microalgae-based carotenoids and their association with the lipid matrix. *Food Research International*, *148*, 110596. DOI: 10.1016/j.foodres.2021.110596

Fresán, U., Núñez, M., Valls, I., & Rosenbaum, R. K. (2025). Quantifying the environmental human health burden of food demand in Spain: A life cycle assessment study. *Environmental Research*, 122147. DOI: 10.1016/j.envres.2025.122147

Funabashi, M. (2018). Human augmentation of ecosystems: Objectives for food production and science by 2045. *NPJ Science of Food*, *2*(1), 16. DOI: 10.1038/s41538-018-0026-4

García-Vaquero, M., Brunton, N., & Lafarga, T. (2021). Microalgae as a source of pigments for food applications. In Lafarga, Tomás, Acién, Gabriel (eds.), *Cultured Microalgae for the Food Industry* (pp. 177–198). Academic Press. DOI: 10.1016/B978-0-12-821080-2.00014-9

Gowder, S. J. (2024). Social aspects of food and nutrition: An overview. *Journal of Ecohumanism*, *3*(7), 2953–2961.

Grossmann, L., Loeffler, M., & Smetana, S. (2022). Production, downstreaming, and utilization of proteins and exopolysaccharides from single cells in food matrices. *Frontiers in Sustainable Food Systems*, *6*, 1091347. DOI: 10.3389/fsufs.2022.1091347

Guerra, I., Torres, C., Cardoso, H., Pereira, H., Trovão, M., Rodrigues, A. M., & Varela, J. (2025). Impact of seasons on industrial cultivation of Limnospira platensis (Spirulina): A year-round case study on biomass, phycocyanin, and protein productivity in Portugal. *Algal Research*, 104335. DOI: 10.1016/j.algal.2025.104335

Herrera, M., & Roca, M. (2023). Microalgal chlorophylls for food/feed applications. In Jacob-Lopes, Eduardo, Queiroz, Maria Isabel, Maroneze, Mariana Manzoni, Zepka, Leila Queiroz (eds.), *Handbook of Food and Feed from Microalgae* (pp. 147–160). Academic Press. DOI: 10.1016/B978-0-323-99196-4.00041-3

Hosseini, S. V., Dastgerdi, H. E., & Tahergorabi, R. (2024). Marine mannitol: Extraction, structures, properties, and applications. *Processes*, *12*(8), 1613. DOI: 10.3390/pr12081613

Iñiguez-Moreno, M., González-Meza, G. M., Araújo, R. G., Flores-Contreras, E. A., & de la Rosa, O. (2025). Microalgae biomass as a sustainable solution for food security: Advances in biotechnology and their role in achieving Sustainable Development Goals. *Bioresource Technology Reports*, *31*, 102183. DOI: 10.1016/j.biteb.2025.102183

Iqbal, B., Alabbosh, K. F., Jalal, A., Suboktagin, S., & Elboughdiri, N. (2025). Sustainable food systems transformation in the face of climate change: Strategies, challenges, and policy implications. *Food Science and Biotechnology*, *34*(4), 871–883. DOI: 10.1007/s10068-024-01712-y

Jaén, M. D. R. S., & Roca, M. (2025). Chlorophylls and their derivatives. In Pérez-Gálvez, Antonio, Jacob-Lopes, Eduardo, Zepka, Leila Queiroz, Roca, María (eds.), *Microalgae and One Health* (pp. 147–158). Academic Press. DOI: 10.1016/B978-0-443-22080-7.00029-6

Katiyar, R., & Arora, A. (2020). Health promoting functional lipids from microalgae pool: A review. *Algal Research*, *46*, 101800. DOI: 10.1016/j.algal.2020.101800

Kaur, S., Sharma, A., Bala, S., Satheesh, N., Nile, A. S., & Nile, S. H. (2025). Microalgae in the food-health nexus: Exploring species diversity, high-value bioproducts, health benefits, and sustainable market potential. *Bioresource Technology*, 132424. DOI: 10.1016/j.biortech.2025.132424

Khanra, S., Mondal, M., Halder, G., Tiwari, O. N., Gayen, K., & Bhowmick, T. K. (2018). Downstream processing of microalgae for pigments, protein and carbohydrate in industrial application: A review. *Food and Bioproducts Processing*, *110*, 60–84. DOI: 10.1016/j.fbp.2018.02.002

Khatri, P., Kumar, P., Shakya, K. S., Kirlas, M. C., & Tiwari, K. K. (2024). Understanding the intertwined nature of rising multiple risks in modern agriculture and food system. *Environment, Development and Sustainability, 26*(9), 24107–24150. DOI: 10.1007/s10668-023-03638-7

Kołodziejczyk, A. M., & Karwowski, B. (2025). Anti-DNA damage mechanisms and the role of carotenoids, vitamin A, and its derivatives. *Nutrients, 17*(17), 2721. DOI: 10.3390/nu17172721

Lafarga, T., Fernández-Sevilla, J. M., González-López, C., & Acién-Fernández, F. G. (2020). Spirulina for the food and functional food industries. *Food Research International, 137*, 109356. DOI: 10.1016/j.foodres.2020.109356

Laroche, C. (2022). Exopolysaccharides from microalgae and cyanobacteria: Diversity of strains, production strategies, and applications. *Marine Drugs, 20*(5), 336. DOI: 10.3390/md20050336

Liberman, G. N., Ochbaum, G., Mejubovsky-Mikhelis, M., Bitton, R., & Arad, S. M. (2020). Physico-chemical characteristics of the sulfated polysaccharides of the red microalgae Dixoniella grisea and Porphyridium aerugineum. *International Journal of Biological Macromolecules, 145*, 1171–1179. DOI: 10.1016/j.ijbiomac.2019.09.205

Lucas, B. F., & Brunner, T. A. (2024). Factors influencing the willingness to purchase and consume microalgae-based foods: An exploratory consumer study. *International Journal of Gastronomy and Food Science*, 37, 100974. DOI: 10.1016/j.ijgfs.2024.100974

Marques, F., Pinho, M., Guerra, I. M., Conde, T. A., Silva, J., Cardoso, H., … Domingues, M. R. (2024). Unlocking functional lipid ingredients from algae by food-grade biosolvents and ultrasound-assisted extraction for nutritional applications. *LWT, 200*, 116136. DOI: 10.1016/j.lwt.2024.116136

Martínez-Ruiz, F. E., Andrade-Bustamante, G., Holguín-Peña, R. J., Renganathan, P., Gaysina, L. A., Sukhanova, N. V., & Puente, E. O. R. (2025). Microalgae as functional food ingredients: Nutritional benefits, challenges, and regulatory considerations for safe consumption. *Biomass, 5*(2), 25. DOI: 10.3390/biomass5020025

Mosibo, O. K., Ferrentino, G., & Udenigwe, C. C. (2024). Microalgae proteins as sustainable ingredients in novel foods: Recent developments and challenges. *Foods, 13*(5), 733. DOI: 10.3390/foods13050733

Puri, M., Gupta, A., McKinnon, R. A., & Abraham, R. E. (2022). Marine bioactives: From energy to nutrition. *Trends in Biotechnology, 40*(3), 271–280. DOI: 10.1016/j.tibtech.2021.08.004

Rajamanickam, R., Das, S., Sharma, S., Ro, R., Prabhu, A. A., Banerjee, S., … Selvasembian, R. (2025). Microalgae-based nutritional supplements: Sustainable applications for high-nutritional-value food production. *Process Biochemistry*. DOI: 10.1016/j.procbio.2025.07.009

Raymundo, A., Fradinho, P., & Nunes, M. C. (2023). Application of microalgae in baked goods and pasta. In Jacob-Lopes, Eduardo, Queiroz, Maria Isabel, Maroneze, Mariana Manzoni, Zepka, Leila Queiroz (eds.), *Handbook of Food and Feed from Microalgae* (pp. 317–334). Academic Press. DOI: 10.1016/B978-0-323-99196-4.00022-X

Reimão, M., Almeida, L., Ramos, C., Eusébio, N., Martins, R., da Silva, M. V., … & Freitas, M. (2025). Colouring applications of microalgae and cyanobacteria photosynthetic pigments: Challenges for industrial and market acceptance. *Journal of Cleaner Production, 520*, 146071. DOI: 10.1016/j.jclepro.2025.146071

Sarkar, N., & Ray, R. R. (2025). β-carotene: algal diversity, biosynthetic pathways, extraction methods and technical aspects, current prospects/applications, market trends. In: Kumar, L., Bharadvaja, N., Kumar, D., & Anand, R. (eds.), *Algae-Derived Biochemicals of Industrial Importance. Sustainable Landscape Planning and Natural Resources Management*. Springer. DOI: 10.1007/978-3-031-85550-4_8

Shirai, M. A., Baú, T. R., Zanela, J., & Pimentel, T. C. (2025). Microalgae as an innovative active ingredient for edible films and coatings for food applications. *Algal Research, 86*, 103959. DOI: 10.1016/j.algal.2025.103959

Sidari, R., & Tofalo, R. (2019). A comprehensive overview on microalgal-fortified/based food and beverages. *Food Reviews International*, *35*(8), 778–805. DOI: 10.1080/87559129.2019.1608557

Sivaramakrishnan, R., Kanwal, S., Incharoensakdi, A., Nirmal, N., & Srimongkol, P. (2025). Exploring the nutraceutical and functional food potential of microalgae: Implications for health and sustainability. *Journal of Agriculture and Food Research*, 102148. DOI: 10.1016/j.jafr.2025.102148

Srivastava, H., Bisht, B., James, J., Malhotra, R. K., Kurbatova, A., Dabral, A., … & Kumar, V. (2025). Advanced extraction technologies and functional applications of algal polysaccharides in modern food systems. *Discover Food*, *5*(1), 272. DOI: 10.1007/s44187-025-00585-2

Timira, V., Meki, K., Li, Z., Lin, H., Xu, M., & Pramod, S. N. (2022). A comprehensive review on the application of novel disruption techniques for proteins release from microalgae. *Critical Reviews in Food Science and Nutrition*, *62*(16), 4309–4325. DOI: 10.1080/10408398.2021.1873734

Torky, A., Saad, S., & Eltanahy, E. (2023). Microalgae as dietary supplements in tablets, capsules, and powder. In Jacob-Lopes, Eduardo, Queiroz, Maria Isabel, Maroneze, Mariana Manzoni, Zepka, Leila Queiroz (eds.), *Handbook of Food and Feed from Microalgae* (pp. 357–369). Academic Press. DOI: 10.1016/B978-0-323-99196-4.00002-4

van de Velde, F., de Hoog, E. H., Oosterveld, A., & Tromp, R. H. (2015). Protein-polysaccharide interactions to alter texture. *Annual Review of Food Science and Technology*, *6*(1), 371–388. DOI: 10.1146/annurev-food-022814-015558

Van Wychen, S., Rowland, S. M., Lesco, K. C., Shanta, P. V., Dong, T., & Laurens, L. M. (2021). Advanced mass balance characterization and fractionation of algal biomass composition. *Journal of Applied Phycology*, *33*(5), 2695–2708. DOI: 10.1007/s10811-021-02508-x

Vendruscolo, R. G., Deprá, M. C., Pinheiro, P. N., Furlan, V. J. M., Barin, J. S., Cichoski, A. J., … Wagner, R. (2022). Food potential of Scenedesmus obliquus biomasses obtained from photosynthetic cultivations associated with carbon dioxide mitigation. *Food Research International*, *160*, 111590. DOI: 10.1016/j.foodres.2022.111590

Vieira, B. M., Silva, G. N. D. M. E., & Silva, M. I. (2025). Nutritional elements I: Nutrients, proteins, carbohydrates, and lipids. In Fajemiroye, James O. (ed.), *Fundamentals of Drug and Non-Drug Interactions: Physiopathological Perspectives and Clinical Approaches* (pp. 35–56). Springer Nature Switzerland. DOI: 10.1007/978-3-031-80107-5_2

Wang, Y., Tibbetts, S. M., & McGinn, P. J. (2021). Microalgae as sources of high-quality protein for human food and protein supplements. *Foods*, *10*(12), 3002. DOI: 10.3390/foods10123002

Wassmann, B., Hartmann, C., & Siegrist, M. (2024). Novel microalgae-based foods: What influences Singaporean consumers' acceptance? *Food Quality and Preference*, *113*, 105068. DOI: 10.1016/j.foodqual.2023.105068

Zakaria, M. F., Zaidi, F. N. A. M., Kamal, A. H. A., Aleng, N. A., Abd Wahid, M. E., Katayama, T., & Jusoh, M. (2025). PUFA from microalgae: Challenges, factors affecting high production and industrial application. *Aquaculture and Fisheries*. DOI: 10.1016/j.aaf.2025.01.004

11 Microalgae-Based Feed
A Promising and Sustainable Alternative Feedstock for Animal Production Systems

Sameh A. Abdelnour and Wael A. Khalil

11.1 INTRODUCTION

The accessibility of microalgae as natural food and feed for various livestock is high due to their widespread availability. Furthermore, certain species are highly valued because they can promptly produce and store large quantities of superior nutritional and bioactive molecules (Lafuente-Rincon et al., 2024). Microalgae are rich in proteins, carbohydrates, carotenoids, vitamins, lipids, and omega-3 or omega-6 fatty acids (Wang et al., 2025), making them attractive as versatile feed additives to improve the productivity and health of livestock (Bature et al., 2022). Microalgae are single-celled organisms, found everywhere from oceans to soil, that use photosynthesis to create their own fuel (Wang et al., 2025). Their adaptability and rapid growth make them metabolic powerhouses, which is why scientists are now extensively studying them as a source of high-value biological products (Sucu, 2020; Song et al., 2025). Microalgae are rapidly gaining recognition as a valuable and novel ingredient for formulated animal diets. They are receiving widespread attention as a feed additive for animals in various production systems, including aquaculture, poultry, and both monogastric and ruminant livestock (Bature et al., 2022; El-Sayed et al., 2024; Ringø et al., 2025). This interest stems from their advantageous traits as aquatic organisms, specifically their rapid growth rate, widespread presence, and inherent ability to adapt to different environments (Kashenye and Zhang, 2025).

Globally, the livestock sector is an essential component of food security and nutrition, providing 17% of total calories and 33% of protein consumed worldwide through animal-derived products, especially those from ruminants (Unicef, 2024). This crucial role faces a significant future challenge as population growth is expected to drive total food demand up by an estimated 35% to 56% from 2010 to 2050 (Mitra et al., 2024). The expanding demand for animal foods is projected to drive the growth of ruminant livestock production, resulting in adverse environments such as heightened methane emissions (Shah et al., 2025). It is crucial to address this dilemma by finding a balance between product demand, nutritional value, sustainability, and health risk mitigation. Urgent strategies are needed to enhance the economic viability of livestock systems while minimizing their environmental impact. The usage of

DOI: 10.1201/9781003611479-11

alternative feed supplements, such as microalgae, shows great promise in achieving both market expansion and environmental sustainability (Ma and Hu, 2024). Microalgae possess a powerful range of bioactive properties incorporating anticancer, antifungal, antioxidant, and anti-inflammatory effects (Kashenye and Zhang, 2025), which are critical for animal physiological health (Tabarzad et al., 2020). Notably, these bioactive components also contribute to environmental sustainability by facilitating the decrease in methane production in ruminants and improving their overall digestibility (Kashenye and Zhang, 2025). Studies confirm that adding macroalgae to feed improves nutrient utilization and supports gastrointestinal health (Emam et al., 2024).

The integration of microalgae into animal diets has been proven to boost health and growth performance (Bature et al., 2022; El-Sayed et al., 2024; Boukrouh et al., 2025). Its balanced nutrition helps animals achieve optimal growth rates, making it a key focus for animal feed over direct human consumption. Critically, microalgae contain bioactive compounds (e.g., polyphenols, flavonoids, and antioxidants) that demonstrate antimicrobial, anti-inflammatory, and immunomodulatory properties (Ma and Hu, 2024; Ringø et al., 2025; Wang et al., 2025). By leveraging these compounds, microalgae help animals mitigate oxidative stress, strengthen their immune response, and enhance their resistance to thermal stress. This chapter reviews the intrinsic biological activities of microalgae and summarizes their potential for use in animal diets. Specifically, we evaluate the impact of microalgal inclusion on livestock performance, health, and reproductive efficiency and assess their utility in reducing methane emissions and improving heat resistance.

11.2 MICROALGA TYPES AND BIOACTIVE COMPOUND MICROALGAE

Microalgae are tiny, single-celled eukaryotic organisms found in both fresh and saltwater. Essentially, they are the unicellular plants of the water with massive potential in science and technology (Ampofo and Abbey, 2022). Microalgae offer a fresh perspective on global challenges like carbon cycling, carbon capture, sustainability, and oxygen production due to their fast growth and carbon-fixation capabilities (Basheer et al., 2020). There are many types of microalgae that have been extensively used in animal diets (Balasubramanian et al., 2021; Bature et al., 2022) and aquaculture due to their ease of cultivation and adaptability to various environments (El-Shall et al., 2023; Song et al., 2025). The blue-green microalgae *Spirulina* (*Arthrospira platensis, SP*) is recognized globally for its high nutritional value and is widely utilized as a nutraceutical food supplement (El-Shall et al., 2023; Abdelfatah et al., 2024). Phycocyanin is the primary active compound in *Spirulina*, responsible for its numerous biological benefits. *Chlorella vulgaris* is a species of green microalgae that is spherical, unicellular, and has a high concentration of chlorophyll, making it bright green. *C. vulgaris* is widely used as a feed additive (Elnady et al., 2024), and its biomass provides a significant protein source for livestock, specifically ruminants (Kholif and Olafadehan, 2021). Spirulina is known for its high protein and iron content, as well as being rich in omega-6 unsaturated fatty acids (Ötleş and Pire, 2001). On the other hand, *Chlorella* is a good source of essential vitamins, especially the B-complex group and beta-carotene, along with significant levels of protein and omega-3 fatty acids (Figure 11.1). The results of Alharthi et al. (2024) show that Spirulina is a rich source of crude protein (58.18%), ash (12.57%), and starch (47.60%). It also contains significant amounts of bioactive compounds such as caffeine (60.41%), amphetamine (23.99%), 4-aminophthalimide (2.70%), 5,8-indolezidine (1.89%), and 9-octadecinamide (1.15%).

Haematococcus pluvialis is essentially a treasure chest of nutrients, containing valuable proteins, fatty acids, and lutein. Its biggest claim to fame, however, is being the richest natural source of astaxanthin (ASTA) (Chen et al., 2024), a potent bioactive compound widely used across different industries for its multifunctional properties (Abdel-Wareth et al., 2024). The microalgae genus *Nannochloropsis*, such as *N. oceanica*, has also been used as animal feed additives. This single-celled organism is characterized by a single chloroplast and a polysaccharide cell wall. Nannochloropsis is commonly used in livestock diets to enhance reproductive health and reduce oxidative stress (El-Sayed et al., 2024). It is only recently that the long-established practice of consumption has been matched by widespread cultivation and harvesting. Microalgae are a nutritionally diverse and valuable source of essential compounds (Wu et al., 2023). They are rich in proteins, vitamins (including C, E, and B complex), minerals (such as iodine, iron, calcium, and magnesium), and dietary fiber (Fox and Zimba, 2018). Beyond basic nutrition, microalgae contain powerful bioactive compounds like polysaccharides, pigments, polyphenols, vitamin B_{12}, peptides, carotenoids, and phycobiliproteins that contribute to their health benefits (Ampofo and Abbey, 2022). These compounds make microalgae promising for functional

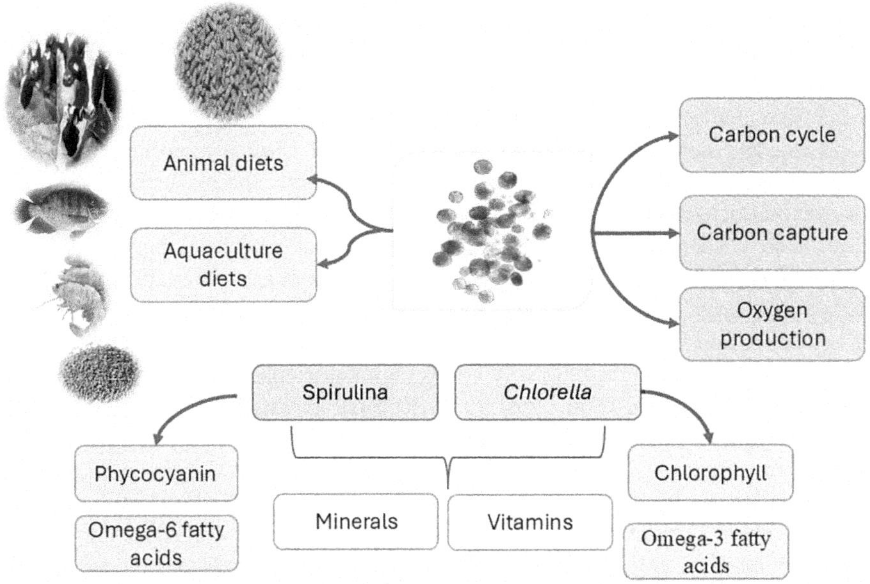

FIGURE 11.1 General uses of microalgae and their main components based on some literature.

foods and nutraceuticals (Abdelnour et al., 2019b). For example, microalgae-derived pigments show immunomodulatory, antioxidant, and antitumor activities (Ampofo and Abbey, 2022). Beyond these fundamentals, macroalgae are also rich sources of vital bioactive compounds, including omega-3 fatty acids and omega-6 fatty acids. The high content of pigments provides anti-inflammatory, antiviral, antibacterial, and cardioprotective effects (Zhu et al., 2024; Wang et al., 2025). Furthermore, certain species have beneficial omega-3 fatty acids, which support cardiovascular health, and their prebiotic activity can perfect gut health by stimulating beneficial bacteria growth (Zhu et al., 2024; Wang et al., 2025). Research confirms that microalgae benefits animal health (in pigs, fish, chickens, and ruminants) through its antibacterial, antioxidant, and anti-inflammatory actions.

11.3 BIOLOGICAL ACTIVITY OF MICROALGAE

Microalgae are rich sources of bioactive compounds, making them valuable food additives. They produce over 400 carotenoids, many of which are used as natural colorants. For instance, ASTA (*from Haematococcus* spp.) is a powerful antioxidant that combats oxidative stress, inflammation, and protects tissues (Basheer et al., 2020). Beta-carotene is a crucial precursor to vitamin A, supporting immune function and offering antitumor and anti-inflammatory benefits. Furthermore, microalgae supply a full range of vitamins (A, B, C, and E), notably which are vital for vegetarian diets (Fiedor and Burda, 2014). Phycocyanin, also found in SP responsible, is for its numerous biological benefits and mitigating oxidative stress induced by thermal stress (Abdelnour et al., 2020). These molecules exhibited antioxidant properties that translate to overall health protection, including enhanced immunity, anti-aging effects, and support for liver function via sustained gut health and enhanced digestion and nutrient absorption (Abdelnour et al., 2024). Microalgae are rich in essential omega-3 and omega-6 fatty acids, which are vital for supporting animal growth and reproductive health. Furthermore, the carotenoids and polysaccharides found in these microalgae exhibit powerful antioxidant and anti-inflammatory (Abdelnour et al., 2019a) properties by regulating antioxidative enzymes (Abd El-Hack et al., 2019) and key cytokines (such as IL-2 and IL-4), while also supporting anti-inflammatory mediators like Interleukin-10 (Abdelnour et al., 2019b).

11.4 MICROALGAE AS A FEED SUPPLEMENT IN MONOGASTRIC

The poultry industry is essential for meeting global demand for animal protein, but it faces key challenges related to sustainability, disease management, and developing alternative antibiotics (Hafez and Attia, 2020). Recent advances point toward novel ingredients, such as macroalgae extracts, which offer possible advantages like a declined environmental impact, improved animal health, and heightened meat quality and shelf life (Matshogo et al., 2020). Despite improving the nutritional quality of the meat, the long-term inclusion of high concentrations of *Spirulina* in broiler diets may negatively impact growth performance, potentially by increasing digesta viscosity (Spínola et al., 2024). The potential efficacy of *Spirulina* when included in broiler diets has been validated by a variety of reviews (El-Shall et al., 2023; Irshad et al.,

2024; Wlaźlak and Biesek, 2025). Additionally, *Spirulina* extract (400 mg/kg diet) as a natural immunomodulatory mediator in conjunction with the H9N2 vaccine to realize the highest possible level of defense against H9N2 Avian Influenza disease (Yehia et al., 2024). Algae function as immunostimulants, offering an effective alternative to conventional veterinary inhibition in poultry flocks. Their health-promoting effects stem from enhancing the host's antimicrobial defenses, specifically by encouraging antibody production, boosting T cell activity, and favorably differentiating the cecum microbiome (Abotaleb et al., 2020). *Spirulina* (2 g/kg) has been shown to reduce Newcastle disease virus excretion by up to 87% and may enhance vaccine efficacy. Furthermore, its anti-inflammatory action helps combat Avian Influenza (H9N2) by increasing phagocytic and lysozyme activity, which reduces viral excretion and liver damage (Yehia et al., 2024). *Chlorella vulgaris* (5 g/kg) strengthens the intestinal barrier in young broilers by boosting the expression of key defense markers and promoting CD4+ T cell growth (Lee et al., 2023). Abdelfatah et al. (2024) recommend supplementing poultry diets with 0.5% *Spirulina*. This inclusion significantly improves the intestinal barrier function, immune response, and digestive capacity (Abdelfatah et al., 2024). *Spirulina platensis* (alone or with garlic powder) is effective at mitigating the negative effects of heat stress in Ross 308 chickens, successfully improving immune parameters, lymphoid organ health, and intestinal *Lactobacillus* populations (Attia et al., 2023). Adding 1 g of *C. vulgaris* improved the growth and overall health of growing rabbits by regulating the synthesis of IgM and IgG, as well as supporting the antioxidative defense system (Abdelnour et al., 2019b). However, while these immune-enhancing effects are consistently reported, the field lacks clarity. Direct comparisons between different algae species and their active compounds remain scarce. Furthermore, many studies simply note changes in immune markers without fully explaining the underlying biological mechanisms, for instance, which specific microalgae compounds activate particular immune pathways in poultry (Abdelnour et al., 2019b). Microalgae show promise for achieving the crucial goal of optimal weight gain and sustainability. For example, supplementing broilers with 3% *Laminaria japonica* and cecropin improved growth performance and increased Newcastle disease antibodies (Bai et al., 2019). Furthermore, this supplementation balanced gut flora by inhibiting *Escherichia coli* while increasing *Lactobacillus* growth (Abdelnour et al., 2019b; Bai et al., 2019). Other studies show that adding 2% to 3.5% of *Ulva* spp. to chicken feed increased weight gain and feed intake but did not affect digestibility (Nhlane et al., 2020). Microalgae also enhance meat quality and health and decrease the water-holding capacity of broiler meat (Abdelnour et al., 2019a; Balasubramanian et al., 2021). Dietary microalgae supplementation improves egg production and quality by reducing lipid and cholesterol levels while increasing beneficial probiotic bacteria and reducing harmful enteric bacteria in the gut (Alders et al., 2021). Microalgae show promising potential in combating Aflatoxin B_1 contamination in poultry (Bature et al., 2022). By both binding to aflatoxins in the gastrointestinal tract and enhancing the poultry immune system, they mitigate toxic effects and improve resilience, all while boosting overall feed quality. Table 11.1 and Figure 11.2 show the summarized effects of dietary inclusion on some productive and physiological responses in poultry.

TABLE 11.1
Impacts of Dietary Inclusion of Various Microalgae on the Monogastric Animals

Species	Dose and Type	Findings	References
Broiler	15% Spirulina powder	• Significant increase in ileum viscosity in SP diets • Birds in SP groups had greater values of yellowness (b*), breast, thigh muscles, pigments, and n-3 PUFA than the control diet	Spínola et al., 2024
Broiler	Spirulina extract (200 or 400 mg/kg diet) in vaccinated broiler	• Spirulina extract displayed an anti-inflammatory impact by downregulating nitric oxide levels	Yehia et al., 2024
Broiler	*Spirulina* (2 g/kg)	• *Spirulina* (2 g/kg) has been shown to reduce Newcastle disease virus excretion by up to 87% and may enhance vaccine efficacy	Abotaleb et al., 2020
Broiler	*Chlorella vulgaris* (5 g/kg)	• *C. vulgaris* strengthens the intestinal barrier in young broilers • It also boosts the expression of key defense markers and promotes CD4+ T cell growth	Lee et al., 2023
Broiler	0.5% *Spirulina*	• SP (0.5%) significantly improved the intestinal barrier function, immune response, and digestive capacity	Abdelfatah et al., 2024
Ross 308 chickens	1 g SP/kg + 200 mg garlic powder/kg diet during heat stress	• SP improved immune parameters, lymphoid organ health, and intestinal Lactobacillus populations	Attia et al., 2023
Broiler and quails	0.5–1 g PS or *C. vulgaris*	• PS or *C. vulgaris* controls aflatoxin B_1 in poultry diets	Abdelnour et al., 2019b
Laying hens	*Haematococcus pluvalis* (astaxanthin at 25, 50, and 100 mg/kg)	• The dietary ASTA enhanced the total antioxidant capacity (TAC), superoxide dismutase (SOD), glutathione peroxidase (GPX), and reduced the malondialdehyde (MDA) levels • ASTA reduced lipid content in eggs by regulating the mRNA expression of scavenger receptor class B type 1 (SCARB1) and very-low-density lipoprotein receptor (VLDLR) genes	Gao et al., 2020
Laying hens	*Haematococcus pluvalis* (astaxanthin at 20, 40, and 80 mg/kg)	• The inclusion of 40 mg/kg ASTA in the diet led to a significant increase in the activity of antioxidant enzymes GPX and SOD and a corresponding reduction in MDA concentration in laying hens	Heng et al., 2021

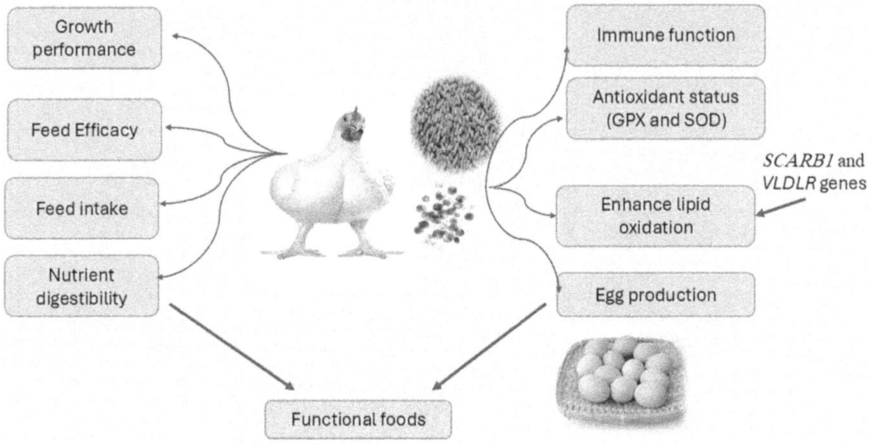

FIGURE 11.2 The role of microalgae in improving the growth and physiology, immune function, and antioxidant status in poultry.

In an earlier study, microalgae *Haematococcus pluvialis*, Gao et al. (2020) evaluated the impacts of ASTA from *Haematococcus pluvialis* at 25, 50, and 100 mg/kg diet on hens' immunity and antioxidants. The dietary ASTA enhanced the total antioxidant capacity (TAC), superoxide dismutase (SOD), glutathione peroxidase (GPX), and reduced the malondialdehyde (MDA) levels, as well as reducing lipid content in eggs by regulating the mRNA expression of scavenger receptor B type 1 (SCARB1) and very-low-density lipoprotein receptor (VLDLR) genes. The inclusion of 40 mg/kg ASTA in the diet led to a significant increase in the activity of antioxidant enzymes GPX and SOD and a corresponding reduction in MDA concentration in laying hens (Heng et al., 2021). Importantly, these health benefits did not compromise production metrics, such as egg weight, feed consumption, feed efficiency, laying rate, Haugh unit, and eggshell strength, which remained unaffected. Mechanistically, the group receiving 40 mg/kg ASTA showed significantly elevated expression of key antioxidant genes, including Cu–Zn SOD (SOD1), Mn SOD (SOD2), and NRF2, in both the liver and kidney (Heng et al., 2021).

11.5 MICROALGAE AS A FEED SUPPLEMENT IN RUMINANTS

Microalgae, the single-celled aquatic organisms, synthesize a diverse range of bioactive compounds. These potent molecules are responsible for the microalgae's efficacy in enhancing the immune function and improving the antioxidant status of animals. As a result, dietary fortification with microalgae has been a subject of extensive research in livestock settings, demonstrating potential benefits such as optimizing growth, increasing milk production, boosting overall health and immunity, and reducing inflammation (Zhu et al., 2024). Numerous microalgal species from prokaryotes like *Spirulina* (*Arthrospira*) to eukaryotes such as *Chlorella*, *Scenedesmus*, and *Schizochytrium* are currently used as ruminant feed additives (Zhu et al., 2024). Their simple, fast-growing cells allow them to rapidly accumulate significant levels of the major nutritional components (proteins, carbohydrates, and lipids) under controlled conditions (Girotto and Scapini, 2025). *Spirulina* spp. and *Chlorella* spp. are the dominant choices for ruminant diets because of their superior nutritional content and suitability for large-scale commercial production. These species typically provide 50–70% protein by dry weight, along with an outstanding profile of essential amino acids required by ruminants (Alharthi et al., 2024). For instance, Alharthi et al. (2024) reported that the addition of *Spirulina* at 0.2–0.8% of lamb diets significantly improved growth performance, enhanced immune function, and modulated the rumen microbiota. This gut modulation was characterized by a reduction in *Butyrivibrio fibrisolvens* and an increase in the levels of *Ruminococcus albus, Ruminococcus flavefaciens*, and *Streptococcus bovis*, which were higher in the *Spirulina* groups (Alharthi et al., 2024). The appeal of using microalgae as ruminant feed additives is significantly boosted by the high levels of bioactive compounds in their biomass, such as unsaturated fatty acids and essential vitamins, which are vital for ruminant health and product quality.

For instance, *Spirulina* species are recognized for their high concentration of protein and iron, in addition to being primarily composed of omega-6 unsaturated fatty acids (Ötleş and Pire, 2001). In contrast, *Chlorella* species are notably rich in essential vitamins (particularly the B-complex group) and beta-carotene, alongside significant levels of protein and omega-3 fatty acids (Ötleş and Pire, 2001; Heo et al., 2026). These compounds notably improve ruminant health by alleviating oxidative stress and reinforcing immunity (Ringø et al., 2025). Other species offer valuable omega-3 fatty acids: *Nannochloropsis* spp. accumulates high levels of eicosapentaenoic acid (EPA) under stress, while *Isochrysis galbana* and *Schizochytrium* spp. are rich sources of docosahexaenoic acid (DHA) (Barcelo-Villalobos et al., 2025; Girotto and Scapini, 2025; Heo et al., 2026). Both EPA and DHA are known to enhance both animal health and the nutritional value of their products.

The effectiveness of microalgae as a ruminant supplement is highly sensitive to the strain, culture conditions, and growth stage, which directly influence the concentration of nutritional and bioactive components (Boukrouh et al., 2025). This makes quality control vital; for instance, the DHA content in *Aurantiochytrium sp.* T66 is highly dependent on culture optimization (Alders et al., 2021). Beyond strain selection and basic optimization, metabolic engineering and traditional mutagenesis offer powerful tools to enhance microalgal strains, boosting key components like omega-3 fatty acids or proteins to better suit ruminant diets (Boukrouh et al., 2025). Critically, because few species have been tested, and the mechanisms behind their benefits are poorly understood, there is a large scope for further research.

Future investments should prioritize evaluating new, fast-growing species and employing omics technologies (genomics, metabolomics) to precisely elucidate how microalgae impact rumen fermentation and nutrient metabolism.

Previous studies have investigated the impact of microalgal feed additives on ruminant growth and performance. Although direct comparisons are challenging due to variations in microalgal species, biomass amount, and feeding protocols, the overall results suggest that selected species hold significant promise as feed supplements. Among those tested, the high-protein species *Spirulina platensis* has received the most attention over the last decade. For instance, including *S. platensis* in steer diets was shown to increase feed intake and the efficiency of microbial protein production, though it simultaneously decreased digesta retention time (Panjaitan et al., 2015).

According to the systematic analysis, Boukrouh et al. (2025) found that including microalgae in the diet can improve DHA and PUFA levels. Supplementing goat diets with microalgae may also enhance milk quality and production. Furthermore, a linear correlation was noted between the nitrogen derived from *S. platensis* supplementation and the animals' average daily gain. *Spirulina platensis* and live yeast (at 2% of the diet) improved rumen fermentation, microbiota, feed efficiency, milk yield, and composition in lactating Shami goats. (Emara Rabee et al., 2025).

The positive effects of *S. platensis* extend to feed components derived from it, as the residue remaining after lipid extraction showed a positive correlation with increased organic matter intake and digestibility (Drewery et al., 2014).

Collectively, these findings suggest that the high-protein *S. platensis* is a highly promising supplement for cattle grazing low-protein forages, potentially boosting feed intake and nutrient utilization efficiency to promote growth and productivity (Panjaitan et al., 2015). In dairy cows, *S. platensis* has also demonstrated benefits. For instance, supplementing Lithuanian black and white cows with 200 g/day of *S. platensis* during early lactation for 90 days improved their body condition (making them 8.5–11% fatter) (Kulpys et al., 2009).

Similarly, its daily inclusion in the diets of Finnish Ayrshire cows increased the digestibility of fiber and nitrogen (Lamminen et al., 2017). However, feed intake was not consistently affected, a variation possibly attributed to differences in individual cow adaptability or the palatability of the microalgae dosage (Lamminen et al., 2017).

Recently, *S. platensis* can be used as a functional food when fortified in cow diets (Barcelo-Villalobos et al., 2025). *S. platensis* (250 g/day, 1% w/w of the feed ration) significantly reduces bacterial content and somatic cell count in milk cows (Barcelo-Villalobos et al., 2025). This study demonstrated the antimicrobial activity of *Spirulina platensis* and its ability to modulate the rumen microbiota (Panjaitan et al., 2015), affecting the microbial composition in milk. The benefits of *Spirulina* supplementation have also been documented in goats and lambs. For example, feeding weaner lambs a diet incorporating 10 (w/v) of *S. platensis* daily for six weeks led to increases in both live weight and body condition score (Holman et al., 2012). Further evidence shows that including 1 g of *S. platensis* per 10 kg body weight daily for 35 days not only boosted live body weight and daily live weight gain but also increased feed intake and the feed conversion ratio in lambs (El-Sabagh et al., 2014). Similarly, in lambs facing drought conditions, supplementing with medium (100 mL) and high (200 mL) levels of 10 (w/v) *S. platensis* per day over nine weeks improved liveweight and average daily gain (Holman et al., 2014).

Nannochloropsis exhibits many biological activities that support its use in animal diets for enhancing immunity and overall health of livestock. In a recent study by El-Sayed et al. (2024), it was discovered that supplementing pregnant ewes with 10 g of *Nannochloropsis oculata* significantly improved their immune function and antioxidant status by regulating gene expression related to these functions. These findings were confirmed by enhanced blood metabolism and reduced levels of TNF-α and MDA, resulting in a decrease in stillbirths among newly born lambs. Additionally, Salehian et al. (2025) indicated that the addition of *N. oculate* (1–5 g/100 g dry matter) in cattle had no significant effects on the nutrient digestibility and kinetics of gas production. Research on animals has demonstrated that species like *N. oculata* are a fantastic addition to the diet. Not only can they substitute conventional protein sources and provide essential EPA (C20:5 n3) (Zahran et al., 2023), but they also enhance animal health. Studies show beneficial effects on palatability, digestion, immunity, and even exhibit antioxidant, anti-inflammatory, and anticancer properties (Sanjeewa et al., 2016).

11.6 MICROALGAE AND METHANE MITIGATION

Given the substantial contribution of ruminant digestion to livestock's methane emissions, microalgal supplementation is being studied as a mitigation strategy, mainly via *in vitro* assays. The impact of microalgal feed additives on rumen-derived methane production is highly inconsistent (Ike et al., 2024), varying significantly between strains. For example, *in vitro* assays assessing species like *Scenedesmus* sp., *C. vulgaris*, and *Nannochloropsis granulate* showed varying results, with only *Tetracystis* sp. hinting at a potential to reduce methane generation (Alders et al., 2021).

These studies often show promise, though results vary by species and preparation method. For instance, supplementing non-lactating Holstein cows' rumen fluid with *Euglena gracilis* led to a 9.1% reduction in methane without harming fermentation (Aemiro et al., 2016). Likewise, 5% *Nannochloropsis oculate* reduced methane production in Barki sheep rumen fluid (Gomaa et al., 2018). Interestingly, *Scenedesmus* sp. biomass reduced methane only after lipid extraction, not as whole cells. Finally, *Chlorella* spp. also demonstrated a minor reduction in methane production at a 25% inclusion rate (Sucu, 2020). The comparative study on reducing bovine methane emissions concluded that *Spirulina* was more effective than *Chlorella* (Lobo et al., 2024). This suggests *Spirulina* is the more promising algae source for CH_4 mitigation. An *in vitro* experiment demonstrated that feeding lactating Shami goats a diet fortified with 2% *Spirulina platensis* and live yeast was effective in lowering methane emissions (Emara Rabee et al., 2025). Adding *N. oculata* (1–5 g/100 g dry matter) to the diet reduced methane production, protozoa population, total volatile fatty acids (VFA), and modulated the rumen microbiota while increasing the partitioning factor (Salehian et al., 2025). Other *in vitro* studies with *C. vulgaris*, *Nannochloropsis oculate* (Gomaa et al., 2018), and *Schizochytrium* sp. failed to reduce emissions and sometimes even increased methane output (Ruiz-González et al., 2018). Given these varied *in vitro* findings, moving forward with in vivo experiments is crucial to validate microalgae's value in mitigating methane. It is important to note that the most effective algal supplement identified to date is not microalgae but the red macroalga *Asparagopsis*, which consistently reduces methane due to its natural synthesis of bromoform. Due to their shared evolutionary path and habitat, it is plausible that microalgae with similar bromoform-synthesizing capabilities could be discovered. Additionally, well-developed genetic engineering techniques could be leveraged to reconstruct bromoform synthesis in certain microalgae, offering another path to methane reduction. However, because bromoform is a suspected carcinogen, further safety research is necessary. Finally, screening microalgae for bioactive compounds that directly inhibit methanogenic archaea, or their processes, represents an alternative, promising strategy for decreasing ruminant methane emissions.

11.7 MICROALGAE AND ENVIRONMENTAL HEAT STRESS

Heat stress is a critical environmental issue negatively impacting animal health, well-being, and homeostasis. In one study, adding *Spirulina* to the diets of stressed broilers at 0.5, 1, and 1.5% of the diet improved growth indices (the 1 % dose was most effective) (Moustafa et al., 2021). Researchers also observed that *Spirulina* lowered lipid accumulation and MDA levels and enhanced antioxidant enzyme activity (SOD and GSH). In broilers, adding Spirulina to water at 15 or 20g/L did not affect the productive performance of broilers under hot conditions. However, these doses enhanced the cell-mediated and humoral immunity against Phytohemagglutinin-P and Newcastle disease vaccination (Kolluri et al., 2022). For laying hens subjected to heat stress (Al-Otaibi et al., 2022), adding 90 g/kg of *Spirulina* to the diet demonstrably improved productive performance, cholesterol profile, redox status (MDA, GPX, and SOD), and pro-inflammatory cytokine levels (IL-1β, IL-6, and TNF-α). *Spirulina* supplementation (1 g/kg diet) significantly improved the performance and health status of Japanese quails subjected to heat stress (Fawaz et al., 2025). This benefit was achieved by regulating the gut microbiota: the supplementation significantly reduced the populations of pathogenic bacteria, specifically *Escherichia coli* (*E. coli*), *Clostridium* spp., and *Enterococcus* spp., while promoting the growth of beneficial *Bacillus* spp. (Fawaz et al., 2025). The same improvements were evident in laying Japanese quails, attributed to the modulation of the gut microbiota and a decrease in oxidative stress (Hajati et al., 2020). In two separate studies, supplementing the feed of heat-stressed broiler chickens with *Spirulina* (10 g/kg) and selenium nanoparticles (0.1 mg/kg) resulted in improved growth performance, carcass yield, immunity, and antioxidant capacity (Abdel-Moneim et al., 2022a; Abdel-Moneim et al., 2022b). Following a similar trend, multiple studies (El-Ratel et al., 2023a; El-Ratel et al., 2023b) demonstrated a synergistic effect between selenium nanoparticles (SeNPs) at 25–50 mg/kg diet and *Spirulina* at 1 g/kg diet in stressed rabbits. This combination was shown to regulate blood metabolic and antioxidant status while reducing oxidative markers. The same authors also reported that the addition of SeNPs and *Spirulina* enhanced semen quality and growth indices in rabbits. Another study (Bashar et al., 2024) found that supplementing stressed rabbits' diets with either *Spirulina* (1 g/kg) or SeNPs (50 mg/kg), or both together, improved their growth, blood picture, and immune-antioxidant responses. These feed additives serve as an effective nutritional tool for reducing the negative effects of summer stress. The integration of nanotechnology has led to the synthesis of Nano-Spirulina (Nano-SP), which is now used to mitigate the negative impact of heat stress (HS) in female rabbits. A study investigating Nano-SP (at 15 mg and 35 mg/kg BW) found that the nano-formulation enhanced oxidative status, boosted reproductive hormone levels, and offered protection to reproductive tissues against HS effects (Hanafi et al., 2025). Notably, *Spirulina* stimulated folliculogenesis earlier compared to the control groups. These findings suggest that Nano-SP successfully enhances puberty in female rabbits experiencing HS. Under HS conditions relevant to global warming, the addition of SP to the diet of stressed sheep was recently shown to be beneficial. SP improved the sheep's resistance to HS by enhancing their immune-antioxidant capacity, adipokines, physio-metabolic, and promoting better growth performance and overall health (Alfaraj et al., 2025). Table 11.2 clarifies the role of microalgae to enhance the heat resistance in animals.

TABLE 11.2
The Role of Microalgae to Enhance the Heat Resistance in Animals

Species	Dose and Type	Finding	References
Laying hens	*Spirulina* (0, 3, 6, and 9% of the diet)	• Improve the productive performance of stressed laying hens. • Enhance the antioxidant enzymes (GSH and MDA) and reduce oxidative stress (MDA). • Decrease the levels of pro-inflammatory cytokines (IL-1β, IL-6, and TNF-α).	Al-Otaibi et al., 2022
Broiler	*Spirulina* (0.5, 1, and 1.5%),	• Reduce lipid and oxidative stress. • Improve antioxidant status and growth performance.	Moustafa et al., 2021
Broiler	*Spirulina* added to drinking water at 15–20g/L	• Enhanced the cell-mediated and humoral immunity against Phytohemagglutinin-P and Newcastle disease vaccination	Kolluri et al., 2022
Japanese quails	*Spirulina* 1 g/kg diet	• Improved the performance and health status of Japanese quails subjected to heat stress. This benefit was achieved by regulating the gut microbiota.	Fawaz et al., 2025
Laying Japanese quails	*Spirulina* 1 g/kg diet	• The same improvements were evident in laying Japanese quails, attributed to the modulation of the gut microbiota and a decrease in oxidative stress	Hajati et al., 2020
Rabbit	Selenium nanoparticles (SeNPs) at 25–50 mg/kg diet and *Spirulina* at 1 g/kg diet	• Improve semen quality and enhance seminal antioxidant enzymes. • Enhancing growth and modulating blood biochemicals.	El-Ratel et al., 2023a, 2023b
Female rabbits	Spirulina nanoparticles (15–35 mg/kg BW)	• Enhance female rabbit reproductive health through the mitigation of inflammation and the strengthening of antioxidant defenses.	Hanafi et al., 2025

In most animal applications, the optimal or most effective dose was found to be 1 g/kg of diet. Overall, SP, a nutrient-dense filamentous cyanobacterium, possesses wide-ranging therapeutic and health benefits due to its rich content of bioactive compounds (e.g., phycocyanin, carotenoids, and phenolics). *C. vulgaris* (60 mg/rabbit/day) and *L. reuteri* (250 units) enhanced rabbit puberty and ameliorated the adverse effects of high environmental temperature by promoting growth rate, ovarian follicular activity, and reducing oxidative stress (Elnady et al., 2024). Its established actions are immunostimulatory, hepatoprotective, antioxidative, and anti-inflammatory. This profile makes SP a potential treatment for various diseases and a useful tool for reducing HS in animals, as studies confirm that it improves the health and immunity of broilers and rabbits by mitigating HS-induced inflammation and oxidative stress.

11.8 ENHANCING REPRODUCTIVE EFFACING USING MICROALGAE IN LIVESTOCK

Microalgae are a promising feed supplement that appears to have positive effects on reproductive efficiency in some livestock, particularly due to their rich profile of essential nutrients and bioactive compounds. The study by Emara Rabee et al. (2025) found that dietary chlorella improved reproductive health by regulating rumen fermentation and nutrient digestibility. Additionally, in a study on pubertal rabbits, Elnady et al. (2024) found that the addition of *Chlorella vulgaris* algae and *Lactobacillus reuteri* reduced the time to puberty in rabbits. Supplementation with *Dunaliella salina* (10 g/head/day) can enhance ovarian follicle development and increase concentrations of plasma metabolites and hormones, leading to improved reproductive performance in Boer goats (Senosy et al., 2017). The study resulted in an improvement in the reproductive efficiency of goats by regulating ovarian activity and hormonal systems with the addition of 10 g of *Chlorella* per day orally (Silva et al., 2023). There is a lack of studies regarding the potential impacts of microalgae on reproductive efficiency in livestock. Further studies are required in this area, especially when using omega-3 or omega-6 isolated from microalgae.

11.9 CHALLENGES OF APPLYING MICROALGAE IN LIVESTOCK DIETS

Despite the unique nutritional value and growth characteristics of microalgae, several challenges hinder their adoption as a ruminant feed additive. The primary hurdle is the high cost of microalgal biomass compared to conventional feed sources. This expense largely stems from current culture methods, particularly autotrophic culture, which exhibit relatively low productivity, and from the energy-intensive and costly processes of harvesting and dewatering (Lafuente-Rincon et al., 2024; Song et al., 2025). To improve commercial viability and ensure consistent quality, optimizing methods for large-scale cultivation, harvesting, dewatering, and contamination control are essential. However, species like *Schizochytrium* and *Chlorella* offer a promising path, as they thrive heterotrophically using well-developed large-scale fermentation (Saadaoui et al., 2021). A final, significant challenge is regulatory approval, as only a limited number of species, such as *Chlorella*, *Spirulina*, and *Schizochytrium*, are currently approved for use in food and feed (Bature et al., 2022).

Obtaining regulatory approvals for microalgal feed additives is a lengthy, costly process that requires robust scientific backing (Lafuente-Rincon et al., 2024). Therefore, a practical initial strategy is to prioritize the development and testing of microalgal species that already hold regulatory clearance for use in ruminant diets. Furthermore, issues with feed palatability and resulting reduced intake must be addressed to ensure effective adoption. Future research is essential to improve the flavor of microalgal supplements. Finally, comprehensive studies are necessary to fully understand how different microalgal species impact ruminant nutrition and health, particularly focusing on the interaction between microalgae and rumen microbes and the underlying mechanisms driving their observed effects.

11.10 CONCLUSIONS AND PERSPECTIVES

The integration of microalgae-based feed additives shows great promise for advancing ruminant production and enhancing the sustainability of the livestock industry. Growing recognition of microalgae's nutritional value has fueled significant research interest. Cumulative studies have confirmed that microalgae serve as unique nutritional supplements, effectively meeting ruminant dietary requirements, boosting physiological performance, and improving the nutritional profile of animal products. However, several critical areas require further investigation: improving microalgal production efficiency, deepening our understanding of the molecular mechanisms behind these beneficial effects, and thoroughly assessing the function of microalgal supplementation in reducing methane emissions. Addressing these areas will facilitate the widespread adoption of microalgae as ruminant feed supplements in the future.

COMPETING INTERESTS

No potential conflict of interest was reported by the author(s).

FUNDING

Not applicable

ACKNOWLEDGMENTS

Not applicable

AVAILABILITY OF DATA AND MATERIAL

The data supporting the findings of this study will be made available upon reasonable request to the corresponding author.

REFERENCES

Abd El-Hack, M.E., Abdelnour, S., Alagawany, M., Abdo, M., Sakr, M.A., Khafaga, A.F., Mahgoub, S.A., Elnesr, S.S., Gebriel, M.G. (2019). Microalgae in modern cancer therapy: Current knowledge. *Biomedicine & Pharmacotherapy* 111, 42–50.

Abdelfatah, S.H., Yassin, A.M., Khattab, M.S., Abdel-Razek, A.S., Saad, A.H. (2024). Spirulina platensis as a growth booster for broiler; Insights into their nutritional, molecular, immunohistopathological, and microbiota modulating effects. *BMC Veterinary Research* 20, 11.

Abdel-Moneim, A.-M.E., Shehata, A.M., Mohamed, N.G., Elbaz, A.M., Ibrahim, N.S. (2022a). Synergistic effect of *Spirulina platensis* and selenium nanoparticles on growth performance, serum metabolites, immune responses, and antioxidant capacity of heat-stressed broiler chickens. *Biological Trace Element Research* 200, 768–779.

Abdel-Moneim, A.-M.E., Shehata, A.M., Selim, D.A., El-Saadony, M.T., Mesalam, N.M., Saleh, A.A. (2022b). *Spirulina platensis* and biosynthesized selenium nanoparticles improve performance, antioxidant status, humoral immunity and dietary and ileal microbial populations of heat-stressed broilers. *Journal of Thermal Biology* 104, 103195.

Abdelnour, S.A., Abd El-Hack, M.E., Arif, M., Khafaga, A.F., Taha, A.E. (2019a). The application of the microalgae Chlorella spp. as a supplement in broiler feed. *World's Poultry Science Journal* 75, 305–318.

Abdelnour, S.A., Mahasneh, Z.M.H., Barakat, R.A., Alkahtani, A.M., Madkour, M. (2024). Microalgae: A promising strategy for aflatoxin control in poultry feeds. *Toxicon* 244, 107770.

Abdelnour, S.A., Sheiha, A.M., Taha, A.E., Swelum, A.A., Alarifi, S., Alkahtani, S., Ali, D., AlBasher, G., Almeer, R., Falodah, F., Almutairi, B., Abdel-Daim, M.M., Abd El-Hack, M.E., Ismail, I.E. (2019b). Impacts of enriching growing rabbit diets with *Chlorella vulgaris* microalgae on growth, blood variables, carcass traits, immunological and antioxidant indices. *Animals* 9, 788.

Abdelnour, S.A., Swelum, A.A., Salama, A., Al-Ghadi, M.Q., Qattan, S.Y.A., Abd El-Hack, M.E., Khafaga, A.F., Alhimaidi, A.R., Almutairi, B.O., Ammari, A.A., El-Saadony, M.T. (2020). The beneficial impacts of dietary phycocyanin supplementation on growing rabbits under high ambient temperature. *Italian Journal of Animal Science* 19, 1046–1056.

Abdel-Wareth, A.A., Williams, A.N., Salahuddin, M., Gadekar, S., Lohakare, J. (2024). Algae as an alternative source of protein in poultry diets for sustainable production and disease resistance: Present status and future considerations. *Frontiers in Veterinary Science* 11, 1382163.

Abotaleb, M.M., Mourad, A., Abousenna, M.S., Helal, A.M., Nassif, S.A., Elsafty, M.M. (2020). The effect of Spirulina algae on the immune response of SPF chickens to commercial inactivated Newcastle vaccine in poultry. *VacciMonitor* 29, 51–57.

Aemiro, A., Watanabe, S., Suzuki, K., Hanada, M., Umetsu, K., Nishida, T. (2016). Effects of Euglena (*Euglena gracilis*) supplemented to diet (forage: concentrate ratios of 60:40) on the basic ruminal fermentation and methane emissions in in vitro condition. *Animal Feed Science and Technology* 212, 129–135.

Alders, R.G., Campbell, A., Costa, R., Guèye, E.F., Ahasanul Hoque, M., Perezgrovas-Garza, R., Rota, A., Wingett, K. (2021). Livestock across the world: Diverse animal species with complex roles in human societies and ecosystem services. *Animal Frontiers* 11, 20–29.

Alfaraj, A.I., Mahmoud, H.K., Reda, F.M., Monem, U.M.A., Almutairi, L.A., Al-Shahari, E.A., Alsulami, R.N., Abdelnour, S.A. (2025). Nanozymes or spirulina platensis: Enhancing sheep thermo-tolerance through physio-metabolic, immune, and antioxidant pathways. *Biological Trace Element Research* 203, 6503–6517. https://doi.org/10.1007/s12011-025-04656-4

Alharthi, A.S., Al-Baadani, H.H., Alghonaim, A.A., Al-Garadi, M.A., Alowaimer, A.N., Alhidary, I.A. (2024). Effects of spirulina platensis addition on performance, immune response, hematological, selected bacteria activity and rumen morphology of lambs. *Italian Journal of Animal Science* 23, 1134–1145.

Al-Otaibi, M.I., Abdellatif, H.A., Al-Huwail, A.K., Abbas, A.O., Mehaisen, G.M., Moustafa, E.S. (2022). Hypocholesterolemic, antioxidative, and anti-inflammatory effects of dietary spirulina platensisis supplementation on laying hens exposed to cyclic heat stress. *Animals* 12, 2759.

Ampofo, J., Abbey, L. (2022). Microalgae: Bioactive composition, health benefits, safety and prospects as potential high-value ingredients for the functional food industry. *Foods* 11, 1744.

Attia, Y.A., Hassan, R.A., Addeo, N.F., Bovera, F., Alhotan, R.A., Al-Qurashi, A.D., Al-Baadani, H.H., Al-Banoby, M.A., Khafaga, A.F., Eisenreich, W. (2023). Effects of Spirulina platensis and/or Allium sativum on antioxidant status, immune response, gut morphology, and intestinal *Lactobacilli* and *Coliforms o*f heat-stressed broiler chicken. *Veterinary Sciences* 10, 678.

Bai, J., Wang, R., Yan, L., Feng, J. (2019). Co-supplementation of dietary seaweed powder and antibacterial peptides improves broiler growth performance and immune function. *Brazilian Journal of Poultry Science* 21, eRBCA-2018-0826.

Balasubramanian, B., Shanmugam, S., Park, S., Recharla, N., Koo, J.S., Andretta, I., Kim, I.H. (2021). Supplemental impact of marine red seaweed (*Halymenia palmata*) on the growth performance, total tract nutrient digestibility, blood profiles, intestine histomorphology, meat quality, fecal gas emission, and microbial counts in broilers. *Animals* 11, 1244.

Barcelo-Villalobos, M., Arrebola, F., Vizcaino, A., Alarcón, F.J., Acién, F.G. (2025). Influence of including Arthrospira platensis as a functional ingredient in dairy cow feed. *Algal Research* 91, 104295.

Bashar, A.M., Abdelnour, S.A., El-Darawany, A.A., Sheiha, A.M. (2024). Dietary supplementation of microalgae and/or nanominerals mitigate the negative effects of heat stress in growing rabbits. *Biological Trace Element Research* 202, 3639–3652.

Basheer, S., Huo, S., Zhu, F., Qian, J., Xu, L., Cui, F., Zou, B. (2020). *Microalgae in Human Health and Medicine. Microalgae Biotechnology for Food, Health and High Value Products*. Springer, pp. 149–174.

Bature, A., Melville, L., Rahman, K.M., Aulak, P. (2022). Microalgae as feed ingredients and a potential source of competitive advantage in livestock production: A review. *Livestock Science* 259, 104907.

Boukrouh, S., Mnaouer, I., Mendes de Souza, P., Hornick, J.-L., Nilahyane, A., El Amiri, B., Hirich, A. (2025). Microalgae supplementation improves goat milk composition and fatty acid profile: A meta-analysis and meta-regression. *Archives Animal Breeding* 68, 223–238.

Chen, L., Wang, L., Li, Y., Wu, X., Long, X. (2024). Comprehensive metabolomics and transcriptomics analyses investigating the regulatory effects of different sources of dietary astaxanthin on the antioxidant and immune functions of commercial-sized rainbow trout. *Frontiers in Immunology* 15, 1408168.

Drewery, M., Sawyer, J., Pinchak, W., Wickersham, T. (2014). Effect of increasing amounts of postextraction algal residue on straw utilization in steers. *Journal of Animal Science* 92, 4642–4649.

Elnady, S., Shehab El-Din, M.I., Khalifa, W.H., Danial, E.N., Aboelmaaty, A., Hanafy, E., Madboli, A.E.-N.A. (2024). The effects of dietary supplementation of *Chlorella vulgaris* alga and *Lactobacillus reuteri* bacteria in female rabbit puberty subjected to heat stress. *Egyptian Journal of Veterinary Sciences* 56(12), 3211–3223.

El-Ratel, I.T., Elbasuny, M.E., El-Nagar, H.A., Abdel-Khalek, A.-K.E., El-Raghi, A.A., El Basuini, M.F., El-Kholy, K.H., Fouda, S.F. (2023b). The synergistic impact of Spirulina and selenium nanoparticles mitigates the adverse effects of heat stress on the physiology of rabbits bucks. *Plos One* 18, e0287644.

El-Ratel, I.T., El-Kholy, K.H., Mousa, N.A., El-Said, E.A. (2023a). Impacts of selenium nanoparticles and spirulina alga to alleviate the deleterious effects of heat stress on reproductive efficiency, oxidative capacity and immunity of doe rabbits. *Animal Biotechnology* 34, 3519–3532.

El-Sabagh, M.R., Abd Eldaim, M.A., Mahboub, D., Abdel-Daim, M. (2014). Effects of Spirulina platensis algae on growth performance, antioxidative status and blood metabolites in fattening lambs. *Journal of Agricultural Science* 6, 92.

El-Sayed, A., Ebissy, E., Ateya, A. (2024). Positive impacts of Nannochloropsis oculata supplementation on gene expression of immune and antioxidant markers and metabolic profile of Barki sheep in the transition period and lipogenic effects on progeny. *Veterinary Research Communications* 48, 2207–2226.

El-Shall, N.A., Jiang, S., Farag, M.R., Azzam, M., Al-Abdullatif, A.A., Alhotan, R., Dhama, K., Hassan, F.-U., Alagawany, M. (2023). Potential of Spirulina platensis as a feed supplement for poultry to enhance growth performance and immune modulation. *Frontiers in Immunology* 14, 1072787.

Emam, K., Ali, S.A., Morsy, A., Fouda, W.A., Elbaz, A.M. (2024). Role of Nannochloropsis oculata supplement in improving performance, antioxidant status, blood metabolites, and egg quality of laying hens under hot environmental conditions. *Scientific Reports* 14, 16884.

Emara Rabee, A., Ghandour, M.M.M., Sallam, A.M., Raef, O., Elwakeel, E.A., Sabra, E.A., Abdel-Wahed, A.M., Abo Bakr, S., Saad ElSamahy, H., Amin Hamed, A. (2025). Milk yield, rumen fermentation, and microbiota of Shami goats fed diets supplemented with spirulina and yeast. *AMB Express* 15, 108.

Fawaz, M.A., Ali, M.A., Khalaphallah, R., Hassan, H.A., Ali, A.H.H. (2025). Impacts of propolis and Spirulina platensis supplementation on growth, nutrient digestibility, and gut microbiota of Japanese quails under heat stress. *Scientific Reports* 15, 31688.

Fiedor, J., Burda, K. (2014). Potential role of carotenoids as antioxidants in human health and disease. *Nutrients* 6, 466–488.

Fox, J.M., Zimba, P.V. (2018). Minerals and trace elements in microalgae. In Ira A. Levine and Joël Fleurence (eds.), *Microalgae in Health and Disease Prevention*. Elsevier, pp. 177–193.

Gao, S., Li, R., Heng, N., Chen, Y., Wang, L., Li, Z., Guo, Y., Sheng, X., Wang, X., Xing, K., Ni, H., Qi, X. (2020). Effects of dietary supplementation of natural astaxanthin from *Haematococcus pluvialis* on antioxidant capacity, lipid metabolism, and accumulation in the egg yolk of laying hens. *Poultry Science* 99, 5874–5882.

Girotto, F., Scapini, A. (2025). Microalgal biomass in the European food industry: Navigating regulation, technological innovation, and consumer acceptance. *Algal Research* 91, 104288.

Gomaa, A.S., Kholif, A.E., Kholif, A.M., Salama, R., El-Alamy, H.A., Olafadehan, O.A. (2018). Sunflower oil and Nannochloropsis oculata microalgae as sources of unsaturated fatty acids for mitigation of methane production and enhancing diets' nutritive value. *Journal Of Agricultural and Food Chemistry* 66, 1751–1759.

Hafez, H.M., Attia, Y.A. (2020). Challenges to the poultry industry: Current perspectives and strategic future after the COVID-19 outbreak. *Frontiers in Veterinary Science* 7, 516.

Hajati, H., Zaghari, M., Oliveira, H. (2020). Arthrospira (*Spirulina*) platensis can be considered as a probiotic alternative to reduce heat stress in laying Japanese quails. *Brazilian Journal of Poultry Science* 22, eRBCA-2018-0977.

Hanafi, E.M., Khalifa, W.H., Elnady, I.A., Danial, E.N., Ramadan, M.M., Ezzo, O.H., Aboelmaaty, A.M., Mdboli, A.E.-N.A. (2025). The role of nano-spirulina platensis in enhancement of female rabbit puberty and amelioration of the heat stress adverse effects. *Egyptian Journal of Chemistry* 68, 515–526.

Heng, N., Gao, S., Chen, Y., Wang, L., Li, Z., Guo, Y., Sheng, X., Wang, X., Xing, K., Xiao, L., Ni, H., Qi, X. (2021). Dietary supplementation with natural astaxanthin from Haematococcus pluvialis improves antioxidant enzyme activity, free radical scavenging ability, and gene expression of antioxidant enzymes in laying hens. *Poultry Science* 100, 101045.

Heo, Y., Kim, M.-Y., Cho, J.Y. (2026). *Chlorella vulgaris*, a representative edible algae as integrative and alternative medicine. *Integrative Medicine Research* 15, 101228.

Holman, B., Kashani, A., Malau-Aduli, A. (2012). Growth and body conformation responses of genetically divergent Australian sheep to Spirulina (*Arthrospira platensis*) supplementation. *American Journal of Experimental Agriculture* 2, 160–173.

Holman, B., Kashani, A., Malau-Aduli, A. (2014). Effects of spirulina (*Arthrospira platensis*) supplementation level and basal diet on liveweight, body conformation and growth traits in genetically divergent Australian dual-purpose lambs during simulated drought and typical pasture grazing. *Small Ruminant Research* 120, 6–14.

Ike, K.A., Okedoyin, D.O., Alabi, J.O., Adelusi, O.O., Wuaku, M., Olagunju, L.K., Anotaenwere, C.C., Gray, D., Dele, P.A., Kholif, A.E., Thomas, M.D., Anele, U.Y. (2024). The combined effect of four nutraceutical-based feed additives on the rumen microbiome, methane gas emission, volatile fatty acids, and dry matter disappearance using an in vitro batch culture technique. *Fermentation* 10, 499.

Irshad, Z., Qasim, M., Hajati, H., Hosseini, S.A. (2024). Functional feed for broiler chickens: Exploring Spirulina platensis as a nutritional supplement. *World's Poultry Science Journal* 80, 511–525.

Kashenye, B.N., Zhang, J. (2025). Gastrointestinal modification based on probiotic feed additive enviro-alleviators to reduce enteric methane production in ruminant and non-ruminant livestock. *Resources, Environment and Sustainability* 22, 100276.

Kholif, A.E., Olafadehan, O.A. (2021). Chlorella vulgaris microalgae in ruminant nutrition: A review of the chemical composition and nutritive value. *Annals of Animal Science* 21, 789–806.

Kolluri, G., Marappan, G., Yadav, A.S., Kumar, A., Mariappan, A.K., Tyagi, J.S., Rokade, J.J., Govinthasamy, P. (2022). Effects of Spirulina (*Arthrospira platensis*) as a drinking water supplement during cyclical chronic heat stress on broiler chickens: Assessing algal composition, production, stress, health and immune-biochemical indices. *Journal of Thermal Biology* 103, 103100.

Kulpys, J., Paulauskas, E., Pilipavicius, V., Stankevicius, R. (2009). Influence of cyanobacteria Arthrospira (*Spirulina*) platensis biomass additive towards the body condition of lactation cows and biochemical milk indexes. *Agronomy Research* 7, 823–835.

Lafuente-Rincon, D.F., Hernández-Beltrán, J.U., Hernández-Terán, F., Balagurusamy, N. (2024). Microalgae-based feed: A sustainable bioproduct for the livestock sector in arid zones. In Cristobal Noé Aguilar, Norma Paola Meléndez-Renteria, and Karen Nathiely Ramirez-Guzman (eds.), *Exploration and Valorization of Natural Resources from Arid Zones*. Apple Academic Press, pp. 233–257.

Lamminen, M., Halmemies-Beauchet-Filleau, A., Kokkonen, T., Simpura, I., Jaakkola, S., Vanhatalo, A. (2017). Comparison of microalgae and rapeseed meal as supplementary protein in the grass silage based nutrition of dairy cows. *Animal Feed Science and Technology* 234, 295–311.

Lee, J.Y., Yoon, J.H., An, S.H., Cho, I.H., Lee, C.W., Jeon, Y.J., Joo, S.S., Ban, B.C., Lee, J.-Y., Jung, H.J. (2023). Intestinal immune cell populations, barrier function, and microbiomes in broilers fed a diet supplemented with Chlorella vulgaris. *Animals* 13, 2380.

Lobo, R.R., Almeida, E., Monteiro, A., da Silva, S.S., Salas-Solis, G., Coronella, C.J., Hiibel, S.R., Faciola, A.P. (2024). Replacing soybean meal with microalgae biomass in diets with contrasting carbohydrate profiles can reduce in vitro methane production and improve short-chain fatty acid production. *Journal of Dairy Science* 107, 5542–5555.

Ma, M., Hu, Q. (2024). Microalgae as feed sources and feed additives for sustainable aquaculture: Prospects and challenges. *Reviews in Aquaculture* 16, 818–835.

Matshogo, T.B., Mnisi, C.M., Mlambo, V. (2020). Dietary green seaweed compromises overall feed conversion efficiency but not blood parameters and meat quality and stability in broiler chickens. *Agriculture* 10, 547.

Mitra, A., Tripathy, A.K., Singh, R.K., Bujarbaruah, K.M. (2024). Transforming the Indian livestock sector. In K. C. Bansal, W. S. Lakra, and Himanshu Pathak (eds.), *Transformation of Agri-Food Systems*. Springer, pp. 173–191.

Moustafa, E.S., Alsanie, W.F., Gaber, A., Kamel, N.N., Alaqil, A.A., Abbas, A.O. (2021). Blue-green algae (*Spirulina platensis*) alleviates the negative impact of heat stress on broiler production performance and redox status. *Animals* 11, 1243.

Nhlane, L.T., Mnisi, C.M., Mlambo, V., Madibana, M.J. (2020). Nutrient digestibility, growth performance, and blood indices of Boschveld chickens fed seaweed-containing diets. *Animals* 10, 1296.

Ötleş, S., Pire, R. (2001). Fatty acid composition of chlorella and spirulina microalgae species. *Journal of Aoac International* 84, 1708–1714.

Panjaitan, T., Quigley, S., McLennan, S.R., Swain, A., Poppi, D.P. (2015). Spirulina (*Spirulina platensis*) algae supplementation increases microbial protein production and feed intake and decreases retention time of digesta in the rumen of cattle. *Animal Production Science* 55, 535–543.

Ringø, E., Ashour, M., Ahmed, S., Sharawy, Z., Goda, A., El-Haroun, E. (2025). *Microalgae and Seaweeds as Feed Additives for Aquatic Animals: Effects on Growth, Immunity, and Disease Resistance*. IntechOpen.

Ruiz-González, A., Debruyne, S., Dewanckele, L., Escobar, M., Vandaele, L., Van Den Broeck, W., Fievez, V. (2018). Supplementation of DHA-Gold pre and/or postnatally to goat kids modifies in vitro methane production and rumen morphology until 6 mo old. *Journal Animal Science* 96, 4845–4858.

Saadaoui, I., Rasheed, R., Aguilar, A., Cherif, M., Al Jabri, H., Sayadi, S., Manning, S.R. (2021). Microalgal-based feed: Promising alternative feedstocks for livestock and poultry production. *Journal of Animal Science and Biotechnology* 12, 76.

Salehian, Z., Khalilvandi-Behroozyar, H., Pirmohammadi, R., Ahmadifard, N., Almasi, H., Rodríguez, G.B., Lackner, M., Salem, A.Z. (2025). The effect of supplementing Nannochloropsis oculata microalgae on ruminal fermentation, methane production and microbial population. *Journal of Agriculture and Food Research* 19, 101663.

Sanjeewa, K.K.A., Fernando, I.P.S., Samarakoon, K.W., Lakmal, H.H.C., Kim, E.-A., Kwon, O.-N., Dilshara, M.G., Lee, J.-B., Jeon, Y.-J., Kalpa, W.S. (2016.) Anti-inflammatory and anti-cancer activities of sterol rich fraction of cultured marine microalga *Nannochloropsis oculata*. *Algae* 31, 277–287.

Senosy, W., Kassab, A.Y., Mohammed, A.A. (2017). Effects of feeding green microalgae on ovarian activity, reproductive hormones and metabolic parameters of Boer goats in arid subtropics. *Theriogenology* 96, 16–22.

Shah, A.M., Zhang, H., Shahid, M., Ghazal, H., Shah, A.R., Niaz, M., Naz, T., Ghimire, K., Goswami, N., Shi, W. (2025). The vital roles of agricultural crop residues and agro-industrial by-products to support sustainable livestock productivity in subtropical regions. *Animals* 15, 1184.

Silva, M.R.L., Alves, J.P.M., Fernandes, C.C.L., Cavalcanti, C.M., Conde, A.J.H., Bezerra, A.F., Soares, A.C.S., Teixeira, D.A., do Rego, A.C., Rondina, D. (2023). Effect of short-term nutritional supplementation of green microalgae on some reproductive indicators of Anglo-Nubian crossbred goats. *Veterinary World* 16, 464–473.

Song, Y., Hu, Z., Liu, S., Luo, S., He, R., Yang, X., Li, S., Yang, X., An, Y., Lu, Y. (2025). Utilization of microalgae and duckweed as sustainable protein sources for food and feed: Nutritional potential and functional applications. *Journal Of Agricultural And Food Chemistry* 73, 4466–4482.

Spínola, M.P., Costa, M.M., Tavares, B., Pestana, J.M., Tavares, J.C., Martins, C.F., Alfaia, C.M., Maciel, V., Carvalho, D.F., Mourato, M.P. (2024). Impact of long-term feeding a high level of Spirulina combined with enzymes on growth performance, carcass traits and meat quality in broiler chickens. *Frontiers in Veterinary Science* 11, 1451516.

Sucu, E., 2020. Effects of microalgae species on in vitro rumen fermentation pattern and methane production. *Annals of Animal Science* 20, 207–218.

Tabarzad, M., Atabaki, V., Hosseinabadi, T. (2020). Anti-inflammatory activity of bioactive compounds from microalgae and cyanobacteria by focusing on the mechanisms of action. *Molecular Biology Reports* 47, 6193–6205.

Unicef. (2024). *The State of Food Security and Nutrition in the World 2024*. https://data.unicef.org/wp-content/uploads/2024/07/SOFI2024_Report_EN_web.pdf

Wang, Y.-H., Chen, H.-H., Jiang, J.-G. (2025). Microalgae biofactories for animal based products. *Trends in Food Science & Technology* 163, 105155.

Wlaźlak, S., Biesek, J. (2025). *Spirulina platensis* and *Chlorella vulgaris* in poultry nutrition - A review of current research and potential opportunities. *Poultry Science* 104, 105456.

Wu, J.Y., Tso, R., Teo, H.S., Haldar, S. (2023). The utility of algae as sources of high value nutritional ingredients, particularly for alternative/complementary proteins to improve human health. *Frontiers in Nutrition* 10, 1277343.

Yehia, N., Mohamed, F.H., Al-Zaban, M.I., Amer, F., Baazaoui, N., Khattab, M.S., Abd Elhalem Mohamed, A., Salem, H.M., El-Saadony, M.T., El-Tarabily, K.A., Omar, D.M. (2024). The influence of Spirulina extract on pathogenicity, immune response, and vaccine efficacy against H9N2 avian influenza virus in specific pathogen free chickens. *Poultry Science* 103, 103194.

Zahran, E., Elbahnaswy, S., Ahmed, F., Ibrahim, I., Khaled, A.A., Eldessouki, E.A. (2023). Nutritional and immunological evaluation of Nannochloropsis oculata as a potential Nile tilapia-aquafeed supplement. *BMC Veterinary Research* 19, 65.

Zhu, M., Singer, S.D., Guan, L.L., Chen, G. (2024). Emerging microalgal feed additives for ruminant production and sustainability. *Advanced Biotechnology* 2, 17.

12 Microalgae for Soil Applications

New Perspectives for Plant Growth Additives

Iara Barbosa Magalhães, Moacir Caetano do Couto Junior, Diego A. R. Hennessey, Tatiana Cardoso e Búfalo, Joyce Dória, Maria Lúcia Calijuri, Eduardo de Aguiar do Couto, and Paula Assemany

12.1 INTRODUCTION

The growing global population, coupled with economic development and dietary shifts associated with higher-income groups, exerts increasing pressure on the food system to expand its boundaries and reconsider conventional agricultural production. This imperative requires not only the expansion of agricultural frontiers but, more critically, a fundamental re-evaluation of traditional agricultural paradigms (Yang et al., 2024). Crop production and livestock activities currently occupy 40% of the world's land area while contributing approximately 25% of global greenhouse gas emissions, accounting for 85% of all freshwater withdrawals for human use, and consuming 80–90% of the world's nitrogen and phosphorus inputs (Campbell et al., 2017). The intensive use of fertilizers, agrochemicals, and pesticides—together with irrigation, livestock farming, and land clearing—has been linked to severe negative impacts, including soil, air, and water pollution, water scarcity, climate change, and biodiversity loss (Campbell et al., 2017). These impacts are not isolated but often interconnected, amplifying environmental degradation on a global scale.

A sustainable future, therefore, necessitates transformative changes in agricultural production, embedding principles of circularity and sustainability within the sector. While sustainability can be advanced through effective strategies that address the economic, environmental, and social dimensions (Bathaei and Štreimikienė, 2023), circular agriculture specifically emphasizes enhancing food production via regenerative farming practices, closing nutrient loops, and minimizing waste (Moreira et al., 2023). The overarching goal is to reduce reliance on nonrenewable resources (e.g., fossil fuels), minimize water and nutrient losses, and maximize the recovery of by-products (Muscat et al., 2021). Within this framework, the Water–Energy–Food

 DOI: 10.1201/9781003611479-12

Nexus provides a valuable approach for addressing the interconnected complexity of global resource systems, serving as a critical tool for achieving social, economic, and environmental goals. By balancing the needs of diverse resource users and interests, it plays a key role in safeguarding ecosystem integrity (Dubois et al., 2014).

Microalgae are a diverse group of microorganisms with the capacity to address multiple objectives simultaneously, making them a promising practical solution for fostering synergies and enabling a more coordinated use of natural resources across various sectors. They can be cultivated in wastewater, thereby contributing to water treatment, CO_2 sequestration, and nutrient recovery while generating value-rich biomass. Several intrinsic characteristics make microalgae particularly attractive for mitigating the impacts of agriculture: they exhibit rapid growth, can be cultivated on non-arable land with non-potable water, and can be harvested year-round. These features position microalgae as a compelling and practical solution to global environmental challenges.

Their biomass contains a broad array of bioactive compounds including phytohormones, carotenoids, polysaccharides, antioxidants, and macro and micronutrients that hold considerable potential for agricultural applications such as biofertilizers, biostimulants, biopesticides, and soil conditioners, among others (Alvarez et al., 2021). Moreover, nanotechnology, widely recognized as a key enabler of sustainable agriculture, integrates seamlessly with microalgae biotechnology. This interdisciplinary convergence is expected to unlock higher biomass productivity and improved biomass quality, substantially enhancing the efficiency and specificity of microalgae-derived products (Alishah Aratboni et al., 2023; Cardoso e Bufalo et al., 2025; Patnaik et al., 2024).

Accordingly, this chapter reviews the most recent advances in microalgae-derived bioproducts with potential agricultural applications. It also explores how the integration of nanotechnology can optimize the production of targeted agronomic metabolites, addressing economic and environmental dimensions while discussing the challenges and future prospects of applying microalgae in soil systems.

12.2 MICROALGAE BIOPRODUCTS WITH AGRONOMIC POTENTIAL

Microalgal biomass holds significant potential for diverse agronomic applications, owing to its unique biochemical composition and metabolic versatility. Among the key attributes underpinning this potential is its macronutrient content, which can reach 1–3% phosphorus (P) and 6–8% nitrogen (N) on a dry weight basis (Thajuddin et al., 2026; Zhang et al., 2024). In addition, due to their high photosynthetic efficiency, these species demonstrate a remarkable ability to convert CO_2 into biomass, thereby contributing to atmospheric carbon sequestration and, consequently, to climate change mitigation. This capacity translates into higher productivity per unit area when compared with terrestrial plants (Zhu et al., 2023). As a result, the biomass obtained can reduce reliance on chemical fertilizers, positioning itself as a sustainable alternative for plant nutrition while simultaneously improving soil quality and supplying both macro- and micronutrients (Prado et al., 2023).

Algal biomass is also rich in lipids, carbohydrates, and proteins, making it an attractive and highly versatile raw material for producing a wide range of bioproducts, including biofertilizers (nutrition), biostimulants (physiological stimulation), and biopesticides (biocontrol) (Nur et al., 2025) (Figure 12.1). Accordingly, the current market already offers various liquid bioinputs derived from microalgal extracts, as well as products based on live cells intended for soil inoculation, foliar application, or seed treatment. These examples illustrate the flexibility and potential of microalgae to provide customized agronomic solutions, whether in the form of cell extracts or as live inocula (Chabili et al., 2024; Shukla et al., 2025). The formulation and mode of application vary according to the crop, as do the effects associated with each type or presentation of the product.

In the case of soil inoculation, for example, the direct effects stem from the presence of living microorganisms that maintain their metabolic activity and produce exudates. These exudates can generate indirect benefits to the soil, such as increasing

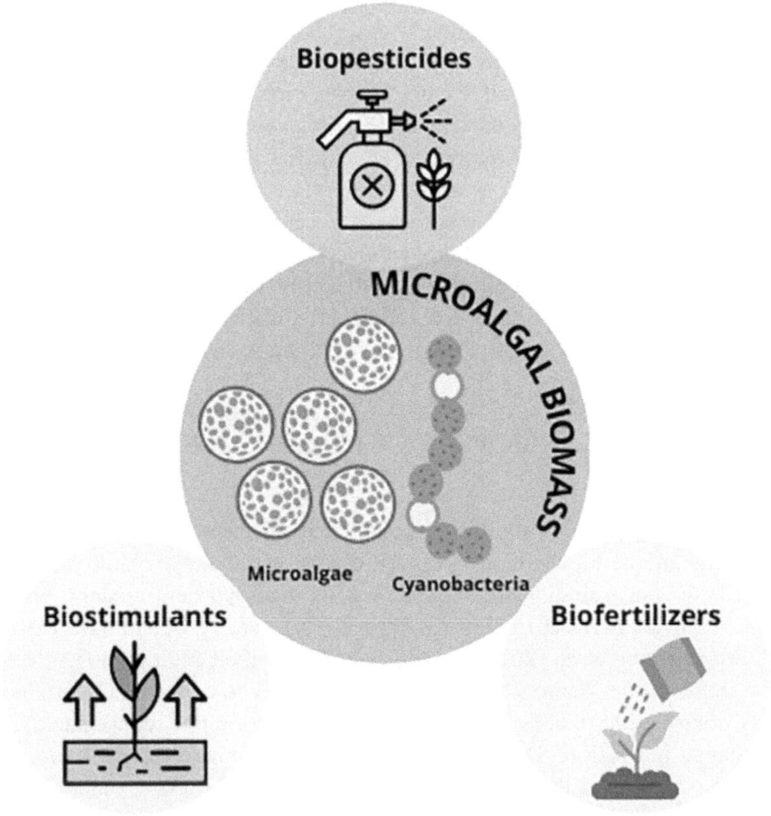

FIGURE 12.1 Different microalgae-based agronomic products.

organic matter content, improving water retention capacity, and promoting biofilm formation, which stimulates soil microbiota, enhances fertility, and supports plant growth. Certain microalgal genera, such as *Spirulina* spp. and *Chlorella* spp., are widely recognized for their potential and ability to produce exudates that exert beneficial effects on both soil and plants, and they have been employed in the formulation of bioinputs (Kholssi et al., 2022; Parmar et al., 2023). Table 12.1 presents selected findings from the literature, followed by a discussion of the characteristics of the three main bioproducts.

12.2.1 Biofertilizers

Algal biomass can be used as a source of macronutrients and as a growth- and nutrition-promoting agent for plants in the form of biofertilizers. Microalgae-based biofertilizers (MBFs) also provide additional benefits, such as improving soil quality and promoting long-term soil health through the input of organic matter and microelements like copper, iron, and manganese, as well as contributing to nitrogen fixation and the solubilization of inorganic phosphorus (P). Furthermore, the polysaccharides and amino acids present in the biomass enhance soil porosity, increase water retention, and stimulate the development of active microbiota, thereby promoting greater degradation of organic compounds that are essential for the resilience and productivity of the soil ecosystem (Arahou et al., 2023; Ramírez Mérida and Rodríguez Padrón, 2023).

The mechanisms of action of MBFs can be classified into two main groups, offering distinct yet complementary modes of action: those directly associated with the application of biomass as a nutrient source and those related to the application of live microalgae. When algal biomass is applied to the soil, compounds such as amino acids, vitamins, and other metabolites are released, enriching the soil and providing essential nutrients for plant development. Conversely, when applied in a living form, microalgae contribute to biological nitrogen fixation, a process influenced by factors such as light availability. In addition, they promote the biosolubilization of phosphorus through the action of phosphatase enzymes, which hydrolyze organophosphates and convert poorly available forms such as calcium and iron phosphates into more plant-assimilable forms (Abinandan et al., 2019).

The application of microalgae-based biofertilizers (MBFs), either as a supplement or in combination with conventional nitrogen fertilizers, enhances nitrogen utilization in the soil and reduces losses through volatilization and gaseous emissions. This represents an intelligent strategy to optimize plant nutrition while minimizing environmental impact. This effect occurs because soluble nitrogen fertilizers release nitrogen rapidly and are highly prone to volatilization in the form of NH_3. In contrast, MBFs release nitrogen more gradually, optimizing natural nutrient cycling processes in the soil while positively interacting with the soil microbiota. This combined effect helps reduce ammonia volatilization and N_2O emissions, thereby increasing nitrogen use efficiency by plants (Castro et al., 2017).

Similarly, MBFs can act in association with phosphate fertilizers by forming an organic coating that regulates phosphorus release, reducing losses through

TABLE 12.1
Literature Results for Different Microalgae-Based Agronomic Products

Microalgae and Cyanobacteria Species	Growth Medium	Application	Crop	Main Results	Reference
Chlorella sp., *Anabaena sp.*	BG-11	Biofertilizer	*Triticum aestivum L.*	Aumentou significativamente os Significantly increased protein levels (40%), chlorophyll A (29.2%), and chlorophyll B (33.5%). Promoted plant height, leaf number, and fresh/dry weight comparable to conventional fertilizers.	Zhang et al. (2024)
Mixed biomass	Sewage	Biofertilizer combined with phosphate fertilization.	*PennisetumglaucumL.*, *ZeamaysL.*	Fertilizer containing 12% MA promoted a higher P recovery rate by plants compared to pure phosphate fertilizer.	Castro et al. (2020)
Chlorella, Scenedesmus, Chlorococcum, Chroococcus	Sewage	Biofertilizer	*Triticum aestivum L. HD2967*	Increased nitrogen, phosphorus, and potassium content in roots, stems, and grains.	Dagnaisser et al. (2022)
Chlorella vulgaris	BG-11	Biostimulant	*Eruca sativa*	Increased germination rate by 7%, hypocotyl length by 20%, and fresh biomass compared to untreated controls.	Del Prete et al. (2025)
Arthrospira paltensis MS001, Dunaleilla salina MS002, Porphyridium sp. MS099	Synthetic medium (Walne's medium and Zarrouk medium)	Biostimulant (polysaccharide extract)	*Solanum lycopersicum*	Increased number of nodes (75%), dry weight (46.6%), and shoot length (25.26%) on average compared to control treatment. Higher carotenoid and chlorophyll content in leaves.	Rachidi et al. (2020)
Tetradesmus obliquus, Chlorella protothecoides, Chlorella vulgaris	Swine effluent	Biopesticide	*Fusarium oxysporum*	*C. vulgaris* (45.6–49.6%), *T. obliquus* (43.5–46.8%), and *C. protothecoides* (35.1–53.2%) inhibited the growth of *Fusarium oxysporum* depending on concentration (1–5 g/L).	Ferreira et al. (2021)

translocation and leaching. Moreover, the organic compounds and micronutrients present in the biomass enrich the soil and stimulate biological processes that enhance nutrient uptake by plants. As a result, higher crop yields can be achieved, along with more efficient utilization of chemical inputs (Castro et al., 2020).

12.2.2 Biostimulants

Biostimulants are organic substances such as phytohormones, fatty acids, polysaccharides, oligosaccharides, peptides, and phenolic compounds that promote plant growth, enhance the efficiency of physiological processes, and stimulate soil microbiota (Brito-Lopez et al., 2025). Biostimulants derived from microalgae can primarily include phytohormones such as gibberellins, auxins, and cytokinins; nitrogenous compounds such as proteins and amino acids; or extracellular polymeric substances (EPS) (Braun and Colla, 2023), in addition to ethylene and abscisic acid.

Among the benefits observed from phytohormone application are their ability to coordinate and optimize a wide range of plant responses. These include stimulation of cell division and elongation, as promoted by the auxins IAA and IBA; enhanced efficiency in nutrient translocation and fruit enlargement, mediated by cytokinins; induction of protein and pigment synthesis, associated with gibberellins; and improved tolerance to biotic and abiotic stresses, linked to ethylene activity (Kang et al., 2021).

The biostimulant potential of amino acids arises from their biochemical structure, which enables multiple functions in both soil and plants, serving as essential building blocks as well as signaling molecules. In addition to stimulating root system development, they promote cell elongation and effective plant growth at low concentrations. Furthermore, they provide a source of organic nitrogen to the soil, primarily in the form of amino groups (Dagnaisser et al., 2022).

The physical responses of the soil and the physiological responses of plants to algal biomass application extend beyond nutrition and growth stimulation. Algal biomass acts as a multifunctional agent that actively contributes to resistance against biotic stresses, including pathogens, insects, and other harmful microorganisms, thereby enhancing plant resilience under adverse environmental conditions (Sarsekeyeva et al., 2024). This effect can be attributed to substances excreted by plants in response to algal biomass application, as well as to microalgal-derived EPS, which may contain phenolic compounds with protective functions by inhibiting attack mechanisms such as the oxidation of plant cell walls (Sarsekeyeva et al., 2024).

These physical effects were demonstrated by Del Prete et al. (2025) who verified the impact of microalgae-based biostimulants by applying liquid extracts in in vitro assays with arugula (Eruca sativa) at a concentration of 0.5 mg/mL. An increase in 7% in germination rate, 20% in hypocotyl development and biomass, and a threefold higher concentration of dihydrophaseic acid 4-O-β-D-hexoside compared with control seedlings were observed. These results indicate the activation of metabolic pathways related to stress responses, leading to more resilient plants.

12.2.3 Biopesticides

Microalgae-derived biopesticides are compounds that inhibit the growth and proliferation of pathogens, bacteria, fungi, and nematodes in the soil. These compounds act against phytopathogenic organisms and weeds through biochemical mechanisms such as the inhibition of lipid peroxidation, the action of allelochemicals, and the regulation of stomatal size by enzymes, among others (Parmar et al., 2023). In certain situations, microalgae excrete these compounds as an adaptive response to hostile environments and as a defense mechanism, thereby generating an indirect effect resulting from biomass application (Casanova et al., 2023).

Some enzymes produced by microalgae, such as hydrolytic enzymes, along with antibacterial compounds including carbamidocyclophane A, benzoic acid, and ambigol A, contribute to the suppression or elimination of soil-borne pathogens. These compounds act through mechanisms such as plasma membrane disruption and the inhibition of enzymatic processes in pathogenic organisms. Moreover, they enhance plant immunity and natural defense responses against pathogen attacks (Renuka et al., 2018).

In this context, the application of microalgae to the soil stimulates the synthesis of enzymes such as polyphenol oxidase and chitinase in plants, which are essential for strengthening plant immunity against a wide range of pathogens. Furthermore, the formation of biofilms by the microbial community in soils inoculated with microalgae can function as a protective barrier, suppressing diseases and insect pests through the release of insecticidal compounds such as amines, alkaloids, and sesquiterpenes (Mol and Purushothaman, 2020).

The effects vary according to the mode of application (live biomass or extracts). In the case of applying live cells, the continuous production and release of secondary metabolites can simultaneously promote biofertilizing, biostimulant, and biopesticidal effects. A similar situation occurs with extracts, representing an integrated outcome of the metabolic activity of microalgae, which are often regarded as living biorefineries. This is because they produce both primary metabolites (lipids, carbohydrates, and proteins), with fertilizing and biostimulant potential, and secondary metabolites (fatty acids, esters, alkaloids, glucosinolates, and amino acids), which act in a complementary and synergistic manner (Costa et al., 2019; Gonçalves, 2021).

Among other perspectives, the application of microalgae for biopesticidal purposes is still under development, with promising results. Ferreira et al. (2021) demonstrated that the application of a microalgae solution cultivated in swine effluent exhibited high biological control potential, reducing the incidence of Panama disease (Fusarium oxysporum) by more than 40%. Therefore, this strategy may hold significant relevance for the fruit production sector, highlighting its potential use as a biopesticide.

12.3 MICROALGAE PRODUCTION TARGETING AGRONOMIC METABOLITES

As previously discussed (Section 12.2), the agronomic potential of microalgae lies in their capacity to improve soil quality, stimulate microbial and enzymatic activity, and release a wide range of bioactive compounds that enhance crop productivity and resilience. To achieve these outcomes, the entire production chain—from strain selection and cultivation to harvesting and downstream processing—must be carefully optimized.

Strain selection plays a decisive role, as it determines both productivity and the biochemical composition of the biomass. Species such as *Chlorella vulgaris*, *Scenedesmus* spp., *Dunaliella tertiolecta*, and *Haematococcus* spp. are frequently employed due to their rapid growth, adaptability to fluctuating conditions, and metabolic versatility. Beyond growth rates, strains are selected for specific functional traits such as efficient CO_2 fixation, nitrogen fixation in cyanobacteria, and beneficial interactions with soil microorganisms that stimulate the synthesis of phytohormones, amino acids, and exopolysaccharides (Chabili et al., 2024). These characteristics make them particularly suitable for biofertilizer and biostimulant applications.

The cultivation system is another key determinant of biomass yield and quality. For instance, open raceway ponds, although cost-effective and widely adopted, are prone to contamination, whereas closed photobioreactors ensure higher productivity and tighter control, albeit at greater operational costs (Magalhães et al., 2024). The choice of cultivation medium is equally critical. While synthetic media provide reproducibility, the use of wastewater represents a sustainable strategy, simultaneously lowering production costs and enabling nutrient recycling. This approach not only removes nitrogen, phosphorus, and organic matter from effluents but also generates biomass enriched in proteins, lipids, and amino acids with strong potential as biofertilizers, although careful monitoring is required to avoid heavy metal, organic, or pathogenic contamination (Gama et al., 2025; Parmar et al., 2023). Cultivation strategies also differ according to trophic modes. Autotrophy enhances photosynthetic activity and carbon fixation, heterotrophy relies on organic carbon sources to boost biomass accumulation, and mixotrophy combines both pathways to balance productivity, costs, and metabolite yield. Increasingly, these approaches are aligned with the principles of the circular bioeconomy, reinforcing the sustainability of microalgae-based agricultural products (Chabili et al., 2024).

The concentration and composition of biostimulant metabolites vary considerably depending on strain type and cultivation conditions. For instance, extracts of *Micractinium pusillum* and *Scenedesmus* sp. have exhibited high levels of auxins and cytokinins, which promote seed germination and seedling development (Rupawalla et al., 2022). Recent studies emphasize that cultivation parameters such as light, salinity, and photoperiod can be strategically manipulated to direct the synthesis of specific compounds. Zuorro et al. (2025) demonstrated that tailoring LED spectra for thermotolerant cyanobacteria enabled the selective enhancement of EPS and indole-3-acetic acid (IAA), while response surface methodology confirmed that photoperiod length directly influences metabolite yield. Likewise, Li et al. (2020) reported that cultivating *Porphyridium purpureum* under both low (5‰) and high (50‰) salinity conditions enhanced the antioxidant activity of algal EPS, indicating that salinity modulation can improve the functionality of biostimulant compounds.

Downstream processing is equally critical to the final quality of algal bioproducts. For biofertilizers, drying and dewatering are less demanding, as the biomass primarily serves as a nutrient source. In contrast, the formulation of biostimulants and biopesticides depends on the preservation of bioactive compounds that are often heat-labile, such as phytohormones, antioxidants, and phycobiliproteins. Excessive heating during drying, particularly at temperatures above 70 °C, can compromise their bioactivity, highlighting the importance of optimized or mild processing techniques (Parmar et al., 2023).

In summary, the production of microalgae for plant growth applications is a complex, multistage process in which strain selection, cultivation strategies, and downstream processing are tightly interconnected and directly influence both biomass yield and metabolite composition. Collectively, these strategies position microalgae as promising biofactories for next-generation agricultural inputs.

12.4 USE OF NANOTECHNOLOGY TO ENHANCE MICROALGAE-BASED BIOPRODUCTS FOR SOIL APPLICATIONS

Nanotechnology is a multidisciplinary field of science that investigates the applications and properties of materials and structures with dimensions ranging from 1 to 100 nm (Wu et al., 2025). Due to their high surface area-to-volume ratio, nanostructures exhibit unique reactivity and properties, making them attractive for a wide range of applications (Hashem et al., 2025; Yadwade et al., 2025). Nanoparticles and nanomaterials display structural, optical, biological, physical, and chemical characteristics that differ significantly from their macroscale counterparts (Periakaruppan et al., 2025; Tank et al., 2025).

Nanoparticles and nanomaterials can be synthesized through physical, chemical, or biological routes. Physical and chemical methods generally require high energy input and involve the use of chemical reagents that may generate toxic residues and by-products harmful to humans and the environment, in addition to being costly. In contrast, biological methods utilize sources such as plants and their extracts, fungi, yeasts, bacteria, viruses, and microalgae, which are more readily accessible, cost-effective, and environmentally sustainable (Goswami et al., 2024b; Tank et al., 2025).

The biosynthesis of nanomaterials using biological agents offers multiple advantages compared with physical and chemical approaches (Arteaga-Castrejón et al., 2024). Consequently, integrating nanotechnology with microalgae biotechnology can optimize biomass utilization and the production of agricultural bioinputs (Sumathi et al., 2025). In agriculture, nanotechnology enables the development of innovative solutions capable of mitigating the environmental damage caused by conventional farming practices while simultaneously enhancing crop productivity (Rodríguez-Seijo et al., 2025).

Despite the previously mentioned potential of microalgae for obtaining different agricultural bioproducts, many of these compounds are found only in small amounts within the cells, making biomass accumulation and the production of such substances critical parameters of utmost importance in algal cultivation. Nanoparticles, whether composed of macro- or micronutrients, can play an important role as nutritional supplements for microalgae when added to the growth medium. They can enhance photosynthetic and growth rates by acting as cofactors for enzymes involved in CO_2 fixation, nitrogen assimilation, nitrate reduction, phosphorus uptake, DNA transcription, biomass production, and the accumulation of bioactive compounds (Alishah

Aratboni et al., 2023; Parveen et al., 2023). However, high concentrations of nanoparticles may cause damage to the cell membrane, alteration of active sites, reduced photosynthesis, nutrient loss, and even algal cell death.

On the other hand, a particularly innovative development in this field is the integration of microalgal cultivation with nanobiotechnology. Extracellular metabolites such as polysaccharides, proteins, and phenolic compounds can serve as reducing and stabilizing agents in the green synthesis of nanoparticles, thereby broadening their applications in agriculture. For instance, the wide variety of biological compounds produced by microalgae enables the biosynthesis of nanofertilizers based on metal oxide nanoparticles, which have attracted growing interest in the agricultural sector. Metabolites produced by microalgae can reduce metal ions to zero valence, allowing their nucleation and, ultimately, their organization and stabilization into biosynthesized nanoparticles (Lawrence et al., 2025). Nanoparticles can be synthesized using live microalgal cells, dried biomass, cell extracts, or cell-free supernatants (Goswami et al., 2024a).

Graesiella emersonii KNUA204 has been shown to produce EPS-rich supernatants capable of mediating silver nanoparticle (AgNP) biosynthesis under light and alkaline conditions without biomass pretreatment (Do et al., 2025). AgNPs were synthesized using a soluble polysaccharide solution from *Chlorella vulgaris*. This AgNP solution was used for seed priming in wheat and common bean, resulting in a significant increase in seedling growth compared with the control. Increases were observed in root length (33% in wheat and 46% in bean), leaf area (75% in wheat and 120% in bean), shoot length (29% in wheat and 41% in bean), as well as dry biomass (40% in wheat and 43% in bean) and fresh biomass (40% in wheat and 61% in bean). Additionally, increases in photosynthetic pigments, protein content, and carbohydrate levels were reported. This overall growth response was attributed to enhanced antioxidant activity (El-Naggar et al., 2020). Iron oxide (Fe_3O_4) nanoparticles were also synthesized using extracts of *Chlorella* K01 as reducing agents. The resulting solution was applied to seeds of rice, maize, mustard, green bean, and watermelon, significantly contributing to germination and plant vigor (Win et al., 2021).

Munaro et al. (2024) nanoencapsulated amino acids, macro- and micronutrients, and polyamines present in a protein hydrolysate derived from the microalga *Arthrospira platensis* (commonly known as Spirulina) using chitosan and soybean lecithin nanoparticles. The nanopolysomal system, employing lecithin and chitosan as carriers of the hydrolyzed protein, was applied to tomato plants, promoting a 49.5% increase in plant height during the vegetative stage and enhancing chlorophyll, carotenoid, flavonoid, and epigallocatechin contents in the leaves, in addition to increasing lycopene levels in the fruits.

Advanced harvesting and extraction methods are essential to improve the recovery of products from algal biomass. Magnetic nanoparticles (Fe_3O_4) have been employed to harvest biomass from *Haematococcus pluvialis*, an astaxanthin-producing (carotenoid) microalga, achieving more than 99% collection efficiency (Sumathi et al., 2025).

In addition to direct agronomic benefits, interactions between nanoparticles and algal cell walls may facilitate the release of lipids, carotenoids, and proteins, although excessive nanoparticle exposure can also induce oxidative stress. Factors such as nutrient availability, light (intensity and quality), CO_2 concentration, pH, temperature, and the specific algal species used can influence biomass production as well as the synthesis and accumulation of bioproducts. Within this context, nanotechnology holds great potential as a tool to improve growth rates, production, and accumulation of bioresources (Alishah Aratboni et al., 2023), as well as to increase biomass harvesting efficiency (Sumathi et al., 2025).

12.5 ENVIRONMENTAL, ECONOMIC ASPECTS AND COMMERCIAL DEVELOPMENT

Efforts to scale microalgae-based agronomic products are increasingly focused on optimizing both environmental performance and financial viability. Life cycle assessments and techno-economic studies identify chemical supply and energy consumption (e.g., for mixing and aeration) as key emission drivers (Gurreri et al., 2024), while overall expenses are mainly attributed to operational costs, including labor, supervision and management, maintenance, taxes, and contingencies (Acién Fernández et al., 2019). Integration with wastewater and agricultural effluents provides a circular pathway: it supplies nutrients, reduces production footprints, and yields biomass with demonstrable biostimulant effects (Alling et al., 2023). Nonetheless, strict quality control for pathogens, heavy metals, and emerging contaminants is essential before agricultural use, as previously discussed.

Despite this promise, production remains relatively costly. Techno-economic assessments of wastewater-fed cultivation systems estimate biomass production costs ranging from approximately €1.9 to €9.7 per kilogram dry weight, depending on productivity and system configuration (Dias et al., 2025). Business models that adopt biorefinery concepts—valorizing the entire microalgal biomass (e.g., pigments, lipids, EPS, and biostimulants)—are widely recommended to improve economic feasibility.

At the market level, the global biostimulants sector was valued at US$ 4.3 billion in 2024 and is projected to reach US$ 4.47 billion by 2032 (Fortune Business Insights, 2025a), with Europe accounting for ~39% of the market share (Markets&Markets, 2025). Although microalgae-derived products currently represent a niche segment, the global microalgae market across all sectors is expected to grow from US$ 782.6 million in 2024 to US$ 1.38 billion by 2032 (Fortune Business Insights, 2025b), providing industrial momentum for agronomic applications.

Regulatory progress is also advancing. Under the EU Fertilising Products Regulation (FPR) 2019/1009, the European Commission issued updated guidance in 2025 clarifying CE marking requirements. Notably, the company AlgaEnergy has secured CE certification for four microalgae-based biostimulants—including a novel product specifically designed for olive cultivation—marking a key milestone in early commercial deployment, with technology readiness levels advancing beyond pilot scale (Agropages, 2025).

In summary, microalgae-based agronomic products are transitioning from proof-of-concept to market reality, supported by environmental integration strategies, emerging cost-effective processes, strong market momentum, and growing regulatory recognition. Strategic emphasis on energy optimization, circular biorefinery integration, crop validation, and regulatory compliance will be pivotal in driving future growth.

12.6 CHALLENGES AND FUTURE PERSPECTIVES

The potential of microalgae application in soils is clear and well-documented in recent literature. However, several gaps and challenges persist, requiring deeper investigation to enable large-scale commercial diffusion of this biotechnology. Key challenges and future directions include:

- Wastewater-based cultivation: Obtaining microalgae-derived agronomic products from wastewater ensures greater sustainability and circularity. However, this step can represent the primary driver of environmental and economic impacts across the production chain. Therefore, production arrangements must be designed under more realistic and scalable scenarios, with wastewater treatment playing a necessary and beneficial role.
- Reliance on synthetic cultivation: Most published results are based on synthetic culture media, which entails unsustainable natural resource consumption and further increases associated environmental burdens.
- Target metabolite accumulation: Developing simple and low-cost strategies that enable higher accumulation of agronomic target metabolites in algal biomass during wastewater treatment is challenging, as it requires reconciling different objectives—effluent purification, biomass production, and metabolite recovery.
- Contaminant risks: Although wastewater use is more sustainable, attention must be paid to biomass contamination with advanced pollutants (e.g., microplastics and pharmaceuticals). These contaminants can be transferred to soils and potentially bioaccumulate within the food chain.
- Nanoparticle biosynthesis: The biosynthesis of nanoparticles using microalgae is a promising strategy that can enhance both biomass productivity and quality while ensuring more sustainable nanoparticle synthesis. However, further studies are needed to better understand potential toxic effects on algal growth, as well as the complex interactions among biomass, nanoparticles, and plants following agricultural application.
- Processing and valorization: The processing and application of algal bioproducts may vary depending on context; however, full biomass utilization may reduce energy and resource consumption. Alternatively, biomass valorization should be approached under a biorefinery framework, in which multiple processes generate multiple value-added products, ensuring comprehensive use of the raw material.

ACKNOWLEDGMENTS

The authors would like to thank the Research Support Foundation of the State of Minas Gerais (FAPEMIG – APQ-06816-24).

REFERENCES

Abinandan, S., Subashchandrabose, S.R., Venkateswarlu, K., Megharaj, M. 2019. Soil microalgae and cyanobacteria: The biotechnological potential in the maintenance of soil fertility and health. *Critical Reviews in Biotechnology* 39, 981–998. https://doi.org/10.1080/07388551.2019.1654972

Acién Fernández, F.G., Fernández Sevilla, J.M., Molina Grima, E. 2019. Costs analysis of microalgae production. In: Ashok Pandey, Jo-Shu Chang, Carlos Ricardo Soccol, Duu-Jong Lee, and Yusuf Chisti (eds.), *Biofuels from Algae*. Elsevier, pp. 551–566. https://doi.org/10.1016/B978-0-444-64192-2.00021-4

Agropages. 2025. *AlgaEnergy Receives CE Marking for Four New Biostimulants, Including a Groundbreaking Specific Solution for Olive Farming* [WWW Document]. https://news.agropages.com/News/NewsDetail---53598.htm (accessed September 10, 2025).

Alishah Aratboni, H., Rafiei, N., Mehdizadeh Allaf, M., Abedini, S., Naseema Rasheed, R., Seif, A., Barati, B., Wang, S., Morones-Ramírez, J.R. 2023. Nanotechnology: An outstanding tool for increasing and better exploitation of microalgae valuable compounds. *Algal Research* 71, 103019. https://doi.org/10.1016/j.algal.2023.103019

Alling, T., Funk, C., Gentili, F.G. 2023. Nordic microalgae produce biostimulant for the germination of tomato and barley seeds. *Scientific Reports* 13, 3509. https://doi.org/10.1038/s41598-023-30707-8

Alvarez, A.L., Weyers, S.L., Goemann, H.M., Peyton, B.M., Gardner, R.D. 2021. Microalgae, soil and plants: A critical review of microalgae as renewable resources for agriculture. *Algal Research* 54, 102200. https://doi.org/10.1016/j.algal.2021.102200

Arahou, F., Lijassi, I., Wahby, A., Rhazi, L., Arahou, M., Wahby, I. 2023. Spirulina-based biostimulants for sustainable agriculture: Yield improvement and market trends. *Bioenergy Research* 16, 1401–1416. https://doi.org/10.1007/s12155-022-10537-8

Arteaga-Castrejón, A.A., Agarwal, V., Khandual, S. 2024. Microalgae as a potential natural source for the green synthesis of nanoparticles. *Chemical Communications* 60, 3874–3890. https://doi.org/10.1039/D3CC05767D

Bathaei, A., Štreimikienė, D. 2023. A systematic review of agricultural sustainability indicators. *Agriculture* 13, 241. https://doi.org/10.3390/agriculture13020241

Braun, J.C.A., Colla, L.M. 2023. Use of microalgae for the development of biofertilizers and biostimulants. *Bioenergy Research* 16, 289–310. https://doi.org/10.1007/s12155-022-10456-8

Brito-Lopez, C., Van Der Wielen, N., Barbosa, M., Karlova, R. 2025. Plant growth-promoting microbes and microalgae-based biostimulants: Sustainable strategy for agriculture and abiotic stress resilience. *Philosophical Transactions of the Royal Society B: Biological Sciences* 380. https://doi.org/10.1098/rstb.2024.0251

Campbell, B.M., Beare, D.J., Bennett, E.M., Hall-Spencer, J.M., Ingram, J.S.I., Jaramillo, F., Ortiz, R., Ramankutty, N., Sayer, J.A., Shindell, D. 2017. Agriculture production as a major driver of the Earth system exceeding planetary boundaries. *Ecology and Society* 22, art8. https://doi.org/10.5751/ES-09595-220408

Cardoso e Bufalo, T., Buttrós, V.H., de Paiva, A.B., de Oliveira, D.D., Ribeiro, C.S.F., Dória, J. 2025. Nanopesticides in Brazilian crops: Classes, mechanisms, efficacy, risks, and photocatalytic remediation. *Plants* 14, 2880. https://doi.org/10.3390/plants14182880

Casanova, L.M., Macrae, A., de Souza, J.E., Neves Junior, A., Vermelho, A.B. 2023. The potential of allelochemicals from microalgae for biopesticides. *Plants* 12. https://doi.org/10.3390/plants12091896

Castro, J.D.S., Calijuri, M.L., Assemany, P.P., Cecon, P.R., de Assis, I.R., Ribeiro, V.J. 2017. Microalgae biofilm in soil: Greenhouse gas emissions, ammonia volatilization and plant growth. *Science of the Total Environment* 574, 1640–1648. https://doi.org/10.1016/j.scitotenv.2016.08.205

Castro, J.D.S., Calijuri, M.L., Mattiello, E.M., Ribeiro, V.J., Assemany, P.P. 2020. Algal biomass from wastewater: Soil phosphorus bioavailability and plants productivity. *Science of the Total Environment* 711, 135088. https://doi.org/10.1016/j.scitotenv.2019.135088

Chabili, A., Minaoui, F., Hakkoum, Z., Douma, M., Meddich, A., Loudiki, M. 2024. A comprehensive review of microalgae and cyanobacteria-based biostimulants for agriculture uses. *Plants* 13, 159. https://doi.org/10.3390/plants13020159

Costa, J.A.V., Freitas, B.C.B., Cruz, C.G., Silveira, J., Morais, M.G. 2019. Potential of microalgae as biopesticides to contribute to sustainable agriculture and environmental development. *Journal of Environmental Science and Health - Part B Pesticides, Food Contaminants, and Agricultural Wastes* 54, 366–375. https://doi.org/10.1080/03601234.2019.1571366

Dagnaisser, L.S., dos Santos, M.G.B., Rita, A.V.S., Chaves Cardoso, J., de Carvalho, D.F., de Mendonça, H.V. 2022. Microalgae as bio-fertilizer: A new strategy for advancing modern agriculture, wastewater bioremediation, and atmospheric carbon mitigation. *Water, Air, and Soil Pollution* 233. https://doi.org/10.1007/s11270-022-05917-x

Del Prete, F., Esposito, T., Pane, C., Manganiello, G., Pepe, G., Salviati, E., Campiglia, P., Mencherini, T., Sansone, F., Aquino, R.P. 2025. Extract from Chlorella vulgaris: Production, characterization, and effects on the germination, growth and metabolite profile of Eruca sativa microgreens. *Industrial Crops and Products* 233, 121490. https://doi.org/10.1016/j.indcrop.2025.121490

Dias, R.R., Deprá, M.C., de Menezes, C.R., Zepka, L.Q., Jacob-Lopes, E. 2025. Microalgae cultivation in wastewater: How realistic is this approach for value-added product production? *Processes* 13, 2052. https://doi.org/10.3390/pr13072052

Do, J.-M., Hong, J.W., Yoon, H.-S. 2025. Microalgae-mediated green synthesis of silver nanoparticles: A sustainable approach using extracellular polymeric substances from Graesiella emersonii KNUA204. *Frontiers in Microbiology* 16. https://doi.org/10.3389/fmicb.2025.1589285

Dubois, O., Faurès, J.M., Felix, E., Flammini, A., Hoogeveen, J., Pluschke, L., Puri, M., Ünver, O. 2014. *The Water-Energy-Food Nexus: A New Approach in Support of Food Security and Sustainable Agriculture*. Food and Agriculture Organization of the United Nations.

El-Naggar, N.E.-A., Hussein, M.H., Shaaban-Dessuuki, S.A., Dalal, S.R. 2020. Production, extraction and characterization of Chlorella vulgaris soluble polysaccharides and their applications in AgNPs biosynthesis and biostimulation of plant growth. *Scientific Reports* 10, 3011. https://doi.org/10.1038/s41598-020-59945-w

Ferreira, A., Melkonyan, L., Carapinha, S., Ribeiro, B., Figueiredo, D., Avetisova, G., Gouveia, L. 2021. Biostimulant and biopesticide potential of microalgae growing in piggery wastewater. *Environmental Advances* 4, 100062. https://doi.org/10.1016/j.envadv.2021.100062

Fortune Business Insights. 2025a. *Biostimulant Market Overview* [WWW Document]. https://www.fortunebusinessinsights.com/industry-reports/biostimulants-market-100414 (accessed September 10, 2025).

Fortune Business Insights. 2025b. *Microalgae Market Overview* [WWW Document]. https://www.fortunebusinessinsights.com/microalgae-market-110314 (accessed September 10, 2025).

Gama, R.C.N., Castro, J.D.S., Marangon, B.B., de Jesus Junior, M.M., Ribeiro, V.J., Silva, J.D., Calijuri, M.L. 2025. Microalgae bioinputs as disruptive technology for a sustainable agriculture: A systematic and bibliometric review. *Journal of Environmental Chemical Engineering* 13, 116034. https://doi.org/10.1016/j.jece.2025.116034

Gonçalves, A.L. 2021. The use of microalgae and cyanobacteria in the improvement of agricultural practices: A review on their biofertilising, biostimulating and biopesticide roles. *Applied Sciences (Switzerland)* 11, 1–21. https://doi.org/10.3390/app11020871

Goswami, R.K., Monika, Agrawal, K., Alferov, S.V., Verma, P. 2024a. Bio fabrication of microalgae mediated nanoparticles: An insight into its mechanism, characterization and applications. *Biocatalysis and Agricultural Biotechnology* 60, 103330. https://doi.org/10.1016/j.bcab.2024.103330

Goswami, S., Bishnoi, A., Tank, D., Patel, P., Chahar, M., Khaturia, S., Modi, N., Khalid, M., Alam, M.W., Kumar Yadav, V., Alreshidi, M.A., Yadav, K.K. 2024b. Recent trends in the synthesis, characterization and commercial applications of zinc oxide nanoparticles - A review. *Inorganica Chimica Acta* 573, 122350. https://doi.org/10.1016/j.ica.2024.122350

Gurreri, L., Calanni Rindina, M., Luciano, A., Falqui, L., Fino, D., Mancini, G. 2024. Microalgae production in an industrial-scale photobioreactors plant: A comprehensive Life Cycle assessment. *Sustainable Chemistry and Pharmacy* 39, 101598. https://doi.org/10.1016/j.scp.2024.101598

Hashem, A.H., Saied, E., Badr, B.M., Dora, M.S., Diab, M.A., Abdelaziz, A.M., Elkady, F.M., Ali, M.A., Issa, N.I., Hamdy, Z.A., Nafea, M.E., Khalifa, A.N., Adel, A., Hasib, A., Hawela, A.M., El-Gazzar, M.M., Nouh, M.A., Nahool, A.A., Attia, M.S. 2025. Biosynthesis of trimetallic nanoparticles and their biological applications: A recent review. *Archives of Microbiology* 207, 50. https://doi.org/10.1007/s00203-025-04237-y

Kang, Y., Kim, M., Shim, C., Bae, S., Jang, S. 2021. Potential of algae–bacteria synergistic effects on vegetable production. *Frontiers in Plant Science* 12, 1–13. https://doi.org/10.3389/fpls.2021.656662

Kholssi, R., Lougraimzi, H., Grina, F., Lorentz, J.F., Silva, I., Castaño-Sánchez, O., Marks, E.A.N. 2022. Green agriculture: A review of the application of micro- and macroalgae and their impact on crop production on soil quality. *Journal of Soil Science and Plant Nutrition* 22, 4627–4641. https://doi.org/10.1007/s42729-022-00944-3

Lawrence, L.V., Rewanwar, C., Shown, O., Vishnu, D. 2025. Impact of nanoparticles in microalgal quantification: Advancements in enhancing biomass valorization and evaluating toxicity effects. *Bioresource Technology Reports* 32, 102354. https://doi.org/10.1016/j.biteb.2025.102354

Li, S., Ji, L., Chen, C., Zhao, S., Sun, M., Gao, Z., Wu, H., Fan, J. 2020. Efficient accumulation of high-value bioactive substances by carbon to nitrogen ratio regulation in marine microalgae Porphyridium purpureum. *Bioresource Technology* 309, 123362. https://doi.org/10.1016/j.biortech.2020.123362

Magalhães, I.B., de Paula Pereira, A.S.A., Silva, T.A., Ferreira, J., Braga, M.Q., Couto, E.A., Assemany, P.P., Calijuri, M.L. 2024. Advancements in high-rate algal pond technology for enhanced wastewater treatment and biomass production: A review. *Journal of Water Process Engineering* 66, 105929. https://doi.org/10.1016/j.jwpe.2024.105929

Markets&Markets. 2025. *Biostimulant Market Overview* [WWW Document]. https://www.marketsandmarkets.com/Market-Reports/biostimulant-market-1081.html?utm_source=chatgpt.com (accessed September 10, 2025).

Mol, I., Purushothaman, T. 2020. Micro-algae as bio-pesticides for the development of sustainable agriculture. *Wide Spectrum* 8, 35–48.

Moreira, S.G., Hoogenboom, G., Nunes, M.R., Martin-Ryals, A.D., Sanchez, P.A. 2023. Circular agriculture increases food production and can reduce N fertilizer use of commercial farms for tropical environments. *Science of the Total Environment* 879, 163031. https://doi.org/10.1016/j.scitotenv.2023.163031

Munaro, D., Mazo, C.H., Bauer, C.M., da Silva Gomes, L., Teodoro, E.B., Mazzarino, L., da Rocha Veleirinho, M.B., Moura e Silva, S., Maraschin, M. 2024. A novel biostimulant from chitosan nanoparticles and microalgae-based protein hydrolysate: Improving crop performance in tomato. *Scientia Horticulturae* 323, 112491. https://doi.org/10.1016/j.scienta.2023.112491

Muscat, A., de Olde, E.M., Ripoll-Bosch, R., Van Zanten, H.H.E., Metze, T.A.P., Termeer, C.J.A.M., van Ittersum, M.K., de Boer, I.J.M. 2021. Principles, drivers and opportunities of a circular bioeconomy. *Nature Food* 2, 561–566. https://doi.org/10.1038/s43016-021-00340-7

Nur, M.M.A., Mahreni, Murni, S.W., Setyoningrum, T.M., Hadi, F., Widayati, T.W., Jaya, D., Sulistyawati, R.R.E., Puspitaningrum, D.A., Dewi, R.N., Hadiyanto, Hasanuzzaman, M. 2025. Innovative strategies for utilizing microalgae as dual-purpose biofertilizers and phycoremediators in agroecosystems. *Biotechnology Reports* 45, e00870. https://doi.org/10.1016/j.btre.2024.e00870

Parmar, P., Kumar, R., Neha, Y., Srivatsan, V. 2023. Microalgae as next generation plant growth additives: Functions, applications, challenges and circular bioeconomy based solutions. *Frontiers in Plant Science* 14. https://doi.org/10.3389/fpls.2023.1073546

Parveen, A., Bhatnagar, P., Bisht, B., Kumar, S., Joshi, S., Gautam, P., Kumar, V., Irina, A., Anna, K. 2023. Algal nanobionics to enhance value added products – A review. *Current Research in Biotechnology* 5, 100124. https://doi.org/10.1016/j.crbiot.2023.100124

Patnaik, R., Kumar Bagchi, S., Rawat, I., Bux, F. 2024. Nanotechnology for the enhancement of algal cultivation and bioprocessing: Bridging gaps and unlocking potential. *Bioresource Technology* 406, 131025. https://doi.org/10.1016/j.biortech.2024.131025

Periakaruppan, R., Govindharaj, K., Martin, J.A., Vijai Selvaraj, K.S., Al-Dayan, N. 2025. Utilization of Ulva rigida for fabrication of iron oxide nanoparticles and its physicochemical characterization. *Applied Biochemistry and Biotechnology* 197, 4850–4860. https://doi.org/10.1007/s12010-025-05253-w

Prado, L.D.O., Bolzani, H.R., Souza, H.H.D.S., Ruas, G., da Silva, G.H.R. 2023. Microalgal cultivation in open and closed systems under a tropical climate: A life cycle comparison. *Journal of Cleaner Production* 422, 138631. https://doi.org/10.1016/j.jclepro.2023.138631

Rachidi, F., Benhima, R., Sbabou, L., El Arroussi, H. 2020. Microalgae polysaccharides biostimulating effect on tomato plants: Growth and metabolic distribution. *Biotechnology Reports* 25, e00426. https://doi.org/10.1016/j.btre.2020.e00426

Ramírez Mérida, L.G., Rodríguez Padrón, R.A. 2023. Application of microalgae in wastewater: Opportunity for sustainable development. *Frontiers in Environmental Science* 11, 1–12. https://doi.org/10.3389/fenvs.2023.1238640

Renuka, N., Guldhe, A., Prasanna, R., Singh, P., Bux, F. 2018. Microalgae as multi-functional options in modern agriculture: Current trends, prospects and challenges. *Biotechnology Advances* 36, 1255–1273. https://doi.org/10.1016/j.biotechadv.2018.04.004

Rodríguez-Seijo, A., Santás-Miguel, V., Arenas-Lago, D., Arias-Estévez, M., Pérez-Rodríguez, P. 2025. Use of nanotechnology for safe agriculture and food production: Challenges and limitations. *Pedosphere* 35, 20–32. https://doi.org/10.1016/j.pedsph.2024.09.005

Rupawalla, Z., Shaw, L., Ross, I.L., Schmidt, S., Hankamer, B., Wolf, J. 2022. Germination screen for microalgae-generated plant growth biostimulants. *Algal Research* 66, 102784. https://doi.org/10.1016/j.algal.2022.102784

Sarsekeyeva, F.K., Sadvakasova, A.K., Sandybayeva, S.K., Kossalbayev, B.D., Huang, Z., Zayadan, B.K., Akmukhanova, N.R., Leong, Y.K., Chang, J.S., Allakhverdiev, S.I. 2024. Microalgae- and cyanobacteria-derived phytostimulants for mitigation of salt stress and improved agriculture. *Algal Research* 82, 103686. https://doi.org/10.1016/j.algal.2024.103686

Shukla, P.S., Michalak, I., Burlot, A.S., Bourgougnon, N., Daridon, B., Ronga, D., Hurtado, A., Critchley, A.T. 2025. Phycobiostimulants: Next-generation sustainable agricultural inputs. *ACS Agricultural Science and Technology* 5, 424–453. https://doi.org/10.1021/acsagscitech.4c00777

Sumathi, Y., Kumar, P., Singhania, R.R., Chen, C.-W., Gurunathan, B., Dong, C.-D., Patel, A.K. 2025. Harnessing Fe3O4 nanoparticles for sustainable harvesting of astaxanthin-producing microalgae: Advancing industrial-scale biorefinery. *Separation and Purification Technology* 353, 128408. https://doi.org/10.1016/j.seppur.2024.128408

Tank, D., Bishnoi, A., Goswami, S., Ambegaonkar, N.J., Patel, P., Chahar, M., Khaturia, S., Yadav, V.K. 2025. Recent advances in phyto- and microorganisms-mediated synthesis of copper nanoparticles and their emerging applications in healthcare, environment, agriculture and food industry. *Bioprocess and Biosystems Engineering*. https://doi.org/10.1007/s00449-025-03196-4

Thajuddin, F., Rasheed, A.S., Elumalai, A., Palanivel, P., Nooruddin, T., Dharumadurai, D. 2026. Harnessing microalgae and cyanobacteria for sustainable bioenergy Production: Integration of molecular engineering and AI-driven approaches. *Renewable Energy* 256, 123961. https://doi.org/10.1016/j.renene.2025.123961

Win, T.T., Khan, S., Bo, B., Zada, S., Fu, P. 2021. Green synthesis and characterization of Fe3O4 nanoparticles using Chlorella-K01 extract for potential enhancement of plant growth stimulating and antifungal activity. *Scientific Reports* 11, 21996. https://doi.org/10.1038/s41598-021-01538-2

Wu, C., Shen, X., Feng, W., Li, P., Chen, Y. 2025. Agricultural nanotechnology. *Coordination Chemistry Reviews* 543, 216906. https://doi.org/10.1016/j.ccr.2025.216906

Yadwade, R., Gharpure, S., Lestari, G.A.D., Sur, U.K., Sarma, L.S., Singh, S.C.E., Sarmah, J.K., Mohanta, D., Ankamwar, B. 2025. Seed-mediated synthesis and biosynthesis of iron oxide nanoparticles. *3 Biotech* 15, 258. https://doi.org/10.1007/s13205-025-04431-1

Yang, Y., Tilman, D., Jin, Z., Smith, P., Barrett, C.B., Zhu, Y.-G., Burney, J., D'Odorico, P., Fantke, P., Fargione, J., Finlay, J.C., Rulli, M.C., Sloat, L., Jan van Groenigen, K., West, P.C., Ziska, L., Michalak, A.M., Lobell, D.B., Clark, M., Colquhoun, J., Garg, T., Garrett, K.A., Geels, C., Hernandez, R.R., Herrero, M., Hutchison, W.D., Jain, M., Jungers, J.M., Liu, B., Mueller, N.D., Ortiz-Bobea, A., Schewe, J., Song, J., Verheyen, J., Vitousek, P., Wada, Y., Xia, L., Zhang, X., Zhuang, M. 2024. Climate change exacerbates the environmental impacts of agriculture. *Science* 1979, 385. https://doi.org/10.1126/science.adn3747

Zhang, Z., Xu, M., Fan, Y., Zhang, L., Wang, H. 2024. Using microalgae to reduce the use of conventional fertilizers in hydroponics and soil-based cultivation. *Science of the Total Environment* 912, 169424. https://doi.org/10.1016/j.scitotenv.2023.169424

Zhu, C., Hu, C., Wang, J., Chen, Y., Zhao, Y., Chi, Z. 2023. A precise microalgae farming for CO2 sequestration: A critical review and perspectives. *Science of the Total Environment* 901, 166013. https://doi.org/10.1016/j.scitotenv.2023.166013

Zuorro, A., Lavecchia, R., Moncada-Jacome, K.A., García-Martínez, J.B., Barajas-Solano, A.F. 2025. Light-driven optimization of exopolysaccharide and indole-3-acetic acid production in thermotolerant cyanobacteria. *Science* 7, 108. https://doi.org/10.3390/sci7030108

13 Microalgae-Based Biotechnology as an Alternative for Industrial Wastewater Treatment

Adegoke Isiaka Adetunji and S'fiso Thuthukani Gumbi

13.1 INTRODUCTION

The rise in industrialization, urbanization, and agricultural activities has led to the continuous discharge of large amounts of industrial wastewater into the receiving environment, posing serious hazards to the aquatic and terrestrial ecosystems, as well as human health (Chai et al., 2021; Walters et al., 2025). The wastewater is produced from various industries such as tannery industries, pharmaceutical industries, slaughterhouses, dairy industries, poultry processing industries, acid mine drainage (AMD), pulp and paper mill industries, fish processing industries, textile industries, and petrochemical industries (Pérez et al., 2016; Adetunji and Olaniran, 2018; Patel et al., 2021; Devi et al., 2023; Rathour et al., 2024; Adetunji and Erasmus, 2025). Industrial wastewater consists of toxic organic and inorganic pollutants, including heavy metals, nutrient load, dissolved solids, fats, oils, and grease (FOG), phenolic compounds, and pathogens (Abdelfattah et al., 2023). The concentration of the pollutants varies, depending on the operations and sources of the wastewater. When untreated and improperly disposed of, these wastewaters pose severe hazards to the ecosystems by contaminating surface and underground water and causing air pollution. In addition, poorly treated wastewater discharged into receiving water bodies depletes the amount of dissolved oxygen required for the survival of aquatic organisms, thus leading to the death of aquatic organisms (Singh et al., 2023). The viscous nature of FOG in industrial wastewater blocks sewer lines and drainages, resulting in the corrosion of the sewer lines and release of unpleasant odor and unsightly appearance on the water body surface (Adetunji and Olaniran, 2018). Chemical species (e.g., heavy metals) in wastewaters settle at the bottom of the receiving water bodies, which, upon exposure, threaten human health by causing chronic diseases, including anemia, kidney and liver damage, bronchitis, lung disease, and skin inflammation (Rambabu et al., 2020; Masindi et al., 2022; Moyo et al., 2024). Therefore, to lessen the negative effects of untreated industrial wastewater, proper and effective

DOI: 10.1201/9781003611479-13

treatment technologies are paramount for ensuring long-term environmental sustainability (Mosai et al., 2024).

Over decades, traditional physicochemical approaches, such as membrane separation, chemical coagulation, adsorption, sedimentation, gravity separation, and ion-exchange techniques, have been employed for wastewater treatment (Kushwaha et al., 2011; Alcolea et al., 2012; Pathak et al., 2016; Abuhasel et al., 2021; Azizi et al., 2024; Dube et al., 2024; Mosai et al., 2024). Nevertheless, these methods are susceptible to some constraints, such as the release of residual pollutants, high operational difficulties, operational costs, and low treatment efficacy. In addition, these technologies require huge amounts of energy and large spaces for the reclamation of contaminants from the waste effluent and the release of clean water. Consequently, biological technology involving the use of microalgae is considered an attractive approach for industrial wastewater treatment.

The application of microalgae for the remediation of hazardous pollutants in industrial wastewater is an innovative, cheap, eco-friendly, and sustainable technology with a reduction in greenhouse gas emissions (Rezvani and Sarrafzadeh, 2025). In addition, microalgae are capable of acclimatizing and surviving in extreme environments, thereby making them appropriate for the treatment of different types of wastewaters (Devi et al., 2023). The microalgae detoxify the pollutants in wastewater using various mechanisms, including electrostatic interaction, micro-precipitation, surface complexing, and intracellular accumulation, based on the wastewater sources and pollutant types (Samal et al., 2020; Goswami et al., 2021; Almeida et al., 2023; Li et al., 2023a, 2023b). Microalgae such as *Oscillatoria*, *Scenedesmus*, *Chlorella*, *Chlamydomonas*, *Botryococcus*, *Ankistrodesmus*, *Phormidium*, and *Limnospira* are excellent candidates for the bioremediation of heavy metals, pathogens, dyes, nutrients, and emerging pollutants from industrial wastewaters (Gonçalves et al., 2017). Wastewater treatment using microalgae yields sustainable biomass that can be utilized for the production of valuable products, including bioenergy, animal feed, biochemicals, biofertilizers, and polyunsaturated fatty acids (PUFAs) (Cordeiro et al., 2022; Gumbi et al., 2022; Patel et al., 2022; Nerom et al., 2024; Peng et al., 2025) (Figure 13.1). Factors such as pH, temperature, nutrient and CO_2 supply, light intensity and duration, and salinity affect microalgal growth and biomass production. This chapter provides insights into the applicability of microalgae in the remediation of

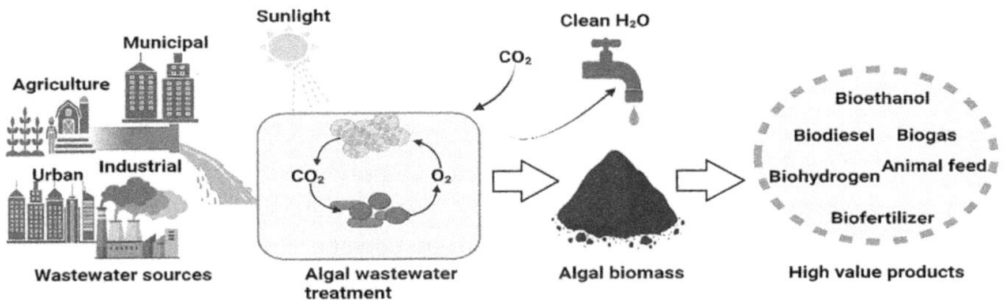

FIGURE 13.1 Microalgae-mediated treatment of various wastewater sources.

pollutants from various industrial wastewaters using different cultivation systems. Additionally, it explores various techniques for recovering microalgae biomass from culture medium, as well as the bio-based products that can be synthesized from microalgal biomass.

13.2 INDUSTRIAL WASTEWATERS: SOURCES AND CHARACTERISTICS

Industrial wastewater is heterogeneous, consisting of large amounts of nutrients, including phosphorus, nitrogen in the form of ammonium, nitrate, and nitrite, as well as dissolved solids, micropollutants, pathogens, and considerable heavy metals (such as lead, zinc, and copper), dyes, and pharmaceuticals (Amin et al., 2022; Anand et al., 2023). The physicochemical characteristics of some of the industrial wastewaters are discussed in detail below.

13.2.1 Dairy Wastewater

Dairy wastewater is concentrated in nature and contains huge quantities of organic compounds, including lipids, lactose, and proteins, contributing significantly to its high levels of chemical oxygen demand (COD), biochemical oxygen demand (BOD), and suspended solids (Adulkar and Rathod, 2015; Adetunji, 2017; Adetunji and Olaniran, 2018). Dairy wastewater has a whitish appearance, accompanied by high-strength black sludge and a strong butanoic acid odor, resulting from the breakdown of casein (Shete and Shinkar, 2013). This wastewater is alkaline or acidic and consists of organic and/or inorganic nitrogen, phosphorus (such as orthophosphate or polyphosphate), and considerable amounts of monovalent, divalent, and trace elements (Demirel et al., 2005; Kushwaha et al., 2011; Pathak et al., 2016; Slavov, 2017).

13.2.2 Poultry Processing Wastewater

Poultry industry wastewater comprises high concentrations of biodegradable organic matter (Valladão et al., 2007; Adetunji and Olaniran, 2018). According to Dors et al. (2013). This wastewater consists of large amounts of COD, suspended solids, and lipids. Furthermore, poultry processing wastewater contains large quantities of nitrogen, phosphorus, heavy metals, organic carbon, etc. (Bustillo-Lecompte and Mehrvar, 2015; Adetunji and Olaniran, 2018). The rise in the levels of organic matter in this waste stream is attributed to the availability of residual blood, excreta, and fat, together with intestinal mucus in the wastewater (Debik and Coskun, 2009).

13.2.3 Distillery Wastewater

The distillery industry is one of the most polluting industries worldwide due to the presence of both biodegradable and nonbiodegradable organic matter in the wastewater, which contributes to its high BOD and COD. In addition, distillery wastewater is characterized by a low pH (4.0–4.5), high ammoniacal nitrogen content, and a dark brown coloration. The dark brown appearance of the wastewater is attributed to the presence of tannic and humic acids, furfural, caramels, and melanoidins (Sankaran et al., 2014). Furthermore, distillery wastewater consists of several endocrine-disrupting chemicals, including 2-hydroxysocaproic acid, benzene propanoic acid, and phthalates. However, the concentration of organic matter in the wastewater is dependent on the process and raw materials employed for ethanol production. The wastewater from distillery industries is referred to as molasses spent wash, vinasses, or stillages (Ravikumar et al., 2021).

13.2.4 Tannery Wastewater

Tanning is a technique that requires substantial amounts of organic and inorganic chemicals and an excessive volume of water, thereby contributing to the release of huge amounts of toxic wastewater to the immediate environment. The tannery wastewater is dark brown in color with a pungent odor and serves as a rich source of nutrients, such as phosphates, nitrogen, and carbon. In addition, the wastewater is characterized with high BOD, COD, phenolics, total dissolved solids, pH, heavy metals (e.g., chromium), formaldehyde, chlorine, detergents, and phthalates (Devi et al., 2023). However, the physicochemical characteristics of the wastewater are influenced by the tanning processes, chemicals, and raw materials used.

13.2.5 Pharmaceutical Wastewater

The substantial usage of pharmaceuticals and personal care products has led to a rapid growth in the pharmaceutical industry, resulting in the discharge of a large volume of wastewater, consisting of toxic and carcinogenic pharmaceutical active compounds, including β-estradiol, carbamazepine, ibuprofen, acetaminophen, estrone, atenolol, gemfibrozil, and sulfamethoxazole. Furthermore, pharmaceutical wastewater is characterized by pungent odor, high BOD and COD, intense color, and low biodegradability (Amin et al., 2022).

13.2.6 Acid Mine Drainage

AMD is acidic water from metal mines or coal mines. It is formed following exposure of mine tailings or deposits to natural environments that consist of air, microorganisms, and water (Oliveira et al., 2024; Wu et al., 2024). This waste effluent has a unique physical appearance ranging from blue, red, black, colorless, or green, depending on the type of dissolved chemical compounds present in the waste stream (Masindi et al., 2022; Adetunji and Erasmus, 2025). However, a typical AMD is reddish in color owing to the oxidation of ferrous ions to ferric ions. Furthermore, AMD is rich in huge concentrations of sulfate, sulfuric acid, dissolved heavy metals, as well as low pH, high hardness, and salinity.

13.3 MODE OF CULTIVATION OF MICROALGAE IN INDUSTRIAL WASTEWATER

Microalgae cultivation systems are categorized into three types: open culture systems, closed culture systems, and hybrid culture systems (Kumar et al., 2024; Ma et al., 2025). These systems differ in terms of design, operational costs, and biomass yield. The advantages and disadvantages of different cultivation systems are summarized in Table 13.1. The different microalgae cultivation systems are discussed in detail below.

13.3.1 Open Systems

This involves the cultivation of microalgae in outdoor facilities under unregulated conditions. Open systems include natural (such as lakes and lagoons) and artificial

TABLE 13.1
Advantages and Disadvantages of Different Microalgae Cultivation Systems

Cultivation Mode	Pros	Cons	Reference
Stabilization pond	Low operating costs, scalability, and high efficiency in removing contaminants	Prone to contamination; affected by seasonal weather changes	Solovchenko et al., 2016
High-rate algal pond	Lower energy consumption, high nutrient removal efficiency, and biomass accumulation	Produce high-solid effluent that requires further treatment, and has inconsistent performance due to changes in environmental factors	Sutherland et al., 2020; Razzak et al., 2024
Tubular PBR	High productivity and minimum pollution	Expensive to operate and has technical complexity	Egbo et al., 2018; Kumar et al., 2024
Column PBR	Offers a highly controlled environment, good mixing, and low power consumption	High capital and cleaning costs, potential for uneven light distribution, and gas gradients	Ranganathan et al., 2022; Penloglou et al., 2024
Flat-panel PBR	High light utilization, good mixing, and low shear stress	High capital costs, limited scalability beyond a certain point, photoinhibition, and fouling	Nuri et al., 2023; Penloglou et al., 2024
Hybrid	Allows separate biomass accumulation, improved contamination control	Higher initial costs and complexity compared to simpler systems like open ponds	Narala et al., 2016; Kumar et al., 2024

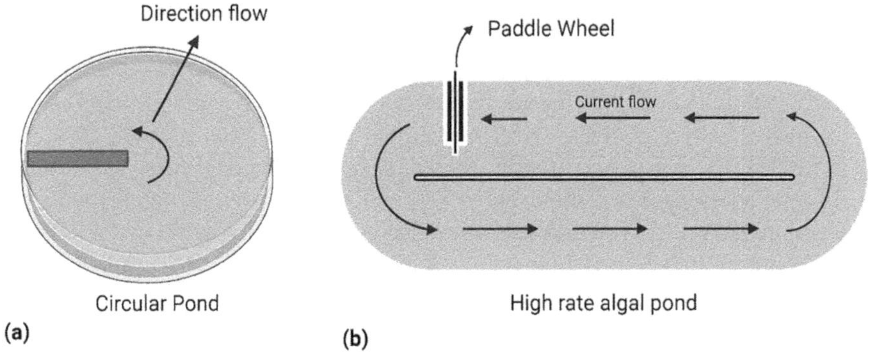

FIGURE 13.2 Schematic diagrams showing different types of ponds for microalgae cultivation: (a) circular pond and (b) high-rate algal pond.

(e.g., circular ponds, artificial tanks, and raceway ponds) pond systems. These systems are characterized by direct exposure to the environment, with raceway ponds being the most widely used for commercial microalgae cultivation and biomass production (Sutherland et al., 2020). An open system is characterized by low construction and operational costs, as well as easy setup (Adetunji et al., 2025). However, open systems are liable to microbial contamination, insufficient light utilization, poor agitation, oxygen accumulation, and less biomass productivity (Razzak et al., 2013). Some open pond systems that are employed for wastewater treatment are discussed in detail below and shown in Figure 13.2.

13.3.1.1 Stabilization Ponds

Stabilization pond systems incorporate a variety of pond types, including anaerobic, facultative, oxidative, and aerobic ponds, each fostering unique microbial, chemical, and biochemical properties influenced by the oxygen levels and nutrient content of their environments (Oberholster et al., 2021). These are inexpensive shallow ponds frequently employed in tropical and rural areas to treat wastewater rich in nutrients and organic pollutants from municipal, domestic, and industrial sources (Razaviarani et al., 2023). Globally, stabilization ponds serve as an alternative biological method for wastewater treatment. In colder regions like Europe, they are utilized in small rural communities of about 2000 people, whereas in warmer climates such as Africa or Asia, they cater to larger populations of up to a million (Mohammed, 2006; Barceló and Petrovic, 2011).

The primary benefits of stabilization ponds include their low capital and operational costs, minimal energy requirements, and effective removal of BOD, suspended solids, nutrients, and pathogens. Green microalgae species like *Chlamydomonas*, *Scenedesmus*, and *Chlorella* are the most prevalent and dominant in stabilization ponds for wastewater treatment (Tang et al., 2024). However, stabilization ponds are characterized by extended hydraulic retention times of 20–30 d, significant water loss due to evaporation, particularly in sunny seasons, and a high reliance on environmental conditions and seasonal variations, making them susceptible to contamination (Novoveská et al., 2023; Razaviarani et al., 2023). They also require large land areas due to their shallow depth (typically less than 1.5 m) and prolonged retention times of several weeks. Biomass harvesting in open ponds is challenging due to the poor settling characteristics of many algal species, which reduce the potential for biomass valorization (Razaviarani et al., 2023; Kumar et al., 2024).

13.3.1.2 High-Rate Algal Ponds

High-rate algal ponds (HRAPs), also known as raceway ponds, are widely recognized as cost-effective and efficient systems for treating various types of wastewaters, including municipal, agricultural, and industrial effluents (Sutherland et al., 2020; Kaloudas et al., 2021). HRAPs operate through three main processes, including the removal of pollutants from wastewater, aerobic biological treatment, and the harvesting of microalgae (Ranjan et al., 2019; Razaviarani et al., 2023). As an alternative to conventional wastewater treatment systems, HRAPs offer several advantages. They have a low carbon footprint, reduce greenhouse gas emissions, and the algal biomass produced can be utilized to produce valuable bio-based products. The design of HRAPs typically features an open racetrack-style pond that is shallow, with depths ranging from approximately 0.2 to 1 m. This design incorporates vertical mixing, allowing microalgae to proliferate, which results in higher biomass accumulation and improved nutrient removal efficiency compared to stabilization ponds (Sutherland et al., 2020; Magalhães et al., 2024). The shallow depths of HRAPs enhance light penetration, which boosts the photosynthetic ability and performance of the microalgae (Razaviarani et al., 2023).

Mixing is generally achieved using a paddlewheel, which produces a mean horizontal water velocity of about 0.15–0.3 m/s. Depending on soil conditions and local regulations, the pond bottom may be lined or unlined. Carbon dioxide can be introduced into a counter-current gas sparging sump (approximately 1.5 m deep), generating turbulent flow within the pond. However, self-shading can occur at higher algal cell densities due to the shallow design of HRAPs, which may impair light penetration. Algae located below the surface can receive insufficient light, resulting in low nutrient uptake and limiting biomass productivity (Adetunji et al., 2025). HRAPs can also be susceptible to contamination by unwanted microorganisms and are prone to significant evaporation losses and CO_2 stripping in hot climates (Al-Jabri et al., 2021; Geremia et al., 2021). Furthermore, after the treatment of wastewater, the recovery of the suspended microalgal biomass can be costly, and inefficient removal may lead to eutrophication. Additionally, the accumulation of sludge over time can create maintenance challenges and reduce the effective volume of the pond (Ranjan et al., 2019).

13.3.2 Closed Systems

Closed photobioreactor (PBR) cultivation systems have been developed to address the challenges faced by open cultivation systems. PBRs are extensively studied because they offer a controllable environment for microalgae cultivation, enable the maintenance of pure algal strains, and minimize contamination (Kumar et al., 2024). In PBRs, microalgae continuously float through a series of transparent tubes that are irradiated from the outside. The culture is aerated with CO_2, and the oxygen produced is degassed to maximize photosynthetic efficiency and the growth

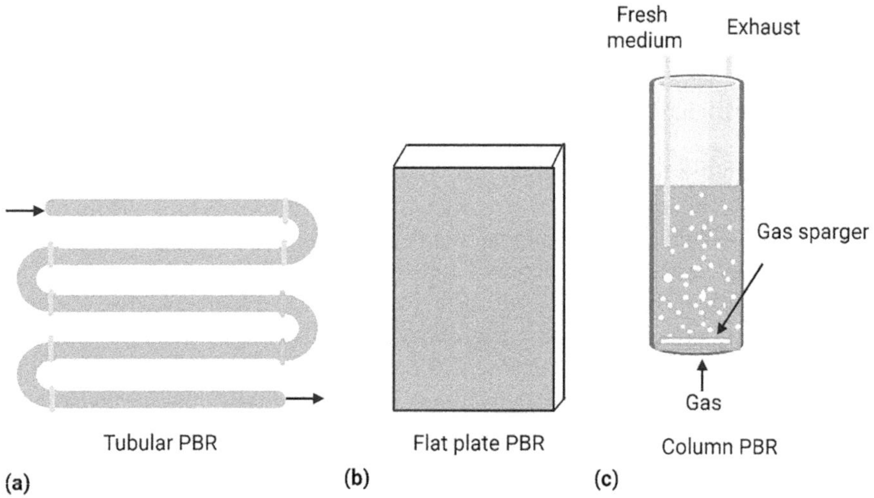

FIGURE 13.3 Schematic diagrams elucidating different closed systems for microalgae cultivation: (a) tubular photobioreactor, (b) flat plate photobioreactor, and (c) column photobioreactor.

of microalgae (Plöhn et al., 2021). Illumination in PBRs is provided by both sunlight and artificial lighting, including warm and cool fluorescent lights. These systems have a high surface-to-volume ratio, which ensures optimal light availability for maximum growth and effective removal of nitrogen and phosphorus (Sen and Mukherjee, 2024). However, this efficiency comes with higher construction and operating costs, as about 40% of the total energy is consumed for aeration and cooling systems needed to maintain temperature (Clippinger and Davis, 2019; Novoveská et al., 2023; Sathinathan et al., 2023). PBRs exist in three common designs, namely tubular, column, and flat-panel PBRs, as illustrated in Figure 13.3.

13.3.2.1 Tubular Photobioreactor

Tubular PBRs are closed-loop microalgae cultivation systems, consisting of a long transparent tube, typically made from glass or plastic, through which microalgae culture is continuously circulated and illuminated using natural sunlight or artificial light. They represent a significant advancement in microalgae cultivation, offering a sophisticated platform for efficiently capturing solar radiation over large areas (Kumar et al., 2024; Yu et al., 2024). Tubular PBRs are known for their high surface-to-volume ratios, which allow for high light efficiencies. Based on the design, tubular PBR are classified into vertical and horizontal tubular PBR. Horizontal tubular PBRs are constructed using clear vertical tubes, which are a parallel set of connected loops of tubes (Ting et al., 2017). This offers an advantage when used in outdoor culture, as they are inherently oriented toward the natural sunlight to allow high light conversion and photosynthetic efficiency (Egbo et al., 2018).

Horizontal tubular PBRs are the most commonly used setup for pilot-scale and full-scale microalgae wastewater treatment systems. They are favored for their ease of land deployment, scalability, and consistent light exposure (García et al., 2018). These systems can handle larger volumes and are easier to operate and expand modularly in real-world wastewater treatment environments (López-Lizárraga and García Cañedo, 2016). Vertical tubular PBRs are constructed using clear vertical tubes, providing a double height and high surface-to-volume ratio with optimal head and mass transfer along the reactor (Cui et al., 2020; Razzak et al., 2024). The mixing of the culture in the tubular PBR is facilitated by passing gas consisting of air and carbon dioxide using a sparger attached to the bottom of the reactor to provide mass transfer of both nutrients and algal cells (Egbo et al., 2018).

13.3.2.2 Column Photobioreactors

Column PBRs are designed similarly to tubular PBRs, featuring a vertical cylindrical shape with a relatively large base and a shorter length compared to tubular models (Xiao and Luo, 2022). In column PBRs, microalgae cultures are circulated through aeration, which enhances mixing efficiency. This effective mixing ensures a uniform dispersion of nutrients throughout the reactor, allowing for even distribution of CO_2 and efficient removal of O_2 (Ranganathan et al., 2022). Column PBRs can be illuminated using natural sunlight or artificial light sources such as light-emitting diodes. When placed outdoors, these reactors can be tilted to optimize the angle of light exposure. Air is usually pumped from the bottom of the reactor through a gas disperser and nozzles in either a bubble column or airlift design (Novoveská et al., 2023).

There are three main types of vertical column PBRs based on their liquid flow and mixing methods: bubble columns, stirred tanks, and airlift reactors. The bubble column PBR is the most common type, featuring a simple design where the height of the reactor is more than twice its diameter. This design allows for a high surface area-to-volume ratio (Egbo et al., 2018). These reactors utilize a sparger at the bottom to inject carbon dioxide, which aids in mixing and aeration. However, column PBRs face challenges regarding limited light penetration in dense cultures, which reduces photosynthetic efficiency and biomass yield, especially as the culture depth increases (Razzak et al., 2024).

13.3.2.3 Flat-Panel Photobioreactor

Flat-panel PBRs are one of the most common types used for both laboratory and large-scale microalgae cultivation, utilizing synthetic media and wastewater. These reactors are constructed from transparent or semi-transparent materials, such as glass, plexiglass, polycarbonate, and plastic bags (Ting et al., 2017). This design allows for the cultivation of algal cultures while ensuring optimal light penetration. Typically, flat panels are short, rectangular reactors that provide a large surface area for illumination, using either natural or artificial light based on the location (Dineshkumar and Sen, 2020; Nuri et al., 2023). When a combination of natural and artificial light is used, the reactor plates can be tilted toward the sun to maximize solar radiation. If solar intensity decreases, artificial lighting is employed, with the tilt angle adjusted accordingly (Ashok et al., 2019). The mixing of the algal culture is primarily achieved through an air sparger, which helps circulate the algae suspension. Exhaust gas emissions occur at the gas–liquid interface, particularly when the reactor is inclined at a specific angle to optimize incident light intensity, especially in outdoor settings (Ting et al., 2017).

Flat plate PBRs are recommended for the mass production of microalgae in both indoor and outdoor systems due to their high surface area for irradiation, minimal accumulation of dissolved oxygen, and convenient modular design for scaling up (Razzak et al., 2024). However, large-scale microalgae cultivation using flat-panel PBRs encounters challenges, such as high operational and maintenance costs, with biomass harvesting logistics, posing a significant obstacle (Razzak et al., 2024). Ongoing advancements in materials, reactor engineering (including cooling and antifouling surfaces), and real-time process control are essential for improving the viability of this system (Ashok et al., 2019; Yu et al., 2024).

13.3.3 Hybrid Systems

The hybrid cultivation system has emerged as a solution to the challenges associated with using open pond systems and PBRs for commercial microalgae cultivation. This two-phase system effectively separates biomass growth needed for downstream applications (Adetunji et al., 2025). It provides improved cultivation conditions that favor monoculture dominance and minimize contamination during the second stage (Andrade et al., 2018). To enhance biomass yield, microalgae are first cultivated in a PBR and then transferred to an open system, which helps maintain culture purity and avoid contamination (Bora et al., 2024). For instance, Yun et al. (2018) cultivated *Parachlorella* sp. JD076, using a hybrid cultivation system that combines a closed PBR with a small open raceway pond fed with wastewater. The PBR served as a continuous source of inoculum for sustaining microalgae growth and biomass production in the open raceway pond. Estrada-Graf et al. (2020) cultivated *Scenedesmus obtusiusculus* in a hybrid PBR, achieving a biomass concentration of 1.2 g/L with a specific growth rate of 0.38/d and biomass productivity of 41.8 mg/L/d using flue gas supplementation. However, the high capital and operating costs, along with the risk of contamination in the open pond, remain significant challenges for the use of hybrid systems in microalgae cultivation. Additionally, the variable biomass density complicates the harvesting and downstream processing of the biomass (Yun et al., 2018; Kumar et al., 2024).

13.4 ROLE OF MICROALGAE IN BIOREMEDIATION OF INDUSTRIAL WASTEWATER

Microalgae-based system is an effective approach for the treatment of pollutant-laden industrial wastewater. The removal of hazardous pollutants in wastewater by microalgae is achieved using three mechanisms, namely passive rapid extracellular adsorption (biosorption), intracellular accumulation of contaminants (bioaccumulation), or biodegradation (Devi et al., 2023; Satya et al., 2023). During biosorption, the positively charged functional groups (e.g., sulfonate, phosphate, hydroxyl, carboxyl, and amino) found on the algal cell wall interact with pollutants present in the wastewater by covalent interactions, van der Waals forces, or ion exchange, thereby leading to the removal of the contaminants from the wastewater to the cell wall. Parameters such as contact time, pollutant concentration, inoculum volume, pH, and temperature influence microalgal biosorption capacity. Conversely, bioaccumulation is an active and slower metabolic process in which the pollutants are taken up and accumulated into algal cells by an energy source (Devi et al., 2023; Agoun and Avci, 2025). Biodegradation utilizes the metabolic activity of microalgae for the breakdown of toxic pollutants, resulting in the formation of low-molecular-weight molecules, which are utilized as a source of nutrients for microalgal growth (Amin et al., 2022; Abdelfattah et al., 2023). The pollutant degradation occurs by complete mineralization or biotransformation in the presence of complex enzymatic processes involving cytochrome P_{450} enzymes, glutathione-*S*-transferases, hydrolases, and transferases, for the formation of less or nontoxic intermediates (Ding et al., 2017; Abdelfattah et al., 2023). Microalgae such as *Anabaena oryzae*, *Chlorella*, and *Oscillatoria* possess excellent biodegradation potentials (Mustafa et al., 2021). The efficacy of microalgae in the removal of pollutants from various industrial wastewaters is shown in Table 13.2.

TABLE 13.2
Efficiencies of Microalgae in the Removal of Pollutants from Various Industrial Wastewaters

Wastewater Type	Microalgae Species	Removal Efficiency	Reference
Tannery wastewater	*Scenedesmus* sp.	BOD 88%; Sulfate 92.68%; P 99.85%; N 90%	Ballén-Segura et al., 2016
Brewery effluent	*Chlorella vulgaris*	TP 54.69%; BOD 91.43%; TN 83.74%; COD 83.11%	Choi, 2016
Petrochemical wastewater	*Chlorella vulgaris*	COD 10.89%; P 69.89%; N 92.59%	Madadi et al., 2021
Pharmaceutical wastewater	*Scenedesmus abundans*	COD 97.24%; P 93.71%; N 97.12%	Nayak and Ghosh, 2019
Acid mine drainage	*Desmodesmus* sp.	Fe^{2+} 90 %	Diale et al., 2023
Dairy wastewater	*Chlamydomonas polypyrenoideum*	COD 80%; NH_4^+ 63%; PO_4^{3-} 69%; NO_3^- 62%; TSS 91%; TDS 89.8%	Umamaheswari and Shanthakumar, 2016
Laundry effluent	*Botryococcus* sp.	N 86.62%; P 78.23%; TOC 76.66%; BOD 61.11%; COD 70.68%	Gani et al., 2015
Textile wastewater	*Chlorella pyrenoidosa*	Cr 80%; Pb 80%; Cd 91%; Cu 85%	Soeprobowati and Hariyati, 2017
Dairy wastewater	*Spirulina* sp.	PO_4^{3-} 72%; NO_3^- 80%; COD 77%	Emparan et al., 2019
Brewery effluent	*Scenedesmus obliquus*	TN 20.8%; COD 57.5%	Mata et al., 2012
Textile wastewater	*Chlorella vulgaris*	COD 99.7%; TN 80%; TP 80%	Fazal et al., 2021
Fish processing wastewater	*Oocystis* sp.	COD 71.1%; phosphate 70%	Riano et al., 2011
Pulp and paper mill wastewater	*Scenedesmus* sp.	COD 75%; BOD 82%; nitrate 65%; phosphate 71.29%	Usha et al., 2016

Note: TOC, total organic carbon; TN, total nitrogen; TP, total phosphate; TSS, total suspended solids; TDS, total dissolved solids.

13.5 METHODS FOR RECOVERY OF MICROALGAL BIOMASS

The extraction of microalgae biomass from the growth medium is a crucial phase in the processing of microalgae. The main techniques employed for harvesting microalgae include flocculation, sedimentation, flotation, and centrifugation (de Morais et al., 2023). The efficiency of microalgae biomass recovery can be improved by integrating two or more of these techniques. However, the high costs associated with biomass harvesting and the absence of a universal harvesting method remain significant obstacles to the financial feasibility and sustainability (Singh and Patidar, 2018). To date, there has been no cost-effective method for harvesting and dewatering microalgae that has been implemented on a commercial scale (Hussian and Ellatif, 2018).

13.5.1 Filtration

The filtration technique uses a membrane for the physical removal of solids from liquid through exclusion, in which solids are trapped while the liquid passes through the membrane. Different types of filtration techniques commonly employed for the recovery of microalgae biomass include microfiltration, ultrafiltration, vacuum filtration, and pressure filtration (Deepa et al., 2023; Zhu et al., 2024). Usually, microfiltration and ultrafiltration are often integrated with a PBR to create the membrane PBR system, allowing simultaneous microalgae cultivation and harvesting (Zhou et al., 2022; Yang et al., 2025). Membrane filtration techniques have several advantages compared to centrifugation, as they offer simplicity, algal cells are recovered gently with low shear stress, scalability, lower energy consumption, and excellent recovery efficiency (de Morais et al., 2023).

Organic materials used for microalgae harvesting are made using materials like polyvinylidene fluoride (PVDF), polyacrylonitrile (PAN), polyether sulfone (PES), polytetrafluoroethylene (PTFE), polyvinylpyrrolidone (PVP), and polyvinyl chloride (PVC). However, ceramic membranes offer better chemical, temperature, and mechanical stability compared to organic membranes (Bamba et al., 2021). However, this method is prone to membrane fouling and clogging based on the membrane properties. As hydrophilic membranes have better anti-fouling capabilities, while hydrophobic membranes are more prone to forming a dense cake layer, causing irreversible fouling (Wang et al., 2025). It may require frequent change of the fresh filter or membrane, which may contribute to increasing its processing cost (Tan et al., 2020). The recovery efficiency of microalgae biomass using filtration in different types of wastewaters is summarized in Table 13.3.

13.5.2 Centrifugation

Centrifugation is a method that separates microalgae based on the differences in particle size and density of the medium components (Najjar and Abu-Shamleh, 2020). As a physical technique, centrifugation utilizes centrifugal forces to speed up the sedimentation and thickening of cell suspensions from highly diluted solutions (Zhu et al., 2024). The separation of algal cells from the suspension also relies on factors such as the settling characteristics of the cells and the retention time of the slurry

TABLE 13.3
Microalgae Biomass Harvesting by Filtration Techniques

Microalgae	Membrane System	Recovery (%)	References
Chlorella vulgaris	Crossflow filtration and ultra-low-pressure filtration	76%	Wan Osman et al., 2021
Desmodesmus sp.	Polysulfone and polyethylene glycol	100%	Zhao et al., 2020
Chlorella sp.	Ultrafiltration membrane	100%	Hwang and Wu, 2015
Chlorococcum sp.	Polyvinylidene fluoride (PVDF) membrane	98.7%	Hashmi et al., 2025
Chlorella vulgaris and *Scenedesmus*	Ceramic microfiltration membrane	80%	Aditya et al., 2023
Chlorella vulgaris	Crossflow membrane filtration, UH050 membrane—hydrophilic polyether sulfone	100%	Elcik and Cakmakci, 2017

in the centrifuge (which is controlled by flow rate and settling depth) (Hussian and Ellatif, 2018; Najjar and Abu-Shamleh, 2020). Centrifugation can achieve high biomass recovery efficiencies, typically greater than 90%, within a short period (Javed et al., 2019). Additionally, this method is suitable for nearly all types of microalgae species and does not require chemical additives (Barros et al., 2015). However, it is important to note that recovery efficiency is linked to high energy consumption, generally ranging from 1 to 8 kWh per cubic meter of culture. This energy consumption constitutes a significant portion of the total harvesting energy in algae-to-biofuel chains (Al Hattab and Ghaly, 2015). Furthermore, the operating and maintenance costs associated with centrifugation make it more expensive compared to other methods such as sedimentation or flocculation (Fasaei et al., 2018).

13.5.3 Flocculation

As an energy-independent method, flocculation harvests microalgae by concentrating the cells into large particles or flocs that can easily be collected by sedimentation or other methods (Al Hattab and Ghaly, 2015; Yuan et al., 2024). This method offers advantages such as high harvesting efficiency, low cost, and scalability for the volume required in wastewater treatment plants (Vasistha et al., 2021; de Morais et al., 2023). Flocculation for microalgae harvesting is mainly performed using chemical flocculation, physical flocculation, and bio-based flocculation, as shown in Figure 13.4. A summary of flocculation techniques utilized in the harvesting of biomass from different microalgae and their recovery efficiencies is given in Table 13.4.

Chemical flocculation is reported to be the most efficient technique for harvesting on a large scale (Matter et al., 2019). This harvesting approach utilizes chemical compounds or synthetic polymers to extract algal cells from the liquid medium. Nevertheless, there have been no commercial large-scale demonstrations confirming their economic viability for the recovery of microalgae (de Morais et al., 2023).

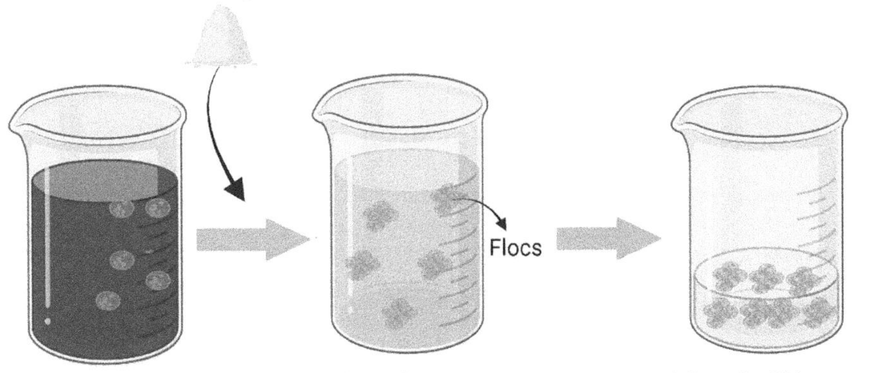

FIGURE 13.4 Schematic diagrams showing the harvesting of microalgae by flocculation techniques.

TABLE 13.4
Microalgae Biomass Harvesting by Flocculation

Harvesting Technique	Microalgae	Procedure	Recovery (%)	Reference
Autoflocculation	*Tribonema* sp. and *Synechocystis* sp.	pH	39.5	Cheng et al., 2020
Autoflocculation	*Ettlia* sp.	pH	91%	Jaiswal et al., 2022
Autoflocculation	*Scenedesmus* sp. NC1	pH	>88.32% 94.95%	He et al., 2022
Flocculation	*Chlorella vulgaris*	Chitosan (biopolymer)	99.1	Elcik et al., 2023
Flocculation	*Chlorella vulgaris*	Aluminum sulfate; ferric sulfate; chitosan	~98–99	Machado et al., 2024
Flocculation	Rhodophyta (*Porphyridium cruentum* and *Galdieria phlegrea*) and Chlorophyta (*Pseudococcomyxa simplex*), and a Cyanobacterium (*Synechococcus bigranulatus*)	$CaCl_2$; chitosan; pH-raising	~96% and ~87%	Imbimbo et al., 2024
Bioflocculation	*Chlorella* sp.	Cyanobacterial (Anabaena)	>80%	Irawan et al., 2023
Bioflocculation	*Picochlorum* sp., *Chlamydomonas reinhardtii*	*Saccharomyces bayanus*	~70–95%	Tripathi et al., 2024
Bioflocculation	*Raphidocelis subcapitata* and *Dunaliella salina*	*Klebsiella* sp.	>90%	Yuan et al., 2024

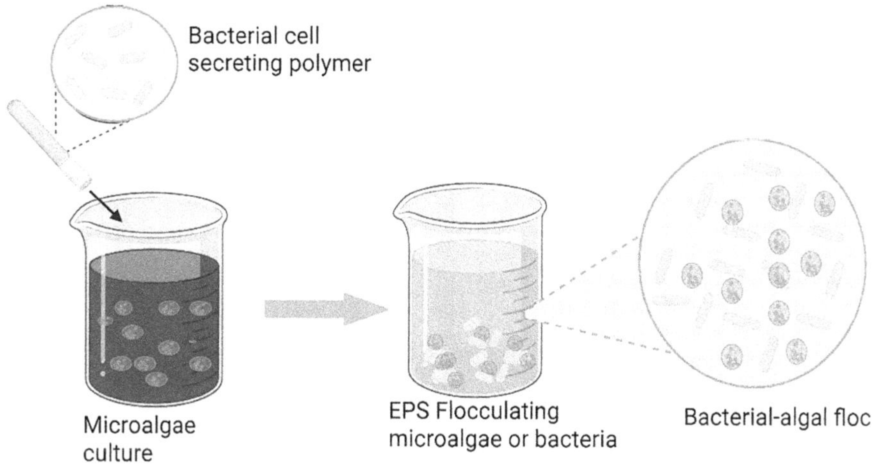

FIGURE 13.5 Schematic diagrams depicting the bioflocculation approach for harvesting of microalgae.

Furthermore, chemical flocculation can result in contamination of microalgal biomass with leftover chemicals, which diminishes its usability. The use of chemical flocculants produces significant amounts of chemical sludge that must be treated separately, thus incurring additional costs and raising the overall expense of the method (Pugazhendhi et al., 2019; Labeeuw et al., 2020). The application of chemicals for coagulation–flocculation might restrict the reuse of both wastewater and the collected biomass.

Microalgae can alternatively be recovered through autoflocculation, a process that occurs when the CO_2 within the algal cells is depleted, causing an increase in pH that leads to the precipitation of carbonate salts along with the coprecipitation of magnesium and calcium ions, which neutralize the cells' negative charge and induce coagulation (Wu et al., 2020; de Morais et al., 2023). Conversely, bioflocculation captures microalgae by utilizing the extracellular polymeric substances (EPS) secreted by microorganisms found in the cultivation medium, as illustrated in Figure 13.5. Flocculants generated by bacteria could play a significant role in advancing the economic feasibility of sustainable biofuel production from microalgae. Bioflocculation negates the need for chemical flocculants, which can be costly, impractical, and toxic (Barros et al., 2015; Alam et al., 2022). When large amounts of microalgae are cultivated in HRAP for wastewater treatment, a considerable quantity of soluble EPS remains in the supernatant post-harvesting, which can be leveraged to enhance the following harvesting stages or extracted and refined for various purposes (de Morais et al., 2023).

13.5.4 Flotation

Flotation separates microalgae by attaching small bubbles to destabilize particles, causing them to float to the surface for recovery (Rao et al., 2023). Dissolved air flotation (DAF) and dispersed air flotation (DiAF) are two common methods promising

TABLE 13.5
Microalgae Biomass Harvesting by Flotation Techniques

Harvesting Technique	Microalgae	Recovery (%)	Reference
Electrocoagulation–flotation	Consortium (*Chlorella* sp., *Spirulina* sp., *Scenedesmus* sp., *Synechocystis* sp.)	~98%	Nageshwari et al., 2022
Dissolved air flotation	*Chlorella vulgaris* *Nannochloropsis oculata*	~99% 71–90%	Rao et al., 2023
Dissolved air flotation—microcystis/coagulation	M. flos-aquae (FACHB 1028)	43.0%	Cheng et al., 2022
Dissolved air flotation	*Scenedesmus obliquus (Tetradesmus obliquus)*	90%	Koley et al., 2017
Microbubble air flotation	*Microcystis* sp.	~63.5%	Ye et al., 2024
Dissolved air flotation, pH modulation	*Chlorella sorokiniana*	96.5–97.9%	de Souza Leite et al., 2020
Nanobubble/foam fractionation	*Microcystis aeruginosa*	69%	Zhang et al., 2025

for harvesting microalgae. The DAF uses small bubbles ranging between 10 and 100 μm, while DiAF uses larger bubbles ranging between 700 and 1500 μm (Singh and Patidar, 2018). Among these methods, DAF utilizes the microalgae's natural tendency to float, and the low-density flocs are formed through coagulation (de Morais et al., 2023). It is known for its faster reaction time, small footprint, moderate operational costs, and high flexibility with lower initial equipment costs (Tan et al., 2020). The biomass recovery efficiency using different flotation techniques is summarized in Table 13.5.

This method has been proven at a large scale, although it generally requires the use of flocculants (Barros et al., 2015). The harvesting efficiency and cost of flotation are improved by combining it with coagulation/flocculation. The surfactants used can increase the probability of bubble particles attached to microalgae cells (Gerardo et al., 2015; Singh and Patidar, 2018). Among the microalgae harvesting methods, few approaches are feasible for generating profits due to the lack of scalability in high-capacity harvesting systems, such as those required for wastewater treatment (Lim et al., 2022).

13.6 VALORIZATION OF MICROALGAE BIOMASS

Microalgae display a great sustainable and commercial advantage as potential feedstock for the production of bio-based products. The algal biomass recovered after wastewater treatment is rich in important macromolecules such as lipids, proteins, and carbohydrates that can be used for the production of biofuels. Biofuel is one of the important types of renewable energy that has received significant attention globally. The main types of biofuels include biodiesel, bioethanol, biogas, and biohydrogen, among others (Pandit et al., 2024; Adetunji et al., 2025). These are discussed in detail below.

13.6.1 Biodiesel

Microalgae, as a third-generation feedstock, have emerged as a promising source for biodiesel production due to their high-lipid content, rapid growth rate, and ability to thrive in arable laboratory conditions using wastewater or seawater (Adetunji et al., 2025). Various microalgal species, such as *Chlorella vulgaris*, *Nanochloropsis oceanica*, *Dunaliella salina*, *Botryococcus*, *Desmodesmus*, *Neochloris*, *Scenedesmus*, and *Tetraselmis*, have been identified as suitable feedstocks for biodiesel production. Certain strains, with lipid content ranging from 15% to 40%, produce biodiesel that meets American Society for Testing and Materials (ASTM) standards (Khan et al., 2023; Wang et al., 2024). The quality and quantity of lipids in microalgae are influenced by environmental factors, including culture conditions and the type of algal strain (Gumbi et al., 2022). Crude lipids extracted from algal biomass are subsequently utilized in the transesterification reaction to produce biodiesel. This process reduces the high viscosity of algal oil, converting it into fatty acid methyl esters with a lower molecular weight. In a study conducted by Abomohra et al. (2021), *Chlorella pyrenoidosa* was cultivated using a transparent polyethylene column PBR. The researchers reported a biomass productivity of 0.182 g $L^{-1}d^{-1}$ and a biodiesel yield of 20.15 % dcw. In another study, Nguyen et al. (2021) extracted lipids from *Chlorella vulgaris* and converted them into biodiesel through in-situ transesterification of wet biomass, achieving a biodiesel yield of 96.9% dcw. For microalgae biodiesel to serve as a viable alternative to fossil fuels, it must comply with established standards such as ASTM D6751, EN 14214:2008, and IS: 15607 (Gumbi et al., 2022; Sharma et al., 2025).

The composition of fatty acid profiles is critical for biodiesel production, as it directly affects the quality of the biodiesel produced. High levels of PUFAs can enhance flow properties, particularly in cold weather, but may adversely affect oxidative stability (Gumbi et al., 2017). Conversely, a high concentration of saturated fatty acids imparts excellent combustion properties to biodiesel but negatively impacts cold flow properties (Jeong et al., 2008). Microalgae diesel exhibits a lower caloric value, ranging from 39 to 41 MJ/kg, compared to petroleum-based diesel, which has a caloric value of 46 MJ/kg (Hoang et al., 2023). Blending microalgae diesel with other diesel types, such as B5, B10, B20, B50, and B100, can enhance thermal efficiency, reduce hydrocarbon and carbon monoxide emissions, and decrease engine smoke emissions (Arun et al., 2021). However, its commercial feasibility remains constrained by high production costs associated with cultivation, harvesting, lipid extraction, and energy-intensive drying and conversion processes (Gaurav et al., 2024). Current research is focused on developing integrated biorefinery systems, employing genetic engineering to produce high-lipid strains, and utilizing wastewater as a nutrient source to improve economic viability and sustainability of biodiesel production from microalgae (Khan et al., 2023).

13.6.2 Bioethanol

Microalgae are considered an excellent feedstock for bioethanol production because certain strains can accumulate carbohydrates that exceed 40% of the dry biomass. The utilization of microalgae for bioethanol production has the potential to address environmental issues, such as climate change (Pandit et al., 2024). As a third-generation feedstock, microalgae are more sustainable than first- and second-generation options, producing up to 60 times more alcohol than agricultural or forestry feedstocks (Joyia et al., 2024). Different from lignocellulosic biomass, microalgae do not contain lignin, which inhibits enzymatic hydrolysis and typically requires extensive pretreatment to remove. Microalgae can yield significant amounts of carbohydrates (approximately 20–40%), which are the primary substrate for bioethanol production. The carbohydrates found in green microalgae are mainly composed of starch and cellulose (Maity and Mallick, 2022). The process of bioethanol production using microalgae consists of several steps, including cultivation, harvesting, pretreatment of the biomass, saccharification (conversion of polysaccharides to sugars), and fermentation (Deshmukh and Pathan, 2026).

Polysaccharides are hydrolyzed into simple sugars before being fermented by microorganisms to produce ethanol and carbon dioxide (Harun et al., 2010). Microalgae can produce bioethanol at rates ranging from 0.07 to 0.5 g/g, depending on the species of microalgae used (Kumari et al., 2024). Commonly selected genera for bioethanol production include *Dunaliella*, *Chlamydomonas*, *Chlorella*, and *Scenedesmus*, as well as cyanobacterial species such as *Synechococcus* and *Spirulina*, all of which are known for their high carbohydrate content (Maity and Mallick, 2022; Khan et al., 2023). In a study by Condor et al. (2022), *Chlorella vulgaris* FSP-E was cultivated using BG-11 at a laboratory scale, resulting in an ethanol production of 0.214 g/g biomass after undergoing acid hydrolysis and fermentation with *Saccharomyces cerevisiae* FAY-1. In a separate study by Constantino et al. (2021), carbohydrates were extracted from 50 g of dry biomass of *Chlorella sorokiniana*, *Tetraselmis* sp., and *Skeletonema* sp. The ethanol yields recorded for these species were high, measuring 14.2 g/L for *Chlorella sorokiniana*, 15.7 g/L for *Tetraselmis* sp., and 12.3 g/L for *Skeletonema* sp. Despite these advantages, the commercialization of bioethanol from microalgae faces several technical and economic challenges. Recent studies have shown that the cultivation and harvesting of microalgal biomass are still costly and energy-intensive. When scaling up production, the pretreatment and hydrolysis of algal biomass require harsh chemicals, significant energy input, and expensive enzymes, all of which can negatively impact yield and profitability (Khan et al., 2023). Additionally, many microalgal strains yield lower carbohydrate content compared to other feedstocks, such as sugarcane and corn (Adetunji et al., 2025).

13.6.3 Biohydrogen

Biohydrogen has garnered attention as an important eco-friendly industrial feedstock due to its high energy density and low environmental impact (Ahmed et al., 2021). Microalgae are excellent candidates for biohydrogen production because of their high photosynthetic efficiency, rapid growth, and ability to utilize atmospheric carbon dioxide (Nagarajan et al., 2021). Various microalgal strains, including *Chlorella*, *Desmodesmus*, *Scenedesmus*, and *Tetraselmis*, have demonstrated significant potential as suitable feedstock for biohydrogen production due to their unique biochemical profiles (Wang et al., 2024). These microalgae can be cultivated either as single or mixed cultures. Single cultures allow for easier cultivation, while mixed cultures provide greater carbohydrate availability for fermentation (Chen et al., 2020; Adamu et al., 2023).

Clean and renewable energy sources, such as light, support the production of biohydrogen from microalgal biomass. Biohydrogen can be produced through either a direct or indirect approach, depending on light availability, fermentation processes, and the thermal conversion of biomass (Ahmed et al., 2021; Ayub et al., 2024).

Microalgae primarily generate biohydrogen through the action of hydrogenase enzymes, which facilitate the conversion of protons into hydrogen. The necessary electrons for this process are supplied either directly from water via the photosynthetic apparatus (Photosystem I and II) during direct biophotolysis or indirectly from stored carbohydrates (such as starch and sugars) during darker or anaerobic conditions in indirect biophotolysis (Samrot et al., 2023). Efficient hydrogen production requires maintaining controlled culture conditions, including sulfur limitation and low-oxygen environments, as hydrogenase enzymes are sensitive to oxygen. Therefore, microalgae offer a renewable and carbon-neutral source for biohydrogen, although their commercial use is restricted due to low production yields and oxygen sensitivity (Mokashi et al., 2023; Singh, 2024). However, the commercialization of algal biohydrogen as fuel faces numerous challenges concerning technical, economic, and environmental feasibility. The hydrogen yield from algal biomass is low due to the inefficiency of hydrogenase enzymes and limited electron flow. These enzymes are highly sensitive to oxygen, necessitating anoxic or sulfur-deprived conditions for optimal activity (Kumar et al., 2024; Singh, 2024).

13.7 CONCLUSIONS AND FUTURE DIRECTIONS

Industrial wastewaters are heterogeneous, consisting of toxic organic and inorganic substances, which, when improperly disposed of, cause serious environmental and health hazards to the immediate environment and human health. Therefore, the utilization of microalgae is an effective, sustainable, eco-friendly, and economical approach for the removal of pollutants from various industrial wastewaters, including pharmaceutical wastewater, AMD, tannery wastewater, dairy wastewater, distillery wastewater, and poultry processing wastewater. The wastewaters serve as a vital medium for the cultivation of microalgae in open, closed, or hybrid systems for the production of biomass, which can be recovered using different techniques such as filtration, centrifugation, flocculation, and flotation. The obtained biomass is utilized as a feedstock for the production of beneficial bio-based products, including biodiesel, bioethanol, and biohydrogen.

Since most studies are limited to laboratory batch cultivation of microalgae for wastewater treatment, the selection of appropriate novel and hyperactive algal strains capable of surviving in high-strength wastewater with high pollutant removal efficiencies and commercial potential is crucial. More so, the design and construction of appropriate PBRs that operate at optimum process conditions is paramount for the optimization of microalgae-based wastewater treatment competence, enhancement of biomass productivity, and shortening of treatment duration. The integration of algae-based wastewater treatment with the circular economy provides an opportunity for the reuse of treated wastewater and conversion of algae biomass into valued-added products (e.g., animal feed and biochemicals) within industrial processes. The use of innovative technologies, including artificial intelligence, machine learning, and genetic modification, for optimization of microalgae growth and pollutant removal efficiencies from industrial wastewaters should be encouraged.

DECLARATION OF INTEREST

The authors declare that they have no known competing financial interests or personal relationships that could have appeared to influence the work reported in this paper.

REFERENCES

Abdelfattah, A., Ali, S. S., Ramadan, H., El-Aswar, E. I., Eltawab, R., Ho, S.-H., Elsamahy, T., Li, S., El-Sheekh, M. M., Schagerl, M., Kornaros, M., Sun, J. (2023). Microalgae-based wastewater treatment: Mechanisms, challenges, recent advances, and future prospects. *Environmental Science and Ecotechnology*, *13*, 100205.

Abomohra, A. E.-F., Zheng, X., Wang, Q., Huang, J., & Ebaid, R. (2021). Enhancement of biodiesel yield and characteristics through in-situ solvo-thermal co-transesterification of wet microalgae with spent coffee grounds. *Bioresource Technology*, *323*, 124640.

Abuhasel, K., Kchaou, M., Alquraish, M., Munusamy, Y., & Jeng, Y. T. (2021). Oily wastewater treatment: Overview of conventional and modern methods, challenges, and future opportunities. *Water*, *13*, 980.

Adamu, H., Bello, U., Yuguda, A. U., Tafida, U. I., Jalam, A. M., Sabo, A., & Qamar, M. (2023). Production processes, techno-economic and policy challenges of bioenergy production from fruit and vegetable wastes. *Renewable and Sustainable Energy Reviews*, *186*, 113686.

Adetunji, A. I. (2017). Treatment of lipid-rich wastewater using free and immobilized bioemulsifer and hydrolytic enzymes from indigenous bacterial isolates. Ph.D. Thesis. University of KwaZulu-Natal, Durban, South Africa.

Adetunji, A. I. & Erasmus, M. (2025). Biological treatment of acid mine drainage: An eco-sustainable strategy for removal of toxic pollutants. *Journal of Hazardous Materials Advances*, *18*, 100659.

Adetunji, A. I., Gumbi, S. F. T., & Erasmus, M. (2025). Harnessing the potential of microalgae in sequestration of CO2 emissions: Removal mechanisms, optimization strategies, and bioenergy production. *Journal of Hazardous Materials Advances*, *18*, 100722.

Adetunji, A. I. & Olaniran, A. O. (2018). Treatment of lipid-rich wastewater using a mixture of free or immobilized bioemulsifier and hydrolytic enzymes from indigenous bacterial isolates. *Desalination and Water Treatment*, *132*, 274–280.

Aditya, L., Vu, H. P., Nguyen, L. N., Mahlia, T. M. I., Hoang, N. B., & Nghiem, L. D. (2023). Microalgae enrichment for biomass harvesting and water reuse by ceramic microfiltration membranes. *Journal of Membrane Science*, *669*, 121287. https://doi.org/10.1016/j.memsci.2022.121287

Adulkar, T. V., & Rathod, V. K. (2015). Pre-treatment of high fat content dairy wastewater using different commercial lipases. *Desalination and Water Treatment 53*, 2450–2455.

Agoun, N. A. & Avci, F. G. (2025). Phycoremediation: A path towards heavy metal bioremediation from wastewater. *Journal of Chemical Technology and Biotechnology*, *100*, 12–23.

Ahmed, S. F., Rafa, N., Mofijur, M., Badruddin, I. A., Inayat, A., Ali, M. S., Farrok, O., & Yunus Khan, T. M. (2021). Biohydrogen production from biomass sources: Metabolic pathways and economic analysis [Mini Review]. *Frontiers in Energy Research*, *9*. https://doi.org/10.3389/fenrg.2021.753878

Al Hattab, M., & Ghaly, A. (2015). Production of biodiesel from marine and freshwater microalgae: A review. *Algae*, *14*, 18.

Alam, M. A., Wan, C., Tran, D. T., Mofijur, M., Ahmed, S. F., Mehmood, M. A., Shaik, F., Vo, D.-V. N., & Xu, J. (2022). Microalgae binary culture for higher biomass production, nutrients recycling, and efficient harvesting: A review. *Environmental Chemistry Letters*, *20*(2), 1153–1168.

Alcolea, A., Vazquez, M., Caparrós, A., Ibarra, I., García, C., Linares, R., & Rodríguez, R. (2012). Heavy metal removal of intermittent acid mine drainage with an open limestone channel. *Mineral Engineering*, *26*, 86–98.

Al-Jabri, H., Das, P., Khan, S., Thaher, M., & AbdulQuadir, M. (2021). Treatment of wastewaters by microalgae and the potential applications of the produced biomass—A review. *Water*, *13*(1), 27. https://www.mdpi.com/2073-4441/13/1/27

Almeida, A., Cotas, J., Pereira, L., & Carvalho, P., 2023. Potential role of Spirogyra sp. and Chlorella sp. in bioremediation of mine drainage: A review. *Phycology 3*, 186–201.

Amin, M., Tahir, F., Ashfaq, H., Akbar, I., Razzaque, N., Haider, M. N., Xu, J., Zhu, H., Wang, N., & Shahid, A. (2022). Decontamination of industrial wastewater using microalgae integrated with biotransformation of the biomass to green products. *Energy Nexus*, *6*, 100089.

Anand, U., Dey, S., Parial, D., Federici, S., Ducoli, S., Bolan, N. S., ... & Bontempi, E. (2023). Algae and bacteria consortia for wastewater decontamination and transformation into biodiesel, bioethanol, biohydrogen, biofertilizers and animal feed: A review. *Environmental Chemistry Letters*, *21*(3), 1585–1609.

Andrade, C. D., Andrade, L. D., Dias, M., Nascimento, C., & Mendes, M. (2018). Microalgae for bioremediation of textile wastewater. An overview. *MOJ Food Process Technol*, *6*(5), 432–433.

Arun, J., Gopinath, K. P., Sivaramakrishnan, R., SundarRajan, P., Malolan, R., & Pugazhendhi, A. (2021). Technical insights into the production of green fuel from CO2 sequestered algal biomass: A conceptual review on green energy. *Science of the Total Environment*, *755*, 142636.

Ashok, V., Gupta, S., & Shriwastav, A. (2019). Photobioreactors for Wastewater Treatment. In Sanjay Kumar Gupta & Faizal Bux (eds.), *Application of Microalgae in Wastewater Treatment* (pp. 383–409). Springer Nature. https://doi.org/10.1007/978-3-030-13913-1_18

Ayub, H. M. U., Nizami, M., Qyyum, M. A., Iqbal, N., Al-Muhtaseb, A., & Hasan, M. (2024). Sustainable hydrogen production via microalgae: Technological advancements, economic indicators, environmental aspects, challenges, and policy implications. *Environmental Research*, *244*, 117815. https://doi.org/10.1016/j.envres.2023.117815

Azizi, S., Beauclair, N., Maaza, M., Mokrani, T., Ambushe, A. A., Seopela, M. P., & Msagati, A. T. M. (2024). Acid mine drainage treatment and metal recovery by means of selective precipitation using magnesium oxide (MgO): An experimental study. *Groundwater for Sustainable Development*, *25*, 101151.

Ballén-Segura, M., Hernández, L., Parra, D., Vega, A., & Pérez, K. (2016). Using Scenedesmus sp. for the phycoremediation of tannery wastewater, *Tecciencia*, *12*, 69–75.

Bamba, B. S. B., Tranchant, C. C., Ouattara, A., & Lozano, P. (2021). Harvesting of microalgae biomass using ceramic microfiltration at high cross-flow velocity. *Applied Biochemistry and Biotechnology*, *193*(4), 1147–1169.

Barceló, D., & Petrovic, M. (2011). *Waste Water Treatment and Reuse in the Mediterranean Region* (Vol. 14). Springer Science & Business Media.

Barros, A. I., Gonçalves, A. L., Simões, M., & Pires, J. C. (2015). Harvesting techniques applied to microalgae: A review. *Renewable and Sustainable Energy Reviews*, *41*, 1489–1500.

Bora, A., Thondi Rajan, A. S., Ponnuchamy, K., Muthusamy, G., & Alagarsamy, A. (2024). Microalgae to bioenergy production: Recent advances, influencing parameters, utilization of wastewater – A critical review. *Science of the Total Environment*, *946*, 174230. https://doi.org/10.1016/j.scitotenv.2024.174230

Bustillo-Lecompte, C. F., & Mehrvar, M. (2015). Slaughterhouse wastewater characteristics, treatment, and management in the meat processing industry: A review on trends and advances. *Journal of Environmental Management*, *161*, 287–302.

Chai, W. S., Tan, W. G., Munawaroh, H. S. H., Gupta, V. K., Ho, S.-H., Show, P. L. (2021). Multifaceted roles of microalgae in the application of wastewater biotreatment: A review. *Environmental Pollution*, 269, 116236.

Chen, S., Qu, D., Xiao, X., & Miao, X. (2020). Biohydrogen production with lipid-extracted Dunaliella biomass and a new strain of hyper-thermophilic archaeon Thermococcus eurythermalis A501. *International Journal of Hydrogen Energy*, *45*(23), 12721–12730.

Cheng, P., Chen, D., Liu, W., Cobb, K., Zhou, N., Liu, Y., Liu, H., Wang, Q., Chen, P., Zhou, C., & Ruan, R. (2020). Auto-flocculation microalgae species Tribonema sp. and Synechocystis sp. with T-IPL pretreatment to improve swine wastewater nutrient removal. *Science of the Total Environment*, *725*, 138263. https://doi.org/10.1016/j.scitotenv.2020.138263

Cheng, S., Zhang, H., Li, L., Yu, T., Wang, Y., Tan, D., & Zhang, X. (2022). Harvesting of Microcystis flos-aquae using dissolved air flotation: The inhibitory effect of carboxyl groups in uronic acid-containing carbohydrates. *Chemosphere*, *300*, 134466. https://doi.org/10.1016/j.chemosphere.2022.134466

Choi, H. J. (2016). Parametric study of brewery wastewater effluent treatment using Chlorella vulgaris microalgae. *Environmental Engineering Research*, *21*, 401–408.

Clippinger, J. N., & Davis, R. E. (2019). *Techno-Economic Analysis for the Production of Algal Biomass Via Closed Photobioreactors: Future Cost Potential Evaluated Across a Range of Cultivation System Designs*. National Renewable Energy Laboratory. NREL/TP-5100-72716. https://www.nrel.gov/docs/fy19osti/72716.pdf

Condor, B. E., de Luna, M. D. G., Chang, Y. H., Chen, J. H., Leong, Y. K., Chen, P. T., Chen, C. Y., Lee, D. J., & Chang, J. S. (2022). Bioethanol production from microalgae biomass at high-solids loadings. *Bioresource Technology*, *363*, 128002. https://doi.org/10.1016/j.biortech.2022.128002

Constantino, A., Rodrigues, B., Leon, R., Barros, R., & Raposo, S. (2021). Alternative chemo-enzymatic hydrolysis strategy applied to different microalgae species for bioethanol production. *Algal Research*, *56*, 102329.

Cordeiro, E. C. N., Mógor, Á. F., Amatussi, J. O., Mógor, G., Marques, H. M. C., & De Lara, G. B. (2022). Microalga biofertilizer improves potato growth and yield, stimulating amino acid metabolism. *Journal of Applied Phycology*, *34*, 385–394.

Cui, X., Yang, J., Feng, Y., & Zhang, W. (2020). Simulation of a novel tubular microalgae photobioreactor with aerated tangent inner tubes: improvements in mixing performance and flashing-light effects. *Archaea*, *2020*(1), 8815263.

de Morais, E. G., Sampaio, I. C. F., Gonzalez-Flo, E., Ferrer, I., Uggetti, E., & García, J. (2023). Microalgae harvesting for wastewater treatment and resources recovery: A review. *New Biotechnology*, *78*, 84–94. https://doi.org/10.1016/j.nbt.2023.10.002

de Souza Leite, L., Dos Santos, P. R., & Daniel, L. A. (2020). Microalgae harvesting from wastewater by pH modulation and flotation: Assessing and optimizing operational parameters. *Journal of Environmental Management*, *254*, 109825.

Debik, E., & Coskun, T. (2009). Use of the static granular bed reactor (SGBR) with anaerobic sludge to treat poultry slaughterhouse wastewater and kinetic modeling. *Bioresource Technology*, *100*, 2777–2782.

Deepa, P., Sowndhararajan, K., & Kim, S. (2023). A review of the harvesting techniques of microalgae. *Water*, *15*(17), 3074.

Demirel, B., Yenigun, O., & Onay, T. T. (2005). Anaerobic treatment of dairy wastewater: A review. *Process Biochemistry*, *40*, 2583–2595.

Deshmukh, M., & Pathan, A. (2026). An eco-friendly solution for bioethanol production from microalgae with optimized process parameters and emission characteristics. *Renewable and Sustainable Energy Reviews*, *226*, 116381.

Devi, A., Verma, M., Saratale, G. D., Saratale, R. G., Ferreira, L. F. R., Mulla, S. I., & Bharagava, R. N. (2023). Microalgae: A green eco-friendly agent for bioremediation of tannery wastewater with simultaneous production of value-added products. *Chemosphere*, *336*, 139192.

Diale, P. P., Hildebrandt, D., Glasser, D., Matambo, T. S., & Makgato, S. S. (2023). An analysis of the processes, kinetics and equilibrium of iron's biosorption on immobilized green microalgae. *South African Journal of Chemical Engineering*, *45*, 210–220.

Dineshkumar, R., & Sen, R. (2020). A sustainable perspective of microalgal biorefinery for co-production and recovery of high-value carotenoid and biofuel with CO2 valorization. *Biofuels, Bioproducts and Biorefining*, *14*(4), 879–897.

Ding, T., Yang, M., Zhang, J., Yang, B., Lin, K., Li, J., & Gan, J. (2017) Toxicity, degradation and metabolic fate of ibuprofen on freshwater diatom Navicula sp. *Journal of Hazardous Materials Advances*, *330*, 127–134.

Dors, G., Mendes, A. A., Pereira, E. B., de Castro, H. F., & Furigo, Jr A. (2013). Simultaneous enzymatic hydrolysis and anaerobic biodegradation of lipid-rich wastewater from poultry industry. *Applied Water Science*, *3*, 343–349.

Dube, V., Phiri, Z., Kuvarega, A. T., Mamba, B. B., & Kock, L.-A. (2024). Exploring acid mine drainage treatment through adsorption: A bibliometric analysis. *Environmental Science and Pollution Research*, *31*, 59659–59680.

Egbo, M., Okoani, A., & Okoh, I. (2018). Photobioreactors for microalgae cultivation-An overview. *International Journal of Scientific Engineering and Research*, *9*(11), 65–74.

Elcik, H., & Cakmakci, M. (2017). Harvesting microalgal biomass using crossflow membrane filtration: Critical flux, filtration performance, and fouling characterization. *Environmental Technology*, *38*(12), 1585–1596. https://doi.org/10.1080/09593330.2016.1237560

Elcik, H., Karadag, D., Kara, A. I., & Cakmakci, M. (2023). Microalgae biomass harvesting using chitosan flocculant: Optimization of operating parameters by response surface methodology. *Separations*, *10*(9), 507. https://www.mdpi.com/2297-8739/10/9/507

Emparan, Q., Harun, R., & Danquah, M. K. (2019). Role of phycoremediation for nutrient removal from wastewaters: A review. *Applied Ecology and Environmental Research Res*, *7*, 889–915.

Estrada-Graf, A., Hernández, S., & Morales, M. (2020). Biomitigation of CO_2 from flue gas by Scenedesmus obtusiusculus AT-UAM using a hybrid photobioreactor coupled to a biomass recovery stage by electro-coagulation-flotation. *Environmental Science and Pollution Research*, *27*(23), 28561–28574.

Fasaei, F., Bitter, J. H., Slegers, P. M., & van Boxtel, A. J. B. (2018). Techno-economic evaluation of microalgae harvesting and dewatering systems. *Algal Research*, *31*, 347–362. https://doi.org/10.1016/j.algal.2017.11.038

Fazal, T., Rehman, M. S. U., Javed, F., Akhtar, M., Mushtaq, A., Hafeez, A., … & Rehman, F. (2021). Integrating bioremediation of textile wastewater with biodiesel production using microalgae (Chlorella vulgaris). *Chemosphere*, *281*, 130758.

Gani, P., Sunar, N. M., Peralta, H. M., Latiff, A., Parjo, U. K., & Razak, A. (2015). Phycoremediation of wastewaters and potential hydrocarbon from microalgae: A review. *Advances in Environmental Biology*, *9*, 1–8.

García, J., Ortiz, A., Álvarez, E., Belohlav, V., García-Galán, M. J., Díez-Montero, R., Álvarez, J. A., & Uggetti, E. (2018). Nutrient removal from agricultural run-off in demonstrative full scale tubular photobioreactors for microalgae growth. *Ecological Engineering*, *120*, 513–521.

Gaurav, K., Neeti, K., & Singh, R. (2024). Microalgae-based biodiesel production and its challenges and future opportunities: A review. *Green Technologies and Sustainability*, *2*, 100060. https://doi.org/10.1016/j.grets.2023.100060

Gerardo, M. L., Van Den Hende, S., Vervaeren, H., Coward, T., & Skill, S. C. (2015). Harvesting of microalgae within a biorefinery approach: A review of the developments and case studies from pilot-plants. *Algal Research*, *11*, 248–262.

Geremia, E., Ripa, M., Catone, C. M., & Ulgiati, S. (2021). A review about microalgae wastewater treatment for bioremediation and biomass production—A new challenge for Europe. *Environments*, *8*(12), 136. https://www.mdpi.com/2076-3298/8/12/136

Gonçalves, A. L., Pires, J. C., & Simoes, M. (2017). A review on the use of microalgal consortia for wastewater treatment. *Algal Research*, *24*, 403–415.

Goswami, R. K., Agrawal, K., Shah, M. P., & Verma, P., 2021. Bioremediation of heavy metals from wastewater: A current perspective on microalgae-based future. *Letters in Applied Microbiology*, *75*, 701–717.

Gumbi, S., Majeke, B., Olaniran, A., & Mutanda, T. (2017). Isolation, identification and high-throughput screening of neutral lipid producing indigenous microalgae from South African aquatic habitats. *Applied Biochemistry and Biotechnology*, *182*(1), 382–399.

Gumbi, S. F. T., Kumar, A., & Olaniran, A. O. (2022). Lipid productivity and biosynthesis gene response of indigenous microalgae Chlorella sp. T4 strain for biodiesel production under different nitrogen and phosphorus load. *BioEnergy Research*, *15*(4), 2090–2101.

Harun, R., Danquah, M. K., & Forde, G. M. (2010). Microalgal biomass as a fermentation feedstock for bioethanol production. *Journal of Chemical Technology & Biotechnology*, *85*(2), 199–203.

Hashmi, Z., Idriss, I. M., Taha, H., Zaini, J., Bakar, M. S. A., Abdullah, R., Nandianto, A. B. D., & Bilad, M. R. (2025). Cultivation of indigenous Chlorococcum sp. in aquaculture wastewater under various light color and biomass harvesting using membrane filtration: A simultaneous wastewater treatment and biomass production. *Cleaner Waste Systems*, *10*, 100209. https://doi.org/10.1016/j.clwas.2025.100209

He, J., Ding, W., Han, W., Chen, Y., Jin, W., & Zhou, X. (2022). A bacterial strain citrobacter W4 facilitates the bio-flocculation of wastewater cultured microalgae Chlorella pyrenoidosa. *Science of the Total Environment*, *806*, 151336.

Hoang, A. T., Sirohi, R., Pandey, A., Nižetić, S., Lam, S. S., Chen, W.-H., Luque, R., Thomas, S., Arıcı, M., & Pham, V. V. (2023). Biofuel production from microalgae: Challenges and chances. *Phytochemistry Reviews*, *22*(4), 1089–1126.

Hussian, M., & Ellatif, A. (2018). The role of microalgae in renewable energy production: Challenges and opportunities. In Muhammet Turkoglu, Umur Önal, & Ali İsmen (eds.), *Marine Ecology - Biotic and Abiotic Interactions*. IntechOpen.

Hwang, K.-J., & Wu, S.-E. (2015). Disk structure on the performance of a rotating-disk dynamic filter: A case study on microalgae microfiltration. *Chemical Engineering Research and Design*, *94*, 44–51. https://doi.org/10.1016/j.cherd.2014.12.009

Imbimbo, P., Ferrara, A., Giustino, E., Liberti, D., & Monti, D. M. (2024). Microalgae Flocculation: Assessment of Extraction Yields and Biological Activity. *International Journal of Molecular Sciences*, *25*(19), 10238. https://www.mdpi.com/1422-0067/25/19/10238

Irawan, A., Rahmawati, A., Fahmi, U., Budiman, A., Maghfiroh, K., Erfianti, T., Andeska, D., Adaranyssa, R., Putri, R., Nurafifah, I., Brilian, R., Ryan Sadewo, B., & Suyono, E. (2023). Studies on bioflocculant exopolysaccharides (EPS) produced by Anabaena sp. and its application as bioflocculant for low cost harvesting of Chlorella sp. *Asian Journal of Agriculture and Biology*, 1–9. https://doi.org/10.35495/ajab.2022.150

Jaiswal, K. K., Kumar, V., Gururani, P., Vlaskin, M. S., Parveen, A., Nanda, M., Kurbatova, A., Gautam, P., & Grigorenko, A. V. (2022). Bio-flocculation of oleaginous microalgae integrated with municipal wastewater treatment and its hydrothermal liquefaction for biofuel production. *Environmental Technology & Innovation*, *26*, 102340.

Javed, F., Aslam, M., Rashid, N., Shamair, Z., Khan, A. L., Yasin, M., Fazal, T., Hafeez, A., Rehman, F., & Rehman, M. S. U. (2019). Microalgae-based biofuels, resource recovery and wastewater treatment: A pathway towards sustainable biorefinery. *Fuel*, *255*, 115826.

Jeong, G. T., Park, J. H., Park, S. H., & Park, D. H. (2008). Estimating and improving cold filter plugging points by blending biodiesels with different fatty acid contents. *Biotechnology and Bioprocess Engineering*, *13*(4), 505–510.

Joyia, M. A. K., Ahmad, M., Chen, Y. F., Mustaqeem, M., Ali, A., Abbas, A., & Gondal, M. A. (2024). Trends and advances in sustainable bioethanol production technologies from first to fourth generation: A critical review. *Energy Conversion and Management 321*, 119037.

Kaloudas, D., Pavlova, N., & Penchovsky, R. (2021). Phycoremediation of wastewater by microalgae: A review. *Environmental Chemistry Letters*, *19*(4), 2905–2920. https://doi.org/10.1007/s10311-021-01203-0

Khan, S., Das, P., Abdul Quadir, M., Thaher, M. I., Mahata, C., Sayadi, S., & Al-Jabri, H. (2023). Microalgal feedstock for biofuel production: Recent advances, challenges, and future perspective. *Fermentation*, *9*(3), 281. https://www.mdpi.com/2311-5637/9/3/281

Koley, S., Prasad, S., Bagchi, S. K., & Mallick, N. (2017). Development of a harvesting technique for large-scale microalgal harvesting for biodiesel production. *RSC Advances*, *7*(12), 7227–7237.

Kumar, C., Sharma, M., Kaur, M., Arya, S., & Khatri, M. (2024). Cultivation of algae: Techniques and challenges. In Shailendra Kumar Arya, Madhu Khatri, & Gursharan Singh (eds.), *Value Added Products From Bioalgae Based Biorefineries: Opportunities and Challenges* (pp. 43–65). Springer Nature. https://doi.org/10.1007/978-981-97-1662-3_3

Kumari, A., Chakraborty, S., Sirotiya, V., Kalita, D., Rai, A., Yadav, K., & Vinayak, V. (2024). A review on economical and impact of bioethanol production from microalgae: Current scenario and future prospect. *Industrial Crops and Products*. https://doi.org/10.1016/j.indcrop.2024.119927

Kushwaha, J. P., Srivastava, V. C., & Mall, I. D. (2011). An overview of various technologies for the treatment of dairy wastewaters. *Critical Reviews in Food Science and Nutrition*, *51*, 442–452.

Labeeuw, L., Commault, A., Kuzhiumparambil, U., Emmerton, B., Nguyen, L., Nghiem, L., & Ralph, P. (2020). A comprehensive analysis of an effective flocculation method for high quality microalgal biomass harvesting. *Science of the Total Environment*, *752*, 141708. https://doi.org/10.1016/j.scitotenv.2020.141708

Li, H., Di, J., Dong, Y., Bao, S., & Fu, S. (2023a). Dynamic experiment on remediation of acid mine drainage by iron-carbon microelectrolysis enhancing sulfate-reducing bacteria. *Environmental Science: Water Research and Technology 9*, 1413–1425.

Li, M., Wang, Y., Zhang, J., Liu, B., Xue, H., Wu, L., & Li, Z. (2023b). Knowledge Mapping of High-Rate Algal Ponds Research. *Water*, *15*(10), 1916. https://www.mdpi.com/2073-4441/15/10/1916

Lim, J. Y., Teng, S. Y., How, B. S., Nam, K., Heo, S., Máša, V., Stehlík, P., & Yoo, C. K. (2022). From microalgae to bioenergy: Identifying optimally integrated biorefinery pathways and harvest scheduling under uncertainties in predicted climate. *Renewable and Sustainable Energy Reviews*, *168*, 112865. https://doi.org/10.1016/j.rser.2022.112865

López-Lizárraga, G. L., & García Cañedo, J. C. (2016). Considerations for photobioreactor design and operation for mass cultivation of microalgae. In N. Thajuddin & D. Dhanasekaran (eds.), *Algae - Organisms for Imminent Biotechnology*. IntechOpen. https://doi.org/10.5772/63069

Ma, J., Cui, Y., Sun, F., Zhou, R., Liu, F., Cao, K., Zhu, B., Zhang, H., Meng, C., & Chen, G. (2025). Strategies for enhancing microalgal carbon sequestration: A review on strain development, culture system optimization, parameter control, and metabolic engineering. *Bioresource Technology*, *435*, 132868.

Machado, C. A., Esteves, A. F., & Pires, J. C. (2024). Chlorella vulgaris Harvesting: Chemical Flocculation with Chitosan, Aluminum Sulfate, and Ferric Sulfate. *Applied Sciences*, *14*(2), 598.

Madadi, R., Zahed, M. A., Pourbabaee, A. A., Tabatabei, M., & Naghavi, M. R. (2021). Simultaneous phycoremediation of petrochemical wastewater and lipid production by Chlorella vulgaris. *SN Applied Sciences*, *3*, 505.

Magalhães, I. B., de Paula Pereira, A. S. A., Silva, T. A., Ferreira, J., Braga, M. Q., Couto, E. A., Assemany, P. P., & Calijuri, M. L. (2024). Advancements in high-rate algal pond technology for enhanced wastewater treatment and biomass production: A review. *Journal of Water Process Engineering*, *66*, 105929. https://doi.org/10.1016/j.jwpe.2024.105929

Maity, S., & Mallick, N. (2022). Trends and advances in sustainable bioethanol production by marine microalgae: A critical review. *Journal of Cleaner Production*, *345*, 131153.

Masindi, V., Foteinis, S., Renforth, P., Ndiritu, J., Maree, J. P., Tekere, M., & Chatzisymeon, E. (2022). Challenges and avenues for acid mine drainage treatment, beneficiation, and valorisation in circular economy: A review. *Ecological Engineering*, *183*, 106740.

Mata, T. M., Melo, A. C., Simões, M., Caetano, N. S. (2012). Parametric study of a brewery effluent treatment by microalgae Scenedesmus obliquus. *Bioresource Technology*, *107*, 151–158.

Matter, I. A., Bui, V. K. H., Jung, M., Seo, J. Y., Kim, Y.-E., Lee, Y.-C., & Oh, Y.-K. (2019). Flocculation harvesting techniques for microalgae: A review. *Applied Sciences*, *9*(15), 3069.

Mohammed, B. (2006). Design and performance evaluation of a wastewater treatment unit. *AU Journal of Technology*, *9*(3), 193–198.

Mokashi, M., Shirsath, A., Lott, P., Müller, H., Tischer, S., Maier, L., & Deutschmann, O. (2023). Understanding of gas-phase methane pyrolysis towards hydrogen and solid carbon with detailed kinetic simulations and experiments. *Chemical Engineering Journal*, *479*, 147556. https://doi.org/10.1016/j.cej.2023.147556

Mosai, A. K., Ndlovu, G., & Tutu, H., 2024. Improving acid mine drainage treatment by combining treatment technologies: A review. *Science of the Total Environment*, *919*, 170806.

Moyo, A., Filho, J. R. D. A., Harrison, S. T. L., & Broadhurst, J. L. (2024). Acid mine drainage and metal(loid) risk potential of South African coal processing wastes. *Mineral Engineering*, *215*, 108825.

Mustafa, S., Bhatti, H. N., Maqbbool, M., & Iqbal, M. (2021). Microalgae biosorption, bioaccumulation and biodegradation efficiency for the remediation of wastewater and carbon dioxide mitigation: Prospects, challenges and opportunities. *Journal of Water Process Engineering*, *41*, 102009.

Nagarajan, D., Dong, C. D., Chen, C. Y., Lee, D. J., & Chang, J. S. (2021). Biohydrogen production from microalgae—Major bottlenecks and future research perspectives. *Biotechnology journal*, *16*(5), 2000124.

Nageshwari, K., Chang, S. X., & Balasubramanian, P. (2022). Integrated electrocoagulation-flotation of microalgae to produce Mg-laden microalgal biochar for seeding struvite crystallization. *Scientific Reports*, *12*(1), 11463. https://doi.org/10.1038/s41598-022-15527-6

Najjar, Y. S. H., & Abu-Shamleh, A. (2020). Harvesting of microalgae by centrifugation for biodiesel production: A review. *Algal Research*, *51*, 102046. https://doi.org/10.1016/j.algal.2020.102046

Narala, R. R., Garg, S., Sharma, K. K., Thomas-Hall, S. R., Deme, M., Li, Y., & Schenk, P. M. (2016). Comparison of microalgae cultivation in photobioreactor, open raceway pond, and a two-stage hybrid system [Original Research]. *Frontiers in Energy Research*, *4*. https://doi.org/10.3389/fenrg.2016.00029

Nayak, J. K., & Ghosh, U. K. (2019). Post treatment of microalgae treated pharmaceutical wastewater in photosynthetic microbial fuel cell (PMFC) and biodiesel production. *Biomass and Bioenergy*, *131*, 105415.

Nerom, S. V., Coleman, B., De Baets, R., Immerseel, F. V., Robbens, J., & Delezie, E. (2024). Microalgae as feed additives in poultry: A review on the health-promoting effects. *Algal Research*, *83*, 103733.

Nguyen, T. T., Lam, M. K., Cheng, Y. W., Uemura, Y., Mansor, N., Lim, J. W., Show, P. L., Tan, I. S., & Lim, S. (2021). Reaction kinetic and thermodynamics studies for in-situ transesterification of wet microalgae paste to biodiesel. *Chemical Engineering Research and Design*, *169*, 250–264. https://doi.org/10.1016/j.cherd.2021.03.021

Novoveská, L., Nielsen, S. L., Eroldoğan, O. T., Haznedaroglu, B. Z., Rinkevich, B., Fazi, S., Robbens, J., Vasquez, M., & Einarsson, H. (2023). Overview and challenges of large-scale cultivation of photosynthetic microalgae and cyanobacteria. *Marine Drugs*, *21*(8), 445. https://www.mdpi.com/1660-3397/21/8/445

Nuri, N. A. S. A., Puad, N. I. M., Azmi, A. S., Ahmad, F., & Rahman, S. N. F. S. A. (2023). Microalgae cultivation in flat panel photobioreactor as a method of carbon capture: A review. *IOP Conference Series: Earth and Environmental Science*, *1281*(1), art.no. 012010.

Oberholster, P. J., Steyn, M., & Botha, A.-M. (2021). A comparative study of improvement of phycoremediation using a consortium of microalgae in municipal wastewater treatment pond systems as an alternative solution to Africa's sanitation challenges. *Processes*, *9*(9), 1677.

Oliveira, M. S. F., Assila, O., Fonseca, A. M., Parpot, P., Valente, T., Rombi, E., & Neves, I. C. (2024). Acid mine drainage precipitates from mining effluents as adsorbents of organic pollutants from water treatment. *Molecules*, *29*, 3521.

Pandit, C., Pandit, S., Kuhad, R. C., Ray, S., Mishra, S. K., Mathuriya, A. S., & Prasad, R. (2024). Microalgal bioethanol production for sustainable development: Current status and future prospects. *Indian Journal of Microbiology*. https://doi.org/10.1007/s12088-024-01364-w

Patel, A. K., Albarico, F. P. J. B., Perumal, P. K., Vadrale, A. P., Nian, C. T., Chau, H. T. B., Anwar, C., Wani, H. M., Pal, A., Saini, R., Ha, L. H., Senthilkumar, B., Tsang, Y.-S., Chen, C.-W., Dong, C.-D., & Singhania, R. R. (2022). Algae as an emerging source of bioactive pigments. *Bioresource Technology*, *351*, 126910.

Patel, K., Patel, N., Vaghamshi, N., Shah, K., Dugdirala, S. M., Dudhagara, P. (2021). Trends and strategies in the effluent treatment of pulp and paper industries: A review highlighting reactor options. *Current Research in Microbial Sciences*, *2*, 100077.

Pathak, U., Das, P., Banerjee, P., & Datta, S. (2016). Treatment of wastewater from a dairy industry using rice husk as adsorbent: Treatment efficiency, isotherm, thermodynamics, and kinetics modelling. *Journal of Thermodynamics*, *2016*, 1–7.

Peng, K.-T., Xiong, B.-H., Cheng, G., Zhong, Y.-H., & Yu, L. (2025). Phosphomolybdic acid boosts polyunsaturated fatty acid and neutral lipid production in *Phaeodactylum tricornutum*. *Frontiers in Marine Science*, *12*, 1531239.

Penloglou, G., Pavlou, A., & Kiparissides, C. (2024). Recent advancements in photo-bioreactors for microalgae cultivation: a brief overview. *Processes*, *12*(6), 1104.

Pérez, L. S., Rodríguez, O. M., Reyna, S., Sánchez-Salas, J. L., Lozada, J. D., Quiroz, M. A., & Bandala, E. R. (2016). Oil refinery wastewater treatment using coupled electrocoagulation and fixed film biological processes. *Physics and Chemistry of the Earth Parts a/b/c*, *91*, 53–60.

Plöhn, M., Spain, O., Sirin, S., Silva, M., Escudero-Oñate, C., Ferrando-Climent, L., Allahverdiyeva, Y., & Funk, C. (2021). Wastewater treatment by microalgae. *Physiologia Plantarum*, *173*(2), 568–578. https://doi.org/10.1111/ppl.13427

Pugazhendhi, A., Shobana, S., Bakonyi, P., Nemestóthy, N., Xia, A., Banu J. R., & Kumar, G. (2019). A review on chemical mechanism of microalgae flocculation via polymers. *Biotechnology Reports*, *21*, e00302. https://doi.org/10.1016/j.btre.2018.e00302

Rambabu, K., Banat, F., Pham, Q. M., Ho, S.-H., Ren, N.-Q., & Show, P. L. (2020). Biological remediation of acid mine drainage: Review of past trends and current outlook. *Environmental Science and Ecotechnology*, *2*, 100024.

Ranganathan, P., Pandey, A. K., Sirohi, R., Tuan Hoang, A., & Kim, S.-H. (2022). Recent advances in computational fluid dynamics (CFD) modelling of photobioreactors: Design and applications. *Bioresource Technology*, *350*, 126920. https://doi.org/10.1016/j.biortech.2022.126920

Ranjan, S., Gupta, P., & Gupta, S. (2019). Comprehensive evaluation of high-rate algal ponds: Wastewater treatment and biomass production. In Sanjay Kumar Gupta & Faizal Bux (eds.), *Application of Microalgae in Wastewater Treatment*. Springer Nature. https://doi.org/10.1007/978-3-030-13909-4_22

Rao, N., Beyer, V., Henderson, R., Thielemans, W., & Muylaert, K. (2023). Microalgae harvesting using flocculation and dissolved air flotation: Selecting the right vessel for lab-scale experiments. *Bioresource Technology*, *374*, 128786. https://doi.org/10.1016/j.biortech.2023.128786

Rathour, R. K., Sharma, D., Ullah, S., Mahmoud, E. M., Sharma, N., Kumar, P., Bhatt, A. K., Ahmad, I., & Bhatia, R. K. (2024). Bacterial-microalgal consortia for bioremediation of textile industry wastewater and resource recovery for circular economy. *Biotechnology for the Environment*, *1*, 6.

Ravikumar, Y., Abdul Razack, S., Yun, J., Zhang, G., Zabed, H. M., & Qi, X. (2021) Recent advances in microalgae-based distillery wastewater treatment. *Environmental Technology and Innovation 24*, 101839.

Razaviarani, V., Arab, G., Lerdwanawattana, N., & Gadia, Y. (2023). Algal biomass dual roles in phycoremediation of wastewater and production of bioenergy and value-added products. *International Journal of Environmental Science and Technology*, *20*(7), 8199–8216. https://doi.org/10.1007/s13762-022-04696-6

Razzak, S., Bahar, K., Islam, K. M. O., Haniffa, A. K., Faruque, M. O., Hossain, S. M. Z., & Hossain, M. M. (2024). Microalgae cultivation in photobioreactors: Sustainable solutions for a greener future. *Green Chemical Engineering*, *5*(4), 418–439. https://doi.org/10.1016/j.gce.2023.10.004

Razzak, S. A., Hossain, M. M., Lucky, R. A., Bassi, A. S., & De Lasa, H. (2013). Integrated CO_2 capture, wastewater treatment and biofuel production by microalgae culturing—A review. *Sustainable Energy Reviews 27*, 622–653.

Rezvani, F., & Sarrafzadeh M.-H. (2025). Innovative strategy for nitrate removal from groundwater: Synergistic interactions of autotrophic microalgae and heterotrophic denitrifiers in sequential batch process. *Water, Air & Soil Pollution*, *236*, 860.

Riano, B., Molinuevo, B., & García-Gonzalez, M. C. (2011). Treatment of fish processing wastewater with microalgae-containing microbiota. *Bioresource Technology*, *102*, 10829–10833

Samal, D. P. K., Sukla, L. B., Pattanaik, A., & Pradhan, D. (2020). Role of microalgae in treatment of acid mine drainage and recovery of valuable metals. *Materials Today: Proceedings*, *30*, 346–350.

Samrot, A. V., Rajalakshmi, D., Sathiyasree, M., Saigeetha, S., Kasipandian, K., Valli, N., Jayshree, N., Prakash, P., & Shobana, N. (2023). A review on biohydrogen sources, production routes, and its application as a fuel cell. *Sustainability*, *15*(16), 12641. https://www.mdpi.com/2071-1050/15/16/12641

Sankaran, K., Premalatha, M., Vijayasekaran, M., & Somasundaram, V. T. (2014) Distillery wastewater treatment through anaerobic digestion and phycoremediation—A green industrial approach. *Renewable and Sustainable Energy Reviews*, *37*, 634–643.

Sathinathan, P., Parab, H. M., Yusoff, R., Ibrahim, S., Vello, V., & Ngoh, G. C. (2023). Photobioreactor design and parameters essential for algal cultivation using industrial wastewater: A review. *Renewable and Sustainable Energy Reviews*, *173*, 113096. https://doi.org/10.1016/j.rser.2022.113096

Satya, A. D. M., Cheah, W. Y., Yazdi, S. K., Cheng, Y.-S., Khoo, K. S., Vo, D.-V. N., Bui, X. D., Vithanage, M., & Show, P. L. (2023). Progress on microalgae cultivation in wastewater for bioremediation and circular bioeconomy. *Environmental Research*, *218*, 114948.

Sen, R., & Mukherjee, S. (2024). Recent advances in microalgal carbon capture and utilization (bio-CCU) process vis-à-vis conventional carbon capture and storage (CCS) technologies. *Critical Reviews in Environmental Science and Technology*, *54*(24), 1777–1802. https://doi.org/10.1080/10643389.2024.2361938

Sharma, A. K., Jaryal, S., Sharma, S., Dhyani, A., Tewari, B. S., & Mahato, N. (2025). Biofuels from microalgae: A review on microalgae cultivation, biodiesel production techniques and storage stability. *Processes*, *13*(2), 488.

Shete, B. S., & Shinkar, N. P. (2013). Dairy industry wastewater sources, characteristics, and its effects on environment. *International Journal of Current Engineering and Technology*, *3*, 1611–1615.

Singh, A. K. (2024). Recent advancements in microalgae-based biofuel production. In Maulin Shah & Deepanwita Deka (eds.), *Emerging Sustainable Technologies for Biofuel Production* (pp. 341–360). Springer.

Singh, B. J., Chakraborty, A., & Sehgal, R. (2023). A systematic review of industrial wastewater management: Evaluating challenges and enablers. *Journal of Environmental Management*, *348*, 119230.

Singh, G., & Patidar, S. (2018). Microalgae harvesting techniques: A review. *Journal of Environmental Management*, *217*, 499–508.

Slavov, A. K. (2017). General characteristics and treatment possibilities of dairy wastewater - A review. *Food Technology and Biotechnology*, *55*, 14–28.

Solovchenko, A., Verschoor, A. M., Jablonowski, N. D., & Nedbal, L. (2016). Phosphorus from wastewater to crops: An alternative path involving microalgae. *Biotechnology advances*, *34*(5), 550–564.

Soeprobowati, T. R., & Hariyati, R. (2017). The phycoremediation of textile wastewater discharge by Chlorella pyrenoidosa H. Chick Arthrospira platensis Gomont and Chaetoceros calcitrans Paulson H Takano. *Aquaculture Aquarium Conservation and Legislation*, *10*, 10640–10651.

Sutherland, D. L., Park, J., Ralph, P. J., & Craggs, R. J. (2020). Improved microalgal productivity and nutrient removal through operating wastewater high rate algal ponds in series. *Algal Research*, *47*, 101850. https://doi.org/10.1016/j.algal.2020.101850

Tan, J. S., Lee, S. Y., Chew, K. W., Lam, M. K., Lim, J. W., Ho, S.-H., & Show, P. L. (2020). A review on microalgae cultivation and harvesting, and their biomass extraction processing using ionic liquids. *Bioengineered*, *11*(1), 116–129.

Tang, C.-C., Hu, Y.-R., He, Z.-W., Li, Z.-H., Tian, Y., & Wang, X. C. (2024). Promoting symbiotic relationship between microalgae and bacteria in wastewater treatment processes: Technic comparison, microbial analysis, and future perspectives. *Chemical Engineering Journal*, *498*, 155703. https://doi.org/10.1016/j.cej.2024.155703

Ting, H., Haifeng, L., Shanshan, M., Zhang, Y., Zhidan, L., & Na, D. (2017). Progress in microalgae cultivation photobioreactors and applications in wastewater treatment: A review. *International Journal of Agricultural and Biological Engineering*, *10*(1), 1–29.

Tripathi, G., Dubey, P., Shamim, A., Farooqui, A., & Mishra, V. (2024). Bio-flocculation: A cost effective and energy efficient harvesting technique for algal biofuel production and wastewater treatment. *Bioresource Technology Reports*, *28*, 101969. https://doi.org/10.1016/j.biteb.2024.101969

Umamaheswari, J., & Shanthakumar, S. (2016). Efficacy of microalgae for industrial wastewater treatment: A review on operating conditions, treatment efficiency and biomass productivity. *Reviews in Environmental Science and Bio/Technology*, *14*, 265–284.

Usha, M. T., Chandra, T. S., Sarada, R., & Chauhan, V. S. (2016). Removal of nutrients and organic pollution load from pulp and paper mill effluent by microalgae in outdoor open pond. *Bioresource Technology*, *214*, 856–860.

Valladão, A. B. G., Freire, D. M. G., & Cammarota, M. C. (2007). Enzymatic pre-hydrolysis applied to the anaerobic treatment of effluents from poultry slaughterhouses. *International Biodeterioration and Biodegradation*, *60*, 219–225.

Vasistha, S., Khanra, A., Clifford, M., & Rai, M. P. (2021). Current advances in microalgae harvesting and lipid extraction processes for improved biodiesel production: A review. *Renewable and Sustainable Energy Reviews*, *137*, 110498.

Walters, C., Steyn, M., Ndlela, L., Nocanda, X., Moloi, M., & Oberholster, P. (2025). Phycoremediation of industrial wastewater: Review of algae consortia. *International Journal of Environmental Science and Technology*, *22*, 6209–6224.

Wan Osman, W. N. A., Mat Nawi, N. I., Samsuri, S., Bilad, M. R., Khan, A. L., Hunaepi, H., Jaafar, J., & Lam, M. K. (2021). Ultra low-pressure filtration system for energy efficient microalgae filtration. *Heliyon*, *7*(6), e07367. https://doi.org/10.1016/j.heliyon.2021.e07367

Wang, M., Ye, X., Bi, H., & Shen, Z. (2024). Microalgae biofuels: Illuminating the path to a sustainable future amidst challenges and opportunities. *Biotechnology for Biofuels and Bioproducts*, *17*(1), 10. https://doi.org/10.1186/s13068-024-02461-0

Wang, Z., Liu, Q., Yan, P., Zhang, J., Liang, X., Ma, X., Li, J., & Cui, Z. (2025). Regulation of the microstructure for hollow fiber ultrafiltration membrane with "flocculation" effect and study on the mechanism of "adsorption flocculation-loose cake layer protection" against humic acid fouling. *Journal of Water Process Engineering*, *69*, 106697. https://doi.org/10.1016/j.jwpe.2024.106697

Wu, M., Li, J., Qin, H., Lei, A., Zhu, H., Hu, Z., & Wang, J. (2020). Pre-concentration of microalga Euglena gracilis by alkalescent pH treatment and flocculation mechanism of Ca3 (PO4) 2, Mg3 (PO4) 2, and derivatives. *Biotechnology for Biofuels*, *13*(1), 98.

Wu, Q., Li, X., Feng, Q., & Li, X. (2024). Source reduction and end treatment of acid mine drainage in closed coal mines of the Yudong river basin. *Water Science and Technology*, *89*, 470–483.

Xiao, Y., & Luo, Y. (2022). Research progress and application of photobioreactor in wastewater treatment. *E3S Web of Conferences*, *352*, 02024. https://doi.org/10.1051/e3sconf/202235202024

Yang, Y., Duan, L., & Wu, H. (2025). Challenges associated with membrane separation of polypeptides and relevant solution strategies. *Separations*, *12*(9), 238.

Ye, J., Zhu, Z., Song, Z., Xu, H., Xu, T., & Liu, H. (2024). Application of microbubble air flotation to harvest Microcystis sp. from agriculture wastewater: The regulation and mechanisms. *Biotechnology and Bioengineering*, *121*(12), 3742–3753. https://doi.org/10.1002/bit.28836

Yu, B. S., Pyo, S., Lee, J., & Han, K. (2024). Microalgae: A multifaceted catalyst for sustainable solutions in renewable energy, food security, and environmental management. *Microbial Cell Factories*, *23*(1), 308. https://doi.org/10.1186/s12934-024-02588-7

Yuan, Y., Lu, J., & Wang, Q. (2024). Evaluating bioflocculation harvesting of freshwater and marine microalgae using exopolysaccharides (EPSs) from Klebsiella sp. *Separations, 11*(12), 355. https://www.mdpi.com/2297-8739/11/12/355

Yun, J.-H., Cho, D.-H., Lee, S., Heo, J., Tran, Q.-G., Chang, Y., & Kim, H.-S. (2018). Hybrid operation of photobioreactor and wastewater-fed open raceway ponds enhances the dominance of target algal species and algal biomass production. *Algal Research, 29*, 319–329. https://doi.org/10.1016/j.algal.2017.11.037

Zhang, W., Zhang, Y., Li, L., Xue, S., Hsu, D., Hurley, A. C., ... & Zhang, X. (2025). Nanobubble-enabled foam fractionation to remove freshwater microalgae and microcystin. *Bioresource Technology*, 133351.

Zhao, Z., Ilyas, A., Muylaert, K., & Vankelecom, I. F. J. (2020). Optimization of patterned polysulfone membranes for microalgae harvesting. *Bioresource Technology, 309*, 123367. https://doi.org/10.1016/j.biortech.2020.123367

Zhou, Q., Sun, H., Jia, L., Wu, W., & Wang, J. (2022). Simultaneous biological removal of nitrogen and phosphorus from secondary effluent of wastewater treatment plants by advanced treatment: A review. *Chemosphere, 296*, 134054. https://doi.org/10.1016/j.chemosphere.2022.134054

Zhu, J., Wakisaka, M., Omura, T., Yang, Z., Yin, Y., & Fang, W. (2024). Advances in industrial harvesting techniques for edible microalgae: Recent insights into sustainable, efficient methods and future directions. *Journal of Cleaner Production, 436*, 140626.

14 Microalgae as Green Filters

Bioremediation and Odor Neutralization in Wastewaters

Nnabueze Darlington Nnaji, Chiemerie Theresa Ekwueme, Job Chinagorom Aleke, Adindu Joy Chinwe, and Helen Onyeaka

14.1 INTRODUCTION

The increasing rate of industrialization and urbanization has led to an ever-growing international problem with wastewater management, where chemical, agricultural, and domestic discharge flood aquatic systems with nutrients, heavy metals, and volatile compounds. Conventional wastewater treatment methods, despite their popularity, often prove to be insufficient to meet sustainability and a circular economy requirement because of the high-energy consumption, reliance on chemicals, and the low effectiveness to eliminate new pollutants, including pharmaceuticals and microplastics (Nagarajan et al., 2020). In this respect, microalgae have become a new, ecologically integrative and multifunctional biotechnological platform, which makes it easy to simultaneously purify pollutants, produce oxygen, capture carbon, and make sure that biomass is used productively. The idea of microalgae as the green filter will not only transform wastewater treatment into a regenerating process but also reinvent waste as a source of bioenergy, bioproducts, and nutrient recovery (Saritas et al., 2025).

As a physiologically and phylogenetically heterogeneous group of photoautotrophic microorganisms, microalgae take a special ecological niche at the border of biology and environmental engineering. These and simple organisms, which include prokaryotic cyanobacteria and eukaryotic Chlorophyta, use sunlight, carbon dioxide, and inorganic nutrients to produce organic compounds exceptionally well (Barsanti & Gualtieri, 2018; Stirk & van Staden, 2022). Their metabolic plasticity, including autotrophic, heterotrophic, and mixotrophic, allows adapting to heterogeneous wastewater conditions containing nitrogen, phosphorus, and organic carbon (Brasil et al., 2017; Gallego et al., 2025). This physiological ability is also vital in their

 DOI: 10.1201/9781003611479-14

involvement in phycoremediation (Queiroz et al., 2020). The microalgal ability to sequester close to half of the atmospheric CO_2 by photosynthesis into biomass also offers a twofold environmental dividend, reducing greenhouse gas emissions and producing renewable feedstock to produce biofuels (Sarıtas et al., 2025). Therefore, microalgal biology and wastewater bioremediation intersectionality represent one of the intersections of environmental protection, renewable energy, and innovation in the circular bioeconomy.

Microalgae are environmentally relevant in the wastewater systems because of their biochemical and cellular adaptability. Various species (*Chlorella vulgaris* and *Scenedesmus obliquus*) are especially effective in absorbing ammonium and nitrate, and *Spirulina platensis* and *Nannochloropsis* sp. could also be grown in wastewater with a high salinity or contaminated with nutrients (Li et al., 2019). By using bioaccumulation and biosorption, these organisms fix heavy metals such as cadmium, lead, and mercury through functional groups on cell surfaces, hence avoiding metal toxicity and facilitating the detoxification of effluents (Rai et al., 2020; Ding et al., 2019). In addition, it is important that microalgae may convert organic pollutants by enzymatic degradation and adsorption (Kong et al., 2019; Subashchandrabose et al., 2011). All these lead to a better reduction in chemical oxygen demand (COD), pathogen control, and nutrient reuse, which underscores microalgae as self-sustaining bioengineers able to close ecological cycles in anthropogenic wastewater systems. The fact that these processes have been integrated into the designs of wastewater treatment units goes further to prove how biological innovation can replace or complement energy-intensive mechanical systems (Rawat et al., 2011; Park et al., 2011).

One of the most crucial but least studied aspects of microalgae use in wastewater management is odor neutralization. Wastewater facilities create major socio-environmental, regulatory, and public health problems due to the emission of odors, which are mostly hydrogen sulfide, ammonia, and volatile organic compounds (VOCs) (Aghalari et al., 2020; Bai et al., 2023). Microalgae reduce these odorants in multiple ways that are interconnected: photosynthetic oxygenation of anaerobic bacteria, which produce hydrogen sulfide, reduces the activity of such bacteria; direct molecular conversion of ammonia into amino acids by the glutamine synthetase–glutamate synthase pathway; the adsorption or biodegradation of VOCs by the algal metabolism (Li et al., 2019; Mubashar et al., 2024). It is this odor control in combination with bioremediation that places microalgae in the category of bio-oxygenators and biofilters, thereby providing an ecologically viable alternative to high-energy aeration or chemical deodorization. Practically, the research on *Chlorella vulgaris* and *Scenedesmus obliquus* in the municipal wastewater has reported significant ammonia dissolved and odor intensity reductions, which coincides with the environmental remediation and the social acceptability (Zeng et al., 2023).

More importantly, the success of microalgal systems is not only determined by the physiology of individual species but also by mutual interactions between microalgae and bacteria. Algal–bacterial consortia are examples of mutualistic ecological engineering: the algae distribute oxygen by photosynthesis, allowing the bacteria to oxidize the organic material, and the bacteria supply the carbon dioxide and mineralized nutrients that feed the algal growth (Qv et al., 2023; Xu & Zhang, 2024). These systems are always more effective than monocultures, with high COD removal, nutrient uptake, and stability of processes with varying loads of effluents (Delrue et al., 2021). This symbiotic metabolic process is used to convert wastewater ponds into micro-ecosystems that replicate the natural biogeochemical cycling and improve resilience, lowering energy expenses, and facilitating the recovery of biomass to be downstream valorized. Therefore, the microalgae–bacteria interface is not only a biotechnological optimization concept but also symbolizes a paradigm shift in wastewater management design toward ecosystem-based management concepts.

Although the idea of microalgae-based treatment systems holds a lot of potential, there are still significant issues that limit the large-scale use of the system. Removal efficiency can be varied by variability in wastewater composition, species-specific growth kinetics, and environmental conditions (light availability, pH, and temperature) (Yuvraj & Padmanabhan, 2017; Barkia et al., 2019). In addition, the high concentration of nutrients can lead to excessive algal growth and cause secondary ecological hazards such as oxygen consumption and the development of toxins by some cyanobacteria (Biswas et al., 2018). These problems cannot be solved only with the help of molecular biology, system engineering, and environmental policy. Developments in omics technologies are enhancing the taxonomy, functional genomics, and metabolic engineering of microalgae, which can be utilized to improve a strain to achieve a desired pollutant removal and bioactive production (Sirohi et al., 2021). Several types of parallel innovations in reactor design, including hybrid photobioreactors (PBRs) and biofilm systems, aim to reduce the disadvantages of the operational process and optimize the distribution of light and nutrients (Anand et al., 2023). Therefore, the future of microalgae-based bioremediation has to be seen in the intersection of biological specifics and environmental systems thinking.

Utilizing microalgae as green filters is the future of the modern science of sustainability, precisely combining ecological processes with technological breakthroughs to solve the mutually dependent environmental problems. Their multidimensional purpose of removing nutrients, sequestration of heavy metals, degradation of organic and elimination of odors puts them at the frontline of bio-based wastewater treatment. In addition to remediation, microalgae represent the transformational possibilities of converting pollution into productivity: wastewater is converted into a nutrient source, and algal biomass is used as an energy, food, and pharmaceutical source. Thus, the study of microalgae in bioremediation and odor neutralization extends beyond an applied environmental inquiry; it represents a broader discourse on how biological intelligence can redefine industrial ecology and contribute to a circular, low-carbon future.

14.2 GROWTH REQUIREMENT (LIGHT, NUTRIENTS, AND CO_2)

Microalgae require certain physicochemical factors and cultivation conditions for luxuriant growth, especially if cultivated in a controlled environment. Conditions such as sunlight, temperature, pH, macronutrients (such as nitrogen, carbon, phosphorus, potassium, sulfur, and silicon for diatoms), minerals, and aeration are primary growth requirements of microalgae. In addition are salinity, alkalinity, and the absence of grazers and contaminants (Brasil et al., 2017; Barkia et al., 2019). The ideal range of these factors fluctuates from species to species, highlighting the variability among the complex microalgae group. Many microalgae are photosynthetic with a photoautotrophic mode of cultivation; therefore, they require abundant sunlight and a CO_2 supply to produce organic compounds (proteins, lipids, and mono or polysaccharides) essential for their growth and metabolic activities. In these photoautotrophic microalgae, the importance of light intensity (daytime and nighttime) and temperature to their growth cannot be overemphasized; nevertheless, equilibrium is needed for an optimum physiological performance to be achieved. Environmental CO_2 is utilized for photosynthesis, but when absorbed in excess, it causes a pH imbalance, which affects cellular growth. In a pilot anaerobic scale, *Dunaliella* and *Spirulina* absorb bicarbonate as their carbon source (Brasil et al., 2017).

Heterotrophic microalgae such as *Chlorella* substitute carbon and energy sources with organic matter. In closed cultivation systems, glycerol, acetic acid, fulvic acid, and sugars (frequently glucose) are introduced into the medium to supply the necessary nutrients. When grown in culture media, microalgae obtain energy from organic carbon and grow in the dark. Alternatively, fermentation technologies like the dark fermentation are employed; in this case, bacteria, which synthesize butyric and acetic acids used as carbon sources by microalgae, are grown together with microalgae (Gallego et al., 2025). The drawback in this method is the probable introduction of bacterial contamination to the medium. Wastewater rich in phosphorus and/or nitrogen from urban or industrial effluents or flue gases rich in CO_2 could also be used as growth additives and sources of organic carbon. However, secondary metabolites produced from such microalgae would be limited for use as pharmaceuticals (Jareonsin & Pumas, 2021).

In aquatic environments, the level of salinity significantly differentiates fresh water from marine water, and this likewise determines the species of microalgae that inhabit such environments, as salinity significantly influences the chlorophyll content in microalgae. It has been reported that aquatic environments with low salinity levels favor the growth of microalgae by facilitating photosynthesis through the seamless absorption of nutrients necessary for growth (Kagan et al., 2014; Sedjati et al., 2019).

Temperature is another significant factor that influences microalgae growth and the synthesis of protein. High temperatures have been shown to favor the production of secondary metabolites from major microalgae such as *Spirulina* to the detriment of the organism. In essence, higher temperature increases photon inactivation rates needed to enhance photosynthesis; thus, as less photons are activated, less photosynthetic reaction occurs, hindering microalgae growth and development (Sarıtas et al., 2025).

The pH of an environment is an essential growth determinant, especially for photosynthetic organisms. It reveals if the microalgae thrive better in an acidic, alkaline, or near-neutral setting, which varies among different species. Inordinate changes in pH result in a drop in photosynthetic rates, leading to retarded microalgae growth or an absolute inhibition of photosynthesis (Del Campo et al., 2000).

14.3 ADAPTABILITY TO DIVERSE WASTEWATER ENVIRONMENT

One major trait of microalgae, which gives them an edge over other plants, is their ability to adapt and bloom in diverse wastewater environments. This water, which could be agricultural runoffs, municipal wastewater, or industrial effluents, is most of the time laden with harmful pollutants such as antibiotics and fertilizer, enriching the water with phosphorus, nitrogen, or CO_2 from industrial exhaust gases. These micropollutants are always difficult to break down in wastewater treatment plants and encourage the spread of multidrug-resistant organisms. In these environments, the ability of microalgae to grow in wastewater becomes a bioremediation strategy and purification technique to eliminate specific pollutants, organic matter, and/or heavy metals from spoilt aquatic systems (Gallego et al., 2025). Microalgae utilize the macronutrients as energy or carbon sources to generate large biomass: This process is termed phycoremediation. Microalgae absorb nitrogen in NH_4^+ and NO_3^- forms efficiently from the wastewater. They also absorb dissolved inorganic carbon in the form of HCO^{3-} and CO_2 and sequester CO_2 in the environment. Microalgae generate exometabolites in the wastewater, which in turn act as allelopathic substances, reducing the microorganisms in the water. They also bioaccumulate these micropollutants in their structure and are later used as biofuels. Through these steps, microalgae purify nitrogen-polluted environments, reduce greenhouse gas emission, and mitigate global climate change (Sarıtas et al., 2025). They also facilitate recycling and limit the use of freshwater for its cultivation. These strategies decrease the waste of valuable resources in waste streams and contribute to removing pollutants in wastewater treatment (Nagarajan et al., 2020). Microalgae like *Desmodesmus* and *Chlorella* have shown considerable abilities to reduce the concentration of pharmaceuticals in culture media. *Scenedesmus* sp., *Tetraselmis* sp., *Spirulina*, *Chlorella* spp., and *Picochlorum* sp. are used in wastewater treatment (Koller et al., 2014; Dvoretsky et al., 2019; Solovchenko et al., 2022; Stirk & van Staden, 2022).

Furthermore, microalgae act as sorbents in wastewater from mining environments, industrial effluents, or power plants polluted with heavy metals such as Zn^{2+}, Cr^{6+}, Pb^{2+}, Co, Cd^{2+}, Cr, Cu^{2+}, and Hg^{2+}. They bioaccumulate these metals and sometimes adsorb them onto the surface of their cell wall, aiding water purification and heavy metal recovery (Warjri & Syiem, 2018).

14.4 MECHANISMS OF BIOREMEDIATION BY MICROALGAE

Wastewater carries nitrogen, phosphorus, heavy metals, and diverse organic compounds that, if not controlled, impair aquatic systems and public health. Microalgae have gained attention as treatment agents because their physiology enables them to assimilate nutrients, immobilize toxic substances, and interact with other microorganisms in ways that improve water quality (Abinandan & Shanthakumar, 2016). This section examines the principal mechanisms through which these organisms contribute to bioremediation. Nutrient removal is considered first, focusing on the assimilation of ammonium, nitrate, and phosphate into cellular metabolism. Attention then shifts to bioaccumulation and biosorption, which highlight the capacity of algal cells and their surfaces to retain metals and organics. A further dimension is the synergy between algae and bacteria, where reciprocal exchanges of oxygen, carbon dioxide, and nutrients strengthen pollutant removal. Together, these perspectives provide a foundation for understanding the role of microalgae in wastewater remediation. These categories and their interactions with microalgal cells are summarized in Figure 14.1.

14.4.1 Nutrient Removal

The ability of microalgae to clean up wastewater of nutrients is also one of the best-studied aspects of algal biotechnology. According to Tan et al. (2023), these organisms use inorganic compounds as their metabolic products and as building blocks, which makes them useful in the reduction of the excess nutrients that normally form eutrophication. These are four general types of pollutants that are of specific interest when it comes to wastewater: nitrogen, phosphorus, heavy metals, and organic pollutants. The effect of microalgal systems in eliminating these contaminants depends on the physiology of the species used, the composition of the wastewater, and the system design.

14.4.1.1 Nitrogen Uptake and Assimilation

Nitrogen is assimilated in microalgae in the form of ammonium, nitrate, and nitrite, and all these enter different metabolic pathways. The energetic benefits of ammonium are that it can be directly inserted into amino acids through the glutamine synthetase–glutamate synthase route (Fernández et al., 2021). There are species like *Chlorella vulgaris* and *Scenedesmus obliquus* that have shown quick ammonium removal and have recorded a removal of above 80 percent within a couple of days in domestic wastewater (Li et al., 2019). Less favorable but reduced by activities of nitrate reductase and nitrite reductase are nitrate and nitrite, which are incorporated into proteins. Wastewaters with changing nitrogen species usually show variability in the rate of uptake, with the example of the higher nitrate form of uptake by a single species, namely *Chlamydomonas reinhardtii*, and a lower form, namely *Scenedesmus* (Chokshi et al., 2016; Silambarasan et al., 2021). Contraindications are nutrient saturation, that is, the maximum rate of uptake is reached when the pools of nitrogen intracellularly have been filled, and competition with bacterial populations, which could absorb nitrogen faster in some circumstances.

14.4.1.2 Phosphorus Removal via Polyphosphate Storage

Phosphorus is a significant nutrient that leads to the degradation of the water quality. Phosphate ions are eliminated by microalgae by attaching them to nucleic acids, phospholipids, and storage granules of polyphosphate. Luxury uptake, whereby the cells absorb phosphorus more than what is immediately needed, is especially beneficial to the wastewater treatment (Powell et al., 2009). Phosphorus has been demonstrated to accumulate in intracellular polyphosphate bodies in both *Chlorella vulgaris* and *Scenedesmus quadricauda*, where batch growth reached over 70% removals (Sutherland et al., 2020; Zeng et al., 2021). But this removal efficiency is affected by the inhibition of cell growth in high phosphate concentration, and fluctuating pH can

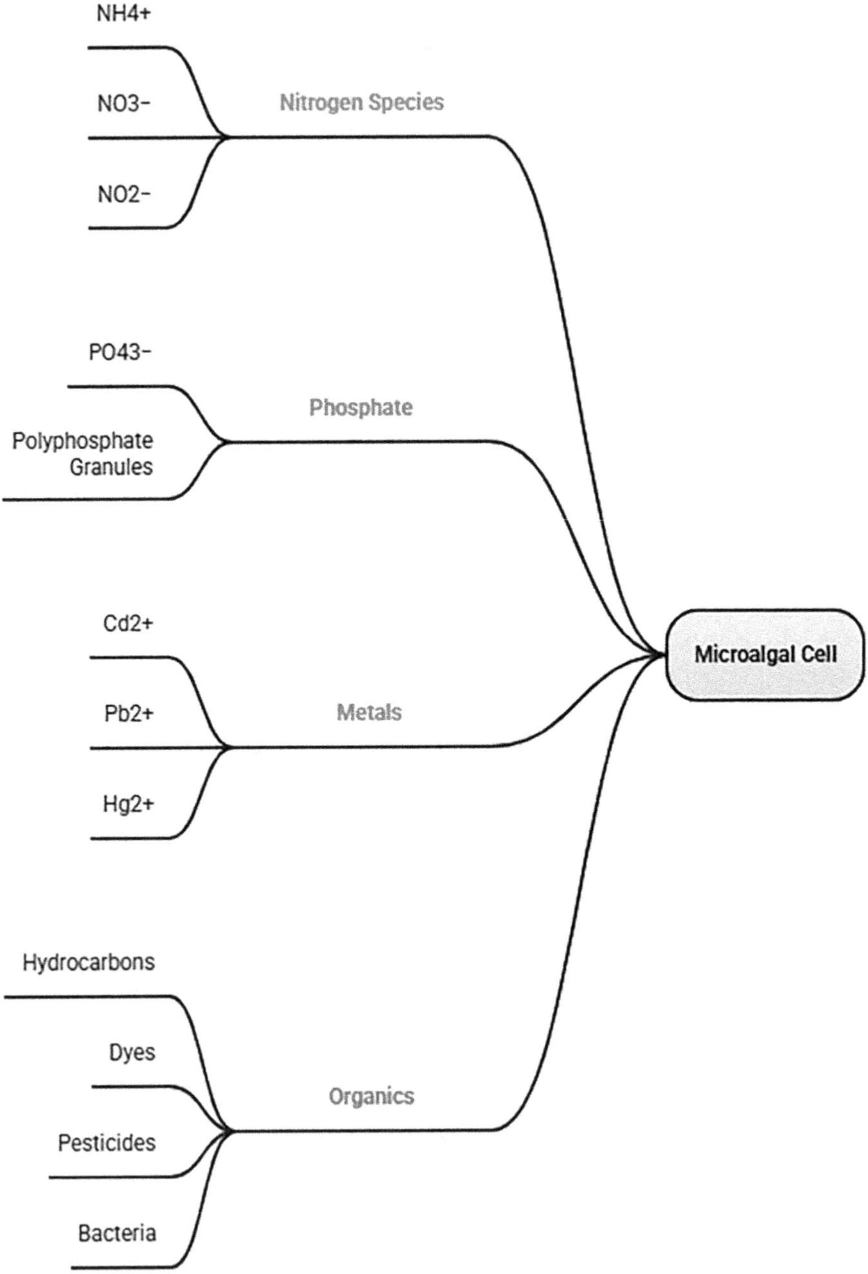

FIGURE 14.1 Microalgal cell interactions with major wastewater contaminants.

The diagram illustrates pathways of nitrogen assimilation (NH_4^+, NO_3^-, and NO_2^-), phosphorus uptake and polyphosphate storage (PO_4^{3-}), heavy metal immobilization (Cd^{2+}, Pb^{2+}, and Hg^{2+}), and interactions with organics (hydrocarbons, dyes, and pesticides) and bacteria.

change phosphate solubility, thus decreasing bioavailability. Besides, light and carbon availability tend to limit the removal of phosphorus, which indirectly influences the rate of algal uptake (Goswami et al., 2023).

14.4.1.3 Heavy Metal Uptake and Immobilization

Microalgae also play a role in immobilizing toxic metals like cadmium (Cd), lead (Pb), and mercury (Hg) that are common in industrial effluents (Baharuddin et al., 2020). It is taken up by active bioaccumulation into the cell and surface-level biosorption through functional groups of carboxyl, hydroxyl, and amino residues on the cell wall of the algae (Rai et al., 2020). The species of *Chlorella* and *Spirulina* have high cadmium tolerance and can significantly decrease the dissolved amounts, but this is highly dependent on the pH and metal speciation. Lead immobilization has been reported in living and dead biomass, thus highlighting the potential of low-cost biosorbent uses (Ding et al., 2019; Pan et al., 2024). The uptake of mercury can also be difficult because of its toxicity and unpredictable chemical forms, which can inhibit growth before it can be removed. Practically, the toxicity threshold limits remediation of heavy metals; beyond a toxicity level, the intracellular systems cannot be replenished, which compromises the viability of cells, which reduces the sustainability of treatment.

14.4.1.4 Organic Pollutant Degradation

In addition to inorganic nutrients, microalgae are involved in the degradation or conversion of organic pollutants, such as hydrocarbons, dyes, and pesticides. The process of hydrocarbon degradation has been reported in species of both the genus *Chlorella protothecoides* and *Nannochloropsis*, which is frequently performed in consortia with bacteria performing the first steps in oxidation (Elumalai et al., 2013). Color intensity and COD in textile wastewater have been reduced by using the biosorption of dye molecules and biodegradation of some enzymes using *Scenedesmus obliquus* that contains synthetic dyes (Kong et al., 2019). The degradation of pesticides has received less attention in research, but it has been shown that some of the green algae decompose the organophosphates and carbamates, thus making them less toxic (Subashchandrabose et al., 2011). Nevertheless, they are not as rapid and comprehensive as bacterial pathways, implying that algal systems can be most useful when incorporated into multi-trophic treatment systems. The difficulty with industrial effluents lies in their complexity, and mixtures of compounds may have synergistic toxicity, thus decreasing algal activity.

As mentioned above, the effectiveness of nutrient removal is not universal across species and systems. Open ponds have low permissiveness to scale-out but can be easily contaminated and unstable in their functioning, contrasting with close PBRs, which have improved scalability to control and elevated removal efficiencies at a higher capital expense (Rawat et al., 2011; Zeng et al., 2021).

14.4.2 Bioaccumulation and Biosorption

Bioaccumulation and biosorption are complementary processes through which microalgae can be used to remove pollutants in wastewater. Bioaccumulation is a metabolically active event of delivering the contaminants across the cell membrane and integrating them into intracellular systems (Musah et al., 2022). Conversely, biosorption is a passive process where the pollutants are attached to the algal cell wall by physicochemical reaction, regardless of cell metabolism (Raval et al., 2020). Collectively, these pathways explain the elimination of metals and dyes among other stubborn compounds in various effluents.

The biosorption mechanisms include ion exchange, complexation with carboxyl, hydroxyl, and phosphate residues, functional groups, as well as surface adsorption due to electrostatic attraction (Gupta & Rastogi, 2008; Zhou et al., 2021). Bioaccumulation, in turn, is based on active transporters and intracellular absorption, and the pollutants are frequently enclosed in vacuoles or connected with proteins and polyphosphates (Kharel et al., 2023). Both mechanisms are useful, as pointed out in case studies. As an example, the application of textile wastewater to the biosorber with the use of *Chlorella vulgaris* and *Scenedesmus obliquus* has been performed, and their cell surfaces have been used to accomplish a large portion of dyes (Kong et al., 2019). On the same note, cadmium and lead have also been reported to be taken up by petrochemical effluents in *Chlorella pyrenoidosa* and bioaccumulation to long-term immobilization in the biomass (Ding et al., 2019; Huang et al., 2024).

Live and dead algal biomass comparative analyses reveal some important differences. The living cells maintain their bioaccumulation ability, therefore, making them able to withstand longer immobilization of the contaminants, but at high levels of pollution, they become highly sensitive to the toxic effect. In comparison, dead biomass provides stable and reproducible biosorption capacity, is often more resilient to extreme effluent conditions, and it is possible to desorb and recover pollutants in some capacity (Bashir et al., 2024; Mahlangu et al., 2024). However, the biosorption process does not have metabolic transformation pathways as the living cells do, so it is relatively less effective with some organic compounds.

As a result, the choice of bioaccumulation and biosorption is dependent on treatment objectives. Live biomass promotes the recycling of nutrients and biomass valorization, and dead biomass is a low-cost, resistant alternative to streams that are heavily contaminated. The two methods are limited by saturation and possible interference of competing ions. By combining these processes with bacterial interactions, some of these limitations can be addressed and more robust treatment systems can be obtained, as discussed in the following section.

14.4.3 Synergistic Interactions with Bacteria

The collaboration of microalgae and bacteria in wastewater treatment is becoming an interesting approach due to the utilization of complementary metabolic activities. The principle behind such consortia is quite straightforward and effective: microalgae release oxygen through the process of photosynthesis and hence promote aerobic decomposition of organic matter by bacteria, and in return obtain carbon dioxide and mineralized nutrients that enhance the growth of algae. This mutual exchange creates a self-sustaining loop that will increase the elimination of the pollutants better than either of the groups could.

14.4.3.1 Enhanced Removal of Organic Load and Nutrients

Algae can be limited by oxygen, but in traditional monocultures, they are frequently the only known species to be able to degrade intricate organics. In combination, consortia enhance COD and biological oxygen demand (BOD) removal owing to the effectiveness of bacteria to oxidize the organics in an oxygen-rich environment, which is developed by algal photosynthesis (Qv et al., 2023; Xu & Zhang, 2024). Carbon dioxide released by bacterial respiration, in its turn, increases the rate at which the algae absorb nitrogen and phosphorus, thereby linking the organic decomposition process to the nutrient-releasing one. Comparative analyses of algal and bacteria system versus the monocultures always indicate greater efficiencies of removal: COD reduction of up to 90 percent in mixed systems versus 60–70 percent in monoculture treatments, accompanied by less lopsided nitrogen and phosphorus uptake (Subashchandrabose et al., 2011).

14.4.3.2 Biofilms and Cocultivation Strategies

Biofilm-based systems are especially effective in synergy. Under conditions of cogrowth of algae and bacteria on carrier surfaces in the form of biofilm, spatial organization promotes intimate contact between metabolites. The outer, light-exposed layers are inhabited by algae, which produce oxygen that diffuses into the interior, and the interior is inhabited by bacteria that degrade organics, thereby releasing carbon dioxide (Anand et al., 2023). This form of structured community provides greater stability to variations in wastewater composition, as well as showing endurance to toxic shocks. The use of cocultivation in suspended systems has also been discussed, such as sequencing batch reactor and membrane PBR systems, which are found to be better than axenic cultures in terms of nutrient uptake and biomass productivity (Sharma et al., 2021).

14.4.3.3 Case Studies and Applications

High-rate algal ponds (HRAPs), which are commonly used in New Zealand and in some regions of the United States, are based on natural algal–bacterial consortia. These systems are also capable of minimizing loads of BOD and pathogens using lower operating expenses than traditional activated sludge systems, and at the same time generate algal biomass that can be utilized in other ways (Liu et al., 2023; Park et al., 2011). In Europe, similar benefits are also attested in pilot-scale tests: the increase in nutrient removal, low cost of aeration, and better quality of effluent (Boelee et al., 2011). At the industrial level, algal–bacterial systems have been utilized to treat agro-industrial wastewater (IWW), where mixed consortia enhance the removal of nitrogen and stabilize the pH and thus, reduce agro-IWW agitation to nitrogen requirements (Delrue et al., 2016).

14.4.3.4 Advantages and Drawbacks

The benefits of these systems are varied. They offer greater total pollutant removal rates, diminish energy usage because of restricted mechanical aeration, and yield biomass that could be utilized in value-added applications. Their environmental stability makes them strong in the face of environmental variations, which is important when it comes to municipal plants that are subjected to variations in loads. However, drawbacks remain. Mixed consortia are harder to manage, and their species composition changes over time depending on the environmental conditions (Abd-Elmaksoud et al., 2021). Infrastructure intrusion by any type of invasive microorganism can cause some community changes, causing performance to decline. Furthermore, the expansion must be done by attentively regulating the light penetration and distribution of carbon dioxide to ensure algal growth in a high population (Razzak et al., 2013).

Algal–bacterial interaction is an example of how biological systems can be optimized by taking advantage of ecological complementarities. These consortia will not operate alone, but they will work together with defined nutrient removal pathways and biosorption processes, as described earlier. Bacterial mineralization of organic matter into bioavailable forms enhances the rate of uptake of nitrogen and phosphorus, and algal surfaces give the organism more biosorption sites in which the heavy metals and organics can be bound (Walters et al., 2025). This way, the synergistic relationships will constitute a holistic remediation mechanism in which the cycling of nutrients, the immobilization of pollutants, and the degradation of organic matter are related and not independent.

14.5 ODOR NEUTRALIZATION POTENTIAL

Wastewater treatment not only produces odors that are unpleasant but also has regulatory, public health, and community implications. The contributors are hydrogen sulfide, ammonia, and various VOCs, which occur under anaerobic or nutrient-enriched conditions in municipal and industrial streams (Aghalari et al., 2020). The traditional control measures are based on the use of chemicals or high-energy aeration that may increase prices and redistribute effects. Microalgae provide an alternative pathway by combining pollutant elimination with odor suppression by photosynthetic oxygenation, nutrient uptake, and surface reactions, which reduce the creation and release of odorants. In the section, evidence has been outlined in three dimensions, namely the sources and chemistry of the most prevailing odorants, the processes through which mitigation of the microalgae is achieved, and the results obtained through operational and pilot systems.

14.5.1 Sources of Odor in Wastewater

The production of odors in wastewater is mostly linked to the degradation of organic compounds by the activity of microbes in an anaerobic environment and the volatilization of the reduced compounds. One of the most noticeable odorants is the hydrogen sulfide (H_2S), which is produced when sulfate-reducing bacteria use sulfur compounds in the absence of oxygen. In addition to the typical rotten egg smell, this gas is also a danger to the safety of workers in high concentrations (Li et al., 2021). The other very common contaminant is ammonia (NH_3), which is formed during the hydrolysis of urea and the degradation of the proteins contained in the municipal sewage and agro-industrial effluents. Ammonia emissions are intensified in nutrient-rich settings where nitrification and denitrification processes are incomplete (Wang et al., 2025; Xu et al., 2021).

VOCs constitute a more diverse group of odorants, encompassing fatty acids, phenols, aldehydes, and solvents derived from both domestic and IWW streams. Jiang et al. (2022) state that industrial plants (like petrochemical plants, textile mills, and food processors) release complicated effluents in which VOCs build up and volatilize, creating long-lasting smells that may go farther than the treatment facilities. At low levels, the compounds are still noticeable and disruptive, and this makes it hard to comply with the odor emission standards.

The significance of odor control extends beyond aesthetic concerns. The unregulated emissions can trigger regulation, with various jurisdictions having strict restrictions on the number of odor units in the air (Bai et al., 2023). Regarding the aspects of public health, chronic exposure to low levels of H_2S or ammonia can worsen breathing issues, but VOCs contain substances with established irritant and toxic characteristics (Capelli et al., 2013; Zhang et al., 2024). Effects on odor also influence community acceptance of wastewater facilities; frequent complaints cause political pressure and potential restrictions in operations. Therefore, odor reduction methods that are effective and sustainable are essential in the management of wastewater. Although chemical scrubbers and aeration are conventional methods, biological processes, especially those based on the mediation of microalgae, have a dual advantage, which is to not only eliminate the pollutant but also the responsible odor.

14.5.2 Microalgal Role in Mitigation

The microalgal communities have effects in altering the odor activities in the wastewater by both physicochemical interactions, as well as the metabolic changes. Their photosynthetic activity, nutrient uptake, and surface-binding capabilities work together to inhibit the release of odor-forming compounds (Mubashar et al., 2024). Compared to traditional aeration or chemical additions, in microalgal systems, oxygenation, nutrient assimilation, and conversion of pollutants are provided in a single process, and hence odor control and improvement of water quality are achieved.

14.5.2.1 Photosynthetic Oxygenation and Suppression of Anaerobic Conditions

The major sites of H_2S and volatile fatty acid production are anaerobic environments. Due to the release of oxygen during photosynthesis, microalgae increase the level of dissolved oxygen in the treatment ponds, which inhibits the activity of sulfate-reducing and methanogenic bacteria (Luo et al., 2021). Increased oxygen prevents the development of H_2S as well as facilitates aerobic degradation, decreasing the precursors of malodorous VOCs. The results of this process are directly connected to the light penetration and mixing: shallow, HRAPs are better at oxygenating and providing more uniform odor reduction compared to basins that are deep and poorly illuminated (Guo et al., 2022).

14.5.2.2 Ammonia Assimilation

Protein-rich effluents contain increased amounts of ammonia emissions and can be significantly removed by algal uptake. Ammonium is a favorite source of nitrogen in most strains, which is directly taken to the glutamine synthetase–glutamate synthase protein synthesis pathway (Li et al., 2019). High rates of assimilation of ammonium are consistently observed in species like those of the *Chlorella vulgaris* and *Scenedesmus obliquus*, and research has indicated that a rate of greater than 70% of ammonia dissolved in municipal wastewater (MWW) is removed (Zeng et al., 2023). It is true that assimilation reduces the supply of volatile ammonia but also cuts off nitrogen to algal biomass that can further be used to generate fertilizer or bio-energy feedstock. Nevertheless, the efficiency of uptake decreases when the concentration of ammonia goes beyond the toxicity levels, which is why load and dilution must be counterbalanced.

14.5.2.3 Sulfide Oxidation and Precipitation

Microalgae also play a role in mitigating sulfide odors. In oxygenated environments, photosynthetically derived oxygen facilitates the biological oxidation of sulfide to sulfate or elemental sulfur. Some algae additionally promote precipitation reactions, where sulfide complexes with metal ions such as iron or manganese present in wastewater (Zhou et al., 2020). This dual pathway (oxidation and precipitation) lowers free sulfide concentrations, thereby reducing H_2S volatilization. The capacity for sulfide management appears enhanced in mixed cultures where algae coexist with bacteria, since bacterial oxidation pathways are stimulated by algal oxygen supply.

14.5.2.4 Volatile Organic Compound Reduction

Although less studied than nitrogen or sulfur compounds, microalgae can reduce VOC emissions. Adsorption of hydrophobic molecules onto algal cell walls provides one removal pathway, while metabolic degradation represents another. Green algae such as *Chlorella* have been shown to metabolize low-molecular-weight aldehydes and phenols, contributing to a measurable decline in odor intensity from industrial effluents (Pradhan et al., 2022). The efficiency of VOC mitigation varies widely depending on compound chemistry and wastewater composition and in many cases is enhanced when algae are cultivated alongside heterotrophic bacteria capable of initial VOC oxidation. The principal pathways by which microalgae mitigate odors are summarized in Figure 14.2.

14.5.2.5 Comparison of Strains and Systems

The efficiency of odor mitigation is strongly influenced by strain physiology and cultivation system design. *Chlorella* and *Scenedesmus* demonstrate good performance in various types of wastewaters because they are tolerant to high levels of nutrient and organic loading. Although its use can be effective, cyanobacteria can also cause odor issues (when the blooms rupture and the intracellular compounds escape) (Mubashar et al., 2024). Open ponds are relatively cheap, and they are well oxygenated under proper conditions, but their functioning is weather- and contamination-sensitive. Closed PBRs provide a greater level of control of odors and nutrient removal but are more expensive to capitalize on, and scaling is a problem (Rawat et al., 2011). Light availability and mixing intensity remain critical design variables; insufficient illumination or stratification reduces photosynthetic oxygenation, weakening the odor mitigation effect.

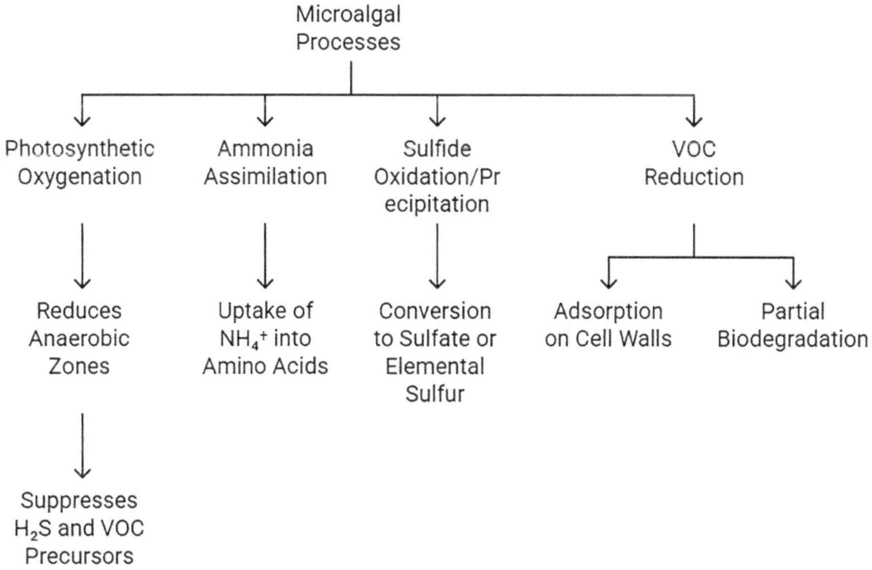

FIGURE 14.2 Microalgal mechanisms of odor mitigation.

The figure illustrates four main processes: photosynthetic oxygenation (reducing anaerobic zones and suppressing H_2S/VOC precursors), ammonia assimilation (incorporation of NH_4^+ into amino acids), sulfide oxidation/precipitation (conversion to sulfate or elemental sulfur), and VOC reduction (adsorption on cell walls and partial biodegradation).

14.5.3 Case Studies Demonstrating Odor Reduction

After decades of growing studies on this topic, empirical evidence of odor control using microalgal systems has now grown to more municipal, industrial, and pilot-scale settings. The case studies will give us an idea about the potential as well as the difficulties with using algal processes to control odors.

14.5.3.1 Municipal Wastewater Systems

HRAPs have been among the most widely studied settings. MWW in New Zealand on hectare-scale HRAPs consistently decreased the concentration of hydrogen sulfide in the headspace and decreased the number of units of detectable odors relative to the traditional stabilization ponds (Craggs et al., 2021). Algal assimilation also reduced the levels of ammonia in the effluent, and the levels were reduced by 6080% under optimal light conditions (Boelee et al., 2020; Park et al., 2011). Seasonal performance, however, revealed limitations: winter months with reduced irradiance correlated with lower dissolved oxygen and increased odor episodes, indicating that climatic variability constrains stability without supplemental mixing or artificial lighting.

14.5.3.2 Industrial Effluents

The industrial case studies reveal the ability of microalgae to treat more complex odor sources. Growth of the *Chlorella vulgaris* in the tannery wastewater, where ammonium and volatile organic sulfur compounds levels are high, achieved high scores of reductions in the intensity of odor, and at the same time, the ammonium concentration was reduced by over 70 percent (Selvaratnam et al., 2020). Food-processing effluents have also been treated using algal–bacterial consortia, where microalgal photosynthesis mitigated odors linked to fatty acids and sulfides while bacterial partners degraded high-COD organics (Singh et al., 2023). Pilot reactors with *Scenedesmus obliquus* were mentioned in petrochemical settings, where they were found to reduce the concentrations of VOCs via adsorption and partial biodegradation, but the extensive data on these systems concerning long-term operation are still unavailable (Delrue et al., 2021).

14.5.3.3 Pilot-Scale Projects

Several pilot-scale projects give quantitative performance results. A MWW treatment plant in Spain cut down H_2S emission by more than 50 percent compared to control basins and kept average dissolved oxygen levels at a level of more than 6 mg/L^{-1} throughout the day (Acién et al., 2020). Agricultural–IWW algal–bacterial PBRs could record significant ammonia odor reduction, with removal efficiency rates of over 75 percent after a few months in China (Zhou et al., 2021). While these studies highlight the operational feasibility of odor control, they also underscore persistent challenges such as maintaining algal activity under variable organic loads and preventing contamination by competing microbial populations.

14.6 TYPES OF WASTEWATERS TREATED

Wastewater comes from a variety of household, commercial, and agricultural processes along with runoff water. With human population growth, the amount of water being utilized also increases significantly, thus leading to higher waste. The nature of wastewater and its composition vary depending on the source of the wastewater. The origins of wastewaters are usually the basis to classify them into municipal, agricultural, and industrial wastewaters.

14.6.1 Municipal Wastewater

Total and absolute growth in the amount of MWW has been triggered by the increased anthropogenic activity worldwide because of the escalation of urbanization and industrialization (Vajargah et al., 2023). MWW is a dilute mixture that is formed as a result of domestic human activities and a minor input of industrial infiltration and rainfall (Gilmour & Zimmerman, 2020). Being the most common wastewater, MWW received much attention during the past decades due to its possible effects on human health and biorefinery possibilities (Sikosana et al., 2019). Each year, MWW amounts to about 360 billion cubic meters (Dai & Wang, 2023). The most significant sources of MWW are residential zones, institutions, recreational zones, and commercial buildings (Bishai, 2021). The composition of MWW varies significantly from one location to another. MWW includes wastewater discharges from toilets, showers, kitchen sinks, etc. Domestic wastewaters may contain household cleaners, detergents, urine, oils, grease, pathogens, salts, pharmaceuticals, and other discharges into the sewer system (Ali et al., 2021). They are typically classified as blackwater and greywater. Blackwater consists of the entire wastewater and excreta collected from toilets, while greywater includes effluents from other areas such as kitchen sink and dishwasher discharge, and wash basins in bathrooms, baths, and showers (Bishai, 2021).

Based on solubility, MWW mainly comprises water with relatively small amounts of organic and inorganic solid waste. The organic constituents include dissolved or suspended compounds such as lignin, proteins, carbohydrates, lipids, and soaps. Inorganic constituents include toxic heavy metals like lead, cadmium, mercury, arsenic, copper, chromium, and zinc (Agarwal et al., 2022). The microbial constituents of MWW are variable in nature, comprising algae, protozoa, bacteria, and viruses. Liu et al. (2021) reported the presence of Adenovirus, Polyomavirus, Norovirus, and Enteroviruses in raw wastewater. Algal composition may include *Chlorella vulgaris*, commonly found in MWW (Pacheco et al., 2021). Variable bacterial communities have also been identified in MWW. For example, Poopedi et al. (2025) identified *Legionella pneumophila, Mycobacterium spp., Arcobacter butzleri*, and *Aeromonas hydrophila from* five MWW sources in South Africa. Along with changing the water's pH, hardness, inorganic components, and pathogen load, the contaminants also raise the levels of COD and BOD in the water (Bishai, 2021). In both developed and developing countries, both treated and untreated MWW have raised serious concerns. This has led to the implementation of WHO guidelines in different regions around MWW discharge and treatment, aimed at protecting the environment and human health (WHO, 2006).

14.6.2 Industrial Wastewater

IWWs are effluents originating from the production activities of trade and industry, characterized by a high level of pollution and complex composition (Erkabaev et al., 2024). IWWs include wastewater from restaurants, service stations, meat processing facilities, food and beverage, paper factories, metal fabrication and finishing, chemical manufacturing, fertilizer production, mills, ceramic industries, oil and gas, hospital wastewater, etc. The wastewater from industries varies greatly in flow, composition, and pollution strength. However, industries typically require large volumes of water for their operations and hence generate large volumes of wastewater. For example, the paper production is the highest consumer of industrial water, utilizing significant volumes of water for processes such as pulping and bleaching (Singh et al., 2023). Other industries, such as petrochemical refining and pharmaceutical production, require water for chemical processes, and coal extraction industries wash and separate impurities from coal using water. Other water-intensive industries include mining, oil and gas, hydropower and geothermal energy production, and food-processing industries (Singh et al., 2023). Wastewater can be discharged from points such as identifiable outlets and nonpoint sources such as runoffs (Kato & Kansha, 2024). The composition varies based on the industry and processes involved, but typical pollutants include heavy metals, organic and inorganic pollutants (e.g., chromium, cadmium, lead, copper, zinc, and nickel), chemicals, oil and grease, and microorganisms (Oladimeji et al., 2024; Suazo et al., 2024; Mansour et al., 2023).

Additionally, IWWs are characterized by high BOD, COD, and total suspended soil (TSS). For example, high concentrations of TSS (2,000–10,000 mg/L), dissolved solids (300–1,000 mg/L), and moderate COD (500–1200 mg/L) have been reported in cement, concrete, and ceramic industries (Kato & Kansha, 2024).

IWW is often more toxic than MWW and generally requires relatively advanced methods of treatment (Nasr, 2022; Häder, 2018). Reusing and recycling are commonly used for efficient utilization of water in certain industries (De Almeida et al., 2016). Large water bodies are becoming exposed to untreated or insufficiently treated wastewater due to the volume of wastewater generated by expanding and newly emerging enterprises (Trianni et al., 2021). Industry noncompliance is a result of regulatory loopholes or insufficient enforcement of current regulations in many regions (Vrzala et al., 2022). Sometimes, there is not enough monitoring of IWW, which results in a lack of awareness of the problem's scope and insufficient response to it (Rana et al., 2022; Zakaria & Michael, 2017).

14.6.3 Agricultural Wastewater

Intensive production of agricultural products requires the use of water primarily from freshwater sources. According to the Food and Agriculture Organisation data in 2017, Egbuikwem et al. (2020) reported an increase by 6.3-fold between the 20th and 21st centuries, accounting for nearly 70% of the global freshwater supply being used in agriculture. With the increase in the production of agricultural products to meet food demands, a large amount of agricultural waste is subsequently produced (Riseh et al., 2023). Agricultural wastewater (AW) includes excess runoffs from surface irrigation composed of fertilizers and manures, livestock activities, agro-processing plants, and chemical applications (pesticides, herbicides, and fungicides) entering watercourses (Jagaba et al., 2023; Peck & Wentworth, 2022). The composition and quantity of AW are dependent on the materials processed, farm-specific operations, and processes. However, most AW contains a significant number of organic materials, with high levels of COD and TSS, which are potentially detrimental to the environment (Azli & Azoddein, 2020). Furthermore, agro-wastes are rapidly increasing with an average annual increase rate of 5–10%, highlighting the need for sustainable waste management (Debnath et al., 2021).

In considering the comprehensive component of AW, irrigation runoff plays a major role in continued AW production. This, in a broader sense, may be classified into agricultural irrigation, which involves the water supplied to crops and farmland, and landscape irrigation runoff, which includes not only agricultural irrigation but also urban landscape irrigation runoffs, such as watering of lawns and ornamental gardens. These runoffs may contain complex waste composition such as pathogens, suspended solids such as eroded soil, nitrates, phosphorus, heavy metals, and other organic pollutants (Xia et al., 2020). This may also include wash-offs from animal dung, over-applied fertilizers, and/or other chemicals. Inland and coastal waters are particularly vulnerable to the impacts of these AW. This is due to associated eutrophication and toxicity to fish, amphibians, invertebrates, and other animals on the food chain (Lan et al., 2024; Newton & Melaram, 2023). Since irrigation runoff water comes into contact with soil pathogens, pesticide residues, and occasionally traces of wild animal intrusion, like excrement, it is frequently regarded as a significant pool of pathogens and chemicals (Rao et al., 2022).

The management of livestock is also a major contributor to AW. The high demand for meat has necessitated the high development of livestock farms. For example, large swine and cattle farms have increased by 70% and 40%, respectively, between 2008 and 2017. Meanwhile, poultry meat and eggs production have increased by 37% and 23%, respectively, highlighting the associated increases in AW (Hu et al., 2020). Due to the nature of livestock farming, high amounts of nutrients, heavy metals, organic materials such as animal dungs, pathogens, and xenobiotics are released from these farms in large amounts, further complicating AW treatments. For instance, zinc (Zn) and copper (Cu), used to boost animals' immune systems and antibiotics, have also been reported to contribute to the emergence of methicillin-resistant *Staphylococcus aureus* (MRSA) (Hejna et al., 2019). Global aquaculture is also rising for human consumption, with about 50 million tons of fish produced annually (Bergheim et al., 2019). Wastewaters from fish farms vary due to the variable composition of feed-derived waste, nutrients, pesticides, pathogens/parasites and therapeutics. However, it has been reported that 1 tonne of frozen fillets releases 740 kg of BOD, 1020 kg of COD, 2050 kg of TSS, 106 kg of nitrogen, and 27 kg of phosphorus. Of this total discharge, up to 60% to 90% comes from wastewater from fishponds, while 3% to 27% comes from sludge from fishponds and wastewater from processing facilities (Thang et al., 2023).

In large agricultural settings, operations and maintenance of processing facilities contribute to AW production. A typical example is dairy processing plants, slaughterhouses, and fruit and vegetable processing. The large discharge of effluent arising from dairy processing is characterized by high COD and BOD levels due to the milk proteins and lactose they contain. Additionally, their rich organic content, particularly high-fat-containing substrates, is a concern in potential treatment strategies. A typical example is the water produced in milk and cheese production. An estimated 18% of milk output is wasted for a variety of reasons (Chang et al., 2017). Presently, the most common method of milk waste management is disposing of large volumes of waste milk into landfills, which raises concerns as runoffs may contaminate recipient waterbodies (Chang et al., 2017). In slaughterhouses, the high amount of water used to maintain a clean and hygienic environment in meat processing operations, blood and meat particles contribute to large AW discharge into the environment (Liu & Haynes, 2011). These wastewaters significantly affect recipient waterbodies depending on the production system of the source AW and over time, result in the contamination and eutrophication of ground and surface water. Considering the high organic contents present in AW, treatment strategies are important for mitigating associated health and environmental concerns.

14.7 CULTIVATION SYSTEMS FOR WASTEWATER TREATMENT

Microalgae can grow and persist in different ecosystems such as ponds and wastewater (Barboza-Rodríguez et al., 2024). They can withstand extreme environmental conditions and persist within distinct microhabitats with little or no light, elevated salt levels, and extreme temperatures (Montuori et al., 2023). The cost of cultivating microalgae for the treatment of wastewater is largely dependent on the method of cultivation. For example, the cost distribution of microalgae cultivation using an open pond indicates that 35–44% of the total production cost is dependent on the facility (Alavianghavanini et al., 2024). Currently, the cultivation of microalgae commonly involves the use of open ponds and closed PBRs.

14.7.1 Open Ponds

Open ponds are common cultivation systems that mimic the natural growth conditions of microalgae. They are simply ground ponds with a mechanical stirring system. They are considered favorable primarily due to comparative cost-effectiveness when used in large-scale industrial production (Barboza-Rodríguez et al., 2024). In an open pond system, microalgae grow on external surfaces and are exposed to sunlight and free oxygen. The ponds can be constructed using plastic or concrete with mechanical paddle wheels and baffles that stir the water and allow for the distribution of nutrients, light, and oxygen (Costa et al., 2019). Common open pond systems include raceway ponds, circular ponds, and unstirred ponds (Table 14.1). An open raceway pond is commonly adopted and accounts for 90% of the global microalgae production (Ren et al., 2025).

The major advantages of open ponds are ease of scalability, low-energy consumption, and low construction and operating costs, and these factors critically influence the choice of open ponds relative to PBRs. Regarding the cost, it has been reported that open pond systems eliminate between 0.3 and 0.4 US dollar/kg microalgae and can be expanded to 0.5 g/L (Trilokesh & Uppuluri, 2021). It has been reported that the financial feasibility of open ponds is significantly greater than that of PBRs (Richardson et al., 2012). Several optimization techniques have also emerged to improve open pond systems for microalgae cultivation and further improve cost-efficiency. For example, Alavianghavanini et al. (2024) reported microalgae cultivation using anaerobically digested effluent. Using a techno-economic approach, microalgae production was reduced by 16 %, being a suitable solution to produce low-cost microalgae (1.7 AUD kg^{-1} dry weight). Again, Zhu et al. (2022) proposed a technology in open ponds that introduced a real-time-based dynamic, which influences the speed of paddle wheels based on the light and temperature variations of the pond.

On the downside, open systems require large areas to scale up and are susceptible to contamination and adverse weather conditions (e.g., adverse temperature and rainfall). These disadvantages affect the production rate of open ponds. For example, biomass yield in raceway pond was shown to be lower, ranging from 0.03 to 0.2 g L^{-1} d^{-1} (6.2–24.5 g $m^{-2}d^{-1}$), compared to 0.15–1.2 g L^{-1} d^{-1} in PBRs (De Vree et al., 2015). This may primarily be due to biological contamination by other microorganisms present in the culture. Representative biological contaminants include other types of microalgae, zooplankton, fungi, bacteria, and viruses (Ding et al., 2019). These contaminants tend to impact biomass yield and/or quality and hence decrease products. Another critical problem is the need for a large area. The location in which the pond is situated is a critical factor in determining the type of pond selected, the algal strain, and the amount of light for photosynthesis. These ponds require large land areas,

TABLE 14.1
Table Showing the Different Types of Open Ponds

Type	Description	Depth	Mixing Mechanism	Advantage	Limitations
Raceway ponds	Typically comprises four long, narrow, and shallow channels about 9 m long and 70 cm wide lined with plastic	0.3 m	Paddle wheel	Cost-effective and environmentally friendly	Limited vertical mixing and uneven light distribution
Circular ponds	This includes a circular pond with a maximum diameter of 50 m	0.3–0.7 m	Centrally stirred	More efficient than unstirred ponds and relatively low land use	Less efficient than raceway ponds, still limited mixing, contamination risk
Cascade ponds	This is a steeped arrangement of surface-level through made of reinforced polyester resins	<0.5 m	Cascade	Cost-effective and environmentally friendly	Limited scalability, possible shear stress, and potential water loss
Unstirred ponds	This is a simple, shallow, open pond without active mixing	<0.1 m	No mixing	Very low operational cost and simple to construct	Least advanced and highly susceptible to contaminants, uneven light, and nutrients

Each open pond is described alongside its advantages and limitations (Ren et al., 2025; Qin et al., 2023; Novoveská et al., 2023; Udayan et al., 2021; Qin et al., 2023).

which may not be available in certain regions. Open ponds are usually built in isolated areas around power plants and away from agricultural lands due to the high-level CO_2 discharge utilized by microalgae in the circular economy. There are high costs of harvesting that are linked to open ponds. Open ponds are the less expensive and scaled-up alternative to cultivating the microalgae; the issues of contamination, environmental reliance, and low yield of biomass are very important impediments. However, new optimization methods are still in the process of making them viable to be used in the treatment of wastewater under a biorefinery strategy.

14.7.2 Closed Ponds

Contrastingly, closed cultivation systems typically refer to closed PBRs and fermenters, which are relatively high in cost due to the controlled environmental conditions and improved microcell biomass production in a laboratory setting (Barboza-Rodríguez et al., 2024). PBR is designed as a transparent culture vessel that allows light penetration and supports photoautotrophic microalgae (Gallego et al., 2025). Closed system utilizes relatively little space and can be situated outdoors or indoors, with light availability and minimal contamination. Generally, using closed systems is advantageous relative to open systems because the environmental factors and parameters to enhance biomass productivity are efficiently controlled. In fact, PBRs have been reported to be 5–10 times more efficient than the open systems (Ahmad et al., 2021). Common PBRs include tubular PRBs, column PBRs, annular PBRs, flat plate PBRs, bags, and stirred tank PBRs (Gallego et al., 2025). Several factors affect the selection of PRBs; however, flat plates are commonly used in wastewater treatment (Ahmad et al., 2021). Air bubbling or mechanical rotation of a motor is used to stir the culture to ensure aeration and distribution of nutrients throughout the culture. Additionally, artificial illumination such as LED can be used to augment sunlight in a PBR (Ahmad et al., 2021).

Based on the literature analysis, PBRs are more effective in wastewater treatment and biomass/biofuel generation due to contamination control and management of major parameters affecting microalgal growth. Overall, PBRs have shown significant advantages over open pond systems, such as minimal use of land area, with the potential to cultivate a single microalgae species over a prolonged period with minimal risk of contamination and high biomass yield. Also, the biomass generated by PBRs is homogeneous and of relatively high quality, which minimizes variation in biorefinery applications and enhances the quality of final products (Valdovinos-Garcida et al., 2021). Although PBRs are widely used, there are still some issues, especially in large-scale applications. The cost of operation is a major dilemma, as energy is forbidden. Dogaris et al. (2015) analyzed laboratory prototypes and were able to find that the mean manufacturing cost of conventional PBRs was four to nine times more than that of open ponds, regardless of the complexity of the design. Other disadvantages include high risk of overheating, maintenance cost, biofouling, and high dissolved oxygen that impacts growth and biomass production (Narala et al., 2016). The design of closed cultivation systems can also be improved to meet specific needs and reduce certain downsides. For example, continuous cultures can be applied in PBRs to improve biomass productivity but cannot be used for lipid induction by stress mechanisms to produce biodiesel (Narala et al., 2016).

Emerging research highlights the potential to minimize the disadvantages of open ponds and PBRs using hybrid systems. In a comparative analysis conducted by Narala et al. (2016), a two-stage cultivation system comprising positive-pressure air-lift-driven bioreactors with a separate synchronized high-lipid induction phase in nutrient-deplete open raceway ponds was used the marine microalga *Tetraselmis sp.* M8. Reports showed relatively high algal lipid production and minimal contamination levels.

14.7.3 Integration with Conventional Wastewater Treatment Infrastructure

The integration of microalgae-based systems with conventional wastewater treatment infrastructure offers a sustainable, circular economy approach to both waste management and resource recovery. Microalgae can be integrated into the conventional wastewater treatment process due to their performance in the removal of nutrients from wastewater. The major phase in conventional wastewater treatment involves the use of biological agents such as activated sludge (Evans et al., 2017). Typically, microalgae can be used in place of activated sludge in secondary wastewater treatment for the reduction in nutrient load, followed by subsequent treatment at the tertiary level to meet regulatory requirements. This will enable the production of microalgae biomass and a reduction in the production of sludges, thus reducing environmental impact.

The traditional way of wastewater treatment includes the combined application of physical, chemical, and biological processes to eliminate suspended solids and dissolved solids and to accomplish biological degradation of organic matter and removal of nutrients (Ranjit et al., 2021). Although the treatment process has been significantly successful in purifying water, it is faced with limitations associated with energy crises, shortage of water, economy, and regulation (Gupta et al., 2021). A conventional wastewater treatment system comprises primary, secondary, and tertiary stages. Primary wastewater treatment involves grit removal, screening, grinding, and sedimentation of raw wastewater. The initial process involves the removal of large amounts of waste such as wood, clothes, and metals. This is followed by decanter centrifugation involving liquid–solid separation, sludge dewatering, and sludge thickening based on density. Coagulant (such as alum, ferric sulfate, ferric chloride, starch, activated silica, and aluminum salt) is then introduced into the water to bind to colloidal particles in the wastewater that are too small to be removed by previous methods. Finally, the water is left undisturbed to allow sedimentation of remaining particles (Altowayti et al., 2022). This process has been reported to remove 10–40% COD and TSS (Oakley, 2019). Some studies have reported the use of microalgae in primary wastewater treatment (AlMomani & Örmeci, 2016). Nonetheless, in certain cases, raw wastewater subjected to primary treatment may not support the growth of microalgae due to high nutrient load, characteristic coloration, and high levels of suspended solids that hinder light availability to support photoautotrophic processes of certain microalgae (Nishshanka et al., 2023).

Recently, microalgae-based wastewater treatment has emerged as a potential alternative to conventional bacteria-based biological treatment at secondary and tertiary treatment levels aimed at the removal of nutrients to nondetectable levels (Nishshanka et al., 2023). Secondary process primarily involves the biological degradation of organic matter present in wastewater by aerobic or anaerobic processes of wastewater microorganisms. Aerobic wastewater treatment systems utilize oxygen to support microbial breakdown of organic matter. Aerobic systems include activated sludge, lagoon, and pond-based treatments, aerobic digestion, and septic tanks (Nguyen et al., 2020). In contrast, anaerobic wastewater treatment occurs in the absence of oxygen, using microbial consortia that generate methane and carbon dioxide as by-products. Common anaerobic processes include anaerobic digesters and upflow anaerobic sludge blanket (UASB) reactors. The two processes yield different end products: aerobic systems produce activated sludge, while anaerobic systems primarily produce biogas and stabilized sludge (Rout et al., 2020). Activated sludge comprises inorganic (e.g., heavy metals) and organic (e.g., bacterial biomass) components, which may differ depending on the source of wastewater (Awasthi et al., 2021). Activated sludge has shown potential as an alternative to chemical fertilizers (Awasthi et al., 2023). In recent years, the use of activated sludge has been reconsidered in favor of microalgae-based systems, which offer multiple advantages. Firstly, there is a reduction in operational costs and energy consumption. Conventional systems require the energy-intensive aeration of wastewater treatment plants to maintain aerobic activity. However, microalgae can produce oxygen as a by-product of photosynthesis that are supplied to aerobic bacteria that promote the oxidization of organic matter (Abdelfattah et al., 2022). This reduces the cost of aeration needed to clean wastewater. Another major advantage of employing microalgae in the treatment of secondary effluents is the uptake of inorganic N and P available in low quantities from the growth media. Amaro et al. (2023) reported 99%, 99%, and 90% of total nitrogen, total phosphorus, and/or organic carbon, respectively, through uptake mechanisms. Thus, microalgae would be able to further treat residual nutrients in the secondary effluent, assisting in limiting the risk of eutrophication when discharged into receiving waterbodies.

In tertiary wastewater treatment, microalgae can be strategically applied to remove residual nutrients, trace metals, and micropollutants that remain after conventional treatment. This tertiary application strengthens the sustainability of treatment systems while enabling biomass recovery for downstream valorization. Again, the use of microalgae can significantly reduce the use of disinfectant against coliform. Certain strains of microalgae have been shown to remove 99.4% and 98.6% of total coliform and *Escherichia coli*, respectively (Amaro et al., 2023).

Similarly, pathogenic strains of *E. coli* O157:H7/H$^-$, *Salmonella* spp., and indicators such as *Bacteroides* spp. and *Enterococcus* spp. were undetectable using molecular tools in secondary effluents of wastewater from PBRs (Bellucci et al., 2019). Furthermore, antibiotics such as sulfamethoxazole, ciprofloxacin, trimethoprim, and azithromycin are reduced to 14%, 35%, 48%, and 7%, respectively, by microalgae. The integration of microalgae for wastewater treatment has shown success in various studies. For example, the European project "Innovative Eco-Technologies for Resource Recovery from Wastewater" (INCOVER) developed an experimental microalgae-based wastewater treatment system for wastewater reuse and added-value products generation. This experimental plant successfully cleaned 6.9 m^3 of urban wastewater and agricultural runoff. A part of the harvested biomass was fed in for anaerobic co-digestion (AcoD) with secondary sludge used to obtain 150 L d^{-1} of biogas, particularly biomethane with 99% concentration. Concomitantly, the remaining biomass was further processed to produce biofertilizers. The treated wastewater was filtered and disinfected and reused in a smart irrigation system in a 125 m^2 sunflowers plantation (Uggetti et al., 2018). Similarly, Kumar et al. (2021) developed a 105-L pilot-scale pond reactor for phyco-mitigation of raw urban wastewater using microalgal strain *Pseudochlorella pringsheimii*-Ind-Jiht-1. This microalgal system provided an 83.2% reduction in COD, and TN and TP removals of 95.1% and 97.2%, respectively, and 66.7% reduction in alkalinity and 69.6% reduction in hardness. The treated wastewater was successfully used to cultivate Catostomus fishes with an average survival rate of 84% when compared to 0% using the raw wastewater, indicating the promising value of treated wastewater using microalgae.

14.7.4 Harvesting and Biomass Recovery Considerations

Harvesting microalgal biomass is a critical step in microalgae processing, which makes up 10–20% of total production cost (Tan et al., 2020). The major reasons for high harvesting cost are the relatively small size (<30 μm) of microalgae and their production in dilute cultures (mass concentration less than 1 g L^{-1}) with similar densities to water. Additionally, their negative charge maintains their dispersed state, making it more difficult to harvest (Barros et al., 2015). Ideally, harvesting methods should be effective for various strains of microalgae, that is, remove as much culture media as possible with high biomass recovery requiring minimal operational cost and maintenance. The harvesting method depends on the physiognomies of the microalgae, density and size of the microalgal cell, final product consideration, and possibility of culture medium reuse (Singh & Patidar, 2018).

Various harvesting and concentrating techniques of microalgae have been reported, including filtration, centrifugation, flocculation, and flotation (Tan et al., 2020). Filtration involves the use of a semipermeable membrane to separate microalgae from the growth media. Filtration techniques include crossflow, axial vibration membrane, polyacrylonitrile-based membrane, tilted membrane panel, and ultrafiltration. Conversely, flocculation utilizes gravity setting to generate large aggregates of cells by sedimentation that can then be easily separated from the liquid (Rao et al., 2023). Techniques used in flocculation involve the use of plants (*Moringa oleifera*), electricity, microorganisms, and chemicals (Tan et al., 2020). Flotation uses small bubbles that attach to microalgae cells in a float layer that can be skimmed off the surface using electricity, ozone, magnesium coagulation-dissolved air, buyo-beads, and foam techniques (Tan et al., 2020). In some cases, multiple techniques can be used to improve harvesting. For example, Rao et al. (2023) utilized flocculation combined with dissolved air flotation (DAF) to harvest microalgae in different benchtop jars.

Regardless of the method used in harvesting biomass, certain factors such as nonspecificity of harvesting method, improved biomass recovery, low operating and maintenance cost, environmental impact, impact on treated water quality, and potential of reuse should be considered (De Morais et al., 2023). Centrifugation is fast and efficient, but its use is limited due to cost and the potential for mechanical damage to microalgae cells (Li et al., 2020a). On the other hand, flocculation is a preferred choice for large-scale harvesting of microalgae for biorefinery use. This method is particularly advantageous due to high harvesting efficiency, low cost, clear-cut operational guidelines, and scalability for the volumes required in wastewater treatment plants (De Morais et al., 2023). Nonetheless, scaling-up may be difficult due to the viscosity of flocculant, and the use of chemicals such as metal-based salts may affect water quality due to their non-biodegradability, which can further affect other applications (De Morais et al., 2023; Singh & Patidar, 2018). This has necessitated the development of organic flocculants such as starch, which are environmentally friendly (He et al., 2022).

Flotation is a fast, cost-effective method with relatively small environmental impact and moderate operational cost (Kyzas & Matis, 2019). Based on recommendations to the largest wastewater treatment plant utilizing microalgae located in

Chiclana (Spain), the use of DAF and a low-energy pressurization pump that consumes 40 Wh/m^3 for the entire system will increase feed for anaerobic digesters and quality standards set by the European Directive 91/271/CE (Morais et al., 2021; Rao et al., 2023). Further optimization recommends that DAF should be combined with coagulation followed by flocculation (Rao et al., 2023). Filtration is a promising method that offers advantages relative to energy usage, biomass recovery rate, cost, and simplicity (Ennaceri et al., 2022; Mkpuma et al., 2021). The use of filtration is majorly limited due to fouling and clogging and potentially increases operational and maintenance costs (Wang et al., 2022). Nonetheless, new filter materials have been developed to improve the efficiency of filtration (Wang et al., 2022; Khairuddin et al., 2018). Based on biomass quantity, biomass quality, cost, processing time, and toxicity, Singh & Patidar (2018) reported flocculation, centrifugation, and filtration as the most preferred harvesting techniques for applications such as biofuel, human and animal food, high-value products, and water quality restoration. Overall, some innovative technologies have emerged to optimize harvesting methods aimed at improving efficiency and reducing costs. Nevertheless, continued research and technological innovation remain essential to enhance efficiency, reduce environmental impact, and make large-scale microalgal harvesting commercially viable.

14.8 VALUE-ADDED PRODUCTS FROM MICROALGAL BIOMASS

The promising concept of microalgal bioremediation also presents the potential for simultaneous conversion of microalgal biomass into useful products. This highlights an innovative approach in the blue economy that promotes the sustainable use of microalgae for improved economy and livelihood (Youssef, 2023). Microalgal biomass obtained from wastewater treatment can be used in the production of value-added products through a biorefinery approach. These products include biofuels such as biodiesel, bioethanol, biohydrogen, biochar, and biomethane. Food products, pharmaceuticals, nutraceuticals, cosmeceuticals, and animal feed can also be obtained (Khaligh & Asoodeh, 2022). Furthermore, it is interesting to note that Gendy and El-Temtamy (2013) reported that 1 kg of algae can fix 1.83 kg of CO_2, and certain algae can utilize other nutrient sources in addition to CO_2, indicating their promising environmental and industrial significance. Moreover, microalgal biomass can ameliorate climate change impact by capturing greenhouse gases, such as nitrous oxide, based on environmental conditions (Cao et al., 2023). Simultaneous and/or sequential extraction of these unprocessed microalgal biomass components will require the development of integrated biorefinery approaches and cost-effective processing technologies. Nonetheless, these products have significant potential for diversifying revenue streams and offering alternative energy sources to replace depleting fossil fuels, thereby contributing to the reduction of CO_2 emissions and improving the economic landscape (Okeke et al., 2022).

The use of microalgae as a renewable feedstock to produce biofuels has shown progress in recent years. In 2021, global primary energy demand increased by 5.8% compared to 2020, surpassing pre-pandemic levels by 1.3%. Fossil fuels continued to dominate the energy mix, accounting for approximately 82% of total primary energy consumption (Wang et al., 2024). Microalgae are adequate feedstocks for third- and fourth-generation biofuels (Wang et al., 2024). Microalgal biofuels have been reported to be physicochemically and functionally like fossil fuels, highlighting their promise as green alternatives to petroleum-based chemicals (Li et al., 2020b). The lipid abundance in microalgae makes it an attractive alternative to fossil fuels. Microalgae contain variable lipids; however, triacylglycerol (TAG) is the preferred lipid component for use as biodiesel feedstocks. *Chromochloris zofingiensis* has been shown to produce relatively high levels of lipid and is considered a particularly suitable option for biodiesel production (Wang et al., 2024). The fatty acids compositional profile is also an important component of microalgal biomass used for biofuel (Wang et al., 2024). Several species, *including Chlorella vulgaris, Scenedesmus bijugatus, Nitzschia recta, Anabaena sp, Nannochloropsis oceanica, Dunaliella salina, Botryococcus, Desmodesmus, Neochloris, Scenedesmus*, and *Tetraselmis*, have been reported as suitable for biofuel among others (Wang et al., 2024; Singh et al., 2023). However, literature has indicated that the sustainable production of microalgae biofuels is only viable with a highly optimized production system (Dasan et al., 2019). To address this, different biofuel productions can be combined to generate valuable coproducts, supporting zero-carbon emissions and minimal waste, thereby promoting a sustainable circular bioeconomy and cost-efficiency. Again, there is also a possibility of combining biofuels and other coproducts, such as bioplastics. In this scenario, optimized production systems may be adopted to convert microalgae biomass from biorefineries into polyhydroxyalkanoate (PHA), which is sustainable as an alternative to synthetic bioplastics (Chong et al., 2021). In some cases, microalgae have been used in other areas of biofuel production. For example, microalgal cultivation technology was adopted in the purification of biogas produced from organic waste from the swine industry to obtain bio methane (Blanes et al., 2024). Further research toward genetically modified microalgae has also gained attention (Saravanan et al., 2022).

In agriculture, microalgae are gaining attention as biofertilizers. Biofertilizers are sustainable biological alternatives to chemical fertilizers that provide land viability and improved soil health (Osorio-Reyes et al., 2023). Naturally, microalgae have a great capacity to assimilate phosphorus and nitrogen in environments with few nutrients, with high rates of specific absorption for their biochemical activities. This is essential in the application of microalgae bioremediation in different wastewater environments with variable concentrations of nutrients that can simultaneously clean the water and improve microalgae biomass for downstream processes (De Souza et al., 2018). Microalgae contain substantial amounts of soil beneficial compounds such as nitrogen, phosphorus, and potassium, along with various plant growth promoters such as Zn, copper (Cu), and Fe, in soil and their translocation inside plants (Gonçalves et al., 2023; Renuka et al., 2018). It has also shown great potential in the restoration of soil health. In fact, some microalgae have been proven to restore degraded soils affected by excess salts (Rocha et al., 2020). Formerly, some microalgae, such as Cyanobacteria, have been restricted to rice crops. However, in the last two decades, there has been a widespread use of microalgal biomass as biofertilizers. Particularly in China, which used to be the highest global producer of chemical fertilizers, accounting for over half of the world's total yield. Current use of biofertilizers has reduced chemical fertilizer usage by 20%–50% without diminishing crop yields (Sun et al., 2025). For example, *Chlorella spp.* and *Anabaena spp.* have been successfully applied as biofertilizers for several crops (Suchithra et al., 2021; Zhang et al., 2023; Puglisi et al., 2020). Nonetheless, environmental and economic viability may require extensive work to develop an optimized production pipeline. De Siqueira Castro et al. (2020) reported a comparative study of microalgae-based phosphate biofertilizer derived from meat processing effluent and a conventional triple superphosphate, using a life cycle assessment (LCA). In an optimized condition, environmental impacts were similar to the potential to further optimize microalgae biofertilizer processing. Aside from direct use in agriculture as biofertilizers, microalgae biomass has also been reported in biocontrol. In a recent study reported by Eckstien et al. (2024), in field conditions, many microalgae strains showed significant inhibitory activity against fungal pathogens. Similarly, allelochemicals produced by microalgae are beneficial biopesticides with a high level of specificity against target pathogens (Casanova et al., 2023). Microalgae have also shown promise as a feed and food source for humans and animals (Kusmayadi et al., 2021).

In other areas, microalgae biomass is considered a rich source of protein and "High-Value Molecules" (HMV) for nutraceutical and pharmaceutical compounds. Microalgae can coproduce other compounds like pigments, proteins, polyunsaturated fatty acids, and antioxidants and are attractive in the food, cosmetics, energy, and pharmaceutical industries. For example, microalgae can produce lipophilic and pigmented compounds such as β-carotene, astaxanthin, lycopene, lutein, and canthaxanthin, called carotenoids, that are useful in industrial applications (Gong & Bassi, 2016). Similarly, microalgal HVM such as β-carotene, vitamins, polysaccharides, peptides, astaxanthin, phycocyanin, and polyunsaturated omega-3 fatty acids (ω-3 PUFAs) have shown potential as food formulations and pharmaceutical applications as anticancer, antiaging, neuroprotective, and antimicrobial activity (Ampofo & Abbey, 2022; Galasso et al., 2019). Over the next ten years, it is projected that these chemicals will propel significant growth in the microalgae biotechnology market due to consumer desire for premium, affordable, and environmentally responsible products.

14.8.1 Opportunities for Circular Economy and Cost Recovery

The integration of microalgal biomass in different industrial processes indicates multiple opportunities within a sustainable circular economy framework. Based on the Ellen MacArthur Foundation's (2013) definition, the circular economy refers to a restorative industrial economy that relies on renewable energy with minimal environmental impacts and eradicates waste through meticulous layout (Santos et al., 2025). The efficient use of microalgal biomass follows the core principle of the circular economy through its lifecycle, beginning from wastewater treatment to the production of valuable products with substantial economic value and environmental benefits. The most significant opportunity for the circular economy is in resource recovery from wastewater in a biorefinery approach. First, the bioremediation process using microalgae provides a waste management strategy that utilizes potentially harmful and toxic waste released to the environment as raw materials for the large-scale culture of microalgae. Aside from being the most effective sustainable method for large-scale wastewater treatment, this option offers a relatively cheap method for wastewater management (Morais et al., 2021). The reuse of treated wastewater in microalgal culture is also an alternative method for water conservation (Kesari et al., 2021). Wastewater provides an alternative source of nutrients for microalgal culture. In an ideal scenario, microalgae require a large volume of freshwater. Hence, wastewater reduces the cost of microalgal production (Amaro et al., 2023). Second, microbial biomass produced from wastewater serves as a raw material to produce several valuable products. This replaces the need for conventional nutrients and allows for nutrient recycling, which has economic and environmental advantages. Technically, this closes the loop between wastewater management and the production of certain products such as biofuels and biofertilizers (Khaligh & Asoodeh, 2022). The biorefinery approach utilized in the production of microalgal biomass-related by-products or coproducts offers potential for cost recovery and industrial diversification. Primarily in an integrated biorefinery approach, a range of value-added products can be produced from microalgal biomass using multiple pathways and conversion technologies. This minimizes the overall raw materials required, energy consumed, and reduces waste (Velvizhi et al., 2021). Furthermore, the multiple production method enhances the economy and spreads supply across different markets.

Microalgae capture and utilization (CCU) is also a promising approach in the circular economy. It is important to note that microalgae-based carbon bio-capture and sequestration is one of the best carbon sequestration mechanisms globally (Alami et al., 2020). Microalgae have the capability to absorb and use atmospheric CO_2 to produce zero/negative emission products (Yang et al., 2024). Microalgae can sequester CO_2 10–50 times faster than land plants (Batista et al., 2015). Some industries with high carbon emissions create potential for the cultivation of microalgae. The price of CO_2 emission influences industrial activities, and most countries have introduced a tax on carbon on companies that use fossil fuels, thus necessitating an economically feasible and environmentally positive solution (Onyeaka et al., 2021). This technology is not very expensive because of less carbon emissions through the fixation of microalgae. In terms of infrastructure, microalgae cultivation units can be developed near an industrial plant, which allows the use of CO_2, wastewater, and other wastes. This closeness makes a circular economy model easy, which helps in saving costs, using waste streams as inputs, and building a symbiotic relationship where the industrial plant seeks a sustainable solution to its products (Haq et al., 2020). Economic viability of microalgal biomass may continue to be improved with the use of genetically modified microalgae to produce a certain compound at a reduced cost, hence increasing biomass yield (Kossalbayev et al., 2024). Although this has its benefits, the implementation of microalgal biomass in the circular economy is limited by scalability, cost, and regulation. However, the present emphasis on sustainability models and environmental technologies can create favorable policies and investment (Omokaro et al., 2025). The potential of microalgae in technology is high and offers a significant chance to develop the circular economy and recover the costs.

14.9 CONCLUSION AND FUTURE DIRECTIONS

Microalgae is an effective and sustainable solution for wastewater treatment, which integrates bioremediation, odor, and resource recovery in one ecological environment. Their remarkable adaptability and photosynthetic efficiency enable them to thrive under diverse environmental conditions while transforming pollutants such as nitrogen, phosphorus, heavy metals, and volatile compounds into valuable biomass. Microalgae transform contaminated effluents to cleaner water and renewable biological resources through nutrient assimilation, bioaccumulation, and biosorption, thus making them an indispensable part of the green infrastructure of the future. The fact that they work in reducing the number of pollutants and in the neutralization of odors contributes further to the sustainability of the wastewater management systems in the environment and social front.

In addition to the existing biological efficiency, microalgae stimulate natural interactions between them and bacteria, creating self-sustainable consortia, which enhance the efficiency of pollutant removal and stabilize ecosystem activity. These interactions make wastewater ponds small ecological systems, in which wastewater oxygen production, nutrient cycling, and organic breakdown are in a continual state of flux. This kind of integration minimizes energy requirements, lowers operating expenses, facilitates carbon recovery, and produces biomass that can be used to produce bioenergy, agriculture, and biotechnology. The ability of microalgae to deliver multiple environmental services in one process is an important indication of changing the traditional approach of wastewater treatment to one of regeneration and circularity.

In perspective, the next stage of developing the utilization of microalgae in large-scale applications necessitates innovation in the development of strains, the growth, and optimization of the process. Future research should focus on improving growth efficiency, enhancing pollutant specificity, and developing hybrid reactor systems that balance productivity with economic feasibility. Policies that support biosafety, standardization, and industrial integration will be equally crucial for mainstream adoption. Ultimately, microalgae exemplify how biological intelligence can be harnessed to achieve sustainable water management, cleaner environments, and a resilient green economy, transforming waste streams into vital resources for future generations.

REFERENCES

Abdelfattah, A., Ali, S. S., Ramadan, H., El-Aswar, E. I., Eltawab, R., Ho, S., Elsamahy, T., Li, S., El Sheekh, M. M., Schagerl, M., Kornaros, M., & Sun, J. (2022). Microalgae-based wastewater treatment: Mechanisms, challenges, recent advances, and future prospects. *Environmental Science and Ecotechnology*, *13*, 100205. https://doi.org/10.1016/j.ese.2022.100205

Abd-Elmaksoud, S., Abdo, S. M., Gad, M., Hu, A., El-Liethy, M. A., Rizk, N., Marouf, M. A., Hamza, I. A., & Doma, H. S. (2021). Pathogens removal in a high-rate algal pond wastewater treatment system. *Sustainability*, *13*(23), 13232. https://doi.org/10.3390/su132313232

Abinandan, S., & Shanthakumar, S. (2016). Evaluation of photosynthetic efficacy and CO2 removal of microalgae grown in an enriched bicarbonate medium. *3 Biotech*, *6*(1), 9. https://doi.org/10.1007/s13205-015-0314-5

Acién, F. G., Gómez-Serrano, C., Morales-Amaral, M. M., Fernández-Sevilla, J. M., & Molina-Grima, E. (2020). Wastewater treatment using microalgae: Recent results from pilot-scale photobioreactors and open ponds. *Water Science and Technology*, *81*(5), 1133–1143. https://doi.org/10.2166/wst.2020.180

Agarwal, S., Darbar, S., & Saha, S. (2022). Challenges in management of domestic wastewater for sustainable development. In Arun Lal Srivastav, Sughosh Madhav, Abhishek Kumar Bhardwaj, and Eugenia Valsami-Jones (eds.), *Current Directions in Water Scarcity Research* (pp. 531–552). Elsevier. https://doi.org/10.1016/b978-0-323-91838-1.00019-1

Aghalari, Z., Dahms, H.-U., Sillanpää, M., & Kazemi, Z. (2020). Odour emission and control in wastewater treatment plants: A review. *Environmental Chemistry Letters*, *18*(5), 1501–1515.

Ahmad, I., Abdullah, N., Koji, I., Yuzir, A., & Muhammad, S. E. (2021). Evolution of photobioreactors: A review based on microalgal perspective. *IOP Conference Series Materials Science and Engineering*, *1142*(1), 012004. https://doi.org/10.1088/1757-899x/1142/1/012004

Alami, A. H., Alasad, S., Ali, M., & Alshamsi, M. (2020). Investigating algae for CO2 capture and accumulation and simultaneous production of biomass for biodiesel production. *The Science of the Total Environment*, *759*, 143529. https://doi.org/10.1016/j.scitotenv.2020.143529

Alavianghavanini, A., Moheimani, N. R., & Bahri, P. A. (2024). Process design and economic analysis for the production of microalgae from anaerobic digestates in open raceway ponds. *The Science of the Total Environment*, *923*, 171554. https://doi.org/10.1016/j.scitotenv.2024.171554

Ali, I., Naz, I., Peng, C., Abd-Elsalam, K. A., Khan, Z. M., Islam, T., Pervez, R., Amjed, M. A., Tehrim, A., Perveen, I., & Sehar, S. (2021). Sources, classifications, constituents, and available treatment technologies for various types of wastewater: An overview. In Kamel A. Abd-Elsalam and Muhammad Zahid (eds.), *Elsevier eBooks* (pp. 11–46). Elsevier. https://doi.org/10.1016/b978-0-12-821141-0.00019-7

AlMomani, F. A., & Örmeci, B. (2016). Performance of *Chlorella vulgaris*, Neochloris oleoabundans, and mixed indigenous microalgae for treatment of primary effluent, secondary effluent and centrate. *Ecological Engineering*, *95*, 280–289. https://doi.org/10.1016/j.ecoleng.2016.06.038

Altowayti, W. A. H., Shahir, S., Othman, N., Eisa, T. A. E., Yafooz, W. M. S., Al-Dhaqm, A., Soon, C. Y., Yahya, I. B., Rahim, N. A. N. B. C., Abaker, M., & Ali, A. (2022). The role of conventional methods and artificial intelligence in the wastewater treatment: A comprehensive review. *Processes*, *10*(9), 1832. https://doi.org/10.3390/pr10091832

Amaro, H. M., Salgado, E. M., Nunes, O. C., Pires, J. C., & Esteves, A. F. (2023). Microalgae systems - Environmental agents for wastewater treatment and further potential biomass valorisation. *Journal of Environmental Management*, *337*, 117678. https://doi.org/10.1016/j.jenvman.2023.117678

Ampofo, J., & Abbey, L. (2022). Microalgae: Bioactive composition, health benefits, safety and prospects as potential high-value ingredients for the functional food industry. *Foods*, *11*(12), 1744. https://doi.org/10.3390/foods11121744

Anand, U., Dey, S., Parial, D., Shukla, V., Jha, N. K., Samanta, A., Karmakar, V., Goswami, R., & Bontempi, E. (2023). Algae and bacteria consortia for wastewater decontamination and transformation into biodiesel, bioethanol, biohydrogen, biofertilizers and animal feed: A review. *Environmental Chemistry Letters*, *21*(3), 1585–1609. https://doi.org/10.1007/s10311-023-01562-w

Awasthi, M. K., Ganeshan, P., Gohil, N., Kumar, V., Singh, V., Rajendran, K., Harirchi, S., Solanki, M. K., Sindhu, R., Binod, P., Zhang, Z., & Taherzadeh, M. J. (2023). Advanced approaches for resource recovery from wastewater and activated sludge: A review. *Bioresource Technology*, *384*, 129250. https://doi.org/10.1016/j.biortech.2023.129250

Awasthi, M. K., Singh, E., Binod, P., Sindhu, R., Sarsaiya, S., Kumar, A., Chen, H., Duan, Y., Pandey, A., Kumar, S., Taherzadeh, M. J., Li, J., & Zhang, Z. (2021). Biotechnological strategies for bio-transforming biosolid into resources toward circular bio-economy: A review. *Renewable and Sustainable Energy Reviews*, *156*, 111987. https://doi.org/10.1016/j.rser.2021.111987

Azli, F. A. M., & Azoddein, A. A. M. (2020). Removal of chemical oxygen demand (COD) and total suspended solid (TSS) using electrocoagulation process for treatment of oleochemical wastewater. *IOP Conference Series Materials Science and Engineering*, *736*(2), 022104. https://doi.org/10.1088/1757-899x/736/2/022104

Baharuddin, N. A., Halim, N. H. A., Yusoff, F. M., & Markou, G. (2020). Current progress, challenges and perspectives in microalgae-based nutrient removal for aquaculture waste: A comprehensive review. *Journal of Cleaner Production*, *277*, 124209. https://doi.org/10.1016/j.jclepro.2020.124209

Bai, X., He, X., Gao, H., & Li, Q. (2023). Odor nuisance and health risk assessment of hydrogen sulfide and ammonia in wastewater facilities. *Chemosphere*, *312*, 136958.

Barboza-Rodríguez, R., Rodríguez-Jasso, R. M., Rosero-Chasoy, G., Aguado, M. L. R., & Ruiz, H. A. (2024). Photobioreactor configurations in cultivating microalgae biomass for biorefinery. *Bioresource Technology*, *394*, 130208. https://doi.org/10.1016/j.biortech.2023.130208

Barkia, I., Saari, N., & Manning, S. R. (2019). Microalgae for high-value products towards human health and nutrition. *Marine Drugs*, *17*(5), 304. https://doi.org/10.3390/md17050304

Barros, A. I., Gonçalves, A. L., Simões, M., & Pires, J. C. (2015). Harvesting techniques applied to microalgae: A review. *Renewable and Sustainable Energy Reviews*, *41*, 1489–1500. https://doi.org/10.1016/j.rser.2014.09.037

Barsanti, L., & Gualtieri, P. (2018). Is exploitation of microalgae economically and energetically sustainable? *Algal Research*, *31*, 107–115. https://doi.org/10.1016/j.algal.2018.02.001

Bashir, M. F., Hu, H., Qadir, F., Iqbal, M., & Liu, H. (2024). A comprehensive review on microalgae-driven heavy metals removal from industrial wastewater using living and nonliving microalgae. *Journal of Hazardous Materials Advances*, *16*, 100492. https://doi.org/10.1016/j.hazadv.2024.100492

Batista, A. P., Ambrosano, L., Graça, S., Sousa, C., Marques, P. A., Ribeiro, B., Botrel, E. P., Neto, P. C., & Gouveia, L. (2015). Combining urban wastewater treatment with biohydrogen production – An integrated microalgae-based approach. *Bioresource Technology*, *184*, 230–235. https://doi.org/10.1016/j.biortech.2014.10.064

Bellucci, M., Marazzi, F., Naddeo, L. S., Piergiacomo, F., Beneduce, L., Ficara, E., & Mezzanotte, V. (2019). Disinfection and nutrient removal in laboratory-scale photobioreactors for wastewater tertiary treatment. *Journal of Chemical Technology & Biotechnology*, *95*(4), 959–966. https://doi.org/10.1002/jctb.6010

Bergheim, A., Schumann, M., & Brinker, A. (2019). Water pollution from fish farms. *Encyclopedia of Water*, 1–10. https://doi.org/10.1002/9781119300762.wsts0101

Bishai, M. (2021). A comprehensive study of COVID-19 in wastewater. In Mohammad Hadi Dehghani, Rama Rao Karri, and Sharmili Roy (eds.), *Elsevier eBooks* (pp. 115–144). Academic Press. https://doi.org/10.1016/b978-0-323-85780-2.00015-9

Biswas, G., Pokkatt, P. P., Ghosh, A., Kamila, B., Adhikari, K., & Dutta, S. (2018). Valorization of waste micro-algal biomass–collected from coke oven effluent treatment plant and evaluation of sorption potential for fluoride removal. *Water Science and Technology*, *78*(1), 132–146. https://doi.org/10.2166/wst.2017.638

Blanes, M. V., Pérez-Hermosillo, I. J., Rueda, A. R., & Sánchez, A. G. (2024). Biogas purification through the use of a microalgae-bacterial system in semi-industrial high-rate algal ponds. *Journal of Visualized Experiments*, *205*. https://doi.org/10.3791/65968

Boelee, N. C., Temmink, H., Janssen, M., Buisman, C. J. N., & Wijffels, R. H. (2011). Scenario analysis of nutrient removal from municipal wastewater by microalgal biofilms. *Water*, *3*(1), 260–276.

Boelee, N. C., Temmink, H., Janssen, M., Buisman, C. J. N., & Wijffels, R. H. (2020). Algal biofilm reactors for wastewater treatment: Effectiveness in nutrient and odor removal. *Water*, *12*(3), 685. https://doi.org/10.3390/w12030685

Brasil, B. D. S. A. F., de Siqueira, F. G., Salum, T. F. C., Zanette, C. M., & Spier, M. R. (2017). Microalgae and cyanobacteria as enzyme biofactories. *Algal Research*, *25*, 76–89. https://doi.org/10.1016/j.algal.2017.04.035

Cao, T. N., Mukhtar, H., Le, L., Tran, D. P., Ngo, M. T. T., Pham, M., Nguyen, T., Vo, T., & Bui, X. (2023). Roles of microalgae-based biofertilizer in sustainability of green agriculture and food-water-energy security nexus. *The Science of the Total Environment*, *870*, 161927. https://doi.org/10.1016/j.scitotenv.2023.161927

Capelli, L., Sironi, S., & Del Rosso, R. (2013). Odor impact assessment in wastewater treatment plants by means of dynamic olfactometry, dispersion modeling and social participation. *Water Research*, *47*(1), 336–349. https://doi.org/10.1016/j.watres.2012.09.048

Casanova, L. M., Macrae, A., De Souza, J. E., Neves, A. Junior, & Vermelho, A. B. (2023). The potential of allelochemicals from microalgae for biopesticides. *Plants*, *12*(9), 1896. https://doi.org/10.3390/plants12091896

Chang, I., Im, J., Chung, M., & Cho, G. (2017). Bovine casein as a new soil strengthening binder from diary wastes. *Construction and Building Materials*, 160, 1–9. https://doi.org/10.1016/j.conbuildmat.2017.11.009

Chokshi, K., Pancha, I., Trivedi, K., George, B., Maurya, R., & Mishra, S. (2016). Biofuel potential of the newly isolated microalgae *Scenedesmus* sp. from India. *Bioresource Technology*, *204*, 222–230. https://doi.org/10.1016/j.biortech.2016.01.046

Chong, J. W. R., Tan, X., Khoo, K. S., Ng, H. S., Jonglertjunya, W., Yew, G. Y., & Show, P. L. (2021). Microalgae-based bioplastics: Future solution towards mitigation of plastic wastes. *Environmental Research*, *206*, 112620. https://doi.org/10.1016/j.envres.2021.112620

Costa, J. A. V., Freitas, B. C. B., Santos, T. D., Mitchell, B. G., & Morais, M. G. (2019). Open pond systems for microalgal culture. In Ashok Pandey, Jo-Shu Chang, Carlos Ricardo Soccol, Duu-Jong Lee, and Yusuf Chisti (eds.), *Elsevier eBooks* (pp. 199–223). Elsevier. https://doi.org/10.1016/b978-0-444-64192-2.00009-3

Craggs, R. J., Sutherland, D. L., & Campbell, H. (2021). Performance of high-rate algal ponds for wastewater treatment in temperate climates: Nutrient removal and odor control. *Journal of Applied Phycology*, *33*(2), 935–947.

Dai, C., & Wang, F. (2023). Potential applications of microalgae–bacteria consortia in wastewater treatment and biorefinery. *Bioresource Technology*, *393*, 130019. https://doi.org/10.1016/j.biortech.2023.130019

Dasan, Y. K., Lam, M. K., Yusup, S., Lim, J. W., & Lee, K. T. (2019). Life cycle evaluation of microalgae biofuels production: Effect of cultivation system on energy, carbon emission and cost balance analysis. *The Science of the Total Environment*, *688*, 112–128. https://doi.org/10.1016/j.scitotenv.2019.06.181

De Almeida, B. A., Almeida, M., Martins, S., Maçarico, V. A., & Da Fonseca, A. T. (2016). Recycling liquid effluents in a ceramic industry. *Boletín De La Sociedad Española De Cerámica Y Vidrio*, *55*(3), 95–104.

De Morais, E. G., Sampaio, I. C. F., Gonzalez-Flo, E., Ferrer, I., Uggetti, E., & García, J. (2023). Microalgae harvesting for wastewater treatment and resources recovery: A review. *New Biotechnology*, *78*, 84–94. https://doi.org/10.1016/j.nbt.2023.10.002

De Siqueira Castro, J., Calijuri, M. L., Ferreira, J., Assemany, P. P., & Ribeiro, V. J. (2020). Microalgae based biofertilizer: A life cycle approach. *The Science of the Total Environment*, *724*, 138138. https://doi.org/10.1016/j.scitotenv.2020.138138

De Souza, M. H. B., Calijuri, M. L., Assemany, P. P., De Siqueira Castro, J., & De Oliveira, A. C. M. (2018). Soil application of microalgae for nitrogen recovery: A life-cycle approach. *Journal of Cleaner Production*, *211*, 342–349. https://doi.org/10.1016/j.jclepro.2018.11.097

De Vree, J. H., Bosma, R., Janssen, M., Barbosa, M. J., & Wijffels, R. H. (2015). Comparison of four outdoor pilot-scale photobioreactors. *Biotechnology for Biofuels*, *8*(1). https://doi.org/10.1186/s13068-015-0400-2

Debnath, B., Haldar, D., & Purkait, M. K. (2021). A critical review on the techniques used for the synthesis and applications of crystalline cellulose derived from agricultural wastes and forest residues. *Carbohydrate Polymers*, 273, 118537. https://doi.org/10.1016/j.carbpol.2021.118537

Del Campo, J. A., Moreno, J., Rodŕiguez, H., Vargas, M. A., Rivas, J., & Guerrero, M. G. (2000). Carotenoid content of Chlorophycean microalgae: Factors determining lutein accumulation in Muriellopsis sp.(Chlorophyta). *Journal of Biotechnology*, *76*(1), 51–59. https://doi.org/10.1016/S0168-1656(99)00178-9

Delrue, F., Álvarez-Díaz, P. D., Fon-Sing, S., Fleury, G., & Sassi, J.F. (2016). The environmental biorefinery: Using microalgae to remediate wastewater, a win-win paradigm. *Energies*, *9*(3), 132.

Delrue, F., Álvarez-Díaz, P. D., Fon-Sing, S., Fleury, G., & Sassi, J.-F. (2021). Environmental biorefinery: Case studies of wastewater phycoremediation with odor reduction. *Energies*, *14*(7), 1972. https://doi.org/10.3390/en14071972

Ding, T., Wang, X., Zhou, Y., & Sun, L. (2019). Biosorption of lead ions from aqueous solutions by microalgae: A review. *Environmental Science and Pollution Research*, *26*(27), 27474–27486.

Dogaris, I., Welch, M., Meiser, A., Walmsley, L., & Philippidis, G. (2015). A novel horizontal photobioreactor for high-density cultivation of microalgae. *Bioresource Technology*, *198*, 316–324. https://doi.org/10.1016/j.biortech.2015.09.030

Dvoretsky, D., Dvoretsky, S., Temnov, M., Markin, I., Akulinin, E., Golubyatnikov, O., … & Eskova, M. (2019). Experimental research into the antibiotic properties of Chlorella vulgaris algal exometabolites. *Chemical Engineering*, *74*. https://doi.org/10.3303/CET1974239

Eckstien, D., Maximov, N., Margolis, N., & Raanan, H. (2024). Towards sustainable biocontrol: Inhibition of soil borne fungi by microalgae from harsh environments. *Frontiers in Microbiology*, *15*. https://doi.org/10.3389/fmicb.2024.1433765

Egbuikwem, P. N., Mierzwa, J. C., & Saroj, D. P. (2020). Assessment of suspended growth biological process for treatment and reuse of mixed wastewater for irrigation of edible crops under hydroponic conditions. *Agricultural Water Management*, *231*, 106034. https://doi.org/10.1016/j.agwat.2020.106034

Elumalai, S., Kalaiselvi, A., & Suresh, V. (2013). Bioremediation of hydrocarbons using microalgae. *International Journal of Pharma and Bio Sciences*, *4*(1), 134–143.

Ennaceri, H., Fischer, K., Schulze, A., & Moheimani, N. R. (2022). Membrane fouling control for sustainable microalgal biodiesel production: A review. *Renewable and Sustainable Energy Reviews*, *161*, 112335. https://doi.org/10.1016/j.rser.2022.112335

Erkabaev, F., Muhammadieva, D., & Rabbimkulova, S. (2024). Composition and properties of industrial wastewater and its electrochemical treatment. *E3S Web of Conferences*, *563*, 01002. https://doi.org/10.1051/e3sconf/202456301002

Evans, L., Hennige, S. J., Willoughby, N., Adeloye, A. J., Skroblin, M., & Gutierrez, T. (2017). Effect of organic carbon enrichment on the treatment efficiency of primary settled wastewater by Chlorella vulgaris. *Algal Research*, *24*, 368–377. https://doi.org/10.1016/j.algal.2017.04.011

Fernández, F. G. A., Gómez-Serrano, C., & Acién, F. G. (2021). The role of microalgae in wastewater treatment: A critical review. *Journal of Applied Phycology*, *33*(1), 289–303.

Galasso, C., Gentile, A., Orefice, I., Ianora, A., Bruno, A., Noonan, D. M., Sansone, C., Albini, A., & Brunet, C. (2019). Microalgal derivatives as potential nutraceutical and food supplements for Human health: A focus on cancer prevention and interception. *Nutrients*, *11*(6), 1226. https://doi.org/10.3390/nu11061226

Gallego, I., Medic, N., Pedersen, J. S., Ramasamy, P. K., Robbens, J., Vereecke, E., & Romeis, J. (2025). The microalgal sector in Europe: Towards a sustainable bioeconomy. *New Biotechnology*. https://doi.org/10.1016/j.nbt.2025.01.002

Gendy, T. S., & El-Temtamy, S. A. (2013). Commercialization potential aspects of microalgae for biofuel production: An overview. *Egyptian Journal of Petroleum*, *22*(1), 43–51. https://doi.org/10.1016/j.ejpe.2012.07.001

Gilmour, D., & Zimmerman, W. (2020). Microbubble intensification of bioprocessing. *Advances in Microbial Physiology/Advances in Microbial Physiology*, 1–35. https://doi.org/10.1016/bs.ampbs.2020.07.001

Gonçalves, J., Freitas, J., Fernandes, I., & Silva, P. (2023). Microalgae as Biofertilizers: A sustainable way to improve soil fertility and plant growth. *Sustainability*, *15*(16), 12413. https://doi.org/10.3390/su151612413

Gong, M., & Bassi, A. (2016). Carotenoids from microalgae: A review of recent developments. *Biotechnology Advances*, *34*(8), 1396–1412. https://doi.org/10.1016/j.biotechadv.2016.10.005

Goswami, R., Ghosh, T., Debsarkar, A., & Pandey, A. (2023). Multifaceted role of microalgae for municipal wastewater treatment: A futuristic outlook toward wastewater management. *CLEAN – Soil, Air, Water*, *51*(6), 2100286. https://doi.org/10.1002/clen.202100286

Guo, R., Li, H., Wang, Y., Zhao, X., & Zhang, W. (2022). Microalgae-based systems for hydrogen sulfide mitigation in wastewater treatment: Insights into mechanisms and applications. *Journal of Hazardous Materials*, *431*, 128607. https://doi.org/10.1016/j.jhazmat.2022.128607

Gupta, S., Mittal, Y., Panja, R., Prajapati, K. B., & Yadav, A. K. (2021). Conventional wastewater treatment technologies. In Sunil Kumar, Rakesh Kumar, and Ashok Pandey (eds.), *Current Developments in Biotechnology and Bioengineering* (pp. 47–75). Elsevier. https://doi.org/10.1016/b978-0-12-821009-3.00012-9

Gupta, V. K., & Rastogi, A. (2008). Biosorption of lead from aqueous solutions by green algae *Spirogyra* species: Kinetics and equilibrium studies. *Journal of Hazardous Materials*, *152*(1), 407–414. https://doi.org/10.1016/j.jhazmat.2007.07.028

Häder, D. (2018). Ecotoxicological monitoring of wastewater. In Donat P. Häder and Gilmar S. Erzinger (eds.), *Bioassays* (pp. 369–386). Elsevier. https://doi.org/10.1016/b978-0-12-811861-0.00018-8

Haq, H., Välisuo, P., Kumpulainen, L., Tuomi, V., & Niemi, S. (2020). A preliminary assessment of industrial symbiosis in Sodankylä. *Current Research in Environmental Sustainability*, *2*, 100018. https://doi.org/10.1016/j.crsust.2020.100018

He, J., Ding, W., Han, W., Chen, Y., Jin, W., & Zhou, X. (2022). A bacterial strain Citrobacter W4 facilitates the bio-flocculation of wastewater cultured microalgae Chlorella pyrenoidosa. *The Science of the Total Environment*, *806*, 151336. https://doi.org/10.1016/j.scitotenv.2021.151336

Hejna, M., Moscatelli, A., Stroppa, N., Onelli, E., Pilu, S., Baldi, A., & Rossi, L. (2019). Bioaccumulation of heavy metals from wastewater through a Typha latifolia and Thelypteris palustris phytoremediation system. *Chemosphere*, *241*, 125018. https://doi.org/10.1016/j.chemosphere.2019.125018

Hu, H., Li, X., Wu, S., & Yang, C. (2020). Sustainable livestock wastewater treatment via phytoremediation: Current status and future perspectives. *Bioresource Technology, 315*, 123809. https://doi.org/10.1016/j.biortech.2020.123809

Huang, J., Wang, Y., Li, M., Wang, Q., Zhang, N., & Liu, Y. (2024). *Chlorella pyrenoidosa* as a potential bioremediator: Tolerance and molecular responses to cadmium and lead. *Science of the Total Environment, 912*, 168712. https://doi.org/10.1016/j.scitotenv.2023.168712

Jagaba, A. H., Bashir, F. M., Lawal, I. M., Usman, A. K., Yaro, N. S. A., Birniwa, A. H., Hamdoun, H. Y., & Shannan, N. M. (2023). Agricultural wastewater treatment using oil palm waste activated hydrochar for reuse in plant irrigation: Synthesis, characterization, and process optimization. *Agriculture, 13*(8), 1531. https://doi.org/10.3390/agriculture13081531

Jareonsin, S., & Pumas, C. (2021). Advantages of heterotrophic microalgae as a host for phytochemicals production. *Frontiers in Bioengineering and Biotechnology, 9*, 628597. https://doi.org/10.3389/fbioe.2021.628597

Jiang, C., Wang, D., He, C., & Xu, Q. (2022). Emissions of ammonia and volatile organic compounds from wastewater treatment plants: A comprehensive review. *Environmental Pollution, 300*, 118993. https://doi.org/10.1016/j.envpol.2022.118993

Kagan, M. L., Sullivan Jr, D. W., Gad, S. C., & Ballou, C. M. (2014). Safety assessment of EPA-rich polar lipid oil produced from the microalgae Nannochloropsis oculata. *International Journal of Toxicology, 33*(6), 459–474. https://doi.org/10.1177/1091581814553453

Kato, S., & Kansha, Y. (2024). Comprehensive review of industrial wastewater treatment techniques. *Environmental Science and Pollution Research, 31*(39), 51064–51097. https://doi.org/10.1007/s11356-024-34584-0

Kesari, K. K., Soni, R., Jamal, Q. M. S., Tripathi, P., Lal, J. A., Jha, N. K., Siddiqui, M. H., Kumar, P., Tripathi, V., & Ruokolainen, J. (2021). Wastewater treatment and reuse: A review of its applications and health implications. *Water Air & Soil Pollution, 232*(5). https://doi.org/10.1007/s11270-021-05154-8

Khairuddin, N. F. M., Idris, A., & Hock, L. W. (2018). Harvesting Nannochloropsis sp. using PES/MWCNT/LiBr membrane with good antifouling properties. *Separation and Purification Technology, 212*, 1–11. https://doi.org/10.1016/j.seppur.2018.11.013

Khaligh, S. F., & Asoodeh, A. (2022). Recent advances in the bio-application of microalgae-derived biochemical metabolites and development trends of photobioreactor-based culture systems. 3 *Biotech, 12*(10). https://doi.org/10.1007/s13205-022-03327-8

Kharel, H. L., Shrestha, I., Tan, M., & Selvaratnam, T. (2023). Removal of cadmium and lead from synthetic wastewater using *Galdieria sulphuraria. Environments, 10*(10), 174. https://doi.org/10.3390/environments10100174

Koller, M., Muhr, A., & Braunegg, G. (2014). Microalgae as versatile cellular factories for valued products. *Algal Research, 6*, 52–63. https://doi.org/10.1016/j.algal.2014.09.002

Kong, Q., Li, L., Martínez, S. S., & Ruan, R. (2019). Removal of dye pollutants from wastewater by microalgae: A mini-review. *Environmental Science and Pollution Research, 26*(28), 28962–28970.

Kossalbayev, B. D., Kakimova, A. B., Sadvakasova, A. K., Bauenova, M. O., Balouch, H., Zaletova, M., Ahmad, F., Kirbayeva, D. K., Ozgul, S., & Allakhverdiev, S. I. (2024). Strategies for genetic modification of microalgae to improve the production efficiency of liquid biofuel. *International Journal of Hydrogen Energy, 100*, 1301–1314. https://doi.org/10.1016/j.ijhydene.2024.12.306

Kumar, V., Jaiswal, K. K., Verma, M., Vlaskin, M. S., Nanda, M., Chauhan, P. K., Singh, A., & Kim, H. (2021). Algae-based sustainable approach for simultaneous removal of micropollutants, and bacteria from urban wastewater and its real-time reuse for aquaculture. *The Science of the Total Environment, 774*, 145556. https://doi.org/10.1016/j.scitotenv.2021.145556

Kusmayadi, A., Leong, Y. K., Yen, H., Huang, C., & Chang, J. (2021). Microalgae as sustainable food and feed sources for animals and humans – Biotechnological and environmental aspects. *Chemosphere*, *271*, 129800. https://doi.org/10.1016/j.chemosphere.2021.129800

Kyzas, G. Z., & Matis, K. A. (2019). The flotation process can go green. *Processes*, *7*(3), 138. https://doi.org/10.3390/pr7030138

Lan, J., Liu, P., Hu, X., & Zhu, S. (2024). Harmful algal blooms in eutrophic marine environments: Causes, monitoring, and treatment. *Water*, *16*(17), 2525. https://doi.org/10.3390/w16172525

Li, S., Hu, T., Xu, Y., Wang, J., Chu, R., Yin, Z., Mo, F., & Zhu, L. (2020a). A review on flocculation as an efficient method to harvest energy microalgae: Mechanisms, performances, influencing factors and perspectives. *Renewable and Sustainable Energy Reviews*, *131*, 110005. https://doi.org/10.1016/j.rser.2020.110005

Li, M., Huang, J., Zhang, N., & Liu, Y. (2021). Odorous volatile organic compounds in wastewater treatment plants: Sources, emission characteristics, and control strategies. *Journal of Hazardous Materials*, *401*, 123349. https://doi.org/10.1016/j.jhazmat.2020.123349

Li, K., Liu, Q., Fang, F., Luo, R., Lu, Q., Zhou, W., & Chen, P. (2019). Microalgae-based wastewater treatment for nutrients recovery: A review. *Bioresource Technology*, *291*, 121934. https://doi.org/10.1016/j.biortech.2019.121934

Li, G., Lu, Z., Zhang, J., Li, H., Zhou, Y., Zayan, A. M. I., & Huang, Z. (2020b). Life cycle assessment of biofuel production from microalgae cultivated in anaerobic digested wastewater. *International Journal of Agricultural and Biological Engineering*, *13*(1), 241–246. https://doi.org/10.25165/j.ijabe.20201301.4178

Liu, Z., Cui, D., Liu, Y., Wang, H., Yang, L., Chen, H., Qiu, G., Xiong, Z., Shao, P., & Luo, X. (2023). New concept in swine wastewater treatment: Development of a self-sustaining synergetic microalgae–bacteria symbiosis system to achieve environmental sustainability. *Bioresource Technology*, *367*, 128304. https://doi.org/10.1016/j.biortech.2022.128304

Liu, Y., & Haynes, R. J. (2011). Origin, nature, and treatment of effluents from dairy and meat processing factories and the effects of their irrigation on the quality of agricultural soils. *Critical Reviews in Environmental Science and Technology*, *41*(17), 1531–1599. https://doi.org/10.1080/10643381003608359

Liu, P., Li, Z., Che, Z., Hu, X., Ying, M., Ren, H., & Zhang, X. (2021). Prevalence of common enteric viruses in municipal wastewater treatment plants and their health risks arising from wastewater reuse. *Blue-Green Systems*, *3*(1), 95–107. https://doi.org/10.2166/bgs.2021.012

Luo, J., Chen, Y., Fang, F., Zhou, W., & Chen, P. (2021). Microalgal ammonium uptake and its role in wastewater treatment: Mechanisms and applications. *Bioresource Technology*, *337*, 125379. https://doi.org/10.1016/j.biortech.2021.125379

Macarthur, Ellen. (2013). Towards the circular economy. *Journal of Industrial Ecology*, *2*(1), 23–44.

Mahlangu, D., Mphahlele, K., De Paola, F., & Mthombeni, N. H. (2024). Microalgae-mediated biosorption for effective heavy metals removal from wastewater: A review. *Water*, *16*(5), 718. https://doi.org/10.3390/w16050718

Mansour, M. S., Abdel-Shafy, H. I., & Ibrahim, A. M. (2023). Petroleum wastewater: Environmental protection, treatment, and safe reuse: An overview. *Journal of Environmental Management*, *351*, 119827. https://doi.org/10.1016/j.jenvman.2023.119827

Mkpuma, V. O., Moheimani, N. R., & Ennaceri, H. (2021). Microalgal dewatering with focus on filtration and antifouling strategies: A review. *Algal Research*, *61*, 102588. https://doi.org/10.1016/j.algal.2021.102588

Montuori, E., Saggiomo, M., & Lauritano, C. (2023). Microalgae from cold environments and their possible biotechnological applications. *Marine Drugs*, *21*(5), 292. https://doi.org/10.3390/md21050292

Morais, E. G., Cristofoli, N. L., Maia, I. B., Magina, T., Cerqueira, P. R., Teixeira, M. R., Varela, J., Barreira, L., & Gouveia, L. (2021). Microalgal systems for wastewater treatment: Technological trends and challenges towards waste recovery. *Energies*, *14*(23), 8112. https://doi.org/10.3390/en14238112

Mubashar, M., Zulekha, R., Cheng, S., Xu, C., Li, J., & Zhang, X. (2024). Carbon-negative and high-rate nutrient recovery from municipal wastewater using mixotrophic microalgae. *Journal of Environmental Management*, *354*, 120360. https://doi.org/10.1016/j.jenvman.2024.120360

Musah, B. I., Xu, Y., Liang, C., Li, Y., Peng, L., & Zhang, S. (2022). Biosorption of chromium(VI), iron(II), copper(II) and nickel(II) ions onto alkaline-modified *Chlorella vulgaris* and *Spirulina platensis* in binary systems. *Environmental Science and Pollution Research*, *29*(43), 62514–62536. https://doi.org/10.1007/s11356-022-19725-7

Nagarajan, D., Chang, J. S., & Lee, D. J. (2020). Pretreatment of microalgal biomass for efficient biohydrogen production – Recent insights and future perspectives. *Bioresource Technology*. https://doi.org/10.1016/j.biortech.2020.122871

Narala, R. R., Garg, S., Sharma, K. K., Thomas-Hall, S. R., Deme, M., Li, Y., & Schenk, P. M. (2016). Comparison of microalgae cultivation in photobioreactor, open raceway pond, and a Two-Stage hybrid system. *Frontiers in Energy Research*, *4*. https://doi.org/10.3389/fenrg.2016.00029

Nasr, M. (2022). Aquatic pollution and wastewater treatment system. In Vimal Chandra Pandey (ed.), *Algae and Aquatic Macrophytes in Cities* (pp. 23–37). Elsevier. https://doi.org/10.1016/b978-0-12-824270-4.00006-7

Newton, A. R., & Melaram, R. (2023). Harmful algal blooms in agricultural irrigation: Risks, benefits, and management. *Frontiers in Water*, *5*. https://doi.org/10.3389/frwa.2023.1325300

Nguyen, P., Carvalho, G., Reis, M. A., & Oehmen, A. (2020). A review of the biotransformations of priority pharmaceuticals in biological wastewater treatment processes. *Water Research*, *188*, 116446. https://doi.org/10.1016/j.watres.2020.116446

Nishshanka, G. K. S. H., Thevarajah, B., Nimarshana, P., Prajapati, S. K., & Ariyadasa, T. U. (2023). Real-time integration of microalgae-based bioremediation in conventional wastewater treatment plants: Current status and prospects. *Journal of Water Process Engineering*, *56*, 104248. https://doi.org/10.1016/j.jwpe.2023.104248

Novoveská, L., Nielsen, S. L., Eroldoğan, O. T., Haznedaroglu, B. Z., Rinkevich, B., Fazi, S., Robbens, J., Vasquez, M., & Einarsson, H. (2023). Overview and challenges of large-scale cultivation of photosynthetic microalgae and cyanobacteria. *Marine Drugs*, *21*(8), 445. https://doi.org/10.3390/md21080445

Oakley, S. (2019). Preliminary treatment and primary sedimentation. In J.B. Rose and B. Jiménez-Cisneros (eds.), *Water and Sanitation for the 21st Century: Health and Microbiological Aspects of Excreta and Wastewater Management* (Global Water Pathogen Project). [J.R. Mihelcic and M.E. Verbyla (eds.), Part 4: Management Of Risk from Excreta and Wastewater - Section: Sanitation System Technologies, Pathogen Reduction in Sewered System Technologies], Michigan State University, UNESCO.

Okeke, E. S., Ejeromedoghene, O., Okoye, C. O., Ezeorba, T. P. C., Nyaruaba, R., Ikechukwu, C. K., Oladipo, A., & Orege, J. I. (2022). Microalgae biorefinery: An integrated route for the sustainable production of high-value-added products. *Energy Conversion and Management X*, *16*, 100323. https://doi.org/10.1016/j.ecmx.2022.100323

Oladimeji, T. E., Oyedemi, M., Emetere, M. E., Agboola, O., Adeoye, J. B., & Odunlami, O. A. (2024). Review on the impact of heavy metals from industrial wastewater effluent and removal technologies. *Heliyon*, *10*(23), e40370. https://doi.org/10.1016/j.heliyon.2024.e40370

Omokaro, G. O., Nafula, Z. S., Iloabuchi, N. E., Chikukula, A. A., Osayogie, O. G., & Nnoli, E. C. (2025). Microalgae as biofactories for sustainable applications: Advancing carbon sequestration, bioenergy, and environmental remediation. *Sustainable Chemistry for Climate Action*, 100098. https://doi.org/10.1016/j.scca.2025.100098

Onyeaka, H., Miri, T., Obileke, K., Hart, A., Anumudu, C., & Al-Sharify, Z. T. (2021). Minimizing carbon footprint via microalgae as a biological capture. *Carbon Capture Science & Technology*, *1*, 100007. https://doi.org/10.1016/j.ccst.2021.100007

Osorio-Reyes, J. G., Valenzuela-Amaro, H. M., Pizaña-Aranda, J. J. P., Ramírez-Gamboa, D., Meléndez-Sánchez, E. R., López-Arellanes, M. E., Castañeda-Antonio, M. D., Coronado-Apodaca, K. G., Araújo, R. G., Sosa-Hernández, J. E., Melchor-Martínez, E. M., Iqbal, H. M. (2023). Microalgae-based biotechnology as alternative biofertilizers for soil enhancement and carbon footprint reduction: Advantages and implications. *Marine Drugs*, *21*(2), 93.

Pacheco, D., Rocha, A. C. S., Garcia, A., Bóia, A., Pereira, L., & Verdelhos, T. (2021). Municipal wastewater: A sustainable source for the green microalgae Chlorella vulgaris biomass production. *Applied Sciences*, *11*(5), 2207. https://doi.org/10.3390/app11052207

Pan, K., Gao, Y., Chen, H., Zhao, Y., Li, Y., & Wang, J. (2024). Using ozone nanobubbles and microalgae to promote the removal of nutrients from aquaculture wastewater: Insights from microbiome changes. *Environmental Research*, *257*, 118186.

Park, J. B. K., Craggs, R. J., & Shilton, A. N. (2011). Wastewater treatment high rate algal ponds for biofuel production. *Bioresource Technology*, *102*(1), 35–42. https://doi.org/10.1016/j.biortech.2010.06.158

Peck, L., & Wentworth, J. (2022). Reducing agricultural pressures on freshwater ecosystems. The Parliamentary Office of Science and Technology. https://doi.org/10.58248/pn661

Poopedi, E., Singh, T., & Gomba, A. (2025). Respiratory and enteric bacterial pathogens in municipal wastewater: A potential risk of infection to workers. *Water*, *17*(2), 268. https://doi.org/10.3390/w17020268

Powell, N., Shilton, A., Pratt, S., & Chisti, Y. (2009). Towards a luxury uptake process via microalgae—Defining the polyphosphate dynamics. *Water Research*, *43*(17), 4207–4213. https://doi.org/10.1016/j.watres.2009.06.011

Pradhan, B., Bhuyan, P. P., Nayak, R., Patra, S., Behera, C. R., Jena, M. R., Panda, S., & Ki, J.-S. (2022). Microalgal phycoremediation: A glimpse into a sustainable environment. *Toxics*, *10*(9), 525. https://doi.org/10.3390/toxics10090525

Puglisi, I., Barone, V., Fragalà, F., Stevanato, P., Baglieri, A., & Vitale, A. (2020). Effect of microalgal extracts from Chlorella vulgaris and Scenedesmus quadricauda on germination of beta vulgaris seeds. *Plants*, *9*(6), 675. https://doi.org/10.3390/plants9060675

Qin, S., Wang, K., Gao, F., Ge, B., Cui, H., & Li, W. (2023). Biotechnologies for bulk production of microalgal biomass: From mass cultivation to dried biomass acquisition. *Biotechnology for Biofuels and Bioproducts*, *16*(1). https://doi.org/10.1186/s13068-023-02382-4

Queiroz, M. I., Vieira, J. G., & Maroneze, M. M. (2020). Morphophysiological, structural, and metabolic aspects of microalgae. In Eduardo Jacob-Lopes, Mariana Manzoni Maroneze, Maria Isabel Queiroz, and Leila Queiroz Zepka (eds.), *Handbook of Microalgae-Based Processes and Products* (pp. 25–48). Academic Press. https://doi.org/10.1016/B978-0-12-818536-0.00002-6

Qv, M., Dai, D., Liu, D., Wu, Q., Tang, C., Li, S., & Zhu, L. (2023). Towards advanced nutrient removal by microalgae-bacteria symbiosis system for wastewater treatment. *Bioresource Technology*, *370*, 128574. https://doi.org/10.1016/j.biortech.2022.128574

Rai, P. K., Lee, S. S., Zhang, M., Tsang, Y. F., & Kim, K. H. (2020). Heavy metals in food crops: Health risks, fate, mechanisms, and management. *Environment International*, *125*, 365–385. https://doi.org/10.1016/j.envint.2019.01.067

Rana, A., Sudhaik, A., Raizada, P., Nguyen, V., Xia, C., Khan, A. A. P., Thakur, S., Nguyen-Tri, P., Nguyen, C. C., Kim, S. Y., Van Le, Q., & Singh, P. (2022). Graphitic carbon nitride based immobilized and non-immobilized floating photocatalysts for environmental remediation. *Chemosphere*, *297*, 134229. https://doi.org/10.1016/j.chemosphere.2022.134229

Ranjit, P., Jhansi, V., & Reddy, K. V. (2021). Conventional wastewater treatment processes. In Naga Raju Maddela, Luz C García Cruzatty, and Sagnik Chakraborty (eds.), *Environmental and Microbial Biotechnology* (pp. 455–479). Springer. https://doi.org/10.1007/978-981-15-8999-7_17

Rao, N., Beyer, V., Henderson, R., Thielemans, W., & Muylaert, K. (2023). Microalgae harvesting using flocculation and dissolved air flotation: Selecting the right vessel for lab-scale experiments. *Bioresource Technology*, *374*, 128786. https://doi.org/10.1016/j.biortech.2023.128786

Rao, A. P., Patel, J., & Pradhan, A. K. (2022). Application of alternative sources of water in agricultural food production — current trends and future prospects. *Current Opinion in Food Science*, *47*, 100877. https://doi.org/10.1016/j.cofs.2022.100877

Raval, N. H., Shah, P. P., Shah, J. N., & Shah, M. B. (2020). Bioremediation of heavy metals using microalgae: Recent advances and mechanisms. *Bioresource Technology*, *303*, 122886. https://doi.org/10.1016/j.biortech.2020.122886

Rawat, I., Ranjith Kumar, R., Mutanda, T., & Bux, F. (2011). Dual role of microalgae: Phycoremediation of domestic wastewater and biomass production for sustainable biofuels. *Applied Energy*, *88*(10), 3411–3424. https://doi.org/10.1016/j.apenergy.2010.11.025

Razzak, S. A., Hossain, M. M., Lucky, R. A., Bassi, A. S., & de Lasa, H. (2013). Integrated CO_2 capture, wastewater treatment and biofuel production by microalgae culturing—A review. *Renewable and Sustainable Energy Reviews*, *27*, 622–653. https://doi.org/10.1016/j.rser.2013.05.063

Ren, S., Shao, C., Zhu, F., Schagerl, M., Hu, X., Sobhi, M., Xu, L., Qian, J., & Huo, S. (2025). Optimization and synergistic enhancement of microalgae productivity in laboratory raceway ponds via co-regulation of automated light-supplemented mixers and electric field system. *Biotechnology for Biofuels and Bioproducts*, *18*(1). https://doi.org/10.1186/s13068-025-02658-x

Renuka, N., Guldhe, A., Prasanna, R., Singh, P., & Bux, F. (2018). Microalgae as multi-functional options in modern agriculture: Current trends, prospects and challenges. *Biotechnology Advances*, *36*(4), 1255–1273. https://doi.org/10.1016/j.biotechadv.2018.04.004

Richardson, J. W., Johnson, M. D., & Outlaw, J. L. (2012). Economic comparison of open pond raceways to photo bio-reactors for profitable production of algae for transportation fuels in the southwest. *Algal Research*, *1*(1), 93–100. https://doi.org/10.1016/j.algal.2012.04.001

Riseh, R. S., Vazvani, M. G., Hassanisaadi, M., & Thakur, V. K. (2023). Agricultural wastes: A practical and potential source for the isolation and preparation of cellulose and application in agriculture and different industries. *Industrial Crops and Products*, *208*, 117904. https://doi.org/10.1016/j.indcrop.2023.117904

Rocha, F., Lucas-Borja, M. E., Pereira, P., & Muñoz-Rojas, M. (2020). Cyanobacteria as a nature-based biotechnological tool for restoring salt-affected soils. *Agronomy*, *10*(9), 1321. https://doi.org/10.3390/agronomy10091321

Rout, P. R., Zhang, T. C., Bhunia, P., & Surampalli, R. Y. (2020). Treatment technologies for emerging contaminants in wastewater treatment plants: A review. *The Science of the Total Environment*, *753*, 141990. https://doi.org/10.1016/j.scitotenv.2020.141990

Santos, B., Freitas, F., Sobral, A. J., & Encarnação, T. (2025). Microalgae and circular economy: Unlocking waste to resource pathways for sustainable development. *International Journal of Sustainable Engineering*, *18*(1). https://doi.org/10.1080/19397038.2025.2501488

Saravanan, A., Deivayanai, V., Kumar, P. S., Rangasamy, G., & Varjani, S. (2022). CO2 biomitigation using genetically modified algae and biofuel production towards a carbon net-zero society. *Bioresource Technology*, *363*, 127982. https://doi.org/10.1016/j.biortech.2022.127982

Sarıtas, S., Kalkan, A. E., Yılmaz, K., Gurdal, S., Göksan, T., Witkowska, A. M., Lombardo, M., Karav, S. (2025). Biological and nutritional applications of microalgae. *Nutrients*, *17*, 93. https://doi.org/10.3390/nu17010093

Sedjati, S., Santosa, G. W., Yudiati, E., Supriyantini, E., Ridlo, A., & Kimberly, F. D. (2019, March). Chlorophyll and carotenoid content of Dunaliella salina at various salinity stress and harvesting time. In *IOP Conference Series: Earth and Environmental Science* (Vol. 246, No. 1, p. 012025). IOP Publishing. https://doi.org/10.1088/1755-1315/246/1/012025

Selvaratnam, T., Pegallapati, A., Montelya, F., Rodriguez, G., Nirmalakhandan, N., & Lammers, P. (2020). Evaluation of microalgae for the treatment of tannery wastewater and associated odor mitigation. *Applied Water Science*, *10*(4), 103.

Sharma, N., Kumar, S., Kumar, R., & Verma, N. (2021). A review on algal–bacterial symbiotic system for effective treatment of wastewater. *Chemosphere*, *271*, 129540. https://doi.org/10.1016/j.chemosphere.2021.129540

Sikosana, M. L., Sikhwivhilu, K., Moutloali, R., & Madyira, D. M. (2019). Municipal wastewater treatment technologies: A review. *Procedia Manufacturing*, *35*, 1018–1024. https://doi.org/10.1016/j.promfg.2019.06.051

Silambarasan, S., Logeswari, P., Sivaramakrishnan, R., Incharoensakdi, A., Cornejo, P., & Kamaraj, B. (2021). Removal of nutrients from domestic wastewater by microalgae coupled to lipid augmentation for biodiesel production and influence of deoiled algal biomass as biofertilizer for *Solanum lycopersicum* cultivation. *Chemosphere*, *268*, 129323. https://doi.org/10.1016/j.chemosphere.2020.129323

Singh, B. J., Chakraborty, A., & Sehgal, R. (2023). A systematic review of industrial wastewater management: Evaluating challenges and enablers. *Journal of Environmental Management*, *348*, 119230. https://doi.org/10.1016/j.jenvman.2023.119230

Singh, V., Ikram, S. F., & Tripathi, B. N. (2023). Exploring the potential of freshwater algal species for biofuel production. *Frontiers in Energy Research*, *11*. https://doi.org/10.3389/fenrg.2023.1271660

Singh, G., & Patidar, S. (2018). Microalgae harvesting techniques: A review. *Journal of Environmental Management*, *217*, 499–508. https://doi.org/10.1016/j.jenvman.2018.04.010

Singh, R., Srivastava, P., & Dhar, D. W. (2023). Utilization of microalgae in wastewater treatment: Recent progress and odor mitigation case studies. *Bioengineered*, *14*(1), 2240655.

Sirohi, R., Joun, J., Choi, H., Gaur, V. K., and Sim, S. J. (2021). Algal glycobiotechnology: Omics approaches for strain improvement. *Microbial Cell Factories*, *20*, 163. https://doi.org/10.1186/s12934-021-01656-6

Solovchenko, A. E., Vasilieva, S. G., Zaitsev, P., Lukyanov, A. A., Skripnikova, E. V., & Antal, T. K. (2022). Approaches to rapid screening of pharmaceutical xenobiotic effects on microalgae via monitoring of photosynthetic apparatus condition. *Journal of Applied Phycology*, *34*(1), 353–361. https://doi.org/10.1007/s10811-021-02660-4

Stirk, W. A., & van Staden, J. (2022). Bioprospecting for bioactive compounds in microalgae: Antimicrobial compounds. *Biotechnology Advances*, *59*, 107977. https://doi.org/10.1016/j.biotechadv.2022.107977

Suazo, K. S., Dobbeleers, T., & Dries, J. (2024). Bacterial community and filamentous population of industrial wastewater treatment plants in Belgium. *Applied Microbiology and Biotechnology*, *108*(1). https://doi.org/10.1007/s00253-023-12822-8

Subashchandrabose, S. R., Ramakrishnan, B., Megharaj, M., Venkateswarlu, K., & Naidu, R. (2011). Consortia of microalgae and bacteria: Potential bioremediators of textile dye wastewater. *Bioresource Technology*, *102*(8), 5238–5243. https://doi.org/10.1016/j.biortech.2011.02.030

Suchithra, M. R., Muniswami, D. M., Sri, M. S., Usha, R., Rasheeq, A. A., Preethi, B. A., & Dineshkumar, R. (2021). Effectiveness of green microalgae as biostimulants and biofertilizer through foliar spray and soil drench method for tomato cultivation. *South African Journal of Botany*, *146*, 740–750. https://doi.org/10.1016/j.sajb.2021.12.022

Sun, X., Xu, Z., Zhang, N., Miao, Y., Zhang, C., Ma, X., Shen, Q., & Zhang, R. (2025). Biofertilizer industry and research developments in China: A mini-review. *Microbial Biotechnology*, *18*(5). https://doi.org/10.1111/1751-7915.70163

Sutherland, D. L., Craggs, R. J., & Shilton, A. N. (2020). Bioremediation of aquaculture wastewater with the microalga *Tetraselmis suecica*: Semi-continuous experiments, simulation and photo-respirometric tests. *Science of the Total Environment*, *738*, 139859. https://doi.org/10.1016/j.scitotenv.2020.139859

Tan, Y. H., Chai, M. K., Na, J. Y., & Wong, L. S. (2023). Microalgal growth and nutrient removal efficiency in non-sterilised primary domestic wastewater. *Sustainability*, *15*(8), 6601. https://doi.org/10.3390/su15086601

Tan, J. S., Lee, S. Y., Chew, K. W., Lam, M. K., Lim, J. W., Ho, S., & Show, P. L. (2020). A review on microalgae cultivation and harvesting, and their biomass extraction processing using ionic liquids. *Bioengineered*, *11*(1), 116–129. https://doi.org/10.1080/21655979.2020.1711626

Thang, H. S., Thao, N. M., Nguyen, V. T., Man, N. Q., & Khieu, D. Q. (2023). The treatment of wastewater from pangasius catfish ponds using peat-derived adsorbent materials. *Materials Research Express*, *10*(12), 125506. https://doi.org/10.1088/2053-1591/ad14c0

Trianni, A., Negri, M., & Cagno, E. (2021). What factors affect the selection of industrial wastewater treatment configuration? *Journal of Environmental Management*, *285*, 112099. https://doi.org/10.1016/j.jenvman.2021.112099

Trilokesh, C., & Uppuluri, K. B. (2021). Biobutanol from lignocellulosic biomass and microalgae: Scope, technology, and economics. In Ramesh C. Ray (ed.), *Elsevier eBooks* (pp. 163–223). Academic Press. https://doi.org/10.1016/b978-0-12-820297-5.00008-6

Udayan, A., Sirohi, R., Sreekumar, N., Sang, B., & Sim, S. J. (2021). Mass cultivation and harvesting of microalgal biomass: Current trends and future perspectives. *Bioresource Technology*, *344*, 126406. https://doi.org/10.1016/j.biortech.2021.126406

Uggetti, E., García, J., Álvarez, J. A., & García-Galán, M. J. (2018). Start-up of a microalgae-based treatment system within the biorefinery concept: From wastewater to bioproducts. *Water Science & Technology*, *78*(1), 114–124. https://doi.org/10.2166/wst.2018.195

Vajargah, M. F., Ramzanipour, M. M., & Mousavi, S. P. (2023). An overview of municipal wastewater and sludge treatment process. *Ecology & Conservation Science Open Access*, *2*(5). https://doi.org/10.19080/ecoa.2023.02.555596

Valdovinos-García, E. M., Petriz-Prieto, M. A., De Los Ángeles Olán-Acosta, M., Barajas-Fernández, J., Guzmán-López, A., & Bravo-Sánchez, M. G. (2021). Production of microalgal biomass in photobioreactors as feedstock for bioenergy and other uses: A techno-economic study of harvesting stage. *Applied Sciences*, *11*(10), 4386. https://doi.org/10.3390/app11104386

Velvizhi, G., Balakumar, K., Shetti, N. P., Ahmad, E., Pant, K. K., & Aminabhavi, T. M. (2021). Integrated biorefinery processes for conversion of lignocellulosic biomass to value added materials: Paving a path towards circular economy. *Bioresource Technology*, *343*, 126151. https://doi.org/10.1016/j.biortech.2021.126151

Vrzala, M., Goňo, M., Goňo, R., Kotulla, M., Wzorek, M., & Leonowicz, Z. (2022). Distributed generation power systems in wastewater management. *Energies*, *15*(17), 6283. https://doi.org/10.3390/en15176283

Walters, C., Steyn, M., Ndlela, L. L., Olaniran, A. O., & Moloi, M. (2025). Phycoremediation of industrial wastewater: Review of algae consortia. *International Journal of Environmental Science and Technology*, *22*(4), 6209–6224. https://doi.org/10.1007/s13762-024-06130-5

Wang, Y., Jiao, Z., Li, W., Zeng, S., Deng, J., Wang, M., & Ren, L. (2022). Superhydrophilic membrane with photo-Fenton self-cleaning property for effective microalgae anti-fouling. *Chinese Chemical Letters*, *34*(8), 108020. https://doi.org/10.1016/j.cclet.2022.108020

Wang, M., Ye, X., Bi, H., & Shen, Z. (2024). Microalgae biofuels: Illuminating the path to a sustainable future amidst challenges and opportunities. *Biotechnology for Biofuels and Bioproducts*, *17*(1). https://doi.org/10.1186/s13068-024-02461-0

Wang, F., Yu, L., Liu, Y., & Zhao, L. (2025). Sources, transformation, and mitigation of volatile sulfur compounds in wastewater systems. *Water Research*, *245*, 120693.

Warjri, S. M., & Syiem, M. B. (2018). Analysis of biosorption parameters, equilibrium isotherms and thermodynamic studies of chromium (VI) uptake by a Nostoc sp. isolated from a coal mining site in Meghalaya, India. *Mine Water and the Environment*, *37*(4), 713–723. https://doi.org/10.1007/s10230-018-0523-3

World Health Organization. (2006). *WHO Guidelines for the Safe Use of Wastewater, excreta and Greywater, Volume II: Wastewater Use in Agriculture (Vol. 2)* [Guidelines]. https://www.who.int/water_sanitation_health/wastewater/guidelines/en

Xia, Y., Zhang, M., Tsang, D. C. W., Geng, N., Lu, D., Zhu, L., Igalavithana, A. D., Dissanayake, P. D., Rinklebe, J., Yang, X., & Ok, Y. S. (2020). Recent advances in control technologies for non-point source pollution with nitrogen and phosphorous from agricultural runoff: Current practices and future prospects. *Applied Biological Chemistry*, *63*(1). https://doi.org/10.1186/s13765-020-0493-6

Xu, G., Li, X., & Wang, J. (2021). Hydrogen sulfide formation and control in sewer systems: A review. *Science of the Total Environment*, *756*, 143775. https://doi.org/10.1016/j.scitotenv.2020.143775

Xu, G., & Zhang, C. (2024). Carbon capture and resource utilization by algal–bacterial consortium in wastewater treatment: A mini-review. *Water*, *16*(15), 2197. https://doi.org/10.3390/w16152197

Yang, Y., Tang, S., & Chen, J. P. (2024). Carbon capture and utilization by algae with high concentration CO2 or bicarbonate as carbon source. *The Science of the Total Environment*, 170325. https://doi.org/10.1016/j.scitotenv.2024.170325

Youssef, M. (2023). Blue economy literature review. *International Journal of Business and Management*, *18*(3), 12. https://doi.org/10.5539/ijbm.v18n3p12

Yuvraj, & Padmanabhan, P. (2017). Technical insight on the requirements for CO2-saturated growth of microalgae in photobioreactors. *3 Biotech*, *7*(2), 119. https://doi.org/10.1007/s13205-017-0778-6

Zakaria, Y., & Michael, K. (2017). An integrated cloud-based wireless sensor network for monitoring industrial wastewater discharged into water sources. *Wireless Sensor Network*, *9*(8), 290–301. https://doi.org/10.4236/wsn.2017.98016

Zeng, X., Danquah, M. K., Chen, X., Chen, G. Q., Lu, Y., & Naqvi, S. R. (2021). Revisiting carbon, nitrogen and phosphorus metabolisms in microalgae for wastewater treatment. *Science of the Total Environment*, *762*, 144590. https://doi.org/10.1016/j.scitotenv.2020.144590

Zeng, Y., Qian, X., Wang, L., & Zhou, J. (2023). Light-driven sulfide oxidation and precipitation by microalgae: Implications for odor management. *Chemosphere*, *330*, 138762. https://doi.org/10.1016/j.chemosphere.2023.138762

Zhang, Z., Xu, M., Fan, Y., Zhang, L., & Wang, H. (2023). Using microalgae to reduce the use of conventional fertilizers in hydroponics and soil-based cultivation. *The Science of the Total Environment*, *912*, 169424. https://doi.org/10.1016/j.scitotenv.2023.169424

Zhang, Y., Zhang, X., Wu, M., & Chen, W. (2024). Characteristics and management of odorous compounds in urban wastewater treatment: Current progress and future perspectives. *Journal of Environmental Management*, *349*, 119540. https://doi.org/10.1016/j.jenvman.2023.119540

Zhou, W., Hu, B., Li, R., Chen, L., & Wang, J. (2020). Oxygenation and pollutant degradation by microalgae in wastewater: Contributions to odor mitigation. *Science of the Total Environment*, *713*, 136633. https://doi.org/10.1016/j.scitotenv.2020.136633

Zhou, L., Li, M., Chen, Y., Zhang, Y., & Wang, L. (2021). Algal–bacterial photobioreactors for agro-industrial wastewater treatment: Case study on ammonia and odor removal. *Bioresource Technology*, *341*, 125798. https://doi.org/10.1016/j.biortech.2021.125798

Zhou, R., Li, C., Wu, X., Luo, Y., Zhang, X., & Hu, Z. (2021). Study on the mechanism of biochar-loaded microalgae *Chlorella* removal of cadmium. *Chemosphere*, *279*, 130499. https://doi.org/10.1016/j.chemosphere.2021.130499

Zhu, C., Ji, Y., Du, X., Kong, F., Chi, Z., & Zhao, Y. (2022). A smart and precise mixing strategy for efficient and cost-effective microalgae production in open ponds. *The Science of the Total Environment*, *852*, 158515. https://doi.org/10.1016/j.scitotenv.2022.158515

15 Microalgae in Carbon Capture and Storage

A Clean and Sustainable Approach for Controlling Environmental Pollution

Zeyuan Wang, Yang Zhou, and Yu Hong

15.1 INTRODUCTION

With the rapid development of industry and human society, CO_2 emissions are significantly higher than their removal rate, leading to a continuous increase in CO_2 concentration in the atmosphere (Li et al., 2025). It is reported that the concentration of CO_2 in the atmosphere has reached 428.1 ppm, nearly 50% higher than pre-industrial levels. The total global CO_2 emissions in 2023 are 377.9 billion t, where China, the United States, and India are the top three countries in terms of emissions (Ma et al., 2025). CO_2 is the largest contributor to the greenhouse effect, accounting for over 68% of the total greenhouse gas emissions in the atmosphere; Therefore, effective measures must be taken for carbon capture and storage (CCS), otherwise it will lead to serious environmental problems such as global warming, melting of polar glaciers, and rising sea levels (Wang et al., 2025).

Although the use of post-combustion separation technology for CCS has been applied and commercialized to some extent, this technology has high energy consumption and cost. Due to the complexity of catalysts and the immaturity of efficient catalyst and reaction medium systems, CCS based on photochemical and electrochemical catalysis has not yet been widely applied (He et al., 2024). Solid adsorption, membrane separation, and other methods have also been used in CCS, followed by compression and storage of CO_2. However, these methods have issues such as long-term leakage, high operating costs, pollutant generation, and low CCS efficiency (Adetunji et al., 2025). Therefore, there is an urgent need for low-cost, low-energy, high-efficiency CCS and environmentally friendly methods.

The use of microalgae for CCS has great application prospects, and this technology has significant and unique advantages, such as high CCS efficiency (Table 15.1), small footprint, high added value (production of lipids, pigments, and proteins, etc.), strong adaptability, and low risk of secondary pollution (Ma et al., 2025). For instance, microalga *Chlorella sorokiniana* NBU-3 efficiently fixed 0.612 g/L/d CO_2,

DOI: 10.1201/9781003611479-15

TABLE 15.1
The CCS Performance of Microalgae in Different Studies

Microalgae	Reactor	CCS Performance	Reference
Chlorella fusca LEB 111	Discontinuous vertical tubular photobioreactors	141.5 mg/L/d[a]	Comitre et al., 2021
Chlorella pyrenoidosa	Six-stage serial algal reactors	92%[b]	Yang et al., 2020
Scenedesmus obliquus	Internal loop photo-bioreactor	0.27 g/L/d[a]	Liu et al., 2020
Chlorella vulgaris	Bubble-column bioreactors	0.52 g/L/d[a]	Wei et al., 2024
Chlorella sorokiniana	Photobioreactor	81.57 mg/L/d[a]	Yang et al., 2024
Chlorella strain (LAMB 31)	Photobioreactor	0.24 g/L/d[a]	Li et al., 2024
Coccomyxa subellipsoidea	Shake flasks	0.74 g/L/d[a]	Liu and Wei, 2023
Micractinium pusillum KMC8	Bubble-column photobioreactor	1.1 g/L/d[a]	Chauhan and Mohanty, 2024
Coccomyxa subellipsoidea	Tubular glass reactor	1065.9 mg/L/d[a]	Li et al., 2023a
Chlorella sorokiniana NBU-3	Round-bottom flasks	0.612 g/L/d[a]	Qin et al., 2023
Spirulina sp. DUT001	Transparent silicate glass bubble-column photobioreactor	0.81 g/L/d[c]	Zhu et al., 2020
Desmodesmus sp.	Photobioreactor	0.21 g/L/d[a]	Premaratne et al., 2021

[a] CO_2 biofixation rate.
[b] CO_2 capture efficiency.
[c] Carbon-capture rate.

as well as accumulating 27.84% lipid and 32.65% protein (Qin et al., 2023). The total CO_2 fixation of *Chlorella sorokiniana* TU5 was 12.7 g/L, and its carbohydrate productivity was as high as 618 mg/L/d, this carbohydrate-rich microalgae biomass produces a maximum H_2 productivity of 312 mL/L/h through dark fermentation (Leong et al., 2025). *Chlorella* sp. L166 not only fixed 28.60% CO_2 but also removed 96.07% total nitrogen (TN), 95.55% total phosphorus (TP), and 78.20% chemical oxygen demand (COD) from soybean processing wastewater (Hu et al., 2020).

The primary objective of this chapter is to systematically elucidate the role of microalgae in CCS as a sustainable solution for mitigating environmental pollution. This chapter first comprehensively summarizes the CCS mechanisms of microalgae and the advanced strategies to enhance CCS performance. Then, this chapter introduced the dual role of microalgae in pollution control and resource reuse, as well as the economic feasibility of microalgal CCS for practical applications. This chapter provides novel insights by synthesizing cutting-edge studies on microalgal CCS, such as using CFD simulation and machine learning to optimize performance.

Additionally, it innovatively frames microalgae as a nexus solution for the water–soil–air system, addressing comprehensive environmental issues. Through the summary and future perspective, this chapter provides certain reference information for advancing microalgal CCS toward industrial implementation, which is conducive to achieving global carbon reduction and carbon neutrality.

15.2 MECHANISMS OF CCS BY MICROALGAE

15.2.1 Carbon Capture

The mechanism diagram of microalgal CCS is listed in Figure 15.1. CO_2 and HCO_3^- enter the cell through the inorganic carbon pumps and the transport proteins (like BicA, BCT1, and SbtA) on the cell membranes of microalgae (Li et al., 2025; Wang et al., 2025). In addition, CO_2 can also enter the cytoplasm of microalgae through extracellular passive diffusion (Adetunji et al., 2025). CO_2 is regenerated through the catalytic conversion process of carbonic anhydrase and then enters the photosynthetic carbon sequestration process (Li et al., 2025). When the concentration of CO_2 in the environment is insufficient, extracellular carbonic anhydrase can also convert HCO_3^- into CO_2, which diffuses into the microalgal cytoplasm and is then absorbed and utilized (Wang et al., 2024a). In addition, many microalgae have developed carbon-concentrating mechanisms, which avoid carbon leakage and maintain high CO_2 concentration around the important enzyme Rubisco in photosynthesis (Tarafdar et al., 2023). At high CO_2 concentrations, the key enzyme carbonic anhydrase in the carbon-concentrating mechanisms is inhibited, and the conversion process between CO_2 and HCO_3^- is weakened. CO_2 is directly transported into microalgal

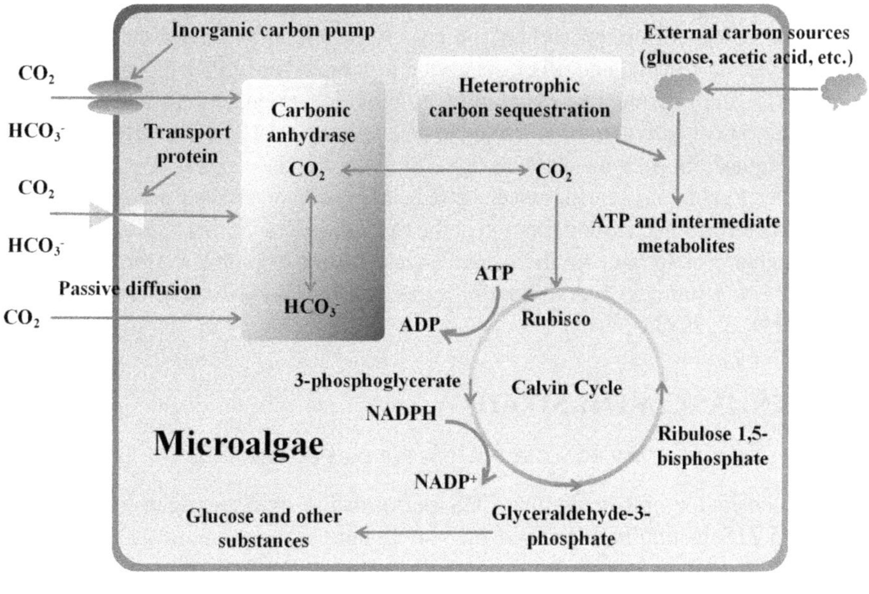

FIGURE 15.1 The mechanism diagram of microalgal CCS.

cells through passive transport under osmotic pressure, simplifying the conversion and transport process of CO_2 and ultimately improving the efficiency of CO_2 supply (Wang et al., 2025).

15.2.2 Carbon Sequestration

Microalgae primarily utilize photosynthesis (including light reaction and the dark reaction) for carbon sequestration, wherein the light reaction is responsible for providing ATP and NADPH for the dark reaction (Miller et al., 2020). The dark reaction consists of carboxylation stage, reduction stage, and regeneration stage: (I) in the carboxylation stage, CO_2 and ribulose-1,5-bisphosphate form 3-phosphoglycerate under the catalysis of ribulose-1,5-bisphosphate carboxylase; (II) in the reduction stage, 3-phosphoglycerate is phosphorylated by ATP to 1,3-diphosphoglycerate under the action of 3-phosphoglycerate kinase and then reduced to glyceraldehyde-3-phosphate by NADPH under the action of glyceraldehyde phosphate dehydrogenase; finally, (III) in the regeneration stage, glyceraldehyde-3-phosphate is acidified by ATP and enzymes to form ribonuclease-1,5-bisphosphate. Then, CO_2 is fixed by the generated ribose 1,5-bisphosphate, converted into products such as starch and glucose, and subsequently converted into amino acids and lipids through a series of metabolic processes (He et al., 2024). Through the above process, microalgae ultimately achieve photosynthetic carbon sequestration of CO_2.

Besides photosynthetic carbon sequestration, microalgae can also absorb external carbon sources (such as glucose and acetic acid) through the pentose phosphate pathway under heterotrophic mode to synthesize and metabolize energy and organic matter; This heterotrophic ability enables microalgae to survive under dark, low light intensity or other adverse environmental conditions and maintain cellular physiological activity and growth (Tarafdar et al., 2023; Wang et al., 2024b). In heterotrophic mode, the absorbed foreign carbon sources are degraded and enter the glycolysis, tricarboxylic acid cycle, and other metabolic processes for further metabolism, producing ATP and intermediate metabolites to provide carbon sources and energy for the physiological activities of microalgae cells (Wang et al., 2024b). Heterotrophic carbon sequestration enables microalgae to carry on the carbon sequestration in wastewater treatment (Tarafdar et al., 2023). Microalgae can also perform autotrophic and heterotrophic carbon fixation at the same time. Under this mode, microalgae have a higher growth rate, but there are also issues such as high cost (requiring more organic carbon sources) and susceptibility to heterotrophic bacterial contamination (Tarafdar et al., 2023).

15.3 ENHANCEMENT STRATEGY

15.3.1 Selecting Microalgae with Superior Performance

The strategies for enhancing the CCS performance of microalgae are listed in Figure 15.2. Selecting high-performance microalgae is the prerequisite for efficient CCS, and there exist significant differences between the CCS abilities of different kinds of microalgae. For example, under the condition of 20% CO_2 aeration, the

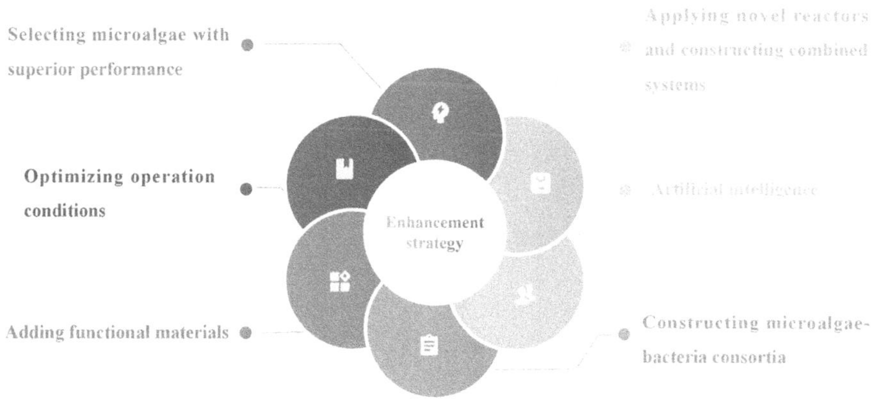

FIGURE 15.2 The strategies for enhancing the CCS performance of microalgae.

CO_2 capture rates of microalgae *Chlorella* sp. AARL G049, *Tetradesmus obliquus* AARL G090, and *Desmodesmus opoliensis* AARL G085 were 0.068 g-CO_2/L/day, 0.072 g-CO_2/L/day, and 0.043 g-CO_2/L/day, respectively (Pekkoh et al., 2024). The CO_2 uptake rates of microalgae *Chlorella sorokinian*a UKM2, *Coelastralla* sp. UKM4, and *Chlorella pyrenoidosa* UKM7 were separately 567.2 mg/L/day, 395.9 mg/L/day, and 485.7 mg/L/day. Genus *Chlorella* was reported to have excellent CO_2 fixation ability (Ding et al., 2020). In another study, the CO_2 fixation rates of *Chlorella vulgaris* and *Chlorella sorokiniana* were separately 0.416 g/L/d and 0.486 g/L/d, which were all higher than that of *Nannochloropsis oculate* (0.330 g/L/d), which once again confirmed that the genus *Chlorella* had good CCS capability (Saranya and Shanthakumar, 2021).

15.3.2 Optimizing Operation Conditions

15.3.2.1 Carbon Source Concentration

Carbon source concentration significantly affects the CCS abilities of microalgae (Figure 15.3). When the total CO_2 increased from 0.12 mol CO_2/L absorbent to 0.49 mol CO_2/L absorbent, the dissolved inorganic carbon fixation rate and dissolved organic carbon fixation rate of microalga *Chlorella* sp. FACHB-8 increased from 7.98 mmol/L/d and 6.21 mmol/L/d to 14.19 mmol/L/d and 12.72 mmol/L/d, respectively (Yan et al., 2024). Similarly, the carbon fixation rate of microalga *Parachlorella kessleri* increased from 31.5 mgC/L/d to 57.7 mgC/L/d when the investigated $NaHCO_3$ concentrations increased from 1 g/L to 10 g/L (Beigbeder et al., 2021). It should be noted that excessive carbon source concentration may actually inhibit the CCS performance of microalgae. Therefore, experiments need to be conducted to explore the optimal range of carbon source concentration, rather than simply increasing the carbon source concentration. For example, the CO_2 capture efficiency of *Chlorella pyrenoidosa* was 82.15% at the condition of 2.0% CO_2, while it decreased to 42.10% when increasing the concentration of CO_2 to 15% (Yang et al., 2020). Similarly, under the CO_2 concentrations of 5%, 15%, and 25%, the maximum CO_2 biofixation

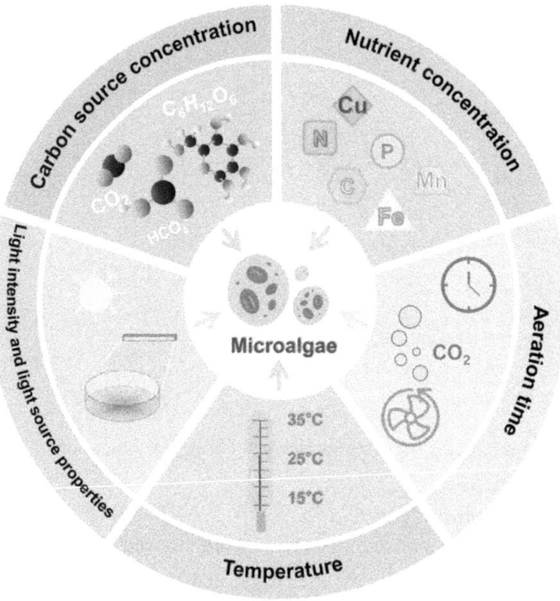

FIGURE 15.3 The key operation conditions affecting microalgal CCS performance.

rates of microalga *Chlorella sorokiniana* NBU-3 were 0.589 g/L/d, 0.612 g/L/d, and 0.445 g/L/d, respectively (Qin et al., 2023).

15.3.2.2 Nutrient Concentration

The CCS performance of microalgae can be optimized by adjusting nutrient concentration (Figure 15.3). Microalgae shown different carbon sequestration ability under the conditions of wastewater with different nutrient characteristics, the CO_2 fixation rate of *Chlorella sorokinana* JD1-1 under the condition of domestic wastewater 2 (DWW2, TN: 57.1 mg/L, TP: 6.8 mg/L) was 0.0646 g CO_2/L/d, while its CO_2 fixation rate increased to 0.1043 g CO_2/L/d under the condition of domestic wastewater 3 (DWW3, TN: 71.3 mg/L, TP: 12.0 mg/L) (Lee et al., 2022). Similarly, *Chlorella* sp. was found to achieve the highest CO_2 fixation rate of 0.204 g/L/d under the wastewater medium with higher phosphate (202.79 mg/L) and nitrate (333.26 mg/L) concentrations compared with other groups (0.112–0.178 g/L/d) (Ramirez et al., 2024). The CO_2 fixation rate of microalga *Tetradesmus obliquus* grown in 75RW + 25RE (75% raw wastewater + 25% recycled effluent) was 0.19 gCO_2/L/d, which was approximately 79% higher than that grown in 100FE (100% wastewater final effluent). This might be due to the low nutrient content in 100FE, resulting in lower biomass productivity and lower CO_2 fixation rate (Hassan et al., 2024). When the initial nitrate concentration increased from 0.075 g/L to 0.375 g/L, the total CO_2 fixation of *Chlorella sorokiniana* TU5 increased from 3.4 to 9.3 g/L (Leong et al., 2025). After adding $CaCl_2$, the CO_2 fixation rate of *Chlorella* sp. HS2 increased by 10.9-fold compared with the group without adding $CaCl_2$ (Yu et al., 2022).

15.3.2.3 Aeration Time

Appropriately increasing the aeration time of CO_2 is beneficial for increasing the solubility of CO_2 in water, thereby increasing the available carbon source and improving the CCS ability of microalgae. For instance, the CO_2 fixation rates of microalgae increased from 14.80 mg/L/d to 34.98 mg/L/d when the aeration time increased from 1 h/d to 9 h/d (Han et al., 2021).

15.3.2.4 Temperature

The growth and CCS activity of microalgae were significantly affected by temperature (Figure 15.3). For example, the carbon fixation rates of microalga *Micractinium pusillum* increased from 43.5 mg/L/d to 99.3 mg/L/d as the temperature increased from 15 °C to 35 °C (Chauhan et al., 2023).

15.3.2.5 Light Intensity and Light Source Properties

There is a close relationship between light intensity and photosynthesis rate, which thereby significantly affects the CCS performance of microalgae (Figure 15.3). In one study, the carbon fixation rates of microalga *Micractinium pusillum* were separately 38.21 mg/L/d, 102 mg/L/d, and 82.8 mg/L/d under the light intensity of 50, 200, and 750 μmol m^{-2} s^{-1}, respectively (Chauhan et al., 2023). As for light source properties, among the different light source (white light, red light, blue light, and the mixed red: blue light), the highest CO_2 fixation rate (0.74 g/L/d) of microalga *Coccomyxa subellipsoidea* was obtained under the condition of red light. This is because red light conditions improved the photosynthetic activity and photoprotection and eventually enhanced the CO_2 fixation performance of microalgae (Liu and Wei, 2023). Another study also showed that microalgae had a higher CO_2 fixation rate under red light conditions compared to white light conditions. Red light promoted carbonic anhydrase activity by regulating COP1/SPA1 complex to maintain pH stability and increased the photosynthetic carbon fixation rate of *Haematococcus pluvialis* (Li et al., 2023b).

15.3.3 Adding Functional Materials

Adding functional materials can efficiently improve the CCS abilities of microalgae. For example, under the role of aggregation-induced emission light-converting films, the carbon sequestration capacity of microalgae *Scenedesmus obliquus*, *Phaeodactylum tricornutum*, and *Chlamydomonas reinhardtii* increased by 17.18%, 38.89%, and 35.71%, respectively. This is because this functional material improved the quality and utilization of light, significantly enhancing the carbon sequestration efficiencies of these microalgae (Song et al., 2024). Oxidative phosphorylation was significantly upregulated after adding graphene oxide quantum dots, which indicated that *Chlorella pyrenoidosa* chloroplasts could get more energy to fix CO_2, and its CO_2 fixation effect increased by 20% (Yang et al., 2022). Ultraviolet (UV) conversion biofilm substrate developed by Wang et al. (2024c) could convert UV light into red light, thereby significantly increasing the carbon sequestration rate of microalga *Chlorella vulgaris* by 381.25%.

When adding 10 mg/L silicon carbide nanoparticles, the carbon fixation rates of microalga *Chlorella sorokiniana* increased from 50.96 mg/L/d to 81.57 mg/L/d. This photocatalytic material captured photons under illumination and transferred electrons to cells for photosynthesis, expanding the ability of microalgae to capture and transport photons, ultimately promoting CO_2 fixation (Yang et al., 2024). Carbonic anhydrase could also enhance the CCS abilities of microalgae. 0.5 mg/L carbonic anhydrase increased the maximum CO_2 fixation rate of microalgae *Chlorella* sp. (FACHB-5) from 1.39 g/L/d to 1.85 g/L/d (Yao et al., 2024). Similarly, under the stress of Hg, the CO_2 fixation rates (170.98–220.92 mg CO_2/L/d) of *Chlorella vulgaris* were strengthened after inoculating indole-3-acetic acid (Kong et al., 2022).

15.3.4 Applying Novel Reactors and Constructing Combined Systems

The application of new reactors is beneficial for the practical engineering application of microalgal CCS. Yang and Xin (2023) developed the novel reactor, S-shaped photobioreactor, which efficiently increased the CO_2 capture rate (0.25–0.43 g/L/d) compared with the traditional photobioreactor. Fu et al. (2021) developed a novel photobioreactor, that was an airfoil-shaped deflector photobioreactor. Compared to a conventional airlift flat plate photobioreactor, this new reactor provided microalgae cells with more opportunities to come into contact with light sources and effectively improved the mass transfer coefficient and decreased mixing time, ultimately increasing the CO_2 fixation rate of microalgae by 10.9%. On the basis of developing novel reactors, the application of CFD simulation can further optimize the CCS performance of microalgae. Yaqoubnejad et al. (2021) developed the novel reactor hexagonal airlift flat plate photobioreactor, and the optimal aeration rate (0.6 vvm) was determined through CFD simulation. The fluid performance of the reactor and the dissolution effect of CO_2 are effectively improved, ultimately promoting the growth of microalgae and the biofixation of CO_2 (from 0.46 g/L/d to 0.85 g/L/d).

Combining microalgae with other treatment systems can achieve good CCS performance. The microalgae-duckweed combined system sequestered CO_2 as highly as 1.65 g/L/day, which surpassed the single microalgae system and the single duckweed system. This is mainly attributed to the synergistic effect of microalgae and duckweed, and the presence of duckweed can promote the growth of microalgae by providing attachment space and a stable microenvironment that supports nutrient cycling (Song et al., 2025). The sequential electrochemical oxidation–algal photobioreactor system constructed by Johnson et al. (2023) also exhibited good CCS performance, and its biomass carbon capture and aqueous carbon capture were 564.24–733.99 mg/L and 245–284 mg/L, respectively. Electrochemical oxidation increased the pH through adjusting nitrates in the culture media, thereby promoting the dissolution of CO_2 in the aqueous phase.

15.3.5 Artificial Intelligence

The use of artificial intelligence can avoid a large number of complex and costly experiments and endow microalgae with outstanding CCS performance. Hossain et al. (2022) combined the crow search algorithm with support vector regression for multiobjective optimization (temperature, light-dark cycles, and nitrogen–phosphorus ratios) to determine the optimal conditions and eventually optimized the CO_2 biofixation rate of *Chlorella vulgaris* to 0.1408 g/L/d. Liu et al. (2024) used a machine learning method, Back Propagation neural network, to optimize the culture conditions of *Chlorella* sp., and the maximum carbon fixation rate of *Chlorella* sp. was 50.86 mg/L/d, which was 2.46-fold that of the control group.

15.3.6 Constructing Microalgae–Bacteria Consortia

Compared with the pure microalgae system, microalgae–bacteria consortia often have better CCS performance. For instance, compared to the pure microalgae *Chlorella* sp. system, *Chlorella* sp.–*Mesorhizobium loti* consortia exhibited better photosynthetic carbon fixation ability (total fixed CO_2 amount: 3.08 g/L, increased by 34.1% compared with the pure microalgae system) (Yao et al., 2025). Similarly, the CO_2 fixation rate of *Chlorella vulgaris* HL02–*Aeromonas* GS-HL01 consortia was 517.68 mg/L/d, 87.10% higher than that of pure *Chlorella vulgaris* HL02 group (276.69 mg/L/d). The large amount of extracellular organic matter secreted by symbiotic bacteria can promote the growth of microalgae through nutrient exchange, biological stimulation, and enhance their fixation of CO_2 (Chen et al., 2024). Optimizing the inoculation ratios of microalgae and bacteria can further enhance the CCS effect. For example, the CO_2 removal efficiency of microalgae–endophytic bacteria consortia increased from 51.82% to 59.33% when the bacterial–algal cell ratios were adjusted from 1:10 to 10:1 (Xu et al., 2020).

15.4 ENVIRONMENTAL POLLUTION CONTROL AND RESOURCE REUSE

15.4.1 Air: Flue Gas CO_2 Mitigation

Microalgae play multiple roles in flue gas CO_2 mitigation, wastewater treatment, and soil carbon sequestration (Figure 15.4). The carbon source cost of microalgae cultivation accounts for a large proportion; Flue gas is an important source of atmospheric CO_2, using CO_2 in flue gas as a carbon source to cultivate microalgae can not only reduce cultivation costs but also fix CO_2, achieving dual economic and environmental benefits (Wang et al., 2025). The excellent CCS performance of microalgae, combined with the reinforcement strategy, has enabled microalgae to play an important role in flue gas CO_2 mitigation. For example, the dual-strain system consisted of *Chlorella* sp. ABC-001 and *Galdieria sulphuraria* 5587.1 efficiently fixed 793 mg/L/d CO_2 in the flue gas, as well as accumulating 113 mg/L/d of high-value substance, that is, lipids (Cho et al., 2024). This indicated that microalgae-based systems can achieve the dual goal of flue gas purification and resource reuse. *Chlorella vulgaris* FACHB-31 exhibited good growth status and CCS performance under the environment of flue gas; its growth rate and average CO_2 fixation rate were 0.40 g/L/day and 0.73 g/L/day, respectively. In addition, it also accumulated abundant carbohydrate (14.0%), demonstrating that *Chlorella vulgaris* FACHB-31 had good carbon assimilation ability (Huang et al., 2023). Similarly, *Chlorella* strain (LAMB 31) grew

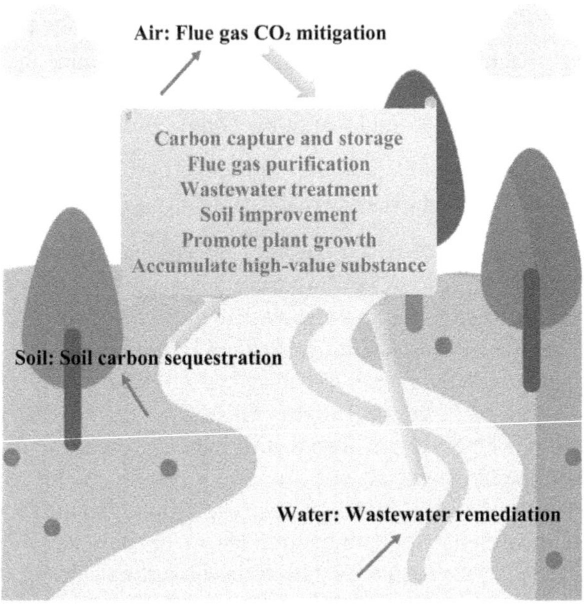

FIGURE 15.4 The multiple roles of microalgae in flue gas CO_2 mitigation, wastewater treatment, and soil carbon sequestration.

well (specific growth rate: 0.15/d) in the flue gas discharged by coal-fired power plants, and its maximum carbon fixation rate and total fatty acid content were 0.24 g/L/d and 9.859 g/100 g DW, respectively (Li et al., 2024).

15.4.2 Water: Wastewater Remediation

Many studies have shown that microalgae can simultaneously treat wastewater and conduct CCS. For instance, the airlift-type photosynthetic microbial fuel cell inoculated with *Chlorella vulgaris* removed 96.3% COD, 95.1% total organic carbon (TOC), 99.1% NH_4^+–N, and 98.9% TP from swine wastewater, as well as fixing 1149 mg/L/d of CO_2 and accumulating 164 mg/L/d of lipids (Li et al., 2021). Similarly, *Chlorella vulgaris* MBFJNU-1 in the 5-m^3 outdoor open raceway ponds removed 82.1% TN, 28.4% TP, and 37.1% COD from real swine wastewater at the pilot scale. In addition, the total fatty acids content, CO_2 fixation rate, and lipid productivity were 21.3%, 63.2 mg/L/d, and 9.1 mg/L/d, respectively (Xie et al., 2022). Besides *Chlorella vulgaris*, *Tetradesmus obliquus* also exhibited good wastewater purification, CCS, and high-value substance accumulation performance. *Tetradesmus obliquus* removed 76.99–100% nitrogen, 97.64–100% phosphorus, and 32.57–100% heavy metals in the municipal wastewater. It also fixed 106.6–275.1 mg/L/d of CO_2 and accumulated 102,812.8–134,232.4 μg/g of total fatty acids (Kong et al., 2021). The above studies indicate that microalgae can achieve the triple objective of wastewater remediation, CCS, and high-value substance accumulation, which has good potential for practical engineering applications.

15.4.3 Soil: Soil Carbon Sequestration

Microalgae can achieve the triple goals of soil improvement, CO_2 reduction, and increased crop yields (Pei and Yu, 2023). Microalgae directly enhance soil carbon sequestration through carbon fixation, extracellular polymeric substance secretion, extracellular enzyme production, and the formation of stable soil aggregates (De Silva et al., 2024). Microalgae also indirectly promote soil carbon sequestration through the following pathway: microalgae promote plant growth, nutrient mineralization, plant hormone synthesis, soil aggregation, and the integration of beneficial soil microorganisms to improve plants to release organic compounds into the soil (De Silva et al., 2024). In addition to promoting soil carbon sequestration, microalgae can also increase the content of nutrients in the soil, enhance soil nitrogen fixation, control erosion, and reduce the use of herbicides (Gurau et al., 2025; Ramanna et al., 2025). Microalgae applied to soil as biofertilizers can effectively improve plant growth status, including germination rate, root development, grain yield, and plant height. (Gurau et al., 2025).

15.5 ECONOMIC FEASIBILITY ANALYSIS

It is necessary to conduct an economic analysis of the CCS performance of microalgae to evaluate their potential for practical applications. Hosseinifard et al. (2025) evaluated that the microalgae system showed a high exergy efficiency of approximately 72.64% and a low cost of approximately 7.17 USD/hour, which had significant advantages such as low-energy consumption, high exergy efficiency, and reduced operational cost compared to the optimized post-combustion carbon capture configuration. Llamas et al. (2021) conducted techno-economic analysis on the process of microalgal CCS through the techno-economic analysis tool from the aspects of land area, biomass productivity, working days, wastewater treatment, CO_2 cost, CO_2 allowances, and raceway reactor cost, etc. The final assessment of the production cost of biomass was 0.41–13.06 €/kg, with the possibility of recovering 0.03–0.1 €/kg of CO_2. In addition, if the emission rights market reached an expected cost of about 50 €/t of CO_2 by 2030, microalgae technology might have the competitive potential (Llamas et al., 2021). Vajdi and Aslani (2023) performed techno-economic analysis of CCS by microalgae-based photobioreactors from capital costs (the cost of equipment, materials, and supplies required to operate the system; installation, engineering, supervision, etc.) and annual operating costs (labor, electricity, consumables, maintenance, administrative, etc.). The result of techno-economic analysis indicated that microalgae-based photobioreactors can reduce 3.9 t CO_2 and produce 2.2 t microalgal biomass (price: 10 $/kg) yearly (Vajdi and Aslani, 2023). Gowd et al. (2025) evaluated that cultivating microalgae with wastewater can reduce the average global warming potential by 164–330% through life cycle assessment. This once again proves that using wastewater to cultivate microalgae for CCS has great potential for large-scale application.

15.6 SUMMARY AND FUTURE PERSPECTIVE

Microalgae-based CCS demonstrated an efficient and sustainable strategy to relieve the greenhouse effect, mitigate environmental pollution, and accumulate high-value substances. This chapter has systematically summarized the mechanisms underlying microalgal CCS, including carbon capture via inorganic carbon pumps, passive diffusion, and carbon-concentrating mechanisms, as well as carbon sequestration through photosynthetic and heterotrophic pathways. Microalgal strain selection, optimization of operation conditions, adding functional materials, artificial intelligence, and other enhancement strategies have demonstrated their efficacy in improving the CCS performance of microalgae. The dual role of microalgae in environmental pollution purification (like flue gas CO_2 mitigation and wastewater remediation) and resource reuse (like biofertilizers and bioenergy production) underscores their potential as a nexus solution for the water–soil–air system. Economic analysis further highlights the feasibility of microalgal CCS.

Despite having significant advantages, some critical challenges must be solved to realize the full potential of microalgal CCS:

1. While *Chlorella* and other kinds of microalgae show good CCS performance, future research should focus on developing new types of microalgae with enhanced CCS, stress tolerance, and high-value substance accumulation. Random mutagenesis, genetic engineering, and metabolic engineering can be applied to obtain novel microalgae that meet the above requirements. In addition, multi-omics approaches (like transcriptomics and metabolomics, etc.) can be used to further explore the reinforcement mechanisms of novel high-performance microalgae. It should be noted that biosafety testing and ecological risk assessment of new strains are required to ensure the safety of microalgae CCS applications.
2. Although many studies have shown that adding functional materials can enhance the CCS performance of microalgae, functional materials may have negative effects on certain physiological indicators of microalgae. Therefore, future research needs to comprehensively consider the comprehensive impact of functional materials on the physiological status of microalgae. In addition, there is a lack of exploration on the long-term impact of functional materials on the ecological environment. It is necessary to strengthen the long-term impact of functional materials on the ecosystem of the application site (such as microbial community structure) to avoid environmental pollution and ecological disorder.
3. At present, research often remains at the laboratory scale. In order to truly achieve the practical engineering application of microalgal CCS, it is necessary to verify the CCS performance of microalgae at a large scale. Translating lab-scale successes to industrial-scale operations requires overcoming barriers such as large-scale reactor design, land use calculation, and energy consumption evaluation. Machine learning, CFD simulation, and economic feasibility analysis may play important roles in reactor design

(improve mass transfer performance, light contact effect, nutrient utilization efficiency, etc.) and process scalability (assess the required operational scale and cost, as well as the economic value generated).

4. Subsidies for applying microalgae-based CCS, emission trading schemes, lawmaking, publicity expansion, and other supportive policies are critical to incentivize the application of microalgal CCS. In addition, it is necessary to promote the collaboration between governments, industries, and research institutes to accelerate the transformation and implementation of technology, establish standardized protocols, and perform actual operation supervision. Strengthening international cooperation and exchanges, such as sharing high-performance algae species, collaborating on the development of microalgae photobioreactors, and economic and technological assessment models, will be more conducive to achieving global carbon reduction and carbon neutrality.

15.7 CONCLUSION

Utilizing microalgae for CCS is a novel and efficient approach to mitigating environmental pollution and global warming. This chapter has systematically elucidated the CCS mechanisms of microalgae, providing theoretical references for enhancing their performance. Strategies to enhance microalgal CCS performance, such as strain selection, optimization of operational conditions, application of functional materials, and artificial intelligence, have shown great potential for improving efficiency. The multiple roles of microalgae in flue gas CO_2 mitigation, wastewater treatment, and soil carbon sequestration underscore microalgal potential as a nexus solution for the water–soil–air system. Economic analyses reveal that microalgal CCS has good cost benefits. Despite the excellent CCS performance demonstrated by microalgae, there are still challenges, such as the lack of high-performance strain development and biosafety assessment. In the future, strategies such as strengthening international cooperation should be adopted to further promote the application of microalgae CCS. In conclusion, microalgal CCS offers a transformative method for relieving global warming. By addressing current limitations and fostering innovation, microalgal CCS will play a greater role in materializing global carbon reduction and carbon neutrality.

DECLARATION OF COMPETING INTEREST

The authors declare that they have no known competing financial interests or personal relationships that could have appeared to influence the work reported in this paper.

ACKNOWLEDGMENTS

This research project was financially supported by the National Natural Science Foundation of China (No. 52270021).

REFERENCES

Adetunji, A.I., Gumbi, S.T., Erasmus, M., 2025. Harnessing the potential of microalgae in sequestration of CO_2 emissions: Removal mechanisms, optimization strategies, and bioenergy production. *J. Hazard. Mater. Adv.* 18, 100722. https://doi.org/10.1016/j.hazadv.2025.100722

Beigbeder, J.-B., Sanglier, M., de Medeiros Dantas, J.M., Lavoie, J.-M., 2021. CO_2 capture and inorganic carbon assimilation of gaseous fermentation effluents using *Parachlorella kessleri* microalgae. *J. CO2 Util.* 50, 101581. https://doi.org/10.1016/j.jcou.2021.101581

Chauhan, D.S., Mohanty, K., 2024. Exploring microalgal nutrient-light synergy to enhance CO_2 utilization and lipid productivity in sustainable long-term water recycling cultivation. *J. Environ. Manage.* 356, 120631. https://doi.org/10.1016/j.jenvman.2024.120631

Chauhan, D.S., Sahoo, L., Mohanty, K., 2023. Acclimation-driven microalgal cultivation improved temperature and light stress tolerance, CO_2 sequestration and metabolite regulation for bioenergy production. *Bioresour. Technol.* 385, 129386. https://doi.org/10.1016/j.biortech.2023.129386

Chen, C., Wang, Y., Dai, Q., Du, W., Deng, X., Zhao, Y., Duan, Q., Liu, H., 2024. Fixation of high concentration CO_2 using *Chlorella* – Bacteria symbiosis system. *J. CO2 Util.* 83, 102803. https://doi.org/10.1016/j.jcou.2024.102803

Cho, J.M., Oh, Y.-K., Lee, J., Chang, Y.K., Park, W.-K., 2024. Development of dual strain microalgae cultivation system for the direct carbon dioxide utilization of power plant flue gas. *Bioresour. Technol.* 393, 130051. https://doi.org/10.1016/j.biortech.2023.130051

Comitre, A.A., Vaz, B.D.S., Costa, J.A.V., Morais, M.G.D., 2021. Renewal of nanofibers in *Chlorella fusca* microalgae cultivation to increase CO_2 fixation. *Bioresour. Technol.* 321, 124452. https://doi.org/10.1016/j.biortech.2020.124452

De Silva, A.G.S.D., Hashim, Z.K., Solomon, W., Zhao, J.-B., Kovács, G., Kulmány, I.M., Molnár, Z., 2024. Unveiling the role of edaphic microalgae in soil carbon sequestration: Potential for agricultural inoculants in climate change mitigation. *Agriculture* 14 (11), 2065. https://doi.org/10.3390/agriculture14112065

Ding, G.T., Mohd Yasin, N.H., Takriff, M.S., Kamarudin, K.F., Salihon, J., Yaakob, Z., Mohd Hakimi, N.I.N., 2020. Phycoremediation of palm oil mill effluent (POME) and CO_2 fixation by locally isolated microalgae: *Chlorella sorokiniana* UKM2, *Coelastrella* sp. UKM4 and *Chlorella pyrenoidosa* UKM7. *J. Water Process Eng.* 35, 101202. https://doi.org/10.1016/j.jwpe.2020.101202

Fu, J., Huang, Y., Liao, Q., Zhu, X., Xia, A., Zhu, X., Chang, J.-S., 2021. Boosting photo-biochemical conversion and carbon dioxide bio-fixation of *Chlorella vulgaris* in an optimized photobioreactor with airfoil-shaped deflectors. *Bioresour. Technol.* 337, 125355. https://doi.org/10.1016/j.biortech.2021.125355

Gowd, S.C., Barathi, S., Lee, J., Rajendran, K., 2025. Sustainability performance of microalgae as a negative emission technology for wastewater treatment. *J. Water Process Eng.* 71, 107393. https://doi.org/10.1016/j.jwpe.2025.107393

Gurau, S., Imran, M., Ray, R.L., 2025. Algae: A cutting-edge solution for enhancing soil health and accelerating carbon sequestration – A review. *Environ. Technol. Innovation* 37, 103980. https://doi.org/10.1016/j.eti.2024.103980

Han, X., Hu, X., Yin, Q., Li, S., Song, C., 2021. Intensification of brewery wastewater purification integrated with CO_2 fixation via microalgae co-cultivation. *J. Environ. Chem. Eng.* 9, 105710. https://doi.org/10.1016/j.jece.2021.105710

Hassan, H., Ansari, F.A., Rawat, I., Bux, F., 2024. Unlocking the potential of microalgae: Cultivation in algae recycled effluent with domestic wastewater for enhancing biomass, bioenergy production and CO_2 sequestration. *J. Water Process Eng.* 68, 106499. https://doi.org/10.1016/j.jwpe.2024.106499

He, D., Yin, Z., Qin, L., Li, Y., Tian, L., Li, J., Zhu, L., 2024. The application of magical microalgae in carbon sequestration and emission reduction: Removal mechanisms and potential analysis. *Renew. Sustain. Energy Rev.* 197, 114417. https://doi.org/10.1016/j.rser.2024.114417

Hossain, S.M.Z., Sultana, N., Razzak, S.A., Hossain, M.M., 2022. Modeling and multi-objective optimization of microalgae biomass production and CO_2 biofixation using hybrid intelligence approaches. *Renew. Sustain. Energy Rev.* 157, 112016. https://doi.org/10.1016/j.rser.2021.112016

Hosseinifard, F., Ebadollahi, M., Amidpour, M., 2025. Sustainable pathways for CO_2 mitigation: A comparative energy, exergy, and economic analysis of optimized post-combustion capture and microalgae-based sequestration. *Clean. Environ. Syst.* 18, 100304. https://doi.org/10.1016/j.cesys.2025.100304

Hu, X., Song, C., Mu, H., Liu, Z., Kitamura, Y., 2020. Optimization of simultaneous soybean processing wastewater treatment and flue gas CO_2 fixation via *chlorella* sp. L166 cultivation. *J. Environ. Chem. Eng.* 8 (4), 103960. https://doi.org/10.1016/j.jece.2020.103960

Huang, Y., Fu, J., Xia, A., Zhu, X., Zhu, X., Liao, Q., 2023. Step-wise SO_2-feeding strategies for microalgae-based CO_2 fixation from flue gas and bioenergy production. *Chem. Eng. J.* 468, 143646. https://doi.org/10.1016/j.cej.2023.143646

Johnson, I., Krishnan, C., Kumar, M., 2023. A sequential electrochemical oxidation – algal photobioreactor system for the treatment of distillery wastewater. *J. Environ. Chem. Eng.* 11, 110208. https://doi.org/10.1016/j.jece.2023.110208

Kong, W., Kong, J., Lyu, H., Ma, J., Wang, Z., Zhang, Z., Feng, S., Shen, B., 2022. Application of indole-3-acetic acid in microalgae cultivation to improve the feasibility of simultaneously purifying wastewater, fixing CO_2 and producing fatty acids under Hg stress. *J. Clean. Prod.* 358, 132028. https://doi.org/10.1016/j.jclepro.2022.132028

Kong, W., Kong, J., Lyu, H., Ma, J., Yuan, P., Wang, Z., Shen, B., Feng, S., 2021. Integrating municipal wastewater treatment with CO_2 fixation and fatty acid production by cultivating *Tetradesmus obliquus*. *J. Clean. Prod.* 320, 128916. https://doi.org/10.1016/j.jclepro.2021.128916

Lee, J.-C., Joo, J.-H., Chun, B.H., Moon, K., Song, S.H., Kim, Y.J., Lee, S.M., Lee, A.H., 2022. Isolation and screening of indigenous microalgae species for domestic and livestock wastewater treatment, biodiesel production, and carbon sequestration. *J. Environ. Manage.* 318, 115648. https://doi.org/10.1016/j.jenvman.2022.115648

Leong, Y.K., Huang, Y.-J., Chang, J.-S., 2025. Boosting CO_2 capture and conversion to biohydrogen through enhanced microalgal biomass yield. *Int. J. Hydrogen Energy* 139, 998–1007. https://doi.org/10.1016/j.ijhydene.2024.10.250

Li, J., Tang, X., Pan, K., Zhu, B., Wang, Z., Zhao, Y., 2024. Application study for the high CO_2 tolerant *Chlorella* strain by flue gas culture: Evaluation of growth performance and adaptive mechanisms. *Chem. Eng. J.* 479, 147700. https://doi.org/10.1016/j.cej.2023.147700

Li, K.-Y., Zhou, J.-L., Guo, S.-Y., Dou, X.-X., Gu, J.-J., Gao, F., 2025. Advances of microalgae-based enhancement strategies in industrial flue gas treatment: From carbon sequestration to lipid production. *Bioresour. Technol.* 423, 132250. https://doi.org/10.1016/j.biortech.2025.132250

Li, L., Wu, Y., Acheampong, A., Huang, Q., 2023b. Red light promotes autotrophic growth of *Haematococcus pluvialis* with improved carbonic anhydrase activity and CO_2 utilization. *Aquaculture* 571, 739462. https://doi.org/10.1016/j.aquaculture.2023.739462

Li, M., Zhou, M., Tian, X., Tan, C., Gu, T., 2021. Enhanced bioenergy recovery and nutrient removal from swine wastewater using an airlift-type photosynthetic microbial fuel cell. *Energy* 226, 120422. https://doi.org/10.1016/j.energy.2021.120422

Li, Y., Luo, Y., Wang, Z., Zou, S., Meng, X., Liu, X., 2023a. Enhancement of carbon biofixation and lipid accumulation in *Coccomyxa subellipsoidea* with triethanolamine CO_2 absorbent manipulation. *Biochem. Eng. J.* 198, 109018. https://doi.org/10.1016/j.bej.2023.109018

Liu, T.-J., Yu, Q., Wang, Y.-T., Li, J.-G., Wang, X.-M., Kang, L.-L., Ji, R., Wang, F.-P., Zeng, Y.-N., Cai, S., 2024. A novel approach for improving carbon fixation of *Chlorella* sp. by elements in converter steel slag using machine learning. *Alex. Eng. J.* 107, 799–818. https://doi.org/10.1016/j.aej.2024.08.112

Liu, X., Chen, G., Tao, Y., Wang, J., 2020. Application of effluent from WWTP in cultivation of four microalgae for nutrients removal and lipid production under the supply of CO_2. *Renew. Energy* 149, 708–715. https://doi.org/10.1016/j.renene.2019.12.092

Liu, Y., Wei, D., 2023. High-efficient CO_2-to-protein bioconversion by oleaginous *Coccomyxa subellipsoidea* using light quality shift and nitrogen supplementation strategy. *Chem. Eng. J.* 473, 145166. https://doi.org/10.1016/j.cej.2023.145166

Llamas, B., Suárez-Rodríguez, M.C., González-López, C.V., Mora, P., Acién, F.G., 2021. Techno-economic analysis of microalgae related processes for CO_2 bio-fixation. *Algal Res.* 57, 102339. https://doi.org/10.1016/j.algal.2021.102339

Ma, J., Cui, Y., Sun, F., Zhou, R., Liu, F., Cao, K., Zhu, B., Zhang, H., Meng, C., Chen, G., Gao, Z., 2025. Strategies for enhancing microalgal carbon sequestration: A review on strain development, culture system optimization, parameter control, and metabolic engineering. *Bioresour. Technol.* 435, 132868. https://doi.org/10.1016/j.biortech.2025.132868

Miller, T.E., Beneyton, T., Schwander, T., Diehl, C., Girault, M., McLean, R., Chotel, T., Claus, P., Cortina, N.S., Baret, J.-C., Erb, T.J., 2020. Light-powered CO_2 fixation in a chloroplast mimic with natural and synthetic parts. *Science* 368, 649–654. https://doi.org/10.1126/science.aaz6802

Pei, H., Yu, Z., 2023. Microalgae: A revolution for salt-affected soil remediation. *Trends Biotechnol.* 41, 147–149. https://doi.org/10.1016/j.tibtech.2022.08.009

Pekkoh, J., Ruangrit, K., Aurepatipan, N., Duangjana, K., Sensupa, S., Pumas, C., Chaichana, C., Pathom-Aree, W., Kato, Y., Srinuanpan, S., 2024. CO_2 to green fuel converter: Photoautotrophic-cultivation of microalgae and its lipids conversion to biodiesel. *Renew. Energy* 222, 119919. https://doi.org/10.1016/j.renene.2023.119919

Premaratne, M., Liyanaarachchi, V.C., Nishshanka, G.K.S.H., Nimarshana, P.H.V., Ariyadasa, T.U., 2021. Nitrogen-limited cultivation of locally isolated *Desmodesmus* sp. for sequestration of CO_2 from simulated cement flue gas and generation of feedstock for biofuel production. *J. Environ. Chem. Eng.* 9 (4), 105765. https://doi.org/10.1016/j.jece.2021.105765

Qin, Y., Wang, X., Lian, J., Zhao, Q., Jiang, H., 2023. Combination of non-sterilized wastewater purification and high-level CO_2 bio-capture with substantial biomass yield of an indigenous *Chlorella* strain. *Sci. Total Environ.* 873, 162442. https://doi.org/10.1016/j.scitotenv.2023.162442

Ramanna, L., Ansari, F.A., Rawat, I., Bux, F., 2025. Microalgae-driven carbon sequestration and bio-fertiliser: Steps towards a sustainable future. *Chem. Eng. J.* 519, 164892. https://doi.org/10.1016/j.cej.2025.164892

Ramirez, K.D.R., Ñañez, K.B., Gomez, C.L.G., Moreira, Í.T.A., 2024. Efficient PAHs removal and CO_2 fixation by marine microalgae in wastewater using an airlift photobioreactor for biofuel production. *Environ. Res.* 261, 119672. https://doi.org/10.1016/j.envres.2024.119672

Saranya, D., Shanthakumar, S., 2021. Insights into the influence of CO_2 supplement on phycoremediation and lipid accumulation potential of microalgae: An exploration for biodiesel production. *Environ. Technol. Innovation* 23, 101596. https://doi.org/10.1016/j.eti.2021.101596

Song, Y., Hu, Z., Yang, X., An, Y., Lu, Y., 2025. Synergistic microalgae-duckweed systems for enhanced aquaculture wastewater treatment, biomass recovery, and CO_2 sequestration: A novel approach for sustainable resource recovery. *Environ. Res.* 274, 121271. https://doi.org/10.1016/j.envres.2025.121271

Song, Y., Xie, L., Zhang, X., Hu, Z., Li, S., Zhang, P., Yang, X., 2024. Enhancement of biomass, lipid accumulation, and carbon sequestration potential in microalgae via cultivation with Aggregation-Induced emission Light-Conversion films. *Chem. Eng. J.* 483, 149148. https://doi.org/10.1016/j.cej.2024.149148

Tarafdar, A., Sowmya, G., Yogeshwari, K., Rattu, G., Negi, T., Awasthi, M.K., Hoang, A., Sindhu, R., Sirohi, R., 2023. Environmental pollution mitigation through utilization of carbon dioxide by microalgae. *Environ. Pollut.* 328, 121623. https://doi.org/10.1016/j.envpol.2023.121623

Vajdi, S., Aslani, A., 2023. Design and techno-economic analysis of direct CO_2 capturing with integrated photobioreactors as a building façade. *Sustain. Energy Technol. Assess.* 56, 103068. https://doi.org/10.1016/j.seta.2023.103068

Wang, H., Liu, J., Phyu, K., Cao, Y., Xu, X., Liang, J., Chang, C.-C., Zhang, K., Zhi, S., 2024b. Microalgae create a highway for carbon sequestration in livestock wastewater: Carbon sequestration capacity, sequestration mechanisms, influencing factors, and prospects. *Sci. Total Environ.* 956, 177282. https://doi.org/10.1016/j.scitotenv.2024.177282

Wang, J., Ma, S., Huang, Y., Xia, A., Zhu, X., Zhu, X., Liao, Q., 2024c. Realizing UV driven microalgae photosynthesis to bio-fix CO_2 by spectrum conversion. *Chem. Eng. J.* 501, 157543. https://doi.org/10.1016/j.cej.2024.157543

Wang, P., Wu, Y., Yang, L., Zheng, X., Long, M., Chen, Y., 2024a. Inorganic carbon metabolism enhanced hydrogen-driven denitrification: Evaluation of carbon fixation pathways and microbial traits. *Chem. Eng. J.* 497, 154528. https://doi.org/10.1016/j.cej.2024.154528

Wang, R., Wang, X., Zhu, T., 2025. Research progress and application of carbon sequestration in industrial flue gas by microalgae: A review. *J. Environ. Sci.* 152, 14–28. https://doi.org/10.1016/j.jes.2024.04.018

Wei, X., Yu, G., Cao, W., Feng, M., Xu, Y., Jin, M., Zhang, Y., Li, T., Guo, L., 2024. Biomass producing and CO_2 capturing simultaneously by *Chlorella vulgaris*: Effect of CO_2 concentration and aeration rate. *Energy* 306, 132321. https://doi.org/10.1016/j.energy.2024.132321

Xie, D., Ji, X., Zhou, Y., Dai, J., He, Y., Sun, H., Guo, Z., Yang, Y., Zheng, X., Chen, B., 2022. *Chlorella vulgaris* cultivation in pilot-scale to treat real swine wastewater and mitigate carbon dioxide for sustainable biodiesel production by direct enzymatic transesterification. *Bioresour. Technol.* 349, 126886. https://doi.org/10.1016/j.biortech.2022.126886

Xu, M., Xue, Z., Sun, S., Zhao, C., Liu, J., Liu, J., Zhao, Y., 2020. Co-culturing microalgae with endophytic bacteria increases nutrient removal efficiency for biogas purification. *Bioresour. Technol.* 314, 123766. https://doi.org/10.1016/j.biortech.2020.123766

Yan, L., Xue, S., Cha, J., Wen, X., Wang, B., Shi, J., Li, P., Zhang, Y., Xie, T., 2024. Effect of excess CO_2 on semi-continuous microalgae systems: Carbon biofixation. *Environ. Technol. Innovation* 35, 103645. https://doi.org/10.1016/j.eti.2024.103645

Yang, H., Xin, X., 2023. CO_2 capture and lipid production performance of microalgae in the S-shaped photobioreactor under different culture modes. *Enzyme Microb. Technol.* 165, 110194. https://doi.org/10.1016/j.enzmictec.2023.110194

Yang, K., Xia, J.-L., Chen, Y.-X., Zhu, R.-M., Jian, Y., Xiang, C., Nie, Z.-Y., Liu, H.-C., 2024. SiC nanoparticles promote CO_2 fixation and biomass production of *Chlorella sorokiniana* via expanding the abilities of capturing and transmitting photons. *Algal Res.* 84, 103814. https://doi.org/10.1016/j.algal.2024.103814

Yang, L., Su, Q., Si, B., Zhang, Y., Zhang, Y., Yang, H., Zhou, X., 2022. Enhancing bioenergy production with carbon capture of microalgae by ultraviolet spectrum conversion via graphene oxide quantum dots. *Chem. Eng. J.* 429, 132230. https://doi.org/10.1016/j.cej.2021.132230

Yang, Q., Li, H., Wang, D., Zhang, X., Guo, X., Pu, S., Guo, R., Chen, J., 2020. Utilization of chemical wastewater for CO_2 emission reduction: Purified terephthalic acid (PTA) wastewater-mediated culture of microalgae for CO_2 bio-capture. *Appl. Energy* 276, 115502. https://doi.org/10.1016/j.apenergy.2020.115502

Yao, D., Li, Y., Tian, K., Wang, M., Tan, D., Fu, X., Wu, Y., Liu, Y., Wang, H., 2025. Enhanced photosynthetic CO_2 fixation in mutualistic *Chlorella* sp. and *Mesorhizobium loti* co-cultures. *Chem. Eng. J.* 516, 163671. https://doi.org/10.1016/j.cej.2025.163671

Yao, D., Wu, L., Tan, D., Yu, Y., Jiang, Q., Wu, Y., Wang, H., Liu, Y., 2024. Enhancing CO_2 fixation by microalgae in a Photobioreactor: Molecular mechanisms with exogenous carbonic anhydrase. *Bioresour. Technol.* 408, 131176. https://doi.org/10.1016/j.biortech.2024.131176

Yaqoubnejad, P., Rad, H.A., Taghavijeloudar, M., 2021. Development a novel hexagonal airlift flat plate photobioreactor for the improvement of microalgae growth that simultaneously enhance CO_2 bio-fixation and wastewater treatment. *J. Environ. Manage.* 298, 113482. https://doi.org/10.1016/j.jenvman.2021.113482

Yu, B.S., Yang, H.E., Sirohi, R., Sim, S.J., 2022. Novel effective bioprocess for optimal CO_2 fixation via microalgae-based biomineralization under semi-continuous culture. *Bioresour. Technol.* 364, 128063. https://doi.org/10.1016/j.biortech.2022.128063

Zhu, C., Zhai, X., Xi, Y., Wang, J., Kong, F., Zhao, Y., Chi, Z., 2020. Efficient CO_2 capture from the air for high microalgal biomass production by a bicarbonate pool. *J. CO2 Util.* 37, 320–327. https://doi.org/10.1016/j.jcou.2019.12.023

16 From Water to Sky
Integrating Microalgal Biomass into Sustainable Aviation Fuel Development

Gabriela F. Ferreira, Fernanda P. Beatriz, and Ingrid L. Motta

16.1 INTRODUCTION

In the context of the global climate emergency and mounting commitments to reducing greenhouse gas emissions, decarbonizing aviation has emerged as a particularly urgent priority. In 2021, aviation contributed approximately 4% to global warming, and it is expected to keep growing (Gössling et al., 2024; Klöwer et al., 2021). Forecasts from the International Civil Aviation Organization suggest that, without significant intervention, international aviation emissions could triple by 2050 compared to 2015 levels (ICAO, 2025). However, unlike other sectors where decarbonization options are rapidly scaling, aviation's dependence on carbon-intensive fuels makes the transition more difficult. Key challenges include high costs, slow adoption of more efficient technologies, and limited government support for sustainable aviation fuels (SAFs) (Bergero et al., 2023; Gössling & Humpe, 2024). These difficulties are further compounded by growing air traffic, limitations in emission trading schemes, and insufficient climate financing (Hasan et al., 2021). These figures underscore the urgency of pursuing ambitious mitigation strategies that align with both global climate targets and the sector's projected growth trajectory.

Aviation decarbonization can proceed along multiple pathways. Improvements in aerodynamic design and propulsion efficiency, as well as optimizing air traffic management, can soften inefficiencies (Kale et al., 2021; Ranasinghe et al., 2019). However, SAF emerges as the most viable near- and medium-term option for achieving significant emission reductions on a large scale. SAF offers the unique advantage of "drop-in" compatibility with existing aircraft engines, fuel distribution networks, and airport infrastructure. Life-cycle assessments demonstrate that SAF can reduce greenhouse gas emissions by 70–93% relative to fossil kerosene, depending on feedstock and production pathway (Borrill et al., 2024; Chagas et al., 2023). Thus, widespread SAF adoption is increasingly recognized as indispensable for achieving the CORSIA (Carbon Offsetting and Reduction Scheme for International Aviation) framework and national net-zero strategies.

DOI: 10.1201/9781003611479-16

Nevertheless, the widespread adoption of SAF is fundamentally constrained by economics. Current production costs of SAF remain two to seven times higher than conventional jet fuel, with feedstock supply representing the largest cost determinant (Hellmann et al., 2025; Capaz et al., 2021). Conventional options include first-generation crop oils (such as soybean or palm), waste lipids (used cooking oils and animal fats), and lignocellulosic residues (Chagas et al., 2023). Each option presents clear trade-offs. Food-based crops are abundant and relatively easy to process but raise sustainability concerns due to land-use change, biodiversity impacts, and food-fuel competition (Ameen et al., 2024). Waste-derived lipids and residues avoid such controversies but suffer from limited global availability, preventing them from meeting projected demand (European Court of Auditors, 2023). This bottleneck illustrates that relying exclusively on conventional feedstocks is unlikely to deliver the volumes required to decarbonize aviation on a large scale. Therefore, identifying next-generation feedstocks with higher productivity and broader scalability is essential, and this is where microalgal biomass has attracted considerable interest.

Microalgae possess several distinctive characteristics that are promising as a SAF feedstock. Their rapid growth rates allow for productivity levels far exceeding those of terrestrial oil crops. For instance, some strains can produce up to 30 times more oil per acre than soybeans (Ameen et al., 2024). Unlike land crops, microalgae can be cultivated on non-arable land, in seawater, brackish water, or even wastewater, thereby avoiding competition with agriculture and freshwater resources. Additionally, microalgal cultivation systems can be integrated with industrial CO_2 point sources, such as power plants or fermentation facilities, simultaneously enhancing biomass yield and contributing to carbon capture (Ferreira et al., 2019). Beyond lipids for hydroprocessed SAF pathways, microalgae also produce carbohydrates and proteins, which can be valorized through alternative conversion routes or as coproducts in biorefineries, further improving economic feasibility (Sharma et al., 2023; Ahamefule et al., 2023). Despite these advantages, realizing their potential requires overcoming persistent technical, economic, and policy challenges that have thus far prevented large-scale deployment.

The aim of this chapter is to explore how microalgal biomass can be integrated into the SAF landscape as a viable feedstock. We frame the discussion around critical topics that shape feasibility and deployment potential. It begins by considering how the intrinsic characteristics of microalgae align with SAF requirements, highlighting the features that make these microorganisms suitable for aviation applications. The discussion then addresses the main conversion routes that can transform microalgae biomass into fuels compatible with aircraft operations. Subsequent sections examine upgrading strategies, certification considerations, and deployment compatibility, before situating these elements within the broader context of feasibility and scale-up challenges. By organizing the chapter around these complementary themes, the discussion underscores the promise and the practical barriers that remain, providing a comprehensive perspective on the opportunities and constraints of this feedstock. In doing so, this chapter situates microalgal SAF within ongoing debates about aviation decarbonization and the evolving bioeconomy.

16.2 ALIGNING MICROALGAE BIOMASS WITH SAF REQUIREMENTS

16.2.1 Lipid Content and Fatty Acid Profile

Microalgae are renowned for accumulating lipids, often in the form of structural polar lipids (such as phospholipids and glycolipids) and storage nonpolar lipids (primarily triacylglycerols or TAGs), which can serve as feedstock for HEFA (hydroprocessed esters and fatty acids) jet fuel. Many species inherently contain 20–50% of their dry weight as lipids under favorable conditions (Cecchin et al., 2020; Maltsev et al., 2023). Certain oleaginous strains can far exceed this: for example, *Nannochloropsis* can reach 65–70% lipid under nitrogen starvation (Latsos et al., 2020). In general, microalgal lipids are rich in C16 and C18 fatty acids (e.g., palmitic, palmitoleic, and oleic acids) suitable for conversion to jet-range hydrocarbons. Highly unsaturated long-chain fatty acids (e.g., eicosapentaenoic acid or EPA, C20:5) are also produced by some (especially marine) algae such as *Nannochloropsis*, though these may be less ideal for fuels due to lower oxidative stability.

Under nutrient-replete conditions, many fast-growing microalgae allocate less carbon to storage lipids (often <20 wt%) and more to proteins and carbohydrates. Imposing stress, usually nitrogen or phosphorus deprivation, is a well-established strategy to boost lipid biosynthesis at the expense of cell growth (Yaakob et al., 2021). During N-starvation, the cessation of protein synthesis diverts carbon into TAG accumulation, often doubling or tripling lipid content after a few days (Cheng et al., 2017). However, this comes with trade-offs: extremely high lipid yields are achieved at the cost of biomass productivity. The fatty acid profile can also shift under stress, sometimes favoring more saturated lipids (desirable for fuel) as unsaturation can be partially reduced (Khan et al., 2023). Overall, the ability to accumulate large quantities of fatty acids makes lipid-rich microalgae excellent candidates for HEFA-based SAF.

16.2.2 Carbohydrates and ATJ Pathway Potential

Carbohydrates (including storage polysaccharides and including starch or other glucans) typically make up 10–40% of algal biomass dry weight under normal conditions. Some microalgae (especially green algae) can be induced to become carbohydrate-rich under certain stresses. For instance, *Chlorella* species under high-CO_2, high-light, or nutrient limitation can accumulate >50 wt% as starch and soluble sugars. In a study, a two-stage process with CO_2 and light stress yielded *Chlorella* sp. biomass with 77.6% carbohydrates (60.3% as starch) (Wiatrowski et al., 2025). These carbohydrates are a potential feedstock for the ATJ (alcohol-to-jet) route, wherein sugars are released (via hydrolysis) and fermented to ethanol or other alcohols, then catalytically converted to jet fuel. Indeed, carbohydrate-rich algae can serve analogously to terrestrial starch crops for ethanol fermentation.

Nitrogen deprivation often triggers lipid accumulation in many algae, but, in some cases, it preferentially induces carbohydrate buildup instead. For example, *Isochrysis* and certain *Chlorella* strains rapidly accumulate starch when N is removed, before later channeling carbon into lipids. Phosphorus or sulfur limitation can also elevate carbohydrate levels (as an energy storage response). The accumulated starch can be converted to fermentable sugars for bioethanol, which is then catalytically upgraded to SAF. *Chlorella* and *Scenedesmus* have both been studied for such processes, with pretreatment and yeast fermentation achieving yields of ~0.2–0.4 g ethanol per g algae in trials (Lorenzo et al., 2023). While carbohydrates are usually not the dominant fraction of biomass unless specifically cultivated for, the flexibility of directing algal biomass toward ATJ (via carbohydrate extraction) provides a useful complement to lipid-to-jet routes. In practice, an integrated biorefinery might extract lipids for HEFA and ferment the residual algal starch to ethanol (for ATJ), maximizing overall fuel yield (Karatzos et al., 2017).

16.2.3 Protein Content and Biorefinery Considerations

Protein is a major component of microalgae, often comprising 30–50% of dry weight in unstressed cultures. For example, *Chlorella vulgaris* can contain over 60% protein in optimal growth conditions (du Preez et al., 2021), making microalgal biomass comparable to soybean meals or other high-protein feeds. Under nitrogen-replete conditions, many algae invest heavily in protein (enzymes and photosynthetic machinery), whereas nitrogen limitation causes protein content to drop substantially. Protein itself is not directly converted to SAF, but it plays an important role in a biorefinery context: high-protein coproducts can be valued (e.g., as animal/fish feed or fertilizer) to improve overall sustainability and economics.

In SAF production pathways, protein can influence processing. In hydrothermal liquefaction (HTL) of whole algae, proteins contribute nitrogen heteroatoms that must be removed during upgrading (Liska et al., 2004). High-protein algae (e.g., *Spirulina*) might be less ideal for direct fuel conversion but are excellent for producing feed coproducts. One strategy in combined algae processing is to extract lipids for fuel, ferment carbohydrates to fuel, and isolate residual proteins for feed or other uses. This integrated approach addresses the "food vs fuel" concern and leverages the full biomass. For instance, after lipid extraction and sugar fermentation, *Nannochloropsis* protein hydrolysate has been explored as a feed additive (Suarez Ruiz et al., 2020). In summary, while proteins are not converted to jet fuel, their presence is important for biorefining. Tuning cultivation to moderate protein (sufficient for growth but not excessive) is beneficial when the goal is maximizing fuels. Any protein-rich residuals can contribute to the economic viability by serving as nutrient-rich coproducts in animal nutrition or soil amendment, thereby improving the sustainability profile of microalgae-derived SAF.

16.2.4 Environmental and Cultivation Effects on Composition

The biochemical makeup of microalgae is highly dynamic and responsive to cultivation conditions. Nutrient availability is the dominant factor: ample nitrogen favors protein synthesis, whereas N-starvation or limitation triggers a shift to storage compounds (Cecchin et al., 2020; du Preez et al., 2021). This inverse relationship is exploited in two-stage cultivation strategies: first grow algae with sufficient nutrients, then induce stress to accumulate lipids/carbohydrates prior to harvest. Light intensity also plays a role: high light (beyond that needed for growth) can lead to excess photosynthate that is stored as TAGs or starch. For example, exposing *Chlorella* to very high irradiance has been shown to raise its TAG content substantially (up to 50 wt% in some cases). On the other hand, under low light or dense cultures, algae allocate more toward chlorophyll and protein to capture light, often reducing their lipid fraction (Latsos et al., 2020).

Carbon supply (e.g., CO_2 level) can influence whether carbon is directed to lipids or carbohydrates. Elevated CO_2 often boosts overall growth; some studies noted that *Chlorella* produced slightly more carbohydrates under very high CO_2 levels (Wiatrowski et al., 2025). Temperature affects enzyme activities and membrane composition; algae grown in cooler conditions may increase unsaturated fatty acids for membrane fluidity, whereas warmer temperatures can favor saturated or storage lipids up to an optimum. Each species has a temperature optimum for growth that also maximizes balanced composition. Salinity is another factor for marine strains: salinity stress (e.g., cultivating a freshwater strain in moderate salt, or a marine strain in hypersaline conditions) can induce accumulation of compatible solutes (glycerol and sugars). Notably, in *Dunaliella*, high salinity leads to glycerol accumulation (a carbohydrate), not lipids (de Jaeger et al., 2018). However, some halotolerant algae such as *N. oleoabundans* can thrive in saltwater with similar lipid productivity (Roostaei et al., 2018; de Vree et al., 2015).

In practice, optimizing biomass composition involves balancing these parameters. For lipid production, a common recipe is to grow under nutrient-replete conditions and then apply N-starvation plus high light to maximize lipid content. If targeting carbohydrates for fermentation, one might impose nutrient stress earlier or use CO_2 enrichment and specific light regimes to maximize starch rather than lipids. It is noteworthy that extreme stress can decrease biomass productivity, so there is a trade-off between composition and growth rate. Overall, understanding and controlling environmental factors allows tuning microalgal biomass toward the desired fuel precursor profile.

16.2.5 Biomass Productivity and Yields

Any evaluation of microalgae for SAF must consider not only composition but also productivity, typically measured in grams biomass per square meter per day ($g/m^2/day$) in outdoor systems. Microalgae are capable of very high areal productivities compared to terrestrial crops. Under real-world outdoor conditions, 4–20 $g/m^2/day$ is a common range for well-managed open ponds, with a few reports exceeding 30 $g\ m^2/day$ (Long et al., 2022). Closed photobioreactors (PBRs) can achieve even higher short-term productivities (10–48 $g/m^2/day$ in some studies) by providing optimum light and CO_2 delivery. For instance, flat-panel PBRs have demonstrated sustained outputs around 24 $g/m^2/day$ (Klein & Davis, 2023). In experimental setups with specialized biofilm or mixotrophic cultivation, microalgae have reached productivity up to ~40 $g/m^2/day$ or more (Khodadadianzaghmari et al., 2024), but such results are often over short periods or require supplementary organic carbon (mixotrophy).

It is important to note that productivity and composition are interlinked. A strain of *Nannochloropsis*, for instance, might grow fast (15–20 $g/m^2/day$) with moderate lipid content (~20%), or slower (10 $g/m^2/day$) with very high lipid (>50%) under stress. Seasonal and daylight variations also impact yields, and real outdoor farms see dips in winter or in cloudy weather and peaks at midday sun (Long et al., 2022). The annual average productivity for current pilot algae farms is often around 18 $g/m^2/day$ average over the year, as reported by NREL for a 3,900-acre hypothetical farm (Wang et al., 2024). Reaching higher averages (>30 $g/m^2/day$ year-round) remains a research and development target, potentially via strain improvements and clever cultivation (e.g., rotating biofilms and waste heat usage to maintain temperature).

From an SAF perspective, higher productivity directly translates to greater fuel output per land area. As a rough comparison, at 20 $g/m^2/day$ and 50% lipids, one hectare of algae could yield on the order of 40–50 barrels of bio-oil per year (Cecchin et al., 2020)vastly outperforming oil crops. Even moderate yields (10 $g/m^2/day$, 30% lipids) are competitive with palm oil on a per-area basis. Generally, *Chlorella* and *Scenedesmus* can be grown very rapidly but typically with moderate oil content, whereas *Nannochloropsis* or *Neochloris* might grow slightly slower but accumulate more energy-dense lipids. Continued innovation, such as breeding strains with truncated light-harvesting to reduce saturation effects (Latsos et al., 2020) or utilizing wastewater nutrients and CO_2, is improving the net productivity and sustainability of microalgal biomass for fuel.

16.2.5.1 Composition and Fuel Pathway Suitability of Selected Strains

Several microalgae species have emerged as model feedstocks in SAF research. Table 16.1 summarizes several representative strains, highlighting their typical composition traits and the SAF conversion routes for which they are best suited. These examples illustrate the diversity of microalgal biomass and the rationale for pairing certain strains with certain fuel pathways:

- *Nannochloropsis* (marine *Eustigmatophyte*): Noted for high lipid content (especially under stress) and relatively lower carbohydrate. Its fatty acid profile (rich in C16:0, C16:1, C18:1) is ideal for HEFA conversion to jet-range alkanes (Latsos et al., 2020). Moderate protein (~30%) allows coproduct use. With fast growth in saline water and lipid levels reaching 50–60% in starved cells, *Nannochloropsis* is often cited as a top candidate for algal jet fuel.
- *Chlorella* (green alga): A robust genus capable of both high protein (up to 60% in nutrient-rich culture) and high starch or lipid under stress (du Preez et al., 2021). *Chlorella* is versatile: it can serve ATJ pathways when cultivated for starch (e.g., >50% carbs achieved in some strains (Wiatrowski et al., 2025)) or HEFA when lipids are induced (e.g., ~30% lipids under N-limited conditions). It also thrives in wastewater, offering integration with waste remediation. Its high protein in residues can be tapped for animal feed, making it attractive for biorefinery approaches.

TABLE 16.1
Examples of Microalgae Species with Typical Biomass Composition and Recommended SAF Conversion Routes

Microalga (Strain)	Lipid (wt%)	Carbohydrate (% DW)	Protein (wt%)	Notable Traits	Suggested SAF Route(s)
Nannochloropsis sp.	15–25 (up to 50+ under N-starvation)	~30	~30	High C16/C18 fatty acids; moderate PUFA (EPA) content	HEFA (lipid-to-jet); HTL (whole biomass)
Chlorella vulgaris	10–20 (can reach ~30–40 under stress)	10–20 (up to ~50 as starch under stress)	40–60	Very high protein normally; flexible lipid vs starch accumulation	ATJ (starch-to-ethanol-to-jet) + HEFA (if lipid-rich); HTL for whole biomass
Scenedesmus sp.	10–20 (up to ~30–40 with stress)	10–15	30–45	Fast-growing, hardy; moderate oils mostly C16/C18; tolerates wastewater	HEFA (moderate lipid); HTL (robust for whole biomass); some ATJ potential
Neochloris oleoabundans	20–30 (up to >50% under N-starve)	20–30 (high when N-replete)	20–35	Switches between lipid and starch storage; salt-tolerant "dual" feedstock	HEFA (when cultivated under lipid mode); ATJ (when cultivated for carbs); HTL
Botryococcus braunii	30–60 (as hydrocarbon oils)	~10 (as cellulosic sugars)	~5–15	Secretes long-chain hydrocarbons; very slow-growing colonies	Direct upgrading of secreted oil (analogous to HEFA); HTL (for whole biomass)

Ranges given are illustrative of literature values for batch cultures; actual composition can vary with conditions.

Routes: HEFA, lipid to hydrocarbon; ATJ, fermentable carbs to jet; HTL, hydrothermal liquefaction of whole biomass; bold indicates a particularly suitable route for that strain.

- *Scenedesmus* (green alga): Often found in wastewater and outdoor pond studies, *Scenedesmus* grows quickly and typically contains roughly 30–45% protein, 10–20% lipids, and 10–20% carbs in balanced growth (Goh et al., 2022). Under nutrient stress, lipids can rise to ~30–40%. Its fatty acids include C16:0, C18:1, and linolenic (C18:3). *Scenedesmus* is thus a good biodiesel/HEFA feedstock, while its modest starch could contribute to ATJ as well. It's often considered in HTL routes too, since whole-biomass processing can bypass the moderate lipid fraction issue.

- *Neochloris oleoabundans* (green alga): Named for being oil-abundant, this strain is known to accumulate both high lipids and high carbohydrates depending on cultivation mode. Under N-starvation, it can exceed 50% lipid (Usman et al., 2023), whereas under N-replete it can accumulate ~50% starch with high protein. This dual capability means *N. oleoabundans* can produce feedstock for HEFA and ATJ simultaneously. Studies have demonstrated >70% lipid in certain conditions. Its salt tolerance adds flexibility (can be grown in seawater) (Farooqui et al., 2021). The ability to obtain oil and starch makes it a promising candidate for a combined HEFA + ATJ biorefinery concept.
- *Botryococcus braunii* (colonial green alga): Uniquely, *Botryococcus* produces long-chain hydrocarbon oils (C30–C40) that resemble petroleum. It can have an extremely high oil fraction (up to 60–75% of dry weight as hydrocarbons) (Cecchin et al., 2020). Those oils can be directly catalytically cracked or hydrotreated to SAF. However, *Botryococcus* grows very slowly and forms dense colonies, resulting in low productivity (often <1 g/m^2/day). Thus, while composition-wise it is excellent for fuel (even eliminating the need for transesterification), its practicality is limited. It remains of research interest as a "best case" for oil content and a natural model of hydrocarbon biosynthesis.

Other notable mentions include *Schizochytrium* (a heterotrophic protist often lumped with microalgae), which can accumulate ~70–77% docosahexaenoic acid or DHA-rich oil (Cecchin et al., 2020), useful for biodiesel/HEFA, though cultivation requires organic carbon. Cyanobacteria like *Spirulina* have over 60% protein but minimal lipids, making them unsuitable for fuel directly but useful in an integrated biorefinery (e.g., producing protein supplements alongside fuel from other fractions). Overall, each strain presents a different profile, and the choice of species for SAF production will depend on whether the process focuses on lipid (favoring high-oil strains for HEFA), sugar (high-carbohydrate strains for ATJ), or a whole-biomass thermochemical route (where total biocrude yield is key, favoring high overall C content and fast growth).

16.3 SAF-ORIENTED CONVERSION ROUTES FROM MICROALGAL BIOMASS

16.3.1 Hydroprocessed Esters and Fatty Acids (HEFA)

The HEFA route is the most developed pathway for commercialized biokerosene production, certified by ASTM D7566 since 2009 and rated Fuel Readiness Level (FRL) 9 by the Commercial Aviation Alternative Fuels Initiative (CAAFI) since 2011 (Ng et al., 2021).

HEFA may use a variety of feedstocks, including vegetable oils, used or waste cooking oils, animal fats, pyrolysis oils, and algal oils (Neuling & Kaltschmitt, 2018), with the oil composition depending on the feedstock type. For instance, most oils and fats are composed of TAGs and unsaturated fatty acids of different chain lengths and degrees of unsaturation (COCPIT, 2025). Alkali metals and impurities may also be present in all feedstocks, requiring removal prior to the HEFA process to avoid catalytic deactivation and coke formation on catalyst surfaces (COCPIT, 2025). The typical HEFA process, shown in Figure 16.1, involves a series of steps (Ng et al., 2021; BIOFAT, 2025), described in the following.

HEFA starts with oil extraction, generating algal oil containing unsaturated fatty acids and triglycerides, which are processed in a catalytic hydrogenation step to remove double bonds by using Ni or NiMo catalysts, pressures of 0.7–4 bar, and temperatures of 150–220°C. The saturated oil then undergoes thermal hydrolysis, in which each triglyceride molecule is decomposed into one molecule of glycerol and three molecules of free fatty acids (FFA) (Neuling & Kaltschmitt, 2018; BIOFAT, 2025). Glycerol is further converted into propane (BIOFAT, 2025), while FFAs proceed to catalytic conversion steps. As FFAs have high oxygen contents, oxygen is later removed via hydrodeoxygenation (HDO) or decarboxylation (DCO) steps, generating straight-chain paraffins (Neuling & Kaltschmitt, 2018; BIOFAT, 2025), such as propane and linear alkanes in the C8–C20 range (Marangon et al., 2024). HDO is a catalytic process in which excess hydrogen attacks the remaining unsaturations, producing octadecane ($C_{18}H_{38}$) and water as a byproduct. HDO happens at high pressures, high temperatures (300–600°C), and under the presence of catalysts such as sulfided NiMo and CoMo supported on alumina (BIOFAT, 2025). DCO, on the other hand, has lower hydrogen demands and uses lower pressure, producing heptadecane ($C_{17}H_{36}$) and CO_2 as a byproduct (Neuling & Kaltschmitt, 2018). However, HDO is usually preferred over DCO as the latter results in increased coking over catalysts (COCPIT, 2025), carbon loss in the form of coke (Marangon et al., 2024), and CO_2 emissions, as well as losses in energy content (Ng et al., 2021).

As the HDO/DCO products do not match the ASTM D7566 cold flow standards for aviation fuels (COCPIT, 2025) such as flash point, freezing point, and cloud point (BIOFAT, 2025), further processing via hydroisomerization and hydrocracking is usually necessary (Neuling & Kaltschmitt, 2018). In the hydroisomerization reactor, the straight-chain molecules are rearranged into branched-chain paraffins of lower freezing point (BIOFAT, 2025), with the highest yields occurring at temperatures of 280–400°C and pressures of 30–100 bar (Marangon et al., 2024). In the hydrocracking reactor, the resulting hydrocarbons are cracked and saturated to form synthetic paraffinic kerosene (SPK) (iso-alkanes in the C9–C15 range) (Ng et al., 2021; BIOFAT, 2025). Both steps occur in the presence of acidic catalysts, such as Pt, Ni, or other metals supported on activated carbon, Al_2O_3, and ZSM-5 zeolites (Ng et al., 2021). Finally, product separation similar to classic refining processes, including distillation and rectification operations, is conducted to generate SPK (Marangon et al., 2024).

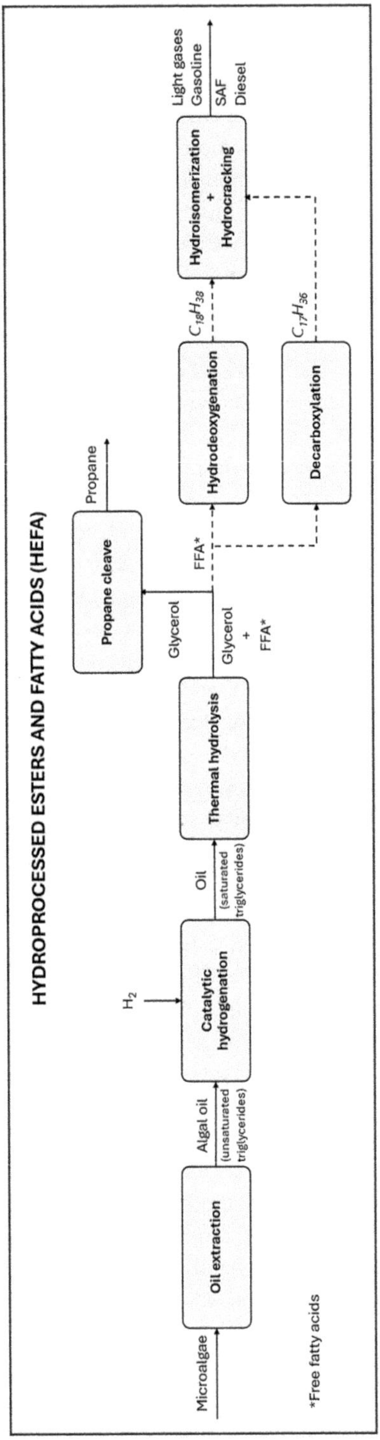

FIGURE 16.1 Schematic flowsheet of HEFA process.

Based on (BIOFAT, 2025).

Process-wise, 1.2 t of oil is processed for each ton of HEFA fuel produced (BIOFAT, 2025). The HEFA yields comprise up to 60% of SPK, with the remaining consisting of naphtha (C5–C10), diesel (C12–C20), and fuel gas/liquefied petroleum gas fractions (C3–C4). Cumulative energy demands range from 1.2 to 1.5 MJ_{input}/MJ_{output} with different oils as feeds, leading to overall SPK efficiencies on an energy basis above 70% (Marangon et al., 2024).

Among the advantages, microalgae are known for their large oil and lipid contents per mass (e.g., Jatropha and Camelina), being suitable for the HEFA routes (BIOFAT, 2025). Also, HEFA may process a variety of oils, increasing process flexibility (Neuling & Kaltschmitt, 2018). HEFA fuels have higher cetane number, higher heating value, and lower oxygen content compared to transesterification, and more than 50% lower auto-ignition delay in comparison to traditional fossil kerosenes. Also, a 50% HEFA fuel blend can reduce fuel consumption by 1% compared to neat Jet A-1 fuel, along with lower particulate and soot emissions (Ng et al., 2021).

Disadvantages of the HEFA route include the high hydrogen consumption (Marangon et al., 2024), typically requiring 300–400 m^3 of H_2/m^3 of vegetable oil to obtain desirable hydrocarbons (Ng et al., 2021; COCPIT, 2025). Recent initiatives exploring the potential of the HEFA technology processing algal oils include the COCPIT project, focused on the microalgae-based fuels for shipping and aviation (Biller & Roth, 2018), and the BIOFAT program, dedicated to the production of biofuels from several algae technologies (Fortier et al., 2014).

16.3.2 Hydrothermal Liquefaction (HTL)

HTL is the thermochemical conversion route most suitable for the processing of wet feedstocks, such as lignocellulosic biomass, sewage sludge, and microalgae (Chong et al., 2023; Behrend et al., 2018). Both the ability to process wet feedstocks and the entire algal biomass are some of the advantages of the HTL of microalgae, avoiding or reducing pretreatment efforts (Kiatkittipong et al., 2022). However, HTL biofuels do not meet aviation fuel standards (not certified by ASTM to produce SAF), being currently in the R&D phase with a technology readiness level (TRL) of 4 (Ng et al., 2021; BIOFAT, 2025).

HTL converts the wet feedstocks, in the presence of a solvent, into a range of products: an organic phase known as biocrude (yields ~20–60 wt%), an aqueous phase (30–50 wt%), solid char and ash (~3 wt%), and a gas phase rich in CO_2 (1–8 wt%) (Zhang, 2016; He et al., 2019). Biocrude is the main HTL product, which is a viscous, heavy-oil-like, and hydrophobic mixture (Zhang, 2016; Fuelgae, 2025) of high carbon content (50–75 wt% of the original feed), intermediate oxygen content (7–20 wt%), and low water concentrations (1–5 wt%) (Behrend et al., 2018), resulting in a product of increased heating value and stability (Neuling & Kaltschmitt, 2018). HTL converts feedstocks of low-to-intermediate solid contents (5–25%) at mild temperatures (200–400°C), high pressures (50–250 bar) (SUNFUSION, 2025; Adnan et al., 2020), and under a wide range of residence times (5–120 min) (Chong et al., 2023). HTL may be conducted in the absence of catalysts, with supercritical water as the main reaction medium (Behrend et al., 2018), adding several benefits due to its simplicity, low costs, and being environmentally friendly (Fuelgae, 2025).

However, cosolvents (ethanol, methanol, isopropanol, and glycerol) and homogeneous catalysts (organic and inorganic acids and organic and alkaline salts) or heterogeneous catalysts (such as metals and metal oxides) have been studied, aiming at reducing aqueous phase formation, increasing biocrude yield, or reducing nitrogen and oxygen contents (Chong et al., 2023). A typical HTL process is shown in Figure 16.2.

Although feedstock drying is not required for HTL, other pretreatment steps may be necessary in the case of microalgae to reduce nitrogen and oxygen contents and avoid transferring N- and O-containing compounds to the biocrude. Hydrothermal pretreatment is the main pretreatment step under development for microalgae HTL, consisting of a first hydrotreatment step under aqueous or acidic media (e.g., using H_2SO_4). Other pretreatment options are also investigated, which are based on cell rupture and ash/protein removal (Chong et al., 2023). Solvents may also be added to adjust the reaction media and produce a pumpable slurry to be pressurized and fed to the HTL reactor (Behrend et al., 2018). In the HTL reactor, the feedstock is converted into a range of HTL products at yields and distributions that depend on the feedstock composition and operational conditions. For example, biomass sources with higher lipid contents result in higher biocrude yields, while feeds with higher protein contents yield biocrudes of higher N content. For each feedstock, specific operational conditions must be determined (Chong et al., 2023). The HTL products are then separated: the aqueous phase is recirculated to the HTL reactor; the char is collected, stored, and directed to specific applications; and the biocrude may be used to produce long-chain hydrocarbons under the desired fuel specifications, either via direct upgrading and hydrotreating, or co-processing with crude oil in oil refineries (Behrend et al., 2018).

The advantages in the microalgae HTL, as mentioned, refer to the possibility of processing wet feedstocks (Kiatkittipong et al., 2022), with supercritical water serving as a simple, low-cost, and environmentally friendly solvent (Fuelgae, 2025). However, several challenges related to feedstock properties and technical aspects prevent the HTL route from achieving higher TRLs.

Concerning the microalgae composition, the diversity of microalgae species directly impacts the operational conditions. Examples include algae with more carbohydrates requiring higher temperatures for conversion; and algae with higher N contents requiring milder HTL conditions and more severe pretreatment methods to reduce N contents (Chong et al., 2023). Other feedstock-related issues include the presence of contaminants, which may lead to fouling, plugging, coke formation, and catalyst deactivation (Neuling & Kaltschmitt, 2018; SUNFUSION, 2025).

Technical challenges comprise the high operating pressures, corrosive nature of biocrude, use of catalysts, and biocrude upgrading. The high operating pressures pose challenges concerning the scale-up of continuous processes and high-pressure feeding (SUNFUSION, 2025). Also, high pressures, along with the corrosive nature of the biocrude, increase the capital costs with the need to develop pressurized vessels that endure corrosive media (Adnan et al., 2020). The use of metallic catalysts also affects the economic production of biocrude via HTL, as the former are expensive and must be recovered via filtration, decantation, or distillation (Chong et al., 2023). On the biocrude upgrading issues, HTL has low yields of middle distillates

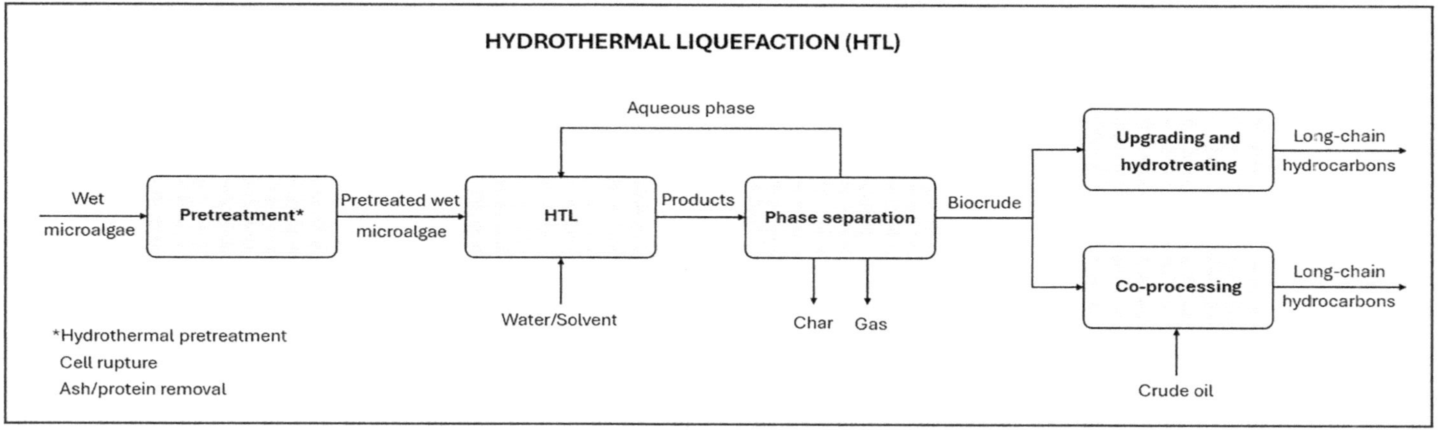

FIGURE 16.2 HTL route for the conversion of wet feedstocks into hydrocarbons.

Based on (Behrend et al., 2018).

over the gasoline carbon range, which may require blending with different feedstocks to increase the share of middle distillates (Behrend et al., 2018). Also, most biocrude upgrading efforts have focused on the production of sustainable diesel and gasoline, with fewer efforts on aviation fuels (Neuling & Kaltschmitt, 2018). Finally, there is still no consensus on which is the best option to obtain HTL-derived kerosene, either via complete refining of biocrude or use as a "drop-in" fuel in refineries (Behrend et al., 2018).

Recent initiatives exploring the potential of the HTL route processing microalgae include several programs, such as the COCPIT project (Biller & Roth, 2018); the FUELGAE project, focused on the use of microalgae for CO_2 capture and production of advanced biofuels (Barroso Soares et al., 2020); and the SUNFUSION program, aiming to convert microalgae and oleaginous yeasts into high energy, sustainable shipping and aviation fuels (Motta et al., 2018).

16.3.3 Gasification–Fischer–Tropsch (GFT) Synthesis

The GFT pathway is the most robust route to convert a wide range of carbon-containing feedstocks into long-chain synthetic hydrocarbons (Ng et al., 2021). The GFT route using coal, natural gas, and biomass sources has been certified by ASTM D7566 since 2009 to produce SPK (Ng et al., 2021; Marangon et al., 2024), with blend limits of 50%. However, GFT using algae as feedstocks is still under laboratory scale (TRL of 4) (Bain & Broer, 2011; Motta et al., 2025). The main steps of the GFT route are shown in Figure 16.3.

Different from HTL, which accepts wet feeds such as microalgae, GFT feedstocks must have moisture contents of 10 wt% or less, requiring a first drying stage. Pretreatment may also include pelletization to ensure uniform particle size and feeding (Dayton et al., 2011), as well as pyrolysis or torrefaction to increase the energy density of the feedstocks (Marangon et al., 2024).

The pretreated algae are then fed to the gasifier, in which the carbohydrate fractions are decomposed at high temperatures (800–1600°C), and sub-stoichiometric oxygen feeds into three main products, namely a gas stream known as synthesis gas or syngas, an undesired liquid stream known as tar, and solid char (Bukur et al., 2016). At industrial scales, autothermal gasifiers (COCPIT, 2025) such as air-steam dual fluidized beds are preferred (Elkelawy et al., 2021). Syngas is the main gasification product (yields up to 60%), composed of H_2, CO, CO_2, CH_4, C_2H_4, C_2H_6, and N-, S-, and Cl- impurities (Bukur et al., 2016).

To ensure a gas stream solely containing H_2 and CO in a ratio of ~2.0 (COCPIT, 2025), syngas undergoes several cleaning and conditioning steps (Michaga et al., 2021). The clean and conditioned syngas is then fed to the Fischer–Tropsch (FT) synthesis reactor, in which H_2 and CO are converted into different-chain-length hydrocarbons, including light gases (C1–C4), naphtha (C5–C10), kerosene (C10–C16), distillate (C14–C20), and waxes (C20+) (BIOFAT, 2025). The FT synthesis involves polymerization reactions (R1 and R2), which follows the Anderson–Schulz–Flory (ASF) statistical distribution. In addition, the water-gas shift reaction (R3) provides additional H_2 (Neves et al., 2020). Since the FT process is highly exothermic, heat removal is necessary to avoid catalyst deactivation and overheating (Li et al.,

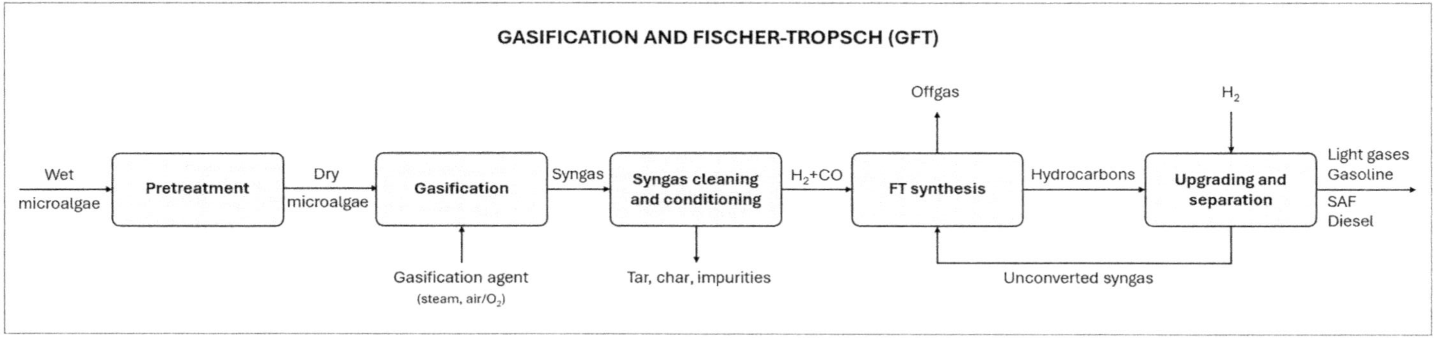

FIGURE 16.3 GFT route for the conversion of wet feedstocks into hydrocarbons.

Based on (Marangon et al., 2024).

2010). To produce more alkanes in the kerosene range, lower FT temperatures are required to increase the value of α in the ASF distribution. Therefore, low-temperature FT (LTFT) at 170–270°C, pressures of 10–40 bar, and using Co catalysts in slurry reactors are preferred (Li et al., 2010; Vela-García et al., 2022). The gases, naphtha, and wax produced via FT are finally upgraded and hydrocracked to increase the kerosene share of the products and separated via distillation to obtain the different hydrocarbon fractions (IEA, 2025a). Typical GFT processes yield 1 ton of FT fuels for every 5–6 tons of biomass (BIOFAT, 2025).

$$(2n+1)H_2 + nCO \rightarrow C_nH_{2n+2} + nH_2O \tag{R1}$$

$$2H_2 + nCO \rightarrow C_nH_{2n} + nH_2O \tag{R2}$$

$$CO + H_2O \rightleftarrows CO_2 + H_2 \tag{R3}$$

Besides the possibility of processing a variety of carbonaceous feedstocks (Ng et al., 2021), the GFT fuels are known for their environmental advantages over fossil fuels. Among the qualities of GFT SAF, one can mention the lower contents of sulfur and aromatics, which result in lower emissions of CO, NO_x, and particulates in jet engines (SUNFUSION, 2025; Redda et al., 2024; Pasa et al., 2022). Despite the advantages, some challenges related to algae composition and syngas quality, complexity in all stages of the GFT route, and biofuel quality persist.

Similar to HTL, the diversity of algae composition greatly affects the syngas quality, differing significantly among various feedstocks. For instance, varying microalgae moisture, heating values, density, and biomass particle size prevent stable feeding rates, while high oxygen and moisture contents result in higher methane content and lower heating value and hydrogen production (Ng et al., 2021). To account for variability in algae composition, specific operational conditions must be defined for different microalgae species.

Syngas cleaning and conditioning are one of the major issues in integrating biomass gasification and FT synthesis, requiring significant development (BIOFAT, 2025). The syngas cleaning stages typically remove a high number of impurities at different temperature levels, increasing capital costs and reducing energy efficiency (Marangon et al., 2024). Special attention is given to tar which must be removed to prevent clogging and catalyst deactivation in the downstream processes (IEA, 2025a). Sulfur and nitrogen must also be completely removed to prevent FT catalyst poisoning (Neuling & Kaltschmitt, 2018). Also, common syngas streams have H_2/CO ratios below 1. Such compositions are unsuitable for FT synthesis and, therefore, require additional conditioning steps. In the FT area, the need for high pressures and expensive catalysts in the synthesis reactor, as well as high H_2 demands in the hydrocracking stages, also increase costs (SUNFUSION, 2025; IEA, 2025a).

According to the database of the Task 39 (Biofuels to decarbonize transport) of the International Energy Agency (IEA), there are no planned or current projects and initiatives to develop microalgae-fed gasification systems aimed at SAF fuels (Mussatto et al., 2022).

16.3.4 Sugar-to-Jet (STJ) and Alcohol-to-Jet (ATJ)

The sugar-to-jet (STJ) route is another option to produce SAF from microalgae (Li et al., 2010). STJ consists of the conversion of starch, carbohydrate, and cellulose-rich fractions into alcohols, followed by the so-called ATJ processes. ATJ is certified by ASTM D7566 for biokerosene production from the conversion of isobutanol since 2016 (blend ratios of 30%, developed by Gevo) and ethanol since 2018 (blend ratios of 50%, developed by Lanzatech) (Ng et al., 2021). An STJ process is typically composed of alcohol synthesis and ATJ areas, as shown in Figure 16.4 and described in the following.

The alcohol synthesis area may be composed of biochemical or thermochemical conversion routes (BIOFAT, 2025; Li et al., 2010). In the biochemical route, microalgae are first hydrolyzed to obtain a solution containing fermentable sugars (Zhu et al., 2023). The sugars are then converted to alcohols (mainly ethanol and n-butanol) by anaerobic fermentation. In the case of ethanol production, *Saccharomyces cerevisiae* is usually selected on an industrial scale due to its high ethanol productivity, tolerance, and ability to ferment a range of sugars. Butanol may be produced via the acetone–butanol–ethanol process using strains of *Clostridium* bacteria. The alcohols are finally separated and purified for the downstream ATJ steps (Honeywell, 2021).

Alternatively, alcohols may be produced thermochemically via microalgae gasification followed by syngas fermentation or catalytic conversion. Syngas fermentation, under development by companies such as Swedish Biofuels and Lanzatech, is an anaerobic process in which acetogenic bacteria ferment CO and H_2 into acetyl-coenzyme A, which is further converted into higher alcohols (ethanol, propanol, and n-butanol) at mild conditions (30–66°C and pH of 5.4–8.5). The catalytic pathway, on the other hand, consists of the catalytic synthesis of methanol (reactions R4 and R5) and higher alcohols (R6) at severe conditions (220–280°C; 50–100 bar) in the presence of Cu- and Zn-based catalysts (IATA, n.d.).

$$CO + 2H_2 \rightarrow CH_3OH \tag{R4}$$

$$CO_2 + 3H_2 \rightarrow CH_3OH + 2H_2O \tag{R5}$$

$$nCO + 2nH_2 \rightarrow C_nH_{2n+1}OH + (n-1)H_2O \tag{R6}$$

The ATJ area is responsible for converting shorter-chain alcohols (C1–C4) into longer-chain hydrocarbons (C8–C16 alkanes) (BIOFAT, 2025; Pasa et al., 2022). The alcohols first undergo a dehydration step (R7), in which an oxygen atom is removed from the alcohol in the presence of acid catalysts (such as ZSM-5 and Amberlyst acidic resins) and at high pressures (<14 bar) and temperatures (170–540°C), producing short-chain alkenes (C_nH_{2n}) and water. The alkenes then undergo oligomerization reactions to produce long-chain alkenes, a process that takes place in a stirred-tank reactor using catalysts such as Amberlyst-35 or Nafion, which favor oligomerization over other reactions. Reaction (R8) shows an example of the oligomerization of butene. The alkenes are later hydrogenated (R9) in the presence of Ni,

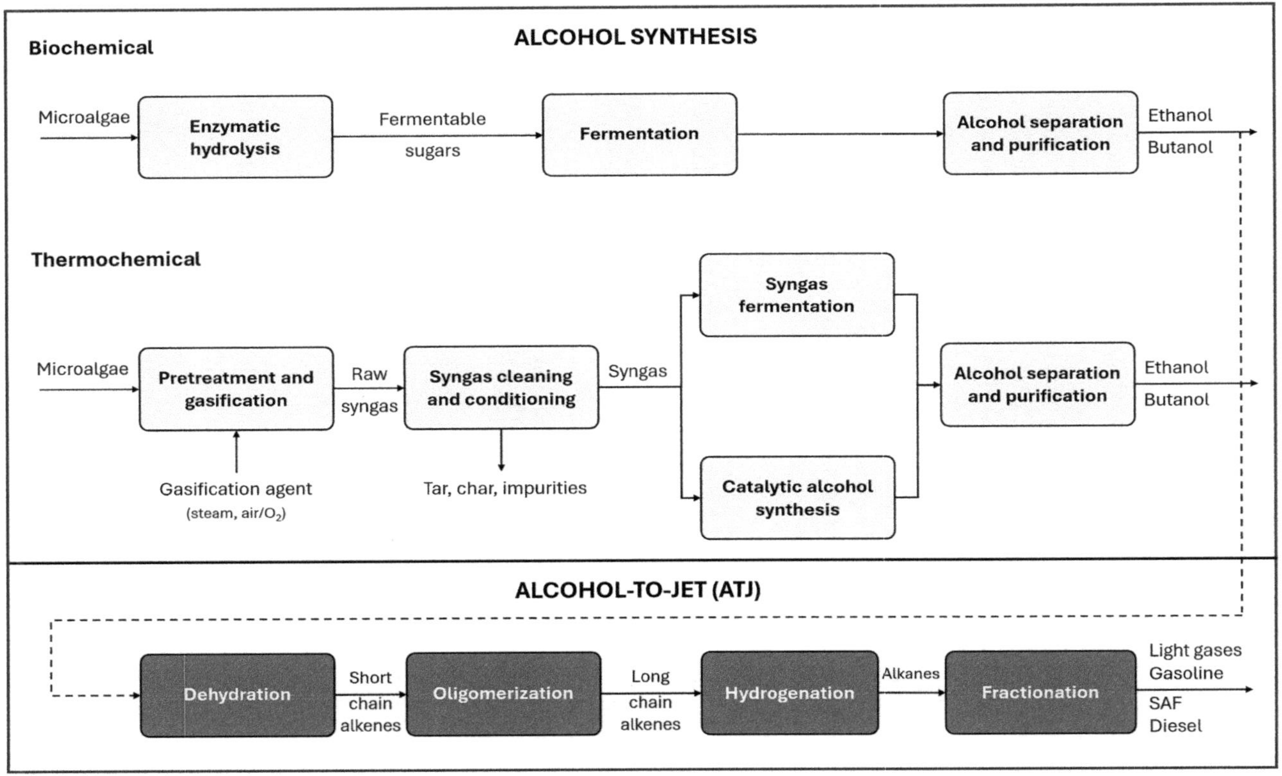

FIGURE 16.4 STJ and ATJ routes to produce hydrocarbons from microalgae.

Based on (BIOFAT, 2025).

Pt, or Pd catalysts at pressures above 20 bar and temperatures of 200–350°C, producing stable SPK (BIOFAT, 2025; IATA, n.d.) The biofuels are finally fractionated in distillation columns to obtain naphtha, SAF, and green diesel (Ng et al., 2021; Zhu et al., 2023), with reported yields of 35%, 48%, and 17%, respectively (Marangon et al., 2024).

$$C_nH_{2n+1}OH \rightarrow C_nH_{2n} + H_2O \quad \text{(R7)}$$

$$nC_4H_8 \rightarrow C_{4n}H_{8n}, \quad n = 2,3,4 \quad \text{(R8)}$$

$$C_nH_{2n} + H_2 \rightarrow C_nH_{2n+2}, \quad n = 8,12,16 \quad \text{(R9)}$$

Challenges in the development of the STJ/ATJ routes relate to feedstock composition, process conditions, and fuel quality. High raw material prices negatively affect the economic feasibility of the process. The composition of ATJ fuels is subject to the composition of the feeds, which may be a concern considering the diversity in microalgae species. Also, the chemical composition of ATJ kerosene differs from its fossil counterpart, influencing the performance of the former in conventional jet engines (Ng et al., 2021). Process-wise, low conversion yields (Ng et al., 2021) and high demands for hydrogen and catalysts in the ATJ area raise the production costs (Neuling & Kaltschmitt, 2018; Marangon et al., 2024). Certification of ATJ routes with different alcohols is still pending (Marangon et al., 2024).

Although each of the ATJ stages has been fully implemented on commercial scales, an entire feedstock-to-fuel process chain is still under pilot and demonstration stages (TRL of 6–7) (Ng et al., 2021). Examples of pilot and demonstration ATJ plants include those of Swedish Biofuels (total production capacity of 10 ton/y) and Gevo (total capacity of ~290 ton/y), respectively (Marangon et al., 2024). Recent initiatives exploring STJ/ATJ with microalgae carbohydrates include the COCPIT (Biller & Roth, 2018) and SUNFUSION (Motta et al., 2018) programs.

16.4 UPGRADING, CERTIFICATION, AND DEPLOYMENT COMPATIBILITY

16.4.1 Upgrading Requirements for Microalgal SAF

Microalgal biomass can be converted to jet-range hydrocarbons through several pathways, as already mentioned, and each route requires upgrading steps to refine the raw bio-oil or intermediates into on-spec jet fuel. Key steps typically include hydrotreating to remove oxygen and saturate double bonds, hydrocracking/isomerization to adjust molecule size and improve cold flow (freezing point), and removal of heteroatoms (primarily nitrogen in algae) via processes such as hydrodenitrogenation (HDN). Finally, fractional distillation isolates the jet-fuel range (C8–C16) hydrocarbons (Advanced Biofuels USA, 2020; Surgenor, 2024). For example, hydroprocessing of heterotrophically grown *Schizochytrium* algae oil (rich in C14–C22 lipids) involved catalytic deoxygenation and isomerization, yielding ~20 wt% jet fuel that satisfied ASTM D7566 specifications after distillation (Lorenzo et al., 2023). This highlights that multistep upgrading can produce a drop-in kerosene from algal oils.

Microalgal feedstocks pose specific upgrading challenges. Whole algae HTL biocrudes contain high levels of nitrogen (from proteins) and oxygen, which require severe hydrotreating. Dedicated HDN steps are needed to remove >99% of nitrogen (as ammonia) from algal biocrudes. In one design, an algal HTL oil's jet fraction contained ~3200 ppm nitrogen, necessitating an HDN stage to reach trace-level N for SAF-quality fuel (Surgenor, 2024). Such heteroatoms can poison catalysts and increase hydrogen demand. Algal oils may also carry metals or sulfur from growth media, which are addressed by pretreatment or sorbents in the refining process. Even lipid extracts (HEFA feedstocks) from microalgae can differ from conventional vegetable oils. For instance, some algae (e.g., *Nannochloropsis*) have high unsaturation or unusual lipids that impact processing. Product variability is another concern: different strains and cultivation conditions yield diverse fatty acid profiles, affecting the cracking and isomerization severity needed. If upgrading is incomplete or lacks sufficient isomerization, the resulting fuel may fail key jet fuel specs such as freezing point and density. In practice, HEFA refining of algae oils closely mirrors that of crop oils, hydrodeoxygenation of triglycerides to n-paraffins, followed by hydrocracking/isomerizing to lighter branched alkanes (Kiatkittipong et al., 2022; Michaga et al., 2021). However, extra catalyst guard beds or acid pretreatment may be used to handle algal contaminants (e.g., chlorophyll). Overall, upgrading microalgal bio-oils demands a systems approach combining chemical refining steps and feedstock pretreatment to ensure a consistent, compliant jet fuel output.

16.4.2 Certification and Fuel Specifications (ASTM D7566/D1655)

While Section 16.3 outlined the main SAF conversion pathways from microalgal biomass, this section focuses on how these products align with aviation fuel standards and the additional upgrading steps required to meet ASTM certification. SAFs must meet stringent specifications to ensure safety and compatibility with existing aircraft and infrastructure. ASTM D7566 defines requirements for aviation turbine fuels containing synthesized hydrocarbons and works alongside ASTM D1655, which governs conventional Jet A and Jet A-1 fuels. Under this framework, a bio-derived fuel component must first be certified via D7566 annexes, which specify approved production pathways and fuel criteria. Once the SAF component is blended with fossil jet fuel (up to the allowed percentage) and the mixture meets all D1655 properties, it is redesignated as conventional Jet A or Jet A-1 fuel, ensuring full "drop-in" compatibility with existing aircraft and infrastructure (Vaillant, 2023). D1655 covers parameters such as density (typically 775–840 kg/m^3), freezing point (max −47°C for Jet A-1), flash point (min 38°C), distillation profile, energy content, and chemical composition (for example, aromatic content up to ~25% volume). Meeting these properties is challenging for algal fuels, which tend to be highly paraffinic. Algal HEFA-SPK contains mostly linear and iso-paraffins and virtually zero aromatics, which improves combustion emissions but can reduce fuel density and affect elastomer swelling in engines. Currently, a minimum aromatics content is not explicitly required by D1655, but, in practice, some aromatics (or cycloparaffins) are needed for seal compatibility. Thus, SAF is used in blends (typically 30–50%) with petroleum jet, so that the final blend retains enough aromatics and meets density limits. Another specification challenge is the freezing point: synthetic paraffinic fuels often have poor cold flow if molecules are mostly linear alkanes. Isomerization is crucial to lower the freezing point, and branching the molecules can ensure that the SAF's freeze point is at or below −47°C (Advanced Biofuels USA, 2020). In summary, algal-derived SAF must be refined and blended to achieve fit-for-purpose properties: cold flow, volatility, stability, and composition on par with fossil jet fuel.

ASTM D7566 has multiple approved production pathways (detailed in its annexes), including those relevant to microalgae feedstocks (AlfaFuels, n.d.):

- HEFA-SPK (hydroprocessed esters and fatty acids): up to 50% blending. This pathway covers oils/fats from plants, algae, animal tallow, and waste greases, converted via hydrogenation into paraffinic kerosene. An algal oil can thus be used in an HEFA process, and the resulting bio-SPK blended into jet fuel up to 50%. HEFA-SPK fuels are very pure (free of sulfur and aromatics) and require blending with conventional fuel for final use.
- FT-SPK and FT-SPK/A (Fischer–Tropsch): up to 50% blending. These cover fuels made from gasification of biomass (or other feedstocks) to syngas and catalytic synthesis. While not specific to algae, microalgal biomass could be gasified to produce FT fuel. FT-SPK/A is unique in that it includes some synthetic aromatics to allow full 50% blending without aromatic shortfall.
- SIP (synthesized iso-paraffins from fermented hydrocarbons): up to 10% blending. This pathway (e.g., Amyris's farnesane from sugar) produces a single branched C15 hydrocarbon. It is not directly an algal route (typically uses sugarcane or corn syrup as feedstock), but it established precedent for fermentation-derived hydrocarbons in jet fuel.
- ATJ-SPK (alcohol-to-jet): up to 50% blending. The ATJ process involves converting alcohols (via dehydration, oligomerization, and hydrogenation) into hydrocarbons. In principle, alcohols from algae could be used (e.g., if algae are processed into fermentable sugars or if engineered algae produce ethanol). While currently ATJ is implemented with corn or cellulosic alcohols, an algae-based ATJ route would follow the same certification once the fuel meets criteria.
- CHJ (catalytic hydrothermolysis jet): This pathway (developed by ARA and Chevron Lummus, branded "CHJ") uses a high-pressure, supercritical water reactor to crack and partially upgrade biogenic oils (free fatty acid oils or whole algal oils) into a hydrocarbon mixture, which is then finished by hydrotreating. CHJ produces a fuel containing normal paraffins, isoparaffins, cycloparaffins, and some aromatics, enabling up to 50% blend. It essentially provides a route for HTL-derived SAF to be used and is very relevant to algae that are processed via hydrothermal methods.
- HC-HEFA-SPK (hydroprocessed hydrocarbons, esters and fatty acids): up to 10% blending (SusAlgaeFuel, n.d.). This pathway was developed by Japan's IHI Corp specifically for microalgae-derived oils. Uniquely, it

covers cases where the feedstock includes hydrocarbon-producing algae (such as *Botryococcus braunii*, which naturally contains long-chain hydrocarbon oils) in addition to traditional lipids. The HC-HEFA process and fuel are similar to HEFA-SPK but account for the algal hydrocarbon fraction.

As of 2025, most major routes to SAF are covered by existing ASTM annexes. Future annexes may address pyrolysis oils or other novel conversions, but those are still in research and development. Microalgae can thus already be utilized under current approvals (via HEFA, CHJ, ATJ, etc.), though each new algal process needs to demonstrate fuel compliance. A noteworthy challenge is qualifying 100% SAF (unblended) for commercial use. Research is underway on formulating fully synthetic jet fuel (for example, combining synthetic aromatics with paraffins) so that one day aircraft could fly on 100% renewable fuel. Currently, blends up to 50% are the norm, and certification ensures that algal SAF, once blended, is indistinguishable from conventional jet fuel in performance and safety.

16.4.3 Real-World Projects and Application Readiness

Microalgae-based jet fuel has moved from early experiments to contemporary pilot projects around the world. Over a decade ago, algae were mentioned as a promising biofuel feedstock; notably in 2009, Continental Airlines flew a test flight partially powered by algae-derived fuel (blended with Jatropha oil) (Campbell, 2022). Several startups (e.g., Solazyme, Sapphire Energy) attracted major investment and even supplied algae-based fuel for demonstration flights and U.S. Navy trials in the early 2010s. However, those efforts struggled with scalability and economics; many programs only reached pilot scale and eventually pivoted away from fuels. By the mid-2010s, algal biofuel research slowed as focus shifted to other feedstocks.

In recent years, a resurgence of projects is targeting microalgal SAF with improved technologies and a systems approach. Notably, the European Union has funded multi-partner initiatives to overcome past barriers. One example is the FUELGAE project (2023–2027), a €5 million Horizon Europe effort to reinvest in algae as a source for sustainable aviation and maritime fuels. Coordinated by the Spanish CSIC, FUELGAE aims to integrate algae cultivation with industrial CO_2 sources (flue gases) and develop energy-efficient harvesting and biorefinery processes. A pilot photobioreactor installed at an ArcelorMittal steel plant in Romania and at a biorefinery in Spain will utilize captured CO_2 for algae growth. The project explores both lipid and polysaccharide production from selected microalgae, HTL conversion of whole biomass, and innovative catalytic upgrading of the resulting lipids/biocrude. By focusing on energy-efficient dewatering, component separation, and catalyst development, FUELGAE seeks to reduce the traditionally high-energy input of algae fuels and reach a fuel production process at TRL 5 (continuous pilot scale) by the end of the project. This includes investigating nutrient recycling (e.g., using HTL biochar as soil amendment) and performing life-cycle assessments to prove a lower carbon footprint. FUELGAE's holistic approach addresses economic and environmental sustainability in parallel (Fuelgae, 2025).

Another EU project, COCPIT ("Circular and Innovative Production of Microalgae-Based Fuels for Aviation and Shipping"), launched in late 2023 with ~€5 million funding over 4 years (Bioenergy Technologies Office. U.S. Department of Energy (DOE), 2024). Led by IMT Atlantique in France, COCPIT's goal is to develop a certified biofuel from lipid-rich microalgae while ensuring circular resource use and economic viability. It acknowledges that jet biofuels require high-energy density and excellent cold properties, criteria unmet by first-gen biofuels such as FAME biodiesel or ethanol. Thus, COCPIT focuses on innovative processes across the supply chain: improving algal strains and photobioreactor yields, low-energy cell separation (for lipids, proteins, and carbohydrates), advanced conversion of lipids to bio-jet, and purification. A key output will be a decision-support tool for investors and refiners, which simulates the techno-economic performance of algae-to-fuel value chains under different configurations. By inputting factors that include geography (sunlight and climate), CO_2 sources, water availability, and process choices, stakeholders can assess the most favorable scenarios for scaling up. This strategic tool aims to "de-risk" investment in algae fuel facilities by providing data-driven projections. COCPIT involves 10 partners (including DLR, the German Aerospace Center, and Hellenic Petroleum) and underscores the importance of industry partnerships. It will demonstrate pilot-scale production and blending of algal SAF and explore the logistics of integrating that fuel into existing refinery or airport systems. Ultimately, COCPIT is expected to deliver not just technical advances but also a clearer roadmap for commercial deployment of algae-based jet fuel (COCPIT, 2025).

The SusAlgaeFuel project (2024–2028) further exemplifies international collaboration on algae SAF. This Horizon Europe project (based in Ireland, with global partners) explicitly addresses the EU's aggressive aviation climate targets, for example, 2% SAF by 2025, increasing to ~64% by 2050 (EASA, 2025). SusAlgaeFuel's strategy is to improve cost-effectiveness through innovations such as waste nutrient reuse, digital sensors for optimizing algae cultivation, cascading biorefinery concepts, and multifunctional catalysts for upgrading (Cruce et al., 2021). By treating waste streams (e.g., wastewater or flue gases) as inputs for algae growth and by coproducing value-added products, the project aims to offset costs and reduce the carbon footprint. SusAlgaeFuel joins the "SUSTAFUELS" cluster alongside FUELGAE and COCPIT, indicating a coordinated effort in Europe to pool knowledge on algal- and CO_2-derived fuels. Such clustering maximizes the impact of research by sharing findings on cultivation, processing, and sustainability across projects. The combined expertise, from microbial technology to solar photochemistry and refining, is hoped to accelerate the application readiness of algae SAF.

Beyond research projects, there have been tangible demonstration achievements in recent years. In 2021, Japan achieved the world's first commercial flights using microalgae-based SAF. Japan Airlines (JAL) and All Nippon Airways each flew a commercial flight using an algal oil-derived jet fuel blend on a Boeing 787 and Airbus A350, respectively. This fuel was produced by IHI Corporation from *Botryococcus braunii* algae and refined using Honeywell UOP's Ecofining™ process, and it met the new ASTM D7566 annex specifications as previously described. These flights proved the fuel compatibility in real airline operations but also validated the end-to-end supply chain: outdoor algae cultivation in open ponds (in Thailand), extraction and hydroprocessing to fuel, rigorous fuel testing, and integration into existing airport fueling systems. In 2018, ExxonMobil and Synthetic Genomics operated a vehicle on algae biofuel, but the JAL flights marked a milestone for aviation. Likewise, in Turkey, the EU-backed "Integrated Biorefinery" project (independent) opened a 2,500 m^2 algae biorefinery in 2022, the first of its kind in Europe. This facility, located at Boğaziçi University, Istanbul, can process ~1,200 tons of wet algal biomass per year and produces various products including a bio-jet fuel fraction (Lim et al., 2021). The algae (both microalgae and macroalgae) are cultivated using industrial CO_2 emissions, and the process is designed to be carbon negative with near-zero waste. By late 2022, the project aimed to conduct a domestic demo flight with Turkish Airlines using a 1–10% algae SAF blend. These real-world pilots and demos, albeit at limited scale, show that microalgal SAF is transitioning from the lab to the field. They also provide invaluable data on fuel performance, life-cycle emissions, and logistics, informing TRL advancement. As of 2025, most algae-to-jet technologies are around TRL 4–5 (lab to pilot scale), with a few reaching TRL 6–7 demonstration (e.g., the Honeywell-IHI fuel). Achieving commercial readiness will require further scaling (to reduce costs) and continued partnerships between biotech firms, refiners, airlines, and government programs. Initiatives such as the U.S. SAF Grand Challenge (to produce 3 billion gallons of SAF by 2030) explicitly include algae as a potential feedstock, ensuring that ongoing R&D will be leveraged toward future deployment (McGowen et al., 2023). The trajectory of microalgal SAF development, from ambitious test flights a decade ago to integrated biorefineries and EU consortia now, underscores a growing confidence that algae could contribute to aviation decarbonization, provided the remaining hurdles of cost and scale can be overcome.

16.4.4 Blending and Infrastructure Compatibility

Blending mandates and policies are emerging as key drivers for SAF uptake. Governments and industry bodies recognize that only a small fraction of jet fuel will be SAF in the near term, but that fraction must steadily increase to meet climate goals. The European Union's ReFuelEU initiative, for instance, sets a binding SAF blending target starting at 2% in 2025, ramping up to ~6% by 2030 and progressively to 70% by 2050 (He et al., 2025). Similar targets are echoed elsewhere (e.g., a 30% SAF goal by 2030 in the U.S. military or proposed mandates in Japan).

In summary, microalgae-based SAF can be integrated seamlessly into today's aviation fuel system, provided it meets the established 'drop-in' fuel criteria. Blending and co-use with existing jet fuel is not just a convenience but a strategic requirement: it allows gradual introduction of SAF at scale without waiting for new aircraft or infrastructure. This compatibility, combined with increasing policy support, means that algae-derived jet fuel (once produced in volume economically) could immediately contribute to airline decarbonization.

16.5 PATH TO FEASIBILITY AND SCALE-UP

Despite the considerable potential of microalgae as a feedstock for SAF production, the transition from laboratory research to deployment has been far slower than anticipated, largely due to a combination of technical, economic, and policy barriers (Ahamefule et al., 2023; Farooq et al., 2025). These interrelated challenges continue to constrain progress, even as evolving policy frameworks, technological synergies, and growing demand from airlines begin to create more favorable conditions. The future of microalgae-based SAF therefore depends on navigating these persistent obstacles while leveraging emerging opportunities that could enhance feasibility soon.

16.5.1 Technical Barriers

Besides the route-related challenges explained throughout Section 16.3, the SAF-oriented conversion routes have other technical barriers associated with the cultivation and processing of microalgae that complicate scale-up. Laboratory systems can achieve impressive biomass yields under controlled conditions, but replicating these outcomes in outdoor or industrial environments has proven difficult (Ahamefule et al., 2023; Farooq et al., 2025). Open pond cultivation, often cited as the most economical approach, is vulnerable to contamination by other microorganisms, to significant water losses through evaporation, and to productivity fluctuations linked to variable light and temperature conditions (Falfushynska, 2024; Davis et al., 2024). These vulnerabilities limit the consistency of biomass yields across seasons and geographies. On the other hand, photobioreactors offer greater control of overgrowth parameters and generally achieve higher productivity, but their operation requires substantial energy inputs and high capital investments. This trade-off between cost and productivity remains unresolved, and large-scale projects often struggle to find the right balance (Davis et al., 2024; IEA, 2025b).

Beyond cultivation, downstream processing adds another layer of technical complexity. Harvesting microalgal biomass is inherently challenging due to the small cell size and dilute concentrations in culture media. Current methods, such as centrifugation, filtration, or flocculation, are energy-intensive and expensive; and although progress has been made in optimizing these techniques, they continue to represent a significant cost component (Ahamefule et al., 2023).

For lipid-based pathways such as HEFA, one of the main technical hurdles lies in consistently achieving high lipid yields from microalgal cultures while minimizing downstream energy inputs. The reliance on drying and solvent-based extraction creates significant operational costs. GFT synthesis also faces challenges from the high moisture leading to high-energy demands for drying (Horizon Europe, 2025).

HTL can process wet biomass, avoiding energy-intensive drying, yet it requires extensive biocrude upgrading, while reactor stability and catalyst durability can complicate scale-up (Khan et al., 2023). The ATJ route, in turn, is limited by the recalcitrant cell walls and variable polysaccharide composition of microalgal carbohydrates, reducing yields and complicating process design (Saccardo et al., 2025).

Taken together, converting microalgal biomass or its metabolites into hydrocarbons suitable for jet fuel adds complexity to the production chain. Moreover, across all pathways, microalgae-derived fuels must demonstrate full compatibility with existing aviation fuel infrastructure and meet stringent ASTM standards for safety and performance. Producing consistent, high-quality output on a large scale under these conditions remains one of the central technical bottlenecks.

16.5.2 Economic Barriers

Closely tied to these technical difficulties are the economic barriers that undermine the commercial viability of algae-based SAF. Production costs are currently several times higher than those of fossil-derived jet fuel and remain well above the costs of more mature SAF pathways (Sharma et al., 2023; Ahamefule et al., 2023; IEA, 2025b). The capital expenditures required for large-scale algae cultivation systems are substantial (IEA, 2025b), while operating costs are further inflated by the energy demands of harvesting, drying, and extraction (Davis et al., 2024). Unlike other feedstocks, microalgae do not benefit from pre-existing large-scale supply chains; instead, new infrastructure must be built from scratch, raising both investment requirements and financial risks.

Economies of scale, which are often critical for reducing unit costs, have been difficult to achieve for algae-based systems (Sharma et al., 2023). Few facilities have advanced beyond the pilot or demonstration stage and, as a result, the industry lacks operational experience and cost reductions that typically accompany scaling. Investors are understandably cautious: the high costs, combined with long payback periods and technological uncertainties, make financing microalgae SAF projects risky. The often-cited solution of a biorefinery approach, in which fuels are produced alongside high-value coproducts such as proteins, pigments, or fertilizers, has yet to deliver consistent economic relief (Sharma et al., 2023). While these coproducts could in theory provide supplementary revenue, their markets are either too small, already competitive, or subject to volatility. Without stronger market pull or dedicated financial support, algae-based SAF continues to face a structural financing gap that hampers its progression toward scale.

16.5.3 Policy Barriers

Policy frameworks play a decisive role in shaping the competitiveness of different SAF pathways (Ameen et al., 2024), and here again microalgae encounter significant obstacles. Most existing incentives are technology-neutral, designed to stimulate SAF production broadly rather than to support specific feedstocks. While this ensures flexibility, it also means that support tends to flow toward the least-cost options, primarily waste oil-based HEFA fuels (IEA, 2025b). Microalgae, with their higher cost structure, are therefore left at a disadvantage. In addition, SAF certification under ASTM standards is an essential requirement for commercial deployment, but the process is both lengthy and expensive, requiring extensive testing and large volumes of fuel for validation (Ahamefule et al., 2023; Davis et al., 2024). Smaller companies or research groups working on algae often lack the resources to navigate these requirements, creating a barrier to entry that further slows progress.

Broader renewable energy policies have also historically overlooked aviation. Many subsidy schemes have focused on electricity generation or biofuels for road transport, sectors where decarbonization pathways are more mature. Aviation, by comparison, has only recently received increased policy attention, despite its growing share of global emissions (Gössling et al., 2024). ReFuelEU Aviation (EU) and Fuel of the Future (Brazil), and research funding bodies, such as Horizon Europe, mark important steps forward (93, 94) as highlighted in this chapter. However, targeted measures for advanced and costly feedstocks like microalgae remain scarce. Without mechanisms recognizing its specific challenges and long-term potential, algae-based SAF risks being sidelined despite its strategic environmental value.

16.5.4 Strategic Enablers

Yet, while barriers remain significant, there are also strategic enablers that could reshape the feasibility landscape for algae-based SAF. One of the most promising involves integrating algae cultivation with waste management and industrial processes. Algae require large amounts of carbon dioxide and nutrients for growth, both of which can be sourced from waste streams (Khan et al., 2023; Kiatkittipong et al., 2022; Elkelawy et al., 2021). Industrial flue gases provide concentrated CO_2 that can be redirected into algal cultivation systems, simultaneously lowering input costs and contributing to emission reductions. Likewise, nutrient-rich wastewater from municipal or agricultural sources can serve as a growth medium, offsetting fertilizer costs while improving wastewater treatment outcomes (Ferreira et al., 2019; Sharma et al., 2023). These integrated systems not only reduce operational expenses but also enhance the environmental profile of algae-based fuels, an important consideration given the emphasis on life-cycle emissions in aviation fuel certification.

Another enabling factor is modular, distributed production. Instead of relying on large, centralized facilities, smaller production units could be colocated near CO_2 or wastewater sources (Sharma et al., 2023; Ahamefule et al., 2023), thereby reducing transportation costs and allowing for gradual, flexible scaling. This approach also reduces financial risk by spreading investment across multiple sites rather than concentrating on a single large facility. At the same time, the biorefinery model remains a key strategy. By valorizing proteins, pigments, or biofertilizers alongside SAF, producers can diversify their revenue streams, improve economic resilience, and make projects more attractive to investors (Ahamefule et al., 2023). Although challenges remain in matching coproduct outputs to market demand, this integrated approach represents one of the few viable pathways toward cost competitiveness.

Perhaps most critically, pilot and demonstration projects serve as indispensable building blocks toward algae-based SAF commercialization. They provide operational data under real-world conditions, enable the production of sufficient fuel for certification testing, and optimize operation conditions. Demonstration projects also reveal practical challenges that laboratory studies cannot capture, such as contamination issues, supply chain logistics, or equipment durability (Falfushynska, 2024). By addressing these challenges early, pilot facilities help de-risk future commercial projects and build a foundation for scale-up. When successful, they act as reference points that signal technical feasibility, attracting both public funding and private capital.

The pathway to feasibility and scale-up of microalgae-based SAF is therefore shaped by a complex interplay of barriers and opportunities. Technical, economic, and policy challenges have thus far hindered commercialization, but prospects for progress exist. Moving forward, the success of this pathway will depend on coordinated efforts that combine technological innovation with supportive policy frameworks. If these conditions align, microalgae could transition from a promising research prospect to a viable contributor within the global SAF portfolio, reinforcing the long-term sustainability of the aviation sector.

16.6 CONCLUSIONS

This chapter examined the potential of microalgal biomass as a feedstock for SAF, following the entire trajectory from "water to sky." The analysis shows that while significant challenges remain, multiple viable conversion pathways exist to transform microalgae into drop-in aviation fuels compatible with ASTM specifications. Lipid-rich strains such as *Nannochloropsis* are particularly well suited for the HEFA route, which is the most technologically mature and already approved for up to 50% blending with Jet A/A-1. HTL of whole wet biomass provides another promising pathway, circumventing energy-intensive drying steps and producing a biocrude that can be upgraded to aviation kerosene via catalytic hydrotreating and fractionation. Carbohydrate-rich species, particularly under stress-induced starch accumulation, offer feedstocks for ATJ processes, while gasification followed by Fischer–Tropsch synthesis provides a route for whole-biomass utilization. Recent certification advances, such as the ASTM Annex A7 approval for HC-HEFA from microalgae-derived hydrocarbons, further demonstrate that algae-based fuels are not only technically feasible but increasingly recognized within regulatory frameworks.

Yet, across these routes, common obstacles persist. Microalgal biomass is compositionally versatile but remains costly to cultivate and harvest at large scale. Balancing growth and stress-induced accumulation of lipids or carbohydrates highlights the fundamental trade-off between productivity and fuel precursor content. Downstream, algal-derived oils and biocrudes often contain high levels of nitrogen, oxygen, and contaminants, demanding severe upgrading and increasing hydrogen consumption. From an economic standpoint, current algae-to-jet fuels cannot yet compete with fossil jet or even other SAF feedstocks, primarily due to cultivation and processing costs. Policy frameworks and certification timelines further constrain deployment, with blending mandates driving demand but not yet tailored to incentivize advanced feedstocks like microalgae.

Looking ahead, strategic priorities for advancing microalgal SAF fall into four key areas. First, integration: positioning algae within broader biorefinery and circular economy systems will be essential to spread costs and valorize coproducts such as proteins and high-value compounds. Utilizing waste CO_2, wastewater, or industrial heat can improve both sustainability and economics. Second, policy support: stronger, feedstock-specific incentives are needed to accelerate algae-based SAF, alongside continued streamlining of certification procedures. Third, infrastructure readiness: ensuring that algae-derived fuels can seamlessly enter existing refinery and airport systems through drop-in compatibility is crucial, and progress with blending and distribution has already demonstrated feasibility. Finally, R&D focus: further work is needed on strain improvement, low-energy harvesting, catalytic upgrading of nitrogen-rich biocrudes, and techno-economic optimization. Coordinated programs, such as recent EU initiatives (FUELGAE, COCPIT, SusAlgaeFuel), highlight the importance of international collaboration in tackling these challenges.

In summary, microalgal biomass remains one of the most promising long-term feedstocks for SAF due to its high productivity potential, non-competition with food, and flexibility across conversion pathways. While its commercial role in aviation decarbonization is still emerging, continued innovation, supportive policies, and integration with industrial systems could enable algae to move from demonstration to deployment. Realizing this vision will require aligning scientific progress with infrastructure readiness and regulatory confidence.

REFERENCES

Adnan MA, Xiong Q, Muraza O, Hossain MM. Gasification of wet microalgae to produce H2-rich syngas and electricity: A thermodynamic study considering exergy analysis. *Renew Energy* 2020;147:2195–205.

Advanced Biofuels USA. *ASTM Approves Pathway for Production of Microalgae-Based Sustainable Aviation Fuel Developed by Japan's IHI*; 2020. https://advancedbiofuelsusa.info/tag/astm

Ahamefule CS, Ogbonna CN, Ahamefule BC, Ogbonna I, Ogbonna J. Lipids and fatty acids from microalgae. In: Eduardo Jacob-Lopes, Maria Isabel Queiroz, Mariana Manzoni Maroneze, and Leila Queiroz Zepka (eds.), *Handbook of Food and Feed from Microalgae*. Elsevier; 2023, pp. 73–86.

AlfaFuels. *The Susalgaefuel Project, New Member of the Sustafuels Cluster*.

Ameen M, Ahmad M, Zafar M, Dawood S, Ahmad S, Kazmi SIM, et al. Sustainable innovations in biodiesel plants: From production to environmental impact. *Biodiesel Plants - Fueling the Sustainable Outlooks*; 2024. Available from: https://www.intechopen.com/chapters/1195241

Bain RL, Broer K. Gasification. In: Robert C. Brown (ed.), *Thermochemical Processing of Biomass: Conversion into Fuels, Chemicals and Power*. John Wiley and Sons; 2011, pp. 47–77.

Barroso Soares R, Ferreira Martins M, Franci Gonçalves R. Experimental investigation of wastewater microalgae in a pilot-scale downdraft gasifier. *Algal Res* 2020;51.

Behrend D, Schreiber C, Pfaff C, Müller A, Grobbelaar J, Nedbal L. Algae as a potential source of biokerosene and diesel - Opportunities and challenges. In: Kaltschmitt M, Neuling U, editors. *Biokerosene - Status and Prospects*. Springer; 2018, pp. 303–24.

Bergero C, Gosnell G, Gielen D, Kang S, Bazilian M, Davis SJ. Pathways to net-zero emissions from aviation. *Nat Sustain* 2023;6(4):404–14. Available from: https://www.nature.com/articles/s41893-022-01046-9

Biller P, Roth A. Hydrothermal liquefaction: A promising pathway towards renewable jet fuel. In: Kaltschmitt M, Neuling U, editors. *Biokerosene: Status and Propspects* (1st ed.). Springer; 2018, pp. 607–38.

Bioenergy Technologies Office. U.S. Department of Energy (DOE). *Updated Algae Report Analyzes National-Scale Prospects for Converting Microalgae to Biofuels and Bioproducts*; 2024.

BIOFAT. *Biofuels Made from Microalgae*; 2025. Available from: https://www.biofat-project.eu

Borrill E, Koh SCL, Yuan R. Review of technological developments and LCA applications on biobased SAF conversion processes. *Front Fuels* 2024;2:1397962.

Bukur DB, Todic B, Elbashir N. Role of water-gas-shift reaction in Fischer - Tropsch synthesis on iron. *Catal Today* 2016;275:66–75.

Campbell M. *A Radical, Carbon Negative Project in Turkey Is Turning Algae into Bio-Jet Fuel*. EuroNews; 2022.

Capaz RS, Guida E, Seabra JEA, Osseweijer P, Posada JA. Mitigating carbon emissions through sustainable aviation fuels: costs and potential. *Biofuels Bioprod Biorefin* 2021;15(2):502–24.

Cecchin M, Berteotti S, Paltrinieri S, Vigliante I, Iadarola B, Giovannone B, et al. Improved lipid productivity in Nannochloropsis gaditana in nitrogen-replete conditions by selection of pale green mutants. *Biotechnol Biofuels* 2020;13(1):78.

Chagas MF, Ferreira GF, Gomes LCM, Zanphorlin LM, Martins LZO, Sampaio ILM, et al. From enzyme to cell-factory: Economic and environmental assessment of biobased pathways to unlock the potential of long-haul transportation biofuels. *Chem Eng J* 2023;469:143878. Available from: https://www.sciencedirect.com/science/article/abs/pii/S1385894723026098?via%3Dihub

Cheng D, Li D, Yuan Y, Zhou L, Li X, Wu T, et al. Improving carbohydrate and starch accumulation in Chlorella sp. AE10 by a novel two-stage process with cell dilution. *Biotechnol Biofuels* 2017;10(1):75.

Chong JW, Chemmangattuvalappil NG, Thangalazhy-Gopakumar S. Aviation biofuels: Conversion routes and challenges. In: Foo DCY, Aziz MKTA, Yusup S, editors. *Sustainable Technologies for the Oil Palm Industry*. Springer; 2023.

COCPIT. *Circular Production of Microalgae-Based Fuels for Shipping and Aviation*; 2025. Available from: https://www.cocpit-horizon.eu/

Cruce JR, Beattie A, Chen P, Quiroz D, Somers M, Compton S, et al. Driving toward sustainable algal fuels: A harmonization of techno-economic and life cycle assessments. *Algal Res* 2021;54:102169. Available from: https://www.sciencedirect.com/science/article/pii/S2211926420310377?via%3Dihub

Davis R, Hawkins TR, Coleman A, Gao S, Klein B, Wiatrowski M, et al. *Economic, Greenhouse Gas, and Resource Assessment for Fuel and Protein Production from Microalgae: 2022 Algae Harmonization Update*; 2024 [cited 1 September 2025]. Available from: https://www.osti.gov/servlets/purl/2318964

Dayton DC, Turk B, Gupta R. Syngas cleanup, conditioning, and utilization. In: Brown RC, editor. *Thermochemical Processing of Biomass: Conversion into Fuels, Chemicals and Power* (1st ed.). John Wiley & Sons; 2011, pp. 78–123.

de Jaeger L, Carreres BM, Springer J, Schaap PJ, Eggink G, Martins Dos Santos VAP, et al. Neochloris oleoabundans is worth its salt: Transcriptomic analysis under salt and nitrogen stress. *PLoS One* 2018;13(4):e0194834.

de Vree JH, Bosma R, Janssen M, Barbosa MJ, Wijffels RH. Comparison of four outdoor pilot-scale photobioreactors. *Biotechnol Biofuels* 2015;8(1):215.

du Preez R, Majzoub ME, Thomas T, Panchal SK, Brown L. Nannochloropsis oceanica as a microalgal food intervention in diet-induced metabolic syndrome in rats. *Nutrients* 2021;13(11):3991.

EASA. *Fit for 55 and ReFuelEU Aviation*. European Union Aviation Safety Agency; 2025.

Elkelawy M, Bastawissi HAE, Radwan AM, Ismail MT, El-Sheekh M. Biojet fuels production from algae: conversion technologies, characteristics, performance, and process simulation. In: Mostafa El-Sheekh and Abd El-Fatah Abomohra (eds.), *Handbook of Algal Biofuels: Aspects of Cultivation, Conversion, and Biorefinery*. Elsevier; 2021, pp. 331–61.

European Court of Auditors. *Special Report 29/2023: The EU's Support for Sustainable Biofuels in Transport*. European Union; 2023. Available from: https://www.eca.europa.eu/en/publications?ref=sr-2023-29

Falfushynska H. Advancements and prospects in algal biofuel production: A comprehensive review. *Phycology* 2024;4(4):548–75. Available from: https://www.mdpi.com/2673-9410/4/4/30/htm

Farooq Z, Wetterlund E, Mesfun S, Furusjö E. Uncovering the economic potential of sustainable aviation fuel production pathways: A meta-analysis of techno-economic studies. *Energy Convers Manage* 2025;341:120076. Available from: https://www.sciencedirect.com/science/article/pii/S0196890425006004?via%3Dihub

Farooqui SA, Sinha AK, Ray A. Progress and trends in renewable jet fuels. In: Deepak Tuli, Sangita Kasture, and Arindam Kuila (eds.), *Advanced Biofuel Technologies: Present Status, Challenges and Future Prospects*. Elsevier; 2021, pp. 333–64.

Ferreira GF, Ríos Pinto LF, Maciel Filho R, Fregolente LV. A review on lipid production from microalgae: Association between cultivation using waste streams and fatty acid profiles. *Renew Sustain Energy Rev* 2019;109:448–66. Available from: https://www.sciencedirect.com/science/article/abs/pii/S1364032119302643?via%3Dihub

Fortier MOP, Roberts GW, Stagg-Williams SM, Sturm BSM. Life cycle assessment of bio-jet fuel from hydrothermal liquefaction of microalgae. *Appl Energy* 2014;122:73–82.

Fuelgae. *Innovative Sustainable On-Site Technologies for Using Microalgae to Capture CO_2 and Produce Advanced Biofuels*; 2025 [cited 27 August 2025]. Available from: https://fuelgae.eu

Goh BHH, Chong CT, Ong HC, Seljak T, Katrašnik T, Józsa V, et al. Recent advancements in catalytic conversion pathways for synthetic jet fuel produced from bioresources. *Energy Conversion and Management* 2022;251:114974.

Gössling S, Humpe A. Net-zero aviation: Transition barriers and radical climate policy design implications. *Sci Total Environ* 2024:912:169107.

Gössling S, Humpe A, Leitão JC. Private aviation is making a growing contribution to climate change. *Commun Earth Environ* 2024;5(1):1–11. Available from: https://www.nature.com/articles/s43247-024-01775-z

Hasan MA, Mamun A Al, Rahman SM, Malik K, Al Amran MIU, Khondaker AN, et al. Climate change mitigation pathways for the aviation sector. *Sustainability* 2021;13(7):3656. Available from: https://www.mdpi.com/2071-1050/13/7/3656/htm

He J, Strezov V, Kumar R, Weldekidan H, Jahan S, Dastjerdi BH, et al. Pyrolysis of heavy metal contaminated Avicennia marina biomass from phytoremediation: Characterisation of biomass and pyrolysis products. *J Clean Prod* 2019;234:1235–45.

He L, Luthfi AAI, Saleh NM, Yaakob Z, Salleh MZM, Ahmad KN, et al. Microalgae biodiesel: Advances, costs, and potential. *Waste Biomass Valorization* 2025;1–19. Available from: https://link.springer.com/article/10.1007/s12649-025-03135-0

Hellmann AJG, Barke A, Heinze W, Heimbs S, Spengler TS. Using lightweight airframe materials in aircraft structures and investigating potential reinforcement effects with sustainable aviation fuels: An analysis of environmental and economic impacts. *J Ind Ecol* 2025;29(4):1135–50.

Honeywell. *Honeywell Technology Enables First Jet Flights with Sustainable Aviation Fuel Produced by Microalgae*; 2021. https://uop.honeywell.com/en/news-events/2021/november/honeywell-technology-enables-first-jet-flights-with-sustainable-aviation-fuel-produced-by-microalgae

Horizon Europe. *European Commission*; 2025 [cited 1 September 2025]. Available from: https://research-and-innovation.ec.europa.eu/funding/funding-opportunities/funding-programmes-and-open-calls/horizon-europe_en

IATA. *Fact Sheet 2 Sustainable Aviation Fuel: Technical Certification.*

ICAO. *Trends in Emissions that affect Climate Change* 2025 [cited 1 September 2025]. Available from: https://www.icao.int/environmental-protection/trends-emissions-affect-climate-change

IEA. *Bioenergy Task 39 Biofuels to Decarbonize Transport. Database on Facilities for the Production of Advanced Liquid and Gaseous Biofuels for Transport*; 2025a [cited 30 August 2025]. Available from: https://demoplants.best-research.eu/

IEA. *Tracking Aviation*; 2025b [cited 1 September 2025]. Available from: https://www.iea.org/energy-system/transport/aviation#tracking

Kale U, Jankovics I, Nagy A, Rohács D. Towards sustainability in air traffic management. *Sustainability* 2021;13(10):5451. Available from: https://www.mdpi.com/2071-1050/13/10/5451/htm

Karatzos S, van Dyk JS, McMillan JD, Saddler J. Drop-in biofuel production via conventional (lipid/fatty acid) and advanced (biomass) routes. *Biofuels, Bioproducts and Biorefining* 2017;11(2):344–62.

Khan S, Das P, Abdul Quadir M, Thaher MI, Mahata C, Sayadi S, et al. Microalgal feedstock for biofuel production: Recent advances, challenges, and future perspective. *Fermentation* 2023;9(3):281.

Khodadadianzaghmari F, Jahadi M, Goli M. Biochemical profile of Scenedesmus isolates, with a main focus on the fatty acid profile. *Food Sci Nutr* 2024;12(8):5922–31.

Kiatkittipong W, Pongsiriyakul K, Lim JW, Kiatkittipong K, Wongsurakul P, Yodpetch V, et al. Bioresources and biofuels—From classical to perspectives and trends. In: *A-Z of Biorefinery: A Comprehensive View*. Elsevier; 2022, pp. 165–220.

Klein B, Davis R. *Algal Biomass Production Via Open Pond Algae Farm Cultivation: 2022 State of Technology and Future Research*; 2023. Available from: www.nrel.gov/publications

Klöwer M, Allen MR, Lee DS, Proud SR, Gallagher L, Skowron A. Quantifying aviation's contribution to global warming. *Environ Res Lett* 2021;16(10):104027. Available from: https://iopscience.iop.org/article/10.1088/1748-9326/ac286e

Latsos C, van Houcke J, Timmermans KR. The effect of nitrogen starvation on biomass yield and biochemical constituents of Rhodomonas sp. *Front Mar Sci* 2020;7:563333.

Li YW, Xu J, Yang Y. Diesel from syngas. In: Vertés AA, Qureshi N, Blaschek HP, Yukawa H, editors. *Biomass to Biofuels: Strategies for Global Industries*. John Wiley & Sons; 2010, pp. 123–39.

Lim JHK, Gan YY, Ong HC, Lau BF, Chen WH, Chong CT, et al. Utilization of microalgae for bio-jet fuel production in the aviation sector: Challenges and perspective. *Renew Sustain Energy Rev* 2021;149:111396. Available from: https://www.sciencedirect.com/science/article/abs/pii/S136403212100681X?via%3Dihub

Liska AJ, Shevchenko A, Pick U, Katz A. Enhanced photosynthesis and redox energy production contribute to salinity tolerance in dunaliella as revealed by homology-based proteomics. *Plant Physiol* 2004;136(1):2806–17.

Long B, Fischer B, Zeng Y, Amerigian Z, Li Q, Bryant H, et al. Machine learning-informed and synthetic biology-enabled semi-continuous algal cultivation to unleash renewable fuel productivity. *Nat Commun* 2022;13(1):541.

Lorenzo K, Santocildes G, Torrella JR, Magalhães J, Pagès T, Viscor G, et al. Bioactivity of macronutrients from chlorella in physical exercise. *Nutrients* 2023;15(9):2168.

Maltsev Y, Kulikovskiy M, Maltseva S. Nitrogen and phosphorus stress as a tool to induce lipid production in microalgae. *Microbial Cell Fact* 2023;22(1):239.

Marangon BB, Castro JDS, Calijuri ML. Aviation fuel based on wastewater-grown microalgae: Challenges and opportunities of hydrothermal liquefaction and hydrotreatment. *J Environ Manage* 2024;354:120418.

McGowen J, Knoshaug EP, Laurens LML, Forrester J. Outdoor annual algae productivity improvements at the pre-pilot scale through crop rotation and pond operational management strategies. *Algal Res* 2023;70:102995. Available from: https://www.sciencedirect.com/science/article/pii/S2211926423000280?via%3Dihub

Michaga MFR, Michailos S, Hughes KJ, Ingham D, Pourkashanian M. Techno-economic and life cycle assessment review of sustainable aviation fuel produced via biomass gasification. In: Ramesh C. Ray (ed.), *Sustainable Biofuels: Opportunities and Challenges*. Elsevier; 2021, pp. 269–303.

Motta IL, Marchesan AN, Maciel Filho R, Maciel MRW. Syngas cleaning and conditioning in biomass gasification plants: a review of technologies, bibliometric trends, and future directions. *International Journal of Hydrogen Energy* 2025;154:150234.

Motta IL, Miranda NT, Maciel Filho R, Wolf Maciel MR. Biomass gasification in fluidized beds: A review of biomass moisture content and operating pressure effects. *Renew Sustain Energy Rev* 2018;94(May):998–1023.

Mussatto SI, Motta IL, Filho RM, van der Wielen L, Capaz R, Seabra J, et al. *Sustainable Aviation Fuels: Production, Use and Impact on Decarbonization* (2nd ed.) Comprehensive Renewable Energy. Elsevier Inc.; 2022, pp. 348–371. Available from: http://doi.org/10.1016/B978-0-12-819727-1.00057-1

Neuling U, Kaltschmitt M. Conversion routes from biomass to biokerosene. In: Kaltschmitt M, Neuling U, editors. *Biokerosene: Status and Propspects* (1st ed.). Springer; 2018, pp. 435–74.

Neves RC, Klein BC, da Silva RJ, Rezende MCAF, Funke A, Olivarez-Gómez E, et al. A vision on biomass-to-liquids (BTL) thermochemical routes in integrated sugarcane biorefineries for biojet fuel production. *Renew Sustain Energy Rev* 2020;119(July 2018), 109607.

Ng KS, Farooq D, Yang A. Global biorenewable development strategies for sustainable aviation fuel production. *Renewable and Sustainable Energy Reviews* 2021;150:111502.

Pasa VMD, dos Santos Oliveira H, Scaldadaferri CA. Main feedstock for sustainable alternative fuels for aviation. In: *Sustainable Alternatives for Aviation Fuels*. Elsevier; 2022, pp. 69–102.

Ranasinghe K, Guan K, Gardi A, Sabatini R. Review of advanced low-emission technologies for sustainable aviation. *Energy* 2019;188:115945.

Redda ZT, Manyazewal DE, Gizaw DG, Laß-Seyoum A, Periyasamy S. Biomass Conversion to synthetic 9 aviation fuels. In: Neelancherry R, Gao B, Wisniewski A, editors. *Agricultural Waste to Value-Added Products: Bioproducts and its Applications*. Springer; 2024, pp. 1–275.

Roostaei J, Zhang Y, Gopalakrishnan K, Ochocki AJ. Mixotrophic microalgae biofilm: A novel algae cultivation strategy for improved productivity and cost-efficiency of biofuel feedstock production. *Sci Rep* 2018;8(1):12528.

Saccardo A, Wolf J, Hankamer B, Bezzo F. Scaling-up information from high-throughput pulsed light data to predict microalgae growth dynamics in photobioreactors. *Algal Res.* 2025;89:104073. Available from: https://www.sciencedirect.com/science/article/pii/S2211926425001821?via%3Dihub

Sharma V, Hossain AK, Duraisamy G, Griffiths G. Microalgal Biodiesel: A challenging route toward a sustainable aviation fuel. *Fermentation* 2023;9(10):907.

Suarez Ruiz CA, Kwaijtaal J, Peinado OC, van den Berg C, Wijffels RH, Eppink MHM. Multistep fractionation of microalgal biomolecules using selective aqueous two-phase systems. *ACS Sustain Chem Eng* 2020;8(6):2441–52.

SUNFUSION. *Advanced Sustainable Biofuels Production from Purified Microalgae and Oleaginous Yeasts via Integrated Solar Hydrothermal Liquefaction*; 2025 [cited 30 August 2025]. Available from: https://sunfusion.eu

Surgenor C. *GreenAir Independent Reporting on Aviation & the Environment*; 2024. New EU-funded FUELGAE project aims to resurrect algae as a promising source for aviation biofuels.

SusAlgaeFuel. *Innovative Approaches to Enable Microalgae Aviation Fuels.*

Usman M, Cheng S, Boonyubol S, Cross JS. The future of aviation soars with HTL-based SAFs: exploring potential and overcoming challenges using organic wet feedstocks. In: *Sustainable Energy and Fuels* (Vol. 7). Royal Society of Chemistry; 2023, pp. 4066–87.

Vaillant PH. *COCPIT Piloting Research into a New Biofuel Based on Microalgae*. IMT Atlantique; 2023.

Vela-García N, Bolonio D, García-Martínez MJ, Ortega MF, Canoira L. Thermochemical conversion of agricultural waste to biojet fuel. In: Abu Yousuf and Cristina Gonzalez-Fernandez (eds.), *Sustainable Alternatives for Aviation Fuels*. Elsevier; 2022, pp. 27–48.

Wang M, Ye X, Bi H, Shen Z. Microalgae biofuels: illuminating the path to a sustainable future amidst challenges and opportunities. *Biotechnol Biofuels Bioprod* 2024;17(1):10.

Wiatrowski M, Klein B, Do TN, Ou L, Cai H, Carlson N, et al. *Technology Case Study: Techno-Economic and Life Cycle Analysis for Microalgae Conversion Pathways to Fuels and Products*; 2025. Available from: www.nrel.gov/publications

Yaakob MA, Mohamed RMSR, Al-Gheethi A, Aswathnarayana Gokare R, Ambati RR. Influence of nitrogen and phosphorus on microalgal growth, biomass, lipid, and fatty acid production: An overview. *Cells* 2021;10(2):393.

Zhang X. Essential scientific mapping of the value chain of thermochemically converted second-generation bio-fuels. *Green Chem* 2016;18(19):5086–117.

Zhu Y, Xu Y, Schmidt A, Thorson M, Cronin D, Santosa D, et al. *Microalgae Hydrothermal Liquefaction and Biocrude Upgrading: 2022 State of Technology*; 2023. Available from: https://www.ntis.gov/about

17 Microalgal Biotechnology in Space Missions

A Potential Strategy for Onboard Sustainable Life-Support Systems

Giacomo Fais, Giovanni Perra, Upasana Mohanty, Nicola Lai, Giacomo Cao, and Alessandro Concas

Long-duration space missions rank among the most demanding endeavors ever conceived by humankind. The prospect of establishing permanent outposts on the Moon or sending crewed expeditions to Mars requires overcoming technological and biological hurdles that, until a few decades ago, were deemed insurmountable. In these scenarios, survival must rely on closed-loop, self-sufficient systems capable of recycling air, water, and nutrients, rather than Earth-based resupply (Afshinnekoo et al. 2020; Kovic 2021). Central to this endeavor is the challenge of sustaining human life in extreme environments that lack readily available resources and are characterized by physical conditions radically different from those on Earth (Belz et al. 2014; Alemany et al. 2019; Nangle et al. 2020). Moreover, ensuring the physiological and psychological well-being of astronauts cannot depend on continuous shipments from Earth, as interplanetary distances and prohibitive costs render such an approach logistically untenable (Green and Kleinhenz 2019).

Next-generation missions target longer and more distant stays with greater crew autonomy. Reference architectures for the Moon envision a sustained human presence to develop experience in living from in-situ resources, whereas Mars mission concepts include both short- and long-surface-stay profiles, with total mission durations on the order of 2–3 years (Drake 2009). A crucial operational constraint is the Martian launch window: under a Hohmann transfer, it typically opens every ~26 months, dictating logistics, resupply schedules, and system design margins (Cates et al. 2016). Together, these constraints make autonomous, bioregenerative provision of air, water, food, and ideally propellants indispensable.

DOI: 10.1201/9781003611479-17

Consequently, it is necessary to develop self-sufficient life-support systems (LSS) that are robust, resilient, and capable of regenerating essential resources in situ, minimizing waste and maximizing recycling efficiency (Wheeler et al. 2003; Fahrion et al. 2021). Conventional Environmental Control and Life-Support Systems (ECLSS) have demonstrated their ability to maintain acceptable living conditions during medium-duration missions, such as those aboard the International Space Station (ISS) (Thirsk et al. 2009; Snyder et al. 2019). However, they rely primarily on physico-chemical technologies, which, while enabling partial closure of water and oxygen cycles, demand high energy input, generate by-products that are difficult to manage, and remain dependent on Earth-based resupply chains (Wheeler et al. 2003; Fahrion et al. 2021; Williamson et al. 2024).

While in low Earth orbit, these issues can be mitigated by relatively frequent cargo flights; however, this dependence becomes prohibitive for lunar or Martian scenarios. Hence, the need for a paradigm shift: integrating regenerative biological processes that, unlike purely physico-chemical systems, can simultaneously deliver breathable air, purified water, and nutritious food (Snyder et al. 2019; Fahrion et al. 2021; Liu et al. 2021). As a design anchor, one astronaut uses ~0.84 kg O_2 and emits ~1.04 kg CO_2 per day. These values define the minimum regenerative load that a photosynthetic module must offset in long-duration missions (James et al. 2011; Cates et al. 2016; Fahrion et al. 2021; Jones 2023).

Within this context, microalgae (including cyanobacteria) emerge as key biological agents—photosynthetic microorganisms of microscopic size but remarkable metabolic plasticity (Cao et al. 2021; Mapstone et al. 2022; Fais et al. 2022a; Barone et al. 2023; De Micco et al. 2023; Koehle et al. 2023; Santomartino et al. 2023; Revellame et al. 2024). Their evolutionary history dates back to the earliest epochs of life on Earth: geochemical and fossil evidence indicate that cyanobacteria were already active more than 3.5 billion years ago in anoxic, iron-rich oceans (Schopf 2011; Schirrmeister et al. 2013). With the advent of oxygenic photosynthesis, they contributed to the Great Oxidation Event (~2.4 billion years ago), which increased atmospheric oxygen, favored formation of the ozone layer, and enabled aerobic metabolism and subsequent eukaryotic diversification (Lyons et al. 2014).

This long selection under extreme environmental conditions (high UV radiation, redox oscillations, and nutrient scarcity) has shaped their exceptional physiological resilience and metabolic flexibility, traits that are particularly valuable for closed systems and the "out-of-equilibrium" conditions typical of space missions (Rothschild and Mancinelli 2001; Cockell and Raven 2004; Aguiló-Nicolau et al. 2023). Among cyanobacteria, several extremophilic lineages have evolved notable tolerance to desiccation, radiation, temperature fluctuations, and nutrient deprivation—adaptations refined over billions of years in the hostile environments of early Earth. Species such as *Chroococcidiopsis thermalis*, *Nostoc commune*, and *Gloeocapsa* sp. have been shown to survive prolonged exposure to vacuum, UV-C, and cosmic radiation, maintaining photosynthetic and repair activity upon rehydration. Their resilience makes them ideal biological models for space biotechnology (Billi et al. 2011; Baqué et al. 2013; Verseux et al. 2016; Cockell 2020; Fais et al. 2024; Santos de Sousa et al. 2025; Billi 2026).

Microalgae are capable of converting CO_2 into O_2 and biomass with high efficiency per illuminated surface area or culture volume, particularly when cultivated in photobioreactors (PBRs), and are therefore regarded as prime candidates for bioregenerative life-support systems (BLSS) (Richmond 2003; Fahrion et al. 2021). During oxygenic photosynthesis, the molar O_2/CO_2 ratio is generally close to 1; in practice, this corresponds, as an order of magnitude, to about 1 g of O_2 released per gram of dry biomass accumulated, with variations depending on biomass composition and growth conditions (Eriksen et al. 2006).

In well-illuminated PBRs, volumetric productivities for *Chlorella vulgaris* are typically in the range of 0.5–2.0 $g \cdot L^{-1} \cdot day^{-1}$, while higher values are reported only under highly optimized laboratory conditions; for instance, long-term operation of a BLSS-oriented reactor achieved ~1.3 $g \cdot L^{-1} \cdot day^{-1}$ (Helisch et al. 2020). On an areal basis, flat-panel or thin-layer systems generally reach 20–50 $g \cdot m^{-2} \cdot day^{-1}$ (Doucha and Livansky 1999; Doucha et al. 2005; Grivalský et al. 2019). Assuming ~1 g O_2 per gram of biomass, these ranges correspond to ~20–50 g $O_2 \cdot m^{-2} \cdot day^{-1}$ and 0.5–2.0 g $O_2 \cdot L^{-1} \cdot day^{-1}$, consistent with estimates for bioregenerative systems designed for long-duration missions (Helisch et al. 2020; Detrell et al. 2021).

For comparison, the average human oxygen consumption is about 0.84 kg $O_2 \cdot day^{-1}$ (James et al. 2011; Cates et al. 2016; Jones 2023). Depending on actual areal productivity and illumination conditions, this implies PBR surface areas of approximately 15–40 m^2 per crew equivalent, with additional margins required in practice (Detrell et al. 2021). Recent reviews further emphasize the strong potential of microalgae for integration into BLSS architectures (De Micco et al. 2023).

In a single biological process, these microorganisms can simultaneously perform multiple vital functions: regenerating the atmosphere by producing oxygen, fixing CO_2, providing food rich in essential nutrients, recycling nitrogen and phosphorus from metabolic wastes, purifying wastewater, and generating biomolecules of technological interest (Gòdia et al. 2002; Detrell et al. 2019; Fahrion et al. 2021; Reza-Solis et al. 2023). PBRs should thus be viewed not as mere "green tanks" but as biological processors, essential nodes within self-sufficient ecosystem architectures. Practically, almost all electrical energy supplied to lighting is ultimately dissipated as heat within the habitat; therefore, thermal management and heat recovery/rejection are integral to PBR design.

Multiple experiments over recent decades have confirmed the notable robustness of species such as *Arthrospira platensis* (*Spirulina*) and *Chlorella* spp. (Mills and Pierson 2000; Sarmistha 2001; Popova 2003; Detrell et al. 2021; Fais et al. 2022a, 2022b). Tested under conditions simulating extraterrestrial environments, they maintained photosynthetic efficiency and growth rates comparable to those observed on Earth (Helisch et al. 2020; Detrell et al. 2021). Some setups report improvements in microgravity, but results are heterogeneous and protocol-dependent; systematic studies are still needed. On the ISS, the "Photobioreactor at the Life Support Rack (PBR@LSR)" with *Chlorella vulgaris* operated continuously for weeks, demonstrating growth control, O_2 production, and pumped-loop management in true microgravity (Detrell et al. 2021). The ArtEMISS/*Arthrospira*-B campaign with *Limnospira indica* (hereafter *Spirulina*) validated reactivation from the dry state, growth, and O_2/CO_2

balance in a membrane micro-PBR (~50 mL), supporting the scalability of edible cultures in space (Poughon et al. 2020; Detrell et al. 2021).

Overall, experimental evidence indicates that these microorganisms not only survive, but also maintain, and sometimes improve, their regenerative functions (O_2 production, CO_2 fixation, and nutrient recycling) in microgravity and under moderate ionizing radiation, as shown by PBR@LSR tests on the ISS with *Chlorella vulgaris* and the ArtEMISS/Arthrospira-B campaigns with *Limnospira* (Poughon et al. 2020; Detrell et al. 2021). Reported O_2 evolution rates (~0.38–3.5 $mg{\cdot}L^{-1}{\cdot}h^{-1}$, species- and condition-dependent) directly inform PBR areal/volumetric sizing (Lee and Palsson 1994; Park and Lee 2000; Tang et al. 2014; Sengupta and Wangikar 2020). In microgravity, phase separation can rely on passive capillary geometries or acoustics with piezoelectric transducers; both options are promising for removing dissolved/bubbled oxygen (Fili et al. 2021). For storage, beyond compressors/concentrators and liquid oxygen (LOX) for propulsion, low-pressure solutions derived from resource-limited settings (e.g., FREO2) ensure continuous delivery with blackout resilience and are useful as medical/onboard buffers (Rassool et al. 2017; Peake et al. 2021; Singla and Chowdhury 2021).

Alongside single-species studies, more complex experiments have aimed to reproduce fully closed, self-sufficient ecosystems. In the 1960s, the Soviet BIOS-3 project represented one of the first concrete attempts: in a pressurized facility, three people were sustained for extended periods thanks to *Chlorella* PBRs combined with terrestrial crops, achieving near-complete closure for oxygen and very high closure for water (Gitelson et al. 2003; Nelson et al. 2010). Historical reviews of BIOS-3 indicate that *Chlorella* PBRs substantially covered the crew's O_2 demand, with highly closed air–water cycles (Gitelson et al. 2003).

Subsequent efforts, such as Biosphere 2 in the United States, expanded the experimental scale by attempting to reconstruct complex ecosystems, though they faced difficulties in stabilizing gas cycles (Nelson et al. 2010). Biosphere 2 exhibited CO_2 oscillations and a progressive drop in O_2 to ~14.5%, attributed to soil microbial respiration and concrete carbonation, crucial evidence of nonlinear interactions in closed systems (Nelson et al. 2010).

More recently, the European MELiSSA program developed a modular model of an artificial ecosystem based on microbial and microalgal compartments capable of transforming organic waste, carbon dioxide, and wastewater into oxygen, potable water, and edible biomass (Lasseur et al. 2010). Specifically, the loop comprises (i) degradation/liquefaction (CI–CII), (ii) aerobic nitrification (CIII), (iii) a photosynthetic PBR (CIVa), and (iv) higher plants (CIVb), with the crew as compartment CV; CIVa is colonized by edible cultures (notably *Limnospira indica* PCC 8005) to regenerate air and produce biomass (Gòdia et al. 2002; Hendrickx et al. 2006; Lasseur et al. 2010; Poughon et al. 2020).

At the MELiSSA Pilot Plant (UAB), CIVa is an ~83 L air-lift PBR (illuminated volume ~55 L) operated in continuous mode; the connection to CIII (packed bed ~7–8 L with Nitrosomonas/Nitrobacter) supplies nitrate as a nitrogen source, with near-100% ammonium conversion over periods on the order of hundreds of days, ensuring nitrogen feed to the microalgal PBR (Gòdia et al. 2002; Poughon et al. 2020). Under steady-state conditions, oxygen production by CIVa reached

~1–3 g $O_2 \cdot h^{-1}$ (depending on flow rate and lighting), aligning with demonstration objectives for meeting the demand equivalent to one human (~0.84 kg $O_2 \cdot day^{-1}$) when integrated with the other modules (Poughon et al. 2020; Fahrion et al. 2021). Reported CO_2 biofixation rates for cyanobacteria vary by ~0.10–1.70 $g \cdot L^{-1} \cdot day^{-1}$ depending on light, hydrodynamics, and nutrition (González López et al. 2009; Ryffel et al. 2012; Rosa et al. 2015). Considering ~1 kg $CO_2 \cdot day^{-1} \cdot$person, four astronauts would require continuous volumes on the order of ~2,000 L at the highest rates, with larger volumes if operating with conservative margins or light/dark cycles. This suggests that, for the near term, PBRs should be integrated with physico-chemical scrubbers, while optimization (illuminance, temperature, aeration, and strain selection/carbon concentrating mechanism (CCM)) remains the primary lever (Zhang et al. 2002; Kumar et al. 2011). For CIVa biomass productivity (*Limnospira*), representative values at the MELiSSA Pilot Plant, the ground-based experimental facility of the MELiSSA program, are approximately 0.5–0.7 $g \cdot L^{-1} \cdot day^{-1}$, with cell concentrations reaching up to ~1.6 $g \cdot L^{-1}$ in optimized PBRs (Poughon et al. 2020); complementary ISS trials with *Chlorella vulgaris* demonstrated the feasibility of pumped-loop PBRs in microgravity for O_2/biomass production, supporting the central role of the photosynthetic compartment (Detrell et al. 2019). Operationally, the CIII→CIVa link has run for hundreds of days at ~95–100% ammonium conversion, providing a stable nitrogen supply for continuous cultivation.

Building on the CIII→CIVa linkage, beyond conventional mineral media, pretreated human urine offers a compact, crew-generated in-situ resource utilization (ISRU) feedstock delivering nitrogen and phosphorus to photosynthetic PBRs (Concas et al. 2023). Coupling a nitrifying module (CIII), which oxidizes urea-derived NH_4^+ to NO_3^-, with the microalgal reactor (CIVa) stabilizes the nitrogen supply and enhances loop robustness under power, volume, and mass constraints. Practical implementation hinges on pretreatment (pH buffering, control of organics, and precipitates), anti-fouling design and hydrodynamics to preserve optical performance, and continuous spectroscopic/optical sensing with hazard analysis and critical control points (HACCP)-style safeguards to maintain product quality in closed-loop operation. This ISRU logic is consistent with cultivation strategies experimentally investigated by Fais et al. for growing edible microalgae on Mars by leveraging in-situ resources (Fais et al. 2022b) and aligns with broader ISRU systems frameworks addressing resource loops and microgravity implications (Cao et al. 2021; Fais et al. 2022b). An illustrative scheme of such an ISRU-based BLSS configuration is shown in Figure 17.1, depicting the integration between the ECLSS, microalgal PBR, and Martian resources, inspired by the model proposed by Fais et al. (2022b).

Moreover, the intended downstream uses as food/functional ingredients mirror the nutraceutical perspective, reinforcing the need for rigorous quality control.

Collectively, the MELiSSA architecture has validated: (a) scalability of CIVa (from a few mL to ~80–100 L) with radiative-kinetic modeling for productivity control; (b) liquid integration CIII→CIVa with pH/TOC control and nitrifier containment; (c) functional closure of air–water–nutrient cycles in demonstrator configurations that also include the animal compartment (CV) as a crew mock-up (Gòdia et al. 2002; Lasseur et al. 2010; Poughon et al. 2020). These architectures

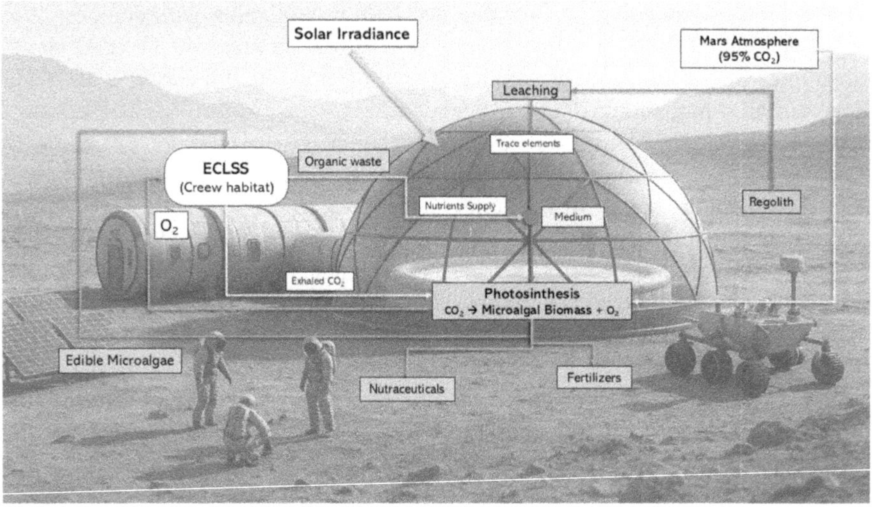

FIGURE 17.1 Example of a conceptual BLSS (bioregenerative life-support system) and ISRU (in-situ resource utilization) integrated scheme for Mars habitats. The diagram illustrates nutrient and gas exchange between the crew habitat (ECLSS) and a microalgal photobioreactor, where exhaled CO_2 and organic waste are converted via photosynthesis into edible biomass and O_2.

bind the water–soil–air nexus: wastewater and metabolic residues provide N/P to the microalgal compartment; atmospheric carbon (CO_2) is fixed into biomass; biomass returns as food/ingredient or biomaterial; oxygen regenerates the atmosphere. The same cycle, replicated on Earth, supports decentralized treatment, nutrient recovery, and low-impact ingredient production.

In China, the Yuegong-1 (Lunar Palace-1) facility conducted the "Lunar Palace-365" experiment (2017–2018), sustaining crews for 370 days in a closed habitat and achieving cycle closure on the order of ~98–98.2% (Liu et al. 2024). In a previous 105-day campaign (2014), the system had already demonstrated near-100% closure for oxygen and water, with ~55% food self-sufficiency (Xie et al. 2017). Despite differing approaches and scales, these results converge on a key point: long-term sustainability requires integrating regenerative biological processes in which microalgae are indispensable components (Table 17.1).

As noted above, the significance of microalgae goes beyond ecological roles in air and water cycle regeneration. They also offer a tangible response to one of the most delicate issues of space missions: maintaining astronaut health under intense physiological and psychological stress. Microgravity and exposure to ionizing radiation drive marked oxidative stress, with the production of reactive oxygen and nitrogen species that damage cell membranes, DNA, and mitochondria (Yatagai et al. 2019; Overbey et al. 2019; Cao et al. 2023; De Micco et al. 2023). This process underlies immunosuppression, cognitive decline, neuroinflammation, and cardiovascular alterations observed in astronauts (Convertino 2005; Aubert et al. 2005; Sonnenfeld 2005; Kalb and Solomon 2007; Morita et al. 2016; Kandarpa et al. 2019; Russomano et al.

TABLE 17.1
Overview of Microalgae as Multifunctional Components of Bioregenerative Life-Support Systems (BLSS) and Terrestrial Circular Bioeconomy Applications

Function in BLSS	Relevant Biological and Technological Parameters	Benefits and Challenges for Space Missions	Terrestrial Applications and Circular Bioeconomy Relevance	Reference Species and Systems	Main References (From Text)
Atmosphere regeneration (O_2/CO_2)	Oxygenic photosynthesis with molar O_2/CO_2 ratio ≈ 1:1; typical productivity 0.5–2.0 $g{\cdot}L^{-1}{\cdot}d^{-1}$ (20–50 $g{\cdot}m^{-2}{\cdot}d^{-1}$ areal); efficiency comparable or superior to terrestrial plants	Supplies O_2 to the crew and fixes emitted CO_2; enables partial closure of air–water cycles; successfully tested under microgravity (PBR@LSR and MELiSSA)	Biological CO_2 capture systems, O_2-producing bioreactors, and closed systems for extreme or urban environments	*Chlorella vulgaris, Limnospira indica*, and *Arthrospira platensis; MELiSSA CIVa, PBR@LSR*	Detrell et al. 2021; Fahrion et al. 2021; Poughon et al. 2020
Nutrition and nutraceuticals	High-protein content (45–70% DW), ω-3 fatty acids, vitamins B and E, and antioxidant pigments (phycocyanin, carotenoids, and chlorophyll); edible and safe biomass (GRAS)	Provides nutrient-dense food and functional compounds counteracting oxidative stress, immunosuppression, and neuroinflammation induced by microgravity and radiation	Production of functional ingredients, supplements, protein flours, and natural pigments; "food-as-medicine" and precision nutrition	*Arthrospira (Spirulina) platensis, Chlorella vulgaris*, and *Haematococcus pluvialis*	Fais, Manca, Concas, et al. 2022b; Karkos et al. 2011; Revellame et al. 2024; Wu et al. 2016
Water and nutrient recycling	Assimilation of N and P from metabolic effluents; $NH_4^+ \rightarrow NO_3^-$ conversion via nitrifying compartment (CIII→CIVa); conversion efficiency >95% in the MELiSSA system	Regeneration of water and nutrient recovery; N–P cycle closure; process stability and microbial containment as key challenges	Biological wastewater treatment, nutrient recovery, circular agriculture, and low-impact purification systems	*MELiSSA Pilot Plant (UAB); Limnospira indica* and *Chlorella vulgaris*	Gòdia et al. 2002; Lasseur et al. 2010; Poughon et al. 2020
Production of biomaterials and bioactive compounds	Production of polysaccharides, pigments, antioxidants, and functional lipids; biomass convertible into pastes, powders, or concentrated extracts	Feedstock for coatings, bioactive ingredients, and functional materials; in-situ production reduces resupply demand	Sustainable production of bioplastics, pigments, functional ingredients, and food additives	*Haematococcus pluvialis, Chlorella vulgaris, Arthrospira platensis*	Fais, Manca, Bolognesi, et al. 2022a; Oslan et al. 2021; Revellame et al. 2024

(Continued)

TABLE 17.1 (Continued)
Overview of Microalgae as Multifunctional Components of Bioregenerative Life-Support Systems (BLSS) and Terrestrial Circular Bioeconomy Applications

Function in BLSS	Relevant Biological and Technological Parameters	Benefits and Challenges for Space Missions	Terrestrial Applications and Circular Bioeconomy Relevance	Reference Species and Systems	Main References (From Text)
Energy and coproduct generation	Biomass conversion into bioH_2 or biogas; thermal energy recovery from lighting systems	Potential auxiliary energy source; waste reduction; integration with physico-chemical subsystems	Microalgal biorefineries; coproduction of energy and fertilizers	*Chlorella sp.*, *Scenedesmus sp.*	Fahrion et al. 2021; Mapstone et al. 2022
Psychological and environmental support	Color chromaticity, living presence, and visual stimulation: perception of "nature" in confined habitats	Enhances psychological well-being and reduces sensory isolation and stress	Bio-design, photosynthetic walls, living walls, and bioreactive architecture	*Chlorella sp.* and *Dunaliella sp.*	Cockell & Raven 2004; Cockell 2020; De Micco et al. 2023
Radiobiological and biomedical protection	Antioxidant pigments (phycobiliproteins and carotenoids) and protective enzymes (SOD and catalase); increased antioxidant activity under moderate radiation doses (~2 kGy)	"Biochemical shield" effect against radiation-induced oxidative damage; prevention of immune and neurodegenerative alterations	Production of radioprotective, antioxidant, and natural cosmetic compounds	*Arthrospira platensis, Haematococcus pluvialis*, and *Chlorella vulgaris*	Badri et al. 2015; Fais, Manca, Bolognesi, et al. 2022a; Shabana et al. 2017)
Advanced biotechnologies (SynBio)	CRISPR, adaptive mutagenesis, and pathway engineering	Targeted production of nutrients or drugs on demand during missions	On-site production of biopharmaceuticals, enzymes, and therapeutic molecules	*Chlamydomonas reinhardtii* and *Synechocystis sp.*	Baqué et al. 2013; Billi 2026; Billi et al. 2011, 2019; Onofri et al. 2025; Santos de Sousa et al. 2025)
Radiobiological and biomedical protection	Accumulation of carotenoids, phycobiliproteins, and antioxidant enzymes (SOD and catalase)	Biochemical shield against ionizing radiation; prevention of DNA damage and oxidative stress	Antioxidant cosmetics and anti-aging and radioprotective supplements	*Haematococcus pluvialis* and *Arthrospira platensis*	Badri et al. 2015; Fais, Manca, Bolognesi, et al. 2022a; Shabana et al. 2017)

2019; Green et al. 2021; Marchal et al. 2024). Leveraging microalgae, due to their biochemical composition, may offer a natural countermeasure (Yang et al. 2012; Neyrinck et al. 2017; Atilgan et al. 2023). *Spirulina*, for example, is characterized by a very high-protein content (60–70% dry weight) together with pigments such as phycocyanin and carotenoids, B-group vitamins, vitamin E, and minerals like iron and magnesium; preclinical models and clinical trials report antioxidant, anti-inflammatory, immunomodulatory, and hypolipidemic activities, with potential neuro- and radioprotective benefits (Karkos et al. 2011; Wu et al. 2016; Fais et al. 2022a). Antioxidant pigments and enzymes (β-carotenes, phycocyanins, superoxide dismutase (SOD), and catalase) in *Arthrospira* increase under moderate stimuli and may help mitigate oxidative stress; gamma doses ≤2.0–2.5 kGy have shown increases in minerals/N-assimilation activities, whereas higher exposures damage PSII (Thaakur and Sravanthi 2010; Badri et al. 2015; Shabana et al. 2017). Edible biomass can thus act as a "biochemical shield" complementing physical protection.

Potential benefits are not limited to the molecular level. Microalgal biomass has been shown to support the immune system by stimulating natural killer cells and modulating cytokine production; it may also aid hematopoietic protection through its iron and vitamin B12 content, mitigating anemia associated with prolonged spaceflight. The central nervous system may likewise benefit from reduced neuroinflammation, with possible positive effects on memory and attention. Added to this is the impact on the gut microbiota: *Spirulina* extracts and pigments such as phycocyanin show prebiotic, probiotic, and antioxidant effects that may support gut-brain axis homeostasis (Fais et al. 2022a). Looking ahead, combining multiple microalgal species could enable the design of "functional" consortia delivering a comprehensive portfolio of nutraceutical and therapeutic benefits.

Nutrition in space is therefore central, and microalgae provide extraordinary added value in this domain. A crucial reference is GRAS status (Generally Recognized As Safe), granted by the FDA and other regulatory authorities to species such as *Arthrospira* (*Spirulina*) and *Chlorella*, attesting to their safety and long terrestrial food tradition (Torres-Tiji et al. 2020; Revellame et al. 2024). This recognition provides the regulatory and scientific basis for their use in extraterrestrial habitats, where tolerability and the absence of toxicological risks are essential criteria. The adoption of pure cultures and dedicated HACCP protocols is essential to avoid contamination by cyanotoxins (microcystins and anatoxins) reported in uncontrolled productions; *Arthrospira* and *Chlorella* are not intrinsically known to produce them, but regular analytical monitoring is needed to guarantee food suitability in closed habitats (Gutiérrez-Salmeán et al. 2015; Furmaniak et al. 2017).

Nutritionally, microalgae are not limited to the role of simple supplements; they can be processed into functional ingredients. Biomass can be converted into rehydratable powders, protein pastes, or extracts enriched in pigments and fatty acids and used in bars, functional beverages, smoothies, doughs for bread and pasta, or high-protein texturized foods. Storage affects the availability of pigments/proteins/antioxidants: in *Arthrospira platensis*, compared with fresh biomass, freezing best preserves phenolics and radical-scavenging capacity; lyophilization maintains high profiles with a lower logistical burden; hot-air drying entails the largest losses; high exposed surface area accelerates degradation of vitamin C/derivatives

(Papalia et al. 2019). Method selection must therefore balance mission energy budgets and nutraceutical requirements. The high nutrient density allows specific needs to be met with small quantities, which is essential where volume and mass are strictly constrained. For example, 10 g of dried *Spirulina* can provide over 6 g of complete protein with markedly lower volume and mass than terrestrial animal sources. In *Arthrospira* (Spirulina), histidine is generally the limiting amino acid (~10 $mg \cdot g^{-1}$): to meet daily needs, a 70-kg astronaut requires ~70 g of dry biomass; other essential amino acids are present in proportions ≥ histidine (National Research Council 1989; Liestianty et al. 2019). *Chlorella* shows a good amino acid profile but with histidine at roughly half the *Spirulina* level, requiring larger doses for equivalent coverage (Shalaby and Yasin 2013). Regarding lipids/ω-3, *Arthrospira* is rich in gamma-linolenic acid (GLA) and linoleic acid (LA) but often contains negligible eicosapentaenoic acid (EPA)/DHA; some *Chlorella* have elevated alpha-linolenic acid (ALA) and, in rare strains, noteworthy EPA, whereas DHA is typical of Dinophyta/Haptophyta/Euglenoids (Ötleş and Pire 2001; Lang et al. 2011). Ensuring ω-3 LC-PUFAs during missions may thus benefit from complementary species consortia or targeted supplementation, given the neuro-cardioprotective role of EPA/DHA (Horrocks and Yeo 1999; Mozaffarian and Wu 2012).

Moreover, the metabolic plasticity of cultures allows tuning of biochemical profiles to mission needs: increasing protein content, enriching carotenoids and antioxidants, or directing production toward specific lipids. This paves the way for genuine precision nutrition in space, where cultivation parameters can be calibrated to deliver biomass with optimized nutritional and functional properties, aligned with dietary needs and preventive medicine protocols. In this sense, PBRs would become "living pharmacies" able to produce customized biomasses according to astronauts' clinical profiles. Beyond serving as nutrient sources, the biomass can be enriched with bioactive compounds, carotenoids, phycobiliproteins, polyphenols, and sterols, exerting antioxidant, immunomodulatory, and neuroprotective actions. Thus, they move beyond the traditional concept of food, assuming the role of "food-as-medicine" that provides targeted protection against key risk factors associated with deep space. *Spirulina*, for instance, via phycocyanin and polyunsaturated fatty acids, may reduce chronic inflammation and oxidative stress, while *Chlorella vulgaris*, rich in chlorophyll and carotenoids, contributes to detoxification and hepatoprotection. *Haematococcus pluvialis*, abundant in astaxanthin, stands out as a potent radioprotective shield (Oslan et al. 2021). Furthermore, combining multiple species in culture consortia, selected and modulated over mission phases, would enable a "dynamic nutraceutical cocktail" that adapts to astronauts' changing physiological conditions.

From an engineering standpoint, these systems hinge on PBRs, which enable controlled cultivation of microalgae. They must be designed to maximize light capture, nutrient distribution, and removal of by-products. In space, microgravity alters fluid behavior, eliminating sedimentation and modifying circulation. This complicates gradient control but can improve light and nutrient uniformity, boosting productivity. Experimental prototypes under simulated microgravity and on the ISS have confirmed that cultures of *Chlorella* and *Spirulina* can not only grow but, under some controlled conditions, also achieve performance comparable to ground controls.

Examples in orbit include: (i) the ArtEMISS/Arthrospira-B PBR, a membrane micro-PBR (≈50 mL) with *Limnospira indica* operated on the ISS in 2017–2018 to verify O_2 production and growth in microgravity and to calibrate a radiative-kinetic model (Detrell et al. 2021); (ii) the "Photobioreactor at the Life Support Rack (PBR@LSR)," launched in 2019 to the Columbus module, with *Chlorella vulgaris* in pumped loops and two reactors adapted to microgravity (total capacity ≈650 mL) to demonstrate long-term cultivation and hybrid integration with ECLSS physico-chemical subsystems (Detrell et al. 2021); (iii) the more recent "Arthrospira-C" campaign, extending flight tests on *Limnospira* with next-generation hardware to validate storage, reactivation, and continuous oxygen/biomass production under space conditions (Poughon et al. 2020).

Despite these milestones, significant challenges remain: preventing biofilm formation that reduces optical efficiency, ensuring long-term sterility, and optimizing energy balances for mixing and lighting, which remains the main technical bottleneck. Long-term analyses of PBR@LSR and MELiSSA benches indicate that biofilms, optical fouling, and process sterility remain primary bottlenecks, requiring in-line (optical/spectroscopic) sensing and advanced control (Poughon et al. 2020; Detrell et al. 2021).

The same properties that make microalgae suitable for long-duration spaceflight translate directly to Earth: CO_2 capture, nutrient recovery, wastewater polishing, and the production of high-value functional ingredients. Commercial deployment is already underway (e.g., EPA/DHA oils for aquafeeds; protein- and pigment-rich ingredients for foods and nutraceuticals), indicating a maturing supply chain from pilot to full scale. In short, space-grade PBR technologies, including high optical efficiency, closed-loop operation, and advanced monitoring, map cleanly onto urban/vertical facilities and decentralized water treatment, accelerating the circular bioeconomy without relying on extensive Earth resupply.

Despite progress, many challenges remain. In space systems, the main difficulty is ensuring long-term culture stability while preventing contamination and mutations that could compromise crew safety. Real-time monitoring of biomass quality is also needed to detect potential accumulation of toxins or heavy metals. PBR engineering must tackle energy balance, maintenance in isolated environments, and process scalability. Finally, the metabolic complexity of microalgae still makes it difficult to predict responses under extreme conditions, necessitating further study in both laboratory settings and experimental missions. To this end, critical steps include the following: (i) developing optical and spectroscopic sensors for real-time detection of pigment variations indicative of stress; (ii) adopting metagenomic and metatranscriptomic approaches for microbial contamination monitoring; (iii) establishing HACCP protocols adapted to closed habitats to guarantee food safety on par with certified terrestrial production.

The evidence gathered in this chapter indicates that microalgae (including cyanobacteria) are a key component of bioregenerative systems in space: they produce oxygen, fix CO_2, recycle nutrients, and provide edible biomass with nutraceutical functions. Demonstrations in microgravity and integrated ground tests (PBR, MELiSSA, and closed habitats) show technical feasibility and pathways to scale, albeit with outstanding bottlenecks (biofilm/fouling, long-term sterility, energy balances for lighting and mixing, and quality monitoring). Hybrid bio + physicochemical integration remains the most robust strategy today, while SynBio and microalgal consortia open avenues for "precision nutrition" and coproduction of functional molecules. In parallel, microalgal biomaterials extend the value of PBRs to non-load-bearing components and functional coatings, reducing dependence on resupply and aligning with ISRU. Looking ahead, operational priorities include in-line sensors for process health, HACCP protocols for closed habitats, radiative-kinetic models for predictive control, and extended-duration demonstration campaigns. The same technologies, transferred to Earth, accelerate the circular bioeconomy (CO_2-to-food/materials, N/P recovery, and water), positioning PBRs as an enabling platform for both space exploration and planetary sustainability.

REFERENCES

Afshinnekoo E, Scott RT, MacKay MJ, et al (2020) Fundamental biological features of spaceflight: Advancing the field to enable deep-space exploration. *Cell* 183:1162–1184. https://doi.org/10.1016/J.CELL.2020.10.050

Aguiló-Nicolau P, Galmés J, Fais G, et al (2023) Singular adaptations in the carbon assimilation mechanism of the polyextremophile cyanobacterium Chroococcidiopsis thermalis. *Photosynth Res* 156:231–245. https://doi.org/10.1007/S11120-023-01008-Y

Alemany L, Peiro E, Arnau C, et al (2019) Continuous controlled long-term operation and modeling of a closed loop connecting an air-lift photobioreactor and an animal compartment for the development of a life support system. *Biochem Eng J* 151. https://doi.org/10.1016/J.BEJ.2019.107323

Atilgan HI, Akbulut A, Yazihan N, et al (2023) The cytokines-directed roles of spirulina for radioprotection of lacrimal gland. *Ocul Immunol Inflamm* 31:271–276. https://doi.org/10.1080/09273948.2022.2026409

Aubert AE, Beckers F, Verheyden B (2005) Cardiovascular function and basics of physiology in microgravity. *Acta Cardiol* 60:129–151. https://doi.org/10.2143/AC.60.2.2005024

Badri H, Monsieurs P, Coninx I, et al (2015) Molecular investigation of the radiation resistance of edible cyanobacterium Arthrospira sp. PCC 8005. *Microbiologyopen* 4:187–207. https://doi.org/10.1002/MBO3.229

Baqué M, De Vera JP, Rettberg P, Billi D (2013) The BOSS and BIOMEX space experiments on the EXPOSE-R2 mission: Endurance of the desert cyanobacterium Chroococcidiopsis under simulated space vacuum, Martian atmosphere, UVC radiation and temperature extremes. *Acta Astronaut* 91:180–186. https://doi.org/10.1016/J.ACTAASTRO.2013.05.015

Barone GD, Cernava T, Ullmann J, et al (2023) Recent developments in the production and utilization of photosynthetic microorganisms for food applications. *Heliyon* 9:e14708. https://doi.org/10.1016/J.HELIYON.2023.E14708

Belz S, Buchert M, Bretschneider J, et al (2014) Physicochemical and biological technologies for future exploration missions. *Acta Astronaut* 101:170–179. https://doi.org/10.1016/J.ACTAASTRO.2014.04.023

Billi D (2026) Desert cyanobacteria under non-Earth conditions: Implications for astrobiology and sustainable life support. *Acta Astronaut* 238:209–214. https://doi.org/10.1016/J.ACTAASTRO.2025.09.022

Billi D, Verseux C, Fagliarone C, et al (2019) A desert cyanobacterium under simulated mars-like conditions in low earth orbit: Implications for the habitability of mars. *Astrobiology* 19:158–169. https://doi.org/10.1089/AST.2017.1807

Billi D, Viaggiu E, Cockell CS, et al (2011) Damage escape and repair in dried chroococcidiopsis spp. from hot and cold deserts exposed to simulated space and Martian conditions. *Astrobiology* 11:65–73. https://doi.org/10.1089/AST.2009.0430

Cao G, Concas A, Orrù R, et al (2023) Recent advances on ISRU technologies and study of microgravity impact on blood cells for deep space exploration. *Front Space Technol* 4:1146461. https://doi.org/10.3389/FRSPT.2023.1146461

Cao G, Pantaleo A, Concas A, et al (2021) Process and kit to investigate microgravity effect on animal/vegetable cells under extraterrestrial cultivation conditions and cultivation process thereof to sustain manned space missions. Google Patents.

Cates G, Stromgren C, Mattfeld B, et al (2016) *The Exploration of Mars Launch & Assembly Simulation*. NASA.

Cockell C (2020) *Astrobiology: Understanding Life in the Universe*. Wiley Blackwell.

Cockell CS, Raven JA (2004) Zones of photosynthetic potential on Mars and the early Earth. *Icarus* 169:300–310. https://doi.org/10.1016/J.ICARUS.2003.12.024

Concas A, Fais G, Enna M, et al (2023) Modeling and experimental assessment of Synechococcus nidulans cultivation using simulated Martian medium and astronauts' urine. *Acta Astronaut* 205:185–198. https://doi.org/10.1016/j.actaastro.2023.01.027

Convertino VA (2005) Consequences of cardiovascular adaptation to spaceflight: Implications for the use of pharmacological countermeasures. *Gravit Space Biol Bull* 18:59–69.

De Micco V, Amitrano C, Mastroleo F, et al (2023) Plant and microbial science and technology as cornerstones to bioregenerative life support systems in space. *NPJ Microgravity* 9:1–12. https://doi.org/10.1038/s41526-023-00317-9

Detrell G, Helisch H, Keppler J, et al (2019) PBR@LSR: The algae-based photobioreactor experiment at the ISS - Operations phase. In: *Proceedings of the International Conference on Environmental Systems*, 1–12.

Detrell G, Verseux C, Trigo-Rodríguez JM (2021) Chlorella vulgaris photobioreactor for oxygen and food production on a moon base-potential and challenges. *Front Astron Space Sci* 8. https://doi.org/10.3389/fspas.2021.700579

Doucha J, Livansky K (1999) Process of outdoor thin-layer cultivation of microalgae and blue-green algae and bioreactor for performing the process. U.S. Patent n. 5,981,271.

Doucha J, Straka F, Lívanský K (2005) Utilization of flue gas for cultivation of microalgae (Chlorella sp.) in an outdoor open thin-layer photobioreactor. *J Appl Phycol* 17:403–412. https://doi.org/10.1007/S10811-005-8701-7

Drake BG (2009) *Human Exploration of Mars Design Reference Architecture 5.0 Mars Architecture*. Steering Group NASA Headquarters.

Eriksen NT, Riisgård FK, Gunther WS, Lønsmann Iversen JJ (2006) On-line estimation of O2 production, CO2 uptake, and growth kinetics of microalgal cultures in a gas-tight photobioreactor. *J Appl Phycol* 19:161. https://doi.org/10.1007/S10811-006-9122-Y

Fahrion J, Mastroleo F, Dussap CG, Leys N (2021) Use of photobioreactors in regenerative life support systems for human space exploration. *Front Microbiol* 12:699525. https://doi.org/10.3389/FMICB.2021.699525

Fais G, Casula M, Sidorowicz A, et al (2024) Cultivation of Chroococcidiopsis thermalis using available in situ resources to sustain life on mars. *Life* 14:251. https://doi.org/10.3390/LIFE14020251

Fais G, Manca A, Bolognesi F, et al (2022a) Wide range applications of spirulina: From earth to space missions. *Marine Drugs* 20:299. https://doi.org/10.3390/MD20050299

Fais G, Manca A, Concas A, et al (2022b) A novel process to grow edible microalgae on Mars by exploiting in situ-available resources: Experimental investigation. *Acta Astronaut* 201:454–463. https://doi.org/10.1016/j.actaastro.2022.09.058

Fili T, Gòdia F, González-Cinca R (2021) Trade-off analysis of phase separation techniques for advanced life support systems in space. *Acta Astronaut* 178:571–583. https://doi.org/10.1016/J.ACTAASTRO.2020.09.052

Furmaniak MA, Misztak AE, Franczuk MD, et al (2017) Edible cyanobacterial genus Arthrospira: Actual state of the art in cultivation methods, genetics, and application in medicine. *Front Microbiol* 8:297783. https://doi.org/10.3389/FMICB.2017.02541

Gitelson II, Degermendzhy AG, Rodicheva EK (2003) Self-restoration as fundamental property of CES providing their sustainability. *Adv Space Res* 31:1641–1648. https://doi.org/10.1016/S0273-1177(03)00111-X

Gòdia F, Albiol J, Montesinos JL, et al (2002) MELISSA: A loop of interconnected bioreactors to develop life support in space. *J Biotechnol* 99:319–330. https://doi.org/10.1016/S0168-1656(02)00222-5

González López CV, Acién Fernández FG, Fernández Sevilla JM, et al (2009) Utilization of the cyanobacteria Anabaena sp. ATCC 33047 in CO2 removal processes. *Bioresour Technol* 100:5904–5910. https://doi.org/10.1016/J.BIORTECH.2009.04.070

Green MJ, Aylott JW, Williams P, et al (2021) Immunity in space: Prokaryote adaptations and immune response in microgravity. *Life* 11:1–17. https://doi.org/10.3390/LIFE11020112

Green RD, Kleinhenz JE (2019) In-situ resource utilization (ISRU) living off the land on the moon and mars. In: *American Chemical Society National Meeting & Exposition*, Orlando, FL.

Grivalský T, Ranglová K, da Câmara Manoel JA, et al (2019) Development of thin-layer cascades for microalgae cultivation: Milestones (review). *Folia Microbiol (Praha)* 64:603–614. https://doi.org/10.1007/S12223-019-00739-7

Gutiérrez-Salmeán G, Fabila-Castillo L, Chamorro-Cevallos G, et al (2015) Aspectos nutricionales y toxicológicos de Spirulina (Arthrospira). *Nutr Hosp* 32:34–40.

Helisch H, Keppler J, Detrell G, et al (2020) High density long-term cultivation of Chlorella vulgaris SAG 211-12 in a novel microgravity-capable membrane raceway photobioreactor for future bioregenerative life support in SPACE. *Life Sci Space Res (Amst)* 24:91–107. https://doi.org/10.1016/J.LSSR.2019.08.001

Hendrickx L, De Wever H, Hermans V, et al (2006) Microbial ecology of the closed artificial ecosystem MELiSSA (micro-ecological life support system alternative): Reinventing and compartmentalizing the Earth's food and oxygen regeneration system for long-haul space exploration missions. *Res Microbiol* 157:77–86. https://doi.org/10.1016/J.RESMIC.2005.06.014

Horrocks LA, Yeo YK (1999) Health benefits of docosahexaenoic acid (DHA). *Pharmacol Res* 40:211–225. https://doi.org/10.1006/PHRS.1999.0495

James JT, Meyers VE, Sipes W, et al (2011) Crew health and performance improvements with reduced carbon dioxide levels and the resource impact to accomplish those reductions. In *41st International Conference on Environmental Systems*, Portland, Oregon (p. 5047). https://doi.org/10.2514/6.2011-5047

Jones HW (2023) *Take Material to Space or Make It There?* NASA.

Kalb R, Solomon D (2007) Space exploration, Mars, and the nervous system. *Arch Neurol* 64:485–490. https://doi.org/10.1001/ARCHNEUR.64.4.485

Kandarpa K, Schneider V, Ganapathy K (2019) Human health during space travel: An overview. *Neurol India* 67:S176–S181. https://doi.org/10.4103/0028-3886.259123

Karkos PD, Leong SC, Karkos CD, et al (2011) Spirulina in clinical practice: Evidence-based human applications. *Evidence-Based Complementary Altern Med* 2011:531053. https://doi.org/10.1093/ECAM/NEN058

Koehle AP, Brumwell SL, Seto EP, et al (2023) Microbial applications for sustainable space exploration beyond low Earth orbit. *NPJ Microgravity* 9. https://doi.org/10.1038/S41526-023-00285-0

Kovic M (2021) Risks of space colonization. *Futures* 126:102638. https://doi.org/10.1016/J.FUTURES.2020.102638

Kumar K, Dasgupta CN, Nayak B, et al (2011) Development of suitable photobioreactors for CO2 sequestration addressing global warming using green algae and cyanobacteria. *Bioresour Technol* 102:4945–4953. https://doi.org/10.1016/j.biortech.2011.01.054

Lang I, Hodac L, Friedl T, Feussner I (2011) Fatty acid profiles and their distribution patterns in microalgae: A comprehensive analysis of more than 2000 strains from the SAG culture collection. *BMC Plant Biol* 11:124. https://doi.org/10.1186/1471-2229-11-124

Lasseur C, Lasseur C, Brunet J, et al (2010) Melissa: The European project of a closed life support system MELiSSA: The European project of closed life support system. *Gravitational Space Biol* 23:3.

Lee C-G, Palsson B (1994) High-density algal photobioreactors using light-emitting diodes. *Biotechnol Bioeng* 44:1161–1167. https://doi.org/10.1002/BIT.260441002

Liestianty D, Rodianawati I, Arfah RA, et al (2019) Nutritional analysis of spirulina sp to promote as superfood candidate. *IOP Conf Ser Mater Sci Eng* 509:012031. https://doi.org/10.1088/1757-899X/509/1/012031

Liu H, Yao Z, Fu Y, Feng J (2021) Review of research into bioregenerative life support system(s) which can support humans living in space. *Life Sci Space Res (Amst)* 31:113–120. https://doi.org/10.1016/J.LSSR.2021.09.003

Liu H, Yao Z, Liu H, et al (2024) Human lunar base: "Lunar Palace 1" team of Beihang University unveils China's "Lunar Palace" plan. *Innovation* 5:100592. https://doi.org/10.1016/J.XINN.2024.100592

Lyons TW, Reinhard CT, Planavsky NJ (2014) The rise of oxygen in Earth's early ocean and atmosphere. *Nature* 506:307–315. https://doi.org/10.1038/NATURE13068

Mapstone LJ, Leite MN, Purton S, et al (2022) Cyanobacteria and microalgae in supporting human habitation on Mars. *Biotechnol Adv* 59. https://doi.org/10.1016/j.biotechadv.2022.107946

Marchal S, Choukér A, Bereiter-Hahn J, et al (2024) Challenges for the human immune system after leaving Earth. *NPJ Microgravity* 10:1–14. https://doi.org/10.1038/s41526-024-00446-9

Mills WR, Pierson DL (2000) Growth and metabolism of the green alga, chlorella pyrenoidosa, in simulated microgravity. National Aeronautics and Space Administration (NASA)/American Society of Engineering Education (ASEE) Summer Faculty Fellowship Program-2000.

Morita H, Abe C, Tanaka K (2016) Long-term exposure to microgravity impairs vestibulo-cardiovascular reflex. *Sci Rep* 6:1–10. https://doi.org/10.1038/SREP33405

Mozaffarian D, Wu JHY (2012) (n-3) Fatty acids and cardiovascular health: Are effects of EPA and DHA shared or complementary? *J Nutr* 142:614S–625S. https://doi.org/10.3945/JN.111.149633

Nangle SN, Wolfson MY, Hartsough L, et al (2020) The case for biotech on mars. *Nat Biotechnol* 38:401–407. https://doi.org/10.1038/S41587-020-0485-4

National Research Council (1989) *Recommended Dietary Allowances*, 10th Edition. National Academies Press.

Nelson M, Pechurkin NS, Allen JP, et al (2010) Closed ecological systems, space life support and biospherics. *Environ Biotechnol*, 517–565. https://doi.org/10.1007/978-1-60327-140-0_11

Neyrinck AM, Taminiau B, Walgrave H, et al (2017) Spirulina protects against hepatic inflammation in aging: An effect related to the modulation of the gut microbiota? *Nutrients* 9. https://doi.org/10.3390/NU9060633

Onofri S, Moeller R, Billi D, et al (2025) Synthetic biology for space exploration. *NPJ Microgravity* 11:1–12. https://doi.org/10.1038/S41526-025-00488-7

Oslan SNH, Tan JS, Oslan SN, et al (2021) Haematococcus pluvialis as a potential source of astaxanthin with diverse applications in industrial sectors: Current research and future directions. *Molecules* 26:6470. https://doi.org/10.3390/MOLECULES26216470

Ötleş S, Pire R (2001) Fatty acid composition of chlorella and spirulina microalgae species. *J AOAC Int* 84:1708–1714. https://doi.org/10.1093/JAOAC/84.6.1708

Overbey EG, da Silveira WA, Stanbouly S, et al (2019) Spaceflight influences gene expression, photoreceptor integrity, and oxidative stress-related damage in the murine retina. *Sci Rep* 9. https://doi.org/10.1038/S41598-019-49453-X

Papalia T, Sidari R, Panuccio MR (2019) Impact of different storage methods on bioactive compounds in Arthrospira platensis biomass. *Molecules* 24. https://doi.org/10.3390/MOLECULES24152810

Park KH, Lee CG (2000) Optimization of algal photobioreactors using flashing lights. *Biotechnol Bioprocess Eng* 5:186–190. https://doi.org/10.1007/BF02936592

Peake D, Black J, Kumbakumba E, et al (2021) Technical results from a trial of the FREO2 low-pressure oxygen storage system, Mbarara Regional Referral Hospital, Uganda. *PLoS One* 16:e0248101. https://doi.org/10.1371/JOURNAL.PONE.0248101

Popova AF (2003) Comparative characteristic of mitochondria ultrastructural organization in chlorella cells under altered gravity conditions. *Adv Space Res* 31:2253–2259. https://doi.org/10.1016/S0273-1177(03)00252-7

Poughon L, Laroche C, Creuly C, et al (2020) Limnospira indica PCC8005 growth in photobioreactor: Model and simulation of the ISS and ground experiments. *Life Sci Space Res (Amst)* 25:53–65. https://doi.org/10.1016/J.LSSR.2020.03.002

Rassool RP, Sobott BA, Peake DJ, et al (2017) A low-pressure oxygen storage system for oxygen supply in low-resource settings. *Respir Care* 62:1582–1587. https://doi.org/10.4187/RESPCARE.05532

Revellame ED, Aguda R, Gatdula KM, et al (2024) Microalgae in bioregenerative life support systems for space applications. *Algal Res* 77:103332. https://doi.org/10.1016/J.ALGAL.2023.103332

Reza-Solis HA, Hernández-Rodríguez OA, Martínez-Rosales AF, Ojeda-Barrios DL (2023) Applications of microalgae in five areas of biotechnology. *Phyton-Int J Exp Bot* 92:2737–2759. https://doi.org/10.32604/PHYTON.2023.029851

Richmond A (2003) *Handbook of Microalgal Culture: Biotechnology and Applied Phycology*. Wiley. https://doi.org/10.1002/9780470995280

Rosa GM, Moraes L, Cardias BB, et al (2015) Chemical absorption and CO2 biofixation via the cultivation of spirulina in semicontinuous mode with nutrient recycle. *Bioresour Technol* 192:321–327. https://doi.org/10.1016/J.BIORTECH.2015.05.020

Rothschild LJ, Mancinelli RL (2001) Life in extreme environments. *Nature* 409:1092–1101. https://doi.org/10.1038/35059215

Russomano T, Da Rosa M, Dos Santos M (2019) Space motion sickness: A common neurovestibular dysfunction in microgravity. *Neurol India* 67:S214–S218. https://doi.org/10.4103/0028-3886.259127

Ryffel GU, Farruggio AP, Chavez CL, et al (2012) Photosynthetic efficiency and rate of CO2 assimilation by Arthrospira (Spirulina) platensis continuously cultivated in a tubular photobioreactor. *Biotechnol J* 7:1412–1417. https://doi.org/10.1002/BIOT.201200177

Santomartino R, Averesch NJH, Bhuiyan M, et al (2023) Toward sustainable space exploration: A roadmap for harnessing the power of microorganisms. *Nat Commun* 14:1–11. https://doi.org/10.1038/S41467-023-37070-2

Santos de Sousa I, Di Stefano G, D'Agostino A, et al (2025) The potential of far-red light-acclimating cyanobacteria to support sustainable outposts on Mars. *Front Astron Space Sci* 12. https://doi.org/10.3389/FSPAS.2025.1658632

Sarmistha N (2001) *Effect of Simulated Microgravity on Growth, Morphological and Biochemical Properties of Chlorella pyrenoidosa*. The University of Houston Clear Lake.

Schirrmeister BE, De Vos JM, Antonelli A, Bagheri HC (2013) Evolution of multicellularity coincided with increased diversification of cyanobacteria and the Great Oxidation Event. *Proc Nat Acad Sci* 110:1791–1796. https://doi.org/10.1073/PNAS.1209927110

Schopf JW (2011) The paleobiological record of photosynthesis. *Photosynth Res* 107:87–101. https://doi.org/10.1007/S11120-010-9577-1

Sengupta A, Wangikar PP (2020) A method to compute instantaneous oxygen evolution rates in cyanobacterial cultures grown in shake flasks. *Eng Rep* 2:e12094. https://doi.org/10.1002/ENG2.12094

Shabana EF, Gabr MA, Moussa HR, et al (2017) Biochemical composition and antioxidant activities of Arthrospira (Spirulina) platensis in response to gamma irradiation. *Food Chem* 214:550–555. https://doi.org/10.1016/J.FOODCHEM.2016.07.109

Shalaby S, Yasin N (2013) Quality characteristics of croissant stuffed with imitation processed cheese containing microalgae Chlorella vulgaris. *Biomass World J Dairy Food Sci* 8, 58–66.

Singla R, Chowdhury K (2021) Enhanced oxygen recovery and energy efficiency in a reconfigured single column air separation unit producing pure and impure oxygen simultaneously. *Chem Eng Process* 162. https://doi.org/10.1016/J.CEP.2021.108354

Snyder JE, Walsh D, Carr PA, Rothschild LJ (2019) A makerspace for life support systems in space. *Trends Biotechnol* 37:1164–1174. https://doi.org/10.1016/j.tibtech.2019.05.003

Sonnenfeld G (2005) The immune system in space, including earth-based benefits of space- based research. *Curr Pharm Biotechnol* 6:343–349. https://doi.org/10.2174/1389201054553699

Tang T, Fadaei H, Hu Z (2014) Rapid evaluation of algal and cyanobacterial activities through specific oxygen production rate measurement. *Ecol Eng* 73:439–445. https://doi.org/10.1016/J.ECOLENG.2014.09.095

Thaakur S, Sravanthi R (2010) Neuroprotective effect of Spirulina in cerebral ischemia–reperfusion injury in rats. *J Neural Transm* 117:1083–1091. https://doi.org/10.1007/S00702-010-0440-5

Thirsk R, Kuipers A, Mukai C, Williams D (2009) The space-flight environment: The international space station and beyond. *CMAJ Can Med Assoc J* 180:1216–1220. https://doi.org/10.1503/CMAJ.081125

Torres-Tiji Y, Fields FJ, Mayfield SP (2020) Microalgae as a future food source. *Biotechnol Adv* 41:107536. https://doi.org/10.1016/J.BIOTECHADV.2020.107536

Verseux C, Baqué M, Lehto K, et al (2016) Sustainable life support on Mars - The potential roles of cyanobacteria. *Int J Astrobiol* 15:65–92. https://doi.org/10.1017/S147355041500021X

Wheeler RM, Sager JC, Prince RP, et al (2003) Crop production for advanced life support systems - Observations from the Kennedy Space Center Breadboard Project. National Aeronautics and Space Administration John F. Kennedy Space Center, Kennedy Space Center.

Williamson J, Wilson JP, Robinson K, Luong H (2024) *Status of ISS Water Management and Recovery*. In *53rd International Conference on Environmental Systems (ICES)* (No. ICES-2024-317).

Wu Q, Liu L, Miron A, et al (2016) The antioxidant, immunomodulatory, and anti-inflammatory activities of Spirulina: An overview. *Arch Toxicol* 90:1817–1840. https://doi.org/10.1007/S00204-016-1744-5

Xie B, Zhu G, Liu B, et al (2017) The water treatment and recycling in 105-day bioregenerative life support experiment in the Lunar Palace 1. *Acta Astronaut* 140:420–426. https://doi.org/10.1016/J.ACTAASTRO.2017.08.026

Yang LL, Zhou QJ, Wang Y, et al (2012) Comparison of the therapeutic effects of extracts from Spirulina platensis and amnion membrane on inflammation-associated corneal neovascularization. *Int J Ophthalmol* 5:32–37. https://doi.org/10.3980/J.ISSN.2222-3959.2012.01.07

Yatagai F, Honma M, Dohmae N, Ishioka N (2019) Biological effects of space environmental factors: A possible interaction between space radiation and microgravity. *Life Sci Space Res (Amst)* 20:113–123. https://doi.org/10.1016/J.LSSR.2018.10.004

Zhang K, Kurano N, Miyachi S (2002) Optimized aeration by carbon dioxide gas for microalgal production and mass transfer characterization in a vertical flat-plate photobioreactor. *Bioprocess Biosyst Eng* 25:97–101.

18 Metrics and Indicators of Sustainability Applied to Microalgal Biotechnology

Rosangela Rodrigues Dias and Eduardo Jacob-Lopes

18.1 INTRODUCTION

Population growth, associated with excessive consumption, has sustained a linear model of extract–make–use–dispose, which disregards the limits imposed by nature. Recent data show that 92.8% of the world economy still operates under this logic, relying almost exclusively on virgin materials (Fraser et al., 2024). This pattern proves unsustainable in the face of the ongoing climate crisis, demanding the construction of alternatives that reconcile economic development and the preservation of the Earth's ecosystems. In this scenario, the circular economy, and more specifically the circular bioeconomy, have emerged as a potential solution (Tan and Lamers, 2021).

Within this perspective, microalgae emerge as strategic agents capable of transforming global challenges into biotechnological opportunities. Their ability to convert sunlight, CO_2, and nutrients into value-added biomass places them at the center of discussions on new paradigms (Biswas, 2024). Microalgal biomass, far from being an isolated resource, can be directed toward various applications that promote the replacement of fossil-based inputs and reduce pressure on finite natural resources. Its biotechnological uses encompass areas as diverse as agriculture, food, health, bioremediation, and biofuels (Reza-Solis et al., 2023). Moreover, microalgae cultivation can be integrated with the utilization of industrial effluents and gases, transforming environmental liabilities into valuable inputs (Abdelfattah et al., 2023; Wang et al., 2025).

Microalgae biotechnology presents itself as an integrative solution capable of linking environmental and economic benefits within the water–soil–air triad. In the water component, microalgae cultivation systems in domestic, agricultural, and industrial wastewater enable the removal of nutrients and pollutants while producing biomass of commercial interest (El Bakraoui et al., 2022). In the soil, products derived from microalgae, such as biofertilizers and biostimulants, promote the recovery of fertility, increase water retention, and substitute chemical inputs, thereby

DOI: 10.1201/9781003611479-18

reducing the environmental pressure of conventional agriculture (Miranda et al., 2024). In the air, microalgal photosynthesis captures CO_2, converting it into biomass and contributing to the mitigation of global warming (Iglina et al., 2022). Moreover, when integrated into biorefinery systems, microalgae generate products and coproducts with a lower environmental footprint. In this sense, microalgae, more than a biological resource, represent a symbol of a break with linear logic (Rodrigues Dias et al., 2023).

However, while all of this is true, technical, economic, environmental, commercial, and political barriers still persist, limiting microalgae-based approaches in circular economy systems. Among the technical challenges, the optimization of biomass productivity; the control of contaminants—especially in bioremediation contexts; the efficiency in capturing and utilizing light and CO_2; and the standardization of processes stand out. Economic barriers include the high cost of production, which, even given the high added value of some products, limits competitiveness compared with conventional inputs. In the commercial sphere, the insertion of microalgae-based products into the market faces obstacles related to consumer acceptance, limited demand, and the complexity of obtaining regulatory certification. On the political plan, the absence of integrated policies supporting biotechnology and the incipient regulation of microalgae cultivation and processing across countries hinders investments, subsidies, and tax incentives. Lastly, regarding environmental barriers, the demand for resources (water and nutrients) and the high energy consumption of bioprocesses can generate direct and indirect emissions sufficient to reduce or even neutralize the potential environmental gains of microalgae cultivation, such as CO_2 capture and effluent treatment (Santos et al., 2025).

In particular, in light of environmental challenges, the relevance of life cycle assessment (LCA) stands out as a technical tool for quantitatively and, ultimately, qualitatively measuring the potential environmental impacts of microalgae-based systems (Carvalho et al., 2022). LCA allows for the quantification of environmental impacts linked to the complete life cycle of products, processes, or services, offering a myriad of information that can guide decisions and, consequently, interventions. This tool enables companies not only to assess the impacts of their processes and products but also to question technological choices, paving the way for business models with concrete environmental benefits (Jacob-Lopes et al., 2021).

In summary, this chapter provides a comprehensive analysis of key environmental sustainability indicators applicable to microalgae biotechnology, organized according to the compartments of the water, soil, and air nexus. Instead of simply describing metrics, this chapter seeks to reflect on how to transform them, questioning the persistence of linear and outdated models.

18.2 FUNDAMENTALS OF ENVIRONMENTAL SUSTAINABILITY ASSESSMENT

Sustainability is a broad concept often structured around three interdependent dimensions: environmental, economic, and social. Together, these three dimensions form what is known as the sustainability tripod, in which sustainable development is achieved if there is a balance between them. Environmental sustainability, the focus of this chapter, refers only to the environmental dimension. In other words, it involves the preservation of ecosystems, the efficient management of natural resources, the reduction of environmental impacts, and the guarantee of intergenerational equity (Vijerathne et al., 2025).

Assessing environmental sustainability is a process aimed at measuring, analyzing, and interpreting the impacts of human activities on the environment. LCA is a tool commonly used to measure the environmental damage of any process, product, or service, considering the flows of materials and energy from cradle to grave. It is structured in four stages, defined by ISO 14040 and 14044 standards, which ensure standardization and comparability between studies. The first stage refers to the definition of objective and scope, in which the purpose of the study, the target audience, the system boundaries, and the functional unit that will serve as the basis for comparison are established. The second stage corresponds to the life cycle inventory, which consists of collecting the input and output flows of the system. Inputs include all resources used, from feedstocks to energy, while outputs encompass products, coproducts, by-products, and emissions. This is one of the most challenging stages of LCA. She creates the information base for subsequent evaluation. The third stage is the environmental life cycle impact assessment (LCIA-E). In this stage, the inventory data are converted into potential environmental impacts, organized into impact categories that can be analyzed at the midpoint level (intermediate points, such as global warming, eutrophication, and acidification) or at the endpoint level (final points, such as damage to human health, ecosystem quality, and climate change). Finally, the fourth stage corresponds to the interpretation, in which the results obtained are critically analyzed in order to generate consistent recommendations to support decision-making (Pacana et al., 2023).

The following sections will discuss in greater detail the midpoint environmental impact categories of LCA, organized according to the water–soil–air nexus compartments, with a special focus on the context of microalgae biotechnology.

18.3 METRICS AND INDICATORS OF ENVIRONMENTAL SUSTAINABILITY IN THE WATER COMPARTMENT

Monitoring environmental indicator metrics related to the water compartment is an essential pillar for identifying critical points and guiding innovations in more circular microalgae cultivation systems. In the context of microalgae biotechnology, whose production depends directly and indirectly on water resources, such indicators become indispensable analytical tools for evaluating water efficiency and the integrity of receiving bodies (de Souza Schneider et al., 2018).

The assessment of environmental sustainability in the water compartment encompasses two intrinsically complementary dimensions: the quantitative, which measures the volumes effectively used throughout the production process, and the qualitative, which reflects the impacts resulting from discharges or transformations in the aquatic ecosystem (Efroymson and Dale, 2015). In LCAs, several impact categories related to water resources have been widely applied, among which the following stand out: Total Water Consumption (TWC), Freshwater Eutrophication (FEU), Freshwater Ecotoxicity (FEC), and Marine Ecotoxicity (MEC) (Dias et al., 2025b).

TWC, generally expressed in cubic meters (m^3), constitutes a quantitative metric that encompasses both direct uses, such as in the culture medium, and indirect uses, for example, the water associated with the generation of inputs such as electricity. Deprá et al. (2020) clearly exemplify direct water consumption. In the cultivation of *Chlorella vulgaris*, water consumption in a raceway pond was estimated at 5 m^3/kg of dry biomass, while in tubular photobioreactors, this value dropped to 1.43 m^3/kg. In summary, closed systems tend to exhibit greater water efficiency, resulting from the combination of higher biomass productivity and a lower evaporation rate—contrasting with the intrinsic characteristics of the technological design of open systems.

The cultivated species also plays a determining role: under similar conditions in a raceway pond, *Chlorella vulgaris* and *Arthrospira platensis* consumed 5 m^3 and 3.12 m^3/kg of dry biomass, respectively (Deprá et al., 2020). In other words, it should be noted that water efficiency is revealed not only as a function of technological design, but also of the biology of the selected species. Furthermore, it is important to highlight that the reuse of the culture medium or the use of domestic, agricultural, and industrial wastewater can significantly reduce water consumption, alleviating pressure on water resources and resulting in better environmental performance, as reported by Herrera et al. (2021). However, it is noteworthy that, although TWC quantifies volumes, it does not distinguish the origin of the water—a gap filled by the water footprint indicator, which differentiates water consumption into blue water footprint (surface or groundwater), green water footprint (rainwater), and gray water footprint (freshwater needed to dilute the pollutant load) (Gerbens-Leenes et al., 2014).

FEU, expressed in phosphorus equivalents (kg P-eq.), is a qualitative indicator that measures the potential of productive processes to promote excessive nutrient enrichment in water bodies. The unbalanced input of nutrients into bodies of water stimulates the excessive growth of algae and aquatic plants, which can lead to reduced water transparency, the occurrence of hypoxia in oceans, rivers, and seas, and compromise the multiple uses of water (de Paula Pereira et al., 2023). In microalgae production, FEU presents a paradoxical character: on the one hand, nutritional supplementation increases the risk of eutrophication; on the other hand, when wastewater is used as an input, the assimilation of nutrients by microalgae can transform this environmental liability into an ecosystem service, mitigating pollutant loads. In LCAs, the results vary substantially depending on the functional unit, system boundaries, and technological routes analyzed. Rodrigues Dias et al. (2023), for example, reported FEU values between ≈1250.00 and 3.69 kg P-eq. for the annual production of 9.38 tons of biomass fractionated into β-carotene, bulk oil, and defatted biomass. According to the authors, the adoption of renewable electricity instead of fossil fuel-based electricity in the upstream and downstream stages led to a significant reduction in the pollutant load that could be released, demonstrating significant environmental gains. This result reveals the centrality of technological choices—notably the energy matrix—in minimizing impacts of this category on the water compartment and in promoting the Sustainable Development Goals, especially SDG 6 (Clean water and sanitation).

FEC and MEC are qualitative indicators that express the ecotoxicological potential associated with chemical emissions throughout the life cycle of a process, product, or service, which can cause adverse effects on aquatic organisms and, consequently, compromise the functional integrity of these ecosystems (Oginah et al., 2023). Both are commonly expressed in 1,4-dichlorobenzene equivalents (kg 1,4-DCB-eq.). Wimmerova et al. (2022), when comparing three microalgae cultivation systems—phototrophic cascade, heterotrophic fermenter, and flat-panel photobioreactor—reported FEC values of 60.26, 67.44, and 1.85 kg 1,4-DCB-eq./kg of biomass and MEC values of 160.8×10^3, 176.5×10^3, and 29.8×10^2 kg 1,4-DCB-eq./kg of biomass, respectively. These results also highlight the impact of technological choices on ecotoxicological responses and, above all, the possibility of making conscious choices in favor of sustainability in the water compartment.

In sum, it is important to emphasize that the inclusion of qualitative indicators in LCAs allows for the revelation of latent risks, often hidden in purely quantitative metrics, ensuring that microalgae-based processes can promote effective net benefits to the water compartment and the integrity of aquatic ecosystems. Finally, microalgae biotechnology is not limited to generating biomass, but to redesigning, in dialogue with water resources, more harmonious pathways for a more circular, regenerative, and integrated production model aligned with natural cycles (Deprá et al., 2022).

18.4 METRICS AND INDICATORS OF ENVIRONMENTAL SUSTAINABILITY IN THE SOIL COMPARTMENT

Soil is a frequently neglected compartment in environmental sustainability assessments. However, when degraded, it compromises the physical and biogeochemical bases that sustain life on Earth (Delgado and Gómez, 2024). Like other productive systems that demand inputs, energy, and the disposal of by-products, microalgae-based processes also exert pressure on the terrestrial environment; therefore, they must be evaluated under this narrative. The understanding of the environmental sustainability of the soil compartment is supported by a critical tripod composed of the land use (LU), terrestrial ecotoxicity (TET), and terrestrial acidification (TA) indicators. Reflecting on these indicators and on how much land can be occupied, how much toxicity can be tolerated, and how much acidification can be allowed before the system fails or loses its regenerative capacity is to get to the core of the resilience of terrestrial ecosystems.

The LU (m^2a crop eq.) indicator represents the relative loss of local species due to various types of LU, such as annual crops and permanent crops (Huijbregts et al., 2017). The two types of LU impact are LU change (transformation) and land occupation. LU change involves the conversion of natural ecosystems, such as forests, into agricultural areas, sometimes irreversibly altering the structure, function, and composition of the landscape. Land occupation, in turn, refers to the period during which the area remains committed to productive purposes, preventing its natural regeneration (Ferreira et al., 2021). In this sense, it is worth highlighting that behind this indicator lies the conversion of complex ecological systems into simplified monocultures, with direct effects on biodiversity loss, decreased soil fertility, reduced carbon sequestration capacity, and microclimatic changes. In evaluations of microalgae-based systems, the LU indicator is essential to demonstrate—or refute—the hypothesis that production from these microorganisms can reduce pressure on LU. This is because microalgae generally exhibit higher production efficiency per unit area and do not require arable land, allowing them to be cultivated in regions with low agricultural suitability, not competing with the traditional agricultural sector, but complementing it (Sanjurjo et al., 2024).

The TEC indicator represents a less visible, but no less alarming, dimension: the chemical load emitted by production processes and their potential adverse effects on organisms living in the soil (Sydow et al., 2020). In LCAs, TEC is expressed in kilograms of 1,4-dichlorobenzene equivalent (kg 1,4-DCB eq), a reference substance representing the standard toxic effect (Cespi, 2023). In microalgal systems, the use of synthetic fertilizers and nutrients; inadequate disposal of residual biomass or cultivation effluents; as well as the use of solvents and reagents in extraction and purification steps, contribute to the ecotoxicological potential. The resulting impacts can lead to a reduction in functional biodiversity and degradation of ecosystem services provided by the soil. Thus, from an environmental sustainability perspective, low TEC values reflect environmentally clean systems capable of protecting the living base of the biosphere—the soil.

Lastly, the TAC indicator reflects the potential of a process, product, or service to contribute to atmospheric deposition of acidifying pollutants, such as nitrogen oxides (NOx), sulfur dioxide (SO_2), and ammonia (NH_3), and their subsequent deposition on the soil. These emissions reverberate through the soil, modifying its pH, altering microbial communities, and compromising vital ecosystem functions. In the long term, these effects reduce soil fertility, biodiversity, and ecological resilience (Roy et al., 2014). TAC, expressed in kilograms of sulfur dioxide equivalent (kg SO_2 eq) per functional unit (e.g., per kg of dry biomass), is not only an impact indicator but also a tool for diagnosing and improving the environmental performance of microalgae-based systems. In these microbiological systems, TAC is mainly associated with indirect emissions derived especially from the consumption of electricity from fossil fuels and the inputs used in the upstream, midstream, and downstream stages of bioprocesses. Low TAC values can reinforce the eco-efficiency of microalgae-based routes compared to conventional ones, contributing to the maintenance of soil quality and chemical balance. Finally, it is important to emphasize that the soil is a faithful mirror of our relationship with the planet, reflecting the balance or imbalance in how we produce, consume, and regenerate the resources that sustain our life on Earth.

18.5 METRICS AND INDICATORS OF ENVIRONMENTAL SUSTAINABILITY IN THE AIR COMPARTMENT

The environmental sustainability assessment of microalgae-based biotechnological systems in the air compartment primarily focuses on atmospheric pollutant emissions throughout the life cycle. Although microalgae production is often considered, empirically, an environmentally promising alternative due to its potential to capture CO_2 and replace fossil feedstocks, many LCA studies have shown that the environmental performance of these systems strongly depends on conditioning factors (Dias et al., 2022). In other words, they essentially depend on the technological and energy context in which they are implemented, being influenced by the type of energy used (renewable or non-renewable), the origin of nutrients, and the cultivation and processing technologies adopted, which can lead to mitigation of impacts or even to outcomes equivalent to or greater than those of conventional processes (Deprá et al., 2020).

In the context of the air compartment, the main environmental impact indicators include: fine particulate matter formation (FPMF), fossil resource scarcity (FRS), global warming potential (GWP), ozone formation affecting terrestrial ecosystems (OFTE), stratospheric ozone depletion (SOD), and ionizing radiation (IR). These indicators allow for a holistic and comparative understanding of potential impacts in the air compartment, reflecting both direct and indirect emissions associated with energy and input consumption (Jacob-Lopes et al., 2021).

The FPMF indicator reflects the potential emission of inhalable particles ($PM_{2.5}$) associated, for example, with the consumption of materials and energy throughout the upstream, midstream, and downstream stages of microalgae-based processes. Fine particulate matter ($PM_{2.5}$), in addition to influencing atmospheric processes such as surface temperature, solar radiation, and the properties of clouds and precipitation, is considered one of the main air pollutants affecting human health, and is associated with approximately 4.1 million premature deaths per year worldwide (Zhang et al., 2019). These particles, with a diameter of 2.5 micrometers, are widely studied for their ability to penetrate and settle in the respiratory system. Furthermore, they not only exacerbate air pollution but also cause adverse effects with regard to nutrient deposition in sensitive ecosystems (Giusti et al., 2022). In the context of microalgae biotechnology, $PM_{2.5}$ emissions can originate mainly from indirect sources, such as fossil fuel electricity generation, the transport of inputs and products, and the production of fertilizers and reagents used in the cultivation and processing of microalgal biomass. Thus, it is worth highlighting that dependence, especially on non-renewable energy, can increase the FPMF, compromising the sustainability of the system. Regarding this, although open systems, such as raceway ponds, have lower energy demands, closed configurations, such as tubular photobioreactors, tend to intensify electricity use, consequently increasing indirect $PM_{2.5}$ emissions (Dias et al., 2023).

In the case of the FRS indicator, products derived from microalgae stand out for their potential to reduce dependence on these resources, replacing petrochemical derivatives with renewable alternatives, such as biofuels. However, on the other hand, the magnitude of this benefit is strongly conditioned by the energy matrix and the inputs used. Systems based on synthetic fertilizers, industrially generated CO_2, and especially fossil fuel-based electricity can increase the FRS, thereby canceling the expected environmental gains. According to Silva et al. (2022), the use of fossil electricity explains the FRS results in microalgae-based processes, where no technology is truly sustainable and green if, in its essence, it still depends on carbon. Adopting integrated strategies, such as the use of liquid and gaseous effluents and renewable and clean energy, significantly reduces this impact, reinforcing the idea that the circular and sustainable nature of microalgae biotechnology should be further explored.

GWP is undoubtedly one of the most emblematic indicators of LCA within the air compartment, representing a true reflection of our relationship with carbon. Although microalgae fix CO_2 during photosynthesis, the net carbon balance depends almost exclusively on the energy that sustains each stage of the process (cultivation, harvesting, and biomass conversion). In fossil fuel-based scenarios, the ideal of neutrality falls apart, and the GWP can equal, or even surpass, that of conventional processes. However, when systems align with clean and renewable sources and feed off residual CO_2 flows, they can transform into carbon sinks, and, ultimately, offsetting remaining emissions is the trampoline for achieving true net climate mitigation, that is, a zero or negative carbon balance (Dias et al., 2022).

The OFTE indicator quantifies the potential for low-altitude photochemical ozone generation resulting from the emission of precursors such as nitrogen oxides (NOx) and volatile organic compounds. This ozone is a secondary pollutant whose potential impact on terrestrial vegetation is assessed by the way it damages plant tissues, reduces photosynthesis, and affects the productivity and biodiversity of terrestrial ecosystems (Khangar & Thangavel, 2024).

The OFTE complements the soil compartment indicators and is commonly expressed in equivalent units of kg NOx eq. In microalgae systems, the main sources of NOx are indirect, associated with the transport of inputs and products and the generation of electrical and thermal energy that powers the bioprocesses. Replacing combustion engine vehicles with electric vehicles charged with renewable electricity, along with the transition from using electricity from fossil fuels to renewable electricity in commercial facilities, can substantially reduce OFTE (Dias et al., 2025a, 2025b; Oliveira et al., 2020). Furthermore, it is worth highlighting that integrated microalgae biorefineries assume a strategic role, transforming environmental liabilities into assets of atmospheric mitigation (Rodrigues Dias et al., 2023).

The SOD allows a greater fraction of ultraviolet B (UV-B) radiation to reach the Earth's surface, resulting in adverse effects not only on human and animal health, but also on ecosystems and biogeochemical cycles. Within the context of LCAs, the SOD impact category is expressed in CFC-11 equivalents (kg CFC-11 eq) and assumes a global scope, as the emission and transport processes of ozone-depleting agents affect the stratospheric layer on a planetary scale (Durante et al., 2015). The most significant role in the destruction of the ozone layer is played by emissions of chlorofluorocarbons (CFCs), which are mainly used in air conditioning and refrigeration equipment, and by NOx emissions, which are produced by the combustion processes of fossil fuels (Kalogirou, 2014).

Although direct unit operations involving microalgae have a low potential for emitting ozone-depleting substances, the reliance on infrastructure with conventional refrigeration systems can contribute to indirect emissions. Adopting environmentally friendly refrigeration technologies and using agents' refrigerants with low ozone depletion potential are essential measures to mitigate this impact. Furthermore, the use of energy from fossil fuels can generate indirect emissions of NOx. Thus, the transition from the use of fossil fuels to renewable energy sources and the improvement of energy efficiency in the cultivation and processing stages also represent fundamental strategies to minimize the indirect contribution of microalgae technologies to the emission of these compounds.

Concerning the IR indicator, it quantifies the potential environmental impact associated with the release of radionuclides, primarily from the nuclear fuel cycle and nuclear medicine (Paulillo et al., 2023). In the context of microalgae production, this impact is indirect and stems mainly from the electricity consumption of national systems that include nuclear energy in their composition. Thus, it is noteworthy that the greater the grid's dependence on nuclear electricity, the higher the IR value tends to be. The adoption of renewable sources, such as wind power or hydroelectric power, in addition to reducing GHG emissions, practically eliminates the impact of IR, becoming a decisive vector for the sustainability of the process (Hosseini et al., 2022).

In general, the analysis of these indicators reveals that the energy matrix is the main determinant of the environmental performance of microalgae-based systems due to their high energy consumption (Deprá et al., 2020). Integration with residual CO_2 and nutrient streams also emerges as a key point for promoting more circular production cycles with reduced atmospheric emissions—provided that biomass productivity is not compromised (Dias et al., 2025a). It is important to highlight that reading each indicator individually only allows one to identify fragments of an interconnected system. In other words, a joint and systemic analysis of these indicators is fundamental to understanding the environmental trade-offs and synergies that shape the contours of sustainability. Recognizing the interdependence between environmental indicators means understanding that none of them, in isolation, is capable of translating the complexity of microalgae-based systems. Furthermore, ignoring this relationship could lead to biased interpretations and conclusions. A holistic approach enables more balanced and effective strategies, guiding the implementation of technologies that maximize environmental benefits while minimizing negative impacts.

18.6 CONCLUSION

The analysis of indicators and some sustainability metrics applicable to microalgae biotechnology shows that the environmentally favorable nature of these microorganisms is not a direct consequence of their biology, but rather the result of the technological and systemic choices that guide their production. Microalgae constitute a unique biotechnological model in which effective sustainability depends on the incorporation of eco-design principles, the use of clean and renewable energy sources, and the processes' integration and intensification to maximize the potential environmental benefits. In this bias, it should be mentioned that LCA has proven to be a powerful tool for measuring and understanding the complex network of direct and indirect impacts of microalgae-based processes. LCA makes it possible to identify critical points and opportunities for improvement that bring microalgae biotechnology closer to a symbiotic model between economy and ecosystem, configuring it as a vector of paradigmatic transformation toward sustainability. Ultimately, it is noteworthy that microalgae biotechnology as a pillar of the circular bioeconomy will depend almost exclusively on our ability to transform knowledge into action, converting metrics into decisions and indicators into effective pathways to sustainability.

REFERENCES

Abdelfattah, A., Ali, S. S., Ramadan, H., El-Aswar, E. I., Eltawab, R., Ho, S. H., ... & Sun, J. (2023). Microalgae-based wastewater treatment: Mechanisms, challenges, recent advances, and future prospects. *Environmental Science and Ecotechnology*, *13*, 100205.

Biswas, T. (2024). Microalgae: Nature's green gold and their potential to meet sustainable development goals. In Shaon Ray Chaudhuri (ed.), *Application of Microbial Technology in Wastewater Treatment and Bioenergy Recovery* (pp. 283–301). Springer Nature Singapore.

Carvalho, J., Nascimento, L., Soares, M., Valério, N., Ribeiro, A., Faria, L., ... & Vilarinho, C. (2022). Life cycle assessment (LCA) of biochar production from a circular economy perspective. *Processes*, *10*(12), 2684.

Cespi, D. (2023). Bio-based chemicals from dedicated or waste biomasses: Life cycle assessment for evaluating the impacts on land. *Sustainable Chemistry*, *4*(2), 184–196.

de Paula Pereira, A. S. A., Magalhães, I. B., Ferreira, J., de Siqueira Castro, J., & Calijuri, M. L. (2023). Microalgae organomineral fertilizer production: A life cycle approach. *Algal Research*, *71*, 103035.

de Souza Schneider, R. D. C., de Moura Lima, M., Hoeltz, M., de Farias Neves, F., John, D. K., & de Azevedo, A. (2018). Life cycle assessment of microalgae production in a raceway pond with alternative culture media. *Algal Research*, *32*, 280–292.

Delgado, A., & Gómez, J. A. (2024). The soil: Physical, chemical, and biological properties. In Francisco J. Villalobos and Elias Fereres (eds.), *Principles of Agronomy for Sustainable Agriculture* (pp. 15–30). Springer International Publishing.

Deprá, M. C., Jacob-Lopes, E., & Zepka, L. Q. (2022). Introductory chapter: Microalgal biotechnology-the gold-green in modern sustainable world. In Leila Queiroz Zepka, Eduardo Jacob-Lopes, and Mariany Costa Deprá (eds.), *Progress in Microalgae Research-A Path for Shaping Sustainable Futures*. IntechOpen.

Deprá, M. C., Severo, I. A., dos Santos, A. M., Zepka, L. Q., & Jacob-Lopes, E. (2020). Environmental impacts on commercial microalgae-based products: Sustainability metrics and indicators. *Algal Research*, *51*, 102056.

Dias, R. R., Deprá, M. C., de Menezes, C. R., Zepka, L. Q., & Jacob-Lopes, E. (2025a). Microalgae cultivation in wastewater: How realistic is this approach for value-added product production? *Processes*, *13*(7), 2052.

Dias, R. R., Deprá, M. C., Rodrigues Carvalho, V. C., de Menezes, C. R., Zepka, L. Q., & Jacob-Lopes, E. (2025b). Decarbonizing the transport of microalgae-based products—The role of E-mobility. *Recent Patents on Biotechnology*, *19*(2), 161–176.

Dias, R. R., Depra, M. C., Sartori, R. B., Schneider, A. T., Zepka, L. Q., & Jacob-Lopes, E. (2023). Integrating solar cell technologies in microalgae facilities: Environmental sustainability metrics and indicators. *Sustainable Energy Technologies and Assessments*, *58*, 103362.

Dias, R. R., Deprá, M. C., Zepka, L. Q., & Jacob-Lopes, E. (2022). Roadmap to net-zero carbon emissions in commercial microalgae-based products: Environmental sustainability and carbon offset costs. *Journal of Applied Phycology*, *34*(3), 1255–1268.

Durante, S., Comoglio, M., & Ridgway, N. (2015). Life cycle assessment in nanotechnology, materials and manufacturing. In Qin Yi (ed.), *Micromanufacturing Engineering and Technology* (Chapter 34, pp. 775–804). Springer Nature Switzerland.

Efroymson, R. A., & Dale, V. H. (2015). Environmental indicators for sustainable production of algal biofuels. *Ecological Indicators*, *49*, 1–13.

El Bakraoui, H., Slaoui, M., Mabrouki, J., Hmouni, D., & Laroche, C. (2022). Recent trends on domestic, agricultural and industrial wastewaters treatment using microalgae biorefinery system. *Applied Sciences*, *13*(1), 68.

Ferreira, J., Esteves, B., Ayata, Ü., Cruz-Lopes, L., & Domingos, I. (2021). Land use impact of maritime pine and eucalypt: A life cycle assessment study. *BioResources*, *16*(2), 3760.

Fraser M, Conde Á, Haigh L (2024). Circularity gap report 2024.

Gerbens-Leenes, P. W., Xu, L., De Vries, G. J., & Hoekstra, A. Y. (2014). The blue water footprint and land use of biofuels from algae. *Water Resources Research*, *50*(11), 8549–8563.

Giusti, G., Vieira, J. G. V., de Souza Tadano, Y., Silva, D. A. L., & Fantke, P. (2022). Health effects of particulate matter formation in life cycle impact assessment: Critical review and recommendation of models for Brazil. *The International Journal of Life Cycle Assessment*, *27*(6), 868–884.

Herrera, A., D'Imporzano, G., Fernandez, F. G. A., & Adani, F. (2021). Sustainable production of microalgae in raceways: Nutrients and water management as key factors influencing environmental impacts. *Journal of Cleaner Production*, *287*, 125005.

Hosseini, S. M., Kanagaraj, N., Sadeghi, S., & Yousefi, H. (2022). Midpoint and endpoint impacts of electricity generation by renewable and nonrenewable technologies: A case study of Alberta, Canada. *Renewable Energy*, *197*, 22–39.

Huijbregts, M. A., Steinmann, Z. J., Elshout, P. M., Stam, G., Verones, F., Vieira, M., … & Van Zelm, R. (2017). ReCiPe2016: A harmonised life cycle impact assessment method at midpoint and endpoint level. *The International Journal of Life Cycle Assessment*, *22*(2), 138–147.

Iglina, T., Iglin, P., & Pashchenko, D. (2022). Industrial CO2 capture by algae: A review and recent advances. *Sustainability*, *14*(7), 3801.

Jacob-Lopes, E., Zepka, L. Q., & Deprá, M. C. (2021). *Sustainability Metrics and Indicators of Environmental Impact: Industrial and Agricultural Life Cycle Assessment*. Elsevier.

Kalogirou, S. A. (2014). *Solar Energy Engineering: Processes and Systems*. AP.

Khangar, N. S., & Thangavel, M. (2024). Assessment of environmental impacts: A life cycle analysis of wheat and rice production in Madhya Pradesh. *Agronomy Research*, *22*(S2), 636–658. https://doi.org/10.15159/AR.24.011

Miranda, A. M., Hernandez-Tenorio, F., Villalta, F., Vargas, G. J., & Sáez, A. A. (2024). Advances in the development of biofertilizers and biostimulants from microalgae. *Biology*, *13*(3), 199.

Oginah, S. A., Posthuma, L., Maltby, L., Hauschild, M., & Fantke, P. (2023). Linking freshwater ecotoxicity to damage on ecosystem services in life cycle assessment. *Environment International*, *171*, 107705.

Oliveira, G. M., Caetano, N., Mata, T. M., & Martins, A. A. (2020). Biofixation of CO_2 emissions from natural gas combined cycle power plant. *Energy Reports*, *6*, 140–146.

Pacana, A., Siwiec, D., Bednárová, L., & Petrovský, J. (2023). Improving the process of product design in a phase of life cycle assessment (LCA). *Processes*, *11*(9), 2579.

Paulillo, A., McKone, T. E., & Fantke, P. (2023). Characterizing human health damage from ionizing radiation in life cycle assessment. *The International Journal of Life Cycle Assessment*, *28*(12), 1723–1734.

Reza-Solis, H. A., Hernández-Rodríguez, O. A., Martínez-Rosales, A. F., & Ojeda-Barrios, D. L. (2023). Applications of microalgae in five areas of biotechnology. *Phyton (0031-9457)*, *92*(10), 2737.

Rodrigues Dias, R., Deprá, M. C., Ragagnin de Menezes, C., Queiroz Zepka, L., & Jacob-Lopes, E. (2023). The high-value product, bio-waste, and eco-friendly energy as the tripod of the microalgae biorefinery: Connecting the dots. *Sustainability*, *15*(15), 11494.

Roy, P. O., Azevedo, L. B., Margni, M., van Zelm, R., Deschênes, L., & Huijbregts, M. A. (2014). Characterization factors for terrestrial acidification at the global scale: A systematic analysis of spatial variability and uncertainty. *Science of the Total Environment*, *500*, 270–276.

Sanjurjo, C., Oulego, P., Bartolomé, M., Rodríguez, E., Gonzalez, R., & Battez, A. H. (2024). Biodiesel production from the microalgae Nannochloropsis gaditana: Optimization of the transesterification reaction and physicochemical characterization. *Biomass and Bioenergy*, *185*, 107240.

Santos, B., Freitas, F., Sobral, A. J., & Encarnação, T. (2025). Microalgae and circular economy: Unlocking waste to resource pathways for sustainable development. *International Journal of Sustainable Engineering*, *18*(1), 2501488.

Silva, T., Ferreira, J., Castro, J., Braga, M., & Calijuri, M. L. (2022). Microalgae from food agro-industrial effluent as a renewable resource for agriculture: A life cycle approach. Available at SSRN 4109357.

Sydow, M., Chrzanowski, Ł., Hauschild, M. Z., & Owsianiak, M. (2020). Influence of metal speciation on soil ecotoxicity impacts in life cycle assessment. *Journal of Environmental Management*, *266*, 110611.

Tan, E. C., & Lamers, P. (2021). Circular bioeconomy concepts—A perspective. *Frontiers in Sustainability*, *2*, 701509.

Vijerathne, D. T., Wahala, W. M. P. S. B., Rathnasiri, M. S. H., De Silva, V., & Dewasiri, N.J. (2025). Literature review on sustainability. *Greening our economy for a sustainable future*, 33–45.

Wang, R., Wang, X., & Zhu, T. (2025). Research progress and application of carbon sequestration in industrial flue gas by microalgae: A review. *Journal of Environmental Sciences*, *152*, 14–28.

Wimmerova, L., Keken, Z., Solcova, O., & Vavrova, K. (2022). A comparative analysis of environmental impacts of operational phases of three selected microalgal cultivation systems. *Sustainability*, *15*(1), 769.

Zhang, H., Li, R., Chen, B., Lin, H., Zhang, Q., Liu, M., … & Wang, X. (2019). Evolution of the life cycle primary PM2. 5 emissions in globalized production systems. *Environment International*, 131, 104996.

19 Life-Cycle Assessment Pathways toward Sustainable Marine Microalgal Biotechnology

Neha Arora, Elizabeth Pyle, and George P. Philippidis

19.1 INTRODUCTION

The marine ecosystem spanning pelagic, coastal, benthic, and extreme environments is a rich source of microbial diversity. Across the microbial communities, microalgae form the base of the marine microbiome, driving the structure and function of the ecosystem. Marine microalgae are among the earth's primary species that convert inorganic carbon (CO_2) to organic carbon (biomass) and produce the majority of atmospheric O_2 in addition to assimilating exogenous nitrogen and phosphorus and serve as the food source for several marine organisms (Nishshanka et al. 2022). Thanks to their metabolic ability to adapt to a wide range of ecosystems, algae are a promising source of sustainable food, energy, and high-value bioactive compounds, such as polyunsaturated fatty acids (PUFAs), pigments, carotenoids, vitamins, and polysaccharides (Figure 19.1). In fact, due to a continuous increase in the demand for natural marine products, 28,000 compounds have been isolated from microalgae to date, with hundreds more being discovered every year (Carroll et al. 2019; Lauritano et al. 2019). According to the global forecast, microalgae-based bioproducts will reach \$1.8 billion with a 10.3% compounded annual growth rate (CAGR) by 2028 (Selvam et al. 2024). Key commercially important algal metabolites include omega-3 fatty acids, β-carotene, astaxanthin, lutein, essential amino acids, whole-cell protein, sulfated polysaccharides, vitamins (B12, E, and ascorbic acid), antimicrobial peptides, and biostimulants with applications in the food, energy, cosmeceuticals, pharmaceutical, nutraceutical, agriculture, and aquaculture industries.

Marine microalgae, such as *Nannochloropsis* sp., *Dunalliella salina*, and *Euglena gracilis*, have been exploited for industrial biofuel and bioproduct (omega-3 fatty acids, animal feed, and aquaculture) production by various companies, including Cellana (Hawaii, United States); Veramaris (DSM-Evonik joint venture, the

DOI: 10.1201/9781003611479-19

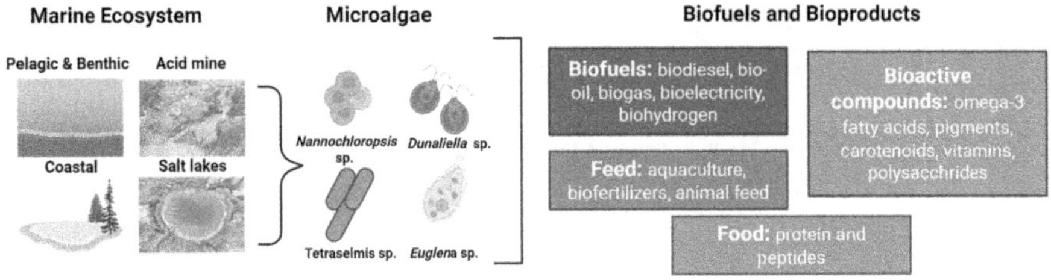

FIGURE 19.1 Marine microalgae-derived biofuels and bioproducts.

TABLE 19.1
Industrially Produced Marine Microalgae Products

Microalgae	Company	Location	Main Product(s)	Website Link
Nannochloropsis oculata + proprietary marine strains	Cellana, Inc.	Hawaii and San Diego, United States	ReNew™ Algae-Omega-3 oils, biofuels, and feed	https://phytosmart.com/cellana
Marine microalgae (EPA/DHA)	Veramaris (DSM/Evonik JV)	Netherlands and United States	EPA/DHA algal oils	https://www.veramaris.com
Nannochloropsis sp.	BlueBioTech GmbH	Germany	Pigments and feed/food ingredients	https://www.bluebiotech-gmbh.de/en
	Arizona Algae Products LLC	United States	EPA	https://www.azalgae.com/contact
Marine microalgae (*Nannochloropsis* sp.)	A4F/Aliga	Denmark/Portugal	Algal oils and biomass	https://a4f.pt/en
Tetraselmis chui	All microalgae	Portugal	Algal biomass	https://www.allmicroalgae.com/en
Dunaliella salina	Now foods	United States	Natural β-carotene softgels	https://www.nowfoods.com
Mixed marine microalgae	AlgaEnergy	Spain	Larval feed	https://www.algaenergy.com/en/index
Euglena gracilis	Euglena Co., Ltd	Japan	Food, biofuels, and paramylon	https://www.euglena.jp/en

EPA, eicosapentaenoic acid; DHA, docosahexaenoic acid.

Netherlands and the United States); BlueBioTech International (Bielefeld, Germany); Algae for Future/Aliga (Denmark/Portugal); and Euglena Co. (Tokyo, Japan) (Table 19.1). Most of these companies target high-value bioproducts instead of fuels, as biofuel production generally requires large economies of scale and therefore significant financing (Maeda et al. 2018). Commercial interest in such high-value algal bioproducts is due to their structural form and stereochemistry, which are not easily found or achievable in terrestrial or synthetic counterparts, enhancing the bioactivity and functional performance of the metabolites. For instance, omega-3 fatty acids derived from fish oils often contain mercury, which is not suitable for consumption, whereas algae-derived eicosapentaenoic acid (EPA, 20:5) and docosahexaenoic acid (DHA, 22:6) have a high purity (Adarme-Vega et al. 2012). Likewise, paramylon is commercially produced from *E. gracilis*, moving away from yeast and cereal grains due to enhanced immunostimulatory properties (Häder and Hemmersbach 2022).

However, despite the extensive potential of marine microalgae, only a few bioproducts have been commercially marketed to date. The limited commercialization is due to a combination of bottlenecks, including biological (low productivity, culture instability, and crashes), technical (cost of nutrients and CO_2, energy-intensive harvesting and downstream processing, and infrastructure cost), economic (operational scale and low market price of biofuels), and regulatory (complex approval pathways, government incentives, and environmental regulations). To address these limitations, life-cycle assessment (LCA) studies are quintessential, as they can help identify crucial process improvements leading to the commercialization of algal bioproducts. This chapter provides a comprehensive review of the LCA studies conducted on marine microalgae for the production of biofuel and bioproducts and the future outlook of their development and commercialization.

19.2 LIFE-CYCLE ASSESSMENT OF MARINE MICROALGAE FOR BIOFUELS AND BIOPRODUCTS

The LCA is a standardized tool that is typically used to evaluate the environmental impacts of a product or a process starting from the acquisition of raw materials and spanning product development and waste management (Finnveden et al. 2009). LCA studies are particularly useful to identify key bottlenecks, such as harvesting, drying, or nutrient supply in algae cultivation and downstream processing, thereby guiding the development of more efficient, cost-effective, and environmentally sustainable algae-based biofuels and bioproducts. A comparison of net energy ratio (NER) and life-cycle-greenhouse gas (LC-GHG) emissions in LCA studies is shown in Table 19.2 to provide a comprehensive overview of the feasibility of marine algae for biofuels and bioproducts. The NER is defined as the ratio of energy contained in the final product(s) to the cumulative energy consumed for product manufacturing. An NER>1 indicates an economically favorable process or product. The LC-GHC is used to quantify the total emissions of GHGs during the product life cycle. Most of the LCAs for algal biofuel studies have used *Nannochloropsis* sp., a model alga for large-scale production due to its high productivity and ease of cultivation (Collet et al. 2014). In a baseline scenario, biodiesel production from microalgae that employs open-raceway pond cultivation, biomass productivity of 10 g/m^2/d, freshwater media, energy-intensive centrifugation, and chemical transesterification to biodiesel often has an NER <1. However, LCA studies examining different scenarios, including improved biomass productivity (10–15 g/m^2/d), use of renewables (solar or wind) for electricity generation, water recycling, and efficiently designed photobioreactors (PBRs) or open-raceway ponds to improve mixing and light penetration

TABLE 19.2
Life-Cycle Assessment Studies of Marine Algal Biofuel and Bioproduct Production

Microalgae	Cultivation System	Biomass Productivity	Functional Unit	System Boundary	Process	Products	NER (Output/ Input)	LC-GHG (Kg CO_2 eq/FU)	Reference
Tetraselmis sp.	Seawater or brackish water in semi-continuous mode in ORP	14.6–19.5	1 GJ	Cradle to grave	Hydrothermal liquefaction	Biocrude oil	2.81	129	Al-Jabri et al. 2022
	Open-raceway pond + wastewater	–	1 vehicle Km travelled		Transesterification + direct combustion of residual algal biomass	Biodiesel	–	0.2	Clarens et al. 2011
Tetraselmis chui	Photobioreactor		–		Transesterification	Biodiesel. Biogas and biochar		0.42	Grierson et al. 2013
Nannochloropsis occulata	Standard open raceways under greenhouses	10–15 g/m²/d	1 MJ					3.5–7.1	Collet et al. 2014
Freeze-dried Nannochloropsis gaditana	-	–	1000 ton/year biodiesel					–	Taher et al. 2020
Nannochloropsis sp.	cultivation with CO_2 transportation and fertilizer supply	330 kg/year	–	Cradle to gate	Hydro processing	Aviation fuel		17.23–51.04	Guo et al. 2016
	Open-raceway pond + wastewater	16 g/m²/d	1 kg of algal oil	Pond to wheels	Transesterification	Biodiesel and biogas	0.48	–	Razon and Tan 2011
	Flat-panel PBR	–	1 MJ		Supercritical CO_2 extraction	Biodiesel and glycerol	–	0.1	Monari et al. 2016

	Integrated photobioreactor + open-raceway pond			Cradle to gate	Transesterification	Biodiesel		0.2	Khoo et al. 2011
Marine microalgae	Seawater pumped into flattened toroidal ponds	30 g/m²/d	–	Cradle to grave				–27.6 to 18.2	Campbell et al. 2011
Nannochloropsis oculata	1-L Erlenmeyer flasks with 500 mL of culture media with nitrogen replete and deplete conditions	1.79 and 9.56 g/L	1MJ				<1	–	Soh et al. 2014
Tetraselmis suecica		1.54 and 0.30 g/L	Per kg of biodiesel						
Dried Nannochloroposis oceanica	–	–	–		Mild oxidative torrefaction	Solid biofuel (biochar)	0.94–1.2		Zhang et al. 2022
Nannochloropsis oceanica	Upscaled Green Wall Panel photobioreactors		1 kg total fatty acid	Cradle to gate	HTL	Total fatty acids	–		Gaber et al. 2022
Nannochloropsis sp.	Polyethylene bags having a 10-L capacity (PBR), exposed to continuous light	1 g total biomass	MJ	Cradle to gate	Soxhlet extraction	Biodiesel, biohydrogen, and pigments		13–15	Ferreira et al. 2013
Nannochloropsis gaditana	Open-raceway photobioreactor and close vertical bioreactor over 4 seasons	–	2,000 hours of machining time	Cradle to gate	Solvent extraction	Metal working fluids (MWF)		210–260	Guiton et al. 2022
Tetraselmis sueccica	Pond with a rotating cylinder biofilm	30 g/m²/d	MJ	Pond to gate	Alkaline solubilization	Algal biomass and protein concentrate (algal meal)		0.18–0.33	Penaranda et al. 2023

(*Continued*)

TABLE 19.2 (Continued)
Life-Cycle Assessment Studies of Marine Algal Biofuel and Bioproduct Production

Microalgae	Cultivation System	Biomass Productivity	Functional Unit	System Boundary	Process	Products	NER (Output/Input)	LC-GHG (Kg CO_2 eq/FU)	Reference
Nannochloropsis salina	Coastal marine water-based algal biorefinery	15–35 g/m²/d	1 kg algal biomass	Cradle to gate	CML	Algal biomass		–1.28	Moradiya and Marathe 2023
Nannochloropsis sp.	Outdoor ponds	0.289 g/L	1 kg crude protein	Cradle to gate	Solvent vs Supercritical CO_2 extraction	Aquaculture feeds		0.456–1.86	McKuin et al. 2023
	Open cascade raceway pond	5.63 tons/year	1 kg of ash-free dry-weight		-	Biomass and EPA		5.6	Lopes et al. 2023
Dunaliella sp.	Tubular photobioreactor	0.70 tons/year				Pigments and glycerol		10.6	
Nannochloropsis sp.	Tubular photobioreactors in flattened toroidal raceway ponds	-	1 kg dry mass	Cradle to gate		Food		2.61	Schade and Meier 2020
	Photobioreactors composed of borosilicate glass and acrylic glass		1 g EPA +DHA	Cradle to store		EPA and DHA		16–26	Schade et al. 2020
Dunaliella salina and *Dunaliella tertiolecta*	*D. salina* cultivated in eight scenarios combining different nitrogen and magnesium concentrations, *D. tertiolecta* cultivated in 5 different scenarios	85 mg/L/d, 113 mg/L/d	1 kg of biomass/bioproduct produced	Gate to gate		Biomass, glycerol, and β-carotene		–157.5, 213.6	De Souza Celente et al. 2024

–, not reported; NER, net energy ratio; LC-GHG, life-cycle-greenhouse gas.

(leading to increased productivity), have been reported to make microalgal biodiesel sustainable (Collet et al. 2014). Such biodiesel has been reported to have a favorable emission profile (–27.6 to 18.2 g CO_2-eq/t-km) compared with fossil diesel (~35.9 gCO_2-eq/t-km) and canola-derived diesel (~81.2 gCO_2-eq/t-km) (Campbell et al. 2011). Moreover, when algae are grown with CO_2 sourced from a nearby industrial flue gas, they have the potential to achieve net negative LC-GHGs. Importantly, the environmental and economic impacts are dominated by cultivation energy needs, dewatering, and lipid extraction. For instance, an LCA study evaluated the feasibility of commercial-scale biodiesel production from freeze-dried *Nannochloropsis gaditana* biomass using supercritical CO_2 lipid extraction and enzymatic transesterification of lipids to biodiesel (Taher et al. 2020). At a plant scale of 1,000 tons/year of biodiesel production, the process was reported to be economically profitable with annual revenues of $27.94 M/year and a 10.5% internal rate of return. However, it was less attractive than compressed natural gas (CNG) due to the latter's lower selling price, but the environmental footprint of algal biodiesel was found to be eight times lower than that of CNG-based biodiesel. It is important to note that freeze drying is an energy-intensive step and may not be a viable option for large-scale algae cultivation. An LCA study on microalgae-based aviation fuel showed that although nitrogen-deprived conditions increased the lipid content in *Nannochloropsis*, leading to lower energy input (2.13 MJ to 3.08 MJ), fossil fuel consumption and GHG emissions increased significantly by 60–200% (Guo et al. 2016). Such results arise as a high lipid content often correlates with reduced biomass productivity, resulting in a higher upstream fuel and material use per unit. The authors highlighted that nitrogen recycling, particularly when using fertilizers, is crucial to reducing the fossil fuel and GHG burdens. Thus, it is essential to take a holistic approach when optimizing microalgae cultivation, rather than simply maximizing lipids, which is often a common simplistic approach.

Another LCA study comparing four different marine algae grown in semi-continuous open-raceway ponds in Qatar to produce biocrude oil reported that *Tetraselmis* sp. outperformed and delivered the most favorable energy balance with a NER of 2.81 and an LC-GHG reduction potential of 129 kg CO_2 equivalent/barrel of biocrude (Al-Jabri et al. 2022). The study assumed the use of seawater brine and recycling of water to reduce the freshwater and pumping burdens, in addition to using hydrothermal liquefaction (HTL) to avoid the high energy costs of drying algal biomass. The reported positive NER for biocrude oil obtained from marine algae is a first requirement for competitiveness. Usage of green wall panel (GWP) flat-panel PBRs for cultivation of *Nannochloropsis oceanica* in Tuscany (central Italy) and Sicily (southern Italy) resulted in 15% lower impacts per kg in Sicily due to more favorable climatic conditions, namely higher solar irradiance and longer productive periods (Gaber et al. 2022). Moreover, nutrient recycling and biocrude energy credits reduced the environmental impact by 11% compared with the baseline scenario, in which neither nutrient recycling nor additional energy credits from conversion residues were taken into account. A further reduction of up to ~36% in environmental impacts could be obtained by integrating the use of renewable electricity into the process. Furthermore, when compared with soybean oil, a common biodiesel source, algal biodiesel production resulted in lower land and water usage. Overall, the LCA study emphasized the importance of proper site selection, nutrient reuse, and renewable electricity incorporation to reduce the environmental burdens.

In addition to biodiesel, marine microalgae *N. oceanica* was evaluated for solid biofuel (biochar) production using mild oxidative torrefaction (Zhang et al. 2022). Conventional torrefaction is a mild thermochemical conversion process operating at 200–300°C for 15–60 min under an inert atmosphere, resulting in the production of a solid fuel with high calorific value (Zhang et al. 2022). However, the long torrefaction duration, high operating cost, and limited biofuel upgrading may hinder the economic feasibility of algal biochar. To address this, the authors used a mild oxidative torrefaction (250°C for 15–30 mins), which proved to be the most energy-efficient process for solid fuel generation, resulting in a higher relative return on investment of 1.2 and lower environmental impact compared with conventional torrefaction. The study highlighted that a selection of optimal thermochemical conversion process conditions could help improve the energy efficiency and environmental outcomes in microalgal biorefineries. Since production of biofuel alone is not competitive, coproduct valorization could help in developing a sustainable industrial algal process. *Nannochloropsis* sp. has been explored for biodiesel production, with valorization of residual algal biomass for biohydrogen and high-value pigments (Ferreira et al. 2013). Although biodiesel obtained via Soxhlet extraction resulted in the lowest energy use (177–245 MJ/MJ) and GHGs (13–15 KgCO_2/MJ), biodiesel extracted using supercritical CO_2, together with biohydrogen and pigment production, was more economically viable. This was mainly due to a lower yield of biohydrogen during Soxhlet extraction, in addition to the recovery of pigments after supercritical CO_2 extraction.

Interestingly, algal oil has also been explored for the generation of metalworking fluids (MWFs) that are typically used as coolants and lubricants (Guiton et al. 2022). Typically, oilseed MWFs are derived from soybean and rapeseed oils and possess only 3% of the total MWF market. Compared with the rapeseed- and mineral oil-based MWFs, microalgal oil (10%) combined with surfactant (Tween 20) and graphene nanoplatelets did not appear to be promising, as it showed higher LC-GHGs (55–95%), water use (20–95%), and fossil resource use (85–95%). The authors reported that 95% of the environmental impact stemmed from electricity consumption during algal cultivation, harvesting, concentration, drying, and fertilizer usage. Indeed, broader integration with wastewater nutrient recovery, renewable electricity, and coproduct allocation could improve the competitiveness of microalgal MWFs.

As mentioned above, marine microalgal biomass has been mainly exploited for the generation of food, aquaculture feed, and high-value bioproducts. With regards to food and feed, algal biomass is increasingly considered a high-quality and sustainable source of protein due to the presence of essential amino acids, such as methionine and lysine, which increase its nutritional value (Selvam et al. 2024). A recent LCA study reported the use of a rotating algal biofilm cultivation system for easy harvesting and potentially higher productivity with reduced dewatering (Penaranda et al. 2023). By optimizing key process parameters, reducing the electricity demand of rotors and blowers, and integrating renewable electricity, the economic and environmental burdens of the algal process can be reduced. Interestingly, ammonia release during algae cultivation in conventional rotating biofilm reactors was identified as a major environmental concern, since it contributes strongly to acidification, eutrophication, and particulate matter formation. The replacement of ammonium-based nitrogen sources with nitrates, pH control, and integration of media with wastewater, in addition to bioreactor modifications, resulted in an impressive 70.8% reduction in environmental impact compared with conventional aquafeeds (fishmeal and soymeal) (Penaranda et al. 2023). The impact of various bioreactor building materials (borosilicate glass and polymethylacrylate (PMMA)) and cultivation season length was assessed to provide a flexible model cultivation system that could be applied to any promising algal strains under any climatic conditions (Schade and Meier 2020). The use of PMMA for building PBRs, instead of glass, could result in reduced LC-GHG emissions over the course of several years. The authors also noted that the practice of using hydrogen peroxide for cleaning PBR tubes, regardless of the material of construction, was the top contributor to the environmental impact, closely followed by the use of nitrogen fertilizer.

Likewise, an LCA study evaluated the potential of defatted *Nannochloropsis* sp. (lipid-free biomass) for fish meal using two harvesting (wet vs. dry) and oil extraction (solvents vs. supercritical CO_2) scenarios (McKuin et al. 2023). The authors highlighted that across all scenarios, cultivation and biomass production were the largest contributors to environmental impact, with the use of urea fertilizer and pure CO_2 sparging being the major bottlenecks. Use of dry biomass with solvent extraction (0.46 Kg CO_2eq/kg protein) gave a better life-cycle performance when using supercritical CO_2 (1.86 Kg CO_2eq/kg protein) (Table 19.2). Notably, the authors reported that the LC-GHGs, water (−0.0005 m^3 water/kg protein), and land (0.0004 m^2 land/kg protein) use were lower for algae than for small pelagic fish (0.61 Kg CO_2eq/kg protein, 0.002 m^3water/kg protein, and 0.005 m^2 land/kg protein). The study highlighted the potential of a biorefinery approach for protein and PUFA coproduction to enhance the sustainability of marine algae-based bioproducts. Another LCA evaluation of a multi-strain microalgae biorefinery showed that strain selection, water management, CO_2 sourcing, and process integration were the main factors affecting the economic viability and sustainability of large-scale algae biomass production (Lopes et al. 2023). Among the three strains tested (*Nannochloropsis* sp., *Dunaliella* sp., and *Spirulina* sp.), *Nannochloropsis* sp. resulted in the lowest LC-GHGs, whereas *Dunaliella* sp. had the highest environmental impact, which is the result of *Nannochloropsis* having high biomass productivity and a reduced requirement for NaCl (Table 19.2). The authors emphasized that water circulation during algae cultivation, along with the use of flue gas CO_2, could significantly minimize water and nutrient use and freshwater ecotoxicity impacts. Additionally, a high biomass productivity (35 g/m^2/d) resulted in lower LC-GHC emissions (−1.34 kgCO_2eq/kg algae) compared with the baseline scenario, in which biomass productivity (15 g/m^2/d) resulted in −1.11 kgCO_2eq/kg algae, signifying the importance of employing superior-performing strains for large-scale algal cultivation (Moradiya and Marathe 2023).

The environmental impact of media components (nitrogen, magnesium, and salt) using two marine strains, *Dunaliella salina* and *Dunaliella tertiolecta*, was evaluated for biomass, glycerol, and β-carotene production (De Souza Celente et al. 2024). Overall, salt and deionized water had a pronounced contribution to the LCA, which could be significantly reduced by recycling brine from desalination plants and reusing the cultivation medium. For *D. salina*, the use of 1 g/L of KNO_3 and 1.1 g/L of $MgCl_2$ resulted in the lowest environmental impact for biomass and glycerol production, whereas 0.1 g/L of KNO_3 and 1.9 g/L of $MgCl_2$ were best for β-carotene synthesis. On the other hand, for *D. tertiolecta*, 1 g/L of KNO_3 with 116.9 g/L salt offered the best approach for glycerol and biomass production (De Souza Celente et al. 2024).

19.3 CONCLUSION AND FUTURE OUTLOOK

Marine microalgae occupy a strategically important position in the development of sustainable biofuels and bioproducts due to their capability to synthesize diverse metabolites while consuming CO_2 and generating O_2. Thanks to their rapid growth rate, flexible metabolic capacity, and ability to grow in a range of environmental conditions, algae are uniquely suited for a wide spectrum of applications ranging from bioenergy, food, feed, and fertilizers to specialty metabolites, such as vitamins, omega-3 fatty acids, and pigments. This chapter reviewed LCA studies conducted on various marine microalgae for biofuels and bioproducts, which potentially could support a circular bioeconomy to alleviate pressure on terrestrial agriculture, freshwater supplies, and wild fisheries. Indeed, the sustainability and competitiveness of marine microalgae are strongly dependent on the efficiency of cultivation and downstream processing. The studies indicate that algae-derived biofuels require high energy input, have marginal NER, and elevated LC-GHGs, which are mainly attributed to high electricity use, nutrient requirements, and energy-intensive harvesting and dewatering. The adoption of HTL or mild oxidative torrefaction could help improve NER and lower LC-GHG emissions, while enabling the removal of eutrophication-causing nitrogen and phosphorus from aqueous and solid residues. However, in the baseline scenario of biofuel production through lipid extraction and biodiesel or aviation fuel upgrading, algal biofuels underperformed relative to fossil- and plant-derived fuels on a life-cycle basis. In contrast, the development of an integrated algae biorefinery that coproduces food, feed, fertilizer, and high-value products leads to improved environmental and economic performance that is superior to that of conventional fuels. Similarly, marine microalgae-derived food and nutraceutical products, such as EPA, DHA, proteins, and pigments, often outperformed fish-derived products. Moreover, algae-derived aquaculture feed showed system-level environmental benefits by replacing fishmeal and fish oil, thereby reducing reliance on marine resources.

Although there is wide variation in key LCA metrics, such as NER and LC-GHGs, as a result of differences in algal processes, system boundaries, sources of fossil energy, and coproduct formation, several cross-cutting determinants across all the LCA studies were identified that repeatedly governed the environmental outcome. These included biomass productivity, electricity usage, nutrient sourcing and recovery, and system integration. Biomass productivity has been reported to consistently outweigh bioproduct yield, emphasizing the importance of strain selection for large-scale algal cultivation. Moreover, cultivation location and climate are equally critical, as regions with moderate, stable temperatures and high solar irradiance achieve high biomass productivities. Importantly, colocation of the algae cultivation with industrial sources of waste CO_2 and wastewater is crucial for reducing the cost of pure CO_2 and fertilizers during algae cultivation, thereby improving NER and lowering LC-GHG emissions.

Collectively, although in the near future marine microalgae will not be able to replace fossil fuels or plant-derived biofuels and bioproducts, they can still serve as a versatile and resilient alternative for several bioproducts. Future scale-up efforts should prioritize the development of an integrated biorefinery, where high-value products constitute a primary value stream that complements commodity biofuels. In addition, the development and adoption of wet-biomass technologies can significantly reduce the energy (drying) burden of algal biofuels and bioproducts. Future LCA studies should incorporate pilot-scale cultivation data, as opposed to just small-scale data, and adopt a standardized approach for calculating NER and LC-GHG emissions to enable more meaningful comparisons across products and technologies. Finally, policy frameworks should incentivize the non-energy benefits of marine microalgae, including nutrient recycling and reduced land and freshwater use, which are essential for achieving environmental and economic viability at a commercial scale.

ACKNOWLEDGEMENTS

The figure was made using Biorender.com

REFERENCES

Adarme-Vega, T. Catalina, David K. Y. Lim, Matthew Timmins, Felicitas Vernen, Yan Li, and Peer M. Schenk. 2012. "Microalgal Biofactories: A Promising Approach Towards Sustainable Omega-3 Fatty Acid Production." *Microbial Cell Factories* 11 (1): 96. https://doi.org/10.1186/1475-2859-11-96

Al-Jabri, Hareb, Probir Das, Shoyeb Khan, et al. 2022. "A Comparison of Bio-Crude Oil Production from Five Marine Microalgae – Using Life Cycle Analysis." *Energy* 251 (July): 123954. https://doi.org/10.1016/j.energy.2022.123954

Campbell, Peter K., Tom Beer, and David Batten. 2011. "Life Cycle Assessment of Biodiesel Production from Microalgae in Ponds." *Bioresource Technology* 102 (1): 50–56. https://doi.org/10.1016/j.biortech.2010.06.048

Carroll, Anthony R., Brent R. Copp, Rohan A. Davis, Robert A. Keyzers, and Michèle R. Prinsep. 2019. "Marine Natural Products." *Natural Product Reports* 36 (1): 122–73. https://doi.org/10.1039/C8NP00092A

Clarens, Andres F., Hagai Nassau, Eleazer P. Resurreccion, Mark A. White, and Lisa M. Colosi. 2011. "Environmental Impacts of Algae-Derived Biodiesel and Bioelectricity for Transportation." *Environmental Science & Technology* 45 (17): 7554–60. https://doi.org/10.1021/es200760n

Collet, Pierre, Laurent Lardon, Arnaud Hélias, et al. 2014. "Biodiesel from Microalgae – Life Cycle Assessment and Recommendations for Potential Improvements." *Renewable Energy* 71 (November): 525–33. https://doi.org/10.1016/j.renene.2014.06.009

De Souza Celente, Gleison, Rosana De Cassia De Souza Schneider, Jennifer Julich, Tiele Medianeira Rizzetti, Eduardo Alcayaga Lobo, and Yixing Sui. 2024. "Life Cycle Assessment of Microalgal Cultivation Medium: Biomass, Glycerol, and Beta-Carotene Production by Dunaliella Salina and Dunaliella Tertiolecta." *The International Journal of Life Cycle Assessment* 29 (12): 2269–82. https://doi.org/10.1007/s11367-023-02209-2

Ferreira, Ana F., Lauro A. Ribeiro, Ana P. Batista, et al. 2013. "A Biorefinery from Nannochloropsis Sp. Microalga – Energy and CO2 Emission and Economic Analyses." *Bioresource Technology* 138 (June): 235–44. https://doi.org/10.1016/j.biortech.2013.03.168

Finnveden, Göran, Michael Z. Hauschild, Tomas Ekvall, et al. 2009. "Recent Developments in Life Cycle Assessment." *Journal of Environmental Management* 91 (1): 1–21. https://doi.org/10.1016/j.jenvman.2009.06.018

Gaber, Kirsten, Christine Rösch, and Natascia Biondi. 2022. "Life Cycle Assessment of Total Fatty Acid (TFA) Production from Microalgae Nannochloropsis Oceanica at Different Sites and Under Different Sustainability Scenarios." *BioEnergy Research* 15 (3): 1595–615. https://doi.org/10.1007/s12155-021-10279-z

Grierson, Scott, Vladimir Strezov, and Jonas Bengtsson. 2013. "Life Cycle Assessment of a Microalgae Biomass Cultivation, Bio-Oil Extraction and Pyrolysis Processing Regime." *Algal Research* 2 (3): 299–311. https://doi.org/10.1016/j.algal.2013.04.004

Guiton, Mélanie, David Suárez-Montes, Rubén Sánchez, et al. 2022. "Comparative Life Cycle Assessment of a Microalgae-Based Oil Metal Working Fluid with Its Petroleum-Based and Vegetable-Based Counterparts." *Journal of Cleaner Production* 338 (March): 130506. https://doi.org/10.1016/j.jclepro.2022.130506

Guo, Fang, Jing Zhao, and Xiaoyi Yang. 2016. "Life Cycle Assessment of Microalgae-Based Aviation Fuel: Influence of Lipid Content with Specific Productivity and Nitrogen Nutrient Effects." *Bioresource Technology* 221 (December): 350–57. https://doi.org/10.1016/j.biortech.2016.09.044

Häder, Donat-P., and Ruth Hemmersbach. 2022. "Euglena, a Gravitactic Flagellate of Multiple Usages." *Life* 12 (10): 1522. https://doi.org/10.3390/life12101522

Khoo, H.H., P.N. Sharratt, P. Das, R.K. Balasubramanian, P.K. Naraharisetti, and S. Shaik. 2011. "Life Cycle Energy and CO2 Analysis of Microalgae-to-Biodiesel: Preliminary Results and Comparisons." *Bioresource Technology* 102 (10): 5800–5807. https://doi.org/10.1016/j.biortech.2011.02.055

Lauritano, Chiara, Maria Immacolata Ferrante, and Alessandra Rogato. 2019. "Marine Natural Products from Microalgae: An -Omics Overview." *Marine Drugs* 17 (5): 269. https://doi.org/10.3390/md17050269

Lopes, Tiago F., Joana Ortigueira, Cristina T. Matos, et al. 2023. "Conceptual Design of an Autotrophic Multi-Strain Microalgae-Based Biorefinery: Preliminary Techno-Economic and Life Cycle Assessments." *Fermentation* 9 (3): 255. https://doi.org/10.3390/fermentation9030255

Maeda, Yoshiaki, Tomoko Yoshino, Tadashi Matsunaga, Mitsufumi Matsumoto, and Tsuyoshi Tanaka. 2018. "Marine Microalgae for Production of Biofuels and Chemicals." *Current Opinion in Biotechnology* 50 (April): 111–20. https://doi.org/10.1016/j.copbio.2017.11.018

McKuin, Brandi, Anne R. Kapuscinski, Pallab K. Sarker, Nicolette Cheek, Jacqueline Lim, and Martin Sabarsky. 2023. "Comparative Life Cycle Assessment of Marine Microalgae, *Nannochloropsis* Sp. and Fishmeal for Sustainable Protein Ingredients in Aquaculture Feeds." *Elementa: Science of the Anthropocene* 11 (1): 00083. https://doi.org/10.1525/elementa.2022.00083

Monari, Chiara, Serena Righi, and Stig Irving Olsen. 2016. "Greenhouse Gas Emissions and Energy Balance of Biodiesel Production from Microalgae Cultivated in Photobioreactors in Denmark: A Life-Cycle Modeling." *Journal of Cleaner Production* 112 (January): 4084–92. https://doi.org/10.1016/j.jclepro.2015.08.112

Moradiya, Keyur K., and Kumudinee V. Marathe. 2023. "Life Cycle Assessment (LCA) of Marine Microalgae Cultivation and Harvesting Process for the Indian Context." *Sustainable Energy Technologies and Assessments* 56 (March): 103063. https://doi.org/10.1016/j.seta.2023.103063

Nishshanka, Gannoru Kankanamalage Sanuji Hasara, Randeniya Arachchige Don Praveen Anthonio, P.H.V. Nimarshana, Thilini U. Ariyadasa, and Jo-Shu Chang. 2022. "Marine Microalgae as Sustainable Feedstock for Multi-Product Biorefineries." *Biochemical Engineering Journal* 187 (November): 108593. https://doi.org/10.1016/j.bej.2022.108593

Penaranda, Diego, Hubert Bonnefond, Freddy Guihéneuf, Marjorie Morales, and Olivier Bernard. 2023. "Life Cycle Assessment of an Innovative Rotating Biofilm Technology for Microalgae Production: An Eco-Design Approach." *Journal of Cleaner Production* 384 (January): 135600. https://doi.org/10.1016/j.jclepro.2022.135600

Razon, Luis F., and Raymond R. Tan. 2011. "Net Energy Analysis of the Production of Biodiesel and Biogas from the Microalgae: Haematococcus Pluvialis and Nannochloropsis." *Applied Energy* 88 (10): 3507–14. https://doi.org/10.1016/j.apenergy.2010.12.052

Schade, S., and T. Meier. 2020. "Distinct Microalgae Species for Food—Part 1: A Methodological (Top-down) Approach for the Life Cycle Assessment of Microalgae Cultivation in Tubular Photobioreactors." *Journal of Applied Phycology* 32 (5): 2977–95. https://doi.org/10.1007/s10811-020-02177-2

Schade, S., G. I. Stangl, and T. Meier. 2020. "Distinct Microalgae Species for Food—Part 2: Comparative Life Cycle Assessment of Microalgae and Fish for Eicosapentaenoic Acid (EPA), Docosahexaenoic Acid (DHA), and Protein." *Journal of Applied Phycology* 32 (5): 2997–3013. https://doi.org/10.1007/s10811-020-02181-6

Selvam, Jairam, Joyabrata Mal, Sukhendra Singh, et al. 2024. "Bioprospecting Marine Microalgae as Sustainable Bio-Factories for Value-Added Compounds." *Algal Research* 79 (April): 103444. https://doi.org/10.1016/j.algal.2024.103444

Soh, Lindsay, Mahdokht Montazeri, Berat Z. Haznedaroglu, et al. 2014. "Evaluating Microalgal Integrated Biorefinery Schemes: Empirical Controlled Growth Studies and Life Cycle Assessment." *Bioresource Technology* 151 (January): 19–27. https://doi.org/10.1016/j.biortech.2013.10.012

Taher, Hanifa, Adewale Giwa, Hana Abusabiekeh, and Sulaiman Al-Zuhair. 2020. "Biodiesel Production from Nannochloropsis Gaditana Using Supercritical CO2 for Lipid Extraction and Immobilized Lipase Transesterification: Economic and Environmental Impact Assessments." *Fuel Processing Technology* 198 (February): 106249. https://doi.org/10.1016/j.fuproc.2019.106249

Zhang, Congyu, Meng Wang, Wei-Hsin Chen, Anelie Pétrissans, Mathieu Pétrissans, and Shih-Hsin Ho. 2022. "A Comparison of Conventional and Oxidative Torrefaction of Microalga Nannochloropsis Oceanica through Energy Efficiency Analysis and Life Cycle Assessment." *Journal of Cleaner Production* 369 (October): 133236. https://doi.org/10.1016/j.jclepro.2022.133236

20 Linking Sustainability and Consumer Perceptions

The Nexus Behind Microalgae-Based Food Acceptance

Ibrahim Abu-Reidah

LIST OF ABBREVIATIONS

CO_2	Carbon dioxide
DHA	Docosahexaenoic acid
EPA	Eicosapentaenoic acid
FAO	Food and Agriculture Organization
GHG	Greenhouse gas
LCA	Life-Cycle assessment
PUFA	Polyunsaturated fatty acids
SDG	Sustainable Development Goals
WTP	Willingness to pay

20.1 INTRODUCTION

20.1.1 Overview of Microalgae-Based Foods

Microalgae consist of photosynthetic microorganisms, which include eukaryotic algae and cyanobacteria that thrive in freshwater, marine, and hypersaline environments. Microalgae cultivation through bioreactors and open pond systems allows for high productivity and consistent nutrient supply (Ennaceri et al., 2025). The food industry uses *Arthrospira platensis* (spirulina), *Chlorella vulgaris*, *Nannochloropsis gaditana*, *Tetraselmis chuii*, and *Dunaliella salina* as their primary microalgae species. The organisms contain more than 50% protein by dry weight, and they are also rich in eicosapentaenoic acid (EPA) and docosahexaenoic acid (DHA) fatty acids, B vitamins, iron, magnesium, selenium, and bioactive compounds that exhibit antioxidant, antimicrobial, and immunomodulatory effects (Oliveira & Bragotto,

 DOI: 10.1201/9781003611479-20

2022). This nutrient density has positioned microalgae as a credible alternative to conventional protein sources and a functional ingredient in foods and supplements (Caporgno & Mathys, 2018). The cultivation of microalgae requires comparatively less land, can utilize saline or wastewater, and captures atmospheric CO_2, making it an attractive solution to address nutritional and environmental challenges simultaneously (Ali et al., 2025).

20.1.2 The Need for Sustainable Food Sources

Global food systems face growing pressures related to climate change, population growth, biodiversity loss, soil degradation, and freshwater scarcity. By 2050, the world's population is expected to reach close to 9 billion, increasing demands for nutrient-dense, affordable, and environmentally sustainable food sources (FAO, 2023). Traditional livestock and crop production are among the largest contributors to greenhouse gas emissions, deforestation, and water overuse, making alternative protein sources an urgent priority. Microalgae have the potential to contribute significantly to more sustainable diets due to their low-input, high-yield model of production that can be well integrated into circular economy frames. Their ability to convert CO_2 into valuable biomass, utilize marginal lands, and integrate with industrial CO_2 capture systems enhances their environmental credentials (Somers et al., 2021). Moreover, with their rapid cycles of growth, they are highly amenable to scalable production in a variety of regions, including those with poor agricultural conditions (Hosny et al., 2025). Their ability to make use of non-potable water, tolerate saline conditions, and be integrated with CO_2 capture technologies enables the development of production systems with lower environmental impacts while producing nutrient-dense biomass (Figure 20.1). These characteristics make microalgae particularly attractive for climate-vulnerable regions and urban environments where land and freshwater are limiting factors. Integrating microalgae into more general food strategies will enhance the resilience of societies to disruptions and help to meet both SDG 2 (Zero Hunger) and SDG 13 (Climate Action) (Ali et al., 2025).

20.2 MICROALGAE AND SUSTAINABILITY

20.2.1 Environmental Benefits

Microalgae cultivation has a significantly lower environmental footprint compared with terrestrial agriculture. These microorganisms can be grown on nonarable land, can utilize either saline water or wastewater instead of fresh water, and require much smaller fertilizer inputs per unit of protein produced. Because of photosynthesis, it can be easily paralleled with direct CO_2 capture. For example, several closed photobioreactor systems have reported sequestration rates that are higher than those of many terrestrial plants on a per-hectare basis (Acién et al., 2016). When the production is integrated with industrial flue gas streams or biogenic CO_2 sources, microalgal growth can function as a carbon sink and thus support climate mitigation strategies. Life-cycle assessments (LCAs) have shown that microalgae reduce greenhouse

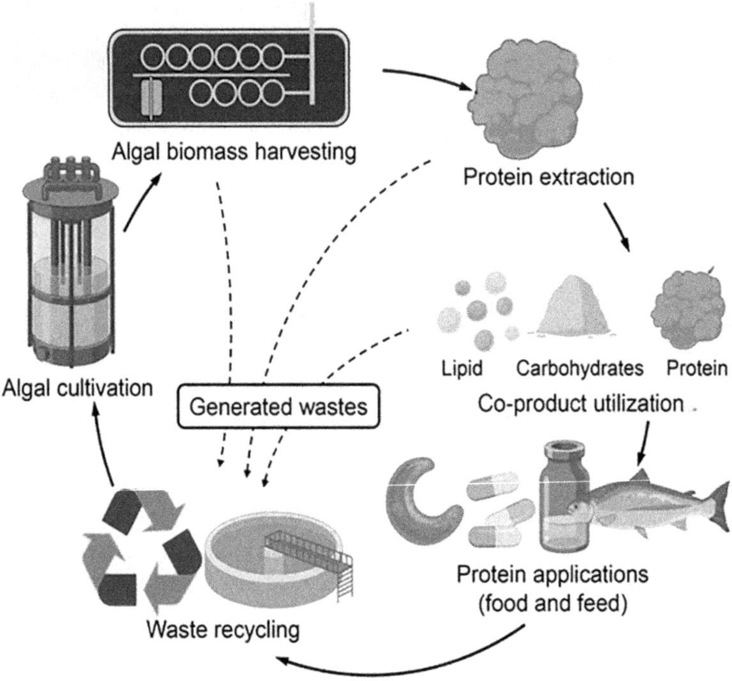

FIGURE 20.1 Integrated microalgae production system schematic.

gas emissions and eutrophication impacts when their cultivation involves renewable energy inputs. For instance, algae cultivation can be combined with nutrient recovery from municipal and agro-industrial wastewater, enabling circular resources with reduced emissions (Santo et al., 2025).

20.2.2 Nutritional Benefits and Potential

Microalgae are among the most nutrient-dense biomass sources available. They contain high levels of complete proteins, essential fatty acids, and micronutrients. Proteins usually make up more than 50% of the dry weight of microalgae and have balanced amino acid profiles comparable with animal proteins (Table 20.1). A number of strains produce long-chain omega-3 fatty acids like EPA and DHA, commonly derived from marine fish, as well as vitamins such as B_{12}, E, K, and carotenoids; minerals including iron, magnesium, and zinc; and antioxidant pigments like chlorophylls and carotenoids, which are able to support metabolic and immune functions (Caporgno & Mathys, 2018; García-Encinas et al., 2025).

These characteristics make them especially suitable for tackling nutritional deficiencies among vulnerable groups and for diversifying diets where access to animal proteins is limited. Very recently, human nutrition trials have shown that

TABLE 20.1
Examples of Some Microalgae Species Main Composition %

Species	Protein (%)	PUFA (%)
Chlorella vulgaris	58	23
Spirulina platensis	60	16
Nannochloropsis sp.	52	27

microalgae-based ingredients may contribute to improving lipid profiles, antioxidant status, and overall dietary quality (Wu et al., 2025).

20.2.3 Technological and Economic Advancements

The scalability of microalgae production has increased significantly because of better photobioreactor design, optimized lighting strategy, and hybrid cultivation systems. Other improvements, like LED-based light management, strain engineering, and adaptive growth control, have increased productivity by reducing energy demand. Economically speaking, integrated biorefinery models are being developed to valorize multiple coproducts such as pigments, proteins, biofuels, and fertilizers, thereby offsetting production costs (Behera et al., 2022; Koyande et al., 2020). In addition, coupling algae cultivation with existing infrastructure, such as wastewater treatment facilities or renewable energy hubs, reduces operational costs while improving environmental performance. Emerging techno-economic analyses point to the fact that food-grade algae can achieve competitiveness with other alternative proteins in less than a decade with policy support. These are in line with broader drives toward decarbonizing food systems and enhancing climate resilience (Sarker & Kaparaju, 2024).

20.3 CONSUMER PERCEPTIONS AND ACCEPTANCE

20.3.1 The Role of Consumer Perceptions in Food Acceptance

In many cases, consumer perception outweighs purely nutritional or environmental attributes in decisions to adopt novel foods. Microalgae, though rich in health-promoting compounds, are still an unfamiliar and unconventional ingredient in many markets. Psychological factors of food neophobia, perceived naturalness, and trust in food production processes mediate acceptance. Cultural context and sensory expectations (flavor, aroma, and texture) also substantially affect willingness to buy (Shan et al., 2025). Positive framing, familiarity through incorporation in more familiar food formats, and transparent communication of benefits significantly enhance acceptance in behavioral nutrition. Trust in labeling, regulatory oversight, and producer reputation further strengthens consumer openness to receiving such foodstuffs (Lucas & Brunner, 2024; Meixner et al., 2023).

20.3.2 Barriers to Acceptance of Microalgae-Based Foods

Major factors that constrain acceptance relate to sensory drawbacks, unfamiliarity, and consumer neophobia—all of which reduce willingness to try algae-forward products. Production and processing costs are high, and cultivation is variable, further limiting scalability and market competitiveness. Ambiguous regulation and incomplete data on safety and bioavailability add further hesitation among manufacturers and consumers alike. The lack of retail visibility and/or unclear product positioning also weakens market uptake. These will be overcome only through the improvement in sensory qualities and integration of microalgae into familiar formats. Safety standards need to be clearly established with better scientific evidence and clearer communication of nutritional and sustainability advantages to build trust. Greater distribution, lower costs, and enhanced bioprocessing can help move microalgae from a niche ingredient into a mainstream food resource (Olsen, 2024). Despite their sustainability and nutritional credentials, several barriers hinder the mainstreaming of microalgae-based foods. At the forefront of these are sensory issues: many microalgal strains have distinct green or blue-green pigmentation, marine or earthy odors, and unfamiliar textures that can be off-putting to unaccustomed consumers (Andrade-Bustamante et al., 2025). Price sensitivity is another big factor, given that many products derived from microalgae currently occupy premium niches in health food markets. Regulatory ambiguity in some jurisdictions can also result in consumer hesitancy. These identified obstacles illustrate the need for a combination of technological, sensory, and communication strategies to normalize algae-based foods.

20.3.3 Drivers of Acceptance

On the other hand, various factors facilitate consumer acceptance of microalgae-based products. Growing awareness of environmental issues such as climate change, resource depletion, and loss of biodiversity has set up a receptive market for innovative food propositions that are sustainable. Health-conscious consumers, especially younger demographics and those on plant-based diets, have a higher level of acceptance of nutrient-dense functional ingredients. The exposure through mainstream products, like smoothies, snack bars, or pasta enriched with microalgae, may lighten the perception of novelty. Label transparency, clear health benefit communication, and resonance with ethical and environmental values all foster a greater level of trust in adopting the product. Furthermore, early adopter segments often catalyze diffusion for wider audiences, particularly if the cuisine is innovative and supported by chef, influencer, or health professional endorsement (Lucas & Brunner, 2024).

20.4 THE NEXUS BETWEEN SUSTAINABILITY AND CONSUMER PERCEPTIONS

20.4.1 How Sustainability Impacts Consumer Preferences

In recent times, sustainability considerations have emerged as an influential driver of consumer choice, including food. A growing share of consumers, especially the younger age groups, report that environmental impact is a consideration in the way they make their purchasing choices. Such an ethically aware customer base is definitely the one that microalgae-based foods can target, with very clear positional values regarding sustainability and climate friendliness. This is further emphasized when such sustainability claims are matched with credible third-party certifications, such as in the form of carbon footprint labels and eco-certification, along with transparent communication. This can also be furthered by developing LCA data and end-to-end traceability systems with a view to building up trust and reducing perceived risk. Several studies report that messages on sustainability have the potential to increase the acceptance of unknown foods, provided they are complemented with information about taste, nutritional value, and safety. Sustainability, however, alone seldom has the power to ensure adoption; it needs to be combined with meeting sensory expectations and price (Lucas & Brunner, 2024).

20.4.2 How Consumer Perceptions Shape the Adoption of Microalgae-Based Foods

Consumer perceptions are a sort of lens through which messages about sustainability are read. Branding, framing, and labeling are central to attitude formation with regard to the new food: microalgae. Terms like "green protein" or "nutrient-rich algae" receive better responses than the term "algae biomass," showing the role of language in the management of perception. Packaging design cues of naturalness, freshness, and transparency enhance consumer trust, while storytelling about production methods may give a human face to the technology-heavy process. Social influence from chefs, influencers, or generally trusted institutions can reduce perceived risk and increase cultural familiarity. In markets where algae-based products had been introduced in mainstream food formats rather than niche supplements, acceptance rates tended to increase (Aschemann-Witzel et al., 2023).

20.4.3 Case Studies

Practical case studies illustrate how sustainability narratives and consumer perceptions interact to shape market outcomes. In Switzerland, segmentation studies identified clusters of consumers—innovators and the eco-conscious—who are particularly receptive to algae-based products when sustainability attributes are saliently framed (Lucas & Brunner, 2024). In Austria, discrete choice experiments with algae crackers indicated higher willingness to pay for organic labels and origin, hinting at the importance of eco-labeling and product origin (Meixner et al., 2023). Similarly, in

TABLE 20.2
Case Studies and Consumer Segmentation Patterns in Microalgae-Based Food Acceptance

Study	Location/ Sample	Product Format	Key Findings	Barriers Identified	Reference
Lucas & Brunner (2024)	Switzerland (n=584)	General survey	Identified receptive and skeptical consumer clusters	Conservative consumers hesitant	Lucas & Brunner, 2024
Meixner et al. (2023)	Austria	Algae crackers	WTP increased with sustainability claims	Price sensitivity	Meixner et al., 2023
UK Study (2021)	United Kingdom (n=34)	Focus groups	Openness when informed	Low familiarity	Bleakley & Hayes, 2021
Spanish survey (2022)	Spain (n=3084)	National survey	Consumers view algae as sustainable, but lack habitual consumption	Low awareness	Bourgade et al., 2022

the United Kingdom, qualitative research demonstrated that baseline knowledge of microalgae is very low, but that willingness to try significantly increased with targeted communication as well as the chance to sample (See Table 20.2). Lessons from other novel foods—insect protein and cultured meat—further indicate that early market acceptance requires the combination of sustainability claims with sensory experience and trusted branding.

20.5 STRATEGIES TO ENHANCE MICROALGAE-BASED FOOD ACCEPTANCE

20.5.1 Improving Product Appeal

In all, the main factors that influence consumer acceptance are the sensory profiles of algae-based products. Minimizing off-flavors, color intensity, and textural deviations is the only way to make these foods palatable for mainstream consumers. Processing technologies such as microencapsulation, fractionation, deodorization, and fermentation may substantially reduce the apparent typical marine odor and bitterness found in some species of algae. Developing or selecting strains that possess low chlorophyll content or characteristically milder sensory attributes will also make it easier to incorporate microalgae into more familiar food matrices. Application in small proportions (1–2%) in common products such as bread, pasta, snack bars, and beverages normalizes its presence and hence minimizes resistance. Examples include algae-fortified bakery goods and smoothies that retain conventional expectations while offering an enhanced nutritional profile.

20.5.2 Consumer Education and Awareness

Education serves as one of the cornerstones in changing perceptions toward new food sources. Most consumers know little about microalgae's nutritional value and environmental benefits, and ignorance begets caution. Transparently communicating cultivation processes, standards of safety, and health benefits could enhance trust in and purchasing of the product. Generally speaking, educational campaigns achieve better results when combined with some experiential strategy, such as tasting sessions or cooking demonstrations, or in cooperation with figures trusted by audiences, including chefs and nutritionists. By telling stories of sustainability and community benefits, it further humanizes the technology, creating an emotional resonance with consumers. Digital media and influencer-driven strategies extend reach, especially among more sustainability-sensitive target groups of younger generations (Duerlund et al., 2019).

20.5.3 Policy and Regulatory Support

Robust policy frameworks are needed to develop and scale up algae-based food. Clear regulatory pathways for species approval, contaminant limits, labeling, and safety standards will be fundamental in gaining the public's trust and reducing market entry barriers. Governments can catalyze this with funding for research, subsidies for pilot facilities, and the integration of algae into public procurement programs, such as school meals or hospital catering. Providing incentives for sustainable production practices, such as using renewable energy and recycling nutrients, would help make algae-based products cost-competitive with conventional proteins. International harmonization of regulatory frameworks could further reduce barriers to global trade and stimulate industry investment in the sector (Acién et al., 2016; Caporgno & Mathys, 2018; Sarker & Kaparaju, 2024).

20.6 FUTURE DIRECTIONS AND RESEARCH NEEDS

20.6.1 Advancing Consumer Research

Future consumer research must go beyond basic attitudinal surveys to capture the evolving dynamics of food system transformation. Longitudinal studies tracking shifts in perceptions over time are essential for understanding how early exposure, sustainability narratives, and cultural context shape acceptance trajectories. Cross-cultural comparative research can reveal regional differences in attitudes toward algae-based foods, which can then inform targeted marketing and education strategies (Habib et al., 2025). Experimental designs—including sensory trials, in-store interventions, and digital campaigns—can help identify effective communication and framing strategies. Segmenting consumers by environmental values, dietary preferences, and familiarity with alternative proteins can enable more precise product positioning. Behavioral economic approaches and nudging strategies have shown particular promise in encouraging trial and adoption of novel foods (Lucas & Brunner, 2024).

20.6.2 Innovations in Microalgae Cultivation and Food Design

Technological developments will be instrumental in making algal-based foods economically viable, as well as sensorially acceptable. Future innovations might pertain to low-energy-consuming photobioreactors, hybrid open-closed cultivation systems, and advanced strain selection for improving sensory and nutritional profiles. Precision fermentation and non-GMO strain engineering might allow the production of tailor-made algal ingredients that exhibit minimal sensory drawbacks. Cultivation integrated with industrial CO_2 sources and wastewater valorization can decrease environmental impact and costs. Innovations in food design, including texture engineering, flavor modulation, and ingredient blending with plant proteins, could enhance palatability and open wider food applications. A multiscale approach considering cultivation, processing, and product design will be crucial for effective scaling of production (Behera et al., 2022; Koyande et al., 2020; Wu et al., 2025).

20.6.3 Long-Term Sustainability and Global Impact

Microalgae can change global food systems by offering a sustainable protein source rich in nutrition without competing for arable land or freshwater. With increasing climate pressures, the integration of algae into food systems offers opportunities to diversify protein sources and enhance the sustainability of food security while lowering environmental impacts related to livestock and intensive crop production (Hosny et al., 2025). LCAs show that when combined with renewable energy, algae-based proteins have the potential to lower the carbon footprint up to 80% compared with conventional animal proteins. Besides food, algae make many contributions to the wider bioeconomy through their possible uses in bioplastics, biofertilizers, and renewable fuels in ways that contribute to a circular and regenerative economy. This will be achieved through comprehensive coordination across policy, industry, science, and civil society to align technological innovation with consumer acceptance and sustainability goals (Acién et al., 2016; Sarker & Kaparaju, 2024).

20.7 CONCLUSION

Microalgae represent a unique convergence of high nutritional value, low environmental impact, and production versatility that places them as perhaps the key component of sustainable future diets. But consumer perceptions remain one of the most important determinants of success. Adoption trajectories will be shaped as much by sensory characteristics, cultural familiarity, trust, and affordability as by technological capability. Bridging this gap will require coordinated strategies across innovation, communication, and policy. To make the transition from niche to mainstream, microalgae require interdisciplinary collaboration. It requires scientists to refine production systems, marketers to shape compelling narratives, policymakers to enable supportive frameworks, and consumers to be engaged as partners rather than passive recipients. This human–technology interface will decide if microalgae meet their potential of contributing to resilient and sustainable food systems.

REFERENCES

Acién, F. G., Gómez-Serrano, C., Morales-Amaral, M. M., Fernández-Sevilla, J. M., & Molina-Grima, E. (2016). Wastewater treatment using microalgae: How realistic a contribution might it be to significant urban wastewater treatment? *Applied Microbiology and Biotechnology*, *100*, 9013–9022. https://doi.org/10.1007/s00253-016-7835-7

Ali, S. S., Al-Tohamy, R., Al-Zahrani, M., Schagerl, M., Kornaros, M., & Sun, J. (2025). Advancements and challenges in microalgal protein production: A sustainable alternative to conventional protein sources. *Microbial Cell Factories*, *24*, 61. https://doi.org/10.1186/s12934-025-02685-1

Andrade-Bustamante, G., Martínez-Ruiz, F. E., Ortega-García, J., Renganathan, P., Gaysina, L. A., Mahendhiran, M., & Puente, E. O. R. (2025). Microalgae-based functional foods: A blue-green revolution in sustainable nutrition and health. *Applied Microbiology*, *5*(2), 39. https://doi.org/10.3390/applmicrobiol5020039

Aschemann-Witzel, J., Asioli, D., Banovic, M., Perito, M. A., & Peschel, A. O. (2023). Consumer understanding of upcycled foods–Exploring consumer-created associations and concept explanations across five countries. *Food Quality and Preference*, *112*, 105033.

Behera, B., Selvam, M. S., & Balasubramanian, P. (2022). Hydrothermal processing of microalgal biomass: Circular bio-economy perspectives for addressing food–water–energy nexus. *Bioresource Technology*, *359*, 127443. https://doi.org/10.1016/j.biortech.2022.127443

Bleakley, S., & Hayes, M. (2021). Algal proteins: Extraction, application, and challenges concerning production. *Foods*, *10*(2), 249. https://doi.org/10.3390/foods10020249

Bourgade, T., Macías-Sánchez, M. D., & Sánchez, J. L. G. (2022). Microalgae as a sustainable source of food and feed. *Trends in Food Science & Technology*, *120*, 182–195. https://doi.org/10.1016/j.tifs.2021.12.010

Caporgno, M. P., & Mathys, A. (2018) Trends in microalgae incorporation into innovative food products with potential health benefits. *Frontiers in Nutrition*, *5*, 58. https://doi.org/10.3389/fnut.2018.00058

Duerlund, M., Andersen, B. V., Grønbeck, M. S., & Byrne, D. V. (2019). Consumer reflections on post-ingestive sensations. A qualitative approach by means of focus group interviews. *Appetite*, *142*, 104350.

Ennaceri, H., Mkpuma, V. O., Amo-Duodu, G., Menkiti, M., Ishika, T., Vadiveloo, A., & Moheimani, N. R. (2025). Microalgae as sustainable bio-factories for nanoparticle biosynthesis: Progress and challenges. *Bioresource Technology Reports*, *30*, 102089. https://doi.org/10.1016/j.biteb.2025.102089

FAO. (2023). *The State of Food Security and Nutrition in the World 2023*. Food and Agriculture Organization. https://doi.org/10.4060/cc3017en

García-Encinas, J. P., Ruiz-Cruz, S., Juárez, J., Ornelas-Paz, J. J., Del Toro-Sánchez, C. L., & Márquez-Ríos, E. (2025). Proteins from microalgae: Nutritional, functional and bioactive properties. *Foods (Basel, Switzerland)*, *14*(6), 921. https://doi.org/10.3390/foods14060921

Habib, M., Singh, S., Jan, S., Jan, K., & Bashir, K. (2025). The future of the future foods: Understandings from the past towards SDG-2. *NPJ Science of Food*, *9*(1), 138. https://doi.org/10.1038/s41538-025-00484-x

Hosny, S, Elshobary, M. E., & El-Sheekh, M. M. (2025). Unleashing the power of microalgae: A pioneering path to sustainability and achieving the sustainable development goals. *Environmental Science and Pollution Research International*, *32*(29), 17312–17342. https://doi.org/10.1007/s11356-025-35885-8

Koyande, A. K., Chew, K. W., Rambabu, K., Tao, Y., Chu, D. T., & Show, P. L. (2020). Microalgae: A potential alternative to health supplementation for humans. *Food Science and Human Wellness*, *9*(1), 55–64. https://doi.org/10.1016/j.fshw.2019.12.001

Lucas, F. B., & Brunner, T. A. (2024). Attitudes and perceptions towards microalgae as an alternative food: A consumer segmentation in Switzerland. *Algal Research*, *78*, 103386.

Meixner, O., Nieschalk, R., & Haas, R. (2023). Microalgae-based food: Consumer perception and willingness to pay in Austria—A discrete choice based experiment. *International Journal on Food System Dynamics*, *14*(4), 381–393. https://doi.org/10.18461/ijfsd.v14i4.H2

Oliveira, A. P. F. D., & Bragotto, A. P. A. (2022). Microalgae-based products: Food and public health. *Future Foods*, *6*, 100157

Olsen, Y. (2024). Consumer acceptance of microalgae as a novel food – Where are we now? *Physiologia Plantarum*, *176*(2), 143–157. https://doi.org/10.1111/ppl.14337

Santos, B., Freitas, F., Sobral, A. J. F. N., & Encarnação, T. (2025). Microalgae and circular economy: Unlocking waste to resource pathways for sustainable development. *International Journal of Sustainable Engineering*, *18*(1). https://doi.org/10.1080/19397038.2025.2501488

Sarker, N. K., & Kaparaju, P. (2024). Microalgal bioeconomy: A green economy approach towards achieving sustainable development goals. *Sustainability*, *16*(24), 11218. https://doi.org/10.3390/su162411218

Shan, Y., Wang, H., & Wang, W. (2025). Food neophobia: Psychological dimensions of consumer perception and emotional sentiment in social media discourse. *Frontiers in Nutrition*, *12*, 1584409. https://doi.org/10.3389/fnut.2025.1584409

Somers, M. D., Chen, P., Clippinger, J., Cruce, J. R., Davis, R., Lammers, P. J., & Quinn, J. C. (2021). Techno-economic and life-cycle assessment of fuel production from mixotrophic *Galdieria sulphuraria* microalgae on hydrolysate. *Algal Research*, *59*, 102419. https://doi.org/10.1016/j.algal.2021.102419

Wu, S., Menut, P., Miao, S., & Turchiuli, C. (2025). Algae protein creates sustainable alternatives for various food matrices: From function to nutrition. *Comprehensive Reviews in Food Science and Food Safety*, *24*(5), e70264. https://doi.org/10.1111/1541-4337.70264

21 Harnessing the Potential of Microalgae Biotechnology through the Water–Energy–Food Nexus

A System Dynamics Approach to Sustainable Innovation

Erika Cristina Francisco

21.1 INTRODUCTION TO THE WATER–ENERGY–FOOD NEXUS

The Water–Energy–Food (WEF) nexus underscores the complex interconnections and trade-offs among these essential resources, highlighting the urgency of adopting integrated management approaches to achieve global sustainability (Francisco et al., 2023; Lens et al., 2023). Because these sectors are interdependent, decisions in one area inevitably influence the others, shaping overall resource security and societal well-being. A rising global population and increasing consumption intensify these linkages, driving the demand for innovative solutions that improve efficiency while reducing environmental pressures across the nexus (Dal Poz et al., 2022; Sutherland et al., 2021). Within this perspective, it becomes imperative to design synergistic strategies that optimize resource flows and minimize ecological damage, particularly through the adoption of emerging biotechnologies such as microalgae cultivation.

Integrating microalgal systems into the WEF framework presents an effective pathway to address resource limitations and foster sustainability by transforming waste streams into valuable bioresources (Khandelwal & Lens, 2023). Microalgae can simultaneously absorb carbon dioxide from the atmosphere, thereby reducing air pollution, and utilize wastewater as a cultivation medium, linking water recovery with energy and food production through biomass generation (Arora et al., 2021; Hasnain et al., 2023). Embedding microalgae into these systems enhances resource

DOI: 10.1201/9781003611479-21

efficiency, lowers environmental burdens, and supports circular economy principles by valorizing waste into high-value products (Lens et al., 2023; Saldarriaga-Hernández et al., 2020). This integration reflects the foundations of a circular bioeconomy, where resource recovery and waste minimization strengthen the resilience of the WEF nexus (Hu et al., 2023).

Despite these opportunities, the interdependencies within the nexus also generate challenges such as resource scarcity, environmental degradation, and geopolitical conflicts, all of which are further aggravated by climate change and population growth (Shu et al., 2021). At the same time, these challenges encourage the development of forward-looking policy frameworks and innovative solutions that enhance sustainability and resilience (Pahl-Wostl et al., 2020; Yaykıran & Ekdal, 2025). A holistic recognition of the tight coupling of water, energy, and food is vital for formulating strategies that balance competition among resources while ensuring mutual benefits (David et al., 2023). The WEF nexus approach has thus gained significant international traction, shaping regional and global agreements that acknowledge the profound interdependence of these sectors (Ansari et al., 2023). Addressing these interdependencies requires deep insight into feedback loops, trade-offs, and synergies in resource allocation, especially in light of biotechnological innovations (García et al., 2018). This calls for robust modeling and analytical tools that can capture the dynamic behavior of these systems and their vulnerability to external uncertainties (Govindan & Al-Ansari, 2019). Such an understanding is essential to optimize resource distribution and reduce risks tied to supply chain disruptions and climate variability (Huang & Han, 2025).

Sustainable innovation, particularly within biotechnology, provides transformative avenues for alleviating resource stress and reinforcing resilience across WEF systems. This involves the development of novel technologies and processes that enhance efficiency, promote circular material use, and minimize waste within these interlinked domains (Ansari et al., 2023). Transitioning toward integrated resource management is therefore central to advancing a circular economy, where resources are maximized and waste is reduced throughout the value chain (Hejnowicz et al., 2022). By merging the principles of the circular economy with bioeconomy, this paradigm—often referred to as the circular bioeconomy—promotes closed-loop systems that recover valuable resources from biological waste streams, with microalgae cultivation serving as a prime example. In this model, biological residues are sustainably converted into reusable products, aligning closely with the United Nations Sustainable Development Goals on clean water, sanitation, and renewable energy (Dāce et al., 2024; Samsatli et al., 2019). This transformative shift enables the decoupling of economic growth from resource depletion by leveraging bio-based resources and processes to enhance efficiency and reduce environmental impacts (Ansari et al., 2023).

21.2 MICROALGAE'S ROLE IN THE WEF NEXUS

Microalgae demonstrate exceptional potential in treating a wide range of wastewater streams—municipal, industrial, and agricultural—by assimilating nutrients such as nitrogen and phosphorus, while also capturing heavy metals and other contaminants.

This dual function enables both water purification and the recovery of valuable resources like clean water and biomass (Ansari et al., 2023). Beyond addressing water treatment needs, the process generates biomass that can be converted into biofuels, biofertilizers, and other high-value products, thereby contributing simultaneously to energy and food security within the WEF nexus (Kilbane, 2022). Embedding microalgae in wastewater treatment transforms waste into a resource, embodying the "waste-to-wealth" concept central to a circular bioeconomy (Sharma et al., 2021). In addition, microalgae efficiently capture atmospheric CO_2, converting it into biomass that supports greenhouse gas mitigation strategies (Ryndin et al., 2023). This integration of wastewater treatment with biomass generation not only reduces environmental pollution but also enhances sustainability by supplying renewable resources (Ansari et al., 2023; Morais et al., 2023). Through photosynthesis, microalgae release oxygen, which lowers biological oxygen demand in wastewater, while their uptake of carbon, phosphorus, and nitrogen helps prevent secondary emissions and provides an energy-efficient waste management strategy (Hawrot-Paw et al., 2019; Hasnain et al., 2023).

Microalgal biomass, characterized by rapid growth rates and high lipid content, represents a promising feedstock for producing biofuels such as biodiesel, bioethanol, and biomethane, offering a renewable substitute for fossil fuels (Harini & Rajkumar, 2022). Their capacity to thrive on nonarable land and utilize wastewater for cultivation further strengthens their role as an environmentally sustainable energy source, avoiding competition with food crops and conserving freshwater resources (Hasnain et al., 2023). With high photosynthetic efficiency and the ability to capture CO_2 from industrial flue gases, microalgae provide additional benefits as an energy crop while contributing to emission reductions (Guo et al., 2015). Unlike traditional bioenergy crops, they do not compete for farmland or freshwater, as they can be cultivated in saline or wastewater environments (Narala et al., 2016). This adaptability makes them a viable candidate for biomass-based energy production while playing a role in mitigating climate change by reducing global carbon emissions (Du et al., 2014). Beyond liquid and gaseous biofuels, microalgae can also support bioelectricity production through bioremediation processes that couple wastewater treatment with energy generation (Rasheed et al., 2023).

Their diverse biochemical composition—including lipids, carbohydrates, and proteins—enables conversion into multiple bioproducts, ranging from functional foods and animal feed to fertilizers, thereby reinforcing their role in the bioeconomy (Harini & Rajkumar, 2022; Razaviarani et al., 2022). However, the economic feasibility of large-scale cultivation remains constrained by high harvesting and dewatering costs, which may represent over 30% of total production expenses (Liu et al., 2017). Overcoming these barriers requires technological innovation in cultivation, processing, and system design to lower costs and unlock the full potential of microalgae as a sustainable resource (Alazaiza et al., 2023).

In addition to energy applications, microalgae represent a highly nutritious and sustainable source of food and feed. They are rich in essential amino acids, vitamins, and polyunsaturated fatty acids and provide a resilient food supply independent of arable land and freshwater resources (Olhoff & Olsen, 2011). Their rapid growth and high photosynthetic efficiency allow for substantial biomass yields within relatively small footprints, making them a promising solution to the rising demand for sustainable protein and lipid sources (Manirafasha et al., 2016). Some strains exhibit exceptional nutritional qualities, with higher oil, carbohydrate, and protein yields than conventional crops such as soybean and palm oil (Inuwa et al., 2022). This superior nutritional value, combined with cultivation flexibility, positions microalgae as a cornerstone of future food systems, serving both human diets and animal feed (Hu et al., 2023).

Historically, algae have been used in food and feed since the 1970s, while modern research has focused on maximizing productivity and optimizing nutritional profiles (Kim et al., 2019; Saad et al., 2019). Widely cultivated species like *Chlorella* and *Spirulina* are well known for their hemoglobin-enhancing properties, their application as biofertilizers, and their ability to provide essential nutrients for both humans and livestock (Harini & Rajkumar, 2022). Today, over 75% of cultivated microalgal biomass is processed into powders, tablets, or capsules for human consumption, reinforcing its role as an alternative food source to combat global food insecurity (Maurya et al., 2021; Vieira et al., 2021). Certain species contain protein levels reaching up to 80%, surpassing soy flour and chicken, while also displaying amino acid profiles with superior Essential Amino Acid Index scores as defined by the Food and Agriculture Organization of the United Nations (FAO) (Stegemüller, 2024).

Microalgae also provide highly efficient land and water use for protein production, outperforming crops like peas, which require significantly more resources per kilogram of protein (Sun et al., 2024). Their exceptional protein and nutrient content make them particularly valuable in addressing protein deficiencies in food-insecure populations and in creating sustainable feed formulations that reduce dependency on conventional protein sources (Ye et al., 2023). With nearly double the protein content of beef, pork, or poultry on a dry weight basis, microalgae biomass provides a competitive advantage over traditional food sources (Ye et al., 2023). Species such as *Arthrospira platensis* (Spirulina), with a protein content of 55–70%, are especially efficient protein sources with substantial growth potential (Gkarane et al., 2020).

21.3 SYSTEM DYNAMICS FOR ANALYZING MICROALGAE BIOTECHNOLOGY

System dynamics provides a valuable framework for examining the complex interactions that shape microalgae biotechnology, encompassing environmental, economic, and social dimensions. By capturing these interdependencies, the approach helps identify leverage points and unintended consequences within the system, thereby supporting more effective policy design and strategic planning for sustainable innovation (Jacob-Lopes et al., 2018; Noüe & Pauw, 1988). This methodology is particularly relevant for analyzing the WEF nexus in the context of microalgae cultivation, as it allows the simulation of multiple scenarios to optimize resource use and reduce ecological burdens (Abdulmajeed & Ibrahim, 2018; Francisco et al., 2023). Through modeling feedback loops and delays inherent in scaling up biotechnology, system dynamics offers a robust tool for evaluating the long-term sustainability and economic feasibility of microalgae production pathways (Gao et al., 2024). It helps map the causal linkages among core variables—such as biomass yields, energy inputs, nutrient recycling, and market demand—showing how changes in one dimension cascade throughout the system (Acién et al., 2012; Pulz, 2001). This allows for quantitative assessments of system behavior under different policy options or technological innovations, providing critical insights into the prospects for sustainable commercialization of microalgae-based solutions (Acién et al., 2017).

Importantly, this approach also accommodates external drivers, including climate change, regulatory measures, and market fluctuations, thus offering a holistic perspective for strategic decision-making. By anticipating risks and opportunities, stakeholders can craft adaptive strategies that strengthen the role of microalgae biotechnology in sustainability transitions (Elsawah et al., 2017). System dynamics has already been applied successfully to other complex systems—such as smallholder agriculture and biomass supply chains—demonstrating its applicability to the multifaceted WEF nexus in microalgae systems (Malard et al., 2015; Zahraee et al., 2020).

The power of system dynamics lies in its ability to represent feedback loops, time delays, and nonlinear interactions, making it especially suitable for analyzing resource allocation and environmental impacts in bio-based industries (Herrera & Kopainsky, 2020; Zahraee et al., 2020). This quasi-continuous modeling technique is particularly effective for understanding interdependent and nonlinear supply chains, such as those characterizing microalgae biorefineries (Habibi et al., 2023). It facilitates a systems-level understanding of how technical parameters (e.g., cultivation strategies, nutrient inputs, and energy demands) intersect with economic and social dimensions, thereby capturing the system evolution over time (Acién et al., 2017; Guzmán et al., 2020). By integrating multiple perspectives, system dynamics enables decision-makers to evaluate both short- and long-term consequences of interventions, which is particularly critical in resource-constrained and rapidly changing environments (Azar, 2012; Nazari et al., 2023).

Given the inherent complexity of the WEF nexus, system dynamics offers a robust means to assess the interactions and dependencies among water, energy, and food within microalgae systems (Mostefaoui et al., 2024). The approach can uncover the feedback mechanisms that drive system performance, thereby clarifying how decisions or policies influence outcomes in interconnected supply chains (Lagarda-Leyva, 2021). Models can explicitly account for water availability, energy requirements for cultivation and processing, and the nutritional value of resulting biomass, offering insights into sustainable management strategies (Zahraee et al., 2020). Scenario testing—such as modifying water recycling rates or transitioning to renewable energy sources—can reveal ripple effects across the entire nexus (Zahraee et al., 2020). Moreover, the capacity to simulate "what-if" scenarios enables researchers and policymakers to anticipate long-term impacts of technological advances or regulatory reforms on the resilience and sustainability of microalgae-based systems (Dal Poz et al., 2022; Francisco et al., 2023).

System dynamics is further strengthened by its ability to incorporate uncertainties and variability, resulting in more realistic representations of real-world conditions (Zahraee et al., 2021). This makes it a powerful tool for optimizing supply chain design and operations, ensuring the long-term viability of microalgae biorefineries while contributing to circular bioeconomy principles (Mohseni & Pishvaee, 2016). Ultimately, integrating system dynamics with WEF nexus modeling helps identify synergies and trade-offs among subsystems, supporting cooperation and mitigating resource competition (Namany et al., 2019).

21.4 FUTURE DIRECTIONS AND RECOMMENDATIONS

Ongoing progress in enhancing microalgal productivity and photosynthetic performance—driven by metabolic engineering, multi-omics approaches, and synthetic biology—remains fundamental to the advancement of sustainable microalgal biorefineries (Hu et al., 2023). Genetic engineering to strengthen carbon sequestration and redirect carbon fluxes is particularly important for improving product yields and light utilization efficiency (Muñoz et al., 2021). These innovations are paving the way for the development of "smart strains" with higher productivity and resilience to environmental fluctuations, positioning microalgae as a key player in future biotechnology (Benedetti et al., 2018). In parallel, research is increasingly focused on integrating microalgal platforms with eco-friendly innovations such as microalgae-derived bioplastics, offering new opportunities for synergistic applications and sustainable technological breakthroughs (Lens et al., 2023). As a result, microalgal biorefineries are emerging as critical elements of the green economy, capable of generating a wide range of biofuels and biochemicals that support environmentally sound solutions for resource recovery from waste and wastewater (Khandelwal & Lens, 2023).

To accelerate these developments, coordinated efforts among policymakers, researchers, and industry stakeholders are required. Policymakers should prioritize funding for scalable pilot projects, researchers should build interdisciplinary collaborations to address persistent technological barriers, and industries must invest in infrastructure that supports large-scale cultivation and processing (D'Este, 2017; Hu et al., 2023). The biorefinery concept—producing multiple coproducts alongside biofuels—plays a decisive role in improving the economic feasibility of algal cultivation (Lens et al., 2023; Lawrence et al., 2023). Such integration enhances profitability while reducing waste generation, thereby reinforcing the transition toward a circular bioeconomy (Kostas et al., 2021; Rasheed et al., 2023). Examples include the production of biogas, bioethanol, and bioelectricity through bioelectrochemical systems, as well as other high-value compounds, thereby transforming waste into valuable resources within sustainable economic frameworks (Lens et al., 2023).

Moreover, situating biorefineries near industrial waste streams and flue gas sources can provide cost-effective cultivation media, improving economic viability while valorizing industrial by-products (Abdullah et al., 2024). This spatial integration exemplifies a transformative pathway that embraces circular economy principles and contributes directly to sustainable development objectives (Khandelwal & Lens, 2023). Flexibility and modularity in biorefinery design are equally essential, allowing adaptation to different biomass feedstocks and shifting market demands (Lens et al., 2023; Paladino & Neviani, 2023).

Adopting a system dynamics perspective provides a valuable framework to capture the complexity of these interconnected systems, enabling identification of optimal leverage points to maximize resource recovery and reduce environmental impacts (Lens et al., 2023). Such an approach supports the creation of adaptive strategies that can withstand global environmental challenges. Life cycle assessments (LCAs) are also indispensable for quantifying environmental impacts and ensuring sustainability, particularly by evaluating energy requirements across cultivation, harvesting, and processing stages, and by assessing co-benefits such as CO_2 mitigation and wastewater treatment (Lens et al., 2023). Furthermore, microalgal-bacterial systems offer great promise, as their synergistic interactions significantly improve both carbon capture and pollutant removal, reinforcing their role as a key environmental biotechnology (Hasnain et al., 2023). Coupling algal cultivation with wastewater treatment further enhances economic feasibility, as it reduces nutrient inputs and creates additional value streams (Javed et al., 2019; Li et al., 2019). This integration not only promotes sustainable wastewater management but also converts waste into a resource, thereby strengthening the circular economy (Thoré et al., 2021).

Continued investment in research and innovation, supported by computational modeling and advanced simulation tools, will be critical to advancing these solutions. In particular, system dynamics modeling can help anticipate long-term outcomes, evaluate trade-offs, and guide the integration of microalgal systems into the broader WEF nexus. These tools provide insights into both environmental and economic performance, thereby supporting informed decisions that align with sustainable development priorities (Lens et al., 2023).

In conclusion, advancing microalgae-based biorefineries requires a coordinated balance of scientific innovation, supportive policy frameworks, and industrial commitment. Cutting-edge tools such as metabolic engineering, system dynamics modeling, and other tools, such as LCAs, are instrumental in optimizing productivity and sustainability, while integration with wastewater treatment and industrial by-products strengthens economic viability and circular economy outcomes. By fostering collaboration across disciplines and sectors, and strategically situating biorefineries within existing resource flows, microalgae can transition from niche applications to a cornerstone of sustainable biotechnology. Ultimately, this multifaceted approach holds the potential to deliver scalable, resilient, and environmentally responsible solutions for energy, water, and food challenges, aligning directly with global sustainability and climate goals.

21.5 CONCLUSION

Microalgae biotechnology exemplifies how sustainable innovation can be achieved within the WEF nexus. By simultaneously addressing water treatment, renewable energy, and food security, microalgae provide integrated solutions to some of the most urgent global sustainability challenges. The use of system dynamics enhances our ability to model and understand these interdependencies, offering pathways for adaptive policies and resilient strategies.

While high costs, technological barriers, and regulatory gaps remain, continued advancements in engineering, interdisciplinary research, and supportive policy frameworks can unlock the full potential of microalgae. When coupled with circular bioeconomy strategies, microalgae biotechnology can drive progress toward climate resilience, food security, and sustainable resource management. Ultimately, its integration across the WEF nexus highlights its role not only as a scientific and industrial innovation but also as a cornerstone of future sustainable development.

REFERENCES

Abdullah, M. A., Nazir, M. S., Hussein, H. A., Shah, S. M. U., Azra, N., Iftikhar, R., Iqbal, M. S., Qamar, Z., Ahmad, Z., Afzaal, M., Daud, A., Shaharah, M. I., Rak, A. E., & Hung, Y. (2024). New perspectives on biomass conversion and circular economy based on Integrated Algal-Oil Palm Biorefinery framework for sustainable energy and bioproducts co-generation. *Industrial Crops and Products*, *213*, 118452. https://doi.org/10.1016/j.indcrop.2024.118452

Abdulmajeed, B. A., & Ibrahim, A. R. (2018). Potential of microalgae cultivation in dairy wastewater as a step in low-cost biofuel production. *Journal of Engineering*, *24*(4), 58. https://doi.org/10.31026/j.eng.2018.04.04

Acién, F. G., Fernández, J. M., Magán, J. J., & Molina, E. (2012). Production cost of a real microalgae production plant and strategies to reduce it. *Biotechnology Advances*, *30*(6), 1344. https://doi.org/10.1016/j.biotechadv.2012.02.005

Acién, F. G., Molina, E., Fernández-Sevilla, J. M., Barbosa, M. J., Gouveia, L., Sepúlveda, C., Bazaes, J., & Arbib, Z. (2017). Economics of microalgae production. In Cristina Gonzalez-Fernandez and Raúl Muñoz (eds.), *Elsevier eBooks* (p. 485). Elsevier BV. https://doi.org/10.1016/b978-0-08-101023-5.00020-0

Alazaiza, M. Y. D., Albahnasawi, A., Maskari, T. A., Abujazar, M. S. S., Bashir, M. J. K., Nassani, D. E., & Amr, S. S. A. (2023). Biofuel production using cultivated algae: Technologies, economics, and its environmental impacts. *Energies*, *16*(3), 1316. https://doi.org/10.3390/en16031316

Ansari, A., Wuryandani, S., Pranesti, A., Telaumbanua, M., Ngadisih, Hardiansyah, M.Y., Alam, T., Supriyanta, S., Martini, T., & Taryono, T. (2023). Optimizing water-energy-food nexus: Achieving economic prosperity and environmental sustainability in agriculture. *Frontiers in Sustainable Food Systems*, *7*. https://doi.org/10.3389/fsufs.2023.1207197

Arora, K., Kumar, P., Bose, D., Li, X., & Kulshrestha, S. (2021). Potential applications of algae in biochemical and bioenergy sector [Review]. *3 Biotech*, *11*(6). https://doi.org/10.1007/s13205-021-02825-5

Azar, A. T. (2012). System dynamics as a useful technique for complex systems. *International Journal of Industrial and Systems Engineering*, *10*(4), 377. https://doi.org/10.1504/ijise.2012.046298

Benedetti, M., Vecchi, V., Barera, S., & Dall'Osto, L. (2018). Biomass from microalgae: The potential of domestication towards sustainable biofactories [Review]. *Microbial Cell Factories*, *17*(1). https://doi.org/10.1186/s12934-018-1019-3

D'Este, M. (2017). Algal biomass for bioenergy and bioproducts production in biorefinery concepts. *Research Portal Denmark*, *58*.

Dāce, E., Cascavilla, A., Bianchi, M., Chioatto, E., Zecca, E., Ladu, L., & Yılan, G. (2024). Barriers to transitioning to a circular bio-based economy: Findings from an industrial perspective. *Sustainable Production and Consumption*, *48*, 407. https://doi.org/10.1016/j.spc.2024.05.029

Dal Poz, M. E. S., Ignácio, P. S. A., Azevedo, A., Francisco, E. F., Piolli, A. L., Silva, G. G., & Ribeiro, T. P. (2022). Food, energy and water nexus: An urban living laboratory development for sustainable systems transition. *Sustainability*, *14*, 1763, 1–22. https://doi.org/10.3390/su14127163

David, L. O., Adepoju, O. O., Nwulu, N., & Aigbavboa, C. (2023). Determining the impact of economic indicators on water, energy and food nexus for sustainable resource security. *Clean Technologies and Environmental Policy*, *26*(3), 803. https://doi.org/10.1007/s10098-023-02651-8

Du, J. J., Zeng, P., Zhang, F., & Wang, C. (2014). The comparison study of four lab scale microalgae collection ways. *Advanced Materials Research*, *1729*. https://doi.org/10.4028/www.scientific.net/amr.875-877.1729

Elsawah, S., Pierce, S. A., Hamilton, S. H., Delden, H. V., Haase, D., Elmahdi, A., & Jakeman, A. J. (2017). An overview of the system dynamics process for integrated modelling of socio-ecological systems: Lessons on good modelling practice from five case studies. *Environmental Modelling & Software*, *93*, 127. https://doi.org/10.1016/j.envsoft.2017.03.001

Francisco, E. F., Piolli, A. L., Ignácio, P. S. A., & Dal Poz, M. E. S. (2023). Food-Energy-Water (FEW) nexus: Sustainable food production governance through system dynamics modelling. *Journal of Cleaner Production*, *386*, 135825. https://doi.org/10.1016/j.jclepro.2022.135825

Gao, Y., Bernard, O., Fanesi, A., Perré, P., & Lopes, F. (2024). The effect of light intensity on microalgae biofilm structures and physiology under continuous illumination. *Scientific Reports*, *14*(1). https://doi.org/10.1038/s41598-023-50432-6

García, I. F., García, A. M., Díaz, J. A. R., Montesinos, P., & Poyato, E. C. (2018). Water–energy nexus in irrigated areas. Lessons from real case studies. In Iván Francisco García Tejero and Víctor Hugo Durán Zuazo (eds.), *Elsevier eBooks* (p. 41). Elsevier BV. https://doi.org/10.1016/b978-0-12-813164-0.00002-8

Gkarane, V., Ciulu, M., Altmann, B. A., Schmitt, A. O., & Mörlein, D. (2020). The effect of algae or insect supplementation as alternative protein sources on the volatile profile of chicken meat. *Foods*, *9*(9), 1235. https://doi.org/10.3390/foods9091235

Govindan, R., & Al-Ansari, T. (2019). Computational decision framework for enhancing resilience of the energy, water and food nexus in risky environments. *Renewable and Sustainable Energy Reviews*, *112*, 653. https://doi.org/10.1016/j.rser.2019.06.015

Guo, Z., Phooi, W. B. A., Lim, Z. J., & Tong, Y. W. (2015). Control of CO2 input conditions during outdoor culture of Chlorella vulgaris in bubble column photobioreactors. *Bioresource Technology*, *186*, 238. https://doi.org/10.1016/j.biortech.2015.03.065

Guzmán, J. L., Acién, F. G., & Berenguel, M. (2020). Modelado y control de la producción de microalgas en fotobiorreactores industriales. *Revista Iberoamericana de Automática e Informática Industrial RIAI*, *18*(1), 1. https://doi.org/10.4995/riai.2020.13604

Habibi, F., Chakrabortty, R. K., & Abbasi, A. (2023). Towards facing uncertainties in biofuel supply chain networks: A systematic literature review [Review]. *Environmental Science and Pollution Research*, *30*(45), 100360. https://doi.org/10.1007/s11356-023-29331-w

Harini, A. B., & Rajkumar, R. (2022). Development of sustainable bioproducts from microalgae biomass: Current status and future perspectives. *BioResources*, *17*(4), 7285. https://doi.org/10.15376/biores.17.4.harini

Hasnain, M., Zainab, R., Ali, F., Abideen, Z., Yong, J. W. H., El-Keblawy, A., Hashmi, S., & Radicetti, E. (2023). Utilization of microalgal-bacterial energy nexus improves CO2 sequestration and remediation of wastewater pollutants for beneficial environmental services [Review]. *Ecotoxicology and Environmental Safety*, *267*, 115646. https://doi.org/10.1016/j.ecoenv.2023.115646

Hawrot-Paw, M., Koniuszy, A., Gałczyńska, M., Zając, G., & Szyszlak-Bargłowicz, J. (2019). Production of microalgal biomass using aquaculture wastewater as growth medium. *Water*, *12*(1), 106. https://doi.org/10.3390/w12010106

Hejnowicz, A. P., Thorn, J., Giraudo, M. E., Sallach, J. B., Hartley, S. E., Grugel, J., Pueppke, S. G., & Emberson, L. (2022). Appraising the water-energy-food nexus from a sustainable development perspective: A maturing paradigm? *Earth's Future*, *10*(12). https://doi.org/10.1029/2021ef002622

Herrera, H., & Kopainsky, B. (2020). Using system dynamics to support a participatory assessment of resilience. *Environment Systems & Decisions*, *40*(3), 342. https://doi.org/10.1007/s10669-020-09760-5

Hu, J., Meng, W., Su, Y., Qian, C., & Fu, W. (2023). Emerging technologies for advancing microalgal photosynthesis and metabolism toward sustainable production. *Frontiers in Marine Science*, *10*. https://doi.org/10.3389/fmars.2023.1260709

Huang, R., & Han, Y. (2025). Differentiated optimization policies for water–energy–food resilience security: Empirical evidence based on Shanxi province and the GWR model. *Water*, *17*(10), 1540. https://doi.org/10.3390/w17101540

Inuwa, A. B., Zeb, I., Mahmood, Q., Irshad, U., Irshad, M., Hafeez, F., Iqbal, A., Pervez, A., & Nazir, R. (2022). Novel microalgae strains from selected lower Himalayan aquatic habitats as potential sources of green products. *PLoS ONE*, *17*(5). https://doi.org/10.1371/journal.pone.0267788

Jacob-Lopes, E., Maroneze, M. M., Deprá, M. C., Sartori, R. B., Dias, R. R., & Zepka, L. Q. (2018). Bioactive food compounds from microalgae: An innovative framework on industrial biorefineries. *Current Opinion in Food Science*, *25*, 1. https://doi.org/10.1016/j.cofs.2018.12.003

Javed, F., Aslam, M., Rashid, N., Shamair, Z., Khan, A. L., Yasin, M., Fazal, T., Hafeez, A., Rehman, F., Rehman, M. S. U., Khan, Z., Iqbal, J., & Bazmi, A. A. (2019). Microalgae-based biofuels, resource recovery and wastewater treatment: A pathway towards sustainable biorefinery. *Fuel*, *255*, 115826. https://doi.org/10.1016/j.fuel.2019.115826

Khandelwal, A., & Lens, P. N. L. (2023). Algal systems for resource recovery from waste and wastewater. In Piet N. L. Lens and Amitap Khandelwal (eds.), *IWA Publishing eBooks* (p. 3). UWA Publishing. https://doi.org/10.2166/9781789063547_0003

Kilbane, J. J. (2022). Shining a light on wastewater treatment with microalgae [Review]. *Arabian Journal for Science and Engineering*, *47*(1), 45. https://doi.org/10.1007/s13369-021-06444-3

Kim, H. S., Park, W.-K., Lee, B., Seon, G., Suh, W. I., Moon, M., & Chang, Y. K. (2019). Optimization of heterotrophic cultivation of Chlorella sp. HS2 using screening, statistical assessment, and validation. *Scientific Reports*, *9*(1). https://doi.org/10.1038/s41598-019-55854-9

Kostas, E. T., Adams, J. M. M., Ruíz, H. A., Durán-Jiménez, G., & Lye, G. J. (2021). Macroalgal biorefinery concepts for the circular bioeconomy: A review on biotechnological developments and future perspectives [Review]. *Renewable and Sustainable Energy Reviews*, *151*, 111553. https://doi.org/10.1016/j.rser.2021.111553

Lagarda-Leyva, E. A. (2021). System dynamics and lean approach: Development of a technological solution in a regional product packaging company. *Applied Sciences*, *11*(17), 7938. https://doi.org/10.3390/app11177938

Lawrence, J. F., Oliva, A., Murphy, J. D., & Lens, P. N. L. (2023). Macroalgae biorefinery and its role in achieving a circular economy. In Piet N. L. Lens and Amitap Khandelwal (eds.), *IWA Publishing eBooks* (p. 53). UWA Publishing. https://doi.org/10.2166/9781789063547_0053

Lens, P. N. L., Khandelwal, A., Khandelwal, A., Lens, P. N. L., Pessi, B. A., Rada-Ariza, A. M., López-Vázquez, C. M., Steen, P., Lens, P. N. L., Lawrence, J. F., Oliva, A., Murphy, J. D., Lens, P. N. L., Acién, F. G., Gómez-Serrano, C., Morillas, A., Arbib, Z., Sanchez, A., Nordio, R., … Hooshmand, M. (2023). Algal systems for resource recovery from waste and wastewater. In Piet N. L. Lens and Amitap Khandelwal (eds.), *IWA Publishing eBooks*. UWA Publishing. https://doi.org/10.2166/9781789063547

Li, S., Zhao, S., Yan, S., Qiu, Y., Song, C., Li, Y., & Kitamura, Y. (2019). Food processing wastewater purification by microalgae cultivation associated with high value-added compounds production — A review [Review]. *Chinese Journal of Chemical Engineering*, *27*(12), 2845. https://doi.org/10.1016/j.cjche.2019.03.028

Liu, C., Hao, Y., Jiang, J., & Liu, W. (2017). Valorization of untreated rice bran towards bioflocculant using a lignocellulose-degrading strain and its use in microalgal biomass harvest. *Biotechnology for Biofuels*, *10*(1). https://doi.org/10.1186/s13068-017-0780-6

Malard, J., Adamowski, J. F., Rojas, D. M., Carrera, J., Gálvez, J., Monadres, H., & Melgar-Quiñonez, H. (2015, July 26). Use of participatory system dynamics modelling to assess the sustainability of smallholder agriculture. *ASABE International Meeting*. https://doi.org/10.13031/aim.20152189490

Manirafasha, E., Ndikubwimana, T., Zeng, X., Lu, Y., & Jing, K. (2016). Phycobiliprotein: Potential microalgae derived pharmaceutical and biological reagent. *Biochemical Engineering Journal*, *109*, 282. https://doi.org/10.1016/j.bej.2016.01.025

Maurya, R., Zhu, X., Valverde-Pérez, B., Kiran, B. R., General, T., Sharma, S., Sharma, A., Thomsen, M., Mohan, S. V., Mohanty, K., & Angelidaki, İ. (2021). Advances in microalgal research for valorization of industrial wastewater [Review]. *Bioresource Technology*, *343*, 126128. https://doi.org/10.1016/j.biortech.2021.126128

Mohseni, S., & Pishvaee, M. S. (2016). A robust programming approach towards design and optimization of microalgae-based biofuel supply chain. *Computers & Industrial Engineering*, *100*, 58. https://doi.org/10.1016/j.cie.2016.08.003

Morais, E. G. D., Silveira, J. T. D., Schüler, L., Freitas, B. C. B. D., Costa, J. A. V., Morais, M. G. D., Ferrer, I., & Barreira, L. (2023). Biomass valorization via pyrolysis in microalgae-based wastewater treatment: Challenges and opportunities for a circular bioeconomy. *Journal of Applied Phycology*, *35*(6), 2689. https://doi.org/10.1007/s10811-023-03104-x

Mostefaoui, L., Sušnik, J., Masia, S., & Jewitt, G. (2024). A water–energy–food nexus analysis of the impact of desalination and irrigated agriculture expansion in the Ain Temouchent region, Algeria. *Environment Development and Sustainability*. https://doi.org/10.1007/s10668-024-05151-x

Muñoz, C. F., Südfeld, C., Naduthodi, M. I. S., Weusthuis, R. A., Barbosa, M. J., Wijffels, R. H., & D'Adamo, S. (2021). Genetic engineering of microalgae for enhanced lipid production [Review]. *Biotechnology Advances*, *52*, 107836. https://doi.org/10.1016/j.biotechadv.2021.107836

Namany, S., Al-Ansari, T., & Govindan, R. (2019). Optimisation of the energy, water, and food nexus for food security scenarios. *Computers & Chemical Engineering*, *129*, 106513. https://doi.org/10.1016/j.compchemeng.2019.106513

Narala, R. R., Garg, S., Sharma, K., Thomas-Hall, S. R., Deme, M., Li, Y., & Schenk, P. M. (2016). Comparison of microalgae cultivation in photobioreactor, open raceway pond, and a two-stage hybrid system. *Frontiers in Energy Research*, *4*. https://doi.org/10.3389/fenrg.2016.00029

Nazari, B., Kanani, E., & Sepehri, S. (2023). Assessment of water productivity improvement strategies using system dynamics approach. *Applied Water Science*, *13*(12). https://doi.org/10.1007/s13201-023-02044-8

Noüe, J., & Pauw, N. D. (1988). The potential of microalgal biotechnology: A review of production and uses of microalgae [Review]. *Biotechnology Advances*, *6*(4), 725. https://doi.org/10.1016/0734-9750(88)91921-0

Olhoff, A., & Olsen, K. H. (2011). Integrating climate change adaptation in energy planning and decision-making - Key challenges and opportunities. *Research Portal Denmark*, *25*.

Pahl-Wostl, C., Gorris, P., Jager, N. W., Koch, L., Lebel, L., Stein, C., Venghaus, S., & Withanachchi, S. S. (2020). Scale-related governance challenges in the water–energy–food nexus: Toward a diagnostic approach. *Sustainability Science*, *16*(2), 615. https://doi.org/10.1007/s11625-020-00888-6

Paladino, O., & Neviani, M. (2023). Interchangeable modular design and operation of photo-bioreactors for Chlorella vulgaris cultivation towards a zero-waste biorefinery. *Enzyme and Microbial Technology*, *173*, 110371. https://doi.org/10.1016/j.enzmictec.2023.110371

Pulz, O. (2001). Photobioreactors: Production systems for phototrophic microorganisms [Review]. *Applied Microbiology and Biotechnology*, *57*(3), 287. https://doi.org/10.1007/s002530100702

Rasheed, R. N., Pourbakhtiar, A., Allaf, M. M., Baharlooeian, M., Rafiei, N., Aratboni, H. A., Morones-Ramírez, J. R., & Winck, F. V. (2023). Microalgal co-cultivation -recent methods, trends in omic-studies, applications, and future challenges [Review]. *Frontiers in Bioengineering and Biotechnology*, *11*. https://doi.org/10.3389/fbioe.2023.1193424

Razaviarani, V., Arab, G., Lerdwanawattana, N., & Gadia, Y. (2022). Algal biomass dual roles in phycoremediation of wastewater and production of bioenergy and value-added products. *International Journal of Environmental Science and Technology*, *20*(7), 8199. https://doi.org/10.1007/s13762-022-04696-6

Ryndin, K. G., Butyrin, A. V., Grigorenko, A. V., Chunzhuk, E. A., Чернова, Н. И., Киселева, С. В., Malaniy, S. Y., Bakumenko, E. A., Slavkina, O. V., Ossipov, К., Maryutina, T. A., Kumar, V., & Vlaskin, M. S. (2023). From the cultivation of Arthrospira platensis at an increased CO2 concentration to the bio-oil production by hydrothermal liquefaction. *Applied Sciences*, *13*(17), 9950. https://doi.org/10.3390/app13179950

Saad, M. G., Dosoky, N. S., Khan, M. S., Zoromba, M. S., Mekki, L., El-Bana, M. I., Nobles, D. R., & Shafik, H. M. (2019). High-throughput screening of chlorella vulgaris growth kinetics inside a droplet-based microfluidic device under irradiance and nitrate stress conditions. *Biomolecules*, *9*(7), 276. https://doi.org/10.3390/biom9070276

Saldarriaga-Hernández, S., Hernández-Vargas, G., Iqbal, H. M. N., Barceló, Damià, & Parra-Saldívar, R. (2020). Bioremediation potential of Sargassum sp. biomass to tackle pollution in coastal ecosystems: Circular economy approach [Review]. *The Science of the Total Environment*, *715*, 136978. https://doi.org/10.1016/j.scitotenv.2020.136978

Samsatli, S., Martínez-Hernández, E., & Ng, K. S. (2019). Towards a sustainable bio-economy: Working in harmony with the environment–food–energy–water nexus. *Food and Bioproducts Processing*, *119*, 371. https://doi.org/10.1016/j.fbp.2019.12.001

Sharma, P., Gaur, V. K., Sirohi, R., Varjani, S., Kim, S., & Wong, J. W. C. (2021). Sustainable processing of food waste for production of bio-based products for circular bioeconomy [Review]. *Bioresource Technology*, *325*, 124684. https://doi.org/10.1016/j.biortech.2021.124684

Shu, Q., Scott, M., Todman, L., & McGrane, S. J. (2021). Development of a prototype composite index for resilience and security of water-energy-food (WEF) systems in industrialised nations. *Environmental and Sustainability Indicators*, *11*, 100124. https://doi.org/10.1016/j.indic.2021.100124

Stegemüller, L. (2024). Unleashing the full potential of haematococcus for astaxanthin production using digital tools. *Research Portal Denmark*, *96*.

Sun, H., Gong, Q., Fan, Y., Wang, Y., Wang, J., Zhu, C., Mou, H., Yang, S., & Liu, J. (2024). Unlocking 3D printing technology for microalgal production and application [Review]. *Advanced Biotechnology*, *2*(4). https://doi.org/10.1007/s44307-024-00044-6

Sutherland, D. L., McCauley, J. I., Labeeuw, L., Ray, P., Kuzhiumparambil, U., Hall, C., Doblin, M. A., Nguyen, L. N., & Ralph, P. J. (2021). How microalgal biotechnology can assist with the UN sustainable development goals for natural resource management. *Current Research in Environmental Sustainability*, *3*, 100050. https://doi.org/10.1016/j.crsust.2021.100050

Thoré, E. S. J., Schoeters, F., Cuyper, A. D., Vleugels, R., Noyens, I., Bleyen, P., & Miert, S. V. (2021). Waste is the new wealth – Recovering resources from poultry wastewater for multifunctional microalgae feedstock. *Frontiers in Environmental Science*, *9*. https://doi.org/10.3389/fenvs.2021.679917

Vieira, M. V., Turkiewicz, I. P., Tkacz, K., Fuentes-Grünewald, C., Pastrana, L., Fuciños, P., Wojdyło, A., & Nowicka, P. (2021). Microalgae as a potential functional ingredient: Evaluation of the phytochemical profile, antioxidant activity and in-vitro enzymatic inhibitory effect of different species. *Molecules*, *26*(24), 7593. https://doi.org/10.3390/molecules26247593

Yaykıran, S., & Ekdal, A. (2025). Sustainable water optimization tool (SUWO): An optimization framework for the water–energy–food–ecosystem nexus. *Water*, *17*(9), 1280. https://doi.org/10.3390/w17091280

Ye, L., Guo, Y., Qiao, L., Xia, H., Neupane, N., Chaudhary, P., Rijal, Y., Ghimire, B., Bhandari, R., Nanyiti, S., Kabaalu, R., Alicai, T., Abidrabo, P., Seal, S., Bouvaine, S., Bailey, A. M., Foster, G. D., Li, J., Yuan, M., … Hlatshwayo, S. A. (2023). Enhancing food production system resilience for food security facing a changing environment. *Frontiers in Sustainable Food Systems*, 7, 1204098. https://doi.org/10.3389/978-2-8325-3995-8

Zahraee, S. M., Golroudbary, S. R., Shiwakoti, N., & Stasinopoulos, P. (2021). Particle-Gaseous pollutant emissions and cost of global biomass supply chain via maritime transportation: Full-scale synergy model. *Applied Energy*, *303*, 117687. https://doi.org/10.1016/j.apenergy.2021.117687

Zahraee, S. M., Golroudbary, S. R., Shiwakoti, N., Stasinopoulos, P., & Krasławski, A. (2020). Water-energy nexus and greenhouse gas–sulfur oxides embodied emissions of biomass supply and production system: A large scale analysis using combined life cycle and dynamic simulation approach. *Energy Conversion and Management*, *220*, 113113. https://doi.org/10.1016/j.enconman.2020.113113

Index

Pages in *italics* refer to figures and pages in **bold** refer to tables.

A

acidification, **116**, 457
adsorption, 310, 318, 341, 352–354
aeration, 99, 104, **116**, 140, 165, 317, 389–390
agricultural wastewater, 356
agronomic potential, 293, 298
Alcohol-to-Jet (ATJ), 417
algae biomass, 53, 328, 483
algae blooms, 39
algal cultivation, 44, 495
algal productivity, 53
ammonia emissions, 350–351
animal health, 275
antagonismo, 49, **52**
antigens, 24–25
Arrhenius model, 95
artificial intelligence, 95, 249, 390
astaxanthin, 25–26, 185–188
aviation decarbonization, 401–402

B

bacterial communities, 42, 44
bead-beating, 166, 170
bioaccessibility, 184
bioaccumulation, 318–319, 347–348
bioactive compounds, 6, 47, 53, 83, 99, 127, 197, 259, 274–275, **441**, 465, 478
bioactive peptides, 189, 261
biofertilizers, 113, 198, 294–295
biofilm-based systems, 349
biogeochemical processes, 38
biomass production, 68–69, **108**, 184–185, 302–303, 473–474
biopesticides, 293–294, 298–299
biopolymer(s), 235–253
bioregenerative life-support systems, 437, **441–442**
biosorption, 318–319, 347–348
biostimulants, 138, 297, 302
biotechnological compounds, 40
bulk commodities, 5–6

C

carbon cycling, 51, 273
carbon footprint(s), 83, **116**, 244, 315, 483
carbon neutrality, 100, 243
carbon sequestration, 367, 385–386, 391–395
carbon source concentration, 387
carbon/nitrogen ratio, 141, **224**
carotenoid(s), 25–27, 175, 187–188
centrifugation, 110, 170, 225, 245–246, 320–321, 363–354
chemical ecology, 46
chemical oxygen demand (COD), 66, 98, 311, 341, 384
chlorophyll(s), 162–175, 188–189, 263
chloroplast endoplasmic reticulum (CER), 236
circular bioeconomy, 83, 453, 490–491
cladocerans, 63–66
climate change, 364
closed systems, **85**, 87, 163, *316*, 360, 456
commensalism, 39, 48
competition, 49
computational fluid dynamics (CFD), 93
consumer acceptance, 199, 454, 482
consumer perception(s), 199, 264, 478, 481, 483
cultivation technologies, 7
cyanobacteria, 62, 71, 140–142, 148, 188

D

dairy wastewater, 311
DISCOVR program, 106
dissolved oxygen, 94, 309, 351–354
distillery wastewater, 311–312
domestic wastewaters, 354
downstream processes, 105, 110
droplet-based microfluidic assays, 8

E

ecological niches, 7
economic analysis, 249, 393–394
edible cyanobacteria, 2
eicosapentaenoic acid (EPA), 195, 263, 403, 444, 478
endosymbiotic events, 3
energy analysis, **116**
engineering bacterial consortia, 50
environmental conditions, 104–105, 127–128, 163, 210, 247, 260, 357, 359, 386
environmental footprint, 454, 471, 479
environmental stressors, 63, 217
environmental Sustainability, 235, 250, 259, 273, 310, 422, 454–455

enzyme-assisted extraction, **169**, 172, **227**
eukaryotic microalgae, 1–3, 10, 18
eutrophic systems, 67
eutrophication, 67–68, **116**, 362, 456, 480
extracellular polymeric substances, 127
extremophilic microalgae, 4

F

feed additives, 272, 278–279, 282
feed supplement, 284
fermenter(s), 116, 165–166, 359
filtration, 245, 320–321, 363–364
fine particulate matter formation (FPMF), 459
flocculation, 245–246, 321–323, 363–364
flotation, 323–324, 363
food additives, 198, 275
food colourants, 198–199
food neophobia, 481
food stability, 199
fossil resource scarcity (FRS), 459
freshwater ecosystems, 63, 68, 75–76
freshwater ecotoxicity (FEC), 456
freshwater eutrophication (FEU), 456

G

genetic engineering, 9–10, 20, 26, 62, 65, 252, 494
genetically modified microalgae, 23, 69
genetically modified organisms (GMOs), 66
genome editing, 9
genome mining, 10–11
genome sequencing, 3, 9, 50
global food systems, 479
global warming potential (GWP), 459
Good Manufacturing Practices (GMP), 106
gravity sedimentation, 226, **246**
greenhouse gases (GHGs), 244, 364

H

hazard analysis and critical control points (HACCP), 106, 439
heat stress, 282
heterotrophic growth, 165
high-rate algal ponds (HRAPs), 164, 315
hybrid systems, **85**, 111, 318

I

industrial wastewater, 309–311
industrial-scale microalgae cultivation, 110
intrinsic transfer efficiency (k_L), 94
ion-exchange techniques, 310
ionic liquids, 172
ionizing radiation (IR), 459

K

kinetic models, 95, 104

L

land use (LU), 457
large-scale cultivation, 2, 200, 238, 494
life-cycle assessment (LCA), 83, 114–115, 454, 467
life-cycle impact assessment (LCIA), 115
life-cycle inventory (LCI), 115
life-support systems (LSS), 436–437
light intensity, **96**, 139, 215–216, 389
lipid accumulation, 209–210, 218, 220–221

M

maceration, **168**, 171
machine learning, 93, 249
marine ecotoxicity (MEC), 456
membrane separation, 143
metabolomics, 40, 53
metagenomics, 40, 49–50
metatranscriptomics, 49
microalgae–bacteria consortia, 39, 51, 391
microalgae-based biofertilizers (MBFs), 295
microalgae-based feed, 272
microalgae-derived biopolymers, 237
microalgal bioprospecting, 7
microalgal exopolysaccharides, 138
microalgal growth, **96**
microalgal microbiome, 39
microenvironment, 40–43, 49
microwave-assisted extraction (MAE), **169**
mixing, 315, 317, **359**
modern biotechnology, 3, 83
modular cloning, 16–19
monounsaturated fatty acids (MUFAs), 213
mutualism, 39, 47, **52**

N

nanomaterials, 300
nanotechnology, 300
natural pigments, 162, 173–174, 185–186
novel foods, 109, 112, 144, 194, 481, 484
nuclear engineering, 11, 19
nuclear transformation, 18, 22, 24
nutritional profile(s), 1, 260, 484
nutritional value, 184, 251, 260, 262, 273

O

odor activities, 351
online flow cytometry, 95
open systems, 84, **85**, 184, 313–314, 358, 360

optimum growth temperature, 139
osmotic stress, 217

P

parasitism, 39, 41, 48–49
patente(s), 174–175
pH, 139–140
pharmaceutical wastewater, 312
phenomics platforms, 7
phenotypic optimization, 9
photobioreactor(s), 93–94, 390
phycoremediation, 344
phycosphere, 38–54
policy frameworks, 425–426, 428, 485, 490, 495
polyunsaturated fatty acids (PUFAs), 64, 183, 195, 209, 262
poultry industry wastewater, 311
precautionary principle, 69, 74
pyrolysis technique, 240

Q

quorum-quenching (QQ), 40
quorum-sensing (QS), 40

R

raceway ponds, 84, **85**, **107–108**, 164, 299, 314–315, 358–359
Raman spectroscopy, 8
regulatory framework(s), 75, 100, 112, 114, 118
rotifers, 63–66, 75–76

S

salinity, 140, 217, 343, 405
saturated fatty acids (SFAs), 213, 325
sensory expectations, 481, 483
social acceptance, 113
soil inoculation, 294
strain isolation, 7
strain selection, 3, 7, 100, 299–300
stratospheric ozone depletion (SOD), 459
sugar-to-jet (STJ), 417
sulfur, 140, 142, 247, 342, 416, 420
  compounds, 46, 350, 352, 354
  emissions, 51
  limitation, 142, 327, 404
sustainability, 367, 402, 454–455, 457–459, 461, 482–485, 489
sustainable aviation fuels (SAFs), 401
Sustainable Development Goals (SDG), 259, 457, 490
sustainable foods, 259–260
sustainable innovation, 490
symbiotic relationships, 3, 40–41, 47
synthetic biology, 1, 5–7, 9–11
synthetic phycosphere, 45

T

TAL effector endonuclease (TALEN), 19
tannery wastewater, 312, 453
techno-economic assessment (TEA), 101
technological maturity, 106, 111, 113, 117
Technological Readiness Level(s), 82
terpenes, 196–197
terrestrial acidification (TA), 457
terrestrial ecotoxicity (TET), 457
thermochemical conversion, 411, 417, 472
total water consumption (TWC), 456
transcription, 50
transcriptomics, 8, 50, 54, 96
transgene expression, *21*, 26
triacylglycerol (TAG), 5, 216, 365, 403
two-step cultivation, 218

U

ultrasound-assisted extraction (UAE), 166, **169**
upstream, 105, 456, 458

V

value-added products (VAPs), 99, 244, 303, 367
volatile organic compounds (VOCs), 341, 460
volumetric mass-transfer coefficient (K_La), 94, **96**
volumetric productivity, 9, 87

W

wastewater treatment technologies, 99
water footprint, 115, **116**, 456
Water–Energy–Food (WEF) nexus, 489

X

xanthophylls, 187, 263
xylose, 128, 133, 221

Y

yeast, 20, 24, 280, 404, 467

Z

zeaxanthin, 26, 187, 263
zinc, 143, 263, 355, 357, 480
zinc-finger nuclease, 19

For Product Safety Concerns and Information please contact our EU representative GPSR@taylorandfrancis.com
Taylor & Francis Verlag GmbH, Kaufingerstraße 24, 80331 München, Germany

www.ingramcontent.com/pod-product-compliance
Ingram Content Group UK Ltd.
Pitfield, Milton Keynes, MK11 3LW, UK
UKHW022028061126
477531UK00002B/91
9781041007500